# PHYSICAL INORGANIC CHEMISTRY

# PHYSICAL INORGANIC CHEMISTRY

## Reactions, Processes, and Applications

Edited by

**Andreja Bakac**

**WILEY**

A JOHN WILEY & SONS, INC., PUBLICATION

Published by John Wiley & Sons, Inc., Hoboken, New Jersey
Published simultaneously in Canada

For general information on our other products and services or for technical support, please contact our Customer Care Department within the United States at (800) 762-2974, outside the United States at (3 17) 572-3993 or fax (317) 572-4002.

Wiley also publishes its books in a variety of electronic formats. Some content that appears in print may not be available in electronic formats. For more information about Wiley products, visit our web site at www.wiley.com.

*Library of Congress Cataloging-in-Publication Data:*

Physical inorganic chemistry : reactions, processes and applications / [edited by] Andreja Bakac.
    p. cm.
  Includes index.
  ISBN 978-0-470-22420-5 (cloth)
 1.  Physical inorganic chemistry.  I.  Bakac, Andreja.
  QD476.P488 2010
  547′.13–dc22

                                      2009051078

Printed in the United States of America

10  9  8  7  6  5  4  3  2  1

*To Jojika*

# CONTENTS

# PREFACE

This book is a natural extension of "Physical Inorganic Chemistry: Principles, Methods, and Models," a 10-chapter volume describing the methods, techniques, and capabilities of physical inorganic chemistry as seen through the eyes of a mechanistic chemist. This book provides an insight into a number of reactions that play critical roles in areas such as solar energy, hydrogen energy, biorenewables, catalysis, environment, atmosphere, and human health. None of the reaction types described here is exclusive to any particular area of chemistry, but it seems that mechanistic inorganic chemists have studied, expanded, and utilized these reactions more consistently and heavily than any other group. The topics include electron transfer (Weinstock and Snir), hydrogen atom and proton-coupled electron transfer (Fukuzumi), oxygen atom transfer (Abu-Omar), ligand substitution at metal centers (Swaddle), inorganic radicals (Stanbury), organometallic radicals (Kégl, Fortman, Temprado, and Hoff), and activation of oxygen (Rybak-Akimova), hydrogen (Kubas and Heinekey), carbon dioxide (Joó), and nitrogen monoxide (Olabe). Finally, the latest developments in carbon–hydrogen bond activation and in solar photochemistry are presented in the respective chapters by Gunnoe and Meyer.

I am grateful to this group of dedicated scientists for their hard work and professionalism as we worked together to bring this difficult project to a successful conclusion. I am also thankful to my family, friends, and colleagues who provided invaluable support and encouragement throughout the project, and to my editor, Anita Lekhwani, who has been a source of ideas and professional advice through the entire publishing process.

ANDREJA BAKAC

# CONTRIBUTORS

MAHDI M. ABU-OMAR, Department of Chemistry, Purdue University, West Lafayette, IN, USA

GEORGE C. FORTMAN, Department of Chemistry, University of Miami, Coral Gables, FL, USA

SHUNICHI FUKUZUMI, Department of Material and Life Science, Graduate School of Engineering, Osaka University, Suita, Osaka, Japan

THOMAS BRENT GUNNOE, Department of Chemistry, University of Virginia, Charlottesville, VA, USA

D. MICHAEL HEINEKEY, Department of Chemistry, University of Washington, Seattle, WA, USA

CARL D. HOFF, Department of Chemistry, University of Miami, Coral Gables, FL, USA

FERENC JOÓ, Institute of Physical Chemistry, Hungarian Academy of Sciences, University of Debrecen and Research Group of Homogeneous Catalysis, Debrecen, Hungary

TAMÁS KÉGL, Department of Organic Chemistry, University of Pannonia, Veszprém, Hungary

GREGORY J. KUBAS, Chemistry Division, Los Alamos National Laboratory, Los Alamos, NM, USA

GERALD J. MEYER, Department of Chemistry, Johns Hopkins University, Baltimore, MD, USA

JOSÉ A. OLABE, Department of Inorganic, Analytical and Physical Chemistry, Facultad de Ciencias Exactas y Naturales, Universidad de Buenos Aires, Buenos Aires, Argentina

ELENA RYBAK-AKIMOVA, Department of Chemistry, Tufts University, Medford, MA, USA

OPHIR SNIR, Department of Chemistry, Ben-Gurion University of theNegev, Beer Sheva, Israel

DAVID M. STANBURY, Department of Chemistry, Auburn University, Auburn, AL, USA

THOMAS W. SWADDLE,    Department of Chemistry, University of Calgary, Calgary, Canada

MANUEL TEMPRADO,    Department of Chemistry, University of Miami, Coral Gables, FL, USA

IRA A. WEINSTOCK,    Department of Chemistry, Ben-Gurion University of the Negev, Beer Sheva, Israel

# 1 Electron Transfer Reactions

OPHIR SNIR and IRA A. WEINSTOCK

## 1.1 INTRODUCTION

Over the past few decades, applications of the Marcus model to inorganic electron transfer reactions have become routine. Despite the many approximations needed to simplify the theoretical descriptions to obtain a simple quadratic model, and the assumptions then needed to apply this model to actual reactions, agreement between calculated and observed rate constants is remarkably common. Because of this, the Marcus model is widely used to assess the nature of electron transfer reactions. The intent of this chapter is to make this useful tool more accessible to practicing chemists. In that sense, it is written from a "reaction chemist's"[1] perspective. In 1987, Eberson published an excellent monograph that provides considerable guidance in the context of organic reactions.[2] In addition to a greater focus on inorganic reactions, this chapter covers electrolyte theory and ion pairing in more detail, and worked examples are presented in a step-by-step fashion to guide the reader from theory to application.

The chapter begins with an introduction to Marcus' theoretical treatment of outer-sphere electron transfer. The emphasis is on communicating the main features of the theory and on bridging the gap between theory and practically useful classical models.

The chapter then includes an introduction to models for collision rates between charged species in solution, and the effects on these of salts and ionic strength, which all predated the Marcus model, but upon which it is an extension. Collision rate and electrolyte models, such as those of Smoluchowski, Debye, Hückel, and others, apply in ideal cases rarely met in practice. The assumptions of the models will be defined, and the common situations in which real reacting systems fail to comply with them will be highlighted. These models will be referred to extensively in the second half of the chapter, where the conditions that must be met in order to use the Marcus model properly, to avoid common pitfalls, and to evaluate situations where calculated values fail to agree with experimental ones will be clarified.

Those who have taught themselves how to apply the Marcus model and cross-relation correctly will appreciate the gap between the familiar "formulas" published in numerous articles and texts and the assumptions, definitions of terms, and physical

*Physical Inorganic Chemistry: Reactions, Processes, and Applications*   Edited by Andreja Bakac
Copyright © 2010 by John Wiley & Sons, Inc.

constants needed to apply them. This chapter will fill that gap in the service of those interested in applying the model to their own chemistry. In addition, the task of choosing compatible units for physical constants and experimental variables will be simplified through worked examples that include dimensional analysis.

The many thousands of articles on outer-sphere electron transfer reactions involving metal ions and their complexes cannot be properly reviewed in a single chapter. From this substantial literature, however, instructive examples will be selected. Importantly, they will be explaining in more detail than is typically found in review articles or treatises on outer-sphere electron transfer. In fact, the analyses provided here are quite different from those typically found in the primary articles themselves. There, in nearly all cases, the objective is to present and discuss calculated results. Here, the goal is to enable readers to carry out the calculations that lead to publishable results, so that they can confidently apply the Marcus model to their own data and research.

## 1.2    THEORETICAL BACKGROUND AND USEFUL MODELS

The importance of Marcus' theoretical work on electron transfer reactions was recognized with a Nobel Prize in Chemistry in 1992, and its historical development is outlined in his Nobel Lecture.[3] The aspects of his theoretical work most widely used by experimentalists concern outer-sphere electron transfer reactions. These are characterized by weak electronic interactions between electron donors and acceptors along the reaction coordinate and are distinct from inner-sphere electron transfer processes that proceed through the formation of chemical bonds between reacting species. Marcus' theoretical work includes intermolecular (often bimolecular) reactions, intramolecular electron transfer, and heterogeneous (electrode) reactions. The background and models presented here are intended to serve as an introduction to bimolecular processes.

The intent here is not to provide a rigorous and comprehensive treatment of the theory, but rather to help researchers understand basic principles, classical models derived from the theory, and the assumptions upon which they are based. This focus is consistent with the goal of this chapter, which is to enable those new to this area to apply the classical forms of the Marcus model to their own science.

For further reading, many excellent review articles and books provide more in-depth information about the theory and more comprehensive coverage of its applications to chemistry, biology, and nanoscience. Several recommended items (among many) are a highly cited review article by Marcus and Sutin,[4] excellent reviews by Endicott,[5] Creutz and Brunschwig,[6] and Stanbury,[7] a five-volume treatise edited by Balzano,[8] and the abovementioned monograph by Eberson[2] that provides an accessible introduction to theory and practice in the context of organic electron transfer reactions.

### 1.2.1    Collision Rates Between Hard Spheres in Solution

In 1942, Debye extended Smoluchowski's method for evaluating fundamental frequency factors, which pertained to collision rates between neutral particles D

$$D^n + A^m \xrightleftharpoons[]{rapid} [D^n, A^m] \xrightleftharpoons[k_{ret}]{k_{fet}} [D^{(n+1)}, A^{(m+1)}] \xrightleftharpoons[]{rapid} D^{(n+1)} + A^{(m+1)}$$

Reactants           Precursor           Successor           Products

**SCHEME 1.1**

and A, randomly diffusing in solution, to include the electrostatic effects of charged reacting species in dielectric media containing dissolved electrolytes.[9–11] Debye's colliding sphere model was derived assuming that collisions between $D^n$ (electron donors with a charge of $n$) and $A^m$ (electron acceptors of charge $m$) resulted in the transient formation of short-lived complexes, $D^n - A^m$. Rate constants for these reactions vary in a nonlinear fashion as functions of ionic strength, and the models are intimately tied to contemporary and later developments in electrolyte theory.

Marcus[12] and others[13] extended this model to include reactions in which electron transfer occurred *during* collisions between the "donor" and "acceptor" species, that is, between the short-lived $D^n - A^m$ complexes. In this context, electron transfer within transient "precursor" complexes ($[D^n - A^m]$ in Scheme 1.1) resulted in the formation of short-lived "successor" complexes ($[D^{(n+1)} - A^{(m-1)}]$ in Scheme 1.1). The Debye–Smoluchowski description of the diffusion-controlled collision frequency between $D^n$ and $A^m$ was retained. This has important implications for application of the Marcus model, particularly where—as is common in inorganic electron transfer reactions—charged donors or acceptors are involved. In these cases, use of the Marcus model to evaluate such reactions is only defensible if the collision rates between the reactants vary with ionic strength as required by the Debye–Smoluchowski model. The requirements of that model, and how electrolyte theory can be used to verify whether a reaction is a defensible candidate for evaluation using the Marcus model, are presented at the end of this section.

After electron transfer (transition along the reaction coordinate from $D^n - A^m$ to $D^{(n+1)-} A^{(m-1)}$ in Scheme 1.1), the successor complex dissociates to give the final products of the electron transfer, $D^{(n+1)}$ and $A^{(m-1)}$. The distinction between the successor complex and final products is important because, as will be shown, the Marcus model describes rate constants as a function of the difference in energy between precursor and successor complexes, rather than between initial and final products.

### 1.2.2 Potential Energy Surfaces

As noted above, outer-sphere electron transfer reactions are characterized by the absence of strong electronic interaction (e.g., bond formation) between atomic or molecular orbitals populated, in the donor and acceptor, by the transferred electron. Nonetheless, as can be appreciated intuitively, outer-sphere reactions must require some type of electronic "communication" between donor and acceptor atomic or molecular orbitals. This is referred to in the literature as "coupling," "electronic interaction," or "electronic overlap" and is usually less than $\sim$1 kcal/mol. Inner-sphere electron transfer reactions, by contrast, frequently involve covalent bond

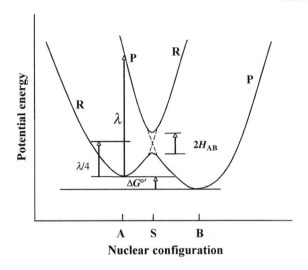

**FIGURE 1.1** Potential energy surfaces for outer-sphere electron transfer. The potential energy surface of reactants plus surrounding medium is labeled R and that of the products plus surrounding medium is labeled P. Dotted lines indicate splitting due to electronic interaction between the reactants. Labels A and B indicates the nuclear coordinates for equilibrium configurations of the reactants and products, respectively, and S indicates the nuclear configuration of the intersection of the two potential energy surfaces.

formation between the reactants and are often characterized by ligand exchange or atom transfer (e.g., of O, H, hydride, chloride, or others).

The two-dimensional representation of the intersection of two $N$-dimensional potential energy surfaces is depicted in Figure 1.1.[4] The curves represent the energies and spatial locations of reactants and products in a many-dimensional ($N$-dimensional) configuration space, and the $x$-axis corresponds to the motions of all atomic nuclei. The two-dimensional profile of the reactants plus the surrounding medium is represented by curve R, and the products plus surrounding medium by curve P. The minima in each curve, that is, points A and B, represent the equilibrium nuclear configurations, and associated energies, of the precursor and successor complexes indicated in Scheme 1.1, rather than of separated reactants or separated products. As a consequence, the difference in energy between reactants and products (i.e., the difference in energy between A and B) is not the Gibbs free energy for the overall reaction, $\Delta G°$, but rather the "corrected" Gibbs free energy, $\Delta G°'$. For reactions of charged species, the difference between $\Delta G°$ and $\Delta G°'$ can be substantial.

The intersection of the two surfaces forms a new surface at point S in Figure 1.1. This ($N - 1$)-dimensional surface has one less degree of freedom than the energy surfaces depicted by curves R and P. Weak electronic interaction between the reactants results in the indicated splitting of the potential surfaces. This gives rise to electronic coupling (resonance energy arising from orbital mixing) of the reactants' electronic state with the products', described by the electronic matrix element, $H_{AB}$. This is equal to one-half the separation of the curves at the intersection of the R and

P surfaces. The dotted lines represent the approach of two reactants with no electronic interaction at all.

This diagram can be used to appreciate the main difference between inner- and outer-sphere processes. The former are associated with a much larger splitting of the surfaces, due to the stronger electronic interaction necessary for the "bonded" transition state. A classical example of this was that recognized by Henry Taube, recipient of the 1983 Nobel Prize in Chemistry for his work on inorganic reaction mechanisms. In a famous experiment, he studied electron transfer from $Cr^{II}(H_2O)_6^{2+}$ (labile, high spin, $d^4$) to the nonlabile complex $(NH_3)_5Co^{III}Cl^{2+}$ (low spin, $d^6$) under acidic conditions in water. Electron transfer was accompanied by a change in color of the solution from a mixture of sky blue $Cr^{II}(H_2O)_6^{2+}$ and purple $(NH_3)_5Co^{III}Cl^{2+}$ to the deep green color of the nonlabile complex $(H_2O)_5Cr^{III}Cl^{2+}$ ($d^3$) and labile $Co^{II}(H_2O)_6^{2+}$ (high spin, $d^7$) (Equation 1.1).[14,15]

$$[Cr^{II}(H_2O)_6]^{2+} + [(NH_3)_5Co^{III}Cl]^{2+} + 5H^+ \longrightarrow [(H_2O)_5Cr^{III}Cl]^{2+} + [Co^{II}(H_2O)_6]^{2+} + 5NH_4^+$$

blue                         purple                                              green

$$(1.1)$$

Using radioactive $Cl^-$ in $(NH_3)_5Co^{III}Cl^{2+}$, he demonstrated that even when $Cl^-$ was present in solution, electron transfer occurred via direct (inner-sphere) $Cl^-$ transfer, such that the radiolabeled $Cl^-$ remained coordinated to the (now) nonlabile $Cr^{III}$ product.

### 1.2.3   Franck–Condon Principle and Outer-Sphere Electron Transfer

The mass of the transferred electron is very small relative to that of the atomic nuclei. As a result, electron transfer is much more rapid than nuclear motion, such that nuclear coordinates are effectively unchanged during the electron transfer event. This is the Franck–Condon principle.

Now, for electron transfer reactions to obey the Franck–Condon principle, while also complying with the first law of thermodynamics (conservation of energy), electron transfer can occur only at nuclear coordinates for which the total potential energy of the reactants and surrounding medium equals that of the products and surrounding medium. The intersection of the two surfaces, S, is the only location in Figure 1.1 at which both these conditions are satisfied. The quantum mechanical treatment allows for additional options such as "nuclear tunneling," which is discussed below.

### 1.2.4   Adiabatic Electron Transfer

The classical form of the Marcus equation requires that the electron transfer be adiabatic. This means that the system passes the intersection slowly enough for the transfer to take place and that the probability of electron transfer per passage is large (near unity). This probability is known as the transmission coefficient, $\kappa$, defined later in this section. In this quantum mechanical context, the term "adiabatic" indicates that nuclear coordinates change sufficiently slowly that the system (effectively) remains at equilibrium as it progresses along the reaction coordinate. The initial eigenstate of the system is modified in a continuous manner to a final eigenstate according to the Schrödinger equation, as shown in Equation 1.2. At the adiabatic limit, the time required for the system to go from initial to final states approaches infinity (i.e., $[t_f - t_i] \rightarrow \infty$).

$$|\psi(x, t_f)|^2 \neq |\psi(x, t_i)|^2 \tag{1.2}$$

When the system passes the intersection at a high velocity, that is, the above condition is not met even approximately, it will usually "jump" from the lower R surface (before S along the reaction coordinate) to the upper R surface (after S). That is, the system behaves in a "nonadiabatic" (or diabatic) fashion, and the probability per passage of electron transfer occurring is small (i.e., $\kappa \ll 1$). The nuclear coordinates of the system change so rapidly that it cannot remain at equilibrium. At the nonadiabatic limit, the time interval for passage between the two states at point S approaches zero, that is, $(t_f - t_i) \rightarrow 0$ (infinitely rapid), and the probability density distribution functions that describe the initial and final states remain unchanged:

$$|\psi(x, t_f)|^2 = |\psi(x, t_i)|^2 \tag{1.3}$$

Another cause of nonadiabicity is very weak electronic interaction between the reactants. This means that $\kappa$ is inherently much smaller than 1, such that the splitting of the potential surfaces is small. In other words, electronic communication between reactants is too small to facilitate a change in electronic states, from reactant to product, at the intersection of the R and P curves. Graphically, this means that the splitting at S is small, and the adiabatic route (passage along the lower surface at S) has little probability of occurring.

The "fast" and "slow" changes described here, which refer to "velocities" of passage through the intersection, S, correspond to "high" and "low" frequencies of nuclear motions. Hence, "nuclear frequencies" play an important role in quantum mechanical treatments of electron transfer.

### 1.2.5   The Marcus Equation

In his theoretical treatment of outer-sphere electron transfer reactions, Marcus related the free energy of activation, $\Delta G^{\ddagger}$, to the corrected Gibbs free energy of the reaction, $\Delta G^{\circ\prime}$, via a quadratic equation (Equation 1.4).[2,4,13]

$$\Delta G^{\ddagger} = \frac{z_1 z_2 e^2}{D r_{12}} \exp(-\chi r_{12}) + \frac{\lambda}{4}\left(1 + \frac{\Delta G^{\circ\prime}}{\lambda}\right)^2 \tag{1.4}$$

The terms $\Delta G^{\circ\prime}$ and $\lambda$ in Equation 1.4 are represented schematically in Figure 1.1, and $\chi$ is the reciprocal Debye radius (Equation 1.5).[11,16]

$$\chi = \left( \frac{4\pi e^2}{DkT} \sum_i n_i z_i^2 \right)^{1/2} \tag{1.5}$$

In Equation 1.5, $D$ is the dielectric constant of the medium, $e$ is the charge of an electron, $k$ is the Boltzmann constant, and $\sum n_i z_i^2 = 2\mu$, where $\mu$ is the total ionic strength of an electrolyte solution containing molar concentrations, $n_i$, of species $i$ of charge $z$ (ionic strength $\mu$ is defined by $\mu \equiv \frac{1}{2} \sum n_i z_i^2$).

The first term in Equation 1.4 was retained from Debye's colliding sphere model: the electron-donor and electron-acceptor species were viewed as spheres of radii $r_1$ and $r_2$ that possessed charges of $z_1$ and $z_2$, respectively. This term is associated with the electrostatic energy (Coulombic work) required to bring the two spheres from an infinite distance to the center-to-center separation distance, $r_{12} = r_1 + r_2$, which is also known as the distance of closest approach (formation of the precursor complex $[D^n-A^m]$ in Scheme 1.1). The magnitude of the Coulombic term is modified by a factor $\exp(-\chi r_{12})$, which accounts for the effects of the dielectric medium (of dielectric constant $D$) and of the total ionic strength $\mu$.

The *corrected* Gibbs free energy, $\Delta G^{\circ\prime}$, in Equation 1.4 is the difference in free energy between the successor and precursor complexes of Scheme 1.1 as shown in Figure 1.1. The more familiar, Gibbs free energy, $\Delta G^{\circ}$, is the difference in free energy between separated reactants and separated products in the prevailing medium. The corrected free energy, $\Delta G^{\circ\prime}$, is a function of the charges of the reactants and products. It is calculated using Equation 1.6, where $z_2$ is the charge of the electron donor and $z_1$ is the charge of the electron acceptor.

$$\Delta G^{\circ\prime} = \Delta G^{\circ} + (z_1 - z_2 - 1) \frac{e^2}{Dr_{12}} \exp(-\chi r_{12}) \tag{1.6}$$

If one of the reactants is neutral (i.e., its formal charge is zero), the work term in Equation 1.4 equals zero. As a consequence of this (and all else being equal), highly negative charged oxidants may react more rapidly with neutral electron donors than with positively charged electron donors. This is somewhat counterintuitive because one might expect negatively charged oxidants to react more rapidly with positively charged donors, to which the oxidant is attracted. In other words, attraction between oppositely charged species is usually viewed as contributing to the favorability of a reaction. For example, the heteropolyanion, $Co^{III}W_{12}O_{40}{}^{5-}$ ($E^{\circ} = +1.0\,V$), can oxidize organic substrates with standard potentials as large as $+2.2\,V$. This is because the attraction between the donor and acceptor in the *successor* complex, generated by electron transfer, leads to a favorable attraction between the negative heteropolyanion and the oxidized (now positively charged) donor. This attraction makes the corrected free energy more favorable, the activation energy smaller, and the electron transfer reaction kinetically possible.[2,17]

In Equation 1.6, the electrostatic correction to $\Delta G^\circ$ vanishes when $z_1 - z_2 = 1$ (e.g., when $z_1$ and $z_2$ are equal, respectively, to 3 and 2, 2 and 1, 1 and 0, 0 and $-1$, $-1$ and $-2$, etc.).[2] In these cases, the difference in Gibbs free energy between the successor and precursor complexes is not significantly different from that between the individual (separated) reactants and final (separated) products (Scheme 1.1).

The relation between $\Delta G^\circ$ and the standard reduction potential of the donor and acceptor, $E^\circ$, is given by

$$\Delta G^\circ = -nFE^\circ \tag{1.7}$$

where $n$ is the number of electrons transferred and $F$ is the Faraday constant. This, combined with Equation 1.6, is often used to calculate $\Delta G^{\circ\prime}$ from electrochemical data.

### 1.2.5.1  Reorganization Energy

The $\lambda$ term in Equation 1.4 is the reorganization energy associated with electron transfer, and more specifically, with the transition from precursor to successor complexes. As noted above, there are two different and separable phenomena, termed "inner-sphere" and "outer-sphere" reorganization energies, commonly indicated by the subscripts "in" and "out." The total reorganization energy is the sum of the inner- and outer-sphere components (Equation 1.8).

$$\lambda = \lambda_{in} + \lambda_{out} \tag{1.8}$$

The inner-sphere reorganization energy refers to changes in bond lengths and angles (in-plane and torsional) of the donor and acceptor molecules or complexes. Due to electron transfer, the electronic properties and charge distribution of the successor complex are different from those of the precursor complex. This causes reorientation or other subtle changes of the solvent molecules in the reaction medium near the reacting pair, and the energetic cost associated with this is the outer-sphere (solvent) reorganization energy.

The inner-sphere reorganization energy can be calculated by treating bonds within the reactants as harmonic oscillators, according to Equation 1.9.

$$\lambda_{in} = \sum_j \frac{f_j^r f_j^p}{f_j^r + f_j^p} (\Delta q_j)^2 \tag{1.9}$$

Here, $f_j^r$ is the $j$th normal mode force constant in the reactants, $f_j^p$ is that in the products, and $\Delta q_j$ is the change in the equilibrium value of the $j$th normal coordinate.

A simplified expression for the outer-sphere reorganization energy, $\lambda_{out}$, was obtained by treating the solvent as a dielectric continuum.[18] For this, it is assumed that the dielectric polarization outside the coordination shell responds linearly to changes in charge distributions, such that the functional dependence of the free energy of the dielectric polarization on charging parameters is quadratic. Marcus then used a two-step thermodynamic cycle to calculate $\lambda_{out}$.[19,20] This treatment allows the individual solvent dipoles to move anharmonically, as indeed they do in the liquid state. The forms of the relationships that describe $\lambda_{out}$ depend on the geometrical model chosen to represent the charge distribution. For spherical reactants, $\lambda_{out}$ is given by

Equation 1.10.

$$\lambda_{\text{out}} = (\Delta e)^2 \left[ \frac{1}{2r_1} + \frac{1}{2r_2} - \frac{1}{r_{12}} \right] \left[ \frac{1}{D_{\text{op}}} - \frac{1}{D_{\text{s}}} \right] \qquad (1.10)$$

Here, $\Delta e$ is the charge transferred from one reactant to the other, $r_1$ and $r_2$ are the radii of the two (spherical) reactants, $r_{12}$ is, as before, the center-to-center distance, often approximated[18] as the sum of $r_1 + r_2$, and $D_{\text{s}}$ and $D_{\text{op}}$ are the static and optical (square of refractive index) dielectric constants of the solvent, respectively. This model for $\lambda_{\text{out}}$ treats both the reactants as hard spheres (i.e., the "hard sphere" model). For other shapes, more complex models are needed, which are rarely used by reaction chemists.[21]

### 1.2.6     Useful Forms of the Marcus Model

*1.2.6.1     The Eyring Equation and Linear Free Energy Relationships*     In principle, one could use nonlinear regression to fit the Marcus equation (Equation 1.4) to a plot of $\Delta G^{\ddagger}$ versus $\Delta G^{\circ\prime}$ values for a series of reactions, with $\lambda$ as an adjustable parameter. To obtain a reasonably good fit, the shapes, sizes, and charges of reactants and products, and their $\lambda$ values, must be similar to one another. A good fit between calculated and experimental curves would be evidence for a common outer-sphere electron transfer mechanism, and the fitted value of $\lambda$ is an approximate value for this parameter. In practice, $\Delta G^{\ddagger}$ values cannot be measured directly. However, bimolecular rate constants, $k$, can be. These are related to $\Delta G^{\ddagger}$ by the Eyring equation (Equation 1.11).

$$k = \kappa Z \exp(-\Delta G^{\ddagger}/RT) \qquad (1.11)$$

Here, $\Delta G^{\ddagger}$ is defined by Equation 1.4, $\kappa$ is the transmission coefficient, and $Z$ is the collision frequency in units of $M^{-1}s^{-1}$. The transmission coefficient is discussed above. In practice, $\kappa$ is often set equal to 1. Although this gives reasonable results in numerous cases, this is one of the many assumptions "embedded" within the familiar, classical form of the Marcus equation (Equation 1.4). Expansion of Equation 1.4 gives Equation 1.12, in which the Coulombic work term (i.e., the first term on the right-hand side of Equation 1.4) is abbreviated as $W(r)$.

$$\Delta G^{\ddagger} = W(r) + \frac{\lambda}{4} + \frac{\Delta G^{\circ\prime}}{2} + \frac{(\Delta G^{\circ\prime})^2}{4\lambda} \qquad (1.12)$$

Substituting Equation 1.12 into Equation 1.11, taking the natural logarithm, and rearranging gives Equation 1.13.

$$RT \ln Z - RT \ln k = W(r) + \frac{\lambda}{4} + \frac{\Delta G^{\circ\prime}}{2} + \frac{(\Delta G^{\circ\prime})^2}{4\lambda} \qquad (1.13)$$

Values of $\ln k$ versus $\Delta G^{\circ\prime}$ can be plotted for a series of reactions and fitted to Equation 1.13 by nonlinear regression using $\lambda$ as an adjustable parameter.

If $|\Delta G^{\circ\prime}| \ll \lambda$, the last term in Equation 1.12 can be ignored, and the Marcus equation can be approximated by a *linear* free energy relationship (LFER) (Equation 1.14).

$$\Delta G^{\ddagger} = W(r) + \frac{\lambda}{4} + \frac{\Delta G^{\circ\prime}}{2} \qquad (1.14)$$

In principle, if the $\lambda$ values for a set of like reactions are similar to one another, and $W(r)$ is small or constant, a plot of $\Delta G^{\ddagger}$ versus $\Delta G^{\circ\prime}$ will be linear and have a slope of 0.5. As noted above, it is rarely possible to measure $\Delta G^{\ddagger}$ values directly. An alternative option for plotting data using a linear relationships is to use Equation 1.15 and the definition of the equilibrium constant, $K = A \exp(-\Delta G^{\circ\prime}/RT)$, in which $A$ is a constant. If the equilibrium constants for a series of reactions can be measured or calculated, one can plot $\ln k$ versus $\ln K$ (Equation 1.15). A linear result with a slope of 0.5 is indicative of a common outer-sphere electron transfer mechanism.

$$\ln k = \ln Z - \frac{W(r)}{RT} - \frac{\lambda}{4RT} + 0.5 \ln K + \ln A \qquad (1.15)$$

Another useful linear relationship is based on electrochemical data and is obtained by recourse to the fact that $\Delta G^{\circ} = -nFE^{\circ}$. For a series of outer-sphere electron transfer reactions that meet the criteria discussed in context with Equation 1.14, a plot of $\ln k$ versus $E^{\circ}$ will have a slope of $0.5(nF/RT)$, and a plot of $\log k$ versus $E^{\circ}$ will have a slope of $0.5(nF)/2.303RT$ or $8.5 \text{ V}^{-1}$ for $n = 1$ at $25^{\circ}\text{C}$.[5] All the above methods can be used to obtain a common (approximate) value of $\lambda$ for a series of similar reactions. For single reactions of interest, however, $\lambda$ values can often be measured directly by electron self-exchange.

***1.2.6.2   Electron Self-Exchange***   In many cases, $\Delta G^{\circ}$ can be easily measured (usually electrochemically), while $\lambda$ is more difficult to determine. The methods discussed above require data for a series of similar reactions. This information is not always accessible, or of interest. An alternative and more direct method is to determine $\lambda$ values from rate constants for electron self-exchange. This requires that a kinetic method be available for measuring the rate of electron exchange between one-electron oxidized and reduced forms of a complex or molecule. One requirement for this is that the oxidation or reduction involved does not lead to rapid, irreversible further reactions of either partner. In this sense, self-exchanging pairs whose $\lambda$ values can be measured kinetically are often reversible or quasi-reversible redox couples.

In self-exchange reactions, such as that between $A^m$ and $A^{m+1}$ (Equation 1.16), $\Delta G^{\circ\prime} = 0$.

$$*A^m + A^{m+1} \rightleftharpoons *A^{m+1} + A^m \qquad (1.16)$$

In this special case, the Marcus equation (Equation 1.4) reduces to Equation 1.17.

$$\Delta G^{\ddagger} = W(r) + \frac{\lambda}{4} \qquad (1.17)$$

Because $\Delta G^{\ddagger}$ is not directly measurable, $\lambda_{11}$ can be calculated from the observed rate constant $k$ for the self-exchange reaction by using Equation 1.18. This is obtained by substituting Equation 1.17 into Equation 1.11 and assuming that $\kappa = 1$.

$$k = Z \exp\left[-\frac{W(r) + \lambda/4}{RT}\right] \tag{1.18}$$

Equation 1.18 can also be converted to a linear form (Equation 1.19) by taking the natural logarithm.

$$RT \ln k = RT \ln Z - W(r) - \frac{\lambda}{4} \tag{1.19}$$

For reactions in solution, $Z$ is often on the order of $10^{11} \, M^{-1} \, s^{-1}$ (values of $Z = 6 \times 10^{11} \, M^{-1} \, s^{-1}$ are also used).[7] To calculate $W(r)$, one must know the charges and radii of the reactants, the dielectric constant of the solvent, and the ionic strength of the solution. The reorganization energy $\lambda_{11}$ can then be calculated from $k$. Worked examples from the literature are included in Section 1.3.

#### 1.2.6.3 The Marcus Cross-Relation

The rate constant, $k_{12}$, for electron transfer between two species, $A^m$ and $B^n$ (Equation 1.20) that are not related to one another by oxidation or reduction, is referred to as the Marcus cross-relation (MCR).

$$A^m + B^{n+1} \rightleftharpoons A^{m+1} + B^n \tag{1.20}$$

It is called the "cross-relation" because it is algebraically derived from expressions for the two related electron self-exchange reactions shown in Equations 1.21 and 1.22. Associated with these reactions are two self-exchange rate constants $k_{11}$ and $k_{22}$ and reorganization energies $\lambda_{11}$ and $\lambda_{22}$.

$$*A^m + A^{m+1} \rightleftharpoons *A^{m+1} + A^m, \quad \text{rate constant} = k_{11} \tag{1.21}$$

$$*B^m + B^{m+1} \rightleftharpoons *B^{m+1} + B^m, \quad \text{rate constant} = k_{22} \tag{1.22}$$

The MCR is derived by first assuming that Equation 1.23 holds. This means that the reorganization energy for the cross-reaction, $\lambda_{12}$, is equal to the mean of the reorganization energies, $\lambda_{11}$ and $\lambda_{22}$, associated with the two related self-exchange reactions.

$$\lambda_{12} \cong \frac{1}{2}(\lambda_{11} + \lambda_{22}) \tag{1.23}$$

The averaging over the outer-sphere components of $\lambda_{11}$ and $\lambda_{22}$, that is, $\lambda_{11\text{out}}$ and $\lambda_{22\text{out}}$, is only valid if $A^m$ and $B^{n+1}$ (Equation 1.20) are of the same size (i.e., $r_1 = r_2$).

As is demonstrated later in this chapter, differences in size between donors and acceptors can lead to large discrepancies between calculated and experimental values.

The assumption in Equation 1.23 is used to derive the MCR (Equations 1.24–1.26)

$$k_{12} = (k_{11}k_{22}K_{12}f_{12})^{1/2}C_{12} \qquad (1.24)$$

where

$$\ln f_{12} = \frac{1}{4}\frac{(\ln K_{12} + (w_{12}-w_{21})/RT)^2}{\ln(k_{11}k_{22}/Z^2) + (w_{11}-w_{22})/RT} \qquad (1.25)$$

and

$$C_{12} = \exp[-(w_{12} + w_{21} - w_{11} - w_{22})/2RT] \qquad (1.26)$$

As mentioned above, $Z$ is the pre-exponential factor and $w_{ij}$ are Coulombic work terms (Equation 1.4) associated with all four combinations of the reacting species. If $k_{22}$ is known, one can use $k_{12}$ and Equations 1.24–1.26 to calculate $k_{11}$, which is related by Equation 1.19 to its reorganization energy $\lambda_{11}$. The $C_{12}$ and $f_{12}$ terms often approach unity for molecules that possess small charges.[4,22] For reactions of charged inorganic complexes, however, these terms can be important.

### 1.2.7   Additional Aspects of the Theory

***1.2.7.1   The Inverted Region***   The Marcus model predicts that as absolute values of $\Delta G^{\circ\prime}$ decrease (i.e., as electron transfer becomes more thermodynamically favorable), electron transfer rate constants should *decrease*. Because energetically more favorable reactions generally occur more rapidly, it is counterintuitive to expect the opposite to occur. However, this is precisely the case in the inverted region where more thermodynamically favorable reactions occur more slowly. The exothermic region in which this occurs is therefore referred to as "inverted." Marcus predicted this behavior in 1960,[13,22] and the first experimental evidence for it was provided more than two decades later.[23]

In a series of related reactions possessing similar $\lambda$ values but differing in $\Delta G^{\circ\prime}$, a plot of the activation free energy $\Delta G^{\ddagger}$ versus $\Delta G^{\circ\prime}$ (from Equation 1.12) can be separated into two regions. In the first region, $\Delta G^{\ddagger}$ decreases, and rates increase, as $\Delta G^{\circ\prime}$ decreases from zero to more negative values. This is the "normal" region. When $\Delta G^{\circ\prime}$ becomes sufficiently negative such that $\Delta G^{\circ\prime} = -\lambda$, $\Delta G^{\ddagger}$ becomes zero. (This is true for reactions in which $W(r) = 0$; for cases where $W(r) \neq 0$, $\Delta G^{\ddagger} = W(r)$.) In the next region (the inverted region), $\Delta G^{\ddagger}$ begins to increase (and rates decrease), as $\Delta G^{\circ\prime}$ becomes even more negative than $-\lambda$.

The dependence of $\ln k$ on $\Delta G^{\circ\prime}$ is depicted in Figure 1.2. The rate first increases as $\Delta G^{\circ\prime}$ becomes more negative (region I, the normal region) and reaches a maximum at $-\Delta G^{\circ\prime} = \lambda$ (point II). Then, as $\Delta G^{\circ\prime}$ becomes even more negative, $\ln k$ begins to decrease (region III, the inverted region).

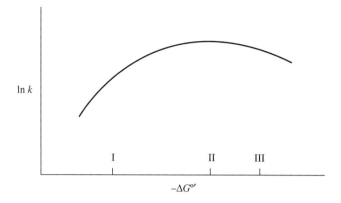

**FIGURE 1.2**    Plot of $\ln k$ versus $-\Delta G^{\circ\prime}$. Region I is the normal region, region III is the inverted region, and at point II ($-\Delta G^{\circ\prime} = \lambda$) $\ln k$ reaches a maximum.

The reason for this can be understood by reference to Figure 1.1. The plots in Figure 1.1 show the locations of the R and P surfaces in the normal region. To reach the inverted region, $\Delta G^{\circ\prime}$ must become more negative. This corresponds to lowering the P surface relative to the R surface in Figure 1.1. As this proceeds, the free energy barrier $\Delta G^{\ddagger}$ decreases until it becomes zero at $\Delta G^{\circ\prime} = -\lambda$. At this point, the intersection of the R and P surfaces occurs at the minimum of the R curve, and the reaction has no activation barrier. Further decrease in $\Delta G^{\circ\prime}$ then *raises* the energy at which the R and P surfaces intersect. This corresponds to an increase in the activation barrier, $\Delta G^{\ddagger}$, and a decrease in rate. This case, that is, the inverted region, is shown in Figure 1.3, which depicts the relative locations of the reactant and product energy surfaces in the inverted region.

Indirect evidence for the inverted region was first provided by observations that some highly exothermic electron transfer reactions resulted in chemiluminescence, an indication that electronically excited products had been formed (surface P* in Figure 1.3). When the ground electronic state potential energy surface of the products, P, intersects the R surface at a point high in energy on the R surface (intersection of curves R and P in the inverted region; Figure 1.3), the reaction is slow. In this case, less thermodynamically favorable electron transfer to a product excited state (P* surface) can occur more rapidly than electron transfer to the ground electronic state of the products (P surface). Electron transfer to a P* state, and chemiluminescence associated with a subsequent loss of energy to give the ground electronic state, was observed by Bard and coworkers.[24] Direct experimental confirmation of the inverted region (electron transfer from the R to P surfaces in Figure 1.3) was provided a few years later.[25,26]

### 1.2.8  Nuclear Tunneling

The Marcus equation (and useful relationships derived from it) is a special case characterized by adiabatic electron transfer at the intersection of the reactant and

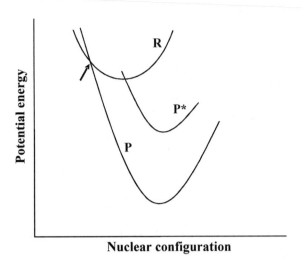

**Nuclear configuration**

**FIGURE 1.3** Potential energy surfaces for the Marcus inverted region. In this highly exothermic reaction, the P surface has dropped in energy to such an extent that further decreases in its energy result in larger activation energies and smaller rate constants. The black arrow indicates the intersection of the R and P surfaces (the splitting of these surfaces due to electronic coupling is not shown). The P* surface depicts an energetically less favorable, yet more rapid transition from the R surface to an electronically excited state of one of the products (see discussion in the text).

product potential energy surfaces. That intersection, S in Figure 1.1, defines the nuclear coordinates and energy of the transition state for electron transfer. This section deals with a quantum mechanical phenomenon in which electron transfer occurs without the nuclear coordinates first reaching the intersection point. Graphically, this means that the system passes from curve R to curve P at a lower (more negative) energy than that of the intersection, S. This is referred to as nuclear "tunneling" from surface R to surface P. The material provided in this section is designed to help the reader understand some basic aspects of this phenomenon.

When nuclear tunneling occurs, the system passes from the R surface to the P surface by crossing horizontally from the first to the second of these surfaces. This is depicted schematically by the horizontal line that extends from "a" to "b" in Figure 1.4. In practice, at room temperature, and for reactions in the "normal" region, nuclear tunneling usually accounts for only minor contributions to rate constants. The cross-relation in the normal region is even less affected by nuclear tunneling, due to partial cancellation of the quantum correction in the ratio $k_{12}/(k_{11}k_{22}K_{12})^{1/2}$.

When it is a viable pathway, nuclear tunneling tends to dominate at low temperatures, at which the probability of the system reaching the intersection point, S, is low. At the same time, nuclear tunneling rates are independent of temperature. This is because, in tunneling, electron transfer occurs at energies near the zero-point

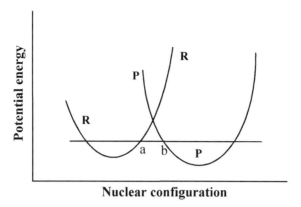

**Nuclear configuration**

**FIGURE 1.4**  Diagrammatic representation of nuclear tunneling. The horizontal line depicts electron transfer via nuclear tunneling from point "a" on surface R to point "b" on surface P.

vibrational energy of the reactants and surrounding media (provided that the energy of the lowest point on the R surface equals or exceeds that of the P surface).

Temperature affects nuclear motion and thereby the Boltzmann probability of attaining the nuclear configuration that corresponds to the intersection between the R and P surfaces. For this reason, temperature determines the contribution that nuclear tunneling makes to overall reaction rates. ($H_{AB}$ is not directly affected by temperature, although it will vary somewhat with nuclear configuration.) Hence, the existence of a temperature-dependent rate constant at high temperatures and an independent one at low temperatures may be a manifestation of nuclear tunneling.

The terms adiabatic and nonadiabatic that were discussed above can be engaged directly to the phenomenon of tunneling. If a system reacts via an adiabatic pathway, the system follows the R surface in the initial stages of the reaction, then remains on the lower surface caused by electronic coupling at the intersection, and continues to the P surface. In a nonadiabatic reaction, the electronic coupling of the reactants is so weak, that is, $2H_{AB}$ is so small, that the probability $\kappa$ of going from the R to the P surface when the system is in proximity to the intersection region in Figure 1.1 is small. In the majority of collisions that result in attaining the energy of the intersection region, the system will stay on the R surface, instead of going on to the P surface. For reactions with intermediate $\kappa$ values, expressions known as the Landau–Zener type are available for calculating $\kappa$.[13] In the "inverted region," a reaction in which the system goes directly from the R to the P surface is necessarily nonadiabatic, and there is no adiabatic path: the system must "jump" from one solid curve to the other in order to form directly the ground-state products.

Nuclear tunneling from the R to the P surfaces is represented by the horizontal line from "a" to "b" in Figure 1.5. Unlike in Figure 1.4, the slopes of the R and P potential energy surfaces have the same sign when approaching the intersection region. Semiclassical models of electron transfer show that in this case nuclear tunneling is much more important.

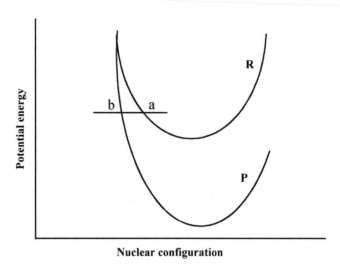

**Nuclear configuration**

**FIGURE 1.5**   Diagrammatic representation of nuclear tunneling from "a" to "b" in the highly exothermic (inverted) reaction.

### 1.2.9    Reactions of Charged Species and the Importance of Electrolyte Theory

*1.2.9.1    Background and Useful Models*    The Marcus equation is an extension of earlier models from collision rate theory. As such, compliance with collision rate models is a prerequisite to defensible use of the Marcus equation. This is particularly important for reactions of charged species, and therefore, for reactions of many inorganic complexes. In these cases, the key question is whether electron transfer rate constants vary with ionic strength as dictated by electrolyte theory, on which the collision rate models are based. When they do not, differences between calculated and experimental values can differ by many orders of magnitude.

The theory of electrolyte behavior in solution is indeed complex, and has been debated since 1923, when Debye and Hückel[27,28] described the behavior of electrolyte solutions at the limit of very low concentration. Subsequently, intense discussions in the literature lasted into the 1980s, by which time a number of quite complex approaches had been promulgated. The latter do give excellent fits up to large ionic strength values, but are generally not used by most kineticists.

The effects of electrolyte concentrations on the rate constants depend on the nature of the interaction between the reacting species. If the reacting species repel one another, rate constants will increase with ionic strength. This is because the electrolyte ions attenuate electrostatic repulsion between the reacting ions. If the reactants have opposite charges, and attract one another, electrolyte ions will attenuate this attraction, resulting in smaller rate constants.

One convenient option is to use the Debye–Hückel equation, also referred to as the Davies equation.[29]

$$\log k = \log k_0 + \frac{2z_1 z_2 \alpha \sqrt{\mu}}{1 + \beta r \sqrt{\mu}} \tag{1.27}$$

In Equation 1.27, $\alpha$ and $\beta$ are the Debye–Hückel constants, equal to 0.509 and 0.329, respectively, at 25°C in aqueous solutions. $k_0$ is the rate constant for the reaction at infinite dilution ($\mu = 0$ M). In this form, $\alpha$ is dimensionless and $\beta$ has units of $\overset{\circ}{A}^{1/2}/mol^{1/2}$.

One fundamental problem is that kineticists almost always deal with solutions of mixed electrolytes. In such cases, the Davies equation is not rigorously correct (based on first principles from thermodynamics). In addition, some workers argue that setting the parameter $r$ equal to the internuclear distance between reacting species is not justified, and that successful results obtained by doing so should be viewed as fortuitous.[30] Nonetheless, as is demonstrated later in this chapter, it can give excellent results in some cases.[31–34]

Alternatively, one may use the Guggenheim equation (Equation 1.28), which is rigorously correct for solutions of mixed electrolytes. In this equation, the specific interaction parameters are moved from the denominator to a second term, in which $b$ is an adjustable parameter.

$$\log k = \log k_0 + \frac{2z_1 z_2 \alpha \sqrt{\mu}}{1 + \sqrt{\mu}} + b\mu \tag{1.28}$$

If one ignores the second term in Equation 1.28, one obtains the "truncated" Guggenheim equation (Equation 1.29), which is identical to the Davies equation, but with $\beta r$ equal to unity. The reader should be aware that many authors refer to Equation 1.29 as the Guggenheim equation.

$$\log k = \log k_0 + \frac{2z_1 z_2 \alpha \sqrt{\mu}}{1 + \sqrt{\mu}} \tag{1.29}$$

Alternatively, one can use more elaborate models.[35] These can yield fine fits to extraordinarily high ionic strengths, but they do not generally provide much additional insight.

In practice, the most common approach[36] is to use the truncated Guggenheim equation (Equation 1.29) and ionic strength no greater than 0.1 M. This is also the method espoused by Espenson.[37] If this fails, the Guggenheim equation (Equation 1.28) is sometimes used. This has more adjustable parameters, and therefore is more likely to produce a linear fit. However, the slopes of the lines obtained often deviate from theoretical values, defined as a function of the charge product, $z_1 z_2$, of the reacting species. Nonetheless, if a good linear fit is obtained (even though the slope is not correct), this might still be used as an argument against the presence of significant ion pairing or other medium effects. A good example of this is provided in an article by Brown and Sutin,[38] who fitted the same data set to a number of models before finally observing a linear relationship between rate constants and ionic strength.

*1.2.9.2 Graphical Demonstration of the Truncated Guggenheim Equation* In Figure 1.6, the truncated Guggenheim equation (Equation 1.29) was used to calculate rate constants as a function of ionic strength. Then, $\log k$ values were plotted as a function of ionic strength. This reveals the effects of ionic strength on rate constants that one might observe experimentally. The curves shown are for reactions with charge products of $z_1 z_2 = -2, -1, 0, 1$, and 2. Charge products of 2 and 1 indicate repulsion between like-charged reactants and those of $-1$ and $-2$ indicate attraction between oppositely charged reactants.

In Figure 1.7, the same rate constants and ionic strengths shown in Figure 1.6 are now plotted according to the truncated Guggenheim equation (Equation 1.29). The horizontal line corresponds to $z_1 z_2 = 0$. The slopes of the lines are equal to $2z_1 z_2 \alpha$ and have values of $-2.036, -1.018, 0, 1.018$, and 2.036, respectively, for $z_1 z_2$ values of $-2, -1, 0, 1$, and 2.

As discussed above, verification that a reaction that involves charged species satisfies the requirements of electrolyte theory is a necessary prerequisite to use of the Marcus model. For this, a plot of $\log k$ versus $\sqrt{\mu}/(1 + \sqrt{\mu})$ (truncated Guggenheim equation) should be linear with a slope of $2z_1 z_2 \alpha$, as shown in Figure 1.7. Deviations from linearity, or, to a lesser extent, slopes that give incorrect charge products, $z_1 z_2$, are indications that the system does not obey this model. When this occurs, other models can be tried. If these fail, ion pairing, other specific medium effects, cation catalysis, or other reaction mechanisms are likely involved.[21] In these cases, the reaction is not a defensible candidate for evaluation by the Marcus model for outer-sphere electron transfer.

Finally, readers should be aware of a comment by Duncan A. MacInnes (in 1939[39]) that "There is no detail of the derivation of the equations of the

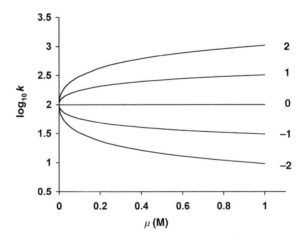

**FIGURE 1.6** Theoretical plots of $\log k$ versus ionic strength $\mu$ for charge products $z_1 z_2 = 2$, $1, 0, -1$, and $-2$. The zero ionic strength rate constant $k_0$ is set equal to $100\,M^{-1}\,s^{-1}$ and $\mu$ is varied from 0.001 to 1 M.

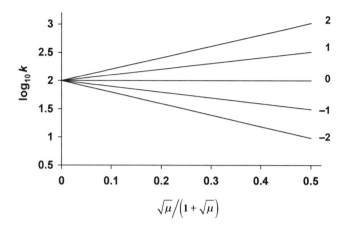

**FIGURE 1.7**    Plots of $\log k$ versus $\sqrt{\mu}/(1 + \sqrt{\mu})$ for charge products $z_1 z_2 = 2, 1, 0, -1$, and $-2$. $k_0$ is taken as $100\,\mathrm{M^{-1}\,s^{-1}}$ and the ionic strength $\mu$ is varied from 0.001 to 1 M.

Debye–Hückel theory that has not been criticized." From this perspective, electrolyte models are simply the best tools available to assess whether the dependence of electron transfer rate constants on ionic strength is sufficiently well behaved to justify use of the Marcus model. For this, and despite their shortcomings, they are indispensable.

## 1.3    GUIDE TO USE OF THE MARCUS MODEL

This section is designed to fill the gap between the familiar "formulas" presented above and the assumptions and definitions of terms and physical constants needed to apply them. Values for all physical constants and needed conversion factors are provided, and dimensional analyses are included to show how the final results and their units are obtained. This close focus on the details and units of the equations themselves is followed by worked examples from the chemical literature. The goal is to provide nearly everything the interested reader may need to evaluate his or her own data, with reasonable confidence that he or she is doing so correctly.

### 1.3.1    Compliance with Models for Collision Rates Between Charged Species

In this section, applications of the Davies and truncated Guggenheim equations are demonstrated through worked examples from the literature.

*1.3.1.1    The Davies Equation*    The Davies equation (introduced earlier in this chapter and reproduced here for convenience in Equation 1.30) is one of several closely related models, derived from electrolyte theory, that describe the functional dependence of rate constants on ionic strength.

$$\log k = \log k_0 + 2\alpha z_1 z_2 \, \mu^{1/2} / (1 + \beta r \, \mu^{1/2}) \tag{1.30}$$

In Equation 1.30, $z_1$ and $z_2$ are the (integer) charges of the reacting ions and $r$ is the hard sphere collision distance (internuclear distance). The latter term, $r$, is approximated as the sum of the radii of the reacting ions, $r_1 + r_2$.[18] The term $\mu$ is the total ionic strength. It is defined as $\mu = \frac{1}{2}\sum n_i z_i^2$ for electrolyte solutions that contain molar concentrations, $n_i$, of species $i$ of charge $z$. The constant $\alpha$ is dimensionless and equal to 0.509, and for $r$ in units of cm, $\beta = 3.29 \times 10^7 \, cm^{1/2}/mol^{1/2}$. Finally, $\log k$ in Equation 1.30 refers to $\log_{10} k$, rather than to the natural logarithm, $\ln k$. This deserves mention because in many published reports, $\log k$ is (inappropriately) used to refer to $\ln k$.

Typically, $\log k$ (i.e., $\log_{10} k$) is plotted ($y$-axis) as a function of $\mu^{1/2}/(1 + \beta r \mu^{1/2})$ ($x$-axis). If the result is a straight line, its slope should be equal to a simple function of the charge product, that is, $2z_1 z_2 \alpha$, and its $y$-intercept gives $\log k_0$, the log of the rate constant at the zero ionic strength limit. The constant $k_0$ is the ionic strength-independent value of the rate constant and can be treated as a fundamental parameter of an electron transfer reaction.

**1.3.1.2  Dimensional Analysis**   The constant $\beta$ in Equation 1.31 is the "reciprocal Debye radius."[11,16] Dimensional analysis of this term is instructive because it involves a number of often needed constants and occurs frequently in a variety of contexts.

$$\beta \equiv 8\pi N e^2 / 1000 D_s kT \tag{1.31}$$

The following values, constants, and conversion factors apply:

Conditions: 298K (25°C) in water
$N$ = Avogadro's number ($6.022 \times 10^{23} \, mol^{-1}$)
$e$ = electron charge ($4.803 \times 10^{-10}$ electrostatic units (esu) or StatC)
$D_s$ = static dielectric constant (78.4) (water at 298K)
$k$ = Boltzmann constant ($1.3807 \times 10^{-16} \, erg/K$)

$$1 \, StatC^2/cm = 1 \, erg \tag{1.32}$$

Evaluation of $\beta$ is as follows:

$$\beta = \left( \frac{8\pi N e^2}{1000 D_s \, kT} \right)^{1/2} \tag{1.33}$$

$$\beta = \left( \frac{8\pi (6.022 \times 10^{23} \, mol^{-1})(4.803 \times 10^{-10} \, StatC)^2}{1000(78.4)(1.3807 \times 10^{-16} erg/K)298K} \right)^{1/2} \tag{1.34}$$

$$\beta = \left( \frac{8\pi(6.022 \times 10^{23}\ \text{mol}^{-1})(4.803 \times 10^{-10}\ \text{StatC})^2}{1000(78.4)(1.3807 \times 10^{-16}\text{erg/K})298\text{K}} \frac{(\text{erg cm})}{\text{StatC}^2} \right)^{1/2} \tag{1.35}$$

$$\beta = 3.29 \times 10^7\ \text{cm}^{1/2}/\text{mol}^{1/2} \tag{1.36}$$

In the Davies equation (Equation 1.30), $\beta$ is multiplied by $r$. If this distance is in cm, one obtains

$$\beta \cdot \text{cm} = (3.29 \times 10^7\ \text{cm}^{1/2}/\text{mol}^{1/2})\text{cm} \tag{1.37}$$

which is equivalent to Equation 1.38.

$$\beta \cdot \text{cm} = 3.29 \times 10^7 \left( \frac{\text{cm}^3}{\text{mol}} \right)^{1/2} \tag{1.38}$$

The units in Equation 1.38 can be viewed as $(\text{vol/mol})^{1/2}$ and are cancelled (i.e., reduced to unity) when multiplied by the units of $\mu^{1/2}$ $(\text{mol/vol})^{1/2}$ in the Davies equation (Equation 1.30).

Note here that the denominator in Equation 1.33 (definition of $\beta$) is multiplied by a factor of 1000. This is needed to "scale up" from $\text{cm}^3$ to L, so that, for $r$ in cm and $\mu$ in mol/L, the units associated with the product $\beta r \mu$ exactly cancel one another. Similarly, for $r$ in units of Å and $\mu$ in units of mol/L, $\beta = 0.329$ Å$^{1/2}/\text{mol}^{1/2}$.

In published articles, $\beta$ is often presented as dimensionless (e.g., as 0.329), or with units of cm$^{-1}$ or Å$^{-1}$. The units of cm$^{-1}$ are obtained if the correction factor of 1000 in the denominator of Equation 1.33 is assigned units of cm$^3$. Once the final units are in cm$^{-1}$, these can be converted to Å$^{-1}$. These options can be disconcerting to those new to the use of these models. For practical purposes, however, one only needs to know that, for $\mu$ in units of molarity (M), $\beta = 0.329$ for $r$ in units of Å and $3.29 \times 10^7$ for $r$ in units of cm.

### *1.3.1.3   Literature Example: Reaction Between* $\alpha$-PW$_{12}$O$_{40}^{4-}$ *and* $\alpha$-PW$_{12}$O$_{40}^{3-}$

In a detailed investigation of electron self-exchange between Keggin heteropolytungstate anions in water, Kozik and Baker used line broadening of $^{31}$P NMR signals to determine the rates of electron exchange between $\alpha$-PW$_{12}$O$_{40}^{3-}$ and one-electron reduced $\alpha$-PW$_{12}$O$_{40}^{4-}$ (Figure 1.8) and between $\alpha$-PW$_{12}$O$_{40}^{4-}$ and the two-electron reduced anion $\alpha$-PW$_{12}$O$_{40}$.[5-31,40] The W ions in the parent anion $\alpha$-PW$_{12}$O$_{40}^{3-}$ are in their highest $+6$ oxidation state (d$^0$ electron configuration).

Structurally, $\alpha$-PW$_{12}$O$_{40}^{3-}$, which is 1.12 nm in diameter,[41] may be viewed as a tetrahedral phosphate anion, P$^{V}$O$_4^{3-}$, encapsulated within a neutral, also tetrahedral, $\alpha$-W$^{VI}_{12}$O$_{36}^{0}$ shell ("clathrate" model[42-44]). According to this model, the P$^{V}$O$_4^{3-}$ anion in the one-electron reduced anion, $\alpha$-PW$_{12}$O$_{40}^{4-}$, is located at the center of a *negatively charged* W$_{12}$O$_{36}^{-}$ shell,[45] which contains a single d (valence) electron. Moreover, the single valence electron is not localized at any single W atom, but is

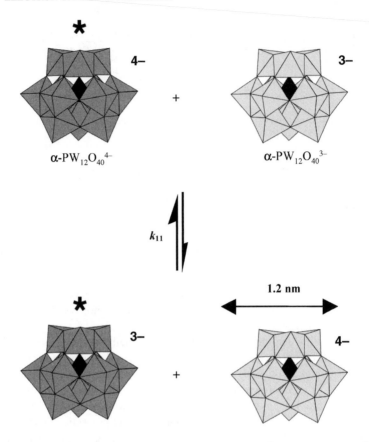

**FIGURE 1.8** Electron self-exchange between $\alpha\text{-PW}_{12}O_{40}{}^{4-}$ and $\alpha\text{-PW}_{12}O_{40}{}^{3-}$. The $\alpha$-Keggin anions are shown in coordination polyhedron notation. Each anion is 1.12 nm in diameter and possesses tetrahedral ($T_d$) symmetry. In each anion, the 12 W addendum atoms are at the center of $WO_6$ polyhedra that each have $C_{4v}$ symmetry. At the center of each cluster (shaded) is a tetrahedral phosphate oxoanion, $PO_4{}^{3-}$.

rapidly exchanged between all 12 chemically equivalent W centers. The rate of intramolecular exchange at 6K is $\sim10^8\,\text{s}^{-1}$ and considerably more rapid[46,47] than most electron transfer reactions carried out at near-ambient temperatures between the reduced anion, $\alpha\text{-PW}_{12}O_{40}{}^{4-}$, and electron acceptors.

Kozik and Baker combined the acid forms of the two anions, $\alpha\text{-H}_3\text{PW}_{12}O_{40}$ and $\alpha\text{-H}_4\text{PW}_{12}O_{40}$ (1.0 mM each), in water using a range of ionic strengths ($\mu$) from 0.026 to 0.616 M.[31] Ionic strength values were adjusted by addition of HCl and NaCl (pH values ranged from 0.98 to 1.8). Under these conditions, the anions are present in their fully deprotonated, "free anion" forms, $\alpha\text{-PW}_{12}O_{40}{}^{3-}$ and $\alpha\text{-PW}_{12}O_{40}{}^{4-}$. Observed rate constants, $k_{obs}$, were fitted to the Davies equation (Equation 1.30). An internuclear distance of $r = 11.2\,\text{Å}$, or twice the ionic radius of the Keggin anions, was used. At 25°C in water, $\alpha = 0.509$ (dimensionless) and $\beta = 3.29 \times 10^7\,\text{cm/mol}^{1/2}$.

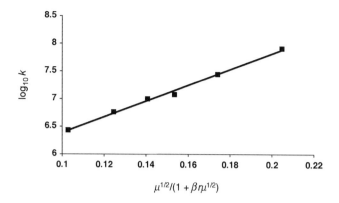

**FIGURE 1.9**   Plot of $\log k$ as a function of $\mu^{1/2}/(1 + \beta r \mu^{1/2})$ (from the Davies equation, Equation 1.30). The slope of the line (equal to $2\alpha z_1 z_2$) gives a charge product $z_1 z_2$ of 14.3.

For the units to "work out," $r$ must be converted to $1.12 \times 10^{-7}$ cm. Using Equation 1.30, $\log k$ was plotted as a function of $\mu^{1/2}/(1 + \beta r \mu^{1/2})$. A linear relationship was observed ($R^2 = 0.998$), whose slope (equal to $2\alpha z_1 z_2$) gave a charge product of $z_1 z_2 = 14.3$ (Figure 1.9). The theoretical charge product is 12. Linearity and comparison to the theoretical charge product are both important considerations in evaluating whether an electron transfer reaction between charged species obeys electrolyte theory to an extent sufficient for proceeding with use of the data in the Marcus model. In the present case, few would argue that the system fails to comply with electrolyte theory.

The linearity and close-to-theoretical slope in Figure 1.9 was surprising because Equation 1.30 was derived for univalent ions at low ionic strengths (up to 0.01 M). Agreement at much higher ionic strengths (greater than 0.5 M) was attributed to the fact that POM anions, "owing to the very pronounced inward polarization of their exterior oxygen atoms, have extremely low solvation energies and very low van der Waals attractions for one another."[48]

In many published examples of careful work, some models fail, while others (including those with empirical corrections) give better fits. In those cases, some judgment is required to assess whether the Marcus model might be used. The most often encountered reasons for failure to comply with these models are the presence of alternative mechanistic pathways and significant ion association between electrolyte ions and the charged species involved in the electron transfer reaction. Ion association is discussed in more detail at the end of this chapter.

### 1.3.2   Self-Exchange Rate Expression

For self-exchange reactions such as that in Figure 1.8, Equation 1.18 applies. This is obtained by setting the free energy terms in Equation 1.12 equal to zero. In most cases, the work term $W(r)$ can be calculated. This is the energy required to bring the reactants from effectively infinite separation to within collision distance $r$, approximated as the sum of the radii of the reacting ions, $r_1 + r_2$. The reorganization energy $\lambda$ is more difficult to calculate. This is because it requires detailed knowledge of all the changes in bond lengths and angles required to reach the transition state for electron transfer (see Equation 1.9) and this information is usually not available. In practice, therefore, self-exchange reactions are most commonly carried out as a means for determining the reorganization energies. This fundamental parameter of the self-exchange reaction can then be used to determine the inherent reorganization barriers associated with other electron transfer reactions (cross-reactions) of interest. In this sense, the species involved in the self-exchange reaction can be deployed as physicochemical "probes."[17,32,33]

The work term $W(r)$, Equation 1.39, is as shown in Equation 1.4, but with the subscript on the collision distance $r$ modified to indicate a self-exchange reaction.

$$W(r) = \frac{z_1 z_2 e^2}{D r_{11}} \exp(-\chi r_{11}) \tag{1.39}$$

Equation 1.39 is commonly written in a more convenient form (Equation 1.40).

$$W(r) = \frac{z_1 z_2 e^2}{D r_{11}(1 + \beta r \,\mu^{1/2})} \tag{1.40}$$

The dimensional analysis provided in Equations 1.33–1.38 applies here as well, and the electron charge, $e$, needed here is equal to $4.803 \times 10^{-10}$ StatC. To use Equation 1.40, a conversion factor must be added, as shown in Equation 1.41.

$$W(r) = \frac{z_1 z_2 (4.803 \times 10^{-10} \text{ StatC})^2}{D r_{11}(1 + \beta r \,\mu^{1/2})} \left(1.439 \times 10^{13} \frac{\text{kcal/mol}}{\text{StatC}^2/\text{cm}}\right) \tag{1.41}$$

To evaluate this expression, $r$ is given in units of cm. Because $z_1$, $z_2$, and the static dielectric constant $D$ are dimensionless, and (as noted above) the units of the term $\beta r \mu^{1/2}$ exactly cancel one another, the units in Equation 1.41 reduce to kcal/mol.

A dramatic example of the effect of charge and $W(r)$ on rate constants involves self-exchange between $\alpha$-AlW$_{12}$O$_{40}{}^{5-}$ and the one-electron reduced anion $\alpha$-AlW$_{12}$O$_{40}{}^{6-}$. At an ionic strength, $\mu$, of 175 mM, the rate constant for this reaction is $3.34 \times 10^2$ M$^{-1}$ s$^{-1}$. By comparison, the rate constant for self-exchange between $\alpha$-PW$_{12}$O$_{40}{}^{3-}$ and the 1e$^-$-reduced anion $\alpha$-PW$_{12}$O$_{40}{}^{4-}$ at $\mu = 175$ mM is $2.28 \times 10^7$ M$^{-1}$ s$^{-1}$. Work terms and reorganization energies for the two reactions at this ionic strength are $W(r) = 4.46$ and 1.78 M$^{-1}$ s$^{-1}$ for charge products, respectively, of 30 and 12. (Also responsible in part for the difference in self-exchange rate constants is the decrease in reorganization energies from 8.8 to 6.1 kcal/mol between the slower and faster reactions.)

By taking the natural logarithm of each side of Equation 1.18, one obtains Equation 1.42 (reproduced here from Section 1.2). The subscript on $k$ is a pair of

1's, used to indicate that this rate constant is for a self-exchange reaction.

$$\ln k_{11} = \ln Z - \frac{W(r)}{RT} - \frac{\lambda}{4RT} \tag{1.42}$$

Hence, by measuring the rate of electron self-exchange, one can readily use the Marcus model to calculate the reorganization energy $\lambda$. To evaluate this, a collision frequency of $Z = 10^{11} \, M^{-1} \, s^{-1}$ is generally used, and the temperature is given in Kelvin.

### 1.3.2.1 Literature Examples: $Ru^{III}(NH_3)_6^{3+} + Ru^{II}(NH_3)_6^{2+}$ and $O_2 + O_2^{-}$

A well-known self-exchange reaction is that between $Ru^{III}(NH_3)_6^{3+}$ and $Ru^{II}(NH_3)_6^{2+}$.[38,49] At 25°C and $\mu = 0.1 \, M$, the experimentally determined self-exchange rate constant $k_{11}$ is $4 \times 10^3 \, M^{-1} \, s^{-1}$.

For $D = 78.4$, the effective radius of the Ru complexes equal to $3.4 \times 10^{-8} \, cm$ (i.e., $r = 6.8 \times 10^{-8} \, cm$), and the other constants defined as shown above, $W(r)$ is calculated as shown in Equation 1.43.

$$W(r) = \frac{3 \cdot 2(4.803 \times 10^{-10} \, \text{StatC})^2}{78.4 \cdot (6.8 \times 10^{-8} \, cm)(1 + 3.29 \times 10^7 \, cm^{1/2}/mol^{1/2}(6.8 \times 10^{-8} \, cm)(0.1 \, mol/L)^{1/2})}$$

$$\cdot \left(1.439 \times 10^{13} \, \frac{kcal/mol}{StatC^2/cm}\right) = 2.19 \, kcal/mol \tag{1.43}$$

The ionic strength $\mu$ in units of mol/L is equivalent to units of mmol/cm$^3$. Note that $\beta$ has units of cm$^{1/2}$/mol$^{1/2}$, such that the units of the product $\beta r \mu$ cancel one another due to the factor of 1000 in the denominator of $\beta$ itself (i.e., as defined in Equation 1.33).

By solving Equation 1.42 for $\lambda$, one obtains

$$\lambda = 4(RT \ln Z - RT \ln k_{11} - W(r)) \tag{1.44}$$

Substitution of $Z = 10^{11}$, $D = 78.4$, $R = 1.987 \times 10^{-3} \, kcal/mol$, $k_{11} = 4 \times 10^3 \, M^{-1} \, s^{-1}$, and $W(r) = 2.19 \, kcal/mol$ into Equation 1.44 gives $\lambda = 31.6 \, kcal/mol$. In this reaction, the work required to bring the charged reactants into close proximity is considerably smaller than the reorganization energy needed to reach the transition state for electron transfer.

In the above example, the reorganization energy $\lambda$ is the *total reorganization energy* and includes both inner-sphere and outer-sphere components, $\lambda_{in}$ and $\lambda_{out}$, discussed earlier in this chapter. Once the total reorganization energy, $\lambda_{total}$, is known, $\lambda_{in}$ and $\lambda_{out}$ can be calculated using Equation 1.45.

$$\lambda_{total} = \lambda_{in} + \lambda_{out} \tag{1.45}$$

As noted earlier, information needed to calculate $\lambda_{in}$ (Equation 1.9) is usually not readily available. However, $\lambda_{out}$ is easily calculated using Equation 1.10. For self-exchange reactions, $r_1 = r_2$ and $r_{12} = 2\,r_1$, and for a one-electron process, $\Delta e = e$. Thus, Equation 1.10 reduces to Equation 1.46.

$$\lambda_{out} = e^2 \left[\frac{1}{r_1}\right] \left[\frac{1}{\eta^2} - \frac{1}{D_s}\right] \tag{1.46}$$

Here, $\eta$ and $D_s$ are the refractive index and static dielectric constant of the solvent. In water at 298K, $\eta = 1.33$ and $D_s = 78.4$.

An important application of Equations 1.45 and 1.46 involves analysis of electron exchange between dioxygen ($O_2$) and the superoxide radical anion ($O_2^-$) in water. Lind and Merényi determined the rate constant for this reaction by reacting $^{32}O_2$ with isotopically labeled $O_2^-$ (Equation 1.47).[50] The labeled superoxide anion was generated by $\gamma$-irradiation of $^{36}O_2$.

$$^{32}O_2 + {}^{36}O_2^- \rightarrow {}^{32}O_2^- + {}^{36}O_2 \tag{1.47}$$

They obtained a rate constant of $450 \pm 150\,\mathrm{M^{-1}\,s^{-1}}$. Because the charge product for this reaction, $z_1 z_2$, is equal to zero (the charge on $O_2$ is zero), $W(r)$ in Equation 1.44 is equal to zero, and that equation reduces to

$$\lambda_{total} = 4RT \left( \ln \frac{Z}{k_{11}} \right) \tag{1.48}$$

The experimentally determined rate constant gave $\lambda_{total} = 45.5\,\mathrm{kcal/mol}$. In this case, it was possible to estimate $\lambda_{in}$ computationally (which involves lengthening of the O–O bond), and then use Equation 1.45 to determine $\lambda_{out}$. They estimated that $\lambda_{in} = 15.9\,\mathrm{kcal/mol}$, which left $29.6\,\mathrm{kcal/mol}$ for $\lambda_{out}$. Next, they used Equation 1.46 to calculate the "effective" radius of $O_2$ (i.e., $r_1/2$) and obtained a value of $3\,\mathring{A}$. More recent work suggests that this seemingly large value might reflect orientational restrictions imposed on collisions between the nonspherical reactants, $O_2$ and $O_2^-$.[51]

### 1.3.3 The Marcus Cross-Relation

Rate constants for outer-sphere electron transfer reactions that involve net changes in Gibbs free energy can be calculated using the Marcus cross-relation (Equations 1.24–1.26). It is referred to as a "cross-relation" because it is derived from expressions for two different self-exchange reactions.

***1.3.3.1  Derivation of the Marcus Cross-Relation***    The cross-relation is derived algebraically by first assuming that the reorganization energy for the "cross" reaction is the average of the reorganization energies associated with the two self-exchange reactions involved. To clarify this, consider the two self-exchange reactions in Equations 1.49 and 1.50. These reactions are, respectively, assigned rate constants of $k_{11}$ and $k_{22}$, which are associated with reorganization energies $\lambda_{11}$ and $\lambda_{22}$. The reorganization energies are inherent properties of these exchanging pairs.

$$A_{red} + A^*_{ox} \rightleftharpoons A_{ox} + A^*_{red} \tag{1.49}$$

$$B_{red} + B^*_{ox} \rightleftharpoons B_{ox} + B^*_{red} \tag{1.50}$$

Associated with these self-exchange reactions are two "cross" reactions (Equations 1.51 and 1.52). Unlike the self-exchange reactions, the cross-reactions are functions of (almost always) nonzero Gibbs free energies $\Delta G^\circ_{12}$ and $\Delta G^\circ_{21}$ (the subscripts indicate a "cross" reaction and are arbitrarily assigned here to one reaction and its reverse). Once the rate constants of the self-exchange reactions, and the Gibbs free energy for the cross-reactions, are known, the MCR can be used to predict the rate constants, $k_{12}$ or $k_{21}$, for each cross-reaction.

$$A_{red} + B_{ox} \rightleftharpoons A_{ox} + B_{red}, \quad \text{rate constant} = k_{12} \tag{1.51}$$

$$A_{ox} + B_{red} \rightleftharpoons A_{red} + B_{ox}, \quad \text{rate constant} = k_{21} \tag{1.52}$$

To derive the MCR, it is necessary to assume that the reorganization energies for the reactions shown in Equations 1.51 and 1.52 are both equal to the mean of $\lambda_{11}$ and $\lambda_{22}$ (Equation 1.53).

$$\lambda_{12} = \frac{1}{2}(\lambda_{11} + \lambda_{22}) \tag{1.53}$$

Next, Equation 1.53 is algebraically combined with equations that describe the dependence of $k_{11}$ and $k_{22}$ on $\lambda_{11}$ and $\lambda_{22}$ (Equations 1.54 and 1.55) and the dependence of $k_{12}$ on $\lambda_{12}$ and $\Delta G^{\circ\prime}_{12}$ (Equation 1.56).

$$\ln k_{11} = w_{11} + \frac{\lambda_{11}}{4} \tag{1.54}$$

$$\ln k_{22} = w_{22} + \frac{\lambda_{22}}{4} \tag{1.55}$$

$$\ln k_{12} = W(r)_{12} + \frac{\lambda_{12}}{4}\left(1 + \frac{\Delta G^{\circ\prime}_{12}}{\lambda_{12}}\right)^2 \tag{1.56}$$

The Gibbs energy term in Equation 1.56 is not the standard Gibbs free energy, $\Delta G^\circ_{12}$, for the cross-reaction in Equation 1.51. Rather, it is the "corrected" Gibbs free energy, $\Delta G^{\circ\prime}_{12}$: the difference in energy between the successor and precursor complexes in Scheme 1.2.

The corrected Gibbs free energy is related to $\Delta G^\circ_{12}$ by the relation in Equation 1.57, where $w_{ij}$ are the Coulombic energies of formation of individual (often short-lived)

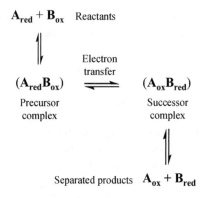

**SCHEME 1.2**

association complexes. The $w_{ij}$ terms are retained in the MCR and will be further clarified below through a worked example.

$$\Delta G_{12}^{o'} = \Delta G_{12}^{o} + w_{21} - w_{12} \tag{1.57}$$

Substituting Equation 1.58 into Equation 1.57 gives Equation 1.59.

$$\Delta G_{12}^{o} = -RT \ln K_{12} \tag{1.58}$$

$$\Delta G_{12}^{o'} = -RT \ln K_{12} + w_{21} - w_{12} \tag{1.59}$$

Using Equations 1.53–1.56 and 1.59, approximately 30 algebraic steps lead to Equations 1.60–1.62 (reproduced here for convenience from Section 1.2). (The equilibrium constant $K_{12}$, rather than a Gibbs free energy term, appears in Equation 1.60, and it is incorporated by substitution of Equation 1.59 into Equation 1.56.)

$$k_{12} = (k_{11}k_{22}K_{12}f_{12})^{1/2} C_{12} \tag{1.60}$$

where

$$\ln f_{12} = \frac{1}{4} \frac{(\ln K_{12} + (w_{12} - w_{21})/RT)^2}{\ln(k_{11}k_{22}/Z^2) + (w_{11} - w_{22})/RT} \tag{1.61}$$

and

$$C_{12} = \exp[-(w_{12} + w_{21} - w_{11} - w_{22}/2RT)] \tag{1.62}$$

The use of this relation requires that Equation 1.53 be valid. Equation 1.53, in turn, is a good approximation for the reorganization energy of the cross-reaction (either

Equation 1.51 or 1.52) only when the reacting species are spherical in shape and identical to one another in size. One of the model's strengths is that reasonable results (within $\sim 1$ order of magnitude between calculated and experimental values) can often be obtained for reactions between electron donors and acceptors that do not meet these criteria. This is fortunate because few reactions do.

### 1.3.3.2 Literature Examples: $\alpha$-AlW$_{12}$O$_{40}$$^{6-}$ + $\alpha$-PW$_{12}$O$_{40}$$^{4-}$ and $\alpha$-PW$_{12}$O$_{40}$$^{4-}$ + O$_2$

The MCR is used here to calculate the rate constant, $k_{12}$, for electron transfer from to $\alpha$-AlW$_{12}$O$_{40}$$^{6-}$ (one-electron reduced) to $\alpha$-PW$_{12}$O$_{40}$$^{4-}$ (one-electron reduced) to give $\alpha$-PW$_{12}$O$_{40}$$^{5-}$ (the two-electron reduced ion) (Equation 1.63).[32]

$$\alpha\text{-AlW}_{12}O_{40}{}^{6-} + \alpha\text{-PW}_{12}O_{40}{}^{4-} \rightarrow \alpha\text{-AlW}_{12}O_{40}{}^{5-} + PW_{12}O_{40}{}^{5-} \qquad (1.63)$$

The MCR will be used below to calculate the rate constant, $k_{12}$(calc), for this reaction and to compare that value with the experimentally determined rate constant, $k_{12}$(exp). For this, $k_{12}$(exp) was determined as shown immediately below. In this example, "zero ionic strength" rate constants are used. These are obtained from plots of rate versus functions of ionic strength derived from electrolyte theory. If these plots give straight lines with slopes near theoretical (i.e., slopes that give the actual charge product, $z_1 z_2$), then extrapolation to zero ionic strength is defensible. As will be shown in a subsequent example, however, the use of zero ionic strength rate constants is not required for use of the MCR.

To determine $k_{12}$(exp), solutions of $\alpha$-AlW$_{12}$O$_{40}$$^{6-}$ and $\alpha$-PW$_{12}$O$_{40}$$^{4-}$ (the latter in large molar excess) were mixed in a stopped-flow apparatus. The change in absorbance with time (determined by UV-Vis spectroscopy) was exponential, and pseudo-first-order rate constants, $k_{obs}$, were determined from each absorbance *versus* time curve (to $\sim 100\%$ completion of reaction). Plots of these rate constants as a function of initial $\alpha$-PW$_{12}$O$_{40}$$^{4-}$ concentration at three ionic strength values gave three straight lines (Figure 1.10). The slopes of these lines are the bimolecular rate constants, $k_{12}$, for the reaction at each of the three ionic strength values used. The range of experimentally useful ionic strength values was limited by practical considerations, but adequate.

The three bimolecular rate constants, $k_{12}$, from Figure 1.10, were plotted as a function of ionic strength using the extended Davies equation (Figure 1.11). The slope ($R^2 = 0.9997$) gave a charge product ($z_1 z_2$) of $23 \pm 1$, within experimental uncertainty of the theoretical value of 24. Extrapolation to zero ionic strength gave $k_{22}^0$(exp) $= 17 \pm 2\,\text{M}^{-1}\,\text{s}^{-1}$ (the superscript "0" is used to indicate that this value refers to zero ionic strength).

The MCR will now be used to obtain $k_{12}$(calc) from the two relevant self-exchange reactions in Equations 1.64 and 1.65:

$$^*\alpha\text{-AlW}_{12}O_{40}{}^{6-} + \alpha\text{-AlW}_{12}O_{40}{}^{5-} \rightarrow {}^*\alpha\text{-AlW}_{12}O_{40}{}^{5-} + \alpha\text{-AlW}_{12}O_{40}{}^{6-} \qquad (1.64)$$

$$^*\alpha\text{-PW}_{12}O_{40}{}^{5-} + \alpha\text{-PW}_{12}O_{40}{}^{4-} \rightarrow {}^*\alpha\text{-PW}_{12}O_{40}{}^{4-} + \alpha\text{-PW}_{12}O_{40}{}^{5-} \qquad (1.65)$$

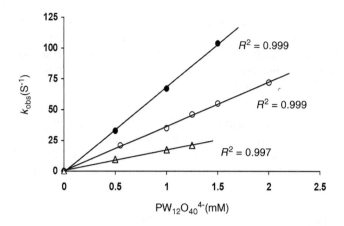

**FIGURE 1.10** Rate constants, $k_{obs}$, for electron transfer from $\alpha$-AlW$_{12}$O$_{40}{}^{6-}$ to $\alpha$-PW$_{12}$O$_{40}{}^{4-}$ (present in large molar excess) at three initial ionic strength values ($\mu$, adjusted by addition of NaCl): 65 mM (($\triangle$), 97 mM (o), and 140 mM ($\bullet$) in 50 mM phosphate buffer (pH 2.15) at 25°C.

Rate constants for both these reactions have been determined, respectively, by [29] Al and [31] P NMR spectroscopy.

The reported zero ionic strength rate constant for self-exchange between $\alpha$-AlW$_{12}$O$_{40}{}^{6-}$ and $\alpha$-AlW$_{12}$O$_{40}{}^{5-}$ (Equation 1.64) is $k_{11}^0 = 6.5 \pm 1.5 \times 10^{-3}\,M^{-1}\,s^{-1}$ and that for the reaction in Equation 1.65 is $k_{22}^0 = 1.6 \pm 0.3 \times 10^2\,M^{-1}\,s^{-1}$.[31] (From here onward in this example, the superscript "0", which indicates values at the zero ionic strength limit, is omitted to avoid confusion, particularly as the equations and general treatment are valid for any ionic strength.) Gibbs free energies for cross-reactions, $\Delta G°$, are most often obtained from electrochemical data. For the cross-

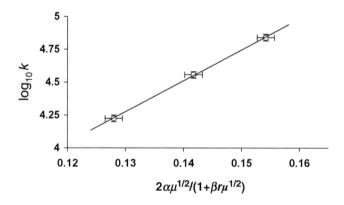

**FIGURE 1.11** Plot of rate constants, $k_{12}$, for electron transfer from $\alpha$-AlW$_{12}$O$_{40}{}^{6-}$ to PW$_{12}$O$_{40}{}^{4-}$ as a function of ionic strength according to the Davies equation (Equation 1.30).

relation, the reduction potentials of the individual self-exchange reactions are used to calculate the equilibrium constant for the cross-reaction, $K_{12}$. Reported reduction potentials (relative to the normal hydrogen electrode, NHE) are $-130 \pm 5$ and $-10 \pm 5$ mV, respectively, for the self-exchanging redox couples $\alpha$-AlW$_{12}$O$_{40}{}^{5-}$/$\alpha$-AlW$_{12}$O$_{40}{}^{6-}$ and $\alpha$-PW$_{12}$O$_{40}{}^{5-}$/$\alpha$-PW$_{12}$O$_{40}{}^{4-}$. The equilibrium constant is calculated by rearranging Equation 1.58 to give Equation 1.66.

$$K_{12} = \exp\left(-\frac{\Delta G_{12}^{\circ}}{RT}\right) \tag{1.66}$$

$\Delta G_{12}^{\circ}$ is calculated from electrochemical data using the definition of the standard potential (Equation 1.67), where $n$ is the charge number of the reaction, that is, the number of electrons involved, and $F$ is the Faraday constant. In convenient units, $F = 23.06$ kcal/(mol V).

$$\Delta G_{12}^{\circ} = -nFE^{\circ} \tag{1.67}$$

Thus, standard potential for the reaction in Equation 1.63 is equal to the reduction potential of the species reduced in the reaction, minus the potential of the electron donor, that is, $-10 - (-130)$ mV, which is equal to $+0.12$ V. Hence, $\Delta G_{12}^{\circ} = 2.77$ kcal/mol (from Equation 1.58). By using Equation 1.66 with $R = 1.987 \times 10^{-3}$ kcal/(mol K) and $T = 298$K, $K_{12} = 107$.

We now have experimentally determined values for $k_{11}$, $k_{22}$, and $K_{12}$. The next step in applying the MCR is to evaluate the terms $\ln f_{12}$ and $C_{12}$. In reactions between species whose charges are small or similar to one another, it is often possible to obtain a reasonably good result by setting both $f_{12}$ and $C_{12}$ equal to 1, such that the MCR reduces to Equation 1.68.

$$k_{12} = (k_{11}k_{22}K_{12})^{1/2} \tag{1.68}$$

Using Equation 1.68, $k_{12}^{0}(\text{calc}) = [(6.5 \times 10^{-3}\text{M}^{-1}\text{s}^{-1})(1.6 \times 10^{2}\,\text{M}^{-1}\,\text{s}^{-1})(107)]^{1/2} = 10.5$ M$^{-1}$ s$^{-1}$. For calculations of this type, the agreement between this value and the experimental one ($17 \pm 2$ M$^{-1}$ s$^{-1}$) is quite good.

This result can be improved upon by including $\ln f_{12}$ and $C_{12}$. For this, the $w_{ij}$ terms in Equations 1.61 and 1.62 must be evaluated, and the subscripts on the $w_{ij}$ terms correctly interpreted. Consider the cross-reaction as written in Equation 1.69:

$$\alpha\text{-AlW}_{12}\text{O}_{40}{}^{6-} + \alpha\text{-PW}_{12}\text{O}_{40}{}^{4-} \rightarrow \alpha\text{-AlW}_{12}\text{O}_{40}{}^{5-} + \text{PW}_{12}\text{O}_{40}{}^{5-} \tag{1.69}$$
$$\quad 1_{\text{red}} \qquad\qquad 2_{\text{ox}} \qquad\qquad\qquad 1_{\text{ox}} \qquad\qquad 2_{\text{red}}$$

The numbers 1 or 2 under each of the four species refer to the two self-exchange reactions (Equations 1.64 and 1.65), associated, respectively, with the rate constants $k_{11}$ and $k_{22}$. For the first self-exchange reaction (involving $\alpha$-AlW$_{12}$O$_{40}{}^{6-}$ and $k_{11}$), the electron donor is $1_{\text{red}}$ and its corresponding oxidized form is $1_{\text{ox}}$. For the

**TABLE 1.1    Notation Guide for Evaluation of $w_{ij}$ Terms in the MCR**

| $w_{ij}$ | Refers to | Evaluation of $z_i z_j$ | $z_i z_j$ |
|---|---|---|---|
| $w_{12}$ | $1_{red}$ and $2_{ox}$ | $(6-)(4-)$ | 24 |
| $w_{21}$ | $1_{ox}$ and $2_{red}$ | $(5-)(5-)$ | 25 |
| $w_{11}$ | $1_{red}$ and $1_{ox}$ | $(6-)(5-)$ | 30 |
| $w_{22}$ | $2_{red}$ and $2_{ox}$ | $(4-)(5-)$ | 20 |

self-exchange reaction involving $\alpha$-$PW_{12}O_{40}{}^{4-}$ and $k_{22}$ (Equation 1.65), the electron acceptor is $2_{ox}$ and the corresponding product of electron transfer is $2_{red}$.

The $w_{ij}$ terms are identical to $W(r)$ as defined in Equations 1.40 and 1.41. To evaluate them, it is necessary to know the charge products, $z_i z_j$, and distances of closest approach, $r_{ij}$. These can be obtained using Table 1.1.

An exactly analogous table could be constructed to assign the $r_{ij}$ values that should be used in the above $w_{ij}$ terms. For the Keggin anions, this is simplified by the fact that all the species in Equation 1.69 possess the same crystallographic radii of 5.6 Å. Therefore, all the $r_{ij}$ values, approximated as the sum of radii, are 11.2 Å or $1.12 \times 10^{-7}$ cm. The individual $w_{ij}$ terms are then evaluated as shown in Equation 1.43.

Use of these terms, and the necessary constants defined earlier in this chapter, in Equations 1.61 and 1.62, gives $f_{12} = 0.80$ and $C_{12} = 1.38$. Using these values in Equation 1.60, and including uncertainties in $k_{11}$ and $k_{22}$, one obtains $k_{12}$(calc) $13.0 \pm \sim 3$ kcal/mol, statistically identical to $k_{22}$(exp). This is unusually good agreement; depending on the reaction(s) involved, results that differ from one another by up to an order of magnitude are often viewed as being in "reasonable" agreement.

The MCR will now be used to calculate the rate of electron transfer, $k_{12}$, from $\alpha$-$PW_{12}O_{40}{}^{4-}$ (one-electron reduced) to $O_2$, the first step in the reaction shown in Equation 1.70.[33]

$$2\alpha\text{-}PW_{12}O_{40}{}^{4-} + O_2 + 2H^+ \rightarrow 2\alpha\text{-}PW_{12}O_{40}{}^{3-} + H_2O_2 \tag{1.70}$$

At pH 2, this reaction occurs via the following steps:

$$\alpha\text{-}PW_{12}O_{40}{}^{4-} + O_2 \rightarrow \alpha\text{-}PW_{12}O_{40}{}^{3-} + O_2^{\bullet-} \qquad (k_{12}, \text{slow}) \tag{1.71}$$

$$O_2^{\bullet-} + H^+ \rightarrow HO_2^{\bullet} \qquad \text{(fast)} \tag{1.72}$$

$$\alpha\text{-}PW_{12}O_{40}{}^{4-} + HO_2^{\bullet} \rightarrow \alpha\text{-}PW_{12}O_{40}{}^{3-} + HO_2^{\bullet-} \qquad \text{(fast)} \tag{1.73}$$

$$HO_2^{\bullet-} + H^+ \rightarrow H_2O_2 \qquad \text{(fast)} \tag{1.74}$$

The one-electron reduced anion $\alpha$-$PW_{12}O_{40}{}^{4-}$ absorbs fairly strongly in the visible region (at $\lambda_{max} = 700$ nm, $\varepsilon \sim 1.8 \times 10^3\,M^{-1}\,cm^{-1}$) and absorbance versus time data were used to determine the rate expression (Equation 1.75) and rate constant for oxidation of $\alpha$-PW$_{12}$O$_{40}{}^{4-}$ by $O_2$ at $\mu = 175$ mM.

$$-d[\alpha\text{-PW}_{12}O_{40}{}^{4-}]/dt = 2k_{12}[\alpha\text{-PW}_{12}O_{40}{}^{4-}][O_2], \quad k_{12}(\exp) = 1.35\text{M}^{-1}\,\text{s}^{-1} \tag{1.75}$$

The MCR will now be used to obtain $k_{12}$(calc). For this, four experimentally determined values are needed: $k_{11} = 2.28 \times 10^7\,\text{M}^{-1}\,\text{s}^{-1}$ for self-exchange between $\alpha$-PW$_{12}$O$_{40}{}^{4-}$ and $\alpha$-PW$_{12}$O$_{40}{}^{3-}$ at $\mu = 175$ mM, $k_{22} = 450\,\text{M}^{-1}\,\text{s}^{-1}$ for self-exchange between $O_2$ and $O_2{}^-$ (Equation 1.47), and the one-electron reduction potentials of PW$_{12}$O$_{40}{}^{3-}$ ($-0.255$ V versus normal hydrogen electrode, NHE) and $O_2$. Because $O_2{}^-$ is not protonated in the rate-determining step (Equation 1.71), the pH-independent reduction potential (i.e., of the $O_2/O_2{}^-$ couple) is used. This value, when based on unit concentrations (i.e., 1 M $O_2$, rather than 1 atm $O_2$ above the solution), is $-0.16$ V.[52] $E°$ for the reaction is $-0.16 - (-0.255) = +0.095$ V, such that Equations 1.66 and 1.67 give $K_{12} = 9.0 \times 10^{-8}$.

Two major problems arise, however, in using the MCR itself and in evaluating the $w_{ij}$ terms in $f_{12}$ and $C_{12}$. First, $O_2$ is much smaller than $\alpha$-PW$_{12}$O$_{40}{}^{4-}$, such that the assumption (Equation 1.53) used to derive the MCR is no longer valid. The second problem is that $O_2$ is far from spherical in shape. Using one-half the O–O bond distance as an approximation for the "radius" of $O_2$, Lind and Merényi found, however, that agreement between calculated and experimental values for electron transfer from many electron donors, with sizes approximately two to three times that of $O_2$, could be obtained if the self-exchange rate constant for electron transfer between $O_2$ and $O_2{}^-$ was set to $\sim 2\,\text{M}^{-1}\,\text{s}^{-1}$ (much smaller than their experimentally determined value of $450\,\text{M}^{-1}\,\text{s}^{-1}$).[53] Using this approximation, Equations 1.60 and 1.61 gave $k_{12}$(calc) $= 1.1\,\text{M}^{-1}\,\text{s}^{-1}$, in close agreement with the experimentally determined value of $k_{12}$(exp) $= 1.35\,\text{M}^{-1}\,\text{s}^{-1}$.[33]

Lind and Merényi's arguments were further confirmed using a MCR modified to allow for differences in size between the reacting species.[51] While detailed analysis of that work is beyond the scope of this chapter, it deserves mention because of the importance of the redox chemistry of $O_2$ in chemistry, biology, and engineering. The modified MCR, along with the experimentally determined value of $450\,\text{M}^{-1}\,\text{s}^{-1}$ for self-exchange between $O_2$ and $O_2{}^-$, gave $k_{12}$(calc) $= 0.96\,\text{M}^{-1}\,\text{s}^{-1}$,[51] quite close to the experimental value of $k_{12}$(exp) $= 1.35\,\text{M}^{-1}\,\text{s}^{-1}$.

### 1.3.4 Ion Pairing between Electrolyte Ions and Electron Donors or Acceptors

For reactions of charged species in solution, ion pairing is probably the single largest impediment to reliable use of the Marcus model. Because of this, ion pairing likely plays an often unrecognized role in thousands of published articles. In many of these cases, disagreements between experimental and calculated values are subjects of extensive discussion, in which simple ion pairing is rarely proposed as the cause of the discrepancy.

Effects of ion pairing are covered in recommended reviews by Wherland[54] and Swaddle,[55] who have both made important experimental contributions in this area. In addition, valuable analyses of ion pairing and electron transfer have been provided by Marcus[56] and Savéant.[57]

Swaddle has noted that effects of ion paring are less pronounced in electron transfer reactions between cationic species and more problematic for reactions between anions. The best example of this is electron transfer by the ferri- and ferrocyanide anions, $Fe^{III}(CN)_6^{3-}$ and $Fe^{II}(CN)_6^{4-}$.

### *1.3.4.1 Solution Chemistry of Ferri- and Ferrocyanide*     In a study of the kinetics of electron self-exchange between $Fe(CN)_6^{3-}$ and $Fe(CN)_6^{4-}$, Shporer observed rate increases in aqueous solution corresponding to the series $H^+$ to $Cs^+$ and $Mg^{2+}$ to $Sr^{2+}$.[58] In alkaline aqueous solutions at constant ionic strength, Wahl[59] observed decreasing rates of self-exchange between $Fe(CN)_6^{3-}$ and $Fe(CN)_6^{4-}$ with increasing size of the tetraalkylammonium cations, $Me_4N^+$, $Et_4N^+$, $n\text{-}Pr_4N^+$, $n\text{-}Bu_4N^+$, and $n\text{-}Pent_4N^+$. Similar effects were observed in acetic acid.[60]

In analyzing the $Fe(CN)_6^{3-/4-}$ self-exchange reaction, Shporer[58] considered the possible effects of ion pairing on electrostatic terms, reorganization energies, and the mechanics of electron transfer itself. Three issues were addressed: (1) the effect of ion pairing on the Coulombic terms associated with bringing together similarly charged species (Coulombic term in Equation 1.4); (2) the effect of ion pairing on the reorganization energies $\lambda$ associated with electron transfer; and (3) the possibility that electron transfer might occur via the ion-paired cation, with the cation serving as a lower energy pathway between donor and acceptor species. Of these, the first and second are undoubtedly important, while the second is more difficult to assess.[59] With regard to the third proposal, work by Kirby and Baker suggests that alkali, alkaline earth, or tetraalkylammonium cations do not serve as conducting bridges for electron transfer between POM anions.[61] In addition, Swaddle has shown that in cation-catalyzed electron transfer reactions between charged species, negative volumes of activation are observed.[55] These are attributed to loss of solvent from the coordination sphere of the associated cations.

Here are some basic facts about the solution chemistries of $Fe^{III}(CN)_6^{3-}$ and $Fe^{II}(CN)_6^{4-}$: In 1935, Kolthoff and Tomsicek[62] showed that even in very dilute aqueous solution, $M_3Fe^{III}(CN)_6$ and $M_4Fe^{II}(CN)_6$ (M = alkali metal cation) are incompletely dissociated. This means that solutions of these salts are mixtures of species with different degrees of cation association and different charges (e.g., solutions of $K_3Fe(CN)_6$ contain substantial amounts of the $-2$ anion, $KFe(CN)_6^{2-}$). This is important for application of the Marcus model because charges determine collision rates, and the sizes of the ion-paired species and the reorganization energies associated with their electron transfer reactions are difficult to assess. Moreover, the chemical potentials of the ion pairs differ from those of the "free" anions, such that the $\Delta G°$ values of the electron transfer reactions are uncertain. Kolthoff and Tomsicek also showed that extents of ion pairing increased in the order: $Li^+ = Na^+ < NH_4^+ < K^+ < Rb^+ < Cs^+$.

In 2002, Swaddle used 18-crown-6 to sequester $K^+$ in solutions of $Fe(CN)_6^{4-}$ and $Fe(CN)_6^{3-}$ and obtained the most (perhaps only) reliable value for the self-exchange rate constant between non-ion-paired $Fe(CN)_6^{4-}$ and $Fe(CN)_6^{3-}$ anions.[63] This highlights another problem in the use of these anions for kinetic studies: the self-exchange rate constant of the $Fe(CN)_6^{3-}/Fe(CN)_6^{4-}$ pair is frequently used to estimate self-exchange rate constants of other species. Swaddle points out, however, that the self-exchange $Fe(CN)_6^{3-}/Fe(CN)_6^{4-}$ rate constants frequently cited in the literature actually describe rate constants for ion-paired species.

In 1935, Kolthoff and Tomsicek also reported that the fourth ionization constant for ferrocyanic acid, $H_4Fe^{II}(CN)_6$, is $5.6 \times 10^{-5}$ at 25°C.[64] This means that $HFe^{II}(CN)_6^{3-}$ is a weak acid. At pH values of $\sim 6$ and below, substantial concentrations of $HFe^{II}(CN)_6^{3-}$ are present. In 1962, Jordan and Ewing showed that $H_2Fe^{II}(CN)_6^{2-}$ is the dominant species at pH 1.[65] They also reported that ferricyanic acid $(H_3Fe^{III}(CN)_6)$ is much more extensively dissociated, such that it is effectively completely deprotonated to the free anion at pH values larger than 1.

Although this information has been available for decades, these complexes continue (at this writing) to be used as kinetic probes in chemistry, colloid, interfacial, and nanoscience and bioinorganic chemistry. Seemingly unaware of the problems involved, many authors find that changes in pH and ionic strength have dramatic effects on electron transfer rates and readily attribute those to the chemistries and electron transfer properties of the materials targeted for study, rather than to ion pairing of the $Fe^{III}(CN)_6^{3-}$ or $Fe^{II}(CN)_6^{4-}$ probes. This highlights how important it is to carefully assess the ionic strength dependences of electron transfer reactions of charged species before using the Marcus model.

## 1.4   CONCLUDING COMMENTS

The information included in this chapter includes an introduction to the Marcus theory for outer-sphere electron transfer and a practical guide to the application of its classical form to bimolecular reactions. The emphasis is on charged species, such as the inorganic and metallo-organic complexes typically encountered in inorganic chemistry. In conjunction with this emphasis on charged species, considerable efforts have been made to explain the effects of electrolytes on reaction rates, and practical guidance is provided on how to assess whether a particular reaction is well behaved in this context. As stated in the introduction, this chapter covers only a very small fraction of the many interesting and important reactions reported in the inorganic literature. However, it is hoped that the detailed step-by-step analysis of the application of the Marcus model to the sequential series of increasingly more complex cases covered here will enable the interested reader to apply this model to his or her own research and to better understand the results of these calculations when they are encountered in the chemical literature.

## REFERENCES

1. Mayer, J. M. *Annu. Rev. Phys. Chem.* **2004**, *55*, 363–390.
2. Eberson, L. E. *Electron Transfer Reactions in Organic Chemistry*; Springer: Berlin, 1987; Vol. 25.
3. Marcus, R. A. *Angew. Chem., Int. Ed.* 1993, *32*, 1111–1121.
4. Marcus, R. A.; Sutin, N. *Biochim. Biophys. Acta* 1985, *811*, 265–322.
5. Endicott, J. F., Ed. *Molecular Electron Transfer*; Elsevier: Oxford, 2004; Vol. 7.
6. Creutz, C. A.; Brunschwig, Eds. *Electron Transfer from the Molecular to the Nanoscale*; Elsevier: Oxford, 2004; Vol. 7.
7. Stanbury, D. M. *Adv. Inorg. Chem.* 2003, *54*, 351–393.
8. Balzano, V., Ed. *Electron Transfer in Chemistry*; Wiley-VCH: Weinheim, 2001; Vols. 1-5.
9. Smoluchowski, M. Z. *Phys. Chem.* 1917, *92*, 129–168.
10. Smoluchowski, M. *Phys. Z.* 1916, *17*, 557–571.
11. Debye, P. *Trans. Electrochem. Soc.* 1942, *82*, 265–272.
12. Marcus, R. A. *J. Phys. Chem.* 1963, *67*, 853–857.
13. Marcus, R. A.; Eyring, H. *Annu. Rev. Phys. Chem.* 1964, *15*, 155–196.
14. Taube, H.; Myers, H.; Rich, R. L. *J. Am. Chem. Soc.* 1953, *75*, 4118–4119.
15. Taube, H. *Angew. Chem., Int. Ed.* 1984, *23*, 329–339.
16. Pelizzetti, E.; Mentasti, E.; Pramauro, E. *Inorg. Chem.* 1978, *17*, 1688–1690.
17. Eberson, L. *J. Am. Chem. Soc.* 1983, *105*, 3192–3199.
18. Marcus, R. A. *J. Phys. Chem.* 1965, *43*, 679–701.
19. Marcus, R. A. *J. Chem. Phys.* 1956, *24*, 966–978.
20. Marcus, R. A. *J. Chem. Phys.* 1956, *24*, 979–988.
21. Chen, P.; Meyer, T. *J. Chem. Rev.* 1998, *98*, 1439–1477.
22. Marcus, R. A. *Discuss. Faraday Soc.* 1960, *29*, 21–31.
23. Miller, J. R.; Calcaterra, L. T.; Closs, G. L. *J. Am. Chem. Soc.* 1984, *106*, 3047–3049.
24. Wallace, W. L.; Bard, A. J. *J. Phys. Chem.* 1979, *83*, 1350–1357.
25. Miller, J. R.; Beitz, J. V.; Huddleston, R. K. *J. Am. Chem. Soc.* 1984, *106*, 5057–5068.
26. Wasielewski, M. R.; Niemczyk, M. P.; Svec, W. A.; Pewitt, E. B. *J. Am. Chem. Soc.* 1985, *107*, 1080–1082.
27. Debye, P.; Hückel, E. *Phys. Z.* 1923, *24*, 185–206.
28. Debye, P.; *Phys. Z.* 1924, *25*, 97–107.
29. Pethybridge, A. D.; Prue, J. E.; In *Inorganic Reaction Mechanisms Part II*; Edwards, J. O., Ed., Wiley: New York, 1972; Vol. *17*, pp 327–390.
30. Czap, A.; Neuman, N. I.; Swaddle, T. W. *Inorg. Chem.* 2006, *45*, 9518–9530.
31. Kozik, M.; Baker, L. C. W. *J. Am. Chem. Soc.* 1990, *112*, 7604–7611.
32. Geletii, Y. V.; Hill, C. L.; Bailey, A. J.; Hardcastle, K. I.; Atalla, R. H.; Weinstock, I. A. *Inorg. Chem.* 2005, *44*, 8955–8966.
33. Geletii, Y. V.; Hill, C. L.; Atalla, R. H.; Weinstock, I. A. *J. Am. Chem. Soc.* 2006, *128*, 17033–17042.
34. Geletii, Y. V.; Weinstock, I. A. *J. Mol. Catal. A: Chem.* 2006, *251*, 255–262.

35. Rubin, E.; Rodriguez, P.; Brandariz, I.; Sastre de Vicente, M. E. *Int. J. Chem. Kinet.* 2004, *36*, 650–660.

36. Stanbury, D. M., private communication.

37. Espenson, J. H. *Chemical Kinetics and Reaction Mechanisms*; 2nd edition, McGraw-Hill: New York, 1995.

38. Brown, G. M.; Sutin, N. *J. Am. Chem. Soc.* 1979, *101*, 883–892.

39. MacInnes, D. A. *The Principles of Electrochemistry*; 1st edition, Dover Publications: New York, 1961.

40. Weinstock, I. A. *Chem. Rev.* 1998, *98*, 113–170.

41. Weinstock, I. A.; Cowan, J. J.; Barbuzzi, E. M. G.; Zeng, H.; Hill, C. L. *J. Am. Chem. Soc.* 1999, *121*, 4608–4617.

42. Day, V. W.; Klemperer, W. G. *Science* 1985, *228*, 533–541.

43. López, X.; Maestre, J. M.; Bo, C.; Poblet, J.-M. *J. Am. Chem. Soc.* 2001, *123*, 9571–9576.

44. Maestre, J. M.; Lopez, X.; Bo, C.; Poblet, J.-M.; Casañ-Pastor, N. *J. Am. Chem. Soc.* 2001, *123*, 3749–3758.

45. López, X.; Poblet, J. M. *Inorg. Chem.* 2004, *43*, 6863–6865.

46. Kwak, W.; Rajkovic, L. M.; Stalick, J. K.; Pope, M. T.; Quicksall, C. O. *Inorg. Chem.* 1976, *15*, 2778–2783.

47. Kazansky, L. P.; McGarvey, B. R. *Coord. Chem. Rev.* 1999, *188*, 157–210.

48. Kozik, M.; Hammer, C. F.; Baker, L. C. W. *J. Am. Chem. Soc.* 1986, *108*, 7627–7630.

49. Meyer, T. J.; Taube, H. *J. Chem. Phys.* 1968, *7*, 2369–2379.

50. Lind, J.; Shen, X.; Merényi, G.; Jonsson, B. Ö. *J. Am. Chem. Soc.* 1989, *111*, 7654–7655.

51. Weinstock, I. A. *Inorg. Chem.* 2008, *47*, 404–406.

52. Sawyer, D. T.; Valentine, J. S. *Acc. Chem. Res.* 1981, *14*, 393–400.

53. Merényi, G.; Lind, J.; Jonsson, M. *J. Am. Chem. Soc.* 1993, *115*, 4945–4946.

54. Wherland, S. *Coord. Chem. Rev.* 1993, *123*, 169–199.

55. Swaddle, T. W. *Chem. Rev.* 2005, *105*, 2573–2608.

56. Marcus, R. A. *J. Phys. Chem. B* 1998, *102*, 10071–10077.

57. Savéant, J.-M. *J. Am. Chem. Soc.* 2008, *130*, 4732–4741.

58. Shporer, M.; Ron, G.; Loewenstein, A.; Navon, G. *Inorg. Chem.* 1965, *4*, 361–364.

59. Campion, R. J.; Deck, C. F.; King, P. J.; Wahl, A. C. *Inorg. Chem.* 1967, *6*, 672–681.

60. Gritzner, G.; Danksagmüller, K.; Gutmann, V. *J. Electroanal. Chem.* 1976, *72*, 177–185.

61. Kirby, J. F.; Baker, L. C. W. *J. Am. Chem. Soc.* 1995, *117*, 10010–10016.

62. Kolthoff, I. M.; Tomsicek, W. J. *J. Phys. Chem.* 1935, *39*, 945–954.

63. Zahl, A.; van Eldik, R.; Swaddle, T. W. *Inorg. Chem.* 2002, *41*, 757–764.

64. Kolthoff, I. M.; Tomsicek, W. J. *J. Phys. Chem.* 1935, *39*, 955–958.

65. Jordan, J.; Ewing, G. J. *Inorg. Chem.* 1962, *1*, 587–591.

# 2 Proton-Coupled Electron Transfer in Hydrogen and Hydride Transfer Reactions

SHUNICHI FUKUZUMI

## 2.1 INTRODUCTION

A number of redox reactions involve cleavage of C–H bonds, which proceeds by hydrogen atom transfer (HAT).[1,2] HAT is an important area of chemistry that has been widely and extensively investigated in the contexts of combustion, halogenation, antioxidant oxidation, and other processes.[1,2] HAT is typically defined as a process in which a hydrogen atom is transferred between two groups. The HAT reactivity has generally been correlated to the bond dissociation energies (BDEs) of substrates from which a hydrogen atom is abstracted. Since a hydrogen atom consists of an electron and a proton, there are two possible reaction pathways in HAT reactions: one is one-step (concerted) HAT, and the other is sequential (stepwise) electron and proton transfer (Scheme 2.1).[3] When the substrate (RH) is a weak acid (e.g., phenols), deprotonation may occur first, followed by electron transfer (ET).[4] The one-step mechanism means that the HAT reaction occurs without an intermediate when an electron and a proton are transferred simultaneously. This is to be contrasted with a sequential pathway that proceeds via mechanistically distinct ET and proton transfer (PT) steps and thereby involving a detectable intermediate $[RH^{\bullet +} A^{\bullet -}]$ in Scheme 2.1. However, the distinction between one-step HAT and sequential ET/PT becomes ambiguous as the lifetime of the intermediate decreases to beyond the detection limit. If ET is the rate-determining step followed by rapid PT, no intermediate would be detected. In such a case, no deuterium kinetic isotope effect (KIE) would be observed. This should be certainly distinguished from one-step HAT that usually exhibits deuterium kinetic isotope effects, even though no intermediates are detected. Thus, no detection of ET intermediates does not necessarily mean that the reaction proceeds via one-step HAT, although detection of ET intermediates provides clear evidence for the sequential ET/PT processes.

*Physical Inorganic Chemistry: Reactions, Processes, and Applications*   Edited by Andreja Bakac
Copyright © 2010 by John Wiley & Sons, Inc.

One-step hydrogen atom transfer

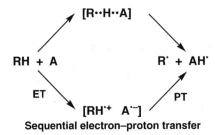

**Sequential electron–proton transfer**

**SCHEME 2.1**

Further complication comes from different ET mechanisms depending on the magnitude of electron coupling element ($H_{DA}$) in ET from a donor (D) to an acceptor (A).[5] ET with $H_{DA} < 200\ cm^{-1}$ is defined as the outer-sphere pathway, which is well analyzed by the Marcus theory of ET,[6] and the substantially greater values pertain to inner-sphere ET.[7–9] The definition of outer-sphere ET originated from ET of metal complexes in which the bimolecular transition state (TS) is traversed with the separate coordination spheres of both the electron donor (D) and the electron acceptor (A) essentially intact,[6] whereas in the inner-sphere ET, the unimolecular (collapsed) transition state typically results from the mutual interpenetration of coordination spheres via a critical bridging ligand.[10,11] There is the same mechanistic dichotomy in one-step inner-sphere ET concerted with a ligand transfer versus sequential outer-sphere ET and a ligand transfer (Scheme 2.2) as in the case of one-step HAT versus sequential ET and PT (Scheme 2.1). The inner-sphere ET is often accompanied by a transfer of the bridging ligand. Such inner-sphere ET reactions are also generally encountered with a variety of organic redox processes, in which the electronic interaction in the transition state can be substantial ($> 1000\ cm^{-1}$),[7–9] as indicated by ubiquitous formation of (preequilibrium) charge transfer (CT) complexes prior to

**One-step inner-sphere ET with ligand transfer**

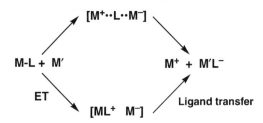

**Sequential outer-sphere ET with ligand transfer**

**SCHEME 2.2**

ET.[7-9,12,13] The inner-sphere character in organic ET reactions is established by their high sensitivity to steric effects.[14]

As the $H_{DA}$ value increases further, the ET process is coupled with the subsequent PT process. This is generally referred to as proton-coupled electron transfer (PCET).[15-24] What distinguishes PCET from classic HAT is that the proton and electron are transferred between two different (noninteracting) orbitals. In particular, PCET plays an important role in enzymatic reactions such as those of lipoxygenases, which are mononuclear nonheme iron enzymes.[25-27] The key step of the lipoxygenase reactions is PCET from a substrate to ferric hydroxide cofactor (Fe(III)-OH) to produce Fe(II)-OH$_2$ and a radical intermediate substrate, when an electron goes to the metal center and a proton is transferred to the OH ligand.[28,29] When PCET is a concerted process whereby a proton and an electron are transferred simultaneously, such a PCET process may be merged into the one-step HAT process.[30] Thus, there has been long-standing ambiguity as to the mechanistic borderline where a sequential PCET pathway is changed to a one-step HAT pathway or vice versa. Understanding HAT reactions certainly requires knowledge of the thermodynamics and kinetics of the overall HAT, ET, and PT steps.

Considering only the two-electron reduction of A, the reduction and protonation give nine species at different oxidation and protonation states as shown in Scheme 2.3. Each species can have an interaction with a variety of metal ions ($M^{n+}$), and such an interaction can control each ET and protonation step, as well as their combined step (hydrogen transfer) in Scheme 2.3.[31-35] The binding of $M^{n+}$ to radical anions of electron acceptors results in a substantial increase in the ET rate.[31-35] This is defined herein as metal ion-coupled electron transfer (MCET) in analogy to PCET.[36] The binding of $M^{n+}$ with $A^{\bullet-}$ can also be combined with other noncovalent interactions such as hydrogen bonding and $\pi-\pi$ interaction.[36] The initial PCET and MCET processes are followed by the second PCET and MCET processes to afford AH$_2$ as the two-electron reduced species of A with two protons.[36]

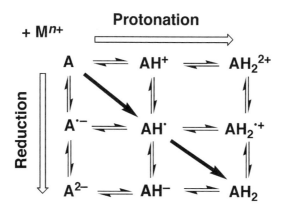

**SCHEME 2.3**

This chapter is intended to focus on the mechanistic borderline between a sequential PCET pathway and a one-step HAT pathway and also the effects of metal ions on HAT reactions as well as overall two-electron and two-proton processes in relation to the borderline between the outer-sphere and inner-sphere ET pathways.

## 2.2   MECHANISTIC BORDERLINE BETWEEN ONE-STEP HAT AND SEQUENTIAL PCET

Among a variety of hydrogen donors, dihydronicotinamide adenine dinucleotide (NADH) and analogues have attracted particular interest, because NADH is the most important source of hydrogen and hydride ion in biological redox reactions.[37–40] If HAT from NADH and analogues to hydrogen atom acceptors (A) occurs in a one-step manner, NAD$^{\bullet}$ and AH$^{\bullet}$ would be the only detectable radical products. In the case of sequential electron and proton transfer, however, NADH$^{\bullet +}$ and A$^{\bullet -}$ would also be detected as the intermediates for the hydrogen transfer reaction. The radical intermediates such as NADH$^{\bullet +}$, NAD$^{\bullet}$, and the corresponding analogues are involved in a variety of thermal and photoinduced ET reactions of NADH and analogues.[39–47] The mechanistic borderline between one-step and sequential pathways in HAT reactions of NADH analogues has been clarified using the triplet excited states of a series of tetrazines ($^3R_2Tz^*$) as hydrogen atom acceptors as described below.[48]

The dynamics of HAT from an NADH analogue, 10-methyl-9,10-dihydroacridine (AcrH$_2$), to $^3Ph_2Tz^*$ was examined by laser flash photolysis measurements. Photoexcitation of a deaerated MeCN solution of [Ru(bpy)$_3$]$^{2+}$ in the presence of Ph$_2$Tz and AcrH$_2$ with 450 nm laser light results in appearance of new absorption bands due to AcrH$^{\bullet}$ ($\lambda_{max} = 360$ and 520 nm)[48] with a concomitant decrease in the absorption band due to $^3Ph_2Tz^*$ ($\lambda_{max} = 535$ nm) as shown in Figure 2.1, whereas no absorption band due to the AcrH$_2^{\bullet +}$ ($\lambda_{max} = 640$ nm)[41] is observed.[48]

Whether HAT from AcrH$_2$ to $^3Ph_2Tz^*$ occurs via a one-step HAT or a rate-determining ET followed by fast proton transfer can be clarified by examining the deuterium kinetic isotope effects. The one-step hydrogen transfer would afford a significant deuterium kinetic isotope effect, whereas the rate-determining ET followed by fast proton transfer would exhibit no deuterium kinetic isotope effect. Comparison of the HAT rated from AcrH$_2$ and the dideuterated compound (AcrD$_2$) to $^3Ph_2Tz^*$ exhibits a significant primary deuterium kinetic isotope effect ($k_H/k_D = 1.80 \pm 0.20$).[48] Thus, HAT from AcrH$_2$ to $^3Ph_2Tz^*$ occurs via a one-step process, which should be faster than ET from AcrH$_2$ to $^3Ph_2Tz^*$.[48]

When $^3Ph_2Tz^*$ ($E^*_{red} = 1.09 \pm 0.04$ V versus SCE) is replaced by a tetrazine derivative that has a slightly higher reduction potential of $^3(ClPh)_2Tz^*$ ($E^*_{red} = 1.11 \pm 0.05$ V versus SCE), AcrH$^{\bullet}$ is also generated by hydrogen transfer from AcrH$_2$ to $^3(ClPh)_2Tz^*$.[48] In contrast to the case of $^3Ph_2Tz^*$, however, only a small primary kinetic isotope effect ($k_H/k_D = 1.11 \pm 0.08$) is observed.[48] No deuterium kinetic isotope effect is observed when $^3(ClPh)_2Tz^*$ ($E^*_{red} = 1.11 \pm 0.05$ V versus SCE) is

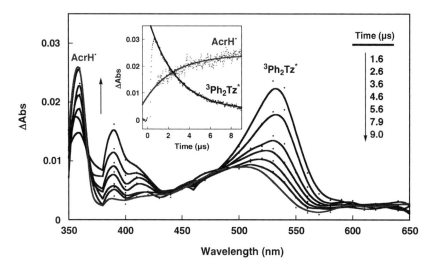

**FIGURE 2.1**  Transient absorption spectra observed by laser flash photolysis of a deaerated MeCN solution of $[Ru(bpy)_3]^{2+}$ ($4.6 \times 10^{-5}$ M) in the presence of $AcrH_2$ ($1.1 \times 10^{-4}$ M) and $Ph_2Tz$ ($9.6 \times 10^{-4}$ M) at 1.6–9.0 µs after laser excitation at $\lambda = 450$ nm at 298K.[48] *Inset:* Time profile of the decay of absorbance at 535 nm due to $^3Ph_2Tz^*$ and the rise of absorbance at 360 nm due to AcrH[^•].[48]

replaced by a tetrazine derivative, $^3Py_2Tz^*$ ($E^*_{red} = 1.25 \pm 0.04$ V versus SCE), which has stronger oxidizing ability than $^3(ClPh)_2Tz^*$.[48] Thus, the mechanism of hydrogen transfer is changed from the one-step hydrogen transfer to the rate-determining ET followed by rapid proton transfer, with increasing $E^*_{red}$ value and decreasing basicity of the nitrogen sites of $^3R_2Tz^*$.

When $AcrH_2$ is replaced by $AcrHPr^i$, the reaction with $^3(ClPh)_2Tz^*$ proceeds via sequential electron–proton transfer, where the formation of radical cation of $AcrHPr^i$ ($\lambda_{max} = 680$ nm)[41] is observed in the laser flash photolysis measurements as shown in Figure 2.2a.[48] The deprotonation from $AcrHPr^{i\bullet+}$ is retarded compared to that from $AcrH_2^{\bullet+}$ because of the steric effect of the $Pr^i$ group than $AcrH_2^{\bullet+}$.[41] The time profiles of the transient absorption at 530 nm due to $^3(ClPh)_2Tz^*$, at 680 nm due to $AcrHPr^{i\bullet+}$, and at 510 nm due to $AcrPr^{i\bullet}$ are shown in Figure 2.2.[48] The absorption at 530 nm due to $^3(ClPh)_2Tz^*$ decays immediately within 2 µs after laser excitation, accompanied by the rise in absorption at 680 nm due to $AcrHPr^{i\bullet+}$. The decay of absorbance at 680 nm due to $AcrHPr^{i\bullet+}$ (Figure 2.2b) coincides with the rise in absorbance at 510 nm due to $AcrPr^{i\bullet}$ (Figure 2.2b). This indicates that ET from $AcrHPr^i$ to $^3(ClPh)_2Tz^*$ occurs rapidly to produce $AcrHPr^{i\bullet+}$ and $(ClPh)_2Tz^{\bullet-}$ within 2 µs, followed by the slower proton transfer from $AcrHPr^{i\bullet+}$ to $(ClPh)_2Tz^{\bullet-}$ to produce $AcrPr^{i\bullet}$. Thus, introduction of isopropyl group at the 9-position of $AcrH_2$ results in change of the mechanism from one-step HAT to sequential ET and PT as shown in Scheme 2.4.[48] ET from another NADH analogue 1-benzyl-1,4-

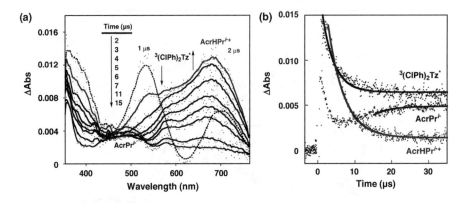

**FIGURE 2.2** (a) Transient absorption spectra observed by laser flash photolysis of a deaerated MeCN solution of $Ru(bpy)_3^{2+}$ ($4.6 \times 10^{-5}$ M) in the presence of $AcrHPr^i$ ($8.8 \times 10^{-4}$ M) and $(ClPh)_2Tz$ ($9.6 \times 10^{-4}$ M) at 1–15 μs after laser excitation at λ = 450 nm at 298K.[48] (b) Time profiles of the decay of absorbance at 530 nm due to $^3(ClPh)_2Tz^*$, the decay of absorbance at 680 nm due to $AcrHPr^{i \bullet +}$, and the rise of absorbance at 510 nm due to $AcrPr^{i \bullet}$.[48]

dihydronicotinamide (BNAH) to $^3Ph_2Tz^*$ also occurs to produce $BNAH^{\bullet +}$ and $Ph_2Tz^{\bullet -}$, followed by slower proton transfer from $BNAH^{\bullet +}$ to $Ph_2Tz^{\bullet -}$ to yield $BNA^{\bullet}$.[48]

The change in the mechanism from one-step HAT to sequential PCET is related to the change from inner-sphere ET to outer-sphere ET. Plots of log $k_{et}$ of outer-sphere ET from various electron donors to $^3R_2Tz^*$ versus the ET Gibbs energy change ($\Delta G_{et}$) are shown in Figure 2.3.[48] The observed rate constant of HAT from $AcrH_2$ to $^3Ph_2Tz^*$ ($k_H = 2.7 \pm 0.1 \times 10^9$ M$^{-1}$ s$^{-1}$) in Figure 2.3a (no. 17) is significantly larger than the expected value from the plot of log $k_{et}$ versus $\Delta G_{et}$. This indicates that the one-step HAT is regarded as inner-sphere ET in which the HDA value is much larger than the value of the outer-sphere limit and the C–H bond of $AcrH_2$ is partially cleaved in the transition state to exhibit the deuterium kinetic isotope effect.[48] In contrast, the observed rate constant of HAT from $AcrH_2$ to $^3(ClPh)_2Tz^*$ ($k_H = 3.1 \pm 0.1 \times 10^9$ M$^{-1}$ s$^{-1}$) in Figure 2.3b (no. 18) is same as the expected value from the correlation for the outer-sphere ET reactions.[48] This indicates that a one-step HAT process and the rate-determining ET followed by rapid PT occur competitively in hydrogen transfer from $AcrH_2$ to $^3(ClPh)_2Tz^*$ as shown in Scheme 2.5.[48]

The rate constants of ET from $AcrHPr^i$ to $^3(ClPh)_2Tz^*$ ($1.0 \times 10^9$ M$^{-1}$ s$^{-1}$) in Figure 2.3b (no. 21) and the rate constant of hydrogen transfer from $AcrH_2$ to $^3Py_2Tz^*$ ($k_H = 3.4 \pm 0.1 \times 10^9$ M$^{-1}$ s$^{-1}$) in Figure 2.3c (no. 19) also agree with the expected values from the plot of log $k_{et}$ versus $\Delta G_{et}$.[48] The Gibbs energy change of ET from $AcrH_2$ ($E_{ox} = 0.81$ V versus SCE) to $^3(ClPh)_2Tz^*$ ($E_{red}^* = 1.11 \pm 0.05$ V versus SCE) ($\Delta G_{et} = -0.30$ eV) is shown in Figure 2.3a–c (dashed line) to emphasize the mechanistic borderline.[48] When $\Delta G_{et}$ becomes smaller than $-0.30$ eV, the reaction

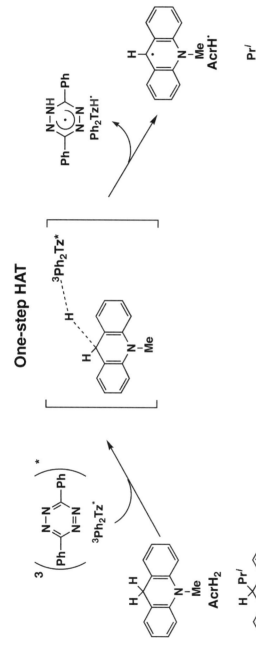

**One-step HAT**

**Sequential electron-proton transfer**

SCHEME 2.4

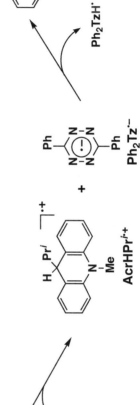

45

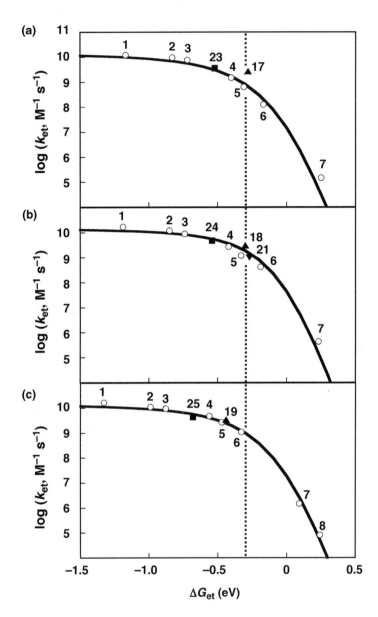

**FIGURE 2.3** (a) Plots of $\log k_{et}$ versus $\Delta G_{et}$ for photoinduced ET from various electron donors to $^3Ph_2Tz^*$ (open circles), including plots of $\log k_H$ versus $\Delta G_{et}$ for quenching of $^3Ph_2Tz^*$ by $AcrH_2$ (no. 17) and BNAH (no. 23) in deaerated MeCN at 298K.[48] (b) Plots of $\log k_{et}$ versus $\Delta G_{et}$ for photoinduced ET from various electron donors to $^3(ClPh)_2Tz^*$ (open circles), including plots of $\log k_H$ versus $\Delta G_{et}$ for quenching of $^3(ClPh)_2Tz^*$ by $AcrH_2$ (no. 18), BNAH (no. 24), and $AcrHPr^i$ (no. 21) in deaerated MeCN at 298K.[48] (c) Plots of $\log k_{et}$ versus $\Delta G_{et}$ for photoinduced ET from various electron donors to $^3Py_2Tz^*$ (open circles), including plots of $\log k_H$ versus $\Delta G_{et}$ for quenching of $^3Py_2Tz^*$ by $AcrH_2$ (no. 19) and BNAH (no. 25) in deaerated MeCN at 298K.[48] Numbers at open circles (numbers 1–8) correspond to electron donors such as ferrocene derivatives.[48]

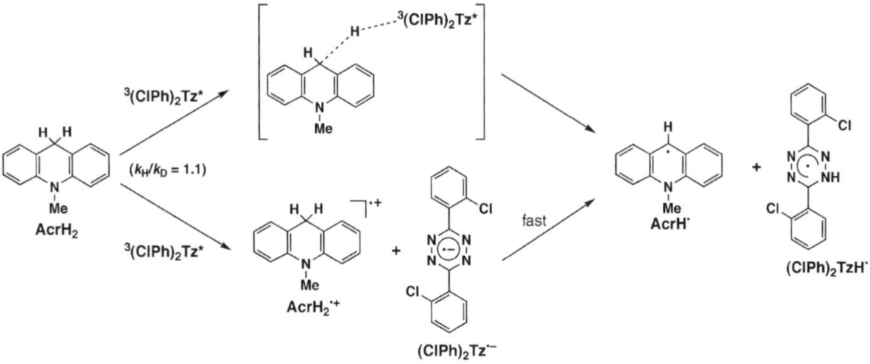

**SCHEME 2.5**

mechanism is changed from the one-step HAT to the rate-determining ET followed by rapid PT, with increasing $E_{red}^*$ value and decreasing basicity of the nitrogen sites of $^3R_2Tz^*$.[48] The one-step HAT is also changed to sequential electron–proton transfer, by replacing AcrH$_2$ with AcrHPr$^i$ and BNAH, when the formation of radical cations of AcrHPr$^i$ and BNAH is observed in the laser flash photolysis measurements.[48] Thus, the reactions of NADH analogues with $^3R_2Tz^*$ occur via one-step HAT, the rate-limiting ET followed by fast PT, or sequential ET and PT depending on the electron-donor ability of NADH analogues as well as the electron-acceptor ability of $^3R_2Tz^*$ and the protonation reactivity of $R_2Tz^{•-}$.[48]

## 2.3 ONE-STEP VERSUS STEPWISE MECHANISM IN INTERPLAY BETWEEN ELECTRON TRANSFER AND HYDROGEN BONDING

### 2.3.1 Hydrogen Bonding by Protonated Amino Acids

In HAT reactions described above, the proton is provided by radical cations of electron donors because the acidity is significantly enhanced by the one-electron oxidation of electron donors. An electron and a proton are transferred by a one-step pathway or a sequential pathway depending on the types of electron donors and acceptors. The same mechanistic dichotomy was reported for PCET when proton is provided externally.[49] ET from an electron donor that has no proton to be transferred to an electron acceptor (A) is coupled with protonation of A$^{•-}$ when the one-electron reduction and protonation of A occur simultaneously (a green arrow in Scheme 2.6a). The binding strength of proton to A$^{•-}$ is regulated by Brønsted bases (:B) such as amino acid residues in the protein environment.[50,51] In such a case, A$^{•-}$ forms a hydrogen bond with H$^+$:B instead of direct protonation when:B acts as the stronger base than A$^{•-}$ (Scheme 2.6b).[52–56] ET coupled with protonation (or hydrogen bond formation) (Scheme 2.6, green arrows) should be thermodynamically more favorable than ET followed by protonation (or hydrogen bond formation) (Scheme 2.6, red and blue arrows).[53,54] However, the one-step ET mechanism may be changed to the

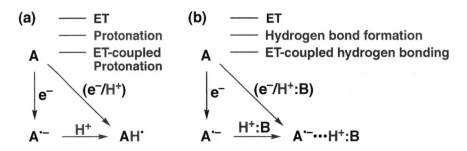

**(a)**  —— ET
—— Protonation
A  —— ET-coupled
Protonation

$$A \xrightarrow{e^-} \quad \diagdown \quad (e^-/H^+)$$

$$A^{\bullet-} \xrightarrow{H^+} AH^\bullet$$

**(b)**  —— ET
—— Hydrogen bond formation
A  —— ET-coupled hydrogen bonding

$$A \xrightarrow{e^-} \quad \diagdown \quad (e^-/H^+:B)$$

$$A^{\bullet-} \xrightarrow{H^+:B} A^{\bullet-}\cdots H^+:B$$

**SCHEME 2.6**    (See the color version of this scheme in Color Plates section.)

stepwise mechanism when the driving force of the initial ET (Scheme 2.6, blue arrows) significantly increases.

With regard to such a mechanistic dichotomy, an important question arises: Are both pathways employed simultaneously? Alternatively, is there a mechanistic continuity? This has been examined in a protonated histidine (His·2H$^+$)-promoted ET reduction of 1-($p$-tolylsulfinyl)-2,5-benzoquinone (TolSQ) by electron donors (see below).[49]

Hydrogen bond formation of semiquinone radical anions with protonated amino acids was examined by ESR in photoinduced ET from 10,10′-dimethyl-9,9′-biacridine [(AcrH)$_2$] to quinones (Scheme 2.7).[49] The (AcrH)$_2$ is known to act as a two-electron donor to produce 2 equivalents of the radical anion of electron acceptor.[59,60] The resulting ESR spectrum of a hydrogen-bonded complex between TolSQ$^{\bullet-}$ and His·2H$^+$ (TolSQ$^{\bullet-}$/His·2H$^+$) is shown in Figure 2.4a.[49] The hyperfine coupling constants (hfc) for three protons of TolSQ$^{\bullet-}$ are $a$(3H) = 0.88, 5.31, and 6.08 G, and for one nitrogen and three equivalent protons of His·2H$^+$ due to superhyperfine splitting are $a$(N) = 1.35 G and $a$(3H) = 2.97 G (Figure 2.4b).[49] The complete agreement of the observed ESR spectrum (Figure 2.4a) with the computer simulation spectrum (Figure 2.4b) clearly indicates formation of the TolSQ$^{\bullet-}$/His·2H$^+$ complex (Scheme 2.7).[49] The optimized structures and the hfc values of TolSQH$^\bullet$ are also obtained by DFT at the BLYP/6-31G$^{**}$ basis (Figure 2.4c).[49]

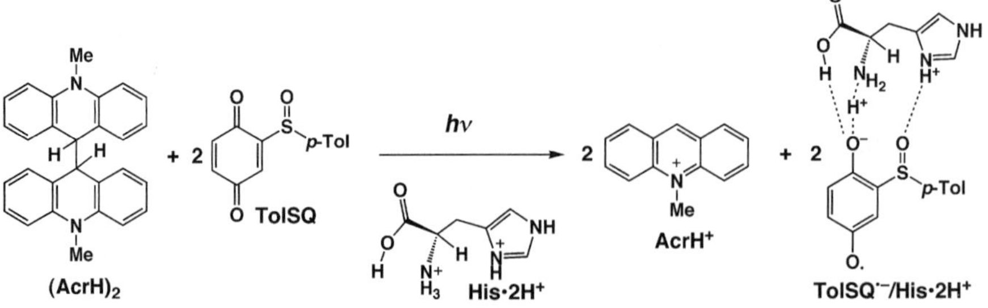

**SCHEME 2.7**

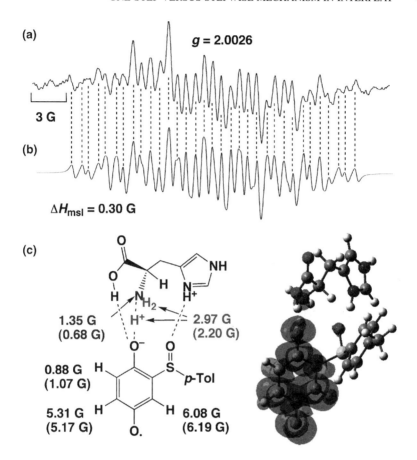

**FIGURE 2.4** (a) ESR spectrum of TolSQ$^{\bullet-}$/His·2H$^+$ produced by photoinduced ET from (AcrH)$_2$ (1.6 × 10$^{-2}$ M) to TolSQ (4.0 × 10$^{-3}$ M) in the presence of His (4.0 × 10$^{-3}$ M) and HClO$_4$ (8.0 × 10$^{-3}$ M) in deaerated MeCN at 298 K and (b) the computer simulation spectrum with the hfc values of TolSQ$^{\bullet-}$/His·2H$^+$.[49] (c) Optimized structure of TolSQ$^{\bullet-}$/His·2H$^+$ calculated by using a density functional theory at the BLYP/6-31G$^{**}$ (the calculated hfc values are given in parentheses).[49]

Strong hydrogen bond formation between TolSQ$^{\bullet-}$ and His·2H$^+$ (TolSQ$^{\bullet-}$/His·2H$^+$), as well as protonation of TolSQ$^{\bullet-}$ (TolSQH$^{\bullet}$), is expected to result in positive shifts of the one-electron reduction potential ($E_{red}$) of TolSQ.[49] Since the free energy change of ET from 1,1'-dimethylferrocene [(C$_5$H$_4$Me)$_2$Fc] ($E_{ox}$ = 0.26 V versus SCE)[58] to TolSQ ($E_{red}$ = −0.26 V versus SCE)[61] in the absence of His·2H$^+$ is highly endergonic ($\Delta G_{et}$ = 0.52 eV), no ET reaction occurs in the absence of His·2H$^+$.[49] In the presence of His·2H$^+$ (5.0 × 10$^{-2}$ M), however, the $E_{red}$ value of TolSQ (−0.26 V versus SCE) is shifted to 0.29 V versus SCE.[49] ET from (C$_5$H$_4$Me)$_2$Fc to TolSQ therefore occurs in the presence of His·2H$^+$ (Equation 2.1), as expected from the negative free energy change of ET ($\Delta G_{et} = -0.03$ eV).[49]

$$R_2Fc + TolSQ \xrightarrow{\hspace{4cm}} R_2Fc^+ + TolSQ^{\cdot-}/His\cdot2H^+ \qquad (2.1)$$

$$R = C_5H_5$$
$$C_5H_4(n\text{-Bu})$$
$$C_5H_4Me$$

The rates of His·2H$^+$-promoted ET from $(C_5H_4Me)_2Fc$ to TolSQ obeyed pseudo-first-order kinetics in the presence of a large excess of TolSQ and His·2H$^+$ relative to the concentration of $(C_5H_4Me)_2Fc$.[49] The observed pseudo-first-order rate constant ($k_{obs}$) increases proportionally with increasing TolSQ concentration.[49] The second-order rate constant ($k_H$) also increases linearly with increasing His·2H$^+$ concentration ([His·2H$^+$]).[49] The rates of ET from $R_2Fc$ to TolSQ exhibit deuterium kinetic isotope effects ($1.3 < k_H/k_D < 1.9$) when His·2H$^+$ is replaced by the deuterated compound (His·2D$^+$-$d_6$) as shown by black circles in Figure 2.5 (see the structure of His·2D$^+$-$d_6$ in Equation 2.1).[49] The observed deuterium kinetic isotope effects may result from partial dissociation of the N–H bond in the hydrogen-bonded NH$_3{}^+$ of His·2H$^+$ at the transition state when the ET is tightly coupled with hydrogen bond formation (Scheme 2.8a).[49]

The plot of $k_H/k_D$ versus the driving force of ET ($-\Delta G_{et}$) is shown in Figure 2.6a, combined with the plot of $\log k_H$ versus $-\Delta G_{et}$ (Figure 2.6b).[49] The $k_H/k_D$ value decreases with increasing $-\Delta G_{et}^0$ value to approach $k_H/k_D = 1.0$ (Figure 2.5a), with a concomitant increase of the $\log k_H$ value (Figure 2.6b). In contrast to the ET coupled with hydrogen bond formation (Scheme 2.8a), the rate-determining ET followed by fast hydrogen bond formation (Scheme 2.8b) would exhibit no deuterium kinetic isotope effect ($k_H/k_D = 1.0$). Thus, the continuous decrease of the deuterium kinetic isotope effect ($k_H/k_D$) with an increase in the ET driving force ($-\Delta G_{et}^0$) (Figure 2.6a) indicates that there is a mechanistic continuity in two reaction pathways; that is, the one-step pathway (Scheme 2.8a) is continuously changed to the stepwise pathway (Scheme 2.8b) with increasing ET driving force ($-\Delta G_{et}^0$).[49] If two reaction pathways were employed simultaneously, the deuterium kinetic isotope effect ($k_H/k_D$) would be constant irrespective of the ET driving force ($-\Delta G_{et}$) above the changeover to the stepwise mechanism (Scheme 2.8b).[49]

### 2.3.2 Intramolecular Hydrogen Bonding

***2.3.2.1 Stepwise Electron Transfer and Hydrogen Bonding***    Photoinduced ET and hydrogen bond formation occurs in a stepwise manner for an electron donor–acceptor dyad with a hydrogen bonding site such as a ferrocene–quinone dyad with an amide spacer (Fc–Q) when the ET process is highly exergonic.[62] Photoexcitation of the Q moiety in Fc–Q in deaerated benzonitrile (PhCN) with 388 nm femtosecond (150 fs width) laser light results in appearance of a new absorption band ($\lambda_{max} = 580$ nm) at 1 ns after the laser excitation.[62] The absorption band at 580 nm is

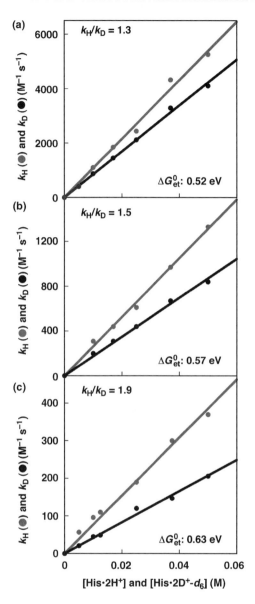

**FIGURE 2.5**   Dependence of $k_H$ (gray circles) and $k_D$ (black circles) on [His·2H$^+$] and [His·2D$^+$-$d_6$] for ET from (a) $(C_5H_4Me)_2$Fc ($1.0 \times 10^{-4}$ M), (b) $[C_5H_4(n\text{-Bu})]_2$Fc ($1.0 \times 10^{-4}$ M), and (c) $(C_5H_5)_2$Fc ($1.0 \times 10^{-4}$ M) to TolSQ in the presence of His·2H$^+$ and His·2D$^+$-$d_6$ in deaerated MeCN at 298K.[49]

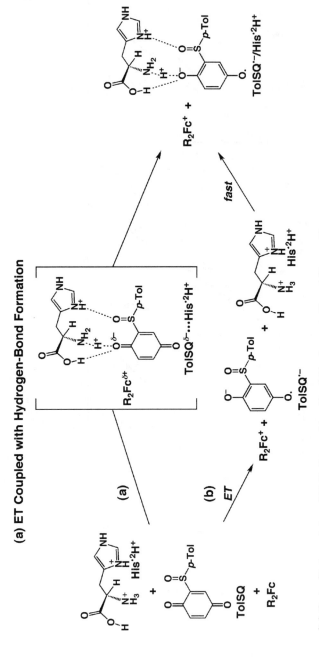

**SCHEME 2.8**

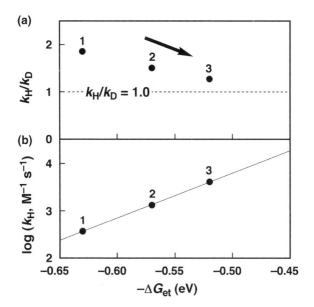

**FIGURE 2.6**    Plots of (a) $k_H/k_D$ and (b) $\log k_H$ versus $-\Delta G_{et}$ for ET from $R_2$Fc to TolSQ in the presence of His·2H$^+$ ($5.0 \times 10^{-2}$ M) in deaerated MeCN at 298K.[49]

significantly redshifted compared to the diagnostic absorption band of semiquinone radical anion at 422 nm, and this is assigned to Q$^{•-}$, which is hydrogen bonded to the amide proton of the spacer.[62] The investigation of the photodynamics revealed that ET from Fc to the singlet excited state of Q occurs rapidly to produce Fc–Q$^{•-}$ without changing the conformation (<1 ps) and then Q$^{•-}$ forms the hydrogen bond with the amide proton of the spacer ($\tau = {\sim}5$ ps) and the resulting radical ion pair decays via a back ET to the ground state as shown in Scheme 2.9.[62] Thus, formation of the hydrogen bond is not coupled with an ET when the ET process is highly exergonic involving the excited state of the Q moiety.[62]

The effect of the hydrogen bonding on the one-electron reduction potential ($E_{red}$) of Q in Fc–Q is shown by the change in the cyclic voltammograms of Fc–Q and Fc–(MeQ) in which the N–H group is replaced by N–Me.[63] The cyclic voltammogram of Fc–Q (Figure 2.7) exhibits two reversible one-electron redox couples at 0.39 and $-0.16$ V (versus SCE) due to the Fc$^+$/Fc couple and the Q/Q$^{•-}$ couple, respectively.[63] The one-electron reduction potential of Q ($E_{red} = -0.16$ V) is significantly shifted to a positive direction compared to $p$-benzoquinone ($-0.50$ V) and this is more positive than the $E_{red}$ value of $p$-benzoquinone possessing an electron-withdrawing substituent ($-0.38$ V for chloro-$p$-benzoquinone).[64] The $E_{red}$ value of Q ($-0.40$ V) in Fc–(Me)Q (Figure 2.7b) is significantly more negative compared to Fc–Q ($-0.16$ V).[63] Such a negative shift in the $E_{red}$ value of Q in Fc–(Me)Q compared to Fc–Q indicates that the radical anion (Q$^{•-}$) is stabilized by the hydrogen bonding with the amide proton of the spacer.[63]

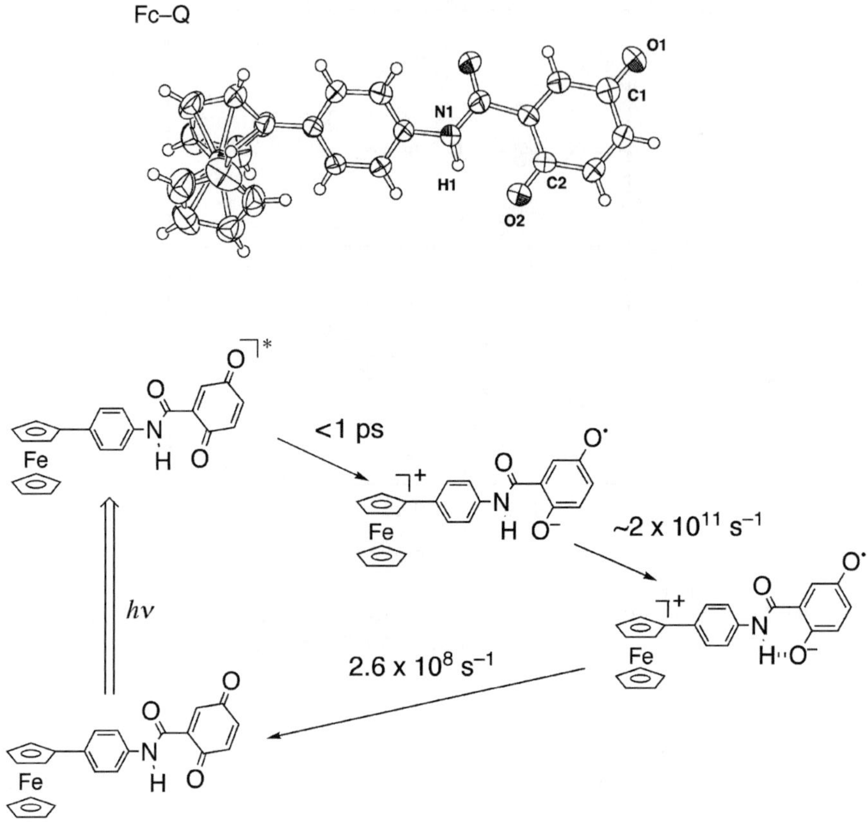

Fc–Q

**SCHEME 2.9**

### 2.3.2.2 Metal Ion-Coupled Electron Transfer and Hydrogen Bonding    The

positive shift in the $E_{red}$ value of Q by the hydrogen bonding with amide proton of
the spacer in Fc–Q (Figure 2.7) is not enough to make thermal ET from Fc to Q
possible. In fact, no ET from Fc to Q occurs in Fc–Q and Fc–(MeQ) thermally in MeCN
at 298K.[62] However, the binding of metal ions to $Q^{\bullet-}$ that further stabilize $Q^{\bullet-}$ makes
the thermal ET possbile.[63] Thus, addition of $Mg^{2+}$ results in the formation of $Fc^+$ as
indicated by the appearance of the absorption band due to $Fc^+$ at 800 nm together with
the absorption band at 420 nm due to $Q^{\bullet-}$ bound to $Mg^{2+}$.[63,65] The ET rates increase
linearly with increasing concentration of $Mg^{2+}$.[63] The second-order rate constant ($k_{et}$)
of MCET is determined to be $1.4 \times 10^3 \, M^{-1} \, s^{-1}$.[63] When Fc–Q is replaced by Fc–(Me)
Q that contains no hydrogen bond acceptor, the $k_{et}$ value ($0.4 \, M^{-1} \, s^{-1}$) of Fc–(Me)Q
becomes much smaller than the $k_{et}$ value ($1.4 \times 10^3 \, M^{-1} \, s^{-1}$) of Fc–Q.[63] This
difference is ascribed to the acceleration of MCET by the hydrogen bonding between
the $Q^{\bullet-}$ moiety and the amide proton. A variety of metal ions ($M^{n+}$: triflate salts) can
also promote ET from Fc to Q in Fc–(Me)Q (Scheme 2.10).[63]

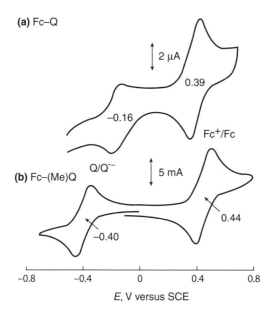

**FIGURE 2.7**   Cyclic voltammograms of (a) Fc–Q (0.5 mM) and (b) Fc–(Me)Q (0.5 mM) in MeCN containing 0.1 M Bu$_4$NPF$_6$.[63]

The promoting effects of metal ions vary significantly depending on the Lewis acidity of metal ions.[63] The Lewis acidity of metal ions has been determined as the $\Delta E$ values of O$_2^{\bullet-}$/M$^{n+}$ derived from the $g_{zz}$ values.[66,67] The log $k_{et}$ values of MCET for Fc–Q and Fc–(Me)Q are linearly correlated with the $\Delta E$ values as shown in Figure 2.8.[63] The $k_{et}$ value of the Sc$^{3+}$-coupled ET in Fc–Q is the largest among metal ions in Figure 2.8 and this is $10^4$ times larger than the corresponding $k_{et}$ value of Fc–(Me)Q. The $10^4$ times difference in the $k_{et}$ values corresponds to the difference in the one-electron reduction potential between Fc–Q ($E_{red}$ versus SCE $= -0.16$ V) and Fc–(Me)Q ($E_{red}$ versus SCE $= -0.40$ V) in Figure 2.7, since the ratio of the rate constant is given by exp(0.24 eV/$k_B T$), which is equal to $1.1 \times 10^4$ at 298K.[63]

The formation of a paramagnetic species, that is, Q$^{\bullet-}$ bound to both Mg$^{2+}$ and the amide proton, was confirmed by the ESR spectrum measured after the addition of Mg$^{2+}$ to an MeCN solution of Fc–Q.[63] The observed ESR spectrum is shown in Figure 2.9a together with the computer simulation spectrum (Figure 2.9b).[63] The largest hfc value (3.95 G) is significantly smaller than the value without Mg$^{2+}$ ion (4.60 G) due to the spin delocalization to the Mg nucleus.[63]

The formation of Fc$^+$ coincides with the formation of Q$^{\bullet-}$ bound to both Mg$^{2+}$ and the amide proton without detection of an intermediate without hydrogen bonding.[63] Thus, in contrast to the stepwise photoinduced ET and formation of the hydrogen bond in Fc–Q (Scheme 2.9), MCET in Fc–Q is coupled with formation of the hydrogen bond as shown in Scheme 2.10.

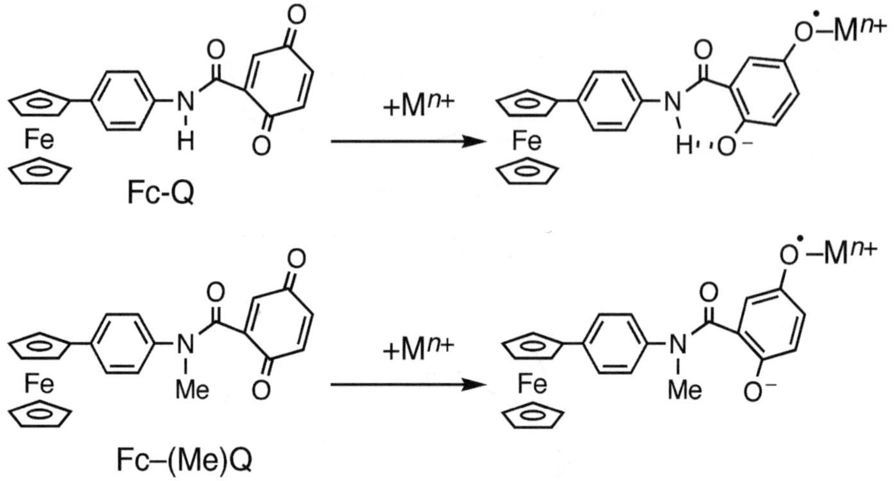

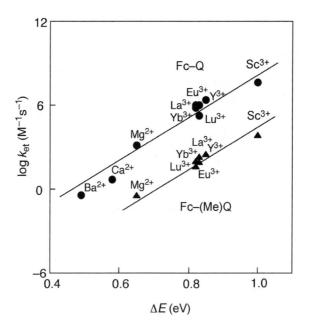

**SCHEME 2.10**

**FIGURE 2.8**  Plots of $\log k_{et}$ versus $\Delta E$ in $M^{n+}$-promoted ET in Fc–Q and Fc–(Me)Q in MeCN at 298K.[63]

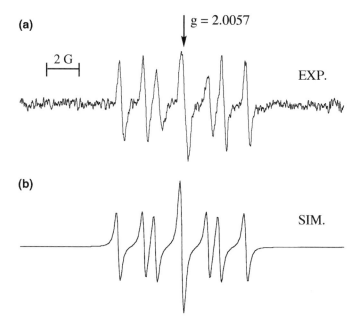

**FIGURE 2.9** (a) ESR spectrum of $Fc^+-Q^{\bullet-}$ ($4.0 \times 10^{-4}$ M) in the presence of $Mg^{2+}$ ($7.5 \times 10^{-2}$ M) in deaerated MeCN at 298 K and (b) the computer simulation spectrum with hfc values of 3.95 (1H), 2.30 (1H), and 1.60 G (1H).[63]

## 2.4 SEQUENTIAL ELECTRON TRANSFER AND PROTON TRANSFER PATHWAYS IN HYDRIDE TRANSFER REACTIONS

### 2.4.1 Hydride Reduction of Quinones by NADH Analogues

The same mechanistic dichotomy for HAT reactions, one-step (concerted) HAT versus sequential (stepwise) electron and proton transfer (Scheme 2.1), is applied to hydride transfer reactions, one-step (concerted) hydride transfer versus sequential (stepwise) ET followed by proton–electron (or hydrogen) transfer.[13,40,64,68] Such one-step versus multistep pathways have been discussed extensively in hydride transfer reactions of dihydronicotinamide coenzyme (NADH) and analogues, particularly including the effect of metal cations and acids,[69–79] because of the essential role of acid catalysis in the enzymatic reduction of carbonyl compounds by NADH.[80] In contrast to the one-step hydride transfer pathway that proceeds without an intermediate, the ET pathway would produce radical cation hydride donors as the reaction intermediates, which have rarely been observed. The ET pathway may become possible if the ET process is thermodynamically feasible.

***2.4.1.1 One-Step Hydride Transfer*** As described above, the ET from $(C_5H_4Me)_2Fc$ to TolSQ becomes possible in the presence of His·2H$^+$

(Equation 2.2).[49] An efficient hydride transfer from AcrH$_2$ to TolSQ to yield AcrH$^+$ and TolSQH$_2$ also becomes possible in the presence of His·2H$^+$ (Equation 2.2).[49]

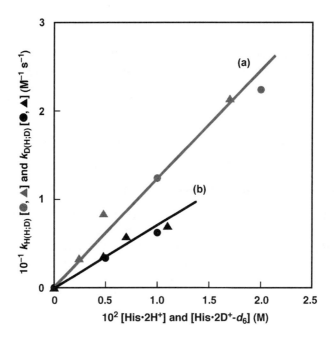

$$(2.2)$$

The second-order rate constant of hydride transfer from AcrH$_2$ to TolSQ with His·2H$^+$ ($k_{HH}$) increases linearly with [His·2H$^+$] (gray circles in Figure 2.10a).[45] The rates of hydride transfer exhibit a deuterium kinetic isotope effect ($k_{HH}/k_{DH}$ 1.7 ± 0.1) when AcrH$_2$ is replaced by the dideuterated compound AcrD$_2$ ($k_{DH}$ denotes the rate constant of hydride transfer from AcrD$_2$ to TolSQ with His·2H$^+$) (black circles in Figure 2.10b).[49] In sharp contrast to this, no deuterium kinetic

**FIGURE 2.10**   (a) Dependence of $k_{HH}$ (gray circles) on [His·2H$^+$] for hydride transfer from AcrH$_2$ (1.0 × 10$^{-4}$ M) to TolSQ in the presence of His·2H$^+$ and that of $k_{HD}$ (gray triangles) on [His·2D$^+$-$d_6$] for hydride transfer from AcrH$_2$ (1.0 × 10$^{-4}$ M) to TolSQ in the presence of His·2D$^+$-$d_6$ in deaerated MeCN at 298K.[49] (b) Dependence of $k_{DH}$ (black circles) on [His·2H$^+$] for hydride transfer from AcrD$_2$ (1.0 × 10$^{-4}$ M) to TolSQ in the presence of His·2H$^+$ and that of $k_{DD}$ (black triangles) on [His·2D$^+$-$d_6$] for hydride transfer from AcrD$_2$ (1.0 × 10$^{-4}$ M) to TolSQ in the presence of His·2D$^+$-$d_6$ in deaerated MeCN at 298K.[49]

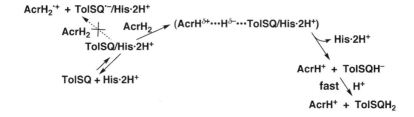

**SCHEME 2.11**

isotope effect ($k_{HH}/k_{HD} = 1.0$ and $k_{DH}/k_{DD} = 1.0$) is observed in hydride transfer from $AcrH_2$ and $AcrD_2$ to TolSQ when His·$2H^+$ is replaced by His·$2D^+$-$d_6$ ($k_{HD}$ denotes the rate constant of hydride transfer from $AcrH_2$ to TolSQ with His·$2D^+$-$d_6$); see gray and black closed triangles in Figure 2.10, respectively.[49] If the hydride transfer proceeds via ET from $AcrH_2$ ($E_{ox} = 0.81$ V versus SCE) to TolSQ ($E_{red} = 0.26$ V versus SCE), as shown by broken arrow in Scheme 2.11, the rates of the formal hydride reactions would exhibit deuterium kinetic isotope effects as in the case of His·$2H^+$-promoted ET from $R_2Fc$ to TolSQ (Figure 2.10), when His·$2H^+$ is replaced by His·$2D^+$-$d_6$. Thus, the observed deuterium kinetic isotope effect ($k_{HH}/k_{DH} = 1.7 \pm 0.1$) by deuterium substitution of $AcrH_2$ by $AcrD_2$ and the absence of the deuterium kinetic isotope effect ($k_{HH}/k_{HD} = 1.0$ and $k_{DH}/k_{DD} = 1.0$) by deuterium substitution of His·$2H^+$ by His·$2D^+$-$d_6$ indicate that the hydride transfer proceeds via the one-step pathway (Scheme 2.11).[49] It should be noted that no absorption band due to $AcrH_2^{\bullet+}$ is observed in the His·$2H^+$-promoted hydride transfer from $AcrH_2$ to TolSQ.[49] The linear correlation between $k_{HH}$ and [His·$2H^+$] (Figure 2.10a) may result from formation of the hydrogen-bonded complex between TolSQ and His·$2H^+$ (TolSQ/His·$2H^+$), which increases with increasing His·$2H^+$ concentration.[49]

**2.4.1.2  Stepwise Electron Transfer Pathway**  A more efficient reduction of TolSQ by $AcrH_2$ occurs to yield $AcrH^+$ and $TolSQH_2$ in the presence of perchloric acid ($HClO_4$), whereas no reaction occurs between $AcrH_2$ and TolSQ in the absence of $HClO_4$.[81] The stoichiometry was confirmed by the spectral titration of TolSQ by $AcrH_2$ in the presence of $HClO_4$ (Figure 2.11a), where all TolSQ molecules are consumed by addition of 1 equivalent of $AcrH_2$ to yield 1 equivalent of $AcrH^+$.[81] The promoting effect of $HClO_4$ on the reduction of TolSQ by $AcrH_2$ should result from protonation of TolSQ (TolSQ + $H^+$ → $TolSQH^+$), which is confirmed by UV-Vis spectral changes of TolSQ in the presence of various concentrations of $HClO_4$.[81]

The dynamics of the reduction of TolSQ by $AcrH_2$ in the presence of $HClO_4$ monitored by using a stopped-flow technique revealed the formation of ET intermediate as shown by a transient absorption band at $\lambda_{max} = 640$ nm (Figure 2.11b), which is ascribed to $AcrH_2^{\bullet+}$.[81] Formation of $AcrH_2^{\bullet+}$ was fully characterized including the ESR detection. The resulting ESR spectrum (Figure 2.11c) agrees with

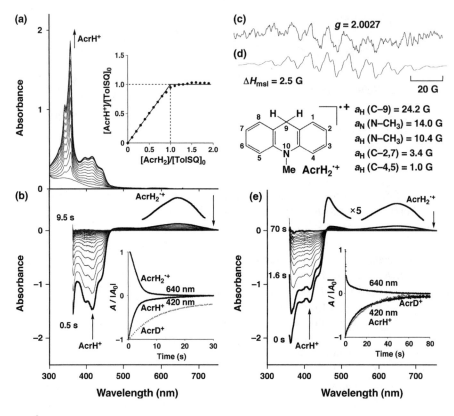

**FIGURE 2.11** (a) Absorption spectral changes observed upon addition of AcrH$_2$ (0–1.9 × 10$^{-4}$ M) to a deaerated MeCN solution of TolSQ (1.0 × 10$^{-4}$ M) in the presence of HClO$_4$ (1.0 × 10$^{-1}$ M) at 298K.[81] (b) Differential spectral changes in the reduction of TolSQ (4.6 × 10$^{-4}$ M) by AcrH$_2$ (6.0 × 10$^{-3}$ M) in the presence of HClO$_4$ (4.9 × 10$^{-2}$ M) in deaerated MeCN at 298K.[81] (c) ESR spectrum of AcrH$_2$$^{•+}$ generated by oxidation of AcrH$_2$ (2.9 × 10$^{-3}$ M) with TolSQ (2.8 × 10$^{-3}$ M) in the presence of HClO$_4$ (7.0 × 10$^{-2}$ M) in deaerated MeCN at 298K and (d) the computer simulation spectrum.[81] (e) Differential spectral changes in the reduction of PQ (4.9 × 10$^{-4}$ M) by AcrH$_2$ (4.8 × 10$^{-3}$ M) in the presence of HClO$_4$ (4.9 × 10$^{-2}$ M) in deaerated MeCN at 298K.[76] *Insets*: (a) Plot of [AcrH$^+$]/[TolSQ]$_0$ versus [AcrH$_2$]/ [TolSQ]$_0$, where [TolSQ]$_0$ is the initial concentration of TolSQ (1.0 × 10$^{-4}$ M).[81] Time course of the absorption change at λ = 640 and 420 nm for the reduction of (b) TolSQ and (e) by AcrH$_2$ and AcrD$_2$; $A_0$ is the initial absorbance.[81]

the computer simulation spectrum (Figure 2.11d) produced using the same hfc values ($a_H$(C-9) = 24.2, $a_N$(N–CH$_3$) = 14.0, $a_H$(N–CH$_3$) = 10.4, $a_H$(C-2,7) = 3.4, and $a_H$(C-4,5) = 1.0 G) of AcrH$_2$$^{•+}$ that was produced by the ET oxidation of AcrH$_2$ by [Fe(bpy)$_3$]$^{3+}$ (bpy = 2,2′-bipyridine).[41,81] The observation of AcrH$_2$$^{•+}$ indicates that ET first occurs from AcrH$_2$ to TolSQH$^+$ (Scheme 2.12).[81]

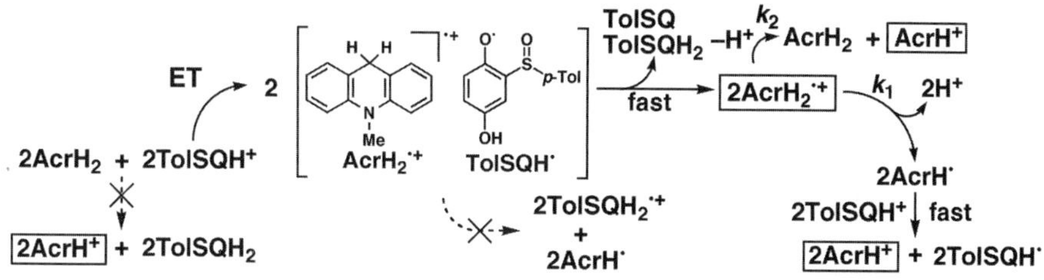

**SCHEME 2.12**

ET from $AcrH_2$ ($E_{ox} = 0.81$ V versus SCE) to TolSQ ($E_{red} = -0.26$ V versus SCE) is highly endergonic because of the highly positive free energy change of ET ($\Delta G_{et} = 1.07$ eV), and thereby no ET reaction occurs in the absence of $HClO_4$.[81] In the presence of $HClO_4$ ($5.0 \times 10^{-2}$ M), however, the one-electron reduction potential of TolSQ is shifted to 0.69 V versus SCE due to protonation of TolSQ.[49] The free energy change of ET from $AcrH_2$ to $TolSQH^+$ is still slightly positive ($\Delta G_{et} = 0.12$ eV). In such a case, the efficient ET from $AcrH_2$ to $TolSQH^+$ is followed by rapid disproportionation of $TolSQH^\bullet$ (Scheme 2.12), which makes the ET undergo to completion.[81]

The absorption at 640 nm due to $AcrH_2^{\bullet+}$ decays accompanied by the rise in absorption at 420 nm due to $AcrH^+$ as shown in Figure 2.11b.[81] The decay dynamics of $AcrH_2^{\bullet+}$ (and rise dynamics of $AcrH^+$) consists of both first-order and second-order processes (Figure 2.11b, inset) due to the deprotonation and disproportionation of $AcrH_2^{\bullet+}$ as shown by solid arrows in Scheme 2.12.[81] Both the first-order and second-order processes exhibit the large primary kinetic isotope effects ($k_H/k_D = 3.2$ and 10, respectively) when $AcrH_2$ is replaced by the dideuterated compound ($AcrD_2$) (Figure 2.11b, inset).[81] The $AcrH^\bullet$, which is produced by deprotonation of $AcrH_2^{\bullet+}$, is a much stronger reductant than $AcrH_2$. Thus, rapid ET from $AcrH^\bullet$ ($E_{ox} = -0.46$ V versus SCE)[41] to $TolSQH^+$ occurs to yield $AcrH^+$ and $TolSQH^\bullet$ (Scheme 2.12).[81] As a consequence, 1 equivalent of $TolSQH^+$ is reduced by 1 equivalent of $AcrH_2$ to yield 1 equivalent of $AcrH^+$ and $TolSQH_2$.[81]

Protonation of TolSQ is also expected to result in enhancement of electrophilicity of TolSQ to accelerate one-step hydride transfer from $AcrH_2$ to $TolSQH^+$, as $His\cdot2H^+$-promoted hydride transfer from $AcrH_2$ to TolSQ (Scheme 2.11).[81] The electrostatic potential map for $TolSQH^+$ (Figure 2.12b) indicates that the positive charges due to protonation of TolSQ are fully delocalized over the entire ring system compared to those of neutral species (Figure 2.12).[81] In such a case, the delocalization of the positive charge (due to $H^+$) in the protonated species ($TolSQH^+$) does not lead to the expected increase of electrophilicity that would promote the ET pathways when the $E_{red}$ value of TolSQ is shifted in the positive direction in the presence of $HClO_4$.[81] This may be the reason why ET from $AcrH_2$ to $TolSQH^+$ occurs instead of the one-step hydride transfer from $AcrH_2$ to $TolSQH^+$.[81]

**(a)**                                                    **(b)**

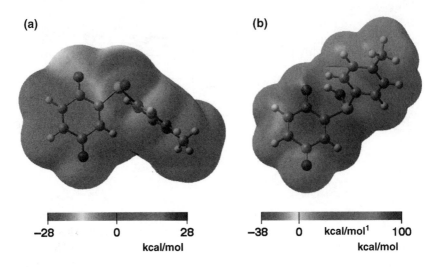

-28              0              28            -38      0    kcal/mol$^1$    100
        kcal/mol                                                    kcal/mol

**FIGURE 2.12**    Electrostatic potential maps for (a) TolSQ and (b) TolSQH$^+$, calculated with density functional theory at the BLYP/6-31G$^{**}$ level.[81] (See the color version of this figure in Color Plates section.)

### 2.4.2    Hydride Reduction of High-Valent Metal-Oxo Complexes by NADH Analogues

*2.4.2.1    One-Step Hydride Transfer via Converted PCET*    Metal-oxo complexes often play important roles as the key species in oxidations of C–H bonds, which are of fundamental importance as biochemical and industrial processes.[82] In particular, high-valent oxoiron(IV) species are frequently invoked as the key intermediates responsible for the oxidation of organic substrates in nonheme iron enzymes.[83–85] Nonheme oxoiron(IV) intermediates have been characterized in the catalytic cycles of *Escherichia coli* taurine: α-ketoglutarate dioxygenase (TauD), prolyl-4-hydroxylase, and halogenase CytC3.[86,87] These results demonstrate unambiguously that nonheme oxoiron(IV) intermediates are capable of abstracting C–H bonds of substrates in biological reactions. In biomimetic studies, a nonheme oxoiron(IV) intermediate was characterized spectroscopically,[88] and then the first crystal structure of an oxoiron(IV) complex, [(TMC)Fe(IV)(O)]$^{2+}$ (TMC = 1,4,8,11-tetramethyl-1,4,8,11-tetraazacyclotetradecane), was obtained in the reaction of [Fe(II)(TMC)]$^{2+}$ and artificial oxidants.[89] Since then, extensive efforts have been devoted to examine the reactivities of mononuclear nonheme oxoiron(IV) complexes bearing tetradentate N4 and pentadentate N5 and N4S ligands in the oxidation of C–H bonds and other substrates, including alkane hydroxylation, olefin epoxidation, alcohol oxidation, N-dealkylation, oxidation of sulfides, and so on.[90–98] There has been the mechanistic discussion with regard to the key initial mechanistic step in C–H bond oxidations, which is ET, HAT, or hydride transfer.[15–30] It has been difficult to understand why one pathway would be preferred over another.

Hydride transfer reactions from NADH analogues to high-valent metal-oxo species provide excellent opportunity to clarify such a mechanistic difference by comparing the hydride transfer reactions with those with $p$-benzoquinone derivatives, which have been discussed in the previous section. A series of NADH analogues, 10-methyl-9,10-dihydroacridine ($AcrH_2$) and its 9-subsituted derivatives (AcrHR: R = H, Ph, Me, and Et), BNAH, and their deuterated compounds, were employed as hydride donors and mononuclear nonheme oxoiron(IV) complexes, $[(L)Fe^{IV}(O)]^{2+}$ (L = N4Py, $N,N$-bis(2-pyridylmethyl)-$N$-bis(2-pyridyl)methylamine; Bn-TPEN, $N$-benzyl-$N,N',N'$-tris(2-pyridylmethyl)ethane-1,2-diamine; TMC, 1,4,8,11-tetramethyl-1,4,8,11-tetraazacyclotetradecane), were used as hydride acceptors (see Chart 2.1).[99]

As a typical example, hydride transfer from an NADH analogue ($AcrH_2$) to $[(N4Py)Fe^{IV}(O)]^{2+}$ (Equation 2.3) occurs efficiently as shown in Figure 2.13a, where the absorption band at 695 nm due to $[(N4Py)Fe^{IV}(O)]^{2+}$ decreases, accompanied by an increase in the absorption band at 357 nm due to 10-methylacridinium ion ($AcrH^+$) and the absorption bands at 380 and 450 nm due to $[(N4Py)Fe^{II}(OH)]^{2+}$.[99] The formation rate of $AcrH^+$ determined from an increase in absorbance at 357 nm coincides with the decay rate of $[(N4Py)Fe^{IV}(O)]^{2+}$ determined from a decrease in absorbance at 695 nm (Figure 2.13a, inset).[99] There was no intermediate observed in the hydride transfer reaction.[99] Thus, the hydride transfer reaction proceeds by one step without intermediates. The pseudo-first-order rate constants ($k_{obs}$) increase linearly with the increase of the $AcrH_2$ concentration (open circles in Figure 2.13c).[99] The second-order rate constant ($k_H$) for the reaction of $[(N4Py)Fe^{IV}(O)]^{2+}$ and $AcrH_2$ was determined to be $1.1 \times 10^2\,M^{-1}\,s^{-1}$ from the linear plot of $k_{obs}$ versus concentration of $AcrH_2$.[99] When $AcrH_2$ was replaced by

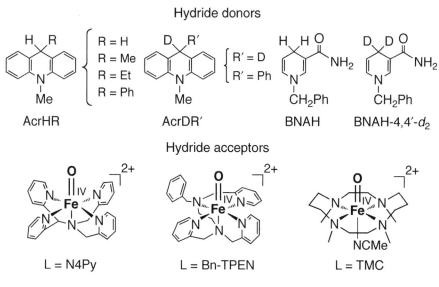

Hydride donors

AcrHR {R = H, R = Me, R = Et, R = Ph}     AcrDR' {R' = D, R' = Ph}     BNAH     BNAH-4,4'-$d_2$

Hydride acceptors

L = N4Py     L = Bn-TPEN     L = TMC

**CHART 2.1**

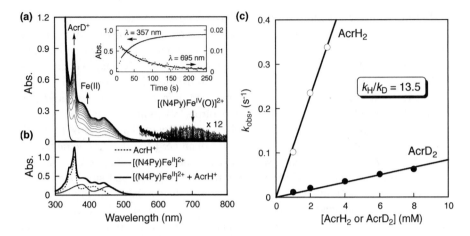

**FIGURE 2.13** (a) Spectral changes observed in the reaction of $[(N4Py)Fe^{IV}(O)]^{2+}$ $(5.0 \times 10^{-5} M)$ and $AcrD_2$ $(2.0 \times 10^{-3} M)$ in deaerated MeCN at 298 K. *Inset*: Time profiles of absorption changes at $\lambda = 357$ and 695 nm for the formation of $AcrH^+$ and decay of $[(N4Py)Fe^{IV}(O)]^{2+}$, respectively. (b) UV-Vis spectra of $AcrH^+$ $(6.0 \times 10^{-5} M)$ and $[(N4Py)Fe^{II}]^{2+}$ $(6.0 \times 10^{-5} M)$ in MeCN. (c) Plots of the pseudo-first-order rate constants $(k_{obs})$ of $[(N4Py)Fe^{IV}(O)]^{2+}$ versus $[AcrH_2$ (○) or $AcrD_2$ (●)].

the dideuterated compound ($AcrD_2$), a large kinetic deuterium isotope (KIE) value of 13.5 was observed (closed circles in Figure 2.13c).[99]

$$AcrH_2 + [(N4Py)Fe^{IV}(O)]^{2+} \xrightarrow{k_H} AcrH^+ + [(N4Py)Fe^{II}(OH)]^+$$

$$\text{AcrH}_2 \qquad\qquad\qquad\qquad\qquad\qquad\qquad \text{AcrH}^+$$

(2.3)

The $k_H$ values determined in the reactions of $[(L)Fe^{IV}(O)]^{2+}$ vary significantly depending on NADH analogues, in particular on the substituent R in AcrHR.[99] Similar reactivity change was observed for hydride transfer reactions from the same series of NADH analogues to *p*-chloranil ($Cl_4Q$) (Equation 2.4).[13] Thus, there are excellent linear correlations between the $k_H$ values of hydride transfer reactions of NADH analogues with $[(L)Fe^{IV}(O)]^{2+}$ and the corresponding values with $Cl_4Q$ as shown in Figure 2.14.[99] The significant decrease in the reactivity by the introduction of a substituent R at the C-9 position can hardly be reconciled by a one-step transfer of a hydride ion, although no intermediates were observed during the reactions.[99] The alkyl or phenyl group at the C-9 position is known to be in a boat axial conformation,[41] and thereby the hydrogen at the C-9 position is located at the

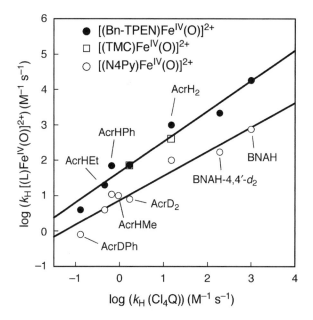

**FIGURE 2.14** Plots of $k_H$ for hydride transfer from NADH analogues to [(Bn-TPEN) Fe$^{IV}$(O)]$^{2+}$ (●), [(TMC)Fe$^{IV}$(O)]$^{2+}$ (□), and [(N4Py)Fe$^{IV}$(O)]$^{2+}$ (○) versus $k_H$ for hydride transfer from the same series of NADH analogues to Cl$_4$Q in MeCN at 298K.[99]

equatorial position, where steric hindrance due to the axial substituent is minimized in the hydride transfer reactions.[13] The introduction of an electron-donating substituent R would activate the release of a negatively charged hydride ion if the concerted hydride transfer should take place. The remarkable decrease in the reactivity with the increasing electron-donor ability of R rather indicates that the reactivity is determined by the process in which a positive charge is released.

$$\text{AcrHR} + \text{Cl}_4\text{Q} \xrightarrow{k_H} \text{AcrR}^+ + \text{Cl}_4\text{QH}^- \qquad (2.4)$$

Linear correlations between hydride transfer reactions of NADH analogues with [(L)Fe$^{IV}$(O)]$^{2+}$ and Cl$_4$Q in Figure 2.14 imply that the hydride transfer mechanism of [(L)Fe$^{IV}$(O)]$^{2+}$ is virtually the same as that of Cl$_4$Q.[99] Although there is still debate on the mechanism(s) of hydride transfer from NADH analogues to hydride acceptors in terms of an ET pathway versus a one-step hydride transfer pathway, the ET pathway is now well accepted for hydride transfer from NADH analogues to hydride acceptors

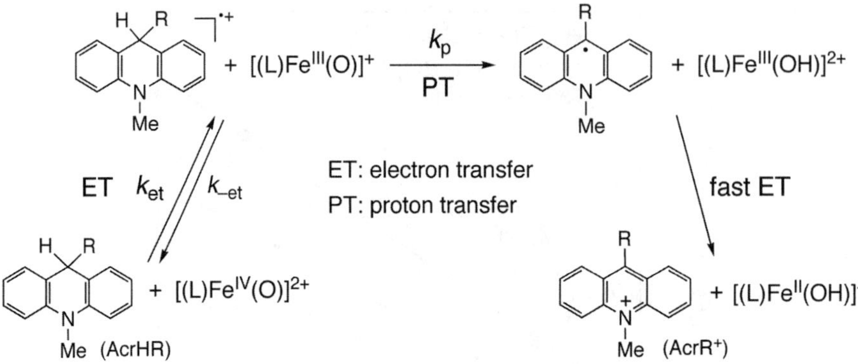

$$[\text{AcrHR}]^{\bullet +} + [(L)Fe^{III}(O)]^+ \xrightarrow[\text{PT}]{k_p} [\text{AcrR}^{\bullet}] + [(L)Fe^{III}(OH)]^{2+}$$

ET $k_{et}$ / $k_{-et}$

ET: electron transfer
PT: proton transfer

fast ET

$\text{AcrHR} + [(L)Fe^{IV}(O)]^{2+}$

$\text{AcrR}^+ + [(L)Fe^{II}(OH)]^+$

**SCHEME 2.13**

that are strong electron acceptors.[13,64,81] It should be noted that the $E_{red}$ values of nonheme oxoiron(IV) complexes ($[(L)Fe^{IV}(O)]^{2+}$; 0.39–0.51 V versus SCE)[100] are more positive than $Cl_4Q$ (0.01 V versus SCE).[64] This means that the $[(L)Fe^{IV}(O)]^{2+}$ complexes are a much stronger electron acceptor than $Cl_4Q$. Thus, the ET from NADH analogues (AcrHR) to $[(L)Fe^{IV}(O)]^{2+}$ is highly likely to occur, and this is followed by rapid proton transfer from $AcrHR^{\bullet +}$ to $[(L)Fe^{III}(O)]^+$ and ET from the resulting $AcrR^{\bullet}$ to $[(L)Fe^{III}(OH)]^{2+}$ in competition with the back ET, affording the final products, $AcrH^+$ and $[(L)Fe^{II}(OH)]^+$, as shown in Scheme 2.13.[99] Since no ET intermediate is observed in this case, the initial ET is coupled with PT, which is a PCET process, followed by fast ET.[99]

A similar mechanism to Scheme 2.13 was proposed for hydride transfer from $AcrH_2$ to a Ru(IV)-oxo species, $cis$-$[Ru^{IV}(bpy)_2(py)(O)]^{2+}$ (bpy = 2,2'-bipyridine; py = pyridine), as shown in Scheme 2.14.[39] The rate-determining step is the PCET (HT) process from $AcrH_2$ to $cis$-$[Ru^{IV}(bpy)_2(py)(O)]^{2+}$, which is followed by fast ET from $AcrH^{\bullet}$ to $cis$-$[Ru^{III}(bpy)_2(py)(OH)]^{2+}$ to yield $AcrH^+$ and $cis$-$[Ru^{II}(bpy)_2(py)(OH)]^+$. The initial HAT process exhibits a large KIE ($k_H/k_D = 12 \pm 1$),[39] which is similar to the case of the PCET process from $AcrH_2$ to $[(N4Py)Fe^{IV}(O)]^{2+}$ (Figure 2.13c). In contrast to the case of $[(N4Py)Fe^{IV}(O)]^{2+}$ in Scheme 2.14, however, nucleophilic attack of $OH^-$ in $cis$-$[Ru^{II}(bpy)_2(py)(OH)]^+$ to $AcrH^+$ occurs slowly in MeCN to give AcrH(OH) and $cis$-$[Ru^{II}(bpy)_2(py)(MeCN)]^{2+}$.[43] This process results from the high $pK_a$ value of $cis$-$[Ru^{II}(bpy)_2(py)(OH_2)]^{2+}$ (10.6) compared to the low $pK_a$ value of $cis$-$[Ru^{III}(bpy)_2(py)(OH_2)]^{2+}$ (0.85).[101] Then, hydride transfer from AcrH(OH) to $AcrH^+$ occurs to yield the final product, 10-methylacridone (AcrO).[102] Thus, the overall reaction is the four-electron oxidation of $AcrH^+$ by 2 equivalents of $cis$-$[Ru^{IV}(bpy)_2(py)(O)]^{2+}$ in MeCN to afford AcrO and $cis$-$[Ru^{II}(bpy)_2(py)(MeCN)]^{2+}$.[43]

Over a series of similar HAT reactions, there is typically a good correlation of rate constants with C–H bond dissociation energies.[103,104] Figure 2.15 shows such a correlation for the oxidations of a series of alkyl aromatic and allylic C–H bonds by

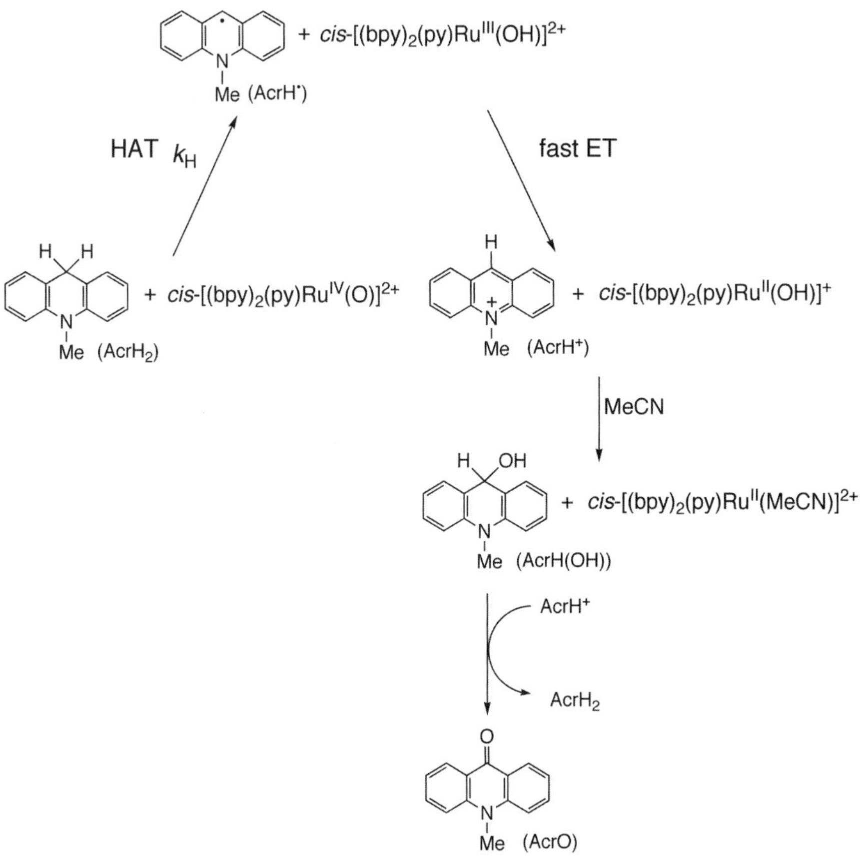

**SCHEME 2.14**

$cis$-$[Ru^{IV}(bpy)_2(py)(O)]^{2+}$.[43] Using the BDEs of $AcrH_2$ (73.7 kcal/mol)[105] and BNAH (67.9 kcal/mol),[105] these rate constants are added to this correlation as shown in Figure 2.15.[43] The rate constants for $AcrH_2$ and BNAH oxidations fit the correlation well. Such a linear correlation in Figure 2.15 strongly indicates that the reactions of all these compounds, including the NADH analogues, proceed via a common HAT mechanism in which a pathway of initial HAT, not hydride transfer from $AcrH_2$, is the rate-determining step. ET from $AcrH^•$ to $cis$-$[(bpy)_2(py)Ru^{III}(OH)]^{2+}$ is highly exergonic when the rate must be very fast.[43]

An alternative mechanism of initial rate-limiting ET from $AcrH_2$ to $cis$-$[Ru^{IV}(bpy)_2(py)(O)]^{2+}$ is ruled out by the large KIE value ($12 \pm 1$).[43] In addition, the ET from $AcrH_2$ ($E_{ox} = 0.81$ V versus SCE)[64] to $cis$-$[Ru^{IV}(bpy)_2(py)(O)]^{2+}$ ($E_{red} < 0.26$ V versus SCE)[101] is thermodynamically infeasible. However, whether the HAT reaction proceeds via one-step transfer of a hydrogen atom or PCET has yet to be clarified.

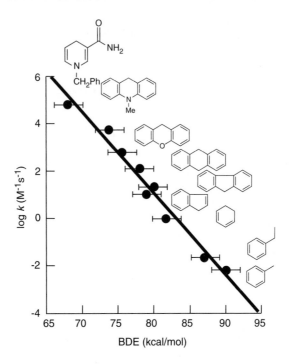

**FIGURE 2.15**   Plot of rate constants for oxidations by $cis\text{-}[Ru^{IV}(bpy)_2(py)(O)]^{2+}$ versus C–H bond energies.[43]

### 2.4.2.2   Stepwise Electron Transfer Pathway   As describe above, hydride transfer from $AcrH_2$ to $[(L)Fe^{IV}(O)]^{2+}$ occurs via PCET without formation of the ET intermediate because ET from $AcrH_2$ to $[(L)Fe^{IV}(O)]^{2+}$ is endergonic (Scheme 2.13). However, the ET process may become exergonic in the presence of $HClO_4$ as in the case of ET from $AcrH_2$ to the protonated hydride acceptor $(TolSQH^+)$ in Scheme 2.12. When the ET process becomes exergonic in the presence of $HClO_4$, the radical cation $(AcrHR^{\bullet +})$ produced in the acid-promoted hydride transfer from AcrHR to $[(L)Fe^{IV}(O)]^{2+}$ can be observed (see below).

ET from AcrDPh to $[(N4Py)Fe^{IV}(O)]^{2+}$ occurs in the presence of $HClO_4$ in MeCN as indicated by instant appearance of a transient absorption band at $\lambda_{max} = 680\,nm$, which is ascribed to the formation of $AcrDPh^{\bullet +}$ as shown in Figure 2.16a.[94] The disappearance of the absorbance at $\lambda_{max} = 680\,nm$ coincides with the appearance of absorbance at 380 nm due to $[(N4Py)Fe^{II}]^{2+}$ and at 360 nm due to $Acr^+\text{–}Ph$ (see inset of Figure 2.16a).[99] Similarly, the formation of $AcrHEt^{\bullet +}$ was observed in the acid-promoted hydride transfer from AcrHEt to $[(N4Py)Fe^{IV}(O)]^{2+}$ as shown in Figure 2.16b, where the disappearance of the absorbance at $\lambda_{max} = 685\,nm$ due to $AcrHEt^{\bullet +}$ also coincides with the appearance of absorbance at 360 nm due to $Acr^+\text{–}Et$ (see inset of Figure 2.16b).[99] These results indicate that an acid-promoted ET from AcrHR to $[(N4Py)Fe^{IV}(O)]^{2+}$ occurs first to produce $AcrHR^{\bullet +}$ and $[(N4Py)Fe^{III}(OH)]^{2+}$, followed by the generation of $AcrR^{\bullet}$ resulting from

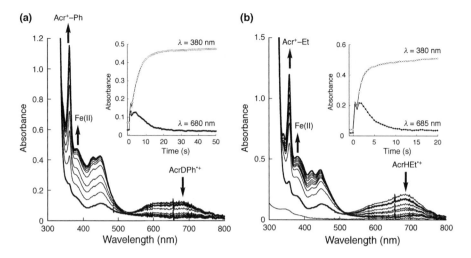

**FIGURE 2.16** Absorption spectral changes observed on addition of (a) AcrDPh ($2.5 \times 10^{-3}$ M) and (b) AcrHEt ($2.5 \times 10^{-3}$ M) to a solution of $[(N4Py)Fe^{IV}(O)]^{2+}$ ($5.0 \times 10^{-5}$ M) in MeCN in the presence of $HClO_4$ ($2.2 \times 10^{-3}$ M) at 298K. *Inset*: Time profiles of the absorption change at (a) $\lambda = 380$ and 680 nm and (b) $\lambda = 380$ and 685 nm.

the deprotonation of AcrHR$^{\bullet+}$.[99] Subsequently, rapid ET from AcrR$^{\bullet}$ to $[(N4Py)Fe^{III}(OH)]^{2+}$ affords the final products, AcrR$^{+}$ and $[(N4Py)Fe^{II}(OH)]^{+}$, as shown in Scheme 2.15.[99]

The ET step from AcrH$_2$ to hydrogen chromate ions (H$_2$CrO$_4$) is also reported to initiate the oxidation of AcrH$_2$ to AcrH$^+$ in H$_2$O/MeCN (4:1, v/v) via a radical chain mechanism that is strongly inhibited by oxygen.[70]

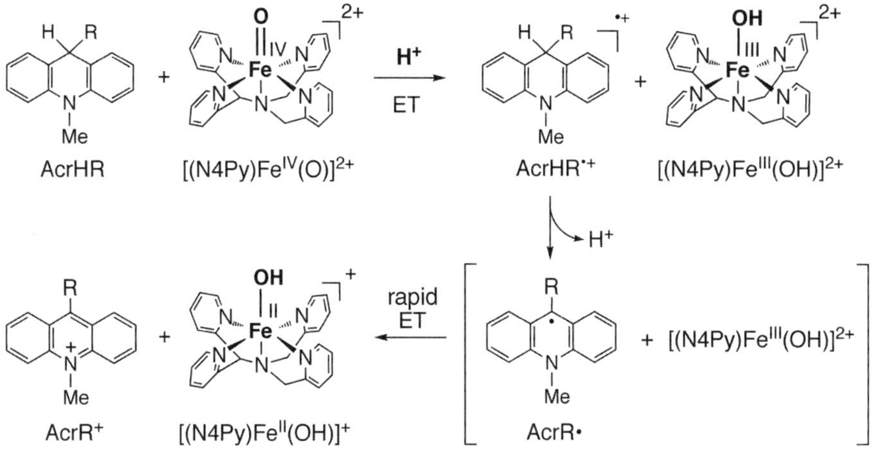

**SCHEME 2.15**

## 2.5 CONCLUSIONS

As demonstrated in this chapter, there have always been the fundamental mechanistic questions in oxidation of C–H bonds whether the rate-determining step is ET, PCET, one-step HAT, or one-step hydride transfer. When the ET step is thermodynamically feasible, ET occurs first, followed by proton transfer for the overall HAT reactions, and the HAT step is followed by subsequent rapid ET for the overall hydride transfer reactions. In such a case, ET products, that is, radical cations of electron donors and radical anions of electron acceptors, can be detected as the intermediates in the overall HAT and hydride transfer reactions. The ET process can be coupled by proton transfer and also by hydrogen bonding or by binding of metal ions to the radical anions produced by ET to control the ET process. The borderline between a sequential PCET pathway and a one-step HAT pathway has been related to the borderline between the outer-sphere and inner-sphere ET pathways. In HAT reactions, the proton is provided by radical cations of electron donors because the acidity is significantly enhanced by the one-electron oxidation of electron donors. An electron and a proton are transferred by a one-step pathway or a sequential pathway depending on the types of electron donors and acceptors. When proton is provided externally, ET from an electron donor that has no proton to be transferred to an electron acceptor (A) is coupled with protonation of $A^{\bullet-}$, when the one-electron reduction and protonation of A occur simultaneously. The mechanistic discussion described in this chapter will provide useful guide to control oxidation of C–H bonds.

## ACKNOWLEDGMENTS

The author gratefully acknowledges the contributions of his collaborators and coworkers mentioned in the references. The author thanks the Ministry of Education, Culture, Sports, Science and Technology, Japan, for the continuous support.

## REFERENCES

1. Dyker, G., Ed.; *Handbook of C–H Transformations*; Wiley-VCH: Weinheim, 2005.
2. Kochi, J. K., Ed.; *Free Radicals*; Wiley: New York, 1973.
3. Tanko, J. M. Reaction mechanisms. Part I. Radical and radical ion reactions. *Annu. Rep. Prog. Chem. B: Org. Chem.* **2008**, *104*, 234–259.
4. Litwinienko, G.; Ingold, K. U. *Acc. Chem. Res.* **2007**, *40*, 222–230.
5. Piotrowiak, P., Ed. *Electron Transfer in Chemistry. Part 1. Principles and Theories*; Balzani, V., Ed.; Wiley-VCH: Weinheim, 2001, Vol. 1.
6. (a) Marcus, R. A. *Discuss. Faraday Soc.* **1960**, *29*, 21–31; (b) Marcus, R. A.; Sutin, N. *Biochim. Biophys. Acta* **1985**, *811*, 265–322.
7. Rosokha, S. V.; Kochi, J. K. *J. Am. Chem. Soc.* **2007**, *129*, 3683–3697.
8. Fukuzumi, S.; Wong, C. L; Kochi, J. K. *J. Am. Chem. Soc.* **1980**, *102*, 2928–2939.

9. Rosokha, S. V.; Kochi, J. K. *Acc. Chem. Res.* **2008**, *41*, 641–653.

10. Taube, H. *Angew. Chem., Int. Ed. Engl.* **1984**, *23*, 329.

11. Taube, H. *Electron-Transfer Reactions of Complex Ions in Solution*; Academic Press: New York, 1970.

12. Fukuzumi, S.; Kochi, J. K. *J. Am. Chem. Soc.* **1981**, *103*, 7240–7252.

13. Fukuzumi, S.; Ohkubo, K.; Tokuda, Y.; Suenobu, T. *J. Am. Chem. Soc.* **2000**, *122*, 4286–4294.

14. Rathore, R.; Lindeman, S.; Kochi, J. K. *J. Am. Chem. Soc.* **1997**, *119*, 9393–9404.

15. Huynh, M. H. V.; Meyer, T. J. *Chem. Rev.* **2007**, *107*, 5004–5064.

16. Cukier, R. I.; Nocera, D. G. *Annu. Rev. Phys. Chem.* **1998**, *49*, 337–369.

17. Chang, C. J.; Chang, M. C. Y.; Damrauer, N. H.; Nocera, D. G. *Biochim. Biophys. Acta* **2004**, *1655*, 13–28.

18. Mayer, J. M.; Rhile, I. J. *Biochim. Biophys. Acta* **2004**, *1655*, 51–58.

19. Mayer, J. M. *Annu. Rev. Phys. Chem.* **2004**, *55*, 363–390.

20. Hammes-Schiffer, S. In *Electron Transfer in Chemistry*; Balzani, V., Ed.; Wiley-VCH: Weinheim, 2001; Vol. 1, pp 189–237.

21. Stubbe, J.; Nocera, D. G.; Yee, C. S.; Chang, M. C. Y. *Chem. Rev.* **2003**, *103*, 2167–2202.

22. Costentin, C. *Chem. Rev.* **2008**, *108*, 2145–2179.

23. Mayer, J. M.; Hrovat, D. A.; Thomas, J. L.; Borden, W. T. *J. Am. Chem. Soc.* **2002**, *124*, 11142–11147.

24. Isborn, C.; Hrovat, D. A.; Borden, W. S.; Mayer, J. M.; Carpenter, B. K. *J. Am. Chem. Soc.* **2005**, *127*, 5794–5795.

25. Brash, A. R. *J. Biol. Chem.* **1999**, *274*, 23679–23682.

26. Boyington, J. C.; Gaffney, B. J.; Amzel, L. M. *Science* **1993**, *260*, 1482.

27. Skrzypczak-Jankun, E.; Bross, R. A.; Carroll, R. T.; Dunham, W. R.; Funk, M. O., Jr. *J. Am. Chem. Soc.* **2001**, *123*, 10814–10820.

28. Lehnert, N.; Solomon, E. I. *J. Biol. Inorg. Chem.* **2003**, *8*, 294–305.

29. Fukuzumi, S. *Helv. Chim. Acta* **2006**, *89*, 2425–2440.

30. Tishchenko, O.; Truhlar, D. G.; Ceulemans, A.; Nguyen, M. T. *J. Am. Chem. Soc.* **2008**, *130*, 7000–7010.

31. Fukuzumi, S. In *Electron Transfer in Chemistry*; Balzani, V., Ed.; Wiley-VCH: Weinheim, 2001; Vol. 4, pp 3–67.

32. Fukuzumi, S. *Bull. Chem. Soc. Jpn.* **1997**, *70*, 1–28.

33. Fukuzumi, S.; Itoh, S. In *Advances in Photochemistry*; Neckers, D. C.; Volman D.H.; von Bünau, G., Eds.; Wiley: New York, 1998; Vol. 25, pp 107–172.

34. Fukuzumi, S.; Itoh, S. *Antioxid. Redox Signal.* **2001**, *3*, 807–824.

35. Fukuzumi, S. *Org. Biomol. Chem.* **2003**, *1*, 609–620.

36. Fukuzumi, S. *Prog. Inorg. Chem.* **2008**, *56*, 49–153.

37. Stryer, L.; *Biochemistry*; 3rd edition; Freeman: New York, 1988, Chapter 17.

38. (a) Eisner, U.; Kuthan, J. *Chem. Rev.* **1972**, *72*, 1–42; (b) Stout, D. M.; Meyers, A. I. *Chem. Rev.* **1982**, *82*, 223–243.

39. Fukuzumi, S.; Tanaka, T. In *Photoinduced Electron Transfer. Part C*; Fox, M. A.; Chanon, M., Eds.; Elsevier: Amsterdam, 1988, Chapter 10.

40. Fukuzumi, S. In *Advances in Electron Transfer Chemistry*; Mariano, P. S., Ed.; JAI Press: Greenwich, CT, 1992, pp 67–175.

41. Fukuzumi, S.; Tokuda, Y.; Kitano, T.; Okamoto, T.; Otera, J. *J. Am. Chem. Soc.* **1993**, *115*, 8960–8968.

42. Gebicki, J.; Marcinek, A.; Zielonka, J. *Acc. Chem. Res.* **2004**, *37*, 379.

43. Matsuo, T.; Mayer, J. M. *Inorg. Chem.* **2005**, *44*, 2150.

44. Fukuzumi, S.; Inada, O.; Suenobu, T. *J. Am. Chem. Soc.* **2003**, *125*, 4808.

45. Pestovsky, O.; Bakac, A.; Espenson, J. H. *J. Am. Chem. Soc.* **1998**, *120*, 13422.

46. Pestovsky, O.; Bakac, A.; Espenson, J. H. *Inorg. Chem.* **1998**, *37*, 1616–1622.

47. Fukuzumi, S.; Ohkubo, K.; Suenobu, T.; Kato, K.; Fujitsuka, M.; Ito, O. *J. Am. Chem. Soc.* **2001**, *123*, 8459–8467.

48. Yuasa, J.; Fukuzumi, S. *J. Am. Chem. Soc.* **2006**, *128*, 14281–14292.

49. Yuasa, J.; Yamada, S.; Fukuzumi, S. *J. Am. Chem. Soc.* **2008**, *130*, 5808–5820.

50. Stowell, M. H. B.; McPhillips, T.; Rees, D. C.; Soltis, S. M.; Abresch, E.; Feher, G. *Science* **1997**, *276*, 812–816.

51. Ädelroth, P.; Paddock, M. L.; Sagle, L. B.; Feher, G.; Okamura, M. Y. *Proc. Natl. Acad. Sci. USA* **2000**, *97*, 13086–13091.

52. Rotello, V. M. In *Electron Transfer in Chemistry*; Balzani, V., Ed.; Wiley-VCH: Weinheim, 2001; Vol. 4, pp 68–87.

53. Okamoto, K.; Ohkubo, K.; Kadish, K. M.; Fukuzumi, S. *J. Phys. Chem. A* **2004**, *108*, 10405–10413.

54. Rhile, I. J.; Markle, T. F.; Nagao, H.; DiPasquale, A. G.; Lam, O. P.; Lockwood, M. A.; Rotter, K.; Mayer, J. M. *J. Am. Chem. Soc.* **2006**, *128*, 6075–6088.

55. Sjödin, M.; Irebo, T.; Utas, J. E.; Lind, J.; Merényi, G.; Åkermark, B.; Hammarström, L. *J. Am. Chem. Soc.* **2006**, *128*, 13076–13083.

56. Costentin, C.; Robert, M.; Savéant, J.-M. *J. Am. Chem. Soc.* **2007**, *129*, 9953–9963.

57. Fukuzumi, S.; Ishikawa, K.; Hironaka, K.; Tanaka, T. *J. Chem. Soc., Perkin Trans.* **1987**, 2, 751–760.

58. Fukuzumi, S.; Mochizuki, S.; Tanaka, T. *J. Am. Chem. Soc.* **1989**, *111*, 1497–1499.

59. Fukuzumi, S.; Kitano, T.; Mochida, K. *J. Am. Chem. Soc.* **1990**, *112*, 3246–3247.

60. Fukuzumi, S.; Tokuda, Y. *J. Phys. Chem.* **1992**, *96*, 8409–8413.

61. Yuasa, J.; Yamada, S.; Fukuzumi, S. *J. Am. Chem. Soc.* **2006**, *128*, 14938–14948.

62. Fukuzumi, S.; Yoshida, Y.; Okamoto, K.; Imahori, H.; Araki, Y.; Ito, O. *J. Am. Chem. Soc.* **2002**, *124*, 6794–6795.

63. Okamoto, K.; Yoshida, Y.; Imahori, H.; Araki, Y.; Ito, O. *J. Am. Chem. Soc.* **2003**, *125*, 1007–1013.

64. Fukuzumi, S.; Koumitsu, S.; Hironaka, K.; Tanaka, T. *J. Am. Chem. Soc.* **1987**, *109*, 305–316.

65. Fukuzumi, S.; Okamoto, T. *J. Am. Chem. Soc.* **1993**, *115*, 11600–11601.

66. Fukuzumi, S.; Ohkubo, K. *Chem. Eur. J.* **2000**, *6*, 4532–4535.

67. Ohkubo, K.; Menon, S. C.; Orita, A.; Otera, J.; Fukuzumi, S. *J. Org. Chem.* **2003**, *68*, 4720–4726.

68. Afanasyeva, M. S.; Taraban, M. B.; Purtov, P. A.; Leshina, T. V.; Grissom, C. B. *J. Am. Chem. Soc.* **2006**, *128*, 8651–8658.

69. Lee, I.-S. H.; Jeoung, E. H.; Kreevoy, M. M. *J. Am. Chem. Soc.* **1997**, *119*, 2722–2728.

70. Pestovsky, O.; Bakac, A.; Espenson, J. H. *J. Am. Chem. Soc.* **1998**, *120*, 13422–13428.

71. Yuasa, J.; Yamada, S.; Fukuzumi, S. *J. Am. Chem. Soc.* **2006**, *128*, 14938–14292.

72. Fukuzumi, S.; Ohkubo, K.; Okamoto, T. *J. Am. Chem. Soc.* **2002**, *124*, 14147–14155.

73. Fukuzumi, S.; Fujii, Y.; Suenobu, T. *J. Am. Chem. Soc.* **2001**, *123*, 10191–10199.

74. Reichenbach-Klinke, R.; Kruppa, M.; König, B. *J. Am. Chem. Soc.* **2002**, *124*, 12999–13007.

75. Fukuzumi, S.; Ishikawa, M.; Tanaka, T. *J. Chem. Soc., Perkin Trans.* **1989**, *2*, 1037–1045.

76. (a) Fukuzumi, S.; Mochizuki, S.; Tanaka, T. *J. Am. Chem. Soc.* **1989**, *111*, 1497–1499; (b) Fukuzumi, S.; Ishikawa, M.; Tanaka, T. *Chem. Lett.* **1989**, 1227–1230.

77. (a) Carlson, B. W.; Miller, L. L. *J. Am. Chem. Soc.* **1985**, *107*, 479–485; (b) Miller, L. L.; Valentine, J. R. *J. Am. Chem. Soc.* **1988**, *110*, 3982–3989.

78. (a) Coleman, C. A.; Rose, J. G.; Murray, C. J. *J. Am. Chem. Soc.* **1992**, *114*, 9755–9762; (b) Murray, C. J.; Webb, T. *J. Am. Chem. Soc.* **1991**, *113*, 7426–7427.

79. Polyansky, D.; Cabelli, D.; Muckerman, J. T.; Fujita, E.; Koizumi, T.; Fukushima, T.; Wada, T.; Tanaka, K. *Angew. Chem., Int. Ed.* **2007**, *46*, 4169–4172.

80. Eklund, H.; Branden, C.-I. In *Zinc Enzymes*; Spiro, T. G., Ed.; Wiley–Interscience: New York, 1983; Chapter 4.

81. Yuasa, J.; Yamada, S.; Fukuzumi, S. *Angew. Chem., Int. Ed.* **2008**, *47*, 1068–1071.

82. Meunier, B., Ed.; *Biomimetic Oxidations Catalyzed by Transition Metal Complexes*; Imperial College Press: London, 2000.

83. Kryatov, S. V.; Rybak-Akimova, E. V.; Schindler, S. *Chem. Rev.* **2005**, *105*, 2175–2226.

84. Abu-Omar, M. M.; Loaiza, A.; Hontzeas, N. *Chem. Rev.* **2005**, *105*, 2227–2252.

85. (a) Borovik, A. S. *Acc. Chem. Res.* **2005**, *38*, 54–61; (b) Costas, M.; Mehn, M. P.; Jensen, M. P.; Que, L. Ć., Jr. *Chem. Rev.* **2004**, *104*, 939–986.

86. (a) Krebs, C.; Galonić Fujimori, D. G.; Walsh, C. T.; Bollinger, J. M., Jr. *Acc. Chem. Res.* **2007**, *40*, 484–492; (b) Hoffart, L. M.; Barr, E. W.; Guyer, R. B.; Bollinger, J. M., Jr.; Krebs, C. *Proc. Natl. Acad. Sci. USA* **2006**, *103*, 14738–14743; (c) Galonić, D. P.; Barr, E. W.; Walsh, C. T.; Bollinger, J. M., Jr.; Krebs, C. *Nat. Chem. Biol.* **2007**, *3*, 113–116.

87. (a) Riggs-Gelasco, P. J.; Price, J. C.; Guyer, R. B.; Brehm, J. H.; Barr, E. W.; Bollinger, J. M., Jr.; Krebs, C. *J. Am. Chem. Soc.* **2004**, *126*, 8108–8109; (b) Price, J. C.; Barr, E. W.; Glass, T. E.; Krebs, C.; Bollinger, J. M., Jr. *J. Am. Chem. Soc.* **2003**, *125*, 13008–13009.

88. Grapperhaus, C. A.; Mienert, B.; Bill, E.; Weyhermüller, T.; Wieghardt, K. *Inorg. Chem.* **2000**, *39*, 5306–5317.

89. Rohde, J.-U.; In, J.-H.; Lim, M. H.; Brennessel, W. W.; Bukowski, M. R.; Stubna, A.; Münck, E.; Nam, W.; Que, L., Jr. *Science* **2003**, *299*, 1037–1039.

90. Nam, W. *Acc. Chem. Res.* **2007**, *40*, 522–531.

91. Que, L., Jr. *Acc. Chem. Res.* **2007**, *40*, 493–500.

92. (a) Kaizer, J.; Klinker, E. J.; Oh, N. Y.; Rohde, J.-U.; Song, W. J.; Stubna, A.; Kim, J.; Münck, E.; Nam, W.; Que, L., Jr. *J. Am. Chem. Soc.* 2004, *126*, 472–473; (b) Oh, N. Y.; Suh, Y.; Park, M. J.; Seo, M. S.; Kim, J.; Nam, W. *Angew. Chem., Int. Ed.* 2005, *44*, 4235–4239; (c) Kim, S. O.; Sastri, C. V.; Seo, M. S.; Kim, J.; Nam, W. *J. Am. Chem. Soc.* 2005, *127*, 4178–4179.

93. Bukowski, M. R.; Koehntop, K. D.; Stubna, A.; Bominaar, E. L.; Halfen, J. A.; Münck, E.; Nam, W.; Que, L., Jr. *Science* **2005**, *310*, 1000–1002.

94. (a) Park, M. J.; Lee, J.; Suh, Y.; Kim, J.; Nam, W. *J. Am. Chem. Soc.* **2006**, *128*, 2630–2634;(b) Sastri, C. V.; Oh, K.; Lee, Y. J.; Seo, M. S.; Shin, W.; Nam, W. *Angew. Chem., Int. Ed.* **2006**, *45*, 3992–3995.

95. (a) Nehru, K.; Seo, M. S.; Kim, J.; Nam, W. *Inorg. Chem.* **2007**, *46*, 293–298; (b) Sastri, C. V.; Lee, J.; Oh, K.; Lee, Y. J.; Lee, J.; Jackson, T. A.; Ray, K.; Hirao, H.; Shin, W.; Halfen, J. A.; Kim, J.; Que, L., Jr.; Shaik, S.; Nam, W. *Proc. Natl. Acad. Sci. USA* **2007**, *104*, 19181–19186.

96. (a) Martinho, M.; Banse, F.; Bartoli, J.-F.; Mattioli, T. A.; Battioni, P.; Horner, O.; Bourcier, S.; Girerd, J.-J. *Inorg. Chem.* **2005**, *44*, 9592–9596; (b) Balland, V.; Charlot, M.-F.; Banse, F.; Girerd, J.-J.; Mattioli, T. A.; Bill, E.; Bartoli, J.-F.; Battioni, P.; Mansuy, D. *Eur. J. Inorg. Chem.* **2004**, 301–308.

97. Bautz, J.; Comba, P.; Lopez de Laorden, C. L.; Menzel, M.; Rajaraman, G. *Angew. Chem., Int. Ed.* **2007**, *46*, 8067–8070.

98. (a) Anastasi, A. E.; Comba, P.; McGrady, J.; Lienke, A.; Rohwer, H. *Inorg. Chem.* **2007**, *46*, 6420–6426; (b) Bautz, J.; Bukowski, M. R.; Kerscher, M.; Stubna, A.; Comba, P.; Lienke, A.; Münck, E.; Que, L., Jr. *Angew. Chem., Int. Ed.* **2006**, *45*, 5681–5684.

99. Fukuzumi, S.; Kotani, H.; Lee, Y.-M.; Nam, W. *J. Am. Chem. Soc.* **2008**, *130*, 15134–15142.

100. Lee, Y.-M.; Kotani, H.; Suenobu, T.; Nam, W.; Fukuzumi, S. *J. Am. Chem. Soc.* **2008**, *130*, 434–435.

101. Moyer, B. A.; Meyer, T. J. *Inorg. Chem.* **1981**, *20*, 436–444.

102. Shinkai, S.; Tsuno, T.; Manabe, O. *J. Chem. Soc., Perkin Trans.* **1984**, *2*, 661–665.

103. Mayer, J. M. *Acc. Chem. Res.* **1998**, *31*, 441–450.

104. Bryant, J. R.; Mayer, J. M. *J. Am. Chem. Soc.* **2003**, *125*, 10351–10361.

105. Zhu, X.-Q.; Li, H.-R.; Li, Q.; Ai, T.; Lu, J.-Y.; Yang, Y.; Cheng, J.-P. *Chem. Eur. J.* **2003**, *9*, 871–880.

# 3 Oxygen Atom Transfer

MAHDI M. ABU-OMAR

## 3.1 INTRODUCTION AND THERMODYNAMIC CONSIDERATIONS

Many reactions in biology and in the chemical industry involve the transfer of an oxygen atom from a transition metal center to a substrate molecule and vice versa. Molecular oxygen ($O_2$) and hydrogen peroxide ($H_2O_2$) are the most common reagents used for inserting oxygen into C–H or C–C bonds and for oxygenation reactions such as epoxidation and sulfoxidation. The enzyme cytochrome P450 is ubiquitous to all forms of life and its various isozymes catalyze the oxidation of various organic substrates. These reactions are vital to biosynthesis, metabolism, and detoxification of drugs. On the industrial front, the success story of the petrochemical industry in the twentieth century has depended largely on the functionalization of hydrocarbons produced from oil cracking to useful products. For example, approximately 10 million metric tons of ethylene oxide (epoxyethylene) is produced annually from the reaction of ethylene with $O_2$ over a heterogeneous silver catalyst (Equation 3.1).[1] Ethylene oxide is used in the production of ethylene glycol and as sterilizing agent for foods and medical supplies. In comparison, propylene oxide, which is used as a monomer for the production of polyether polyols for use in making polyurethane plastics, is still prepared by chlorination of propylene or oxidation with organic peroxides (*t*-butyl hydroperoxide or 1-phenylethyl hydroperoxide) and not directly from propylene and dioxygen. In the new HPPO process, BASF and Dow Chemicals synthesize propylene oxide from the oxidation of propylene with hydrogen peroxide that produces only water as a by-product (Equation 3.2).[2]

$$H_2C{=}CH_2 \ + \ 1/2 \ O_2 \ \xrightarrow[\substack{200\text{-}300\ ^0C \\ 1\text{-}2\ MPa}]{\substack{Ag \\ \text{on alumina}}} \ + H_2\overset{O}{\overset{\wedge}{C}}{-}CH_2 \tag{3.1}$$

$$\diagup\!\!\!\diagdown \ + \ H_2O_2 \ \longrightarrow \ \diagup\!\!\!\triangle\!\!\overset{O}{} \ + \ H_2O \tag{3.2}$$

*Physical Inorganic Chemistry: Reactions, Processes, and Applications* Edited by Andreja Bakac
Copyright © 2010 by John Wiley & Sons, Inc.

The energetics of oxygen transfer is important in determining the suitability of a transition metal catalyst for given oxidant and substrate. Biology circumvented the low reactivity of dioxygen as an oxidant by evolving electron transport chains in the mitochondria and utilizing metalloenzymes. Similarly, chemists have learned to improve the efficiency of catalytic oxidations by designing oxygen transfer chains that take advantage of thermodynamics. A thermodynamic oxygen transfer potential (TOP) scale has been developed for a variety of oxidants and substrates based on calculated reaction free energies ($\Delta G$) for the reaction described in Equation 3.3.[3] TOP values were computed using density functional theory (DFT), and representative examples are illustrated in Figure 3.1. For example, the dihydroxylation of olefin by osmium(VIII) is made "green" by driving the critical oxidation of dioxoosma-2,5-dioxolane (Os$^{VI}$) to trioxoosma-2,5-dioxolane (Os$^{VIII}$) (TOP $= -36$ kcal/mol) with a cascade of higher TOP oxidants (Figure 3.2).[4]

$$X + H_2O_2 \rightarrow XO + H_2O \qquad (3.3)$$

Another large class of oxygen atom transfer (OAT) reactions is between a donor and an acceptor. While those reactions are often thermodynamically favorable, they do not occur at any appreciable rate under ambient conditions. Representative

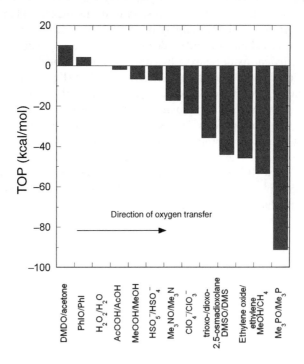

**FIGURE 3.1**  Thermodynamic oxygen transfer potentials. DMDO, dimethyldioxirane; Ac, CH$_3$C(O); DMSO, dimethyl sulfoxide; DMS, dimethyl sulfide.

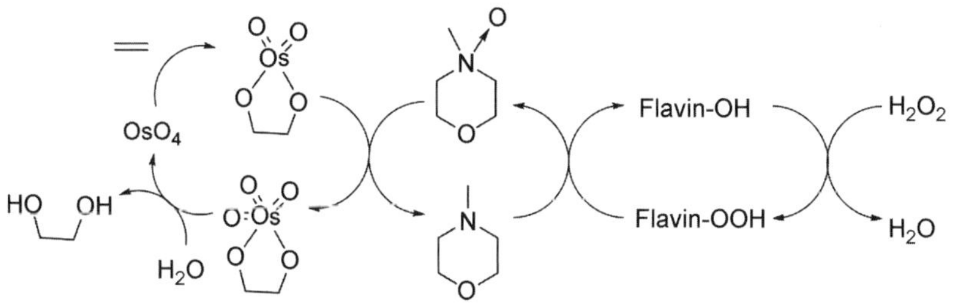

**FIGURE 3.2**  Catalytic cascades for oxidizing alkenes using $OsO_4$ as one of the oxygen transfer catalysts and the environmentally benign $H_2O_2$ oxidant.

examples include disproportionation of organic sulfoxides to sulfide and sulfone, oxidation of sulfides with pyridine-$N$-oxide, perchlorate, or nitrate, and oxidation of organic phosphines with sulfoxides (Table 3.1). Such reactions are catalyzed by transition metal complexes. Biological enzymes use molybdenum and tungsten in the reduction of dimethyl sulfoxide (DMSO), nitrate, trimethylamine $N$-oxide, biotin sulfoxide, perchlorate, and in the oxidation of sulfite and carbon monoxide. In addition to Mo and W model systems, a rich OAT chemistry has been developed for rhenium in the oxidation states $+5$ ($d^2$) and $+7$ ($d^0$). In addition to their rich mechanistic chemistry, OAT inorganic catalysts are important for the environmental remediation of inorganic oxyanions. We will discuss biological OAT and their biomimetic model compounds followed by chemical systems highlighting most recent developments.

## 3.2  BIOLOGICAL OXYGEN ATOM TRANSFER

### 3.2.1  Cytochrome P450

This enzyme is ubiquitous to all forms of life from single cellular organisms to humans, and different classes of this enzyme catalyze a large variety of important reactions including detoxification of toxins. Examples of reactions catalyzed by P450 isozymes are given in Table 3.2. Cytochrome P450 is a heme containing enzyme with a thiolate

**TABLE 3.1   Examples of Oxygen Atom Transfer Reactions Between Closed-Shell Donor and Acceptor Molecules and Their Thermodynamics**

| Reaction | $-\Delta G°$ (kJ/mol) |
|---|---|
| $C_5H_5NO + CH_3SCH_3 \rightarrow C_5H_5N + CH_3S(O)CH_3$ | 63 |
| $ClO_4^- + CH_3SCH_3 \rightarrow ClO_3^- + CH_3S(O)CH_3$ | 84 |
| $2CH_3S(O)CH_3 \rightarrow CH_3SCH_3 + CH_3S(O)_2CH_3$ | 105 |
| $CH_3S(O)CH_3 + Ph_3P \rightarrow CH_3SCH_3 + Ph_3PO$ | 234 |
| $NO_3^- + Ph_3P \rightarrow NO_2^- + Ph_3PO$ | 247 |

**TABLE 3.2   Example of Reactions Catalyzed by Cytochrome P450 Isozymes**

| Entry | Example Reaction | Class Name |
|-------|------------------|------------|
| 1. | Camphor → camphor-OH | Alkane hydroxylation |
| 2. | $CH_3(CH_2)_3CH=CH_2$ → hexene epoxide | Alkene epoxidation |
|    | $C_6H_6$ → $C_6H_6O$ (phenol) | Aromatic hydroxylation |
| 3. | $Me_2S$ → $Me_2SO$ | Sulfoxidation |
| 4. | $Me_3N$ → $Me_3NO$ | Amine oxidation |
| 5. | $Ph(CH_2)Br$ → $Ph(CH_3)$ (toluene) | Reductive halogenation |
| 6. | $Me_2NH$ → $MeNH_2 + H_2CO$ | N-dealkylation |
| 7. | $Ph\text{-}OCH_3$ → $PhOH + H_2CO$ | O-dealkylation |
| 8. | $MeNH_2$ → $NH_3 + H_2CO$ | Oxidative deamination |

($Cys^-$) axial ligand and gets its name from the unique absorption of its $Fe^{II}\text{-}CO$ complex at 450 nm. The resting state of the enzyme is low-spin $Fe^{III}$ (EPR $g$-values 2.41, 2.26, and 1.91). Upon binding of the substrate, the protein loses an axial water ligand, changes the iron's spin to high-spin $Fe^{III}$, and these changes result in a significant perturbation of the iron's reduction potential (low spin $-300$ mV versus high spin ca. $-170$ mV). Once the iron is reduced by the reduction partner (NADPH or NADH), dioxygen binds to $Fe^{II}$ and through several steps that couple electron transfer (ET) with proton transfer (PT), a high-valent $(por^{\bullet+})Fe^{IV}=O$ is generated. This compound is referred to as compound I, and it is responsible for oxidation of C–H bonds through a "rebound" mechanism and for OAT reactions. The catalytic cycle of P450 is summarized in Figure 3.3. The ET–PT reactions needed for activation of dioxygen could be bypassed by using peroxides ($H_2O_2$ or ROOH) through what is known as the peroxide shunt, going to compound I directly from the ferric state. The structures of the intermediates in the P450 catalytic cycle have been arrived at through spectroscopic studies of various P450 enzymes, model compounds, and drawn analogies to similar peroxidase systems.[5] Furthermore, the structure of the $Fe^{II}$ dioxygen adduct and its breakdown to the oxoferryl species (compound I) for camphor P450 have been determined by trapping techniques and single-crystal X-ray crystallography.[6]

   A common tool for probing the rebound mechanism in P450 oxidations is the use of radical clocks.[7] These mechanistic probes provide diagnostic products by undergoing fast reorganization when a radical is formed during the reaction. Many of the radical clocks are based on cyclopropane that rearranges via ring opening once a radical is formed giving rise to ratios of products (an example is give in Figure 3.4). One should be aware that the rate constant for radical clock rearrangement is crucial because they have to be competitive with the rebound rate of the carbon radical substrate and $Fe^{IV}$-OH of the P450 enzyme. Typical enzymatic rebound rates that have been measured for P450 enzymes are in the range of $10^9$–$10^{11}$ $s^{-1}$.[8]

   The identity of active species responsible for oxidation/OAT, whether it is compound I (oxoferryl) or a precursor peroxo complex, has been contentious. Evidence from mutation and model studies has pointed to the possibility of multiple oxidants in which one of the oxidants might very well be a peroxo or an oxidant adduct

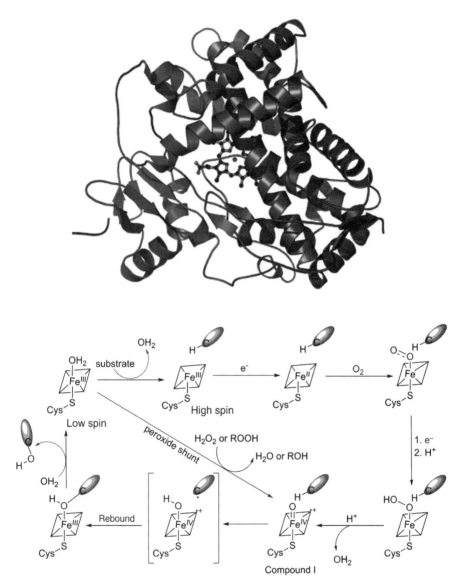

**FIGURE 3.3**   Catalytic cycle for cytochrome P450 and X-ray structure of a human cyto-chrome P450 2D6 that recognizes substrates containing basic nitrogen and an aromatic ring, functional groups found in a large number of central nervous system, and cardiovascular drugs.

of iron in addition to or instead of oxoferryl.[9-11] Recent quantum mechanical/molecular mechanical computations on the mechanism of C–H oxidation in camphor P450 have suggested that indeed the reaction takes place through two different spin states of compound I, doublet (low spin) and quartet (high spin).[12] This has been dubbed "two-state reactivity." The reaction along the doublet state potential was

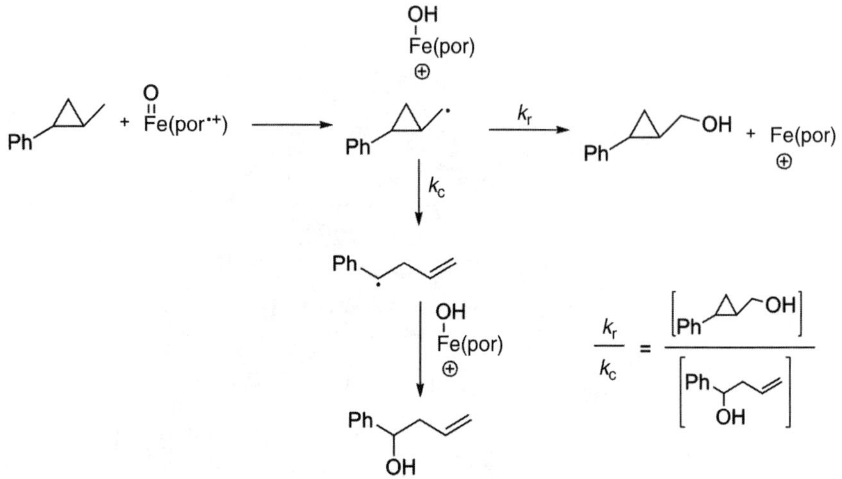

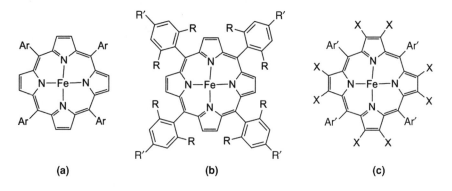

**FIGURE 3.4** Illustration of a radical clock. The ratio of unarranged hydroxylation product and rearranged product reflects the ratios of rebound rate constant ($k_r$) and ring opening of the radical intermediate ($k_c$). In this case, $k_c = 3.4 \times 10^{11}\,\text{s}^{-1}$.

found to be concerted, while the reaction along the quartet state pathway to be stepwise with a distinguishable rebound between the carbon radical substrate and the $Fe^{IV}$-OH.

Iron porphyrins have been studied extensively over the past $30 +$ years as model systems of cytochrome P450.[13] Biomimetic model studies included variants in axial ligands (thiolate and other bases), the oxidation of alkanes, olefins, sulfides, and amines, and utilization of several oxidants such as hypochlorite (bleach), iodosobenzene (ArIO), hydrogen peroxide, and organic peroxides (ROOH). The first-generation models employed the *meso*-tetraarylporphyrins (Figure 3.5). These were

**FIGURE 3.5** Porphyrin structures used as models for P450 enzymes. (a) Earlier models with Ar = aromatic, (b) improved second-generation porphyrin ligands with R = CH$_3$, Cl, or F and R′ = H or CH$_3$, (c) more recent model catalysts with Ar′ as in (b) and X = Br, Cl, or F.

more stable than beta-substituted porphyrins and iron porphyrins derived from heme (protoporphyrin IX). Improvement in catalyst performance and stability was eventually achieved by introducing bulky substituents on the aryl rings such as tetramesityl and dichlorophenyl (Figure 3.5b). More recently, addition of electron-withdrawing halogen atoms in the beta positions increased reactivity and stability toward oxidative degradation (Figure 3.5c). Metal complexes of halogenated porphyrins are capable of oxidizing inert C–H bonds of alkanes. For example, oxidation of heptane to heptanal and heptanone with PhIO proceeds to 38% conversion with Fe(TDCPP)Cl (TDCPP= tetra-2,6-dichlorophenylporphyrin) as catalyst versus 78% conversion with Fe (TDCPCl$_8$P)Cl (TDCPCl$_8$P = tetra-2,6-dichlorophenyl-β-octachloroporphyrin) under the same conditions.[14] Mechanistic studies on alkane oxygenation by halogenated porphyrin complexes suggest a radical chain mechanism in which radicals are generated by oxidation and reduction of the peroxide oxidant.[15]

### 3.2.2  Peroxidases

Hydrogen peroxide is another oxidant of biological significance and a variety of enzymes utilize it as a substrate in three modes of reactions: (1) disproportionation to dioxygen and water (catalase), (2) as an oxidant (peroxidases), and (3) in the halogenation of organic substrates (haloperoxidases). All these enzymes with the exception of the vanadium haloperoxidases contain heme active sites and in many respects are analogous to P450 chemistry. However, the axial ligand and the details of the mechanisms are different. Catalase has a phenolic axial ligand (tyrosine) and an open coordination site in the other axial site for binding H$_2$O$_2$. The structural characterization of bovine liver catalase shows a number of residues in the distal pocket that are essential for function.[16] These include a phenylalanine ring pi-stacked to one of the pyrrole rings of the heme, and His and Asn residues that form hydrogen bonding network. The mechanism proceeds through compound I, which is capable in this environment of oxidizing another molecule of hydrogen peroxide to O$_2$ to complete the catalytic cycle (Equation 3.4). It has been shown that compound I of catalase can oxidize formate and ethanol, as well as hydrogen peroxide. While a common theme for peroxidases is formation of Fe(IV)=O, the position of the radical cation varies. It is either on the prophyrin as in cytochrome P450 or on a protein residue. A recent EPR study on bovine liver catalase showed a broad radical signal at $g \approx 2$ with hyperfine structure that resembles the Tyr$^•$ observed in photosystem II.[17]

$$(\text{enzyme})\text{Fe}^{\text{III}} + \text{H}_2\text{O}_2 \rightarrow \text{compound I}/(\text{enzyme}^•) + \text{Fe}^{\text{IV}} = \text{O} + \text{H}_2\text{O}$$

$$\text{compound I}/(\text{enzyme}^•) + \text{Fe}^{\text{IV}} = \text{O} + \text{H}_2\text{O}_2 \rightarrow (\text{enzyme})\text{Fe}^{\text{III}} + \text{H}_2\text{O} + \text{O}_2 \quad (3.4)$$

Cytochrome $c$ peroxidase (CcP) and horseradish peroxidase (HRP) contain axial histidine ligation and the oxo group from compound I is released as water (Equation 3.5). Cytochrome $c$ peroxidase oxidizes two molar equivalents of cytochrome $c$. HRP reacts with a number of electron donor substrates such as alkylamines and sulfides. While the oxo ligand from compound I is not transferred to the substrates in these reactions, peroxidases have been shown to catalyze a wide variety of synthetically useful OAT reactions with $H_2O_2$ including epoxidation and sulfoxidation.[18] In some instances, the reactions proceed with good to high enantioselectivity. For HRP, compound I, $(por^{\bullet +})Fe^{IV}{=}O$, was characterized by EPR and Mössbauer spectroscopies.[19] A transient tyrosyl radical $(Tyr^{\bullet})$ in equilibrium with $(por^{\bullet +})$ was observed for the Phe172Tyr HRP mutant.[20] In compound I of CcP, the organic radical resides on Trp191.[21] While the active site of ascorbate peroxidase shows structural homology to that of CcP, it stabilizes a $(por^{\bullet +})$ radical.[22]

$$(\text{Substrate-H}_2)_{red} + H_2O_2 \rightarrow (\text{Substrate})_{ox} + 2H_2O \tag{3.5}$$

Chloroperoxidase (CPO) resembles cytochrome P450 in that its axial ligand is a cysteine thiolate. In addition to catalyzing substrate halogenation (Equation 3.6), CPO exhibits peroxidase, catalase, and cytochrome P450-like activities. The catalytic reaction proceeds through compound I intermediate that oxidizes chloride to hypochlorite. The latter in turn halogenates the substrate and releases hydroxide. Thus, the final fate of the oxygen atom from hydrogen peroxide is in water as for the other peroxidases. While the proximal side of CPO resembles cytochrome P450 because of the cysteine ligation, its distal side is very much peroxidase-like featuring polar residues for the binding of peroxide.[23] Access to the active site of CPO is restricted to the distal face that contains a hydrophobic patch above the heme. This allows small organic substrates to access the iron oxo ligand (compound I) accounting for the P450-like activity observed for CPO. Recent X-ray absorption results on oxoiron(IV) (compound II) of CPO were consistent with a protonated $Fe^{IV}$-OH, demonstrating the high basicity of the oxo ligand because of strong donation by the axial thiolate.[24] It has been suggested that the elevated basicity of ferryl accounts for the ability of these enzymes to oxygenate substrates at potentials that can be tolerated by proteins.

$$H_2O_2 + Cl^- + (\text{Substrate-H}) \rightarrow H_2O + (\text{Substrate-Cl}) + OH^- \tag{3.6}$$

Vanadium haloperoxidases are found in marine algae and lichens. They are responsible for the oxidation of bromide (vanadium bromoperoxidases = VBPO) or chloride (vanadium chloroperoxidases = VCPO) with hydrogen peroxide to give $XO^-$, HOX, and $X_2$, either in solution or bound to the protein.[25] The hypohalo species is responsible for halogenating organic substrates such as terpenes in the syntheses of natural products. In this enzyme family, the metal does not undergo a change in oxidation state. Vanadium remains in the $+5$ oxidation state and OAT ensues from a vanadium(V) peroxo complex. It was established recently that VCPO are structurally similar to the (membrane-bound) acid phosphatases.[26] The vanadate ion is held in the active site of enzymes through ligation to one histidine residue and several hydrogen bonding interactions (Figure 3.6).[27] The mechanism for vanadium haloperoxidases is illustrated for VBPO in Figure 3.7.

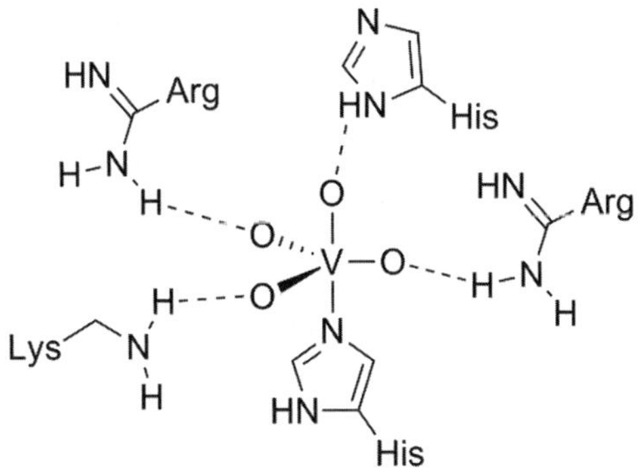

**FIGURE 3.6** Active site of vanadium chloroperoxidase from *C. inaequalis.*

### 3.2.3 Oxygen Evolving Chlorite Dismutase

Chlorite dismutase (Cld) contains a heme b active site, analogous to peroxidases, and occurs in perchlorate and chlorate respiring bacteria.[28] Perchlorate ($ClO_4^-$) is a rocket fuel and used in pyrotechnics and ammunition. It has been found in soil and ground water only recently.[29] Due to its size similarity to iodide, perchlorate inhibits the thyroid gland irreversibly. Microbes have evolved to utilize the oxidizing power of perchlorate. These organisms reduce perchlorate to chlorate and chlorate to chlorite

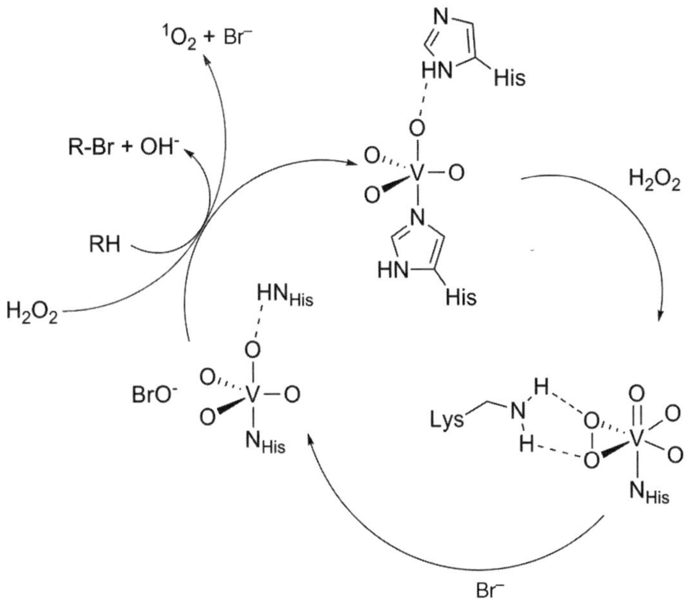

**FIGURE 3.7** Mechanism of vanadium bromoperoxidase.

via a molybdopterin-dependent perchlorate reductase (PerR), a close homologue of bacterial respiratory nitrate reductases. Cld plays a detoxifying role by taking chlorite ($ClO_2^-$) and decomposing it to dioxygen and harmless chloride ($Cl^-$) (Equation 3.7). This enzyme exhibits remarkable selectivity for dioxygen (O=O) formation.[30] In the presence of reducing substrates such as ascorbate and oxygen atom acceptors such as thioanisole (PhSMe), Cld does not catalyze peroxidase chemistry with chlorite but instead produces $O_2$ and $Cl^-$ exclusively. However, with other oxidants such as hydrogen peroxide and peroxy acids, Cld behaves very much as a peroxidase. In contrast, heme peroxidases use chlorite as an oxidant in one-electron oxidation, OAT, and chlorination chemistry.[31]

$$ClO_4^- \xrightarrow{PerR} ClO_3^- \xrightarrow{PerR} ClO_2^- \xrightarrow{Cld} O_2 + Cl^- \tag{3.7}$$

Isotope labeling experiments have shown chlorite to be the sole source of the oxygen atoms in the $O_2$ product.[30] EPR and UV-Vis studies also suggest that Cld forms a compound I-like intermediate upon reaction with $ClO_2^-$. However, the hypochlorite ($ClO^-$) by-product is believed to not diffuse freely into solution, but instead remains protein bound for further reaction with the ferryl to afford $O_2$. The mechanism illustrated in Figure 3.8 has been put forth for this unique heme enzyme.

Water-soluble iron porphyrins catalyze decomposition of chlorite to chlorate and chloride (Equation 3.8) rather than producing $O_2$.[32] An exception has been found when a highly fluorinated iron porphyrin is used (Figure 3.9). [$Fe^{III}(TF_4TMAP)$] catalyzes chlorite decomposition into $O_2$ and $Cl^-$ in addition to the chlorate producing

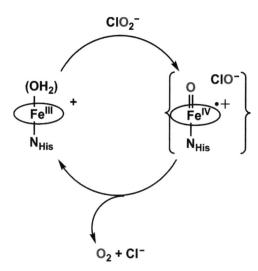

**FIGURE 3.8** Mechanism of chlorite dismutase. (See the color version of this figure in Color Plates section.)

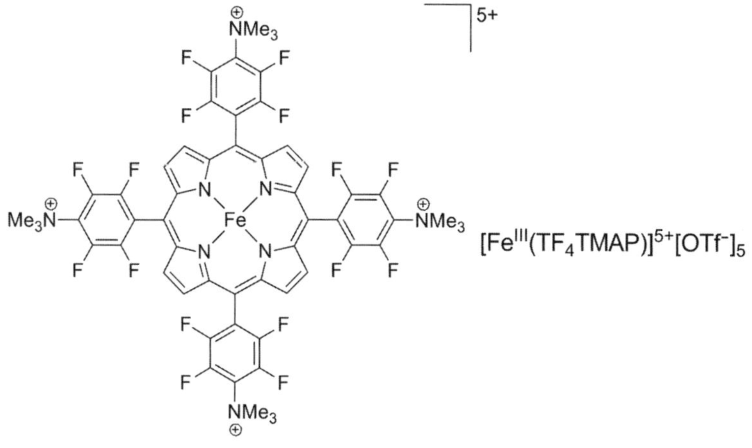

$[Fe^{III}(TF_4TMAP)]^{5+}[OTf^-]_5$

**FIGURE 3.9**    A biomimetic functional model of chlorite dismutase.

pathway. Nevertheless, this biomimetic model remains much less selective than the enzyme itself.

$$3ClO_2^- \rightarrow Cl^- + 2ClO_3^- \tag{3.8}$$

### 3.2.4    OAT with Nonheme Iron

Nonheme iron oxygenases catalyze a wide array of reactions with dioxygen that are as diverse as or more so than those recognized for the heme analogues.[33,34] Some of the nonheme iron enzymes contain a single iron in the active site and require a cofactor. Examples include the aromatic amino acid hydroxylases and α-ketoglutarate-dependent enzymes (Figure 3.10). In other cases, the mononuclear iron enzymes act as dioxygenases, such as catechol dioxygenase (Figure 3.10). Even though cytochrome P450 is not capable of oxidizing methane ($CH_4$), kinetically the most inert hydrocarbon, microorganisms have evolved iron-dependent methane monooxygenase (MMO) that catalyzes the air oxidation of methane to methanol, an extremely difficult reaction to accomplish because of overoxidation: methanol is more reactive toward oxidation than methane.[35] Examples of the different classes of nonheme iron enzymes and their reactions are illustrated in Figure 3.10.

High-valent iron intermediates have been proposed as the active species in OAT and C–H oxidation reactions for nonheme iron enzymes. In some cases, such intermediates have been trapped by rapid freeze-quench studies and characterized. In ribonucleotide reductase from *E. coli* and MMO, intermediates **X** and **Q** with $Fe^{III}$-$(\mu\text{-}O)_2$-$Fe^{IV}$ and $Fe^{IV}$-$(\mu\text{-}O)_2$-$Fe^{IV}$ diamond core, respectively, have been characterized (Figure 3.11).[35] Also, $Fe^{IV}$ oxo intermediates have been observed for mononuclear proteins such as taurine/2-oxoglutarate dioxygenase (TauD) (Figure 3.11).[36]

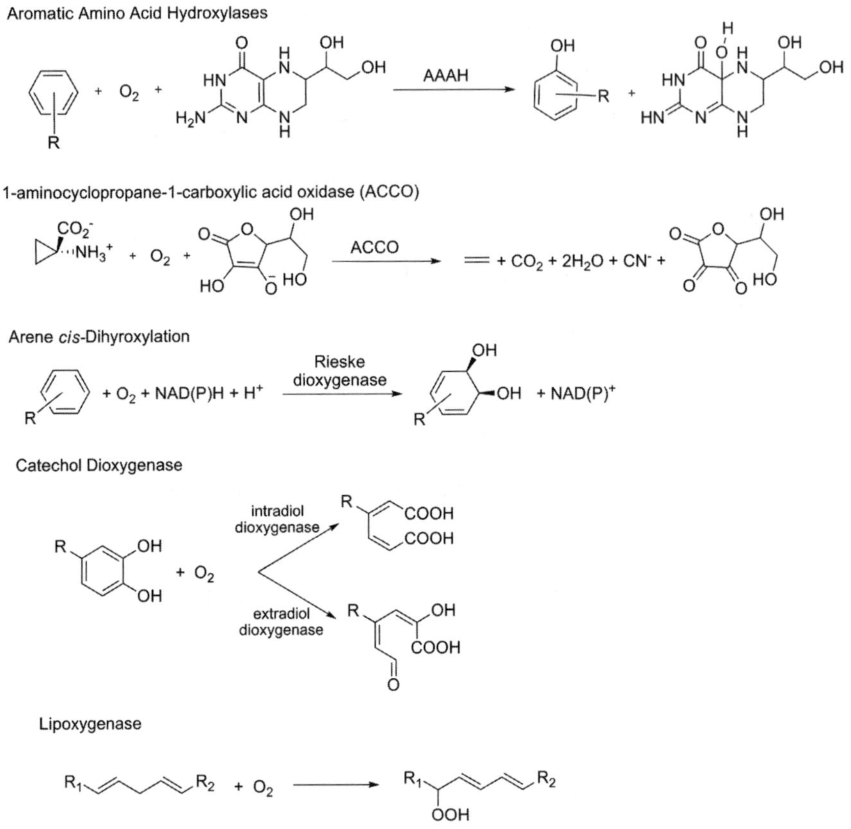
Aromatic Amino Acid Hydroxylases
AAAH
1-aminocyclopropane-1-carboxylic acid oxidase (ACCO)
ACCO
Arene cis-Dihyroxylation
Rieske dioxygenase
Catechol Dioxygenase
intradiol dioxygenase
extradiol dioxygenase
Lipoxygenase

**FIGURE 3.10** Representative reactions of nonheme iron enzymes.

As for synthetic inorganic model complexes, the past few years have witnessed major breakthroughs in the preparation and isolation of high-valent $Fe^{IV}$ oxo coordination complexes. More than a decade ago, tetra-amido macrocyclic ligands (TAML, Figure 3.12) were shown to support high-valent metal centers such as $[Fe^{IV}(TAML)Cl]^-$.[37] More recently, an oxo-bridged dinuclear iron(IV) ($Fe^{IV}$–O–$Fe^{IV}$) has been

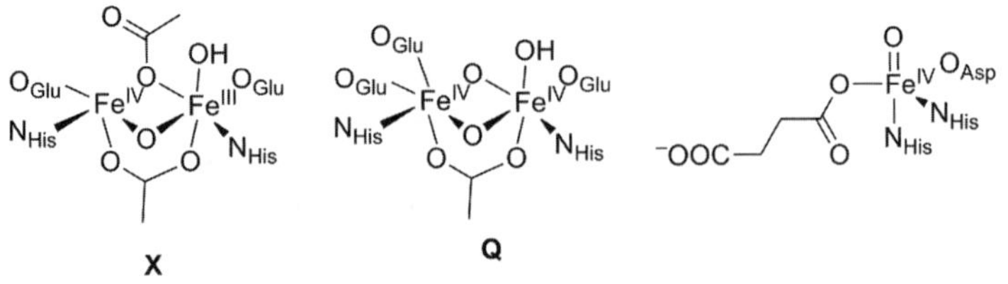
X
Q

**FIGURE 3.11** High-valent intermediates detected in nonheme enzymes.

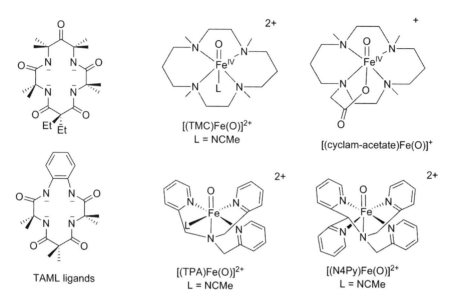

**FIGURE 3.12**    Tetra-amido ligands and nonheme iron(IV) oxo synthetic complexes.

isolated and the first example of Fe$^V$ oxo complex reported.[38] In the past few years, approximately 15 nonheme iron(IV) oxo complexes have appeared in the literature based on tetra-aza macrocyclic ligands and polypyridyl amine ligands (Figure 3.12).[39,40] These ligands support low-spin ($S = 1$) oxoiron(IV) with a short Fe=O bond distance of ~1.65 Å determined by X-ray crystallography and extended X-ray absorption fine structures (EXAFS). Further support for the double bond character of the Fe–O bond comes from IR spectroscopy, v(Fe–O) ~ 830 cm$^{-1}$. The electronic spectra of low-spin nonheme oxoiron(IV) complexes show a characteristic absorption band in the visible–near-IR region between 650 and 1050 nm with low extinction coefficients (200–400 L/(mol cm)). While most synthetic nonheme iron(IV) oxo complexes are low spin, Mössbauer analysis of $[(H_2O)_5Fe^{IV}=O]^{2+}$ indicates a high-spin ($S = 2$) state in acidic aqueous solution. Also, in the enzyme TauD a high-spin iron(IV) oxo intermediate has been identified.

Three synthetic methods have been established for the generation of iron(IV) oxo compounds: (1) reaction with oxygen donors such as PhIO, peracids, ozone, KHSO$_5$, and hypochlorite (Equation 3.9); (2) homolytic cleavage of Fe$^{III}$ organic peroxide complexes (Equation 3.10), and (3) from the reaction of dioxygen with Fe$^{II}$, presumably via a putative μ-peroxo Fe$_2^{III}$ dinuclear species (Equation 3.11). The stability of synthetic nonheme iron(IV) oxo complexes depends on the ancillary ligand.[40] For example, $[(TPA)Fe^{IV}=O]^{2+}$ and $[(cyclam-acetato)Fe^{IV}=O]^+$ are stable only at low temperatures (−40 and −80°C, respectively), and $[(TMC)Fe^{IV}(O)]^{2+}$ is stable at room temperature. The stability of iron(IV) oxo complexes is also pH dependent. While $[(N4Py)Fe^{IV}(O)]^{2+}$ is stable at pH 5–6, it decomposes readily at higher pH.

$$(L)Fe^{II} + PhIO \rightarrow (L)Fe^{IV}(O) + PhI \tag{3.9}$$

$$(L)Fe^{II} + ROOH \rightarrow (L)Fe^{III}OOR \rightarrow (L)Fe^{IV}(O) + RO^{\bullet} \tag{3.10}$$

$$2(L)Fe^{II} + O_2 \rightarrow [(L)Fe^{III}-O-O-Fe^{III}(L)] \rightarrow 2(L)Fe^{IV}(O) \tag{3.11}$$

Synthetic nonheme iron(IV) oxo complexes undergo several OAT and oxidation reactions.[40] They are capable of transferring an oxygen atom to organic phosphines ($PR_3$), organic sulfides ($R_2S$), and alkenes such as cyclooctene and *cis*-stilbene, giving phosphine oxide ($R_3P=O$), sulfoxide ($R_2S=O$), and epoxide, respectively. Reactions of oxoiron(IV) complexes with aryl sulfides showed significant dependence on the *para* substituents with a Hammett reaction constant ($\rho$) values in the range of $-1.4$ to $-2.5$. These values are in agreement with electrophilic oxygen atom transfer from $Fe^{IV}=O$. More relevant to biological systems is the observation that some of the synthetic nonheme iron(IV) oxo complexes such as $[(N4Py)Fe^{IV}(O)]^{2+}$ are capable of oxidizing hydrocarbons such as cyclohexane. These reactions exhibit a significant kinetic isotope effect (KIE) of $>30$, which is consistent with hydrogen atom abstraction in the rate-determining step. The KIE values are similar to those observed for enzymes. The mononuclear iron-dependent TauD shows a KIE of $\sim37$ and the dinuclear MMO a KIE of $>50$.[41,42] Another hydrogen atom transfer (HAT) reaction observed for iron(IV) oxo is that with alcohols to afford aldehydes. Isotope labeling studies using $Fe^{IV}={}^{18}O$ demonstrated that the mechanism of aldehyde formation proceeds via two consecutive HAT reactions rather than a gem-diol intermediate.[40] No ${}^{18}O$ incorporation was observed in the resulting aldehyde.

In sharp contrast to reactions with alkanes and alcohols, synthetic nonheme $Fe^{IV}$ oxo complexes oxidize aromatic compounds such as anthracene to anthraquinone with an inverse KIE $\sim 0.9$.[40] Furthermore, electron-donating substituents influence the reaction rate, giving a large and negative Hammett reaction constant ($\rho = -3.9$). These observations disfavor HAT and are in agreement with electrophilic addition of the iron oxo to $sp^2$ carbon of the aromatic ring. This is analogous to observations made for the iron-dependent aromatic amino acid hydroxylases. There the mechanism is believed to proceed through an arene epoxide followed by 1,2-hydrogen shift that has been termed as the NIH shift.[34]

Nonheme iron(IV) oxo complexes have also been shown to effect oxidative N-dealkylation, a reaction recognized in nature for nonheme iron enzymes.[43] Mechanistic studies on the reaction of synthetic models with *N*,*N*-dimethylaniline demonstrated that rather than HAT the reaction proceeds via electron transfer followed by proton transfer.[40] Finally, nonheme iron(IV) oxo species undergo complete intermetallic OAT to $Fe^{II}$. The oxygen atom transfer depends on the oxidizing power of the starting ferryl. The observation of complete intermetallic oxo transfer in nonheme systems stands in sharp contrast to porphyrin/heme systems in which incomplete OAT occur affording bridged μ-oxo $Fe^{III}$ dinuclear species. A summary of all the reaction types that have been observed for nonheme iron(IV) oxo complexes is shown in Figure 3.13.

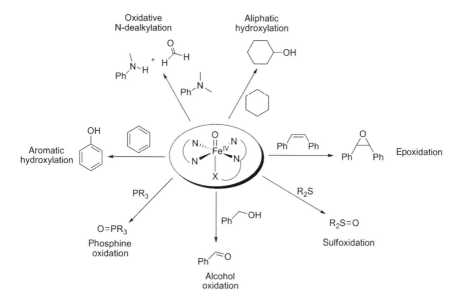

**FIGURE 3.13**   Reactions of nonheme iron(IV) oxo complexes.

### 3.2.5   Molybdenum and Tungsten Oxotransferases

Molybdenum and tungsten enzymes are found in all forms of life from micro-organisms to humans.[44] They catalyze a number of important metabolic reactions involved in the nitrogen and chlorine cycles. Even though the metal site cycles between $Mo^{IV}/W^{IV}$ and $Mo^{VI}/W^{VI}$ with concomitant oxygen atom (oxo) addition to the metal or removal there from, and hence the oxotransferase nomenclature, water is the ultimate sink or source of oxygen in the overall catalytic cycle. The reaction of the substrate (X/XO) is coupled to electron transfer to or from an iron sulfur cluster, a heme protein, or a flavin cofactor. In the past decade, a large number of single-crystal X-ray structures for molybdenum and tungsten enzymes have been reported. These revealed the details of active site coordination in both reduced and oxidized states. Pyranopterindithiolate (also known as molybdopterin or pterindithiolene) is the common ligand for molybdenum and tungsten in these enzymes (Figure 3.14). The molybdenum enzymes have been classified into three families based on their protein sequence and the structures of their active sites (Figure 3.14).[45] Xanthine oxidase and other enzymes in its family contain one pterindithiolene ligand and terminal oxo and sulfido in the $Mo^{VI}$ oxidized state. Members of the sulfite oxidase family also contain a single pterindithiolene ligand, a dioxo in the $Mo^{VI}$ state, and a thiolate (from Cys). Dimethylsulfoxide (DMSO) reductase features a bis-pterindithiolene and a single oxo ligand in the $Mo^{VI}$ oxidized state. In the reduced state, DMSO reductase does not possess a multiply bonded ligand (desoxo) (Figure 3.14). One additional ligand in DMSO reductase is an alkoxide from Ser. In other members of the DMSO reductase family, the serine is replaced with a thiolate from Cys as in nitrate reductase and

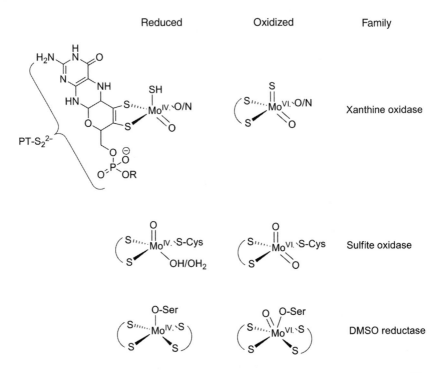

**FIGURE 3.14**    Structures of active sites of molybdenum enzymes in the reduced $Mo^{IV}$ and oxidized $Mo^{VI}$ states alongside their classification into three families.

selenate from Se-Cys as in formate dehydrogenase (FDH). The different reactions catalyzed by molybdenum and tungsten enzymes are summarized in Table 3.3.

Three main families are recognized for the tungsten enzymes.[46] They are aldehyde oxidoreductase, acetylene hydratase, and formate dehydrogenase (Table 3.3). Structures for examples in each of the three families have been determined. The active site tungsten contains two pterindithiolene ($PT-S_2^{2-}$) ligands. Additional ligands are not well identified in all cases but appear to be donor atoms derived from the peptide backbone. The oxidized form has been suggested for aldehyde oxidoreductases to contain oxo and hydroxo groups, $W^{VI}(O)(OH)$. In formate dehydrogenase, a selenocysteinate and one hydroxyl ligand have been reported. The tungsten enzymes also contain one protein-bound $Fe_4S_4$ cluster.

One striking feature of the active site structures of the enzymes in Figure 3.14 is the prevalence of five-coordinate molybdenum and the dithiolene ligation. Prior to structural information on the enzymes, a significant body of modeling work was focused on accomplishing OAT for monooxo molybdenum(IV) and dioxo molybdenum(VI) (Equation 3.12).[47] The earlier model compounds featured a variety of donor ancillary ligands based on O, N, and S, but not dithiolene. Furthermore, these complexes were six-coordinate and none of them mimicked the desoxo observed in the reduced state of DMSO reductase. Nevertheless, important kinetics and

**TABLE 3.3    Examples of Reactions Catalyzed by Molybdenum and Tungsten Enzymes**

| Reaction | Enzyme |
|---|---|
| **Molybdenum (Mo)** | |
| Xanthine $+ H_2O \leftrightarrow$ uric acid $+ 2H^+ + 2e^-$ | Xanthine oxidase |
| $SO_3^{2-} + H_2O \leftrightarrow SO_4^{2-} + 2H^+ + 2e^-$ | Sulfite oxidase |
| $NO_3^- + 2H^+ + 2e^- \leftrightarrow NO_2^- + H_2O$ | Nitrate reductase |
| $Me_2SO + 2H^+ + 2e^- \leftrightarrow Me_2S + H_2O$ | DMSO reductase |
| $RCHO + H_2O \leftrightarrow RCO_2H + 2H^+ + 2e^-$ | Aldehyde oxidoreductase |
| $CO + H_2O \leftrightarrow CO_2 + 2H^+ + 2e^-$ | Carbon monoxide dehydrogenase |
| $H_2AsO_3^- + H_2O \leftrightarrow HAsO_4^{2-} + 3H^+ + 2e^-$ | Arsenite oxidase |
| **Tungsten (W)** | |
| $HCO_2^- \leftrightarrow CO_2 + H^+ + 2e^-$ | Formate dehydrogenase |
| $HC \equiv CH + H_2O \leftrightarrow CH_3CHO$ | Acetylene hydratase |
| $RCHO + H_2O \leftrightarrow RCO_2H + 2H^+ + 2e^-$ | Aldehyde oxidoreductase |

mechanistic chemistry emerged from studying primary OAT reactions with these model systems (Equation 3.12). One of the important lessons learned was the effect of steric congestion on inhibiting the comproportionation of $Mo^{IV}(O)$ and $Mo^{VI}(O)_2$ to give irreversibly $(O)Mo^V$-$\mu$-$O$-$Mo^V(O)$ (Equation 3.13). One of the early nondithiolene systems that demonstrated double oxygen atom transfer reaction is shown in Figure 3.15. From a functional prospective, this complex is a synthetic analogue of the sulfite oxidase family because it involves monooxo and dioxo molybdenum.

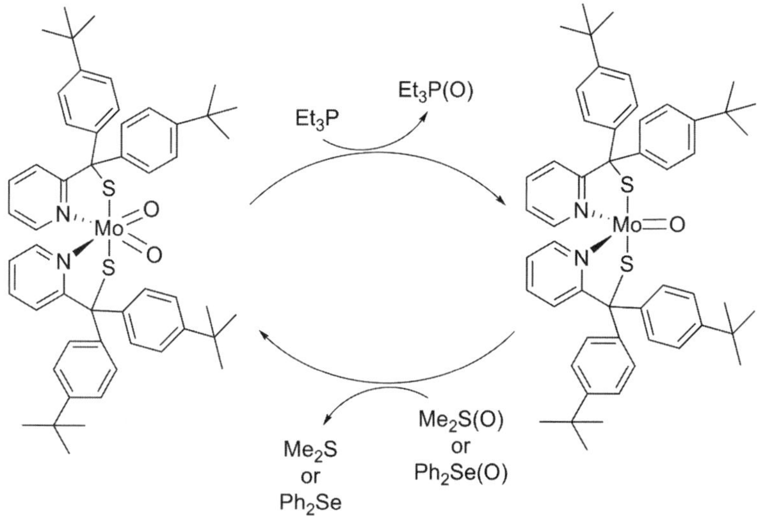

**FIGURE 3.15**    OAT between monooxo and dioxomolybdenum(IV) and (VI), respectively.

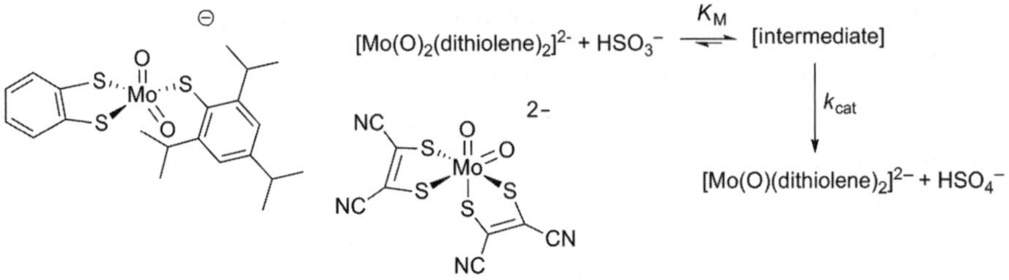

**FIGURE 3.16** Structural mimic of sulfite oxidase family and a functional model based on bis-dithiolene ligation.

$$(L)Mo^{VI}(O)_2 + X \leftrightarrow (L)Mo^{IV}(O) + XO \tag{3.12}$$

$$(L)Mo^{VI}(O)_2 + (L)Mo^{IV}(O) \leftrightarrow (L)(O)Mo^V-O-Mo^V(O)(L) \tag{3.13}$$

Producing a ligand environment that mirrors sulfite oxidase is challenging for the synthetic chemist because it requires a single dithiolene ligation. Bis-dithiolene complexes of molybdenum are easier to access and isolate. Even though dithiolene is a non-innocent ligand and can exist in fully reduced dianion form, partially reduced radical anion and fully oxidized neutral form structural data from the enzymes and synthetic models are consistent with the fully reduced form in molybdenum complexes. Benzene-1,2-dithiolate has been employed successfully to stabilize a structural mimic of the sulfite oxidase family (Figure 3.16). However, in terms of functionality, the bis-dithiolene complex $[Mo(O)_2(mnt)_2]^{2-}$ was found to oxidize bisulfite, following Michaelis–Menten kinetics ($K_M = 10^{-2}$ mol L and $k_{cat} = 0.9\,s^{-1}$) (Figure 3.16). This reaction is inhibited by other anions such as $SO_4^{2-}$, $H_2PO_4^{-}$, and so on. Although the ligand composition does not match that in the enzyme's site, the reaction chemistry parallels that for the enzyme. The observation of saturation kinetics in bisulfite implies that the intermediate is a sulfite adduct of Mo. The xanthine oxidase family also presents challenges on two fronts, dithiolene and mixed oxo sulfido terminal ligands. The latter is particularly difficult because of the high reactivity of the sulfido ligand. In one structural analogue of the xanthine oxidase enzyme family, a trispyrazolylborate ligand was used to generate $[(L)Mo^{VI}(O)(S) (OPh)]$.[47] However, this compound forms a $Mo^V$ dinuclear species bridged by a disulfide ($\mu$-S–S) bond.

The bis-dithiolene complexes of both molybdenum and tungsten have been the best characterized family structurally as well as their kinetics for OAT reactions, making the DMSO reductase family the easiest to mimic chemically. However, well-characterized bis-dithiolene complexes are those featuring $Mo^{IV}(O)$ and $Mo^{VI}(O)_2$ moieties (as illustrated in Figure 3.16). Exact structural analogues of the minimal DMSO reductase active sites, that is desoxo $Mo^{IV}$, have been realized (Figure 3.17). However, the study of their OAT kinetics has been complicated by the fact that monooxo $Mo^{VI}(O)$ complexes are unstable and decompose rapidly to monooxo $Mo^V$

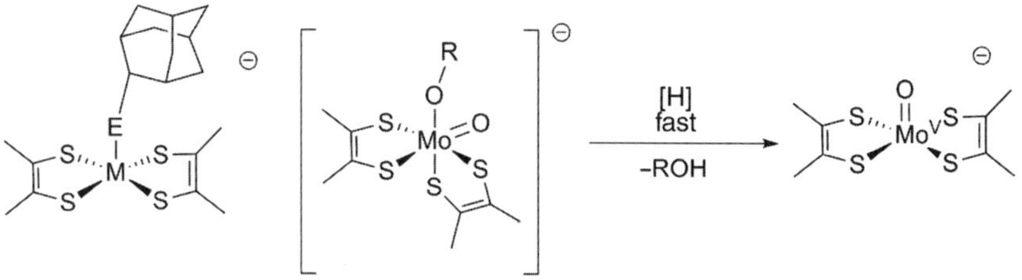

**FIGURE 3.17**    Structural models of desoxo sites of the DMSO reductase enzyme family and decomposition of monooxo $Mo^{VI}$ mimics.

(Figure 3.17). Mechanistic studies of OAT to $M^{IV}(O)$ bis-dithiolene complexes (where M = Mo or W) from amine $N$-oxide or sulfoxide substrates show that the reactions proceed through an adduct intermediate of the oxo donor followed by atom transfer. The kinetics follow trends in oxygen donors that parallel the enthalpy of the XO bond strength in the donor as well as the substrate's ability to act as a Lewis base. In general, alkyl amine $N$-oxides are more reactive than PyO, which is more reactive than organic sulfoxides ($Me_3NO > PyO > R_2S(O)$). The entropy of activation values are large and negative ($\Delta S^{\ddagger} \sim -15$ to $-40 \, cal/(mol \, K)$). These values are consistent with an associative transition state. In the reverse reaction, organic phosphines abstract an oxygen atom from $M^{VI}(O)_2$ with rates that increase with electron–donor groups on the phosphine, consistent with the oxo acceptor acting as a nucleophile. Comparison of Mo versus W reactivity has been possible in systems that contain identical ligands. While the difference in reactivity between the two metals is dependent on the substrate, phosphines exhibiting the largest rate differences, tungsten (W) is more reactive with oxygen atom donors and molybdenum (Mo) more reactive with oxygen acceptors. For $[M^{VI}(O)_2(mnt)]^{2-}$ (mnt = $cis$-1,2-dicyano-1,2-ethylenedithiolene), Mo is 1000× more reactive than W with $(MeO)_2PhP$. In the reaction of $[M^{IV}(O)(bdt)]^{2-}$ (bdt = benzene-1,2-dithiolate) with $Me_3NO$, $k^W/k^{Mo} \sim 3$.

## 3.3    CHEMICAL OXYGEN ATOM TRANSFER

### 3.3.1    High-Valent Corrole and Corrolazine Complexes

While porphyrins serve as models for heme enzyme sites, isolation of high-valent intermediates has been difficult. Several polypyrrolic macrocyclic ligands that are analogues of porphyrin have been studied (Figure 3.18). In the context of OAT and high-valent complexes, the past few years have witnessed substantial advances in metal corrole and corrolazine complexes.[48,49] Both macrocycles are ring-contracted porphyrinoids. This reduction in one core atom makes the fully deprotonated ligands trianionic, creating stronger σ-donation. The outcome of this increase in charge and σ-donation is the ability of ligands (corrole and corrolazine) to stabilize high oxidation states of metal ions. However, one-electron oxidation of the corrole ligand

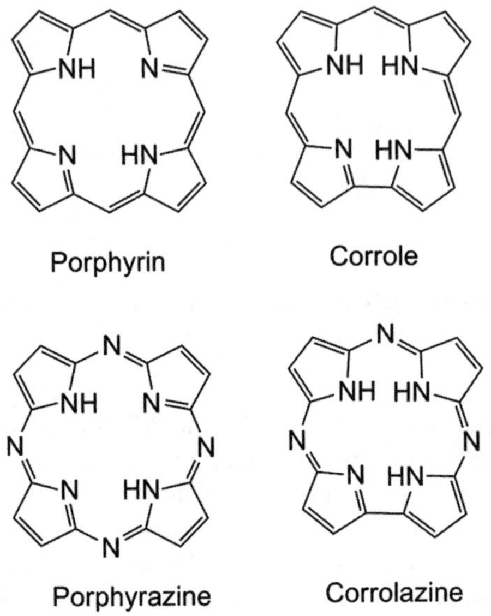

**FIGURE 3.18**    A sampling of polypyrrolic macrocyclic ligands.

to make a π-cation radical is typical, making oxidation state assignment difficult. Corrolazine could be thought of as ring-contracted porphyrazine or aza-substituted corrole. The effect of the contraction is, as discussed above, an increase in charge and σ-donation. Aza-substitution makes the ligand a stronger σ-donor with a larger cavity size and increases the oxidation potential to the π-radical cation by lowering the HOMO ($b_1$ in $C_{2v}$ symmetry).

While the chemistry of corroles has been known for decades, their use has been revived by the recent improvement of their syntheses. Nevertheless, the synthetic methodology is limited to *meso*-substituted derivatives, and β-substituted corroles are still difficult to prepare. In contrast, corrolazines can be prepared by ring contraction of porphyrazines with $PBr_3$, which gives easy access to β-substituted corrolazines. However, the removal of the phosphorus atom to generate the free corrolazine ligand was demonstrated only recently by using $Na/NH_3$ as a reductant. The corrole ligand with penta-fluorophenyl substituents is worthy of special mention because the preparation and isolation of several high-valent metal oxo complexes have been realized (Figure 3.19).[50] In the case of manganese(V), OAT to olefins and other substrates has been demonstrated. The mechanism under catalytic reactions and the role of $Mn^V(O)$ in OAT continue to be a subject of discussion in the literature.[51,49] The $Cr^{III}(tpfc)$ complex reacts with molecular oxygen ($O_2$) to generate $Cr^V(O)(tpfc)$. The latter can transfer an oxygen atom to organic phosphines and norbornene, for example, enabling catalytic oxidations with $O_2$.[52] In addition to OAT, $Mn^V(O)$ corrole and corrolazine complexes have been shown to undergo hydrogen atom transfer reactions with phenolic substrates and allylic C–H bonds (Figure 3.19).[53] Interrogation of OAT from $Mn^V(O)(Cz)$ complex using a double-isotope labeling

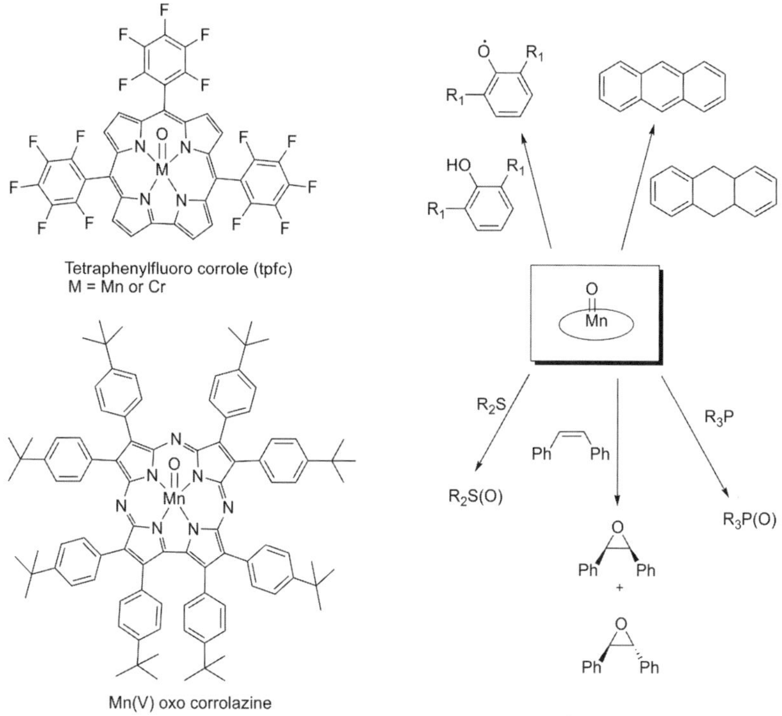

**FIGURE 3.19** High-valent corrole and corrolazine complexes. Oxygen and hydrogen atom transfer reactions of oxomanganese(V) corrole/corrolazine complexes.

experiment provided evidence for oxo transfer from an oxidant adduct of oxomanganese rather than the oxo ligand itself (Figure 3.20).[54] The chemistry of high-valent corrole and corrolazine complexes has been extended to nitrido and imido systems (nitrogen analogues of oxo), which are involved in nitrogen atom transfer reactions.[55,56]

### 3.3.2   OAT with the Element Rhenium

Next to molybdenum the most well-studied element for oxygen atom transfer reactions is rhenium. Oxo complexes of rhenium in the oxidation states of VII and V are plentiful. Their electronic configuration and structure parallel those of Mo and W in oxidation states VI and IV. Simple coordination complexes of rhenium(V) undergo oxygen atom transfer reactions. For example, *mer,trans*-Re(O)Cl$_3$(PPh$_3$)$_2$ reacts with excess DMSO to generate *mer,cis*-Re(O)Cl$_3$(MeS)(OPPh$_3$). The latter catalyzes OAT from sulfoxides to phosphines. Kinetics and mechanistic studies support atom transfer from a Re$^V$ sulfoxide adduct rather than a dioxorhenium(VII).[57] Undoubtedly, the development of organometallic oxorhenium complexes, in particular methyltrioxorhenium(VII) MTO, has invigorated the use of rhenium in OAT

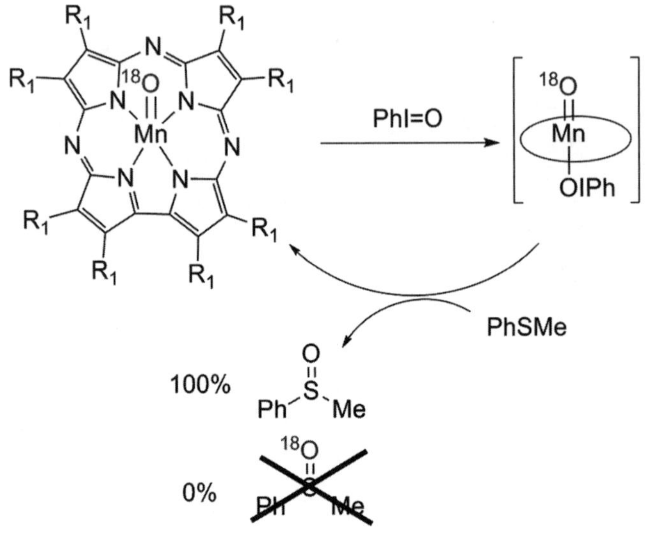

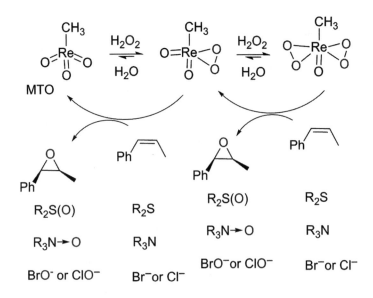

**FIGURE 3.20**    New mechanism for OAT with manganese oxo corrolazine.

reactions.[58–60] MTO catalyzes a wide variety of oxidation reactions with hydrogen peroxide through rhenium peroxo intermediates (Figure 3.21). These rhenium mono- and diperoxo complexes transfer an oxygen atom from the peroxo ligands to substrates such as alkenes, alkynes, sulfides, amine, imines, and so on. The source

**FIGURE 3.21**    OAT transfer reactions with $H_2O_2$ catalyzed by MTO.

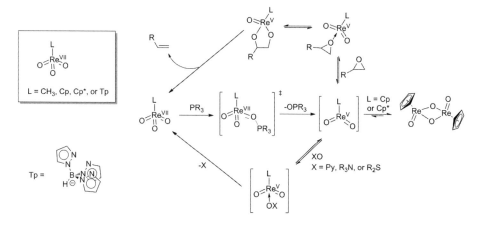

**FIGURE 3.22**    OAT with trioxorhenium complexes such as MTO and CpRe(O)$_3$.

of this peroxo atom is hydrogen peroxide, which is regarded a "green" oxidant because its only by-product is water.

Besides catalytic oxidation reactions with hydrogen peroxide, MTO and other trioxorhenium(VII) complexes such as CpRe(O)$_3$ and TpRe(O)$_3$ undergo OAT reactions with closed-shell organic molecules.[59–61] The trioxorhenium(VII) is reduced by organic phosphines to give reactive dioxorhenium(V) species that is capable of deoxygenating organic sulfoxides, amine N-oxides, PyO, and even epoxides (Figure 3.22). The reaction with epoxides to generate alkenes is believed to proceed via a rhenium diolate intermediate.[62] The kinetics of many of these reactions has been studied in detail and linear free energy relationships developed. Oxo transfer to substrate proceeds via nucleophilic attack of the phosphine substrate onto the electrophilic oxo ligand of rhenium. The transfer of an oxygen atom from the substrate to rhenium(V) follows in some instances Michaelis–Menten kinetics and involves formation of an adduct between the oxo donor substrate and the rhenium followed by unimolecular oxygen atom transfer. An extensive family of thiolato complexes of rhenium(V) containing oxo and methyl ligands has been prepared and investigated in OAT catalytic reactions.[63] The related rhenium(V) catalysts were found to adopt different rate laws, signaling the involvement of different intermediates and chemical steps/mechanisms.

Besides organometallic trioxorhenium(VII) and dioxorhenium(V), cationic monooxorhenium(V) coordination complexes containing oxazoline and salen ancillary ligands have been prepared and studied in the context of oxygen atom transfer reactions (Figure 3.23).[64] Upon reaction of monooxorhenium(V) with oxygen donor substrates such as PyO and organic sulfoxides, a cationic cis-dioxorhenium(VII) is generated. This complex was found to be reactive enough to transfer an oxygen atom to organic phosphines and sulfides. It is worth noting that these coordination complexes are comparable to synthetic models of sulfite oxidase based on molybdenum (Section 3.2.5). However, the reaction kinetics observed with rhenium surpasses

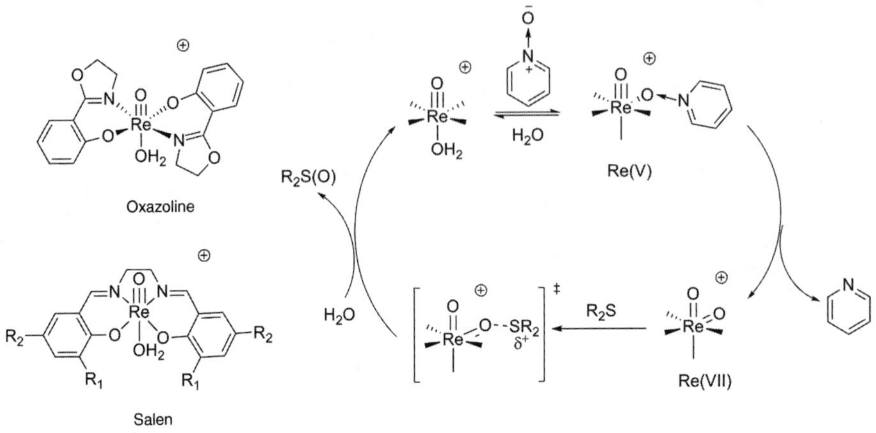

**FIGURE 3.23**  Cationic monooxorhenium(V) oxazoline and salen complexes. Mechanism of OAT for monooxorhenium(V) and dioxorhenium(VII) couple.

the observed rates for molybdenum complexes by $10^3$–$10^4$ times. The reaction kinetics of dioxorhenium(VII) with oxo acceptor substrates is highly dependent on the electronic effect of substituents. The use of several *para*-substituted thioanisole (*p*-X-PhSMe) substrates displayed a Hammett LFER (linear free energy relationship) with a reaction constant $\rho \sim -4$. Activation parameters were also consistent with an associative transition state. Oxo imido analogues of dioxorhenium(VII) salen have been synthesized by the action of organic azides ($RN_3$) on monooxorhenium(V) (Figure 3.24).[65] These compounds were found to undergo oxygen atom transfer with

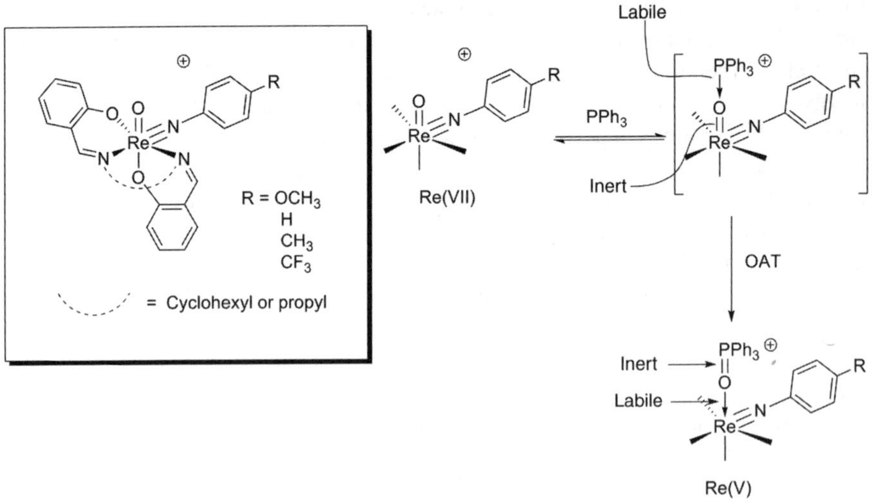

**FIGURE 3.24**  Oxo imido rhenium(VII) complexes and their mechanism of OAT.

substrates such as organic phosphines and sulfides and retain the metal imido ligand. Variation of the electronics of the arylimido ligand had a significant effect on the second-order rate constant of OAT. Electron-withdrawing groups (EWG) on the arylimido ligand were most effective. While inductive effects in metal ligands often translate into subtle perturbation in the rate, substituents on the arylimido ligand of these rhenium(VII) complexes spanned more than three orders of magnitude in OAT reactivity. These findings demonstrate the extent and strength of coupling between the multiply bonded imido ligand and the oxo atom being transferred in the transition state. Furthermore, the kinetics of oxo transfer followed Michaelis–Menten kinetics with substrate-dependent saturation values in support of a previously suggested mechanism in which the reductant forms a prior equilibrium adduct with the rhenium oxo complex ($Re^{VII}=O \leftarrow Y$) (Figure 3.24).

The effect of charge on the rate of oxygen atom transfer has been investigated for two identical monooxorhenium(V) complexes that differed only by one single charge.[66] The cationic monooxorhenium(V) complex [(HCpz$_3$)ReOCl$_2$]$^+$ (HCpz$_3$= tris(pyrazolyl)methane) was compared with the neutral complex [(HBpz$_3$)Re(O)Cl$_2$] (HBpz$_3$ = tris(pyrazolyl)borate) (Figure 3.25). The reaction of each with PPh$_3$ to give the corresponding $Re^{III}$(OPPh$_3$) complexes was studied. The cationic complex was reduced by PPh$_3$ 1000 times faster than its neutral analogue. This finding is consistent with the oxo ligand being more electrophilic in the charged complex, which is expected. However, the magnitude of the effect, three orders of magnitude, is quite remarkable.

### 3.3.3 Perchlorate Reduction by OAT and the Environment

The perchlorate ion $ClO_4^-$ is a recognized contaminant in groundwater as well as soil with more than a hundred contamination sites in 35 states.[67] Perchlorate toxicity arises from its interference with thyroid function. $ClO_4^-$ is unique among the inorganic oxyanions in that it is weakly coordinating and when used as a counteranion with metal complexes, the salts yield high-quality single crystals. Nevertheless, the

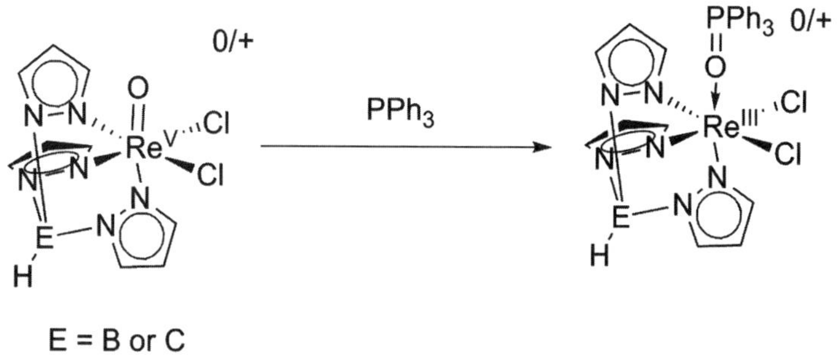

E = B or C

**FIGURE 3.25**    Effect of charge on the rate of OAT. When E = C, charge = + 1, rate of above reaction is $10^3$ that of E = B and charge = 0.

most fascinating aspect of perchlorate chemistry is its dual character. It is a strong oxidant in acidic aqueous solution, $E^0 = +1.23$ V for $ClO_4^- + 2H^+ + 2e^- \rightarrow ClO_3^- + H_2O$, but its reactions in solution are sluggish. This kinetic inertness is attributed to the anion's weak nucleophilicity and basicity. Furthermore, the chlorine (VII) oxidizing center is shielded by four oxygens. The outcome of all these properties is the fact that perchlorate does not react with highly reducing metal ions such as $Cr_{(aq)}^{2+}$.[68] Even the labile ion $Ti_{(aq)}^{3+}$ reduces perchlorate very slowly.[69] It has been argued that reactivity is dependent on transition metal complexes that form stable M=O bonds and have a favorable reaction coordinate to abstracting oxygen atoms from perchlorate.[70] It is interesting to note that a molybdenum-dependent enzyme is known to reduce perchlorate (Section 3.2.3). Therefore, the kinetic inertness of perchlorate should, in principle, be overcome by chemical catalysis.

Methylrhenium dioxide (MDO) when generated in acidic aqueous solution was found to react readily with the perchlorate ion to give chloride as the final product.[70] The rate-determining step is the reduction to chlorate ($k = 7$ L/(mol s) at pH $= 0$ and $T = 25°C$). The subsequent oxygen atom transfer from the chlorate ion, $ClO_3^-$, is four orders of magnitude faster ($k = 4 \times 10^4$ L/(mol s) under the same conditions (pH $= 0$ and $T = 25°C$). The reaction kinetics showed saturation in the oxyanion concentration $[ClO_n^-]$. This finding is consistent with adduct formation prior to atom transfer (Figure 3.26). MDO is generated in aqueous solution by the reduction of MTO with hypophosphorus acid ($H_3PO_2$). Hence, the reaction is catalytic and $H_3PO_2$ serves as the oxygen sink. However, under catalytic conditions the rate-determining step becomes the generation of MDO with a second-order rate constant of 0.03 L/(mol s) at pH $= 0$ and 25°C.[71] The second limitation of this catalytic system is the need to maintain low pH because MDO is presumably coordinated by one or more aqua ligands and at higher pH it forms a polymeric methylrhenium oxide that falls out of

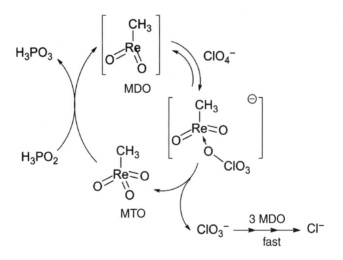

**FIGURE 3.26**   Perchlorate reduction by MDO.

solution. This polymer consists of bridging μ-oxo and hydroxo ligands. In addition to perchlorate, MDO reduces perbromate ($BrO_4^-$) to $Br^-$ ($k = 3 \times 10^5$ L/(mol s) at pH = 0 and $T = 25°C$). It is worth noting that the reaction with $BrO_4^-$ is five orders of magnitude faster than that with perhclorate. Recently, a heterogeneous version of the MTO catalytic system on Pd/C has been shown to effect perchlorate reduction to chloride with dihydrogen ($H_2$) as the stoichiometric reductant (Equation 3.14).[72] This is truly practical for environmental remediation because the by-product is only water and catalyst recovery (being heterogeneous) should be straightforward. Furthermore, rhenium oxides, $Re_2O_7$ and $ReO_4^-$, can be used. The heterogeneous catalysts, however, still exhibit high sensitivity to pH and the reaction requires acidic conditions (pH ≤ 3). Palladium is required for $H_2$ activation. Hydride spillover from the Pd surface to rhenium oxide has been postulated for the mechanism of action with perchlorate reduction rate-determining step.

$$ClO_4^- + 4H_2 \xrightarrow{\text{MTO/Pd/C}} Cl^- + 4H_2O \tag{3.14}$$

Cationic oxorhenium(V) oxazoline (Figure 3.23) was found to be especially reactive toward perchlorate in acetonitrile:water media.[73] The reaction is not pH dependent and proceeds smoothly at neutral pH. The perchlorate ion, $ClO_4^-$, was reduced all the way to the chloride ion, $Cl^-$. The reaction of oxorhenium(V) with perchlorate followed Michaelis–Menten showing saturation in $[ClO_n^-]$. The apparent second-order rate constant for $ClO_4^-$ is ca. 0.5 L/(mol s). The resulting cationic dioxorhenium(VII) species is more reactive toward OAT than MTO and can be turned over with organic sulfides quite readily (Figure 3.27). Therefore, under catalytic

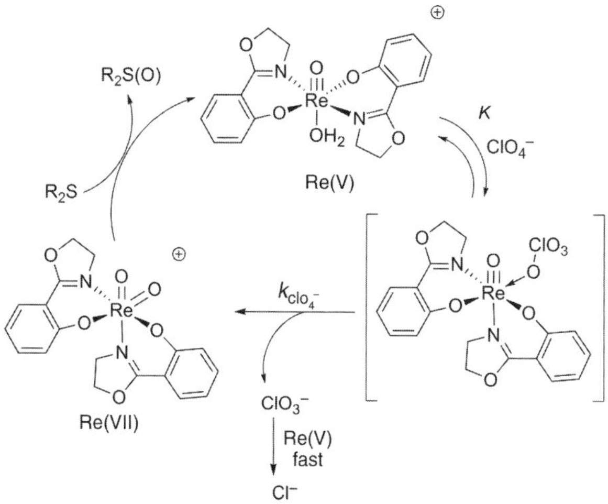

| $ClO_n^-$ | $K$ (L/mol[1]) | $k_{ClO_n}$ (s$^{-1}$) |
|---|---|---|
| $ClO_4^-$ | 5 | 0.09 |
| $ClO_3^-$ | 200 | 0.13 |

**FIGURE 3.27** Mechanism and rate constants for perchlorate and chlorate reduction by oxorhenium(V) oxazoline.

conditions the rate-determining step for the oxazoline system is perchlorate reduction at a second-order rate constant of ca. 0.5 L/(mol s) under ambient conditions and neutral pH. The subsequent reduction of chlorate ($ClO_3^-$) is fast and the kinetics shows saturation in [$ClO_3^-$] giving a first-order rate constant of 0.13 s$^{-1}$ (half-life of ~5 s!). It should be noted that the first-order rate constants measured for $ClO_4^-$ and $ClO_3^-$ at substrate saturation are comparable, and this rate constant corresponds to the OAT step following anion coordination to rhenium(V) (Figure 3.27).[74] Hence, chlorate is a superior oxidant in this instant not because it undergoes faster OAT but because it is a better ligand than perchlorate (see $K$ values in Figure 3.27). The oxazoline catalyst shows product inhibition at high chloride concentrations, which is not surprising. This inhibition can be circumvented by removal of the chloride product by precipitation with Ag(I). Nevertheless, even at high perchlorate concentrations (0.10 M), which translates to significant product inhibition as $Cl^-$ builds up, the oxorhenium(V) oxazoline system gives a turnover number in excess of 10 h$^{-1}$. Another advantage of this catalyst is that it shows very little or no decomposition even after hundreds of turnovers.

### 3.3.4   OAT Reactions with Aqua Metal Ions

Transition metal complexes in intermediary oxidation states with activated oxygen ligands have been studied extensively in aqueous solution.[75] These include $Cr^{III,IV}$, $Co^{III}$, and $Rh^{III}$. In the reduced state $M^{II}$ (where M = Cr, Co, or Rh), the metal ions react with oxygen to give bound superoxide complexes ($M^{III}$-OO). Reduction of the superoxide followed by proton transfer affords the hydroperoxo ($M^{III}$-OOH). In the case of Cr, further electron and proton transfer yields the chromyl ion ($Cr^{IV}=O$). The ligands for Co and Rh are cyclam macrocycles with $H_2O$, $OH^-$, $Cl^-$, or $SCN^-$ as the axial sixth ligand, for Cr aqua ligands, and tetrammonia aqua for Rh (Figure 3.28).

The hydroperoxide complexes react with the halide ions $I^-$ and $Br^-$ to give the OAT products HOI and HOBr in a kinetic step that is first order in [$H_3O^+$], [$M^{III}$-OOH], and [halide]. The HOBr reacts with $M^{III}$-OOH further to give dioxygen and $Br^-$. This reaction resembles $H_2O_2$ disproportionation with HOBr and catalytic disproportionation of $H_2O_2$ by haloperoxidases (Section 3.2.2). Iodide is more reactive than bromide and the rate constants vary over two orders of magnitude depending on the metal and the ligand. For example, $k_I$ (in M$^{-2}$ s$^{-1}$) for oxidation of $I^-$ is as follows for different complexes: $(H_2O)_5Cr(OOH)^{2+} = 990$, (cyclam)($H_2O$) Co(OOH)$^{2+} = 100$, (cyclam)($H_2O$)Rh(OOH)$^{2+} = 540$, and (NH$_3$)$_4$($H_2O$)Rh-(OOH)$^{2+} = 8800$. In all instances, however, the metal hydroperoxide is much more reactive than $H_2O_2$, $k_I(H_2O_2) = 0.17$ M$^{-2}$ s$^{-1}$. Unlike hydrogen peroxide, the metal hydroperoxides do not show acid-independent reactivity. Two mechanisms have been put forth for the OAT to the halide ion (Figure 3.29). Transfer of the remote oxygen seems to be favored for nonlabile metals because protonation of the hydroperoxo in complexes such as (cyclam)($H_2O$)Rh($H_2O_2$)$^{3+}$ and (CN)$_5$Co($H_2O_2$)$^{2-}$ does not result in rapid dissociation of $H_2O_2$. Furthermore, bulky substrates such as PPh$_3$ probably favor nucleophilic attack on the remote oxygen.

Interestingly, the chromyl ion oxidizes bromide via an acid-dependent pathway and all the kinetic data are consistent with a one-electron mechanism. However, there

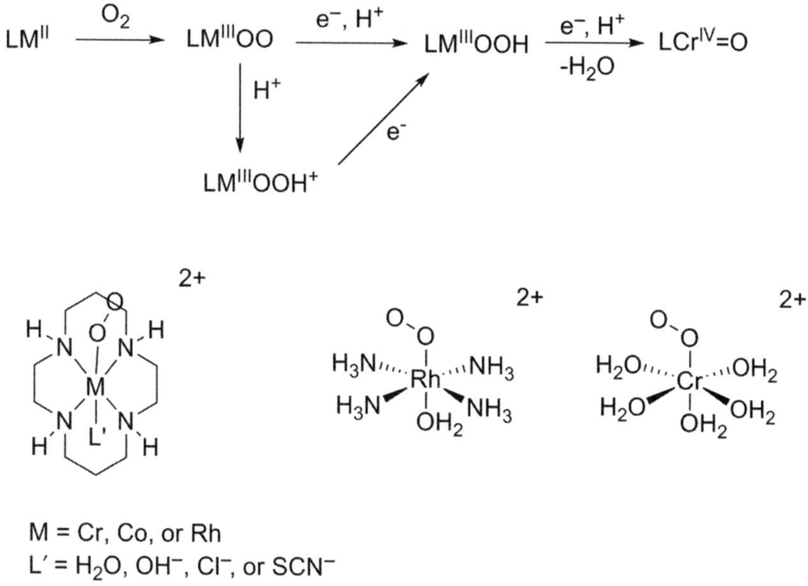

**FIGURE 3.28**   Metal oxygen complexes in intermediary oxidation state and their superoxo, peroxo, and oxo complexes in aqueous solution.

is no spectroscopic evidence for the existence of a protonated chromyl ion ($Cr^{IV}$-$OH^{3+}$). The reaction of chromyl with $I^-$ is too fast to measure even by stopped-flow spectrophotometry. The chromyl ion is stable under most conditions for ca. 1 min. It decomposes through an exceptional disproportionation reaction to give $Cr_{(aq)}^{3+}$ and $HCrO_4^-$. The kinetics of this reaction is second order in chromyl and shows a significant solvent kinetic isotope effect. The rate in water is seven times that in $D_2O$. It is also impossible to envision how $Cr^{IV}=O$ would afford $HCrO_4^-$ in one step. Therefore, multiple steps are likely involved and the rate-determining step must include O–H/D bond breaking from a solvent molecule. An intermediate with a hydroxo bridge has been suggested to give $Cr^{3+}$ and $(OH)Cr^V=O^{2+}$. The latter disproportionates further to Cr(IV) and Cr(VI).

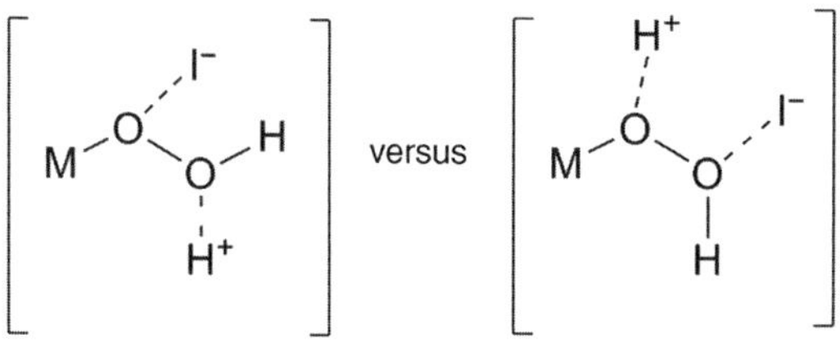

**FIGURE 3.29**   OAT from metal hydroperoxo to halide ion ($I^-$). M = Rh, Co, or Cr.

### 3.3.5  Intermetallic OAT: The Fast and the Slow

Even though oxygen atom transfer from metal complexes to organic substrates and main group species has been extensively studied, studies of intermetallic OAT have been infrequent. Inner-sphere electron transfer reactions that involve atom transfer (one-electron processes) date back to the classical Taube reactions between $Co^{III}$–Cl and $Cr^{2+}$. Kinetic investigations of analogous complete two-electron oxygen transfer reactions, however, have been seldom. In part many OAT reactions between metal centers proceed incompletely to give μ-oxo dinuclear species as in the case of porphyrin ferryl and oxomolybdenum complexes (Equations 3.13 and 3.15). Complete oxygen/chlorine atoms exchange in chromium and titanium porphyrin complexes has been examined (Equation 3.16).[76] However, in these systems the net redox process is one-electron transfer. Nevertheless, the reactions are presumed to involve intermediacy of an oxygen-bridged μ-oxo complexes rather than a chloride bridge. The kinetics showed inhibition by Cl$^-$ ion in support of dissociation of the chloride prior to OAT. The rates span three orders of magnitude with second-order rate constants of 0.14 and 240 L/(mol s) for Cr and Ti, respectively. The reactions in Equation 3.16 involve different porphyrins but the equilibrium constants are close enough to unity and thus the measured kinetics is essentially for degenerate atom transfer and not complicated by driving force.

$$(por)Fe^{IV}(O) + (por)Fe^{II} \rightarrow (por)Fe^{III}\text{–O–}Fe^{III}(por) \tag{3.15}$$

$$(por)M = O + (por')M\text{-Cl} \leftrightarrow (por)M\text{-Cl} + (por')M = O \tag{3.16}$$
$$M = \text{Ti or Cr}$$

Complete intermetallic oxygen atom transfer has been documented between dioxo Mo(VI) and Mo(II), as well as a degenerate example between dioxo Mo(VI) and monooxo Mo(IV) (Equations 3.17 and 3.18).[77,78] In the latter case, the μ-oxo dinuclear Mo(V) intermediate has been observed. The rate constant is faster than that observed for the porphyrin systems, $k = 1470$ L/(mol s). A comprehensive examination of complete OAT between $Mo^{VI}(O)_2$ and $Mo^{IV}(O)$ containing sterically demanding thiolato ligands has been reported.[79] These reactions were spontaneous and they allowed the development of a relative thermodynamic scale for OAT. Also, complete oxygen atom transfer reactions from Mo and Re oxo complexes to W have been documented.[76] In general, the thermodynamic trend in favor of OAT is as follows: Re $\rightarrow$ Mo, Re $\rightarrow$ W, and Mo $\rightarrow$ W.

$$(R_2NCS_2)Mo^{VI}(O)_2 + (R_2NCS_2)Mo^{II}(CO)_2 \rightarrow 2(R_2NCS_2)MoI^{V}(O) + 2CO \tag{3.17}$$

$$(Et_2NCS_2)Mo^{VI}(O)_2 + (Et_2NCS_2)Mo^{IV}(O) \leftrightarrow [Mo^V\text{–O–}Mo^V] \leftrightarrow (Et_2NCS_2)Mo^{VI}(O)_2$$
$$+ (Et_2NCS_2)Mo^{IV}(O) \tag{3.18}$$

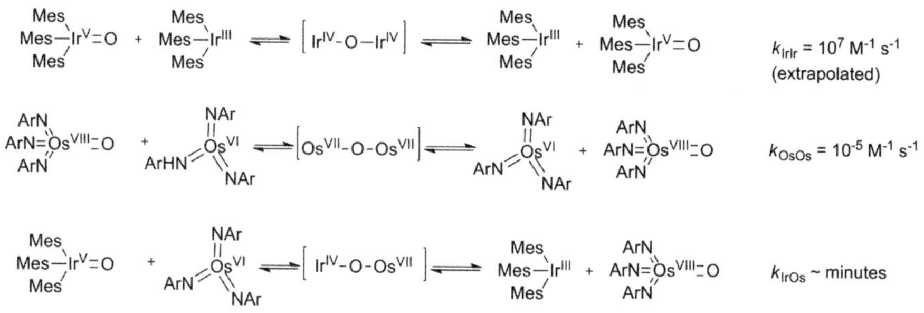

**FIGURE 3.30**    Slow, intermediate, and fast intermetal oxygen atom transfer of the elements Ir and Os.

An intriguing degenerate OAT reactions between $(Mes)_3Ir^V=O$ and $(Mes)_3Ir^{III}$, on the one hand, and $(ArN)_3Os^{VIII}=O$ and $(ArN)_3Os^{VI}$, on the other hand (Figure 3.30), have been investigated in detail recently.[80] The rate constant for complete intermetallic OAT for Ir is 12 orders of magnitude faster than that observed for Os. Both metal oxo display identical reactivity toward $PPh_3$ and the metal oxo strengths are quite comparable. Furthermore, as would be predicted by the Marcus cross relationship for outer-sphere electron transfer, the Ir–Os cross exchange takes place at an intermediate rate (Figure 3.30). This sharp contrast in OAT rates has been attributed to geometrical demands in the reorganization required to proceed through the μ-oxo intermediates. In the case of Ir, both $+3$ and $+5$ complexes are pyramidal and can form a pyramidal Ir $+4$ with little energetic penalty. However, the trisimido Os(VI) is planar and its pyramidalization would require occupancy of an antibonding orbital. This is an elegant example of how subtle changes in metal and geometry would impose a dramatic rate penalty on the order of $10^{12}$.

## 3.4    CONCLUSION

Since we live on an aerobic planet, the chemistry of oxygen takes center stage in all aspects of life. Living organisms use oxygen for respiration and for metabolizing drugs. The use of oxygen, however, comes at a price, namely, oxidative damage. Hence, antioxidant chemistry is also important. Finally, our industrial development in the past 100 years has relied heavily on the utilization of feedstock from petroleum that necessitated the use of oxidation chemistry for making functionalized chemicals. All of these fields rely on OAT reactions, albeit for different reasons. In many of these chemical and biochemical transformations, the transfer of an oxygen atom is mediated by a transition metal. We have detailed in this chapter some examples of transition metal coordination complexes that are involved in oxygenations of C–H bonds, carbon–carbon double bonds, halides, and closed-shell organic and inorganic molecules. The ligands vary quite a bit from being porphyrin/heme to O, N, or sulfur-based monodentate as well as multidentate. The metals are also quite versatile spanning all three rows of the transition element series. The kinetics and mechanisms are very rich and in some instances the observed rates can be rationalized on thermodynamic basis and in other instances effects other than thermodynamics are at play. Much of the OAT chemistry has been dominated by oxygenation, that is, addition of an oxygen atom to make new and interesting molecules. As the demand for renewable feedstock mandates that we move away from using fossil fuels, we will move toward the development of efficient chemistries for the utilization of biomass. This will require advancement of OAT reactions in the reverse direction, that is, deoxygenation of lignocellulosic biomass.

## REFERENCES

1. McClellan, P. P. *Ind. Eng. Chem.* **1950**, *42*, 2402–2407.

2. Tullo, A. *Chem. Eng. News* 2004, *82* (36), 15.

3. Duebel, D. V. *J. Am. Chem. Soc.* 2004, *126*, 996–997.

4. Jonsson, S. Y.; Färnegårdh, K.; Bäckvall, J.-E. *J. Am. Chem. Soc.* 2001, *123*, 1365–1371.

5. Groves, J. T.; Han, Y. In *Cytochrome P450: Structure, Mechanism and Biochemistry*, 2nd edition; Ortiz de Montellano, P. R., Ed.; Plenum Press: New York, 1995; pp 3–48.

6. Schlichting, I.; Berendzen, J.; Chu, K.; Stock, A. M.; Maves, S. A.; Benson, D. E.; Sweet, R. M.; Ringe, D.; Petsko, G. A.; Sligar, S. G. *Science* 2000, *287*, 1615–1622.

7. Griller, D.; Ingold, K. U. *Acc. Chem. Res.* 1980, *13*, 317–323.

8. McLain, J. L.; Lee, J.; Groves, J. T. In *Biomimetic Oxidations Catalyzed by Transition Metal Complexes*; Meunier, B., Ed.; Imperial College Press: London, 2000; pp 91–169.

9. Toy, P. H.; Newcomb, M.; Coon, M. J.; Vaz, A. D. N. *J. Am. Chem. Soc.* 1998, *120*, 9718–9719.

10. Vaz, A. D. N.; McGinnity, D. F.; Coon, M. J. *Proc. Natl. Acad. Sci. USA* 1998, *95*, 3555–3560.

11. Nam, W.; Lim, M. H.; Moon, S. K.; Kim, C. *J. Am. Chem. Soc.* 2000, *122*, 10805–10809.

12. Schönoboom, J. C.; Cohen, S.; Lin, H.; Shaik, S.; Thiel, W. *J. Am. Chem. Soc.* 2004, *126*, 4017–4034.

13. Mansuy, D. *C. R. Chim.* 2007, *10*, 392–413.

14. Bartoli, J. F.; Brigaud, O.; Battioni, P.; Mansuy, D. *J. Chem. Soc., Chem. Commun.* 1991, 440–442.

15. Grinstaff, M. W.; Hill, M. G.; Labinger, J. A.; Gray, H. B. *Science* 1994, *264*, 1311–1313.

16. Fita, I.; Silva, A. M.; Murthy, M. R. N.; Rossmann, M. G. *Acta Crystallogr. B* 1986, *42*, 497–515.

17. Ivancich, A.; Jouve, H. M.; Gaillard, J. *J. Am. Chem. Soc.* 1996, *118*, 12852–12853.

18. Hager, L. P.; Lakner, F. J.; Basavapathruni, A. *J. Mol. Catal. B* 1998, *5*, 95–101.

19. Shulz, C. E.; Devaney, P. W.; Winkler, H.; Debrunner, P. G.; Doan, N.; Chiang, R.; Rutter, R.; Hager, L. P. *FEBS Lett.* 1979, *103*, 102–105.

20. Miller, V. P.; Goodin, D. B.; Friedman, A. E.; Hartmann, C.; Ortiz de Montellano, P. R. *J. Biol. Chem.* 1995, *270*, 18413–18419.

21. Sivaraja, M.; Goodin, D. B.; Smith, M.; Hoffman, B. M. *Science* 1989, *245*, 738–740.

22. Bonagura, C. A.; Sundaramoorthy, M.; Pappa, H. S.; Patterson, W. R.; Poulos, T. L. *Biochemistry* 1996, *35*, 6107–6115.

23. Sundaramoorthy, M.; Terner, J.; Poulos, T. L. *Structure* 1995, *3*, 1367–1378.

24. Green, M. J.; Dawson, J. H.; Gray, H. B. *Science* 2004, *304*, 1653–1656.

25. Butler, A. *Curr. Opin. Chem. Biol.* 1998, *2*, 279–285.

26. Littlechild, J.; Garcia-Rodriguez, E.; Dalby, A.; Isupov, M. *J. Mol. Recognit.* 2002, *15*, 291–296.

27. Messerschmidt, A.; Wever, R. *Proc. Nat. Acad. Sci. USA* 1996, *93*, 392–396.

28. Coates, J. D.; Achenbach, L. A. *Nat. Rev. Microbiol.* 2004, *2*, 569–580.

29. Urbansky, E. T.; Schock, M. R. *J. Environ. Manage.* 1999, *56*, 79–95.

30. Lee, A. Q.; Streit, B. R.; Zdilla, M. J.; Abu-Omar, M. M.; DuBois, J. L. *Proc. Natl. Acad. Sci. USA* 2008, *105*, 15654–15659.

31. Jakopitsch, C.; Spalteholz, H.; Fûrtmuller, P. G.; Arnhold, J.; Obinger, C. *J. Inorg. Biochem.* 2008, *102;* 293–302.

32. Zdilla, M. J.; Lee, A. Q.; Abu-Omar, M. M. *Angew. Chem., Int. Ed.* 2008, *47*, 7697–7700.

33. Costas, M.; Mehn, M. P.; Jensen, M. P.; Que, L., Jr. *Chem. Rev.* 2004, *104*, 939–986.

34. Abu-Omar, M. M.; Loaiza, A.; Hontzeas, N. *Chem. Rev.* 2005, *105*, 2227–2252.

35. Wallar, B. J.; Lipscomb, J. D. *Chem. Rev.* 1996, *96*, 2625–2658.

36. Krebs, C.; Price, J. C.; Baldwin, J.; Saleh, L.; Green, M. T.; Bollinger, J. M., Jr. *Inorg. Chem.* 2005, *44*, 742–757.

37. Collins, T. J. *Acc. Chem. Res.* 2002, *35*, 782–790.

38. Tiago de Oliviera, F.; Chanda, A.; Banerjee, D.; Shan, X.; Mondal, X.; Que, L., Jr.; Bominaar, E. L.; Münck, E.; Collins, T. J. *Science* 2007, *315*, 835–838.

39. Que, L., Jr. *Acc. Chem. Res.* 2007, *40*, 493–500.

40. Nam, W. *Acc. Chem. Res.* 2007, *40*, 522–531.

41. Price, J. C.; Barr, E. W.; Glass, T. E.; Krebs, C.; Bollinger, J. M., Jr. *J. Am. Chem. Soc.* 2003, *125*, 13008–13009.

42. Nesheim, J. C.; Lipscomb, J. D. *Biochemistry* 1996, *35*, 10240–10247.

43. Mishina, Y.; He, C. *J. Inorg. Biochem.* 2006, *100*, 670–678.

44. Sigel, A.; Sigel, H.; Eds. *Molybdenum and Tungsten: Their Roles in Biological Processes. Metals in Biological Systems*; Marcel Dekker: New York, 2002; Vol. 39.

45. Hille, R. *Trends Biochem. Sci.* 2002, *27*, 360–367.

46. Johnson, M. K.; Rees, D. C.; Adams, M. W. W. *Chem. Rev.* 1996, *96*, 2817–2840.

47. Enemark, J. H.; Cooney, J. J. A.; Wang, J.-J.; Holm, R. H. *Chem. Rev.* 2004, *104*, 1175–1200.

48. Aviv, I.; Gross, Z. *Chem. Commun.* 2007, 1987–1999.

49. Kerber, W. D.; Goldberg, D. P. *J. Inorg. Biochem.* 2006, *100*, 838–857.

50. Gross, Z. *J. Biol. Inorg. Chem.* 2001, *6*, 733–738.

51. Zhang, R.; Newcomb, M. *J. Am. Chem. Soc.* 2003, *125*, 12418–12419.

52. Mahammed, A.; Gray, H. B.; Meier-Callahan, A. E.; Gross, Z. *J. Am. Chem. Soc.* 2003, *125*, 1162–1163.

53. Lansky, D. E.; Goldberg, D. P. *Inorg. Chem.* 2006, *45*, 5119–5125.

54. Wang, S. H.; Mandimutsira, B. S.; Todd, R.; Ramdhanie, B.; Fox, J. P.; Goldberg, D. P. *J. Am. Chem. Soc.* 2004, 18–19.

55. Galina, G.; Gross, Z. *J. Am. Chem. Soc.* 2005, *127*, 3258–3259.

56. Zdilla, M. J.; Abu-Omar, M. M. *J. Am. Chem. Soc.* 2006, *128*, 16971–16979.

57. Abu-Omar, M. M.; Khan, S. I. *Inorg. Chem.* 1998, *37*, 4979–4985.

58. Romão, C. C.; Kühn, F. E.; Herrmann, W. A. *Chem. Rev.* 1997, *97*, 3197–3246.

59. Espenson, J. H. *Chem. Commun.* 1999, 479–488.

60. Owens, G. S.; Arias, J.; Abu-Omar, M. M. *Catal. Today* 2000, *55*, 317–363.

61. Gable, K. P. *Adv. Organomet. Chem.* 1997, *41*, 127.

62. Gable, K. P.; Phan, T. N. *J. Am. Chem. Soc.* 1994, *116*, 833–839.

63. Espenson, J. H. *Coord. Chem. Rev.* 2005, *249*, 329–341.

64. Abu-Omar, M. M. *Chem. Commun.* 2003, 2102–2111.

65. Ison, E. A.; Cessarich, J. E.; Travia, N. E.; Fanwick, P. E.; Abu-Omar, M. M. *J. Am. Chem. Soc.* 2007, *129*, 1167–1178.

66. Seymore, S. B.; Brown, S. N. *Inorg. Chem.* 2000, *39*, 325–332.

67. Brown, G. M.; Gu, B. H. *Perchlorate: Environmental Occurrence, Interactions, and Treatments*; Springer: New York, 2006; pp 17–47.

68. Thompson, R. C.; Gordon, G. *Inorg. Chem.* 1966, *5*, 562–569.

69. Early, J. E.; Tofan, D. C.; Amadie, G. A. In *Perchlorate in the Environment*; Urbansky, E. T., Ed.; American Chemical Society: Washington, DC.

70. Abu-Omar, M. M.; Espenson, J. H. *Inorg. Chem.* 1995, *34*, 6239–6240.

71. Abu-Omar, M. M.; Appleman, E. H.; Espenson, J. H. *Inorg. Chem.* 1996, *35*, 7751–7757.

72. Hurley, K. D.; Shapley, J. R. *Environ. Sci. Technol.* 2007, *41*, 2044–2049.

73. Abu-Omar, M. M.; McPherson, L. D.; Arias, J.; Béreau, *Angew. Chem., Int. Ed.* 2000, *39*, 4310–4313.

74. McPherson, L. D.; Drees, M.; Khan, S. I.; Strassner, T.; Abu-Omar, M. M. *Inorg. Chem.* 2004, *43*, 4036–4050.

75. Bakac, A. *Coord. Chem. Rev.* 2006, *250*, 2046–2058.

76. Woo, L. K. *Chem. Rev.* 1993, *93*, 1125–1136.

77. Chen, G. J.-J.; McDonald, J. W.; Newton, W. E. *Inorg. Chim. Acta* 1976, *19*, L67–L68.

78. Matsuda, T.; Tanaka, K.; Tanaka, T. *Inorg. Chem.* 1979, *18*, 454–457.

79. Harlan, E. W.; Berg, J. M.; Holm, R. H. *J. Am. Chem. Soc.* 1986, *108*, 6992–7000.

80. Fortner, K. C.; Laitar, D. S.; Muldoon, J.; Pu, L.; Braun-Sand, S. B.; Wiest, O.; Brown, S. N. *J. Am. Chem. Soc.* 2007, *129*, 588–600.

# 4 Mechanisms of Oxygen Binding and Activation at Transition Metal Centers

ELENA V. RYBAK-AKIMOVA

## 4.1 INTRODUCTION

Understanding the mechanisms of dioxygen activation at transition metal centers is important for unraveling the mechanisms of metal-containing oxidases and oxygenases, synthesizing new selective oxidation catalysts and new drugs analogous to bleomycin, and suppressing free radical pathways of oxidative damage in biological systems. Additional interest to dioxygen chemistry arises from the need to design efficient oxygen electrodes for fuel cells. The problem of producing hydrogen from water is also related to oxygen chemistry: electrocatalytic or photocatalytic water splitting couples hydrogen production to water oxidation, a process that makes dioxygen from water.

Molecular oxygen is an ideal reagent for chemical oxidations, because it is readily available (air contains about 20% of oxygen) and environmentally clean (the only by-product of oxidations with $O_2$ is water). Oxygen is used in nature in a variety of selective chemical reactions occurring every moment in every aerobic organism, including humans. Chemists are, however, far less sophisticated than nature in their ability to use oxygen for the synthesis of complicated molecules. It is easy to activate oxygen under harsh conditions, when it quickly and completely reacts with almost any organic compound, producing carbon dioxide. This process, which is very useful in producing energy via burning fuels, is not applicable to selective syntheses of desired organic products. Selective oxygen activation under mild conditions is a difficult problem.

Not surprisingly, catalysts for oxygen and peroxide activation are under active development, and one of the recently published texts on modern oxidation methods devotes a section in every chapter to environment-friendly oxidations with oxygen and hydrogen peroxide.[1] Despite numerous successes, problems remain. Many oxygen-activating processes generate free radicals, thus giving rise to nonselective oxidations. The majority of selective catalysts for aerobic oxidations, even those

*Physical Inorganic Chemistry: Reactions, Processes, and Applications*  Edited by Andreja Bakac
Copyright © 2010 by John Wiley & Sons, Inc.

developed recently,[2] are based on platinum metals. Developing cheap, nontoxic catalysts for aerobic oxidations remains an important goal. Mimicking oxygen-activating metalloenzymes, which most commonly utilize iron or copper at the active sites, is a popular and promising approach.

In biology, reactions of dioxygen are catalyzed by two classes of metalloenzymes: oxidases and oxygenases. The oxidase chemistry transfers electrons and protons from substrates to $O_2$ molecule, incorporating both oxygen atoms in water molecules. In contrast, oxygenases incorporate one or two oxygen atoms from $O_2$ into organic products, catalyzing atom transfer rather than electron transfer processes. Much-celebrated discovery of oxygenases by Hayaishi in 1955, along with concurrent work by Mason, is often viewed as the origin of the field of biochemical oxygen activation.[3] Development of modern aerobic oxidations is not limited to narrowly defined, oxygenase-like oxygen activation processes, rather it makes profitable use of oxidase-like reactions.[2]

From the thermodynamic standpoint, reactions with dioxygen are usually favorable. However, the majority of these reactions are slow. Therefore, oxygen activation is primarily a kinetic problem. In contrast to some other small molecules, most notably $N_2$, kinetic inertness of dioxygen cannot be attributed to a strong element–element bond in the diatomic molecule. While the triple $N \equiv N$ bond is the second strongest known chemical bond (bond dissociation energy for $N_2$ is 941 kJ/mol), the double $O=O$ bond is relatively weak (bond dissociation energy for $O_2$ is 495 kJ/mol). The major reason for kinetic inertness of dioxygen in its reactions with typical organic molecules is the mismatch in the spin state of reactants. The $O_2$ molecule has two unpaired electrons in its ground state. Concerted reactions between triplet $O_2$ and singlet organic molecules are spin-forbidden and, therefore, slow.[4] This limitation is lifted in reactions between triplet $O_2$ and paramagnetic species such as organic radicals or transition metal ions or complexes. Organic radicals are generally unstable, and they tend to initiate radical chain reactions that do not selectively produce a sole well-defined product. In contrast, transition metal compounds can be indefinitely stable and can react via a variety of nonradical pathways that would selectively generate target products. Not surprisingly, oxygen activation with transition metals received most attention, and it shows great promise from the standpoint of developing practical, environmentally clean oxidations.

The literature on oxygen activation is extensive, including books,[5–9] chapters,[2,10] journal special issues,[11–13] review articles,[14–16] and numerous primary publications; only a handful of representative references are cited above. A comprehensive review of this active area of research is nearly impossible, and definitely beyond the scope of this work. In this chapter, we will focus on mechanistic aspects of oxygen activation at transition metal centers. General principles that govern kinetics and mechanisms of dioxygen coordination to transition metal complexes will be briefly considered and illustrated by selected examples (preference is given to biomimetic systems). Reactions of metal–oxygen intermediates are briefly introduced in this chapter, providing the link between oxygen binding and oxygen activation. Oxygen transfer processes are treated in detail in Chapter 3; hydrogen atom abstraction and proton-coupled electron transfer are considered in Chapter 2.

An alternative approach to matching the spin states of dioxygen and its targets, generating an excited-state singlet oxygen, $^1O_2$, has its merits in both chemistry and biology, but will not be discussed here. Finally, an allotrope of oxygen, ozone ($O_3$), is significantly more reactive than $O_2$. Ozone can be generated from $O_2$ in electric discharge. Although chemistry of ozone is very interesting in its own right, it will not be treated in detail; selected reactions with ozone that generate metal-oxo inter- mediates will be mentioned wherever appropriate.

## 4.2   REDOX PROPERTIES OF DIOXYGEN AND ITS REACTIONS WITH TRANSITION METAL COMPLEXES

Reactions of dioxygen molecule with transition metal compounds always involve redistribution of electron density that leads to partial or complete dioxygen reduction. Usually, dioxygen reduction is accompanied or followed by coordination to the metal center. Electron transfer from the reducing agent (e.g., a metal complex) to the $O_2$ molecule is favorable for thermodynamically downhill reactions. Thermodynamics of the stepwise reduction of free dioxygen, which is briefly considered in Sec- tion 4.2.1, sets the limits of the range of metal complexes that can activate $O_2$. While these thermodynamic limits cannot be violated for outer-sphere reactions, coordi- nation of $O_2$ or partially reduced $O_2$ to the metal centers shifts the redox potentials, increasing the driving force for dioxygen activation. Typical chemical structures of metal–oxygen complexes will be summarized next. Finally, reversible $O_2$ binding to natural dioxygen carriers will be introduced. These topics serve as an introduction to a more detailed treatment of the mechanisms and kinetics of dioxygen reactions with mononuclear metal complexes (Section 4.3) and di- or polynuclear metal complexes (Section 4.4).

### 4.2.1   Thermodynamics of Stepwise Dioxygen Reduction

Complete reduction of dioxygen requires four electrons and four protons:

$$O_2 + 4H^+ + 4e \rightarrow 2H_2O \tag{4.1}$$

The redox potential for this four-electron oxygen reduction (1 atm $O_2$) in 1 M acid in water is $+1.229\,V$ (versus NHE). Of course, the redox potential of a proton-sensitive process, such as Equation 4.1, depends on the concentration of $H^+$ (Figure 4.1): it drops to $+0.815$ at pH 7, and decreases further to $+0.401$ at pH 14.[5]

The electrochemical potential of dioxygen reduction also depends on the nature of solvent. For example, in 1 M acid in acetonitrile, the redox potential of $+1.79\,V$ was reported (Figure 4.2).[5] Different solvation of reactants and products in water versus $CH_3CN$, different $pK_a$'s of acids involved, and different oxygen solubilities all contribute to changes of redox activity of dioxygen in different solvents. It is important to take this solvent-dependent behavior into account, since nearly all

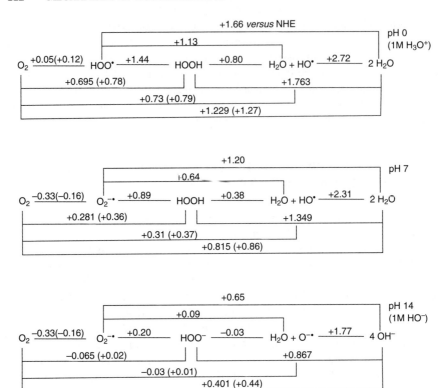

**FIGURE 4.1**    Redox potentials for stepwise $O_2$ reduction in water (Latimer diagrams for $O_2$ at different pH).[5]

studies of oxygen activation with metalloenzymes were performed in water, while the majority of studies with synthetic coordination compounds were carried out in nonaqueous solvents (and acetonitrile appears to be one of the most popular solvents for these model studies). Dioxygen solubility in organic solvents is several times higher than its rather low solubility in water (1.234 M at 25°C and 1 atm of $O_2$); the data for common solvents have been tabulated.[17]

Harnessing full thermodynamic potential of a four-electron oxygen reduction is highly desirable for energy conversion, but managing four electrons and four protons at a time is challenging. Stepwise reduction of dioxygen is much easier and it occurs in the majority of both natural and synthetic oxygen-activating systems. Thermodynamics of stepwise dioxygen reduction is summarized in Figures 4.1 and 4.2. Consecutive one-electron reduction of $O_2$ generates superoxide ($O_2{}^{-\bullet}/HO_2$), peroxide ($O_2{}^{2-}/HO_2{}^-/H_2O_2$), and hydroxyl radicals ($HO^{\bullet}$). Notably, redox potentials depend strongly on pH (in aqueous medium) or effective acidity (in organic solvents). Reduction of $O_2$ places electrons on the antibonding, $\pi^*$ orbitals, thus weakening the $O_2$ $\pi$-bond in superoxide (which has a formal O–O bond order of 1.5) and breaking this $\pi$-bond in peroxide (where two oxygen atoms are connected by one $\sigma$-bond, with a

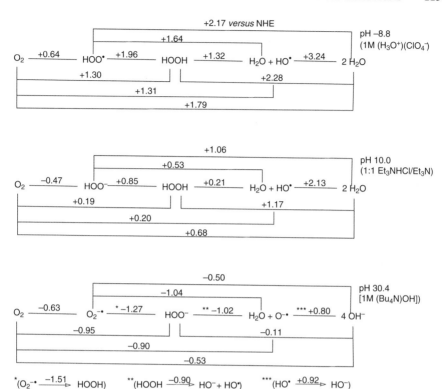

**FIGURE 4.2**  Redox potentials for stepwise $O_2$ reduction in acetonitrile (Latimer diagrams for $O_2$ in $CH_3CN$ at different acidities).[5]

formal bond order of 1). Further reduction of peroxide breaks the remaining O–O bond and separates two oxygen atoms. Addition of the first electron to $O_2$ is thermodynamically less favorable, while subsequent reduction steps are all downhill.[4,5] This trend in sequential redox potentials highlights major challenges in dioxygen activation: initially, it is difficult to add an electron to $O_2$, but once the reduction starts, it is difficult to control the reactivity channels and to stop reactions at a certain oxidation level of oxygen. Ideal oxygen-activating metallocomplexes should be able to accomplish both goals: facilitate the initial steps of dioxygen reduction and allow exquisite control of substrate oxidation pathways by selectively channeling the reactions toward desired metal–oxygen intermediates.

The well-known chemical properties of noncoordinated superoxides, peroxides, and hydroxyl radicals should be taken into account when interpreting mechanisms of metal-based oxygen activation. All these species contain oxygen in intermediate oxidation states. Superoxide is a very poor oxidant, but a reasonable reductant; in simple reactions of inorganic superoxides, reduction chemistry is much more important than oxidation chemistry. Hydrogen peroxide and other peroxides can undergo either an oxidation into $O_2$ (thus acting as reducing agents), or a reduction into $O^{2-}$, $HO^-$, or $H_2O$ (thus acting as oxidizing agents). Both superoxide and hydrogen peroxide easily undergo disproportionation (Figures 4.1 and 4.2).

Both hydrogen peroxide and hydrosuperoxide, $HO_2^\bullet$, are weak acids[18,19]: for $H_2O_2$, $pK_{a1} = 11.65$ and for $HO_2^\bullet$, $pK_a = 4.88$. Consequently, superoxide radical anion, $O_2^{-\bullet}$, acts as a strong Brønsted base; similarly, peroxide or hydroperoxo anions abstract protons in water, forming $H_2O_2$. Superoxide anion is a weak nucleophile, while peroxide is more nucleophilic than $OH^-$. One-electron reduction of $O_2$ is favored in the presence of proton sources; in fact, hydrogen atom abstraction would be more favorable than an electron transfer at the first step of oxygen reduction.[4,5] Notably, hydrogen atom abstraction is also a preferential reactivity mode of hydroxyl radicals, $HO^\bullet$, which are strong, nondiscriminatory oxidants. This radical chemistry has to be avoided in biological systems, where it results in oxidative stress, and in designing selective oxidation reagents and catalysts for synthetic applications.

### 4.2.2 Reactions of Transition Metal Compounds with Dioxygen and Structures of Metal–Oxygen Complexes

Typical reactions of dioxygen with metal ions or complexes can be classified on the basis of the number of electrons transferred. The examples below show changes in the formal oxidation states of dioxygen and the metal center(s). Of course, protonation of metal–oxygen complexes often occurs; for the sake of simplicity, proton transfer events are ignored in the chemical equations shown below.

*One-electron $O_2$ reduction* affords superoxide. Mononuclear or polynuclear metal complexes can act as one-electron reducing agents; the formal oxidation state of one metal center is increased by 1 in the product.

$$M^{n+} + O_2 \rightarrow M^{n+1}(O_2^{-\bullet}) \tag{4.2}$$

One-electron dioxygen reduction may also occur via an outer-sphere electron transfer, affording a noncoordinated superoxide and a one-electron oxidized metal complex. Essentially, all low-valent 3d transition metal ions and numerous 4d and 5d metal ions can react as one-electron reducing agents.

*Two-electron $O_2$ reduction* can occur at one metal center (which changes the oxidation state by 2) or at two metal centers (with each metal changing its oxidation state by 1):

$$M^{n+} + O_2 \rightarrow M^{n+2}(O_2^{2-}) \tag{4.3}$$

or

$$2M^{n+} + O_2 \rightarrow (M^{n+1})_2(O_2^{2-}) \tag{4.4}$$

Typical examples of the two-electron reducing agents that react via Equation 4.3 include (but are not limited to) $Pd^0 \rightarrow Pd^{2+}$; $Fe^{2+} \rightarrow Fe^{IV}$; $Ru^{2+} \rightarrow Ru^{IV}$, and so on.

Most transition metal ions can react via pathways described by Equation 4.4; a few biologically relevant examples include $Fe^{II}Fe^{II} \rightarrow Fe^{III}Fe^{III}$ or $Cu^{I}Cu^{I} \rightarrow Cu^{II}Cu^{II}$ oxidations. Two low-oxidation state metal ions can be a part of a dinuclear complex. Alternatively, a mononuclear complex may undergo oxidation, forming dinuclear complexes in the oxidized form. Two metal ions in Equation 4.4 can be different; in this case, exemplified by $Cu^{I}Fe^{II}$ systems, heterodinuclear oxidation products form.

*Four-electron $O_2$ reduction* rarely occurs at a single metal center, because a jump in four oxidation levels of the metal is required. The most notable exception is ruthenium, where a range of accessible oxidation states (from $+2$ to $+8$) makes four-electron oxidations (e.g., $Ru^{II} \rightarrow Ru^{VI}$) possible. Several scenarios involving two, three, or four metal centers can be envisioned: two metal ions increasing their oxidation states by 2; one metal ion undergoing a two-electron oxidation and two metal ions undergoing concomitant one-electron oxidations; and each of the four metal ions increasing their oxidation state by 1.

In most cases, partially reduced dioxygen molecule remains coordinated to the metal center(s). Typical coordination modes of superoxide and (hydro)peroxide, as well as representative structures of metal-oxo complexes, are shown in Figure 4.3.

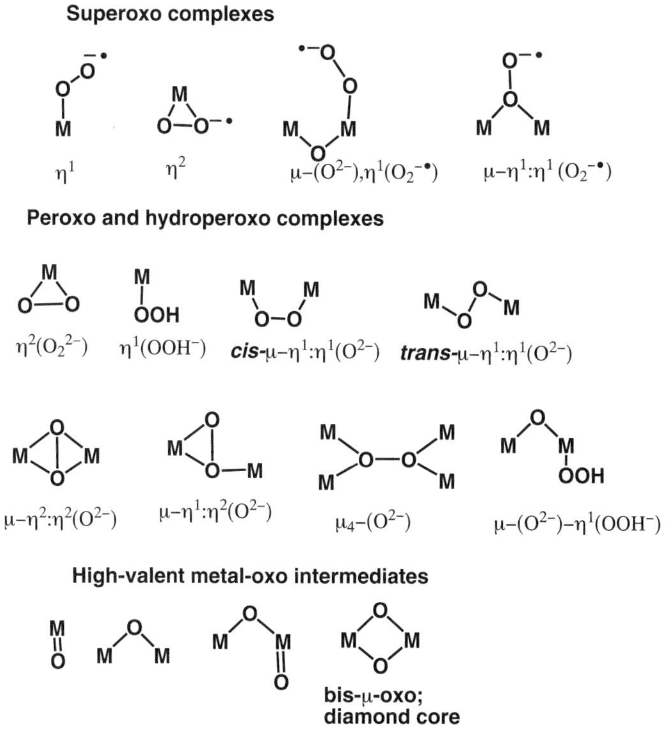

**FIGURE 4.3** Typical oxygen coordination modes in metal–oxygen complexes.

It is not always easy to unambiguously assign a particular dioxygen adduct as a superoxo or a peroxo complex. Equations 4.2 and 4.3 describe limiting scenarios of electron density distribution in the metal–oxygen complexes of the same stoichiometry; in reality, a continuum of states between $M^{n+1}$ ($O_2^{-\bullet}$) and $M^{n+2}(O_2^{2-})$ is possible, and a specific electronic structure of a given complex depends on the nature of the metal and on the particular types of ligands in the complex. Experimentally, spectroscopic and structural data are necessary to classify a given dioxygen adduct as a superoxo or a peroxo complex on the basis of the O–O bond order (judged by O–O vibrational frequencies and O–O bond lengths) and the formal oxidation state of the metal ion. A more precise description of the electronic structure of these dioxygen adducts can be obtained by quantum chemistry calculations, which often show "intermediate" electron density distribution between the metal center and the dioxygen-derived ligand.

Dioxygen can coordinate side-on ($\eta^1$) or end-on ($\eta^2$); when more than one metal ion is present, various bridging coordination geometries ($\mu$-) are possible (Figure 4.3).[18] For example, $\mu$-$\eta^1$: $\eta^1$ structure contains an $O_2$ ligand that coordinates end-on to each of the two metal ions and bridges between them; *cis*- or *trans*-geometry of the bridge are possible. Similarly, $\mu$- $\eta^2$: $\eta^2$ structure means that a bridging $O_2$ ligand coordinates side-on to each of the metal centers, while $\mu$-$\eta^1$: $\eta^2$ structure is nonsymmetric and contains a bridging $O_2$ ligand that binds end-on to one of the metal ions and side-on to another metal ion in the dinuclear complex. For high-valent metal-oxo complexes, terminal or bridging coordination modes for oxo ligands are commonly found; they can also coexist in one complex. Of course, metal-oxo moieties can be protonated, giving metal–hydroxo complexes. Bis-$\mu$-oxo dinuclear complexes are often referred to as "diamond cores."

### 4.2.3 Reversible Oxygen Binding to Biological Oxygen Carriers

Dioxygen coordination to the metal center(s) is the first step in inner-sphere oxygen activation processes. In most practically important oxygen activation reactions, this coordinated $O_2$ molecule undergoes further transformations (hence the term "activation"— initial $O_2$ binding induces subsequent oxidation of substrates with "activated" oxygen). Unraveling the mechanisms of multistep oxidations with dioxygen is necessary for complete understanding of oxygen activation. Extracting quantitative kinetic data for each individual reaction step in these complicated processes, however, is fairly challenging, and sometimes impossible. Fortunately, in many cases, simpler chemical systems can be designed or discovered, giving chemists a chance to study individual reaction steps in detail. An excellent example is provided by reversible oxygen carriers and their models, where dioxygen binding and dissociation is not complicated by any subsequent reactions and can be studied and characterized in detail:

$$P + O_2 \underset{k_{off}}{\overset{k_{on}}{\rightleftharpoons}} P(O_2), \quad K_{eq} = \frac{k_{on}}{k_{off}} \tag{4.5}$$

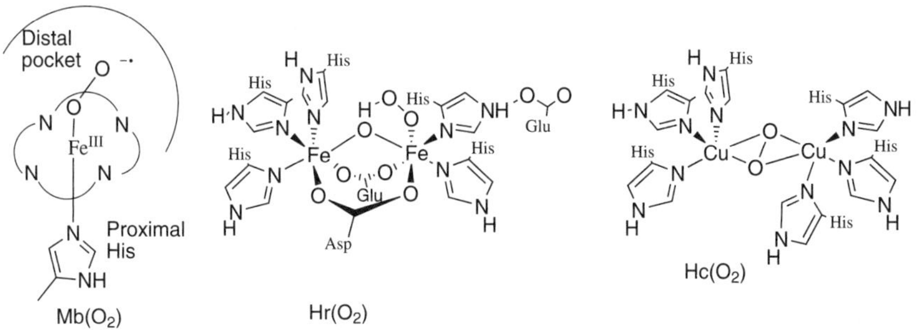

**FIGURE 4.4**  Active sites of natural dioxygen carriers.

In biology, three classes of metalloproteins that reversibly bind and release dioxygen are known: heme proteins (hemoglobin and myoglobin are the best examples), nonheme diiron proteins termed hemerythrins, and dicopper proteins, hemocyanins.[20,21] The active sites of these classes of oxygen carriers are drastically different, leading to different $O_2$ binding modes in their dioxygen adducts (Figure 4.4).

In hemoglobin (a tetrameric oxygen transport protein), myoglobin (a monomeric oxygen storage protein), and related proteins, a single iron(II) is surrounded, in the equatorial plane, by four nitrogen donor atoms from the porphyrin ring, and the fifth, axial nitrogen donor from proximal histidine (F8His, the eighth residue in the helix F of the protein). The incoming dioxygen molecule binds at the vacant sixth coordination site in the distal pocket. This coordination causes the spin state change of iron, from the high-spin state in the five-coordinate deoxy form to the low-spin state in the six-coordinate oxy form. While high-spin Fe(II) is displaced by about 0.6 Å from the porphyrin plane, a smaller low-spin iron ion fits inside the macrocyclic ring and moves into the porphyrin plane. Concomitant motion of coordinated proximal histidine initiates conformational changes in the hemoglobin subunits that are transmitted through their contact area, accounting for cooperativity of dioxygen binding to tetrameric hemoglobins. Dioxygen molecule binds in the bent end-on coordination mode to iron(II) heme in monomeric myoglobin or in hemoglobin subunits, with an Fe–O–O angle of about 125°. Numerous spectroscopic and computational studies showed that electron density distribution in this adduct can be best described as $Fe^{III}(O_2^{-\bullet})$. Detailed studies on myoglobin and hemoglobin variants and mutants uncovered the role of various amino acids in dioxygen affinity of heme proteins. Dioxygen adducts of myoglobin and hemoglobin are stabilized by favorable electrostatic interactions inside the distal pocket, and especially by hydrogen bonding to the distal (noncoordinated) histidine, E7His (the seventh residue in the E helix of the protein). Dioxygen affinity drastically decreases in the presence of excessively bulky groups in the distal pocket: steric constraints prevent efficient $O_2$ binding. The opposite effect was also observed in high-affinity hemoglobins, such as LegHb (a protein from the root modules of legumes), which has a large $K(O_2) = 1.4 \times 10^7 \, M^{-1}$ due to an easy access of $O_2$ to the iron center. Interestingly, steric effects influence primarily dioxygen binding rates, $k_{on}$. In

**TABLE 4.1    Kinetic Parameters of Oxygen Binding to Noncooperative Hemerythrins[17,21]**

| Oxygen Carrier | $k_{on} \times 10^{-6}$ $(M^{-1}s^{-1})$ | $k_{off}$ $(s^{-1})$ | $K_{eq}$ $(M^{-1})$ | $P_{1/2}$ $(O_2)$ (torr) |
|---|---|---|---|---|
| Myoglobin (SW, sperm whale) | 14.3 | 12.2 | $1.2 \times 10^6$ | 0.51 |
| Hemoglobin HbA R ($\alpha$-chain) | 28.9 | 12.3 | $2.39 \times 10^6$ | 0.15–1.5 |
| Hemoglobin HbA T ($\alpha$-chain) | 2.9 | 183 | $1.6 \times 10^4$ | 9–160 |
| Hemerythrin | $78^a$ | 315 | $1.5 \times 10^{5e}$ | |
| *Themiste zostericola* | | | $2.5 \times 10^5$ | |
| (MHr, monomer) | | | | |
| Hemerythrin | 7.5 | 82 | $1.3 \times 10^5$ | 6.0 |
| *Themiste zostericola* | | | | |
| (Hr, octamer, noncooperative) | | | | |
| Hemocyanin | 57 | 100 | $5.7 \times 10^5$ | 9.3 |
| *Panulirus interruptus* | | | | |
| (Hc, monomer) | | | | |
| Hemocyanin | 3.8 | 10 | $3.8 \times 10^5$ | 2.7 |
| *Helix pomatia,* R state | | | | |
| (complex Hc) | | | | |

contrast, various electronic effects, such as strengthening of Fe–$O_2$ bonds due to an increased electron density at the metal center and/or favorable electrostatic interactions of coordinated $O_2$ inside its binding pocket, often have little effect on the rates of oxygen binding, but instead decrease the rates of $O_2$ dissociation from the Fe–$O_2$ adducts. Typical $O_2$ binding rates reach about $10^8 M^{-1} s^{-1}$ (Table 4.1) and are characterized by low activation enthalpies (about 5 kcal/mol).[21]

Replacing iron with cobalt in heme oxygen carriers leads to functional systems that still bind $O_2$ reversibly, albeit with about 100-fold lower affinity (that can be attributed, at least in part, to less negative redox potentials of the $Co^{III}/Co^{II}$ couple compared to $Fe^{III}/Fe^{II}$ couple). Dioxygen binding rates to coboglobins are comparable to the oxygenation rates of myoglobins or hemoglobins, while dioxygen dissociation rates are significantly higher (due to the weaker Co–$O_2$ bond). Functional oxygenation of cobalt-substituted heme proteins inspired successful design and synthesis of numerous synthetic cobalt dioxygen carriers (Section 4.3).

The active site of hemerythrins, buried in a four-helix bundle, contains two nonequivalent iron(II) centers bridged by two carboxylates (one from aspartate and one from glutamate) and one hydroxide anion.[22] The two high-spin iron(II) centers are antiferromagnetically coupled ($J = -14\,cm^{-1}$, $\mathbf{H} = -2J\,\mathbf{S_1S_2}$, as measured by SQUID magnetometry).[23] The coordination sphere at one of the iron(II) centers is completed to six ligands by three histidine nitrogens; the other iron(II) center, however, is bound to only two histidines and remains coordinatively unsaturated, as established by X-ray crystallography[22] and spectroscopy[23] (Figure 4.4). The histidine-rich environment of the iron centers in Hr is different from the carboxylate-rich coordination sphere of diiron oxidative enzymes and presumably helps preventing overstabilization of high oxidation states of iron, thus favoring reversible dioxygen binding. The presence of only one vacant coordination site in Hr is very important for reversible, end-on $O_2$ binding to a single iron center, as opposed to the bridging between both iron centers in dioxygen activating diiron enzymes (e.g., MMO, RNR-R2, or $\Delta$9D).[23] In oxyHr, dioxygen molecule is formally reduced to an end-on bound peroxide, while both irons are oxidized to the $+3$ state. Only one iron atom (initially five-coordinate) directly interacts with $O_2$; the second, six-coordinate iron serves as a charge reservoir. The proton from the hydroxo bridge is transferred to the noncoordinating oxygen atom of the $O_2$ moiety, and a resulting -OOH unit is hydrogen bonded to an oxo bridge (Figure 4.5). This intramolecular hydrogen bonding interaction is likely favored by the hydrophobic nature of Hr oxygen binding pocket, which does not offer alternative H-bonding motifs. Therefore, the reaction between Hr and $O_2$ proceeds through a number of elementary steps, including dioxygen diffusion through solvent and through protein toward the diiron active site, the Fe–O bond formation with a concomitant electron transfer, another one-electron transfer, and a proton transfer (or a proton-coupled electron transfer). In addition, spin changes occur as the $O_2$ diradical interacts with the weakly antiferromagnetically coupled diiron(II) center, yielding a strongly antiferromagnetically coupled ($J = -77\,cm^{-1}$, $\mathbf{H} = -2J\,\mathbf{S_1 S_2}$) diiron(III)-peroxo species.[23]

Detailed kinetic and mechanistic studies were performed in order to determine the mechanism of dioxygen binding to hemerythrin at the molecular level, and quantum

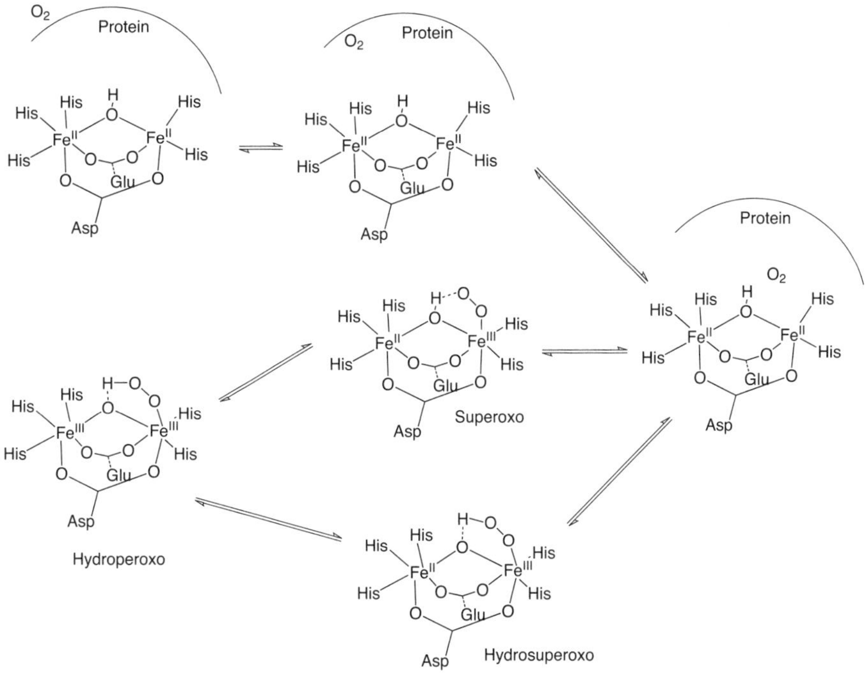

**FIGURE 4.5** Proposed mechanism of dioxygen binding by hemerythrin.[17]

chemical and QM/MM calculations provided additional insight. Despite the multistep nature of hemerythrin oxygenation, kinetics of dioxygen binding to this diiron(II) protein is remarkably simple and quite similar to the oxygenation of other classes of dioxygen carriers (myoglobin and hemocyanin). In most studies (Table 4.1), one-exponential growth of the oxo-Hr was observed under excess $O_2$. The dioxygen binding is a fairly rapid second-order reaction (first order in Hr, and first order in $O_2$), with rate constants (at room temperature) ranging from $10^6$ to $10^8 M^{-1} s^{-1}$. The rate constants of dioxygen binding to hemerythrins are independent on pH (at least in the pH interval from 6 to 9)[24,25] and on the $H_2O/D_2O$ substitution.[25,26] This strongly implies intramolecular, postrate-limiting proton transfer.[17,20]

In hemocyanins, each copper(I) ion is coordinated, in a trigonal geometry, to three nitrogen donors from histidines. The incoming $O_2$ molecule binds side-on to both copper centers, forming a bridge between them; this is a $\mu$-$\eta^2$,$\eta^2$ geometry.[20,21] The distribution of electron density in oxygenated hemocyanin can be best described as $[Cu^{II}(O_2{}^{2-})Cu^{II}]$, with a very strong antiferromagnetic coupling through the peroxo bridge. Hemocyanins are high molecular mass oligomers that often display cooperativity in $O_2$ binding. Thermodynamics and kinetics of $O_2$ binding to these large proteins also depend on media effects (pH, salt concentration, etc.). Nevertheless, oxygenation of each dicopper site occurs via a single observable step, suggesting a concerted mechanism, and appears to be similar to the behavior of other oxygen carriers.

Despite differences in the chemical nature of the active sites, three classes of natural oxygen carriers display amazing similarities in thermodynamic and kinetic parameters that characterize dioxygen binding. All noncooperative proteins or isolated subunits of oligomeric oxygen carriers show reversible $1:1$ $O_2$ binding (Equation 4.5).

Dioxygen affinities of the vast majority of natural oxygen carriers are close to $10^5 - 10^6 M^{-1}$ (Table 4.1). These oxygen affinities correspond to $P_{1/1}(O_2)$ of about $0.5 - 10$ torr ($P_{1/2} = 1/K_P$, an equilibrium constant of oxygen binding expressed in units of pressure). These values of oxygen affinities allow the proteins to scavenge dioxygen at partial pressures comparable to atmospheric conditions and to release dioxygen in hypoxic cells or tissues. Interestingly, all natural oxygen carriers bind $O_2$ extremely rapidly, with second-order rate constants of $10^7 - 10^8 M^{-1} s^{-1}$, thus approaching diffusion-controlled limit (about $10^9 M^{-1} s^{-1}$). These high rates are due to very low values of activation enthalpies for $O_2$ binding, $\Delta H_{on}^{\ddagger} \approx 5$ kcal/mol. It appears that dioxygen binding at a vacant metal coordination site is a rapid, low-barrier process. In many cases, the rates of dioxygen binding are controlled by activation entropies, which are negative (and typically range from $-15$ to $-30$ cal/(K mol)), in agreement with associative nature of the process. The activation volumes for oxygen binding to hemerythrin and to myoglobin, however, were small, but positive, suggesting a significant contribution from protein rearrangement rather than a simple association of the metal center and the oxygen molecule in the transition state.[27]

Dioxygen dissociation from the oxy forms of natural oxygen carriers is characterized by a fairly large activation enthalpy and large positive activation entropy and activation volume. For example, $\Delta H_{off}^{\ddagger} = 92$ kJ/mol, $\Delta S_{off}^{\ddagger} = 117$ J/K mol, and $\Delta V_{off}^{\ddagger} =$

$28\ cm^3/mol$ for a monomeric hemerythrin from *Themiste zostericola*.[26,28] This is typical of reactions that are limited by a bond-breaking event.

While rapid, efficient oxygenation is clearly important for the proper functioning of natural oxygen carriers, the rate constants of $10^8\ M^{-1}\ s^{-1}$ may appear excessive. Indeed, at millimolar concentrations of $O_2$, these rate constants allow the oxygenation process to be completed in microseconds. In most enzymes, including dioxygen-activating enzymes, millisecond to second timescale is sufficient for successful biocatalysis. It is tempting to speculate that reversible nature of $O_2$ binding to oxygen carriers indirectly favors very high oxygenation rates. Indeed, reasonable values of dioxygen affinities are determined by partial pressure and solubility of $O_2$. These rather high $K(O_2)$ values can be achieved by either increasing $O_2$ binding rates or decreasing $O_2$ dissociation rates. However, oxygen release from oxyMb, oxyHr, and oxyHc must be fast for these proteins to able to deliver $O_2$. Therefore, $k_{off}$, the dioxygen dissociation rate constants, cannot drop below $10–100\ s^{-1}$, and $k_{on}$, the dioxygen binding rate constants, must be extremely high. High oxygen affinities coupled with rather high $O_2$ dissociation rates require exceedingly high $O_2$ binding rates.

## 4.3   REACTIONS OF MONONUCLEAR METAL COMPLEXES WITH DIOXYGEN

Coordinatively saturated metal complexes that are kinetically inert with respect to ligand substitution may undergo outer-sphere electron transfer reactions with dioxygen. Typical examples include oxidations of six-coordinate chromium(II) complexes[29] (Equation 4.6) and oxidations of polyoxometallate anions.[30]

$$(L)_3Cr^{2+} + O_2 \rightarrow (L)_3Cr^{3+} + O_2^{-\bullet} \tag{4.6}$$

As discussed in Section 4.1, these reactions are described by Marcus theory and their rates depend on the thermodynamic driving force of the reaction and on the self-exchange rate constants for the oxidant $(O_2/O_2^{-\bullet})$ and the reducing agent (the metal complex).[29,30] Unfortunately, the self-exchange rates for dioxygen/superoxide couple are difficult to determine, and the reported values varied by several orders of magnitude. This controversy and the current state-of-the-art in the field are described in detail in Section 4.1, and will not be further discussed here.

In many cases, oxygenation of the metal complexes yields various metal–oxygen adducts. The structures of typical mononuclear metal–oxygen complexes are shown in Figure 4.3; these complexes contain coordinated $O_2$ as an end-on superoxo or peroxo ligand, coordinated $O_2$ as a side-on superoxo or peroxo ligand, or oxygen as an oxo ligand (protonated forms of these species are also known).[14] In every case, the immediate reaction of dioxygen with any metal complex must be the formation of a bond between one of the two oxygen atoms and the metal ion. Usually this has the consequence that the dioxygen molecule substitutes another ligand (often this is a solvent molecule, however it can be a coordinated anion, an additional ligand, or a

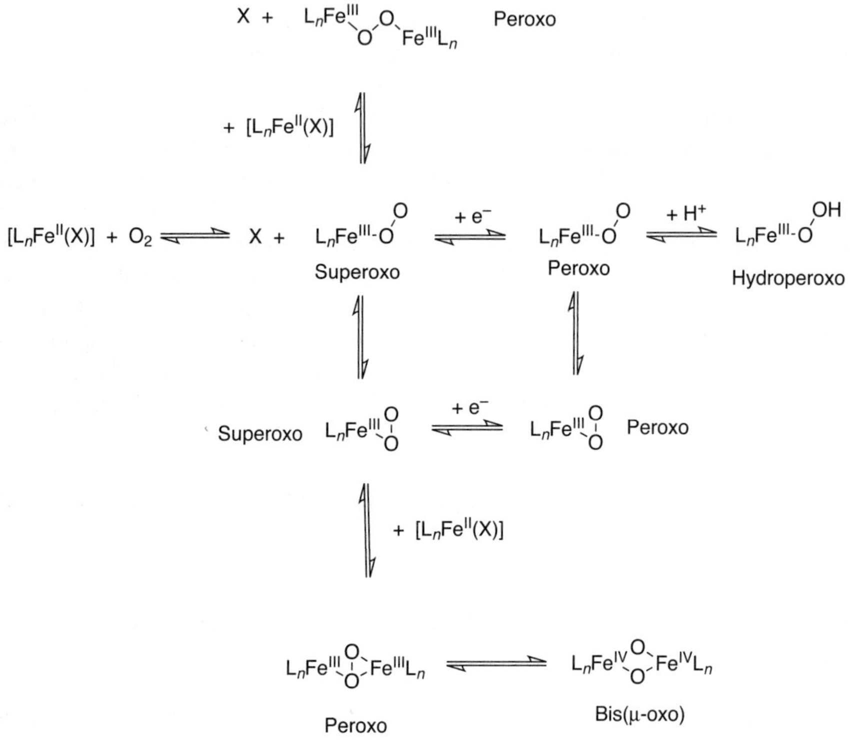

$$X + L_nFe^{III} \underset{O}{\overset{O}{\diagup}} \diagdown _{O} \diagdown Fe^{III}L_n \quad \text{Peroxo}$$

$$+ [L_nFe^{II}(X)]$$

$$[L_nFe^{II}(X)] + O_2 \rightleftharpoons X + L_nFe^{III}\text{-O}^{\diagdown O} \underset{\text{Superoxo}}{\overset{+ e^-}{\rightleftharpoons}} L_nFe^{III}\text{-O}^{\diagdown O} \underset{\text{Peroxo}}{\overset{+ H^+}{\rightleftharpoons}} L_nFe^{III}\text{-O}^{\diagdown OH}$$

Superoxo    Peroxo    Hydroperoxo

$$\text{Superoxo} \quad L_nFe^{III} \overset{O}{\underset{O}{\diagdown}} \overset{+ e^-}{\rightleftharpoons} L_nFe^{III} \overset{O}{\underset{O}{\diagdown}} \quad \text{Peroxo}$$

$$+ [L_nFe^{II}(X)]$$

$$L_nFe^{III} \overset{O}{\underset{O}{\diagdown}} Fe^{III}L_n \rightleftharpoons L_nFe^{IV} \overset{O}{\underset{O}{\diagdown}} Fe^{IV}L_n$$

Peroxo    Bis($\mu$-oxo)

**FIGURE 4.6**   Possible reaction pathways in oxygenation of mononuclear nonheme iron(II) complexes.[17]

ligand arm of the multidentate ligand) followed by an electron transfer reaction leading in a first step to an end-on metal-superoxo species. The extent of this inner-sphere electron transfer can vary. An alternative pathway would be the ligand addition, for example, a five-coordinate $M^{n+}$ complex reacts to form a six-coordinate $M^{n+1}(O_2^{-\bullet})$ complex. After the formation of the first bond, quite different consecutive steps can occur (Figure 4.6 illustrates possible pathways for iron complexes[17]; other metals show analogous chemistry, although their exact oxidation states may differ). It is plausible that a ring-closure step takes place and the end-on superoxo complex converts into a side-on species. Further electron transfer then would lead either to an end-on peroxo or side-on superoxo complex. Reactions with another molecule of a mononuclear iron complexes are possible, leading to dinuclear peroxo complexes discussed in detail in Section 4.4. Formation of 1:1 metal–oxygen complexes is considered below.

### 4.3.1   Formation of End-On 1 : 1 Metal-Superoxo Complexes

The appealing goal of binding dioxygen reversibly inspired numerous studies on mononuclear $\eta^1$ superoxo complexes that resemble $O_2$ coordination mode in natural heme oxygen carriers.[14,21,31,32] Cobalt analogue of myoglobin, coboglobin, is a functional oxygen carrier. Early synthetic work centered on cobalt rather than iron, due to significant stability of oxygenated cobalt complexes with respect to subsequent irreversible oxidations. A variety of complexes with polydentate ligands containing nitrogen donors showed some ability to bind $O_2$ reversibly.[14,33,34] Dioxygen affinity of these systems increases with an increase in electron density at the cobalt center provided by strong electron-donating ligands. These ligands stabilize the oxidized form of the cobalt complex, as can be seen in lower values of the $Co^{III}/Co^{II}$ redox potential. Under all other circumstances being equal, inverse correlation between $E_{1/2}(Co^{III}/Co^{II})$ and $\log K(O_2)$ was often found.[33] To improve relatively low dioxygen affinity of cobalt systems compared to their more easily oxidizable iron analogues, addition of an axial electron-donating ligand (such as pyridine or imidazole derivatives) is often helpful. Electronic factors have little effect on dioxygen binding rates, but change dioxygen dissociation rates (not surprisingly, stronger $Co–O_2$ bonds are harder to break, and $O_2$ dissociation is slower in these systems).

When equatorial ligands do not shield the axial positions, two molecules of an externally added base may coordinate, thus blocking the site for dioxygen binding and shutting down oxygenation:

$$(L)Co^{II} + B \rightarrow (L)(B)Co^{II}(\text{reactive complex}) \tag{4.7}$$

$$(L)(B)Co^{II} + O_2 \rightleftharpoons (L)(B)Co^{III}(O_2)^{-\bullet} \tag{4.8}$$

But

$$(L)(B)Co^{II} + B \rightleftharpoons (L)(B)_2Co^{II}(\text{does not react with } O_2) \tag{4.9}$$

Varying equatorial polyamine ligands (Figure 4.7) influence dioxygen affinity of the cobalt complexes (Table 4.2).[14] In this case, dioxygen binding rates change substantially, increasing as the equatorial ligand field increases for polydentate and especially macrocyclic ligands. The oxygenation rates follow the lability of axial water molecule; the reactions are faster for macrocyclic complexes because of the labilization of the axial positions caused by the strong equatorial field of the macrocyclic ligands. The close similarity between the oxygenation rates and the water substitution rates suggests that ligand substitution is rate limiting in the oxygen binding processes.

The activation enthalpies for these oxygenation reactions are relatively large (they fall in the range between 4 and 20 kcal/mol) (Figure 4.8), while the activation entropies are positive, confirming dissociative nature of the rate-limiting step.[14,35] The activation volumes for the oxygenation of cobalt(II) macrocycles were found to be close to 0, also suggesting that Co-solvent bond breaking occurs along with $Co–O_2$ bond formation.[36] Analysis of the reported kinetic data for $O_2$ binding to cobalt(II) centers revealed an activation enthalpy–activation entropy compensation: a correlation between $\Delta H^{\ddagger}$ and $\Delta S^{\ddagger}$ exists for several groups of complexes (six-coordinate ammine complexes; ammine-aqua complexes; aminocarboxylate complexes;

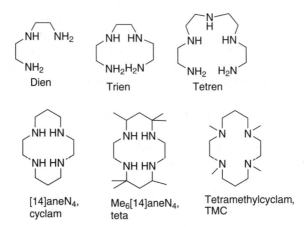

**FIGURE 4.7** Acyclic and macrocyclic polyamine ligands that support dioxygen binding cobalt complexes.

porphyrins embedded into apomyoglobin or apohemoglobin).[35] The dissociative nature of the rate-limiting step explains this enthalpy–entropy compensation. If the sixth ligand dissociation occurs as a rate-limiting step, the activation enthalpy for bond breaking could be high, but the activation entropy would be positive. If the sixth coordination site is vacant prior to oxygenation, Co–$O_2$ bond formation occurs with low barrier, but the activation entropy for the process is negative and unfavorably large. In many systems, intermediate situations are possible, and some of them might provide the highest reaction rates at room temperature.[35]

The bi- or tricyclic systems were designed to create a three-dimensional void about the metal ion, thus allowing for steric discrimination between small and large ligands. These sterically hindered ligands also protect end-on dioxygen adducts from oxidative decomposition via formation of peroxo- or oxo-bridged dimers:

**TABLE 4.2  Rate Constants for the Reaction of Cobalt(II) Complexes with $O_2$ in Aqueous Solution[14]**

| Cobalt Complex[a] | $k_{on}$ ($M^{-1}s^{-1}$) | $k_{off}$ ($s^{-1}$) |
|---|---|---|
| $[Co(NH_3)_5(H_2O)]$ | $2.5 \times 10^4$ | |
| $[Co(dien)_2]^{2+}$ | $1.2 \times 10^3$ | |
| $[Co(trien)(H_2O)_2]^{2+}$ | $2.5 \times 10^4$ | |
| $[Co(tetren)(H_2O)]^{2+}$ | $10^5$ | |
| $[Co([14]aneN_4)(H_2O)_2]^{2+}$ | $1.18 \times 10^7$ | 63 |
| $[Co(Me_6[14]aneN_4)(H_2O)_2]^{2+}$ | $5.0 \times 10^6$ | $2.06 \times 10^4$ |
| $[Co(Me_6[14]aneN_4)(H_2O)Cl]^+$ | $1.80 \times 10^6$ | $3.21 \times 10^3$ |
| $[Co(Me_6[14]aneN_4)(H_2O)(SCN)]^+$ | $7.29 \times 10^6$ | $1.77 \times 10^1$ |
| $[Co(Me_6[14]aneN_4)(H_2O)(OH)]^+$ | $8.9 \times 10^5$ | $2.1 \times 10^{-2}$ |

[a]Structural formulas of the ligands are shown in Figure 4.7.

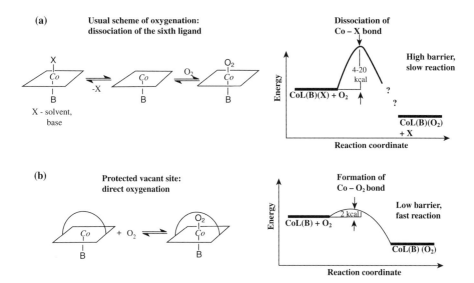

**FIGURE 4.8**    Oxygenation barriers for the five- and six-coordinate cobalt(II) complexes.[35,38]

$$\{(L)M^{II}(O_2) \rightleftharpoons (L)M^{III}(O_2^{-\bullet})\} + (L)M^{II} \rightarrow (L)M^{III}(O_2^{2-})M^{III}(O^{2-}) \qquad (4.10)$$

$$(L)M^{III}(O_2^{2-})M^{III}(O^{2-}) \rightarrow \ \rightarrow (L)M^{III}(\mu\text{-}O^{2-})M^{III}(L) \qquad (4.11)$$

While peroxo dimer formation is sometimes reversible for cobalt complexes, it inevitably leads to further oxidations in case of iron complexes. Steric hindrance about the metal center would prevent dimerization processes.

Extensive efforts to model natural oxygen carriers resulted in impressive collection of sterically hindered porphyrins[21,31,32]; some examples of these molecules are shown in Figure 4.9. Oxygenation of both iron(II) and cobalt(II) complexes was studied, unraveling the factors that control dioxygen binding and release. As expected, strengthening the metal–$O_2$ adducts, for example, by incorporating proximal axial donors or by providing hydrogen bond donors to coordinated $O_2$ in the distal pocket increases dioxygen affinity and decreases $O_2$ dissociation rates.[31] Steric constrains at the $O_2$ binding site decrease dioxygen affinity and $O_2$ binding rates. Extensive reviews discuss this beautiful chemistry in great detail.[21,31,32]

The most thoroughly studied and robust totally synthetic nonporphyrin oxygen carriers are the superstructured Co(II) or Fe(II) cyclidenes, which have been bridged across the cavity, providing sterically protected ligand binding pockets in the resulting lacunar complexes (Figure 4.10).[37] The 16-membered macrocyclic platforms fold because of the conformational requirements of adjacent chelate rings: the low-energy chair or boat conformations of the six-membered saturated rings force the adjacent unsaturated rings to tilt toward the $MN_4$ plane. In the most common cyclidene

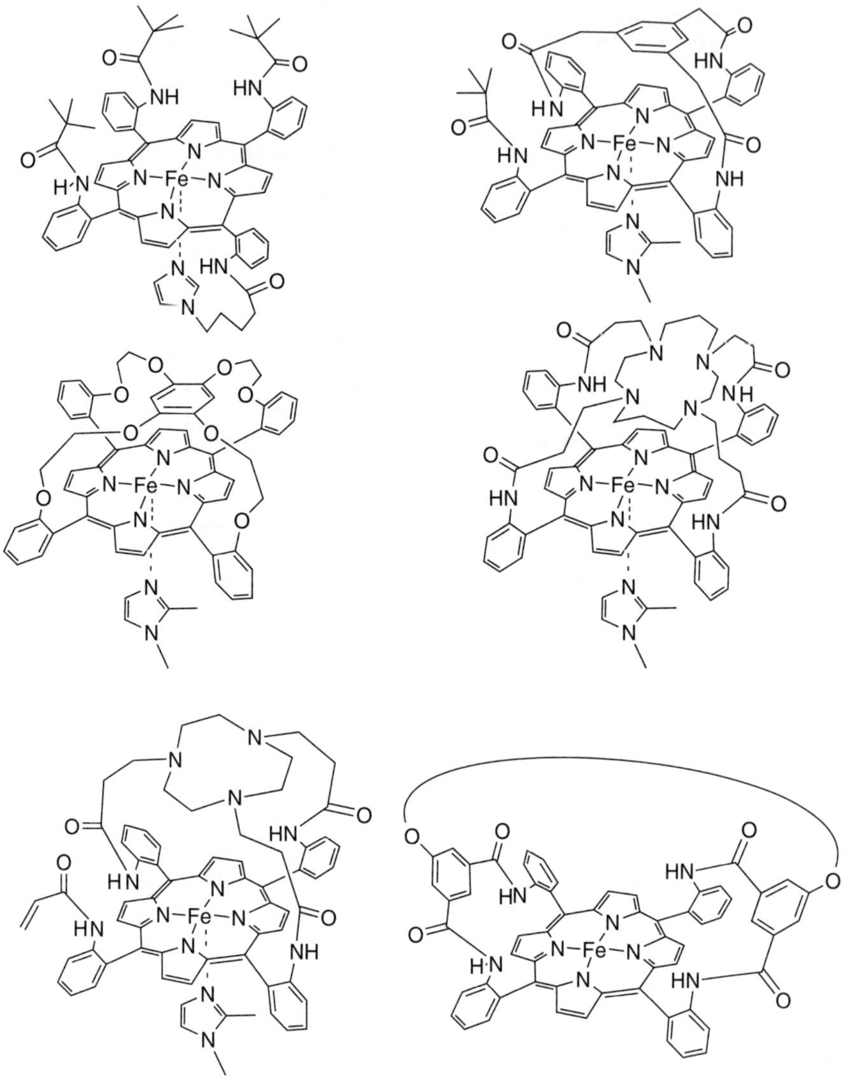

**FIGURE 4.9** Representative sterically constrained porphyrins designed for reversible dioxygen binding to their Fe(II) or Co(II) complexes.[31]

geometry, both unsaturated "wings" are positioned on the same side of the $MN_4$ plane, forming a well-defined cleft.

Dioxygen affinity of the Co(II) and Fe(II) cyclidene complexes is governed by both electronic and steric factors. Coordination of a fifth donor ligand, such as pyridine or imidazole, is required for efficient $O_2$ binding and is usually accomplished by adding the corresponding base to a solution of the macrocyclic complex. The sixth coordination site of the metal should remain vacant or occupied by a labile ligand to allow

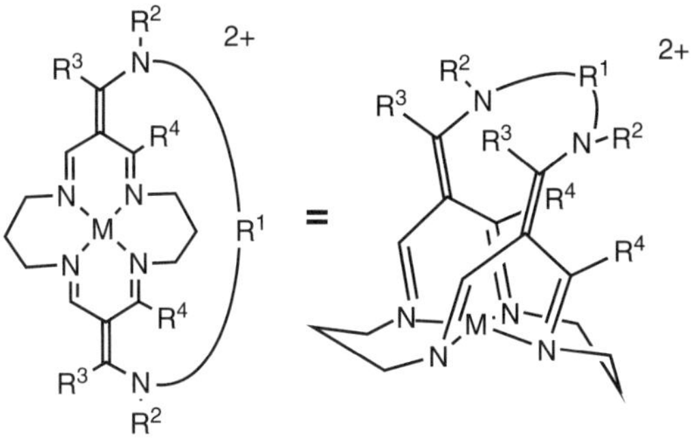

**FIGURE 4.10**  Lacunar cyclidenes.

the reversible binding of the $O_2$ ligand. The lacunar cyclidenes are perfect for selectively binding the small ligands ($O_2$ and CO) inside the cavity, while excluding larger monodentate basic ligands (pyridine and imidazole). The iron(II) complexes with C4 and C5 cyclidenes ($R^1 = (CH_2)_4$ or $(CH_2)_5$, respectively; $R^2 = Ph$; $R^3 = R^4 = Me$) exist in solution (in acetone-pyridine water, or acetonitrile-1-methylimidazole solvent systems) as high-spin, five-coordinate species, while the iron (II) complex of the C6-bridged cyclidene ($R^1 = (CH_2)_6$, $R^2 = Ph$; $R^3 = R^4 = Me$) displays a distinct spin equilibrium between a five-coordinate high-spin form and a six-coordinate low-spin form, which is shifted toward the formation of low-spin six-coordinate complexes at low temperatures ($-40°C$). The C8-bridged complex ($R^1 = (CH_2)_8$, $R^2 = Ph$; $R^3 = R^4 = Me$) exists as a similar equilibrium mixture with a significant fraction of six-coordinate complex present even at room temperature. Molecular modeling studies demonstrate that exclusion of axial base (1-methylimidazole) from the cavity is largely due to van der Waals repulsion between the guest and the bridge (the "roof" of the lacuna), while the interaction between the flat aromatic guest and the walls of the cleft is relatively less important. The behavior of $d^7$ Co(II) ions with respect to axial ligand binding is somewhat different from that of $d^6$ ions, such as cobalt(III) or iron(II): five-coordinate species dominate in solutions of both lacunar and unbridged Co(II) cyclidenes, as evidenced by EPR spectroscopy and electrochemical data. In the case of cobalt(II) cyclidenes (especially for the unbridged complex), the axial base exclusion is due to the electronic factors (destabilization of the $d_{z^2}$ orbital in complexes with strong equatorial ligand field) rather than the superstructure. In the absence of the axial base binding in the sixth position, the walls of the cleft and the bridge shield the cobalt(II) metal center from the noncoordinating solvent (methanol), resulting in facilitated dioxygen binding.[38]

The only crystal structure for a cobalt(II) cyclidene dioxygen adduct, that of $[Co(C6Cyc(O_2)(MeIm)]^{2+}$ ($R^1 = (CH_2)_6$, $R^2 = R^3 = R^4 = Me$), showed that $O_2$ binds at an angle of $121°$, as expected for coordinated superoxide, and that the hexamethylene bridge flips away from the guest.[37]

Dioxygen affinity of the Co(II) complexes parallels the cavity width (Figure 4.11), indicating that there are significant steric interactions between the walls of the

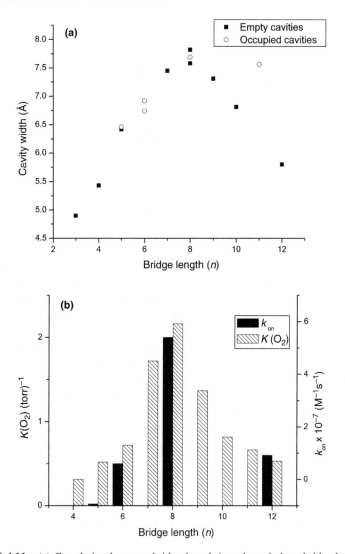

**FIGURE 4.11** (a) Correlation between bridge length in polymethylene-bridged cobalt(II) cyclidenes and cavity width of the complexes. (b) Dioxygen affinities of cobalt(II) complexes measured in $CH_3CN$-1-methylimidazole (MeIm) at 0°C, and dioxygen binding rates of cobalt (II) complexes measured in acetone-MeIm at −70°C.[35,38]

cavity and the guest. The cavity width in cyclidenes, which is defined as the distance between the "edges" of the cleft (the bridgehead nitrogen atoms that are spanned by the bridge $R^1$) (Figure 4.10), depends on the length of the bridge. Short bridges composed of tetra- or pentamethylene chains constrain the distances between the edges of the cleft and enforce narrow cavities. Hexamethylene bridge has an optimal length for spanning the cavity of the 16-membered cyclidene, while longer bridges (C7 or C8) enforce wider separation between the "walls" of the cavity. Even longer

**TABLE 4.3    Kinetic and Thermodynamic Parameters for $O_2$ Binding to Cyclidene Complexes Measured in Acetonitrile/1.5 M 1-MeIm (for Fe(II) Complexes) or in Acetone/1.5 M 1-MeIm (for Co(II) Complexes) Using Stopped-Flow Spectrophotometry**

| $R^1$ | $R^2$ | $R^3$ | $\Delta H_{on}^{\ddagger}$ (kJ/mol) | $\Delta S_{on}^{\ddagger}$ (J/(K mol)) | $\Delta H_{off}^{\ddagger}$ (kJ/mol) | $\Delta S_{off}^{\ddagger}$ (J/(K mol)) |
|---|---|---|---|---|---|---|
| Cobalt complexes | | | | | | |
| $(CH_2)_4$ | Me | Me | 9 | −124 | 66 | 80 |
| $(CH_2)_5$ | Me | Me | 3 | −118 | 77 | 92 |
| $(CH_2)_6$ | Me | Me | 17 | −42 | ND | N/D |
| Iron complexes | | | | | | |
| $(CH_2)_4$ | Me | Ph | 12 | −276 | 95 | 76 |
| $(CH_2)_5$ | Me | Ph | 14 | −229 | 50 | −110 |
| $(CH_2)_6$ | Me | Ph | 41 | −75 | ND | ND |
| $(CH_2)_8$ | Me | Ph | 54 | −40 | ND | ND |
| $m$-Xy | Me | Me | 11.1 | −252 | ND | ND |
| $m$-Xy | Bz | Me | 11.4 | −253 | ND | ND |
| $m$-Xy | Ph | Me | 15.8 | −228 | ND | ND |
| $m$-Xy | Ph | Bz | 9.1 | −243 | ND | ND |

Oxygen dissociation rates were determined from the nonzero intercepts of the plots of $k_{obs}$ versus $[O_2]$, using the equation $k_{obs} = k_{on}[O_2] + k_{off}$. Equilibrium constants were estimated as $K = k_{on}/k_{off}$.[35,38] ND, not determined.

bridges are flexible and can adopt various confirmations that allow the cleft to shrink back to its optimal width (about 7 Å).[37]

The differences in oxygen affinities are determined by the rates of $O_2$ binding rather than $O_2$ dissociation (Figure 4.11, Table 4.3). For a reversible dioxygen binding process depicted in Equation 4.5, the observed oxygenation rate constant depends linearly on the concentration of $O_2$:

$$k_{obs} = k_{on}[O_2] + k_{off} \qquad (4.12)$$

An example of experimental plots that allow the determination of $k_{on}$ and $k_{off}$ (as a slope and an intercept, respectively, of the straight line $k_{obs}$ versus $[O_2]$) is shown in Figure 4.12. Independent measurements of the equilibrium constant, $K(O_2)$, provide the way to check for accuracy of kinetic measurements, which should result in nearly identical values of $K_{kin} = k_{on}/k_{off}$.

Dioxygen binding to cobalt(II) cyclidenes is an entropically controlled, low-barrier process (Figure 4.8).[35] Large negative activation entropies are typical of associative processes. This is consistent with solvent exclusion from the cyclidene cavity that protects a vacant site at cobalt(II). Activation enthalpies for associative oxygen binding at a vacant cobalt(II) site range about 1–4 kcal/mol.[35]

Dioxygen affinity of iron(II) cyclidenes depends on steric constraints for $O_2$ binding, resulting in van der Waals repulsions in dioxygen adducts. This effect is evident from comparison of the complexes with aliphatic bridges of varying length

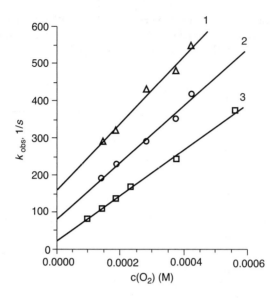

**FIGURE 4.12**    Plot of observed rate constants versus dioxygen concentration during oxygenation of Co(C5Cyc) ($R^1 = (CH_2)_5$, $R^2 = R^3 = R^4 = Me$) in acetone/pyridine, measured at $-35$ (1), $-40$ (2), and $-50°C$ (3).[35]

(Table 4.3). Oxygen binding to five-coordinate C4-, C5-, and $m$-xylene-bridged cobalt (II) or iron(II) cyclidenes is characterized by low activation enthalpies and large negative activation entropies, which are typical of associative processes. The six-coordinate C6- and C8- iron(II) cyclidenes release the solvent molecule in the course of oxygen binding. These ligand substitution reactions have substantially larger activation barriers, but less negative activation entropies (Table 4.3).[38]

Electronic effects on the oxygenation rates are less pronounced, although in some cases still significant. The rates of oxygen binding increase by almost an order of magnitude with an increase in electron withdrawing properties of the $R^2$ and $R^3$ substituents in $m$-xylene-bridged ($R^1 = m$-Xy) iron (II) cyclidenes ($k_{on} = 186\ M^{-1}\ s^{-1}$ for $R^2 = R^3 = Me$, and $1305\ M^{-1}\ s^{-1}$ for $R^2 = Ph$, $R^3 = Bz$ at $-20°C$) (Table 4.3).[38]

Another 3d-metal involved in biological and biomimetic dioxygen activation, copper, can also form end-on superoxo complexes[39–41]:

$$[(L)Cu^I] + O_2 \Leftrightarrow [(L)Cu^{II}(\eta^1\text{-}(O_2^{-\bullet}))] \tag{4.13}$$

In contrast to iron and cobalt, end-on superoxo-copper(II) species do not dominate the field of copper–oxygen chemistry. In 1 : 1 copper–dioxygen adducts, an alternative side-on, $\eta^2$ coordination mode is sometimes observed; these $[(L)Cu^{II}(\eta^2\text{-}(O_2^{-\bullet}))]$ or $[(L)Cu^{III}\text{-}(\eta^2\text{-}(O_2^{2-}))]$ complexes are discussed below. Mononuclear copper–dioxygen complexes easily react with the second molecule of the Cu(I) complex, forming peroxo- or dioxo-bridged dinuclear species (Section 4.4). For sterically unhindered ligands, the rates of bimolecular decay of initial 1 : 1 Cu:$O_2$ adducts often exceed their formation rates; in these cases, the overall kinetics reflects the initial dioxygen binding to a single copper(I) center, but the 1 : 1 intermediates cannot be observed and their structures remain unknown. In another extreme, the formation of the mononuclear Cu–$O_2$ complex is very rapid (often too rapid to be measured), and the subsequent bimolecular decay is rate limiting, accounting for the overall second-order rate law in copper(I). Finally, the rates of two steps may be comparable, so that their individual kinetic parameters can be resolved and the transient 1 : 1 Co–$O_2$ intermediates can be observed.[40] In several instances, spectroscopic data suggested an $\eta^1$ Cu(II)-superoxo formulation for these intermediates.[39,41]

Increasing steric bulk at tetradentate tripodal ligands prevented dimer formation and stabilized end-on Cu–$O_2$ complexes, one of which was crystallographically and spectroscopically characterized (cavity around the copper center was shaped by the ligand (TMG)$_3$TREN) (Figure 4.13).[42] This dramatic result provided additional support for earlier end-on superoxide formulations for the 1 : 1 copper–dioxygen complexes supported by tripodal ligands (such as TPA or TREN derivatives) (Figure 4.13).[39,41]

Mechanisms and rates of dioxygen binding to copper(I) were determined for a series of tripodal complexes that afford end-on superoxo intermediates. Kinetic parameters for a these complexes proved to be rather similar, with activation enthalpies ranging about 30–40 kJ/mol for reactions in coordinating solvents (nitriles). Activation entropies were relatively small for these processes; both small positive and small negative values have been reported.[40] This behavior is indicative of

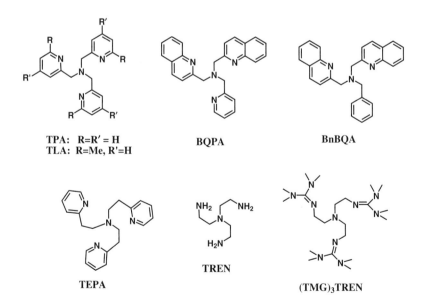

TPA:  R=R′ = H
TLA:  R=Me, R′=H

BQPA

BnBQA

TEPA

TREN

(TMG)$_3$TREN

**FIGURE 4.13**  Representative tripodal ligands forming copper(I) complexes that react with dioxygen.

oxygen binding preceded or accompanied by solvent dissociation. The fifth coordination site in the copper(I) complexes with most studied tetradentate amines or aminopyridines is indeed occupied by the molecule of coordinated nitrile that has to be displaced by the incoming $O_2$.

The topology, electron-donating ability, and steric bulk of polydentate ligands all influence thermodynamic and kinetic parameters of oxygen binding to Cu(I).[39–41] For example, [Cu(TPA)(RCN)]$^+$ rapidly reacts with $O_2$ at low temperatures, while its analogue having ethylpyridine arms in place of methylpyridine ones, [Cu(TEPA)]$^+$, does not react with dioxygen. This effect of chelate ring size can be traced, at least in part, to a large, 450 mV difference in their redox potentials.[41] The opposite effect was observed when [Cu(TPA)(RCN)]$^+$ was compared to a complex with a stronger electron-donating ligand, N-Me$_6$-TREN. A 1000-fold increase in dioxygen affinity for the latter complex is due to about 100-fold decrease in dioxygen dissociation rate, and a smaller, yet sizeable $\sim$10-fold increase in dioxygen binding rate. As expected, an increased steric bulk of the ligand decreases $O_2$ binding and dissociation rates of Cu$^I$(BQPA) compared to [Cu(TPA)(RCN)]$^+$, simultaneously increasing the lifetime of the Cu$^{II}$($\eta^1$-$O_2^{-\bullet}$) intermediate.[40,41]

A more systematic study of the electronic effects took advantage of a series of TPA derivatives bearing different substituents in the 4-position of each pyridine ring of the ligand ($^{R'}$TPA). Dioxygen binding rate constants varied between $1 \times 10^4$ and $3 \times 10^4$ M$^{-1}$ s$^{-1}$ (283 K, EtCN) for R' = H, Me, $^t$Bu, or OMe, while dioxygen dissociation rates decreased by an order of magnitude. These trends are remarkably similar to oxygenation of cobalt(II) or iron(II) complexes: electronic factors have little effect on $O_2$ binding rates, but effect $O_2$ dissociation rates.

A dramatic acceleration of dioxygen binding was caused by switching from a coordinating solvent (EtCN) to a noncoordinating solvent (THF): $k_{on} \approx 10^9$ M$^{-1}$ s$^{-1}$ was measured[43] by laser flash-trap technique for [Cu(TPA)]$^+$. Negative activation entropy of $-45.1$ J/(K mol) corresponds to this associative process. Dioxygen binding at a vacant copper site proved to be a low-barrier process: $\Delta H^{\ddagger} = 7.6$ kJ/mol. These features again resemble kinetic parameters for the oxygenation at vacant sites at cobalt(II) and iron(II) complexes.

The complexes of other metals (chromium, manganese, nickel, etc.) can also form end-on superoxo adducts with dioxygen; examples of this chemistry have been reviewed elsewhere.[10,14]

### 4.3.2 Formation of Side-On 1 : 1 Metal-Superoxo Complexes

A seminal discovery by Vaska[44] of reversible oxygenation of the iridium(I) complex, [IrCl(CO)(PPh$_3$)$_2$], generated excitement in the inorganic community and opened new ways of thinking about oxygen binding and activation. In the resulting dioxygen adduct, [Ir(O$_2$)Cl(CO)(PPh$_3$)$_2$], the $O_2$ molecule binds side-on to the iridium center. Subsequently, a large number of similar $\eta^2$ dioxygen complexes with various transition metals were discovered.[18,45,46] While the exact electronic structure of these dioxygen adducts varies and depends on the nature of the metal and the ligands, the majority of Vaska-like $\eta^2$ metal–$O_2$ complexes can be treated as peroxides. These side-on $O_2$ adducts are easily formed by the low-valent late transition metal ions that tend to undergo two-electron oxidation, such as Pt(0), Pd(0), Ni(0), Ir(I), Rh(I), Co(I), Ru(0), Ru(II), and so on. The O–O bond length in these complexes, 1.4–1.5 Å, is comparable to the O–O bond length in inorganic peroxide (1.49 Å). The oxygenation of Vaska's complex, therefore, can be considered as an oxidative addition reaction (although O–O bond in the $O_2$ molecule is not broken after its addition to the metal):

$$[Ir^I Cl(CO)(PPh_3)_2] + O_2 \rightarrow [Ir^{III}(\eta^2\text{-}O_2^{2-})Cl(CO)(PPh_3)_2] \tag{4.14}$$

Initially four-coordinate iridium center becomes six-coordinate upon oxygen binding, and the formal oxidation state of the metal increases by 2.

Interestingly, in some cases, the incoming oxygen molecule displaces one of the ligands at the metal,[18,45] as can be illustrated by dioxygen binding to a five-coordinate analogue of Vaska's complex or to a three-coordinate platinum(0) complex $Pt(PPh_3)_3$:

$$[Ir^I Cl(CO)(PPh_3)_3] + O_2 \rightarrow [Ir^{III}(\eta^2\text{-}O_2^{2-})Cl(CO)(PPh_3)_2] + PPh_3 \tag{4.15}$$

$$[Pt^0(PPh_3)_3] + O_2 \rightarrow [Pt^{II}(\eta^2\text{-}O_2^{2-})(PPh_3)_2] + PPh_3 \tag{4.16}$$

The tendency of the low-valent late-metal complexes to form adducts with dioxygen is controlled by both electronic and steric factors. Strong $O_2$ binding is irreversible, weaker binding is reversible, and further weakening of putative metal–$O_2$ bonds results in the lack of detectible oxygen binding; complexes that are unreactive with dioxygen are poor reducing agents.[46] Reactivity trends for series of metal complexes in reactions with dioxygen parallel their reactivity trends in other oxidative addition processes.[46,47]

Periodic trends were established for dioxygen binding to the Group 8 complexes: reactivity increases down the group for the second- and third-row elements (Os (0) > Ru(0); Ir(I) > Rh(I)). However, the first-row metals, such as Co(I), react substantially faster than their heavier counterparts. The overall trend, as illustrated in Table 4.4, is Co(I) > Ir(I) > Rh(I).[46,48] The rates correlate with the values of activation enthalpy: $\Delta H_{on}^{\ddagger}$ for the oxygenation of the Co(I) complex is low, but it increases dramatically for the Rh(I) complex. Activation entropies are large negative for all three metal ions, as expected for a bimolecular, associative process. Activation enthalpies appear to follow the changes in crystal field stabilization energies that accompany the transformation of a square planar starting material into a pseudo-octahedral dioxygen adduct.[48]

A clear correlation between the electron density of the metal and the oxygen binding ability of iridium complexes was observed for a large number of compounds of general formula *trans*-Ir(CO)PPh$_3$X.[46,47,49] For X = Me, dioxygen was bound irreversibly and could not be removed; for X = Cl or X = I, dioxygen binding was reversible; and for X = OMe, dioxygen adduct could only be seen at low temperatures. Kinetic parameters of dioxygen binding (Table 4.4) also correlate with the basicity of iridium, which, in turn, depends on the electron-donating ability of the

**TABLE 4.4　Kinetic Parameters of Dioxygen Addition to Low-Valent Complexes of Cobalt, Rhodium, and Iridium[46-49]**

| M | L | X | $k_{on}$ $(M^{-1}s^{-1})$ | $\Delta H^{\ddagger}_{on}$ (kcal/mol) | $\Delta S^{\ddagger}_{on}$ (cal/(K mol)) |
|---|---|---|---|---|---|
| Complex: $[M(Ph_2PCH_2CH_2PPh_2)_2]^+$; solvent: chlorobenzene | | | | | |
| Co | $2\ Ph_2PCH_2CH_2PPh_2$ | — | $1.7 \times 10^4$ | 3.4 | −28 |
| Rh | $2\ Ph_2PCH_2CH_2PPh_2$ | — | 0.12 | 11.6 | −24 |
| Ir | $2\ Ph_2PCH_2CH_2PPh_2$ | — | 0.37 | 6.5 | −38 |
| Complex: $trans$-$[Ir(CO)(PPh_3)_2X$; solvent: chlorobenzene (benzene for X=F) | | | | | |
| Ir | $2\ PPh_2$, CO | F | $1.48 \times 10^{-2}$ | 13.6 | −24 |
| Ir | $2\ PPh_2$, CO | Cl | $3.4 \times 10^{-2}$ | 13.1 | −21 |
| Ir | $2\ PPh_2$, CO | Br | $7.4 \times 10^{-2}$ | 11.8 | −24 |
| Ir | $2\ PPh_2$, CO | I | 0.30 | 10.9 | −24 |

ligands.[47] Furthermore, a linear correlation was found between the activation enthalpies of $O_2$ binding and the energy of the electronic transition in the optical spectra of the square planar Ir(I) complexes (and the corresponding changes in crystal field stabilization energies).[49]

Steric effects also play a role in thermodynamics and kinetics of dioxygen binding to Vaska-type complexes. For example, ortho-substituents on arylphosphine ligands significantly decrease the oxygenation rate or even completely shut down the dioxygen binding.[46]

The formation of a side-on, $\eta^2$ dioxygen adducts may proceed via an initial end-on coordination of $O_2$ followed by the formation of the second oxygen–metal bond. While this two-step mechanism is logical, no intermediates were observed, even at low temperatures, in reactions between square planar iridium complexes and $O_2$. In the absence of spectroscopically observable intermediates, kinetic data may provide additional mechanistic information. An elegant study[24] compared the rates of dioxygen binding to four- and five-coordinate complexes of iridium(I) (Figure 4.14). The four-coordinate complex, $[Ir(cod)(phen)]Cl$, may undergo a concerted, side-on addition of dioxygen. Alternatively, this complex may react with $O_2$ via initial formation of an end-on adduct that subsequently rearranges into the final side-on product. Chloride counterions did not coordinate to iridium in this system and did not interfere with dioxygen binding. However, other anions, such as iodide or thiocyanate, were capable of binding to iridium centers, forming five-coordinate species and blocking (at least temporarily) one dioxygen coordination site. Interestingly, $I^-$ or $SCN^-$ did not suppress dioxygen binding, showing that two vacant sites at iridium are not necessary for this process. Moreover, oxygenation rates increased linearly with an increase in concentration of $I^-$ or $SCN^-$. It was concluded that five-coordinate Ir(I) complexes react with $O_2$ faster than their four-coordinate precursors; dissociation of the $I^-$ or $SCN^-$ anion from this six-coordinate $\eta^1$ dioxygen adduct occurs readily. End-on, $\eta^1$ binding is the most likely coordination mode of $O_2$ in its adducts with five-coordinate iridium complexes (Figure 4.14). By extension, four-coordinate iridium

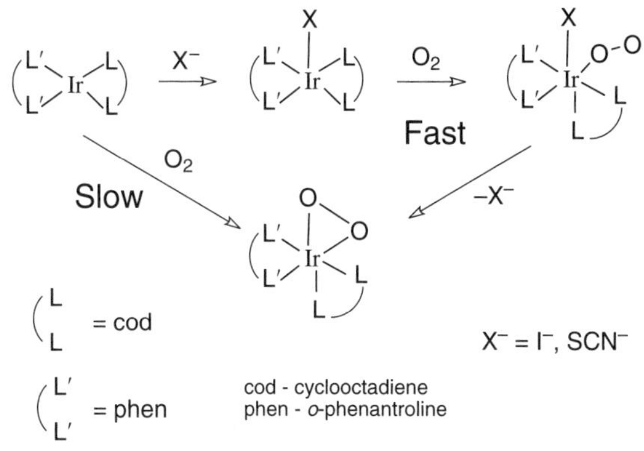

**FIGURE 4.14** Stepwise addition of $O_2$ to four- and five-coordinate complexes of iridium(I).[24]

complexes would likely form, at the first step, an end-on complex with an incoming $O_2$ molecule. This $\eta^1$ complex would rapidly rearrange into the final $\eta^2$ complex, $[Ir^{III}(O_2^{2-})(cod)(phen)]^+X^-$.[24]

Oxygenation of Vaska's complex and related compounds usually occurs much slower and is characterized by higher activation barriers than dioxygen binding to natural oxygen carriers or their synthetic models. However, these highly covalent dioxygen binding complexes generated significant interest due to their potential applications in oxidation catalysis.[47] Renewed interest in catalytic aerobic oxidations will undoubtedly lead to a renaissance of the area of side-on dioxygen complexes with noble metals.[2]

The second- and third-row transition metals, which tend to form highly covalent complexes with dioxygen, resemble, in some sense, chemical behavior of organometallic compounds, but differ from typical biological systems for oxygen binding and activation. Typical bioinorganic approach utilizes classical coordination compounds of 3d metal ions with nitrogen, oxygen, or sulfur donors to mimic biological oxygen activation. Although the first-row transition metals preferentially form end-on superoxo adducts in reactions with $O_2$, examples of side-on complexes are also known. For example, cobalt(I), nickel(I), manganese(II), iron(II), and copper(I) can form $\eta^2$ complexes with dioxygen.[10,50–52] About a dozen complexes from this family were crystallographically characterized; a good correlation was found[53] between the O–O bond distances, $r$, and a function of the O–O vibrational frequencies, $v$ (in agreement with Badger's rule, $r = Cv^{-2/3} + d$). Electronic structure of these complexes was analyzed by rigorous density functional calculations (which were in excellent agreement with experimental geometries and O–O stretching frequencies), showing a "continuum" of charge distribution between the metal ion and the coordinated $O_2$ rather than distinct "islands" of peroxo complexes ($M^{n+2}(O_2^{2-})$) and superoxo complexes ($M^{n+1}(O_2^{-\bullet})$). In most complexes, the precise assignment of the metal oxidation state and the oxidation level of coordinated $O_2$ cannot be made, and the complexes are intermediate between the limiting superoxo and peroxo formulations.

Although mechanistic insights into the formation of side-on $O_2$ adducts with 3d-metals are still limited, a detailed mechanistic picture of the formation of an $\eta^2$-$(O_2)$ copper complexes emerged from a combination of structural, spectroscopic, kinetic, and computational studies of the system depicted in Figure 4.15.[54] Low-coordinate Cu(I) in the starting material was supported by a sterically bulky, bidentate β-ketiminate ligand; the third coordination site was occupied by a weakly bound monodentate solvent molecule (nitrile). Oxygenation of this complex afforded a side-on $O_2$ adduct that displayed significant Cu(III)-peroxo character. This conclusion was confirmed by the O–O stretching frequency ($v(^{16}O_2) = 961$ cm$^{-1}$ and $\Delta v(^{18}O_2) = 49$ cm$^{-1}$), the relatively long O–O distance (1.392(12) Å), and the Cu $K$-edge X-ray absorption spectroscopy data (the pre-edge feature at $-8980.7$ eV is similar to other Cu(III) compounds and exceeds typical values for Cu(II) complexes by about 1.5–2 eV). Irreversible and rapid oxygenation of the Cu(I) complex, which was accompanied by significant changes in the optical spectra, was followed by the low-temperature stopped-flow methodology.

The reaction is first order in copper(I) complex, but the dependence on $[O_2]$ is more complicated and is affected by the presence of MeCN, which moderately decreases the observed rate constant, $k_{obs}$. In neat acetonitrile or acetonitrile-containing solvent mixtures, the $k_{obs}$ depended linearly on the concentration of $O_2$, and the straight line

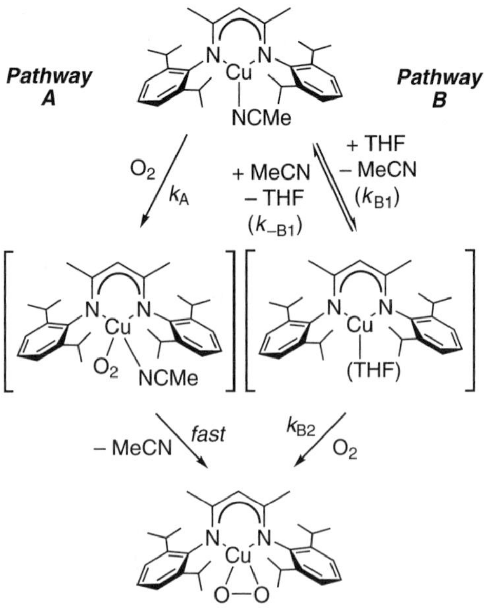

**FIGURE 4.15** Solvent-dependent oxygenation pathways for a mononuclear copper(I) complex with a sterically hindered ketimine ligand.[54]

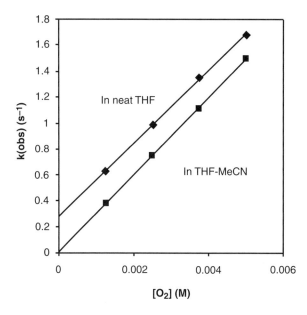

**FIGURE 4.16** Kinetic plot of the observed rate constant of the oxygenation of [Cu(L)(MeCN)] (0.25 mM) in neat THF and THF–MeCN mixture (160:1) at −80°C. Similar parallel straight lines for the two solvents were obtained from the data at −70 and −40°C. The reaction pathways are shown in Figure 4.15.[54]

passed through zero (Figure 4.16). This behavior indicates first order in $O_2$ and the overall second order for the oxygen binding process:

$$k_{\text{obs}}^{\text{THF/MeCN}} = k_A[O_2] \tag{4.17}$$

$$\frac{\partial[\text{LCuO}_2]}{\partial t} = -\frac{\partial[\text{LCu(MeCN)}]}{\partial t} = k_A[\text{LCu(MeCN)}][O_2] \tag{4.18}$$

In neat THF, however, plots of $k_{\text{obs}}$ versus $[O_2]$ also are linear, but with significant positive $y$ intercepts (Figure 4.16). Similar behavior was often observed for reversible oxygen carriers (Section 4.3.1) and could be attributed to the interplay of the oxygen binding rates and the oxygen dissociation rates (with the appropriate rate constants $k_{\text{on}}$ and $k_{\text{off}}$ determined as the slope and the intercept of the $k_{\text{obs}}$ versus $[O_2]$ straight line, respectively) (Figure 4.12). However, this interpretation cannot be applied to irreversible oxygen binding processes, such as the oxygenation of Cu(I) complex with a ketimine ligand.[54] Therefore, a different explanation was proposed for irreversible oxygenation of $\text{LCu}^{\text{I}}$. In this case, a dual reaction pathway was implicated. An additional oxygenation pathway is operative in neat THF that exhibits a zero-order dependence on $[O_2]$, such that the overall rate law has two terms (Equations 4.19 and 4.20):

$$k_{obs}^{THF} = k_A[O_2] + k_B \tag{4.19}$$

$$\frac{\partial[LCuO_2]}{\partial t} = -\frac{\partial[LCu(MeCN)]}{\partial t} = k_A[LCu(MeCN)][O_2] + k_B[LCu(MeCN)] \tag{4.20}$$

The first and rate-limiting step of pathway A (Figure 4.15) is a direct bimolecular reaction of LCu(MeCN) with $O_2$, where adduct $LCu(MeCN)(O_2)$ may be considered as a transition state or an unstable intermediate that releases MeCN and converts to $LCuO_2$ in postrate-limiting step(s). In pathway B, the first and rate-limiting step is solvolysis of LCu(MeCN) to yield a highly reactive intermediate LCu (THF) that is subsequently scavenged rapidly by $O_2$; thus, pathway B is kinetically independent of $[O_2]$. Assuming that $L^1Cu(THF)$ is a steady-state intermediate, the mechanism in Figure 4.15 is described by the rate law (Equation 4.21), which is consistent with the rate laws determined experimentally under the different reaction conditions used.

$$\frac{\partial[LCuO_2]}{\partial t} = -\frac{\partial[LCu(MeCN)]}{\partial t} = k_A[LCu(MeCN)][O_2]$$

$$+ \frac{k_{B1}k_{B2}[LCu(MeCN)][O_2]}{k_{-B1}[MeCN] + k_{B2}[O_2]} \tag{4.21}$$

In the presence of excess MeCN, pathway B is effectively shut down because nearly all LCu(THF) is converted to LCu(MeCN), and the oxygenation process proceeds entirely through pathway A. The rate-determining step for pathway B is exchange of THF solvent for nitrile prior to oxygenation, while for the pathway A, coordination of $O_2$ to the Cu(I)–nitrile complex is rate controlling. Activation parameters for both pathways A ($k_A$) and B ($k_B = k_{B1}$) are consistent with associative mechanisms for their rate-limiting steps: low $\Delta H^{\ddagger}$ and large negative $\Delta S^{\ddagger}$ values were found (Table 4.5).

**TABLE 4.5   Selected Kinetic Parameters for Oxygenation of Mononuclear Cu(I) Complexes Leading to 1 : 1 Cu/O$_2$ Adducts**[54]

| Cu(I) Complex | Solvent | $k_{on}$ (223 K) | $\Delta H^{\ddagger}$ (kJ/mol) | $\Delta S^{\ddagger}$ (J/(mol K)) |
|---|---|---|---|---|
| LCu(MeCN), pathway A | THF-MeCN or THF | $1560 \pm 19\,M^{-1}\,s^{-1}$ | 18 ± 2, 14.9 (*t*) | −100 ± 10, −108 (*t*) |
| LCu(MeCN), pathway B | THF | $3.95 \pm 0.59$ | 30 ± 2, 27.2 (*t*) | −98 ± 10, −101.0 (*t*) |

All experimental values, except those denoted by *t*, which are derived from theoretical calculations.

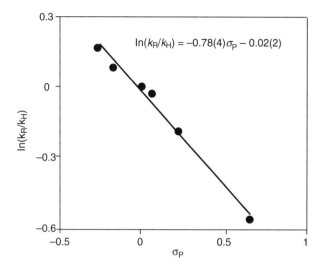

**FIGURE 4.17**    Hammet plot for the rate constants of the reaction between $O_2$ and complexes
CuL($p$-NC-$C_6H_4$-R) {R = OMe, Me, H, F, Cl, CN} in the THF solution at 203 K (Figure 4.15,
pathway A).[54]

Additional insights into the oxygenation mechanisms were obtained by varying the
chemical nature of a nitrile ligand. The excellent Hammet correlation of $k_A$ with
substituent $\sigma_p$ values (Figure 4.17) represents a quantitative measure of the sensitivity
of oxygenation rate constants to electronic effects across a series of metal complexes
that differ only in the nature of coordinated monodentate aromatic nitrile ligand. The
resulting $\rho$ of $-0.34$ reflects a modest buildup of positive charge at the metal center in
the rate-controlling transition state or, in other words, nucleophilic character for the
metal complex in the oxygenation process in pathway A (Figure 4.15). This result is
not surprising considering the degree of reduction of the $O_2$ molecule and corresponding degree of oxidation of the $L^1Cu$ fragment in the product $L^1CuO_2$, shown to
have considerable Cu(III)-peroxide character. Interestingly, electronic effects often
have little influence on the rates of oxygen binding to heme proteins and their
synthetic models, and instead alter oxygen dissociation rates in those systems
(Section 4.2.3).

CASPT2-corrected DFT calculations (including continuum solvation) provided
insights into the structures of the rate-determining transition states and predicted
activation enthalpies and entropies within experimental error for both pathways.
Calculations demonstrated small energy difference between end-on and side-on $O_2$
adducts and suggested that in a solvent-assisted pathway, $O_2$ initially binds end-on
and then rearranges into a side-on final product in a very low-barrier process.[53] It can
be concluded that formation of the $\eta^2$ metal–dioxygen adducts generally occurs in a
stepwise manner, with the initial formation of the end-on ($\eta^1$) complex, which is
similar to (often reversible) processes described in Section 4.3.1. The final side-on
adducts tend to be thermodynamically stable, and their formation is usually (although

not always) irreversible. The oxygenation of first-row transition metals (3d-metals) tends to be much faster than analogous reactions of the second- and third-row metals (4d- and 5d-metals).

### 4.3.3    Formation of Metal-Oxo Complexes

Metalloenzyme-catalyzed biological oxidations with dioxygen take advantage of high-valent metal-oxo intermediates that are highly reactive toward reducing substrates. The best known example of this chemistry is cytochrome P450, a heme enzyme that generates an intermediate three oxidation levels above the initial Fe(II) state:

$$(P^{2-})Fe^{V} = O \rightleftharpoons (P^{-\bullet})Fe^{IV} = O$$

This process requires an additional electron and two protons, as discussed in Section 4.5. In the absence of exquisite control over electron and proton delivery in nonenzymatic systems, chemists were unable to replicate this noncomplementary redox reaction between an Fe(II) porphyrin and dioxygen. Instead, sterically unhindered porphyrins tend to form iron(III) oxo-bridged dimers (see also Equations 4.10 and 4.11):

$$(P^{2-})Fe^{II} + O_2 \Leftrightarrow (P^{2-})Fe^{II}(O_2) \rightleftharpoons (P^{2-})Fe^{III}(O_2^{-\bullet}) \tag{4.22}$$

$$(P^{2-})Fe^{II}(O_2) \rightleftharpoons (P^{2-})Fe^{III}(O_2^{-\bullet}) + (P^{2-})Fe^{II} \rightarrow (P^{2-})Fe^{III}(O_2^{2-})Fe^{III}(P^{2-}) \tag{4.23}$$

$$(P^{2-})Fe^{III}(O_2^{2-})Fe^{III}(P^{2-}) \rightarrow 2(P^{2-})Fe^{IV}(O^{2-}) \tag{4.24}$$

$$(P^{2-})Fe^{IV}(O^{2-}) + (P^{2-})Fe^{II} \rightarrow (P^{2-})Fe^{III}(\mu\text{-}O^{2-})Fe^{III}(P^{2-})$$

(Transient, usually unobserved) $\qquad\qquad$ (4.25)

Sterically protected iron(II) porphyrins bind one molecule of $O_2$, often reversibly, but do not usually form stable, isolable iron(IV)-oxo species (Section 4.3.1). Significant interest in forming high-valent metal-oxo species directly from $O_2$ stems from the goal of developing catalysts for aerobic oxidations. Indeed, high-valent metal-oxo species are assumed to be capable of transferring an oxygen atom, abstracting a hydrogen atom, transferring an electron, and so on. Despite this interest, limited success was achieved in this area, and the mechanisms of direct formation of metal-oxo intermediates from $O_2$ are not completely understood. Important developments in this area include low-temperature trapping of ferryl complexes. Low-temperature studies in organic solvents provided support for the reaction scheme shown in Equations 4.22–4.25. For example, an intermediate in oxygenation of $Fe^{II}(TMP)$, where TMP = tetramesitylporphyrin, was unambiguously detected and spectroscopically characterized as an iron(IV)-oxo complex, $(TMP)Fe^{IV}=O$.[55] This ferryl species formed, via a bimetallic pathway, from a diiron(III) peroxo precursor that was also observed as a transient intermediate. Tetraphenylporphyrin complexes showed similar dioxygen reactivity, and also generated transient ferryl intermediates.

A detailed study of dioxygen reactions with $Fe(F_8TPP)$ (where $F_8TPP$ = tetrakis (2,6-difluorophenyl)porphyrinate(2−), an electron-poor porphyrin) revealed similar individual steps of $O_2$ adduct formation.[56] In this example, oxygenation was solvent dependent. In coordinating solvents (nitriles or THF), typical 1 : 1 end-on $Fe^{III}(O_2^{-\bullet})$ adducts were reversibly formed. The electron-withdrawing fluorine substituents weakened the $Fe–O_2$ bond, thus decreasing dioxygen affinity and increasing dioxygen dissociation rates for $Fe(F_8TPP)$ system; however, kinetic parameters of formation of this 1:1 dioxygen adduct were comparable to those for oxygen binding to other iron (II) porphyrins, with the second-order rate constants of $10^7–10^8\,M^{-1}\,s^{-1}$. Interestingly, in noncoordinating solvents (such as methylene chloride or toluene), reversible formation of diiron(III)-peroxo adducts was reported (Figure 4.18). Addition of a base (DMAP) caused O–O bond cleavage in the $(P)Fe^{III}(O_2^{2-})Fe^{III}(P)$ intermediate, generating a ferryl species, $(F_8TPP)Fe^{IV}=O$. This study[56] further confirms that the formation of Fe(IV)-oxo species from Fe(II) precursors and $O_2$ is a multistep process.

In recent years, numerous parallels between heme chemistry and corresponding nonheme iron chemistry were uncovered. Several nonheme iron(IV)-oxo complexes were characterized spectroscopically and even structurally.[57,58] Although most of these ferryl intermediates were generated from iron(II) or iron(III) precursors and relatively strong oxidants (acting as an oxygen atom donor), it is also possible to

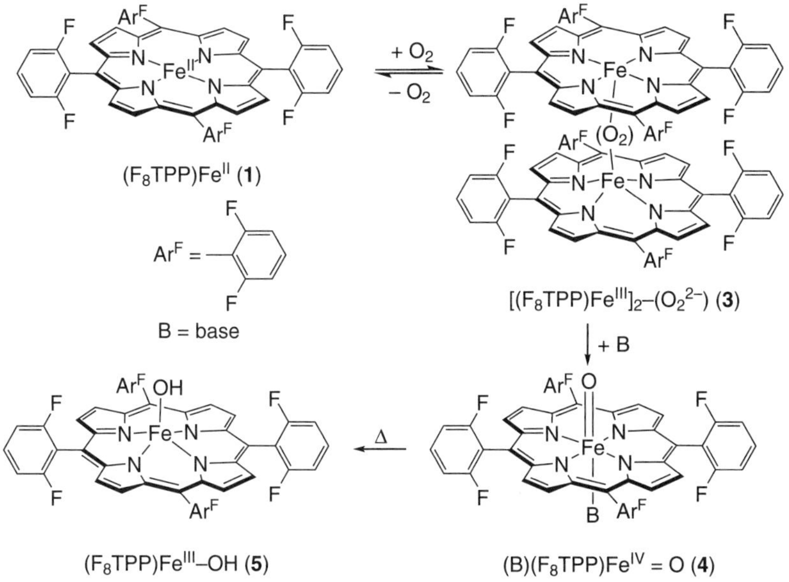

**FIGURE 4.18** Stepwise oxygenation of the $Fe^{II}(F_8TPP)$ affording an iron(IV)-oxo intermediate.[56]

obtain an $(L)Fe^{IV}=O$ complex directly from $(L)Fe(II)$ and dioxygen.[59] In this system, saturated tetraazamacrocycle tetramethylcyclam, TMC (Figure 4.7), was used as an equatorial iron ligand. Oxygenation outcome was again solvent dependent: ferryl species was formed in ethers, THF, and alcohols (solvents that decreased redox potential of the iron complex), but not in $CH_3CN$, acetone, or $CH_2Cl_2$. The 2:1 stoichiometry of the $Fe(TMC)^{2+}/O_2$ reaction suggests a bimolecular mechanism with the intermediacy of diiron(III)-peroxo species, analogous to the mechanism of O–O bond cleavage by iron(II) porphyrins (see above).[59]

Iron is a natural choice for biomimetic studies aimed at modeling ferryl chemistry of natural heme- or nonheme oxygenases. Even though most other transition metals are not found in oxygen-activating enzymes, they can participate in similar reaction pathways and support similar intermediates in their reactions with dioxygen or oxygen atom donors. Metal-oxo complexes are widespread in inorganic chemistry,[60,61] and their ability to transfer an oxygen atom[62] makes them attractive as oxidizing reactants or catalysts.[61] Acid–base properties of the metal-oxo group predict the tendency of low-valent metals to form hydroxo rather than oxo species, while high-valent metals (with the formal oxidation states equal to or exceeding 4) preferentially form oxo compounds even in protic media.[62] The electronic structure of the metal-oxo moiety predicts that the metal ion can possess five or fewer d-electrons; this requirement is sometimes referred to as "The Oxo Wall."[62] Although no late metal-oxo compounds were known until recently,[60] the oxo-wall was broken by latest discoveries of iridium-oxo, palladium-oxo, and platinum-oxo moieties.[63] Still, metal-oxo complexes are most prevalent for the group 6 metals, and their abundance decreases in moving across the period in both directions.[60,62]

Relatively few metal-oxo compounds are prepared in direct reactions of the lower valent metal ion (or its complex) with dioxygen. For 3d-metals, the best studied systems contain chromium(II). Two factors contribute to the desired reactivity mode: on one hand, chromium(II) is highly reducing and its reactions with $O_2$ are thermodynamically favorable and kinetically facile, while on the other hand, $Cr^{IV}=O$ species are sufficiently stable, yet capable of transferring an oxygen atom to a range of substrates.[14,15] For comparison, $TiO^{2+}$ and $VO^{2+}$, which can be obtained from Ti(II) and V(II), respectively, do not display oxidizing properties; $Mn^{IV}=O$ and especially $Mn^V=O$ compounds are potent oxidants, but their plausible Mn(II) precursors are unreactive with $O_2$ (unless the metal is bound to strongly electron-donating ligands, such as corroles[64]).

A variety of chromium(II) compounds, ranging from aqueous $Cr^{2+}$ [14,15] to chromium(II) porphyrins,[65] were shown to produce chromium(IV) oxo species in reactions with $O_2$:

$$2(L)Cr^{II} + O_2 \rightarrow 2(L)Cr^{IV}=O \tag{4.26}$$

The proposed mechanism of oxygenation of Cr(II) porphyrins[65] is identical to the mechanisms of analogous reactions with iron(II) porphyrins (Equations 4.22–4.25), and involves initial coordination of $O_2$ to a single chromium center, followed by the formation of μ-peroxo dimers of Cr(III) and their subsequent transformation into (P)

$Cr^{IV} = O$ via O–O bond homolysis. Similar to iron chemistry, bimetallic activation is implicated to account for the electronic mismatch between a four-electron oxidant $(O_2 \rightarrow 2O^{2-})$ and a two-electron reductant $(Cr^{II} \rightarrow Cr^{IV})$. The oxygenation of $Cr^{2+}_{aq}$ is even more complicated and may include one-electron processes involving odd oxidation states of chromium ($Cr^{III}$ and $Cr^{V}$); a likely source of these intermediates is facile disproportionation of $CrO^{2+}_{aq}$ in water.[66] Excess oxygen also induces side reactions, such as the formation of $CrO_2^{2+}$. Nevertheless, in the absence of excess $O_2$, aquachromyl(IV) intermediate (with a half-life of 30 s in acidic water at room temperature) can be cleanly generated and used in subsequent reactivity studies.

High oxidation states are more stable for second- and third-row metals compared to their first-row counterparts. In view of biological relevance and potential synthetic importance of iron–oxygen chemistry, it is instructive to briefly consider the reactivity of the heavier group 8 metal, ruthenium. Ruthenium chemistry naturally lends itself to multielectron chemistry. Unlike iron, ruthenium can easily access a broad range of oxidation states (from 0 to $+8$), with most stable oxidation states ($Ru^{II}$, $Ru^{IV}$, and $Ru^{VI}$) being separated by two-electron transitions. Much like their iron analogues, sterically protected ruthenium(II) porphyrins react with dioxygen, reversibly forming $O_2$ adducts.[67] Some reactions of sterically unhindered ruthenium(II) porphyrins, however, differ from analogous iron systems. Most interestingly, $[Ru^{VI}(TMP)(O)_2]$ (where TMP = tetramesitylporphyrin) was found to catalyze the aerobic epoxidation of olefins at ambient temperature and pressure.[68] Furthermore, this dioxoruthenium (VI) catalyst is a competent stoichiometric oxidant under anaerobic conditions, affording 1.6 equivalents of epoxide. The oxidant, dioxo(tetramesitylporphyrinato)ruthenium(VI) ($Ru(TMP)(O)_2$), could be obtained in high yield from the $Ru^{II}(TMP)$ and dioxygen. The catalytic cycle involving $(TMP)Ru^{II}$, $(TMP)Ru^{IV} = O$, and $(TMP)Ru^{VI}(O)_2$ was proposed.[67] Experimental studies of the reactive intermediates in this system were somewhat complicated by the instability of $(TMP)Ru^{IV}=O$ with respect to disproportionation. The intermediacy of $Ru^{IV}=O$ species was observed in related porphyrin-catalyzed or -promoted oxidations, including stoichiometric enantioselective alkene epoxidation with a chiral dioxoruthenium(VI) porphyrinato complex.[69] It should be noted, however, that oxygen donors other than $O_2$ (e.g., $m$-chloroperoxobenzoic acid, mCPBA, or iodosobenzene, PhIO) were used in these synthetically appealing applications. Computational studies[70] provided additional insight into possible mechanism of original aerobic epoxidation with $(TMP)Ru^{II}$, and confirmed that $Ru^{IV}=O$ is a likely intermediate in the reaction between $(TMP)Ru^{II}$ and $O_2$ (Figure 4.19).

Bimetallic pathways were implicated in the formation of both $Ru^{IV}=O$ and $Ru^{VI}(O)_2$. There is no doubt that rich chemistry of ruthenium and osmium in their reactions with $O_2$ is ripe for additional exploration, and will provide new insights into the mechanisms of O–O bond breaking with concomitant generation of relatively stable high-valent metal-oxo intermediates. Furthermore, these metals provide the best opportunity for true monometallic activation of dioxygen via a formally four-electron pathway affording the $M^{n+4}(O)_2$ metal-oxo species. This opportunity has to be pursued.

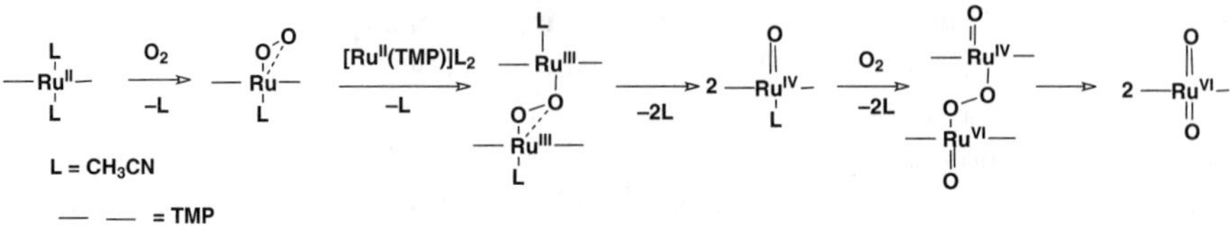

**FIGURE 4.19**   Step-by-step reaction pathway for an oxidation of ruthenium(II) porphyrins, L [Ru$^{II}$P]L (L=CH$_3$CN), by molecular oxygen, supported by computational studies.[69]

An interesting way to bypass the electronic noncomplementarity of typical reactions between transition metal ions or complexes and dioxygen is based on an oxygen atom transfer from an oxygen allotrope, ozone. This approach was very successful in generating aqueous ferryl, an otherwise hardly accessible species[71]:

$$Fe^{2+} + O_3 \rightarrow FeO^{2+} + O_2 \tag{4.27}$$

Such a clean way of producing ferryl intermediates allowed a detailed characterization of subsequent reactivity of aqueous ferryl (see also Section 4.5).

## 4.4   DIOXYGEN BINDING TO TWO METAL CENTERS

Two-electron reduction of dioxygen into coordinated peroxide can be easily performed by two metal centers undergoing concomitant one-electron oxidations, as shown in Equation 4.4 (Section 4.2.2). A variety of transition metal ions (cobalt, nickel, iron, manganese, copper, etc.) can form dinuclear peroxides. These complexes differ in structure (cis-μ-1,2-peroxides, trans-μ-1,2-peroxides, μ-η$^2$:η$^2$-peroxides), in stability and subsequent reactivity modes, and in the protonation state of the peroxo ligands (Figure 4.3). In certain cases, dinuclear μ-η$^2$:η$^2$-peroxides and bis-μ-oxo diamond core complexes interconvert, as discussed below for copper–dioxygen adducts.

Dinuclear peroxides can be assembled from the mononuclear or the dinuclear precursors. The former scenario is more straightforward from the synthetic perspective, because monometallic starting materials are often readily accessible. However, oxygen binding to these systems necessarily goes via at least one step that involves two molecules of the metal complex, giving rise to complicated kinetics. In contrast, mechanistic studies of dioxygen binding to dinuclear metal complexes are somewhat easier, although the synthesis of well-defined bimetallic starting materials is often difficult. Furthermore, two metal centers have to be properly positioned in a dinuclear complex, so that dioxygen molecule can coordinate to both of them. Otherwise, significant rearrangement of the multidentate ligand may be necessary for efficient oxygenation.

### 4.4.1   Formation of the Dinuclear Metal–Oxygen Complexes from Mononuclear Precursors

The formation of dinuclear peroxo-bridged species from the mononuclear complexes and $O_2$ was historically viewed as an undesirable process that prevents fully reversible dioxygen binding to synthetic models of hemoglobin and myoglobin. In the context of oxygen carrier design, formation of dinuclear peroxides is termed "autoxidation." This is indeed the first step in the bimolecular irreversible decomposition pathway of $(P)Fe^{III}(O_2^{-\bullet})$ adducts. As discussed in Section 4.3, sterically hindered, superstructured ligands were designed to protect the metal–$O_2$ binding site in iron porphyrins. Transient diiron(III) peroxides are well documented in sterically unhindered heme systems.[56,72]

Macrocyclic nonheme iron complexes resemble iron porphyrins in their reactivity with dioxygen.[72] Strong electron-donating ligands ensure that iron(II) is susceptible to oxidation and reacts rapidly with an $O_2$ molecule. The initially formed mononuclear iron(III)-superoxo intermediate then attacks the second molecule of iron(II) complex, affording diiron(III)-peroxo species. A recent example[73] of this rather common reactivity takes advantage of amidopyridine macrocycle $H_2pydioneN_5$ (Figure 4.20). Its rigidity results in a predictable (pentagonal–bipyramidal) coordination geometry about the metal and in high thermodynamic and kinetic complex stability, which prevents uncontrolled metal leakage, oxidation, and hydrolysis. Both mono- and dideprotonated versions of the iron(II) complex were isolated in the solid state; the deprotonation state persists in DMSO solvent. Monodeprotonated complex features an intramolecular source of proton(s) that modulates oxygenation chemistry.

Oxygenation of the dideprotonated complex of iron(II), $Fe(pydioneN_5)$, in aprotic solvents proceeds via a path analogous to that of iron(II) porphyrins: via iron(III) superoxo and diiron(III) peroxo species (Figure 4.20), as evidenced by time-resolved spectral changes observed during the reaction. The reaction is second order in the iron (II) complex and shows an inverse dependence of reaction rate on dioxygen concentration. The presence of 1-methylimidazole stabilizes the diiron peroxo intermediate: the half-life increases by about two orders of magnitude. The reaction

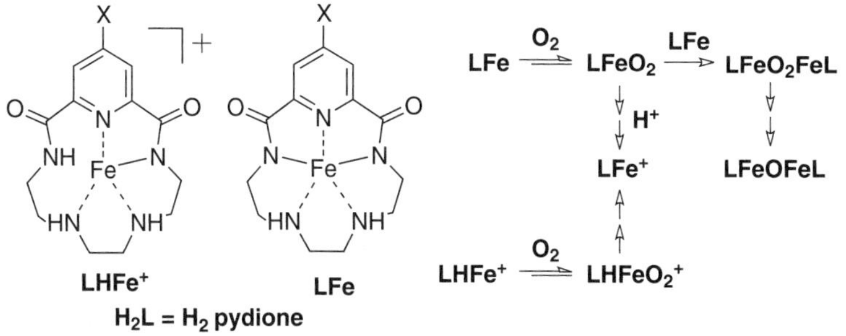

**FIGURE 4.20**   Oxygen binding to a mononuclear macrocyclic iron(II) complex with ligand $H_2$ pydione.[73]

of Fe(pydioneN$_5$) with dioxygen in methanol is distinctly different: it is first order in both iron(II) complex and dioxygen, with no spectroscopically observed intermediate. Similar behavior was observed for the monodeprotonated complex Fe(HpydioneN$_5$) in various solvents. The presence of an accessible proton in the vicinity of the reaction center or from the solvent altered the oxygenation pathway in these macrocyclic systems.[73] Stability of coordinated peroxides often decreases upon protonation in both heme and nonheme complexes.[74]

The final decomposition products of diiron(III) peroxides, μ-oxo-bridged diiron (III) complexes, are often referred to as "bioinorganic rust" because of their thermodynamic and kinetic stability. To achieve oxygen activation, irreversible formation of μ-oxo dimers has to be avoided. In addition to commonly used (but synthetically challenging) steric protection of the metal center, electronic effects may also shut down irreversible decomposition pathways, favoring instead reversible formation of dinuclear peroxides. Decreasing the electron density of the metal causes relative destabilization of high oxidation states and can prevent the irreversible formation of oxidation products. In heme chemistry, electron-withdrawing substituents make iron centers less susceptible toward autoxidation. An interesting example of reversible formation of diiron(III) peroxides from O$_2$ and iron tetrakis-2,6-difluorotetraphenylporphyrin (F$_8$TPP) was recently reported (Figure 4.18); analogous nonfluorinated porphyrins form "bioinorganic rust" in their reactions with O$_2$.[56]

Cobalt(II) compounds are less prone to one-electron oxidation than their iron(II) analogues. This shift in redox potential decreases dioxygen affinity of mononuclear cobalt(II) oxygen carriers, as discussed in Section 4.3.1. The other side of the same coin is increased reversibility of dioxygen binding to cobalt(II). While reversible formation of metal-superoxo adducts is not unusual and can be achieved for both iron and cobalt synthetic oxygen carriers, the tendency of cobalt complexes to reversibly form dinuclear peroxide species is fairly unique. For example, iron porphyrins rarely stop at the stage of diiron(III)-peroxo intermediates in their reactions with O$_2$; electron-poor ligands (such as porphyrins bearing electron-withdrawing substituents) are necessary for reversible formation of diiron(III) peroxides.[56] In contrast, a variety of cobalt(II) complexes that react with O$_2$ proceed, via initial formation of a 1 : 1 Co(III) superoxide, to form dicobalt(III)-peroxides that do not undergo further decomposition.[75] Relative stability with respect to bimolecular autooxidation made cobalt(II) systems very attractive for designing synthetic oxygen carriers. Indeed, reversible dioxygen binding was achieved for complexes with really simple ligands, including cyanides, acyclic and cyclic polyamines, Schiff bases (with various salicylaldehyde derivatives, such as Salen and its numerous relatives, being especially popular), aminocarboxylates, porphyrins, and phthalocyanines, to name a few. Several of these complexes were crystallographically characterized and shown to have *trans*-μ-1,2-geometry.[75]

Thermodynamic and kinetic parameters of the two-step formation of dicobalt(III)-peroxo complexes vary by many orders of magnitude.[14] All four individual rate constants (Equations 4.5 and 4.10) contribute to the overall dynamics of bimolecular oxygenation of Co(II) in most systems. Therefore, complicated kinetics was often observed. In a number of chemical systems, all four individual rate constants could be determined, revealing some trends in two sequential oxygen binding events. Coordination of $O_2$ molecule to a single cobalt(II) center is usually limited by the ligand substitution rates, and the rate constants can be as high as about $10^7$ for certain macrocyclic complexes (Table 4.2, Figure 4.7). The rate constant of the second step, formation of dicobalt(III) peroxide, tends to be somewhat lower. For example, $k_{1,on} = 1.18 \times 10^7 \, M^{-1} s^{-1}$ and $k_{2on} = 4.7 \times 10^5 \, M^{-1} s^{-1}$ for $[Co([14]aneN_4)(H_2O)_2]^{2+}$.[14] This decrease in rate of the second step reflects steric constraints in the process of forming a dinuclear cobalt(III) complex with a relatively short peroxide bridge. Steric effects become apparent when unsubstituted [14]aneN$_4$ ligand is compared to its hexamethylated analogue, Me$_6$[14]aneN$_4$ (Figure 4.7). In the latter case, oxygenation of LCo(II) stops at the first step; no further reaction of cobalt(III) superoxide with the remaining LCo(II) takes place. Methyl substituents on both faces of the macrocyclic ligand disfavor a bimolecular reaction leading to a μ-1,2-peroxide.[14] A popular way of overcoming this obstacle involves design of bimetallic complexes that are preorganized for dioxygen coordination between the two metals.

### 4.4.2 Dioxygen Binding to Dinuclear Metal Complexes Exemplified by Nonheme Iron

Bimetallic oxygen-activating sites in biology include nonheme iron or copper centers. An interest in understanding and modeling biological oxygen binding and oxygen-activating systems fueled numerous studies in nonheme diiron and dicopper chemistry. Detailed picture of dioxygen coordination to biomimetic diiron or dicopper sites is now emerging. The discussion below will start with diiron systems, since many features of their reactions with $O_2$ are generally applicable to other bimetallic complexes.

The nonheme diiron centers in proteins and model complexes with O,N-donors can reach a number of oxidation states spanning from Fe$^{II}$Fe$^{II}$ to Fe$^{IV}$Fe$^{IV}$. The diiron(II) state is reactive with dioxygen yielding different products depending on the nature of ligands and reaction conditions (Figure 4.21). Outer-sphere electron transfer may occur for coordinatively saturated and sterically impeded complexes with sufficiently low redox potential.[17]

Inner-sphere oxygenation seems to be a more common reactivity pathway for diiron(II) complexes.[17,73,76,77] Formation of the first Fe–O bond and transfer of one electron should lead to an iron(II)–iron(III)-superoxo intermediate (Figure 4.21), which has not been observed until very recently.[78] Instead, diiron(III)-peroxo composition is ascribed to the adducts observed in most enzymatic and synthetic systems, suggesting that the second electron transfer and Fe–O bonding usually are postrate-limiting events. Several synthetic complexes with μ-1,2-peroxo-bridged diiron(III) cores were characterized structurally[79–82] and such binding mode is assumed for the peroxo-diiron(III) intermediates of several proteins (soluble methane monooxygenase, stearoyl desaturase, ferritin, and ribonucleotide reductase) and most model compounds.[17] Protonation of the peroxide ligand gives a hydroperoxo

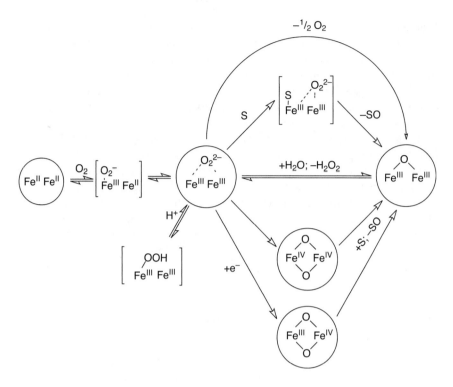

**FIGURE 4.21**   Reactions of diiron(II) complexes with dioxygen and further transformations of diiron–dioxygen adducts.[17]

complex; in hemerythrin such protonation is intramolecular (a $\mu$-OH ligand provides a proton turning into an oxo-bridge) (Section 4.2.3). Several model diiron(III)-hydroperoxo complexes were recently prepared in solution and characterized by spectroscopy.[17]

Peroxo-diiron(III) intermediates vary significantly in their stability and reactivity depending on the nature of supporting ligands and other conditions. Some of them (like oxyhemerythrin) (Section 4.2.3) form reversibly from a diiron(II) complex and dioxygen,[20] but the majority of known peroxo-diiron(III) species form irreversibly and eventually decompose into different types of products. One pathway of the peroxo-diiron(III) species decomposition is the disproportionation into oxo-bridged polyiron(III) complexes and dioxygen.[83] The peroxo intermediate of sMMO undergoes a remarkable transformation into a bis($\mu$-oxo)diiron(IV) species (Section 4.5). Such a reaction has not been achieved in a synthetic system yet, although ($\mu$-oxo)diiron(IV) complexes were recently obtained from an iron(III) precursor and dioxygen, presumably via a ($\mu$-peroxo)diiron(IV) intermediate.[84] Peroxo-diiron(III) species with exchangeable coordination sites can transfer an O-atom to organic phosphines and sulfides, and there is strong evidence that the reactions occur intramolecularly upon the coordination of the substrate.[17] One-electron reduction of a peroxo-diiron(III) complex can yield an oxo-bridged $Fe^{III}Fe^{IV}$

species, a reaction that apparently occur in the R2 subunit of ribunucleotide reductase protein.[76,77] Similar chemistry was observed or suggested in some of the model systems leading to $Fe^{III}Fe^{IV}$ complexes.[77,85] High-valent ($Fe^{III}Fe^{IV}$ and $Fe^{IV}Fe^{IV}$) intermediates are generally much more reactive oxidants than their parent peroxo-iron complexes and can perform even such challenging reactions as alkane hydroxylation (Section 4.5).

Peroxo-diiron(III) complexes can undergo not only redox but also ligand substitution reactions. Liberation of $H_2O_2$ was observed in the reactions with phenols and carboxylic acids leading also to the respective phenolate or carboxylate iron(III) complexes.[86] Hydrolysis of a peroxo-diiron(III) complex results in an oxo-diiron(III) species and hydrogen peroxide. Such reaction is responsible for the "autoxidation" of hemerythrin, but is very slow for the native protein due to hydrophobic shielding of the active site (Section 4.2.3).[20] The hydrolysis of iron(III) peroxides is reversible, and the reverse reaction, the formation of peroxo intermediates from $H_2O_2$ and the (di)iron(III), is often referred to as "peroxide shunt" and is much better studied for model complexes.

Mixed-valent $Fe^{II}Fe^{III}$ intermediates were observed in several cases during the oxygenation of diiron(II) complexes, most probably due to a one-electron outer-sphere oxidation. Oxo- or hydroxo-bridged diiron(III) complexes were the final products of these oxygenation reactions. The overall mechanism of these transformations is apparently complex and not well understood yet.[77]

The reactions of dioxygen with the nonheme diiron(II) complexes often lead to oxo-bridged diiron(III) complexes without observable intermediates. The four-electron reduction of the $O_2$ molecule obviously cannot proceed in one step. It is probable that superoxo, peroxo, and high-valent iron intermediates form in such systems, but in very small steady-state concentrations, while the initial step of $O_2$ binding is rate limiting (Figure 4.21). Kinetic and isotope labeling studies allowed insights into the mechanisms of several such reactions.[73,87,88] Compared to the formation of peroxo- or high-valent diiron intermediate, the generation of oxo-diiron(III) complexes is less interesting from the viewpoint of dioxygen activation, because in the this case all four oxidative equivalents of $O_2$ have been wasted on oxidizing iron, and nothing is left for the oxidation of a cosubstrate (unless the $Fe^{III}/Fe^{II}$ potential is sufficient for such a reaction).

Two stable ($\mu$-oxo)diiron(IV) complexes with tetraamidato ligands were recently obtained in a direct reaction of mononuclear iron(III) precursors and $O_2$.[84] This unprecedented transformation is possible due to very strong electron-donating properties of the ligands that stabilize the high-valent state of iron and bring the $Fe^{IV}/Fe^{III}$ redox potential down.

Nonheme iron enzymes and oxygen carriers contain imidazole nitrogen donors and carboxylate oxygens in their active sites; additional (hydr)oxo-bridges or coordinated water molecules are also commonly found. The majority of biomimetic diiron complexes also contain N,O-ligands. Nitrogen donors are usually provided by amino groups, pyridine rings, or other heterocycles; oxygen donors come from carboxylates, alkoxides, hydroxides, and others. Dinuclear complexes may be supported by polydentate dicompartmental ligands, which often incorporate bridging

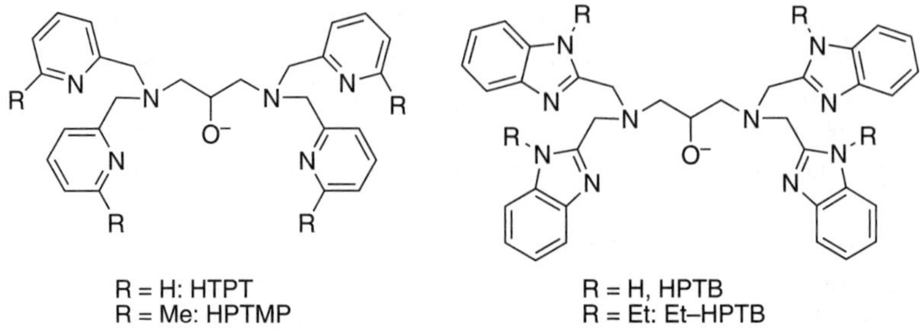

R = H: HTPT
R = Me: HPTMP

R = H, HPTB
R = Et: Et–HPTB

**FIGURE 4.22** Dinucleating aminopyridine ligands supporting alkoxy-bridged diiron–oxygen complexes.

groups (such as alkoxides or phenoxides). Alternatively, bi-, tri-, or tetradentate ligands support single iron centers that are further bridged by externally added carboxylates, hydroxides, or similar monodentate ligands.[17]

The first detailed kinetic study of the oxygenation of diiron(II) species was done with a series of complexes of dinucleating ligands HPTP, Et-HPTB, and HPTMP (Figure 4.22).[83]

Starting complexes $[Fe^{II}_2(\mu\text{-L})(\mu\text{-O}_2CPh)]^{2+}$ have similar ligand environment around the two five-coordinate iron(II) centers bridged by an alkoxide and a carboxylate residues and also ligated by three N-donor atoms each (Figure 4.23). However, there is a significant difference in the steric accessibility of the diiron(II) centers (HPTP > Et-HPTB >> HPTMP), which is controlled by the structure of the dinucleating ligands.

Oxygenation of the complexes in propionitrile solution at low temperatures ($-70$ to $-20°C$) gives high yields of dioxygen adducts formulated as $[Fe^{III}_2(\mu\text{-L})(\mu\text{-}1,2\text{-peroxo})(\mu\text{-O}_2CPh)]^{2+}$. The formation of the HPTP and Et-HPTB peroxides is irreversible, while the HPTMP dioxygen adduct forms reversibly. Oxygenation of

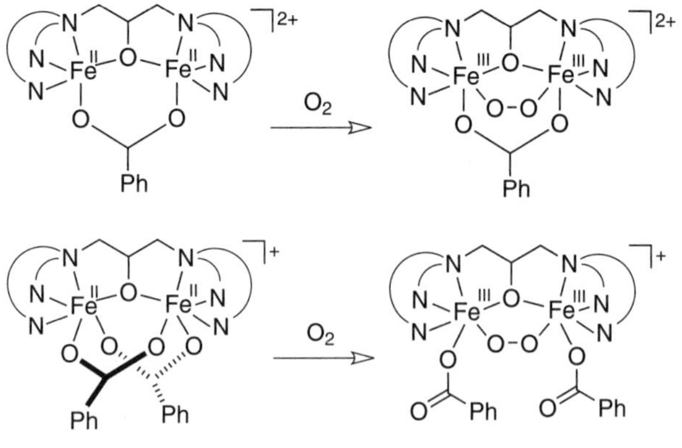

**FIGURE 4.23** Oxygenation of alkoxo-, carboxylato-bridged diiron(II) complexes with dinucleating aminopyridine ligands.[17]

the less sterically hindered diiron(II) complexes (L = HPTP and Et-HPTB) follows the mixed second-order kinetics, $v = k[\text{Fe}_{\text{II}}{}^2][\text{O}_2]$, with almost identical activation parameters ($\Delta H^{\ddagger} = 16\,\text{kJ/mol}$ and $\Delta S^{\ddagger} = -120\,\text{J/(mol K)}$). The reaction of the Et-HPTB complex was also characterized by high-pressure stopped-flow technique (the only such study for the oxygenation of a synthetic diiron(II) complex so far), which gave the volume of activation $\Delta V^{\ddagger} = -13\,\text{cm}^3/\text{mol}$. The low activation barrier and negative values of $\Delta S^{\ddagger}$ and $\Delta V^{\ddagger}$ are indicative of an associative mechanism of the oxygenation reaction, which is consistent with the unsaturated coordination sphere of the starting diiron(II) complexes $[\text{Fe}^{\text{II}}{}_2(\text{L})(\text{O}_2\text{CPh})]^{2+}$ (L = HPTP or Et-HPTB) and the accessibility of the iron centers in them.[83] Oxygenation of the more sterically constrained diiron(II)-HPTMP complex proceeds $10^2$–$10^3$ times slower in the temperature range studied because of a much higher enthalpy of activation ($\Delta H^{\ddagger} = 42\,\text{kJ/mol}$) only partially compensated by a less unfavorable entropy of activation ($\Delta S^{\ddagger} = -63\,\text{J/(mol K)}$), and displayed a more complicated kinetics consistent with reversible $\text{O}_2$ binding.[17,83] The reactions of the less sterically hindered complexes of HPTP and Et-HPTB appear to have highly ordered transition states with dominating bond formation as seen from the low enthalpies of activation and large negative entropy of activation (Table 4.6). A much higher value of $\Delta H^{\ddagger}$ and more favorable value of $\Delta S^{\ddagger}$ for the oxygenation of the diiron(II)-HPTMP complex suggests that this reaction involves a more significant degree of bond breaking between the iron(II) atoms and bridging ligands needed to give way to the incoming $\text{O}_2$ molecule.[83]

Oxygenation of the diiron(II) complexes with HPTP and two additional carboxylate ligands was directly compared to the reactivity of similar complexes with one carboxylate bridge (Figure 4.23).[89] The bis-carboxylate complex $[\text{Fe}^{\text{II}}{}_2(\mu\text{-HPTP})(\mu\text{-O}_2\text{CPh})_2]^+$ has two six-coordinated iron(II) centers in the solid state. In solution (MeCN or $\text{CH}_2\text{Cl}_2$) at low temperatures, the complex reacts with $\text{O}_2$ following the mixed second-order kinetics to give an adduct formulated as $[\text{Fe}^{\text{III}}{}_2(\mu\text{-HPTP})(\mu\text{-peroxo})(\eta^1\text{-O}_2\text{CPh})_2]^+$. Formation of the peroxo complex from the bis-carboxylate precursor is about an order of magnitude slower than the analogous reaction of the monocarboxylate due to a more negative entropy of activation, while the enthalpies of activation are nearly the same (Table 4.6). Therefore, the additional steric bulk in the bis-carboxylate complex and the requirement for a shift from bridging to terminal mode of the carboxylate coordination give rise to a more restricted and highly organized transition state of the dioxygen binding process.

Oxygenation of complexes $[\text{Fe}^{\text{II}}{}_2(\text{Et-HPTB})(\text{O}_2\text{CPh})]^{2+}$, $[\text{Fe}^{\text{II}}{}_2(\text{HPTP})(\text{O}_2\text{CPh})]^{2+}$, and $[\text{Fe}^{\text{II}}{}_2(\text{HPTP})(\text{O}_2\text{CPh})_2]^+$ was studied in several weakly coordinating solvents (MeCN, EtCN, $\text{CHCl}_3$, $\text{CH}_2\text{Cl}_2$, the MeCN/$\text{CH}_2\text{Cl}_2$ mixtures, and others) and only minor kinetic effects were observed, indicating a passive role of the solvents. There was a general tendency for somewhat higher reaction rates in solvents of higher polarity, in agreement with a polar transition state and the electron transfer during the coordination of the $\text{O}_2$ molecule. However, the oxygenation of $[\text{Fe}^{\text{II}}{}_2(\text{HPTP})(\text{O}_2\text{CPh})]^{2+}$ in the mixture MeCN/DMSO (9 : 1 v/v) was characterized by significantly different activation parameters, suggesting that a more strongly coordinating solvent (like DMSO) may change the speciation of the starting diiron(II) complex or otherwise interfere with the reaction.

TABLE 4.6  Kinetic Parameters for Reactions of Diiron(II) Complexes with Dioxygen[73]

| Complex | Solvent | $T$ range (°C) | $k$ (−40°C, $M^{-1}s^{-1}$) | $\Delta H^{\ddagger}$ (kJ/mol) | $\Delta S^{\ddagger}$ (J/(K mol)) |
|---|---|---|---|---|---|
| $[Fe_2(HPTP)(O_2CPh)]^{2+}$ | MeCN | −40 to 0 | 7300 | 15.8(4) | −101(10) |
| $[Fe_2(HPTP)(O_2CPh)]^{2+}$ | MeCN/DMSO (9:1 v/v) | −40 to 0 | 2800 | 8.0(3) | −143(10) |
| $[Fe_2(HPTP)(O_2CPh)]^{2+}$ | $CH_2Cl_2$ | −80 to 0 | 67 | 16.7(2) | −132(8) |
| $[Fe_2(dxlCO_2)_4(Py)_2]$ | $CH_2Cl_2$ | −80 to −30 | 215 | 4.7(5) | −178(10) |
| $[Fe_2(dxlCO_2)_4(MeIm)_2]$ | $CH_2Cl_2$ | −80 to −30 | 300 | 10.1(10) | −153(10) |
| $[Fe_2(dxlCO_2)_4(THF)_2]$ | $CH_2Cl_2$ | −80 to −30 | 3.20 | 14(1) | −135(10) |
| $[Fe_2(OH)_2(TLA)_2]^{2+}$ | $CH_2Cl_2$ | −80 to −40 | 0.67 | 17(2) | −175(10) |
| $[Fe_2(OH)_2(TLA)_2]^{2+}$ | MeCN | −40 to −5 | 1.94 | 16(2) | −167(10) |
| $[Fe_2(OH)_2(TPA)_2]^{2+}$ | $CH_2Cl_2$ | −40 to −15 | 12 | 30(4) | −94(10) |
| $[Fe_2(OH)_2(BQPA)_2]^{2+}$ | $CH_2Cl_2$ | −40 to −15 | 3.2 | 36(4) | −80(10) |
| $[Fe_2(OH)_2(BQPA)_2]^{2+}$ | $CH_2Cl_2/NEt_3$ | −70 to −40 | 2.6 | 36(4) | −81(10) |
| $[Fe_2(OH)_2(BnBQA)_2]^{2+}$ | MeCN | −65 to −25 | 2670 | 16(2) | −108(10) |
| $[Fe_2(OH)(OH_2)(TPA)_2]^{3+}$ | $CH_2Cl_2$ | −30 to +20 | 0.32 | 19(2) | −170(10) |

Mononuclear iron(II) complexes supported by aminopyridine ligands (TPA and its derivatives) self-assemble into bis-hydroxo-bridged diamond core-like structures under basic conditions. The donor atom sets in these diiron complexes are similar to the coordination environment created by dinucleating ligands such as HPTB. The differences in observed oxygenation rates, detailed below, can be attributed to significant steric hindrance about coordinatively saturated iron centers.[73]

Dinuclear complexes $[Fe^{II}_2(\mu\text{-}OH)_2(L)_2]^{2+}$ (L = TLA, TPA, BQPA, and BnBQA) (Figure 4.13) react with $O_2$.[17,73,88] In selected solvents, the oxygenation of complexes with TLA, BQPA, and BnBQA (but not TPA itself) at low temperatures gives diiron (III) peroxo species; the yield of these species could often be increased in the presence of organic bases. Under other conditions, the oxo-bridged iron(III) oligomers $[Fe^{III}_n(O)_n(L)_n]^{n+}$ (n = 2 or 3) formed.[90]

In all cases, the reaction was first order in both diiron(II) complex and in $O_2$ (second order overall) and characterized by a low enthalpy of activation and a strongly negative entropy of activation (Table 4.6), in agreement with expectations for an associative process.[90] Both water extrusion from the diiron-bis-hydroxo-core and the O–H bond breaking (hydrogen atom transfer or proton-coupled electron transfer) are postrate-limiting steps, because oxygenation rates did not depend on the concentration of $H_2O$ or $D_2O$.[91] A similar behavior was reported for oxygen binding to hemerythrin.[17,20]

Kinetic evidence points to $O_2$ coordination to one of the iron(II) centers in $Fe_2(OH)_2$ cores as being the rate-limiting step in the overall oxygenation process and implies the initial formation of a diiron-superoxo intermediate (Figure 4.24). Indeed, such an intermediate was recently observed in the oxygenation of $[Fe_2(OH)_2(TLA)_2]^{2+}$ at low temperature ($-80°C$),[78] and the activation parameters

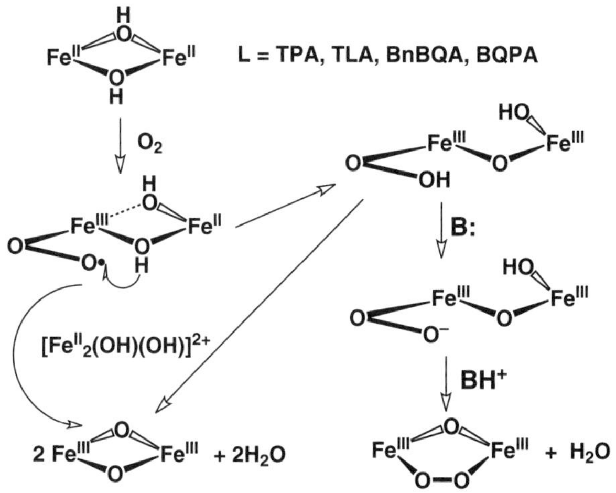

**FIGURE 4.24**  Mechanism of oxygenation of bis-hydroxo-bridged diiron(II) complexes with aminopyridine ligands.[88]

of its formation were essentially identical to the activation parameters of the formation of a more stable μ-1,2-peroxo intermediate as described above (Figure 4.24, Table 4.6).

Binding of the $O_2$ molecule to the diiron(II) core is limited by the rates of ligand substitution rather than by electron transfer rates. An outer-sphere one-electron transfer is unlikely, because the potential for the $O_2/O_2^-$ redox pair in organic solvents is about 1 V more negative than those determined for the $Fe^{II}Fe^{III}/Fe^{II}Fe^{II}$ redox pairs.[90] Furthermore, the oxygenation rates for a series of diiron(II) complexes (Table 4.6) do not correlate with their redox potentials.[90] Instead, the oxygenation rates correlate best with the length of the weakest Fe–donor atom bond in the diiron(II) complex, suggesting that dissociation of one (or more) of the donor atoms initially bound to iron(II) is involved in $O_2$ binding.

Extreme steric hindrance about iron coordination sites in complexes with TLA protects the peroxo intermediate from oxidative decomposition (a desirable effect) but simultaneously makes iron(II) centers much less accessible to even a small incoming $O_2$ molecule, resulting in slow oxygenation ($t_{1/2} \approx 30$ min at $-40°C$).[91] A decrease in steric bulk of the tetradentate ligands failed to significantly improve oxygenation rates, presumably because of a more "compact" and symmetric structure of the $Fe^{II}_2(OH)_2$ cores: the less negative activation entropy for these complexes was compensated for by the higher activation enthalpies (Table 4.6) required to break the shorter Fe–O(H) and/or Fe–N(py) bonds in the course of oxygen binding.

Since ligand substitution limits $O_2$ binding rates, coordinatively unsaturated iron (II) complexes featuring a vacant or labile site were predicted to undergo facile oxygenation. Indeed, a remarkable 1000-fold acceleration of dioxygen binding to $[Fe^{II}_2(OH)_2(BQPA)_2]^{2+}$ was accomplished by replacing one pyridine arm in the polydentate ligand BQPA with a noncoordinating benzyl group in BnBQA.[17]

The addition of a base (NEt$_3$ or another noncoordinating amine) does not influence the oxygenation rate, but in some cases (especially for $[Fe_2(OH)_2(BQPA)_2]^{2+}$) increases substantially the yield of peroxo intermediates, revealing the involvement of a proton-sensitive step that controls reactivity of the initial $O_2$ adduct (Figure 4.24). In the absence of base, the intermediate $O_2$ adduct can oxidize residual diiron(II) complex, yielding oxo-bridged diiron(III) products.[88] Alternatively, free base can deprotonate the hydroxo bridge in the diiron–$O_2$ adduct, promoting formation of a diiron(III)-peroxo species.[91] Electrophilic activation of the diiron(III)-peroxo species may be useful in substrate oxidations if competing, unproductive reactions with the starting Fe(II) species can be prevented.

Carboxylate ligands that can bind to the metal centers in a monodentate, bidentate, or bridging coordination modes were often incorporated in diiron complexes and sometimes altered their reactivity with dioxygen.[17] For example, diiron complexes with polyamine ligands and simple formate- or acetate-bridging ligands showed unusual third-order kinetics (second order in diiron complex and first order in $O_2$) (Figure 4.25).[87] This low rate agrees with the rate-limiting association of two molecules of the diiron(II) complex and a molecule of $O_2$. The transition state for the oxidation of diiron(II) complexes was proposed to contain a side-on peroxo bridge that is connected to both diiron units (Figure 4.25). Molecular mechanics calculations

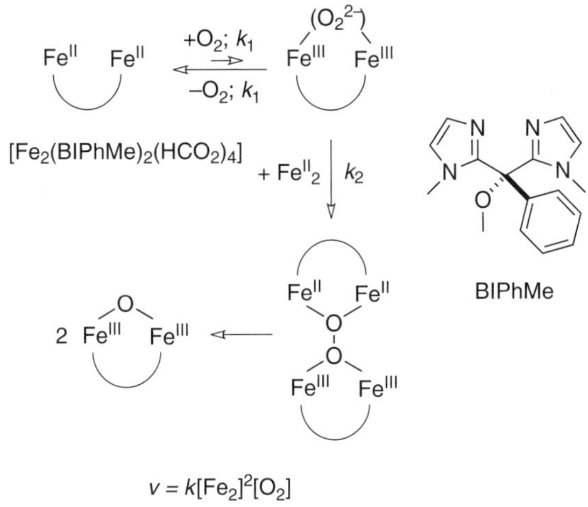

**FIGURE 4.25** Third-order oxygenation of carboxylate-containing diiron complex with PIPhMe.[87]

indicated that such tetranuclear intermediates (or transition states) are not highly strained. The first order with respect to dioxygen rules out intramolecular carboxylate rearrangements as a possible rate-limiting step, although carboxylate rearrangement may occur concurrently with $O_2$ binding. A small value of the activation enthalpy (33 kJ/mol) and a significant negative value of the activation entropy ($-39$ J/(mol K)) obtained for the oxygenation of $[Fe_2(BIPhMe)_2(HCO_2)_4]$ are consistent with rate-limiting oxygenation.

Dinucleating carboxylates support diiron(II) complexes that bind $O_2$ rapidly in a simple bimolecular process (first order in the complex and first order in $O_2$).[92] The activation parameters estimated for the oxygenation of the XDK–Im complex, approximately $\Delta H^{\ddagger} = 16$ kJ/mol and $\Delta S^{\ddagger} = -120$ J/(mol K) are in accord with the values obtained for other diiron(II) complexes and correspond to a low-barrier $O_2$ coordination at a vacant iron(II) center. A mechanism proposed for the oxygenation reaction includes the attack of the $O_2$ molecule on the unsaturated (five-coordinate) iron(II) center concomitant with the carboxylate shift at this center followed by the coordination to the second, six-coordinate iron(II) center and additional ligand rearrangement (Figure 4.26).

The rates of formation of diiron(III)-peroxo intermediates depended primarily on the nature of the nitrogenous base $L_N$. For $L_N = $ Im and $N$-alkylimidazoles, the reaction was very fast (complete within seconds at about $-70°C$) and required stopped-flow instrumentation, while for $L_N = $ Py, it was about five orders of magnitude slower and could be followed by conventional UV-Vis spectrophotometry. This effect was explained by higher electron-donating ability of imidazole compared to pyridine that increases the reducing power of the $Fe^{II}$ centers.[92] It is reasonable to assume, however, that the monodentate base effect is due to labilization of carboxylate ligands and/or to some structural changes in diiron(II)-carboxylate core.

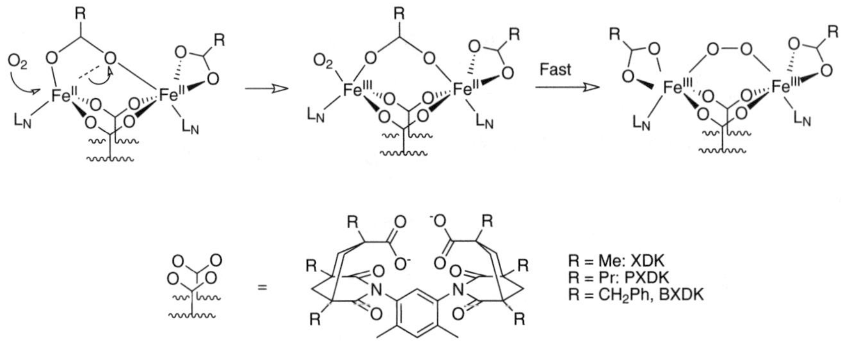

**FIGURE 4.26** Proposed mechanism for oxygenation of diiron(II) complexes with dinucleating dicarboxylate ligands.[92]

Sterically hindered monocarboxylates (e.g., $O_2CAr^{Tol}$, $O_2CAr^{Mes2}$, and DXL) (Figures 4.27 and 4.28) also proved to be very useful in modeling the active centers of nonheme diiron enzymes.[73,77,93] Oxygenation of the paddlewheel complexes $[Fe^{II}_2(DXL)_4(L)_2]$ (L = Py, MeIm, THF) at low temperatures in scrupulously dry noncoordinating solvents (e.g., $CH_2Cl_2$ or toluene) cleanly produced an intermediate containing coordinated $O_2$ (Figure 4.27).[73,90] Oxygenation of these coordinatively unsaturated complexes (both iron(II) centers are five coordinate) occurs rapidly and is complete within seconds even at $T = -80°C$. The dioxygen binding was found to be an irreversible second-order process (first order in diiron complex and first order in dioxygen), with low activation enthalpies and large negative activation entropies (Table 4.6). Capping ligands L (Py, MeIm, and THF) exerted little effect on the kinetic parameters of the oxygenation step, indicating that electron-donating properties of the monodentate ligands do not alter significantly the dioxygen binding rates. Furthermore, oxygenation of the three complexes with different capping ligands proceeded in an isokinetic regime, suggesting a similar rate-determining step for the series of complexes. An isokinetic temperature $T_{iso}$ for the series is seen as a common intersection point of the linear Eyring plot ($T_{iso} = 216 \pm 4$ K), as well as from the linear correlation between activation enthalpy and activation entropy (the slope of the linear fit gave $T_{iso} = 215 \pm 4$ K).[90]

A more sterically constrained complex $[Fe^{II}_2(O_2CAr^{Tol})_4(4\text{-}^tBupy)_2]$ (Figure 4.28) reacts with $O_2$ much slower ($t_{1/2} \approx 10$ min at $-78°C$) than $[Fe^{II}_2(DXL)_4(L)_2]$ (L = Py, MeIm, and THF) ($t_{1/2} \approx 10$ s at $-80°C$).[94] These observations support the importance of steric control in the oxygenation rates of dinuclear iron(II) complexes.[17] Interestingly, dioxygen adduct is not observed for diiron complexes with $O_2CAr^{Tol}$; instead, this elusive intermediate reacts, via a one-electron transfer, with another molecule of diiron(II) starting material, yielding an Fe(II)Fe(III) and an Fe(III)Fe(IV) species (Figure 4.28).[94] Mixed-valent Fe(III)Fe(IV) intermediate in this system is a competent oxidant for selected organic substrates (e.g., phenols).

An elegant approach was developed for investigating the role of structural rearrangements in diiron carboxylates.[95] Absorption bands of these complexes shifted

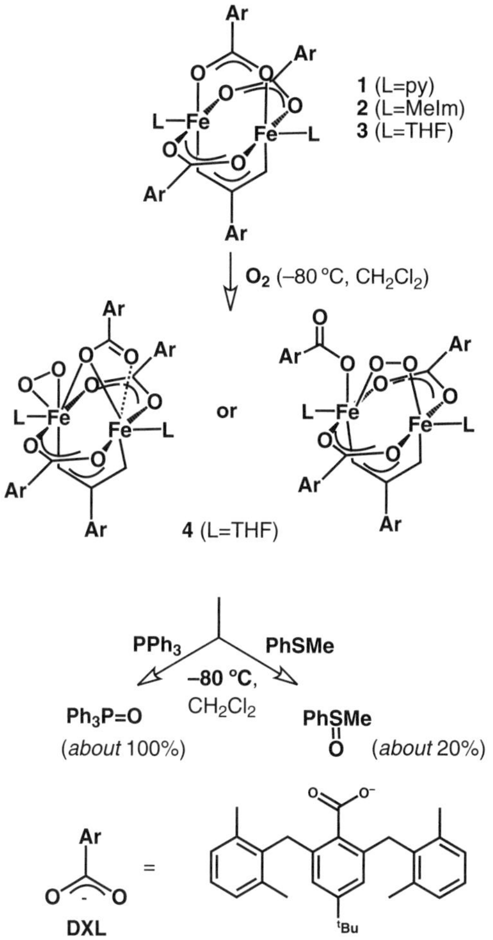

**FIGURE 4.27** Mechanism of oxygenation of diiron complexes with spherically hindered carboxylate ligand DXL.[90]

into the visible range by introducing electron-withdrawing substituents into the molecules of monodentate nitrogen heterocycles, thus allowing the direct observation of paddlewheel–windmill interconversions and subsequent reactions with $O_2$ (Figure 4.29). Iron centers are significantly more accessible to the incoming oxygen molecule in a windmill complex, and faster oxygenation is observed, as expected. Perhaps even more interestingly, addition of water molecules to a paddlewheel diiron (II) complex causes its structural rearrangement into a windmill geometry. As a result, dioxygen binding to a paddlewheel complex can be accelerated by a factor of 40 in the presence of water. The addition of the first water molecule to a paddlewheel complex is the rate-limiting step for the overall oxygenation reaction (Figure 4.29); the second water molecule rapidly adds to the second iron(II) center, and dioxygen molecule

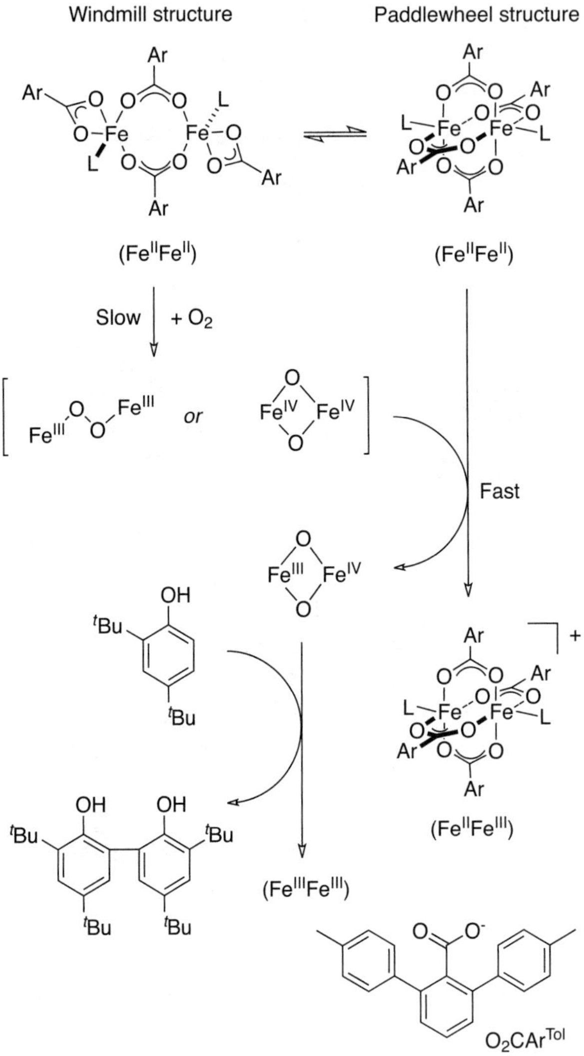

**FIGURE 4.28** Dioxygen activation by the diiron(II) complexes of the ligand $O_2CAr^{Tol}$ (e.g., $[Fe_2(O_2CAr^{Tol})_4(4\text{-}^tBuPy)_2])$.[94]

attacks sterically accessible sites at iron(II). Substitution of a labile water molecule occurs readily, leading to oxygenated products. These results support potential importance of water coordination in carboxylate-rich diiron enzymes.[95]

Dioxygen binding rates to diiron-carboxylate cores can be retarded and oxygenated intermediate can be trapped by embedding the metal complex into dendrimers.[96] A third-generation dendrimer assembled around a paddlewheel diiron complex with $O_2CAr^{Tol}$ reacts with $O_2$ 300 times slower than its nondendritic counterpart.

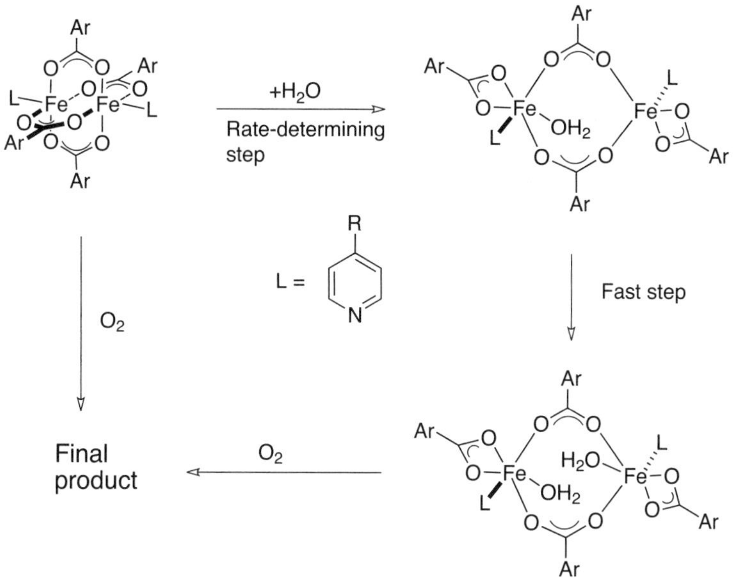

**FIGURE 4.29**  Proposed mechanism of water-accelerated dioxygen binding to carboxylato-bridged diiron(II) complexes.[95]

Furthermore, subsequent bimolecular reaction with the second molecule of diiron(II) complex is suppressed in dendrimers, thus allowing the direct observation and spectroscopic characterization of the Fe(II)Fe(III) intermediate (presumably, a superoxo complex).[96] Oxygen-activating metallodendrimers open a new way toward designing artificial analogues of metalloenzymes.

In summary, dioxygen binding to diiron(II,II) complexes is an inner-sphere process that is usually limited by steric effects. Oxygenation at a vacant site is rapid and low-barrier; oxygenation at six-coordinate iron(II) sites is limited by ligand substitution.

### 4.4.3  Dioxygen Binding to Dicopper Complexes: O–O Bond Breaking That Yields Bis-μ-Oxo Species

Rich dioxygen chemistry of biomimetic copper complexes provides important insights into factors that govern the chemical structure of dicopper–oxygen intermediates and the mechanisms of their formation and reactivity.[40,41,97–100] Similar to other low-valent transition metals, copper(I) reacts with dioxygen in a stepwise manner. The initially formed mononuclear Cu–$O_2$ adduct (end-on or side-on Cu(II)-superoxo or copper(III)-peroxo) reacts with another molecule of copper(I) complex, affording dinuclear $Cu_2(O_2)$ species (in certain cases, trinuclear or tetranuclear adducts form instead). The structures of these dioxygen complexes, some of which were crystallographically characterized, are dictated by the chemical nature of the ligands coordinated to copper centers.

Tetradentate N-donor ligands tend to support end-on, *trans*-μ-1,2-peroxo dicopper (II) complexes.[41] The first crystal structure of a $Cu_2(O_2)$ adduct, obtained by Karlin and coworkers, showed this binding mode in a complex supported by tris-picolylamine ligand (TPA) (Figure 4.13).[101,102] The end-on structure of dioxygen adduct is determined by coordination preferences of Cu(II). This $d^9$ metal ion usually forms five-coordinate complexes; when four sites are occupied by nitrogen donors from the tetradentate ligand, one site remains available for dioxygen binding.[41,97]

Tridentate and bidentate ligands leave at least two coordination sites at each metal available for interaction with dioxygen. As a result, bridging side-on $O_2$ binding becomes dominant in dicopper complexes supported by these ligands. The first example of a crystallographically characterized side-on dicopper–$O_2$ complex, reported by Kitajima and coworkers, took advantage of tris-pyrazolylborate ligands.[103,104] Remarkably, spectroscopic properties of this model complex were extremely similar to those of oxyhemocyanin, thus allowing a correct prediction of the side-on structure of the protein long before it was crystallized. While dicopper–dioxygen core was essentially planar in the first hemocyanine models,[103,104] various degrees of distortion were observed later[97]; a recent example of a butterfly core contains a sterically constrained bidentate ligand sparteine at each copper and a bridging carboxylate.[105]

The O–O bond in side-on dicopper–dioxygen complexes can be easily cleaved, affording dicopper(III) bis-μ-oxo diamond core-type structures.[40,85,106] The first crystallographically crystallized complex of this type, described by Tolman and coworkers, contained N-alkylated triazacyclononane ligands.[106] Subsequently, a large number of bis-μ-oxo dicopper(III) intermediates were observed in various chemical systems.[97] In many cases, dicopper(II) $\mu$-$\eta^2$:$\eta^2$ peroxo species and dicopper (III) bis-μ-oxo species easily interconvert (Figure 4.30).[40]

The mechanisms of oxygenation of mononuclear copper(I) complexes were investigated in detail by stopped-flow methods, revealing three main kinetic scenarios for a two-step process.[40,99] In some cases, the rates of the first step (reversible formation of a 1 : 1 $O_2$ adduct) and the second step (interaction between $Cu^{II}(O_2^{-\bullet})$ and another molecule of the $Cu^I$ starting material) are comparable.[99] This fortunate situation was initially observed for the complexes forming end-on-bridged peroxides.[99] Global analysis of the time-resolved optical spectra, facilitated by significant difference in the absorption maxima of superoxo and peroxo intermediates, allows accurate determination of all the four rate constants in this case. For nonsterically hindered ligands, such as 4-substituted TPA derivatives (Figure 4.13), the oxygenation rates at both reaction steps seem to be limited by ligand substitution (displacement of a coordinated solvent molecule with an incoming $O_2$ or $Cu^{II}$-$O_2^{-\bullet}$); relatively small electronic effects were found for both steps.[107] Furthermore, the rate of the formation of peroxo complex is nearly identical for TPA and for N-alkylated TREN, although the latter is believed to be a more electron-donating ligand.[108] However, the equilibrium constants are higher for oxygen binding to $Cu(Me_6tren)^+$: $K_1$ is higher by two–four orders of magnitude. Interestingly, the values of $K_2$ only differ by a factor of 5, and they are slightly higher for TPA. As is often the case for oxygenation

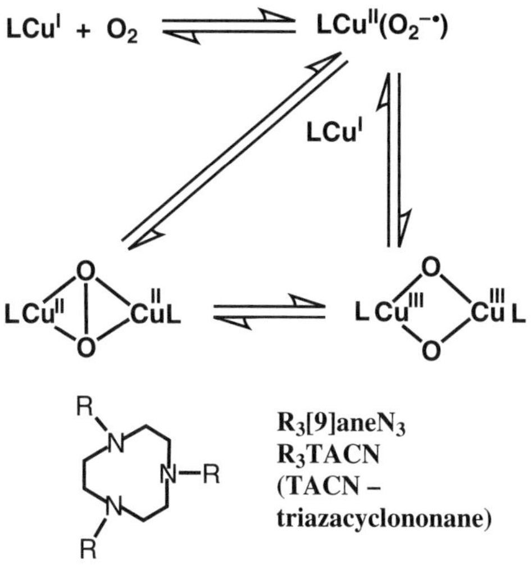

$$LCu^I + O_2 \rightleftharpoons LCu^{II}(O_2^{-\bullet})$$

$LCu^I$

$$LCu^{II} \overset{O}{\underset{O}{<>}} Cu^{II} L \rightleftharpoons L Cu^{III} \overset{O}{\underset{O}{<>}} Cu^{III} L$$

$R_3[9]aneN_3$
$R_3TACN$
(TACN –
triazacyclononane)

**FIGURE 4.30**  Formation of dicopper–oxygen complexes and interconversions of dicopper (II) $\eta^2$ peroxo and dicopper(III) bis-$\mu$-oxo species.[40]

equilibria, electronic effects are reflected primarily in dioxygen dissociation rates. As expected, sterically hindered ligands slow down or even completely suppress the formation of dinuclear copper(II)-peroxo species.[40]

Formation of dicopper–$O_2$ complexes is usually entropically unfavorable. One possible solution of this problem is to covalently link two copper binding sites, thus preorganizing the complex for dioxygen coordination.[41,99] For this "preorganization" to be successful, the geometry of the dicopper(I,I) complex has to match the geometry of the resulting dioxygen adduct. Flexible linkers, such as polymethylene chains of varying length, allow significant adjustments of geometry upon oxygenation. However, entropy gains with such nonrigid complexes are minimal and are often compensated by unfavorable reorganization enthalpy.[41] These trends were demonstrated for a series of aminopyridine ligands connected by $(CH_2)_n$ spacers.[99] Interestingly, bimolecular (in dicopper complex) oxygenation competes with monomolecular oxygenation in these flexible systems: the initially formed $Cu^{II}-O_2^{-\bullet}$ adduct can attack the second copper(I) site within the second compartment of the ligand, or another copper(I) site in a different molecule of a dicopper(I) complex. Nonideal geometry of dicopper–$O_2$ adducts supported by these dinucleating ligands often resulted in nonplanar, "butterfly" geometry of $Cu_2$–$O_2$ moieties.[41]

When rigid aromatic spacers are used instead of flexible aliphatic tethers, very high rates of dicopper(II) peroxide formation can be achieved.[41,99] For example, *meta*-xylene containing dicompartmental ligands were effective in preorganizing two copper ions for dioxygen binding (Figure 4.31). Even more interesting was the finding of intramolecular regioselective aromatic hydroxylation of these xylene-bridged complexes with coordinated $O_2$. Furthermore, independently prepared

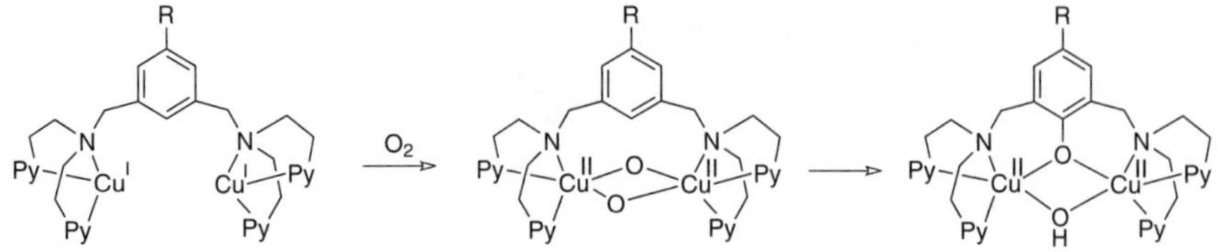

**FIGURE 4.31** Rapid oxygenation of xylene-bridged dicopper(I) complexes and intramolecular aromatic hydroxylation of the resulting side-on peroxo adduct.[41]

phenoxy-bridged dicopper(I) complexes reacted with $O_2$ extremely rapidly (the reaction rates were too fast to be accurately measured by the stopped-flow method even at 173 K). The dioxygen molecule is coordinated nonsymmetrically in the resulting adduct.[41]

In many cases, dicopper–$O_2$ complexes formed without observable intermediates. This behavior can be attributed to one of the two limiting scenarios: (1) the first step, formation of a 1 : 1 Cu–$O_2$ adduct, is rate limiting, while all subsequent reaction steps are fast; (2) a 1 : 1 Cu–$O_2$ adduct forms in a rapid, left-lying equilibrium, and subsequent formation of a $Cu_2(O_2)$ complex is rate limiting.[40] The first situation results in a simple mixed second-order rate law (first order in copper(I) complex and first order in dioxygen); the activation parameters reflect coordination of $O_2$ to a single copper(I) site (usually, relatively small activation enthalpies and negative activation entropies are observed). A copper(I) complex with tris-$N$-isopropyl-triazacyclononane serves as a good example of this kinetic behavior.[40]

The second situation can be identified by an overall third-order kinetics of oxygenation (second order in copper complex and first order in $O_2$).[40]

The rate constant for the overall process, $k$, can be expressed as follows:

$$k = K_1 k_2 \tag{4.28}$$

The effective activation parameters reflect the contributions of the reaction enthalpy of the $Cu^{II}(O_2^{-\bullet})$ formation and of the activation enthalpy of the subsequent step, reaction between $Cu^{II}(O_2^{-\bullet})$ with $Cu^I$ yielding $Cu_2(O_2)$.:

$$\Delta H^{\ddagger} = \Delta H_1^{\circ} + \Delta H_2^{\ddagger} \tag{4.29}$$

Because of the composite nature of the observed activation parameters, negative values of effective activation enthalpy are possible, and were indeed observed, for example, for the oxygenation of copper(I) complexes with 6-PhTPA[109] and some other TPA derivatives.[107]

A striking feature of dicopper–oxygen systems is their ability to undergo rapid interconversion between a side-on peroxo ($Cu^{II}$) and a bis(μ-oxo) ($Cu^{III}$) isomeric forms (Figure 4.30).[40,41,97] These two forms are close in energy, and often coexist at equilibrium that in many cases can be easily shifted by changing reaction conditions (solvents, counterions, or temperature). Dicopper(III) bis-oxo form is enthalpically stabilized but entropically destabilized. For example, for the original system containing *tris-N*-isopropyltriazacyclononane, $\Delta H° = -3.8$ kJ/mol and $\Delta S° = -25$ J/(K mol) for the equilibrium shown in Figure 4.30.[110]

Relative stability of the bis($\mu$-oxo)dicopper(III) complexes depends on the structure of ligands (denticity, steric bulk, and electron-donating ability). Bidentate ligands tend to stabilize bis($\mu$-oxo)dicopper(III) cores[97,111]; this trend can be attributed to the preference of Cu(III), a $d^8$ metal ion, to adopt a square planar coordination geometry.[97] Nevertheless, a large number of tridentate ligands also form bis($\mu$-oxo) dicopper(III) species. DFT calculations showed that dicopper(III) bis-oxo structure is inherently more stable than dicopper(II) side-on peroxo structure. However, many side-on peroxo complexes are also known for dicopper systems. Steric hindrance destabilizes a more compact dicopper(III) diamond core complexes, which have a Cu–Cu separation of about 2.8 Å (as opposed to about 3.6 Å for side-on peroxo-bridged compounds).[40,41,97] Therefore, excessive steric bulk should be avoided to obtain dicopper(III) complexes and intermediates. Electronic effects can be somewhat useful in stabilizing dicopper(III) oxo-bridged species: electron-donating substituents would favor high oxidation state of the metal. However, coordinating anions generally favor dicopper(II) peroxides, since binding of an additional ligand increases the coordination number of the metal (thus destabilizing compact dicopper (III) diamond cores).

With very few exceptions, side-on peroxo bis-$\mu$-oxo rearrangement occurs very rapidly, so $O_2$ binding to copper(I) complexes affords the most stable isomer (or their mixture in cases of their similar stabilities). When two isomers are in a rapid equilibrium, they appear to form from Cu(I) precursors and $O_2$ with equal rates. When dicopper(III) bis-$\mu$-oxo form is favored, it forms directly from two molecules of copper(I) complex and a molecule of $O_2$ (or, in some cases, from one molecule of dicopper(I) complex with a dicompartmental ligand and a molecule of $O_2$). As usual, oxygenation rates are fast, even at low temperatures, and activation barriers are low.[40] The overall dioxygen binding process can be described by Equation 4.30:

$$2(L)Cu^{I} + O_2 \rightarrow (L)Cu^{III}(\mu\text{-}O)_2Cu^{III}(L) \qquad (4.30)$$

The net result of this reaction is a complete cleavage of O–O bond accompanied by an increase of the oxidation state of each copper center by 2. This O–O bond breaking occurs seamlessly and often reversibly. Dicopper systems offer fascinating opportunity for dioxygen activation via O–O bond breaking; resulting copper(III) species can act as competent oxidants for a variety of substrates.[40]

### 4.4.4 Toward Four-Electron Dioxygen Reduction

Dioxygen reduction with dicopper(I) complexes that affords dicopper(III) bis-$\mu$-oxo species (Equation 4.30) can be considered formally as a four-electron process. The four-electron dioxygen reduction with bimetallic systems affording high-valent diamond core products or intermediates is also known for transition metals other than copper. For example, mononuclear Ni(I) complexes with tri- or tetradentate ligands react with $O_2$ and form dinickel(III) bis-$\mu$-oxo species.[51] Unlike dicopper systems, dinickel(III)-oxo diamond cores are thermodynamically favored over the side-on peroxo isomers, and no equilibrium between the two isomeric forms could be observed. While several oxygenation intermediates derived from Ni(I) complexes have been described, kinetics of these reactions were not reported.[51] A similar four-electron reaction with $O_2$ occurs for manganese complexes[14]; however, the "low" oxidation state of the metal is now Mn(II), and the "high," $(n + 2)$ oxidation state of the metal is Mn(IV):

$$2Mn(II) + O_2 \rightarrow Mn^{IV}(\mu\text{-}O)_2Mn^{IV} \tag{4.31}$$

Strong electron-donating ligands are used to shift the redox potentials of the $Mn^{III}/Mn^{II}$ and $Mn^{IV}/Mn^{III}$ couples, thus dramatically increasing susceptibility of the Mn(II) complexes to oxidations. Diamond core structure of the final product cannot be accommodated by rigid, planar ligands such as porphyrins or phthalocyanins, but is readily stabilized by relatively flexible Schiff bases derived from substituted salicylaldehyde and 1,3-diaminopropane (SALPRN).[112] While it is clear that the formation of dimanganese(IV) bis-oxo diamond core is a multistep process, the exact mechanism of this reaction was not established.

In biology, four-electron $O_2$ reduction can be efficiently catalyzed by polynuclear centers in certain well-known metalloproteins: cytochrome c oxidase and multi-copper oxidases.[113] A reverse process, four-electron water oxidation, can be carried out by manganese clusters of photosystem II or by multicopper oxidases.[114,115] While water oxidation can hardly be viewed as dioxygen activation, certain mechanistic features can be shared by these reverse reactions. Interestingly, biomimetic hetero-metallic complexes containing a heme iron center and a copper center were designed to model the active site of cyctochrome c oxidase.[116-118] Initial steps of dioxygen binding to these complexes is rather similar to $O_2$ coordination to individual components (copper(I) sites of iron(II)-porphyrin sites). Subsequent intermediates, however, differ from heme-only or copper-only systems and include asymmetric $\mu$-$\eta^1$:$\eta^2$ peroxo species[116] and Fe(IV)-$Cu^{II}$(OH) species.[119] Electrocatalysis of water oxidation was recently observed with these heterometallic complexes.[120]

## 4.5 REACTIONS OF METAL–OXYGEN INTERMEDIATES

### 4.5.1 Oxygen Activation with Metalloenzymes

Selective oxidation of organic substrates with dioxygen is efficiently catalyzed by metalloenzymes. Although full treatment of this topic is well beyond the scope of this chapter, a glimpse into enzymatic oxygen activation provides important insights into the role of possible reactive intermediates and illustrates the challenges of regulating multistep dioxygen reduction. Cytochromes P450 provide a textbook example of oxygen atom transfer with heme monooxygenases (often referred to as "the heme

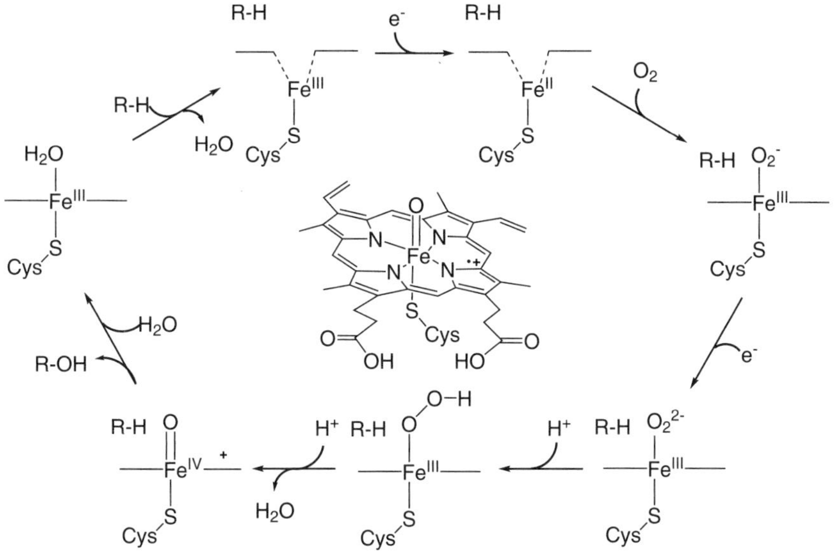

**FIGURE 4.32** Mechanism of substrate oxidation with dioxygen catalyzed by cytochrome P450.[58,124]

paradigm"), although remaining controversial mechanistic issues are still being actively debated.[58,121]

The consensus mechanism depicted in Figure 4.32 includes one-electron reduction of the resting state, $(P)Fe^{III}$, followed by $O_2$ binding to a vacant site at $(P)Fe^{II}$ affording an Fe(III)-superoxo adduct, $(P)Fe^{III}(O_2^{-\bullet})$. This dioxygen complex is converted into an iron(III) hydroperoxide via further one-electron reduction and protonation. Electrons are supplied by NADPH and transferred via reductase-mediated electron transfer pathway. The same $(P)Fe^{III}(OOH)$ intermediate can be accessed by the reaction of $(P)Fe^{III}$ with hydrogen peroxide ("peroxide shunt" pathway). Proton-assisted O–O bond heterolysis generates a highly reactive, formally iron(V)-oxo $(PFe^V=O \leftrightarrow (P^{+\bullet})Fe^{IV}=O)$ intermediate that rapidly hydroxylates the molecule of substrate bound in the enzyme pocket, closing the catalytic cycle and regenerating ferric state of the enzyme. Notably, two electron transfer events as well as proton transfers are involved in dioxygen activation with cytochrome P450.

The ferryl–porphyrin radical intermediate in P450 chemistry was postulated by analogy with peroxidase and catalase Compound I (CpdI). Until recently,[122] no direct experimental observations of this very unstable intermediate were reported. The proposed mechanism of alkane hydroxylation with CpdI intermediate of cytochrome P450 includes a hydrogen atom abstraction followed by an oxygen rebound.[121] Although other intermediates in P450 cycle, such as Fe(III)-OOH and Fe(IV)=O, are sometimes also proposed as reactive species,[123] their oxidizing ability appears to be low.[124]

The electronic structure of hemes is well suited for the stabilization of high-valent intermediates. Cytochrome P450 and related heme enzymes take advantage of this

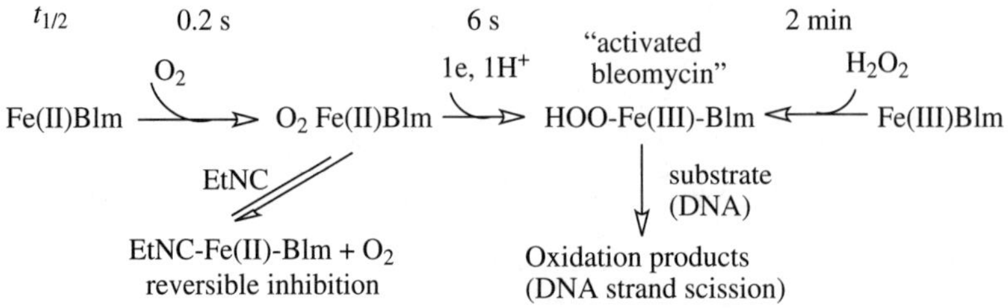

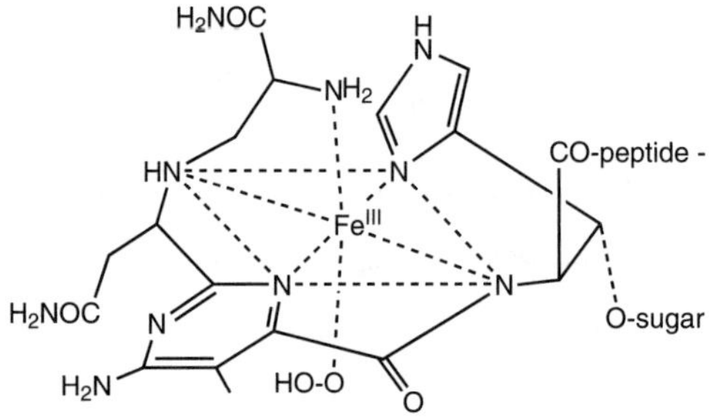

Iron coordination sphere in "activated bleomycin"

**FIGURE 4.33**    Mechanism of substrate oxidations in the presence of iron bleomycin, Fe (Blm).[17,126]

feature and utilize an Fe(V) → Fe(III) transformation for two-electron substrate oxidations. Of course, an additional electron has to be supplied to regenerate oxygen-sensitive Fe(II) and to close the catalytic cycle. In the absence of a porphyrin ligand, different intermediates may carry out catalytic oxidations. For example, a mononuclear system, iron bleomycin (Fe-Blm), causes oxidative DNA cleavage, which is considered to be responsible for the anticancer activity of this drug.[23,125,126] The sequence of events at the iron center in Fe-Blm leading to DNA cleavage (Figure 4.33) includes $O_2$ binding to the Fe(II) center, one-electron reduction of the dioxygen adduct leading to the formation of a so called "activated bleomycin," and an interaction of "activated bleomycin" with the substrate (in case of DNA cleavage, the 4'-hydrogen abstraction from the sugar occurs). Another one-electron reduction regenerates Fe(II); the exact source of electrons in biological oxygen activation with iron bleomycin was not identified. The peroxide activation of Fe(III)Blm (peroxide shunt) is also possible. The "activated bleomycin," $Fe^{III}(OOH)$, may undergo homo- or heterolytical O–O bond cleavage, or it may attack its substrate directly. Spectroscopic and kinetic evidence in favor of a direct reaction of DNA with "activated bleomycin" was recently reported.[127]

Many mononuclear nonheme iron oxygenases require a reducing cofactor (pterin or alpha-keto acid) for dioxygen activation.[23,128] These enzymes utilize Fe(IV) → Fe (II) reduction by two-electron substrates. A simplified catalytic cycle for 2-keto-glutarate-dependent enzymes is shown in Figure 4.34. Keto-glutarate cofactors assist

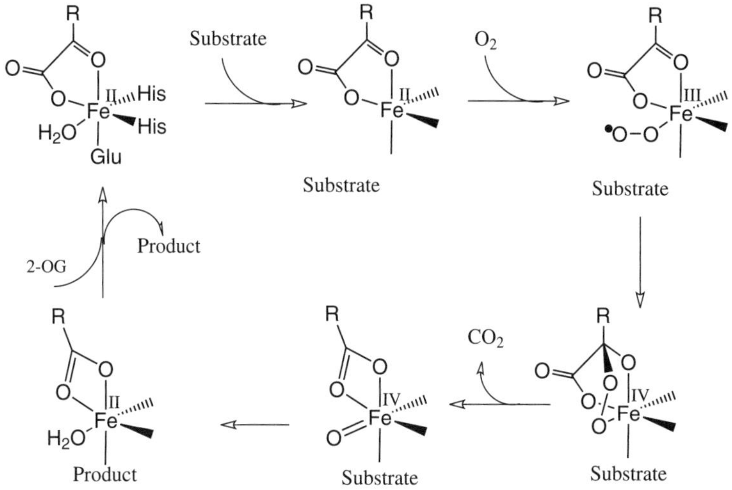

**FIGURE 4.34**    Substrate oxidation catalyzed by 2-keto-glutarate-dependent enzymes.[128]

the iron(II) oxidation process by donating electron density to the metal center and displacing water ligands to allow $O_2$ binding at the remaining vacant or labile site. They also provide another electrophilic center for $O_2$ binding and undergo a formally two-electron oxidation with concomitant formation of Fe(IV). Succinate product binds as a monodentate ligand and can be easily replaced by a bidentate keto-glutarate cofactor in the next catalytic cycle.[23,128]

The dinuclear systems, such as extensively studied methane monooxygenase (MMO) that utilizes a $Fe_2^{IV}/Fe_2^{III}$ redox couple for a formally two-electron oxidation of methane into methanol with an intermediate Q of its hydroxylase component, take advantage of both iron centers in catalysis.[76] Diiron(II) form of MMO reacts with $O_2$ yielding a peroxo-diiron(III) species P as the first spectroscopically observable intermediate (Figure 4.35). Although intermediate P itself does not react with a

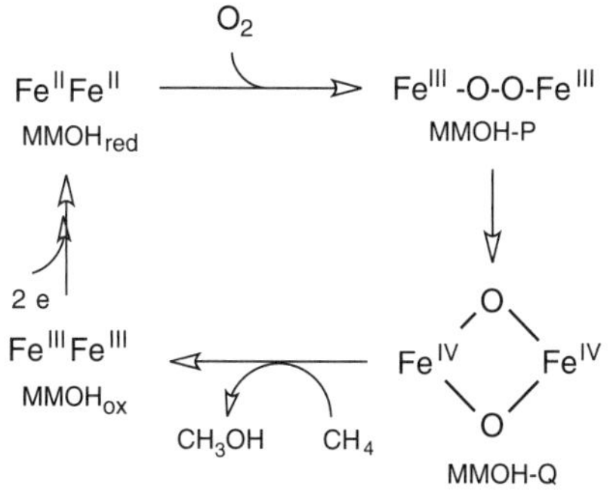

**FIGURE 4.35**    Mechanistic scheme for substrate hydroxylation catalyzed by soluble methane monooxygenase (hydroxylase component) (sMMOH).[17,76]

native substrate of MMO (methane), there is some evidence of its ability to transfer an oxygen atom to olefins. Intermediate P undergoes further transformation, via proton-induced O–O bond cleavage, into a high-valent diiron(IV) species, which most likely has a $Fe_2^{IV}(\mu\text{-O})_2$ core. This high-valent intermediate, termed intermediate Q, is a kinetically competent oxidant for methane (native substrate) and other aliphatic substrates. Substrate hydroxylation yields diiron(III) form of the enzyme. The catalytic cycle is closed by injecting two electrons from NADH, which is mediated by MMO reductase.

A brief description of several catalytic cycles for oxygen-activating enzymes does not even start unraveling the beauty and complexity of metalloprotein structures, reactivities, and their interplay. A closer look at these mechanisms reveals some common features that are sometimes overlooked in designing synthetic catalysts. Initial dioxygen binding to a low-valent metal, which occurs readily at a vacant coordination site, initiates a sequence of reactions generating several metal–oxygen intermediates. Superoxo, (hydro)peroxo, and high-valent metal-oxo intermediates were observed and multiple intermediates may be involved in the oxidation of native and nonnative substrates.[123] Generating reactive high-valent metal-oxo intermediates and/or closing the catalytic cycle by regenerating oxygen-sensitive low-valent metal centers require additional electrons and protons. In other words, oxygenases require sacrificial reducing agents. Temporal and spatial control of electron and proton delivery is really challenging in synthetic systems that would mimic oxygenase catalytic activity.

### 4.5.2    Generating Metal–Oxygen Intermediates in Synthetic Systems

As discussed in previous sections, direct reactions of low-valent metal compounds with $O_2$ afford various metal–oxygen intermediates, including superoxo, peroxo, and high-valent metal-oxo species. However, with the exception of superoxides and dinuclear peroxides, alternative methods of generating these intermediates are more generally applicable, and often produce better results.[15,16]

Metal superoxides, $(L)M^{(n+1)}(O_2^{-\bullet})$, are most commonly prepared from $(L)M^{n+}$ and $O_2$. These dioxygen adducts are rarely stable, and normally undergo subsequent reactions (formation of $\mu$-peroxo dimers is the most common decomposition pathway). Alternatively, direct reactions with inorganic superoxide (e.g., $KO_2$) may yield superoxo complexes:

$$(L)M^{(n+1)} + O_2^- \rightarrow (L)M^{(n+1)}(O_2^{-\bullet}) \tag{4.32}$$

If low-valent $M^{n+}$ compounds are used, peroxides form instead (see below). Most transition metal ions catalyze, to some degree, catalytic decomposition of superoxides (display superoxide dismutase activity), thus limiting applicability of reaction (4.32) for the synthesis or *in situ* generation of metal superoxo complexes.

Metal peroxides can be prepared directly from the low-valent metal complexes and $O_2$ in some cases, including (1) mononuclear precursors with the metal ions that easily undergo two-electron oxidation; complexes of Mn(II) are good examples[14]:

$(L)Mn^{2+} + O_2 \rightarrow (L)Mn^{IV}(O_2^{2-})$; and (2) mono- or dinuclear precursors that yield dinuclear peroxides with both metal ions in $(n+1)$ oxidation state: cobalt(II), iron(II), and copper(I) complexes offer numerous examples. Another highly popular approach to the synthesis of peroxo complexes is a direct reaction of $(L)M^{(n+1)}$ with hydrogen peroxide. This method works well for medium- and high-oxidation state metals. A variation of the method starts with a low-oxidation state metal complex and generates a higher oxidation state peroxide; $H_2O_2$ acts both as an oxidant and as a ligand in this case. A good example is provided by iron chemistry:

$$(L)Fe^{2+} + H_2O_2 \rightarrow (L)Fe^{3+} + H_2O \tag{4.33}$$

$$(L)Fe^{3+} + H_2O_2 \rightarrow (L)Fe^{III}(OOH) \tag{4.34}$$

Deprotonation of coordinated hydroperoxides would yield peroxides.

Hydroperoxo complexes can also be obtained by a one-electron reduction of the superoxo intermediates; $[Ru(NH_3)_6]^{2+}$ is one of the most convenient reducing agents for this chemistry.[15] An interesting way of preparing hydroperoxides via dioxygen activation utilizes a formal $O_2$ insertion into M–H bond in metal hydrides; examples are known for platinum, rhodium, and cobalt complexes, to name just a few.[14,15]

High-valent metal-oxo species, $M^{(n+2)}=O$, are perhaps the most interesting for oxidizing organic substrates. In certain cases, they can be accessed directly from $M^{n+}$ and $O_2$. For example, aqueous $Cr^{2+}$ reacts with $O_2$, affording $Cr_{aq}O^{2+}$[15]; other examples from iron chemistry and ruthenium chemistry have been introduced in Section 4.3.3. If ozone, $O_3$, is used in place of $O_2$, aqueous $Fe^{IV}=O$ can be cleanly generated.[71] For dinuclear complexes, reactions with $O_2$ may directly generate high-valent species (examples for $Cu^ICu^I$ complexes affording bis-$\mu$-oxo $Cu^{III}(\mu\text{-}O)_2Cu^{III}$ have been provided in Section 4.4.3). For the majority of transition metal complexes, however, the best way of obtaining high-valent metal-oxo species is based on reactions between $(L)M^{n+}$ and various oxygen atom donors (amine N-oxides, $R_3NO$; iodosobenzene and its better soluble substituted analogues, ArIO; sodium periodate, $NaIO_4$; peroxoacids, $RC(O)OOH$, etc.). This chemistry, which was initially developed for metalloporphyrins, is also applicable to nonporphyrin transition metal complexes. While the formal oxidation state of the metal ion increases by 2 as a result of an oxygen atom transfer, the initial and final oxidation states depend on the electronic properties of the ligand. Strong electron-donating porphyrinates allow the Fe(III) $\rightarrow$ Fe(V) oxidation, but neutral aminopyridine ligands support Fe(II) $\rightarrow$ Fe(IV) oxidations[57,58]:

$$(P)Fe^{III} + ArC(O)OOH \rightarrow (P)Fe^V = O \rightleftharpoons (P^{+\bullet})Fe^{IV} = O + ArCOOH \tag{4.35}$$

$$(TMC)Fe^{II} + ArC(O)OOH \rightarrow (TMC)Fe^{IV} = O + ArCOOH \tag{4.36}$$

Formally, one-electron oxidations of $M^{(n+1)}$ compounds can also yield high-valent metal-oxo intermediates, $M^{(n+2)}=O$. For example, O–O bond homolysis of iron(III) hydroperoxides or alkylperoxides supported by aminopyridine ligands results in the formation of $Fe^{IV}=O$ species[57]:

$$(L)Fe^{III}(OOR) \rightarrow (L)Fe^{IV} = O + RO^\bullet \qquad (4.37)$$

Similar reactions of dinuclear complexes sometimes give rise to mixed-valent intermediates, for example, diamond core $Fe^{III}(\mu\text{-}O)_2Fe^{IV}$ species obtained from diiron(III) TPA complexes and $H_2O_2$.[85]

### 4.5.3  Reactivity of Synthetic Superoxo, Peroxo, and Metal-Oxo Complexes

Metal–oxygen intermediates react with inorganic or organic substrates via various reaction pathways, such as oxygen atom transfer, hydrogen atom transfer, hydride transfer, electron transfer, proton-coupled electron transfer, free radical reactions, and others.[14–16] The preferential reactivity pathways depend on the nature and oxidation state of the metal, the nuclearity of the complex, and the coordination mode and protonation state of coordinated oxygen-derived ligand(s).

Superoxo complexes can be accessed directly from $M^{n+}$ and $O_2$. However, these dioxygen adducts are usually rather sluggish oxidants. Typical superoxides of mononuclear biological and biomimetic oxygen carriers (myoglobin, coboglobin, cobalt(II) macrocycles, etc.) may undergo radical coupling reactions (e.g., form peroxonitrites in reactions with NO)[15]

$$M^{(n+1)}(O_2^{-\bullet}) + NO \rightarrow M^{n+1}(OON(O))^- \qquad (4.38)$$

and sometimes oxidize organic compounds with weak O–H bonds, such as phenols.[15,37,38] Detailed mechanistic information for these types of reactions was obtained for relatively stable model systems, including superoxo complexes of chromium and rhodium.[14–16] For example, a hydrogen atom abstraction pathway was unambiguously shown for the reactions of $Cr_{aq}OO^{2+}$ with rhodium hydrides; these reactions were thermodynamically favorable and were characterized by a fairly large H/D kinetic isotope effect (in some cases, $k_{RhH}/k_{RhD} \approx 7$ was observed, suggesting possible role for tunneling).[15] Hydrogen atom abstraction also dominates the oxidation of 2,4,6-trialkylphenols with $Cr_{aq}OO^{2+}$.

The reactivity of metal-superoxo complexes depends on the coordination mode of the $O_2^{-\bullet}$ ligand, as can be illustrated by copper–oxygen chemistry. The end-on copper(II) superoxides, such as $[Cu^{II}(TMG)_3TREN(\eta^1\text{-}O^{-\bullet})]$ (Figure 4.13),[42] readily oxidize externally added phenols (thus resembling other metal-superoxo intermediates described above) and promote more challenging intramolecular oxidations (e.g., aliphatic hydroxylation of the $CH_3$ group of the coordinated tripodal ligand in the presence of H-atom donor, TEMPOH).[129] In contrast, side-on superoxides supported by ketimidate ligands (Figure 4.15) were essentially unreactive with respect to reducing substrates: no phenol oxidation was observed and triphenylphosphine did not form phosphine oxide, but instead displaced the $O_2$ ligand.[53] Strong electron-donating ability of the ketimidate ligands also contributes to the lack of oxidizing activity of their Cu(II)-$\eta^2$ superoxo complexes. In the future, it would be interesting to compare complexes with identical (or at least similar) polydentate ligands that differ only in the coordination modes of the metal-bound superoxide.

Rare examples of dinuclear metal complexes with bound superoxide are known; their reactivity is similar to the reactivity of mononuclear superoxo complexes. A recent study on oxygenation of hydroxo-bridged diiron complex with (6-Me)$_3$TPA (TLA) (Figure 4.13) detected a relatively unstable intermediate that was identified by resonance Raman spectroscopy as an Fe(II)Fe(III) complex with the end-on superoxide coordinated to one iron.[78] This species oxidized 2,4-di-*tert*-butylphenol at $-80°C$. Small yet definitive H/D kinetic isotope effect of 1.5 suggested a proton-coupled electron transfer mechanism for this process. Interestingly, $\mu$-1,2-peroxo intermediate in the same chemical system (Figure 4.24) did not react with phenols or other substrates. The lack of oxidizing reactivity of diiron-peroxo intermediate may be unexpected, in view of rather high reactivity of hydrogen peroxide itself. A brief discussion of metal peroxide reactivity follows.

Metal peroxides vary dramatically in their structure and reactivity. Two extremes are represented by peroxo complexes of high- and low-valent metals. Of course, most complexes cannot be adequately described as one of the extremes and fall somewhere in the middle. High-valent metals (Re$^{VII}$, Cr$^{VI}$, Mo$^{VI}$, W$^{VI}$, V$^V$, etc.) preferentially form oxo-anions, where one or more oxygen atoms can be replaced with a peroxo moiety. Unless the metal center is strongly oxidizing and tends to react with coordinated peroxide (producing dioxygen), these high-valent peroxides are fairly stable and often crystallized. Coordination to a high-valent metal ion increases the acidity of hydrogen peroxide, so that doubly deprotonated peroxide anions preferentially form under most practical conditions; side-on coordination mode is common. Despite the lack of protons, peroxides coordinated to high-valent metals are electrophilically activated and are capable of transferring an oxygen atom to substrates.[16]

Similar electrophilic activation of coordinated peroxides or alkylperoxides can be observed for the metal ions in intermediate oxidation states. To give just one example, Sharpless epoxidation takes advantage of an electrophilic activation of alkyl hydroperoxides at titanium(IV). Notably, efficient epoxidation requires substrate binding in the vicinity of coordinated alkylperoxide, thus limiting the substrate scope of this reaction to allylic alcohols (alkoxy group acts as an anchor).[1,45]

Metal-peroxo intermediates of different structures were often implicated in the catalytic cycles of oxidative enzymes (see Figures 4.32–4.35 for representative examples). In most cases, however, these peroxo complexes are incompetent oxidants for native substrates. Instead, they undergo further reactions (usually, O–O bond cleavage that generates high-valent metal-oxo species) yielding more reactive, kinetically competent oxidants. Nevertheless, some peroxides can react with substrates in both enzymatic and synthetic systems.

Deprotonated peroxide coordinates to metals in low- and intermediate-oxidation states tend to bind side-on and display nucleophilic properties. For example, iron(III) porphyrin–peroxo complexes are nucleophilic: they did not transfer an oxygen atom to electron-rich substrates (such as electron-rich olefins), but brought about epoxidation of electron-poor olefins or oxidative deformylation of aldehydes.[130] Dinuclear iron(III) complexes containing deprotonated peroxide are also sluggish oxidants that are nucleophilic in character.[17,77,86] Complexes with sterically hindered carboxylates were investigated in some detail, revealing critical role of substrate proximity to the coordinated peroxide. For example, substrate coordination to one of the iron centers in $[Fe_2^{III}(O_2)(DXL)_4(THF)_2]$ (Figure 4.27) is necessary for productive phosphine oxidation, as the saturation kinetics suggested. The formation of phosphine oxide was controlled by the rate of substitution of the capping monodentate ligand (THF) with the incoming substrate. Similar coordinatively saturated complexes with strongly bound capping pyridine or imidazole did not oxidize triarylphosphines.[73,90] The observed greater reactivity of coordinated phosphines indirectly supports a nucleophilic nature of these deprotonated, coordinated peroxo moieties. However, even nucleophilic peroxides can transfer an oxygen atom to a closely positioned substrate. Similar "proximity effects" were also reported by Tshuva and Lippard for the oxidation of covalently tethered substrates.[77]

A different class of biomimetic dinuclear peroxides, $\mu$-$\eta^2$:$\eta^2$ dicopper-dioxygen adducts $Cu_2^{II}(O_2^{2-})$, are significantly more reactive than $cis$-$\mu$-1,2-peroxides of iron (III) introduced above.[40,41] These complexes participate in various intramolecular oxidations (aromatic hydroxylation and benzylic or aliphatic hydroxylation of the ligand) and intermolecular oxidations (in reactions with phenols or phenolates, phosphines, sulfides, etc.). Interestingly, end-on dicopper(II) peroxides (with either bridging or terminal peroxo ligand) displayed nucleophilic reactivity: they produced phenols from phenolates (via a proton transfer reaction), peroxycarbonates from $CO_2$, and did not transfer an oxygen atom to phosphines (undergoing ligand substitution of $O_2^{2-}$ with $PAr_3$ instead). In contrast, side-on dicopper(II) peroxides supported by very similar aminopyridine ligands transferred an oxygen atom to phosphines and oxidized phenols, displaying electrophilic reactivity.[41] Oxidation rates for a series of substituted phenols correlate with substrate redox potentials via Marcus equation; these results suggested a proton-coupled electron transfer rather than a simple hydrogen atom abstraction mechanism for this reaction.[39,131] Low-barrier O–O bond cleavage in side-on dicopper(II) peroxo complexes may generate dicopper(III) bis-$\mu$-oxo species (Figure 4.30) that are believed to be even more potent oxidants. While one can argue that dicopper(II) peroxides convert into dicopper(III)-oxo species before attacking the substrate, and in certain cases this mechanism appears to operate, in other cases peroxo complexes themselves act as oxidants, and O–O bond cleavage occurs concurrently with an oxygen atom transfer to substrates.[39–41] In general, dicopper(II) peroxo complexes preferentially carry out oxygen atom transfer, while dicopper(III) bis-$\mu$-oxo intermediates are most reactive in hydrogen atom abstraction. Similar to diiron(III) peroxides, both types of copper–dioxygen adducts tend to oxidize coordinated substrates.

Electrophilic activation of coordinated peroxides can be achieved by protonation, which typically yields end-on coordinated hydroperoxides.[16] For example, the well-defined hydroperoxo complexes of rhodium and chromium efficiently oxidize inorganic substrates such as halide anions; these reactions are acid catalyzed.[16] Hydroperoxo intermediates were implicated in some enzymatic oxidations, such as hydroxylation catalyzed by cytochrome P450 (Figure 4.32). Although significant evidence accumulates in favor of multiple reaction pathways operating in P450 chemistry, the idea of (P)Fe$^{III}$-OOH species being one of the truly reactive intermediates remains controversial.[132] On the other hand, a nonheme complex, activated bleomycin, which is believed to have an Fe(III)-hydroperoxo structure (Figure 4.33), appears to directly react with at least some substrates, including its native substrate, DNA.[127] Interestingly, iron-bleomycin may be a rather unusual case of an iron(III)-hydroperoxo complex kinetically competent to directly oxidize substrates. While many examples of similar species in nonheme iron chemistry were observed and characterized, and were proposed to be involved in catalytic cycles of substrate oxidation with H$_2$O$_2$,[90,128] recent direct experiments indicate that pure Fe$^{III}$-OOH compounds are sluggish oxidants that do not transfer an oxygen atom to olefins or even such nonchallenging substrates as phosphines.[132] It appears that excess oxidant (H$_2$O$_2$ in oxidations with hydrogen peroxide) often decreases the lifetime of metal-hydroperoxo intermediates, while increasing their apparent reactivity. This implies that intermediates other than M(OOH) may form, especially in the presence of extra H$_2$O$_2$, and these species would act as competent oxidants. Often-discussed possibilities include high-valent metal-oxo complexes (e.g., Fe$^{IV}$=O or Fe$^{V}$=O, derived from the parent Fe$^{III}$-OOH by O–O bond homolysis or heterolysis, respectively). An interesting alternative may be seen in oxidant coordination to a high-valent metal species. This pathway was proposed for olefin epoxidation with H$_2$O$_2$ catalyzed by a manganese(IV)-oxo complex with a crossed-bridged cyclam ligand, where neither (L)Mn$^{IV}$=O nor (L)Mn$^{V}$=O were plausible (the former was unreactive, and the latter was never observed); Lewis acid activation of H$_2$O$_2$ coordinated to a Mn$^{IV}$=O moiety accounted for the observed reactivity and $^{18}$O labeling results.[133] Similar pathway seems to operate in oxygen atom transfer from PhIO catalyzed by Mn$^{V}$=O complex with corrolazine: while high-valent Mn(V)-oxo complex is stable and unreactive by itself, its adduct with PhIO, (L)Mn$^{V}$(O)(OIPh), is a potent oxidant.[64]

To oxidize the challenging substrates, really reactive intermediates must be generated. Lessons from redox enzymes suggest high-valent metal-oxo species as the best candidates for efficient oxidants. Not surprisingly, a vast literature is devoted to the chemistry of these species, which differ rather dramatically in structure and reactivity and range from simple aqua complexes, such as Cr$_{aq}$O$^{2+}$, to elaborate iron-porphyrin models of the active site of cytochromes P450. The O–O bond is no longer present in these species: it was cleaved prior to their formation. Most of the nonbiological metal-oxo intermediates were prepared from oxygen donors other than dioxygen itself. Substrate oxidation with metal-oxo compounds is always accompanied by a decrease in the metal oxidation state, and oxidizing ability of metal-oxos depends primarily on the redox properties of a particular metal in its specific ligand environment. The term "high-valent," when applied to metal-oxo species, has to be defined in context: the oxidation state of $+3$ is "high" for copper, but usual for iron, and "low" for molybdenum, to name just a few examples. Electronic structure of the ligands plays a crucial role in modulating the reactivity of metal-oxo species. In iron chemistry, strong electron-donating porphyrin ligands favor the formally Fe(V)/Fe(III) redox couple, while Fe(IV) porphyrins are less potent oxidants, and preferentially undergo one-electron reduction into thermodynamically and kinetically stable $(P)Fe^{III}$. On the other hand, the oxidation state $+4$ is "high" in nonheme iron chemistry, where two-electron redox processes often utilize an $Fe(IV)/Fe(II)$ couple.[57,58]

Typical reactions of high-valent metal-oxo species formally include (but are not limited to) formally one-electron processes (hydrogen atom abstraction, electron transfer, and proton-coupled electron transfer) and formally two-electron processes (hydride abstraction and oxygen atom transfer). Electron transfer happens at the metal, hydrogen atom abstraction or hydride abstraction involves adding $H^{\bullet}$ or $H^{-}$ to the terminal oxygen atom of the metal-oxo moiety, a proton-coupled electron transfer combines an electron transfer to the metal and a proton transfer elsewhere (e.g., to the oxo group), and an oxygen atom transfer involved complete cleavage of a metal–oxygen bond.

Electron transfer rates with metal-oxo species as oxidants are described by Marcus theory, similar to the other electron transfer processes. The rates of hydrogen atom abstraction by metal-oxos were analyzed in detail by Mayer, and were shown to follow Polanyi correlations.[134] These free energy relationships link the oxidation rates to the substrate O–H (or C–H) bond energies and the OH bond energies of the metal-containing product ($M^{n+}$-OH derived from $M^{n+1}$=O). The latter term depends on the $M^{n+1}$=O/$M^{n+}$-OH redox potential and on the acidity ($pK_a$) of the metal-hydroxy product. For a given oxidant, linear correlations were found for the rates of substrate oxidation as a function of the C–H bond energies (or O–H bond energies for alcohol or phenol substrates). For a given substrate, the oxidation rates correlate reasonably well with the $M^{n+1}$=O/$M^{n+}$-OH redox potential.[134] However, exceptions from these simple correlations are also known and could often be attributed to steric hindrance and/or excessive geometric reorganization in the course of redox process. A more recent analysis showed applicability of Marcus crossed-relation not only to electron transfer but also to hydrogen atom transfer reactions[135]:

$$k_{xy} = \left(k_{xx}k_{yy}K_{xy}f_{xy}\right)^{1/2} \tag{4.39}$$

Metal-oxo complexes abstract hydrogen atom from substrates faster than their protonated metal-hydroxo counterparts, as was shown for Mn(IV) compounds with crossed-bridged cyclam.[136] Hydrogen atom abstraction is characterized by a fairly large H/D kinetic isotope effects: KIE usually exceeds 3, and sometimes reaches values as high as 30–40.[58] In contrast, a modest kinetic isotope effect of about 1.3–1.5 is usually observed for proton-coupled electron transfer reactions.[137,138]

In many cases, metal-oxo complexes participate in either hydrogen atom abstraction or hydride abstraction reactions; sometimes these two pathways occur concurrently. The preference for one of these pathways over the other depends on the nature of substrate: substrates that preferentially undergo homolytic cleavage of their C–H bonds forming radicals react via hydrogen atom abstraction.[15] The two pathways are often close in energy, as can be exemplified by the reactions of $Cr_{aq}O^{2+}$. This simple metal-oxo species can be prepared directly from $Cr^{2+}$ and $O_2$, and it forms spectroscopically distinguishable products in one-electron (H-atom transfer) versus two-electron (hydride transfer) reductions ($Cr^{2+}$ that quickly reacts with excess $O_2$, affording $CrOO^{2+}$ or $Cr^{III}(OH)$, respectively). Interestingly, methanol undergoes $H^-$ transfer, while aldehydes (such as pivaldehyde) are oxidized via H-atom transfer[15]:

$$Cr_{aq}O^{2+} + CH_3OH \rightarrow CH_2O + H_2O + Cr_{aq}^{2+} \xrightarrow{O^2} Cr_{aq}OO^{2+} \qquad (4.40)$$

$$Cr_{aq}O^{2+} + (CH_3)_3C\text{-}CHO \rightarrow (CH_3)_3C^{\bullet}O + Cr_{aq}OH^{2+} \qquad (4.41)$$

Oxygen atom transfer reactions from metal-oxo species to phosphines (forming phosphine oxides), alkylsulfides (forming sulfoxides), olefins (forming epoxides), and other substrates depend on the relative X=O bond dissociation energies in the metal-oxo starting material and the oxidized product; the reactivity scale based on bond dissociation energies was proposed.[62]

While it is impossible to review individual metal-oxo complexes here, one recent dramatic development, discovery of ferryl(IV) nonheme complexes,[57,58] deserves to be mentioned. Iron(IV)-oxo intermediates were known in heme chemistry for quite some time, but ferryl(IV) in nonheme systems became readily available and well characterized only very recently, when the X-ray structure of the $Fe^{IV}$=O complex with tetramethylcyclam was determined,[139] and several similar complexes were identified spectroscopically and crystallographically (Figure 4.36).[57,58]

This recent breakthrough provides an access to an entirely new class of intermediates in mononuclear nonheme iron systems. Nonheme iron(IV)-oxo complexes can oxidize a variety of substrates (Figure 4.37), their most challenging reaction is alkane hydroxylation. These oxidants are electrophilic in nature, as was demonstrated by Hammet correlations in oxygen atom transfer to aryl sulfides. Nonheme ferryls abstract hydrogen atom from dihydroanthracene and related compounds with a huge (about 30–50) H/D kinetic isotope effect. In contrast, aromatic hydroxylation with (L) $Fe^{IV}$=O is accompanied by small inverse isotope effect, indicative of an electrophilic attack of the aromatic ring with an $sp^2$–$sp^3$ rehybridization at the rate-limiting step. In addition to ferryl species supported by polyamine or aminopyridine ligands, aqueous $Fe_{aq}^{IV}$=O was cleanly generated from Fe(II) and ozone and its reactivity with organic substrates was characterized, showing competing hydrogen atom and hydride transfer pathways.[71]

In view of these successes in preparation and characterization of nonheme ferryl species, it is tempting to invoke Fe(IV)-oxo intermediates in a variety of iron-catalyzed oxidations with dioxygen, peroxides, and other oxygen atom donors. However, direct experiments question this interpretation in several cases. Efficient olefin epoxidation with hydrogen peroxide catalyzed by $Fe(TPA)^{2+}$ or Fe $(BPMEN)^{2+}$ in the presence of acetic acid proceeds under conditions that favor the formation of (L)$Fe^{IV}$=O. This intermediate was independently generated and subjected to reactions with olefins. Surprisingly, no epoxide (or an alternative product, *cis*-diol) formed in this experiment, clearly demonstrating that ferryl(IV) is not a competent intermediate in these epoxidations (or related dihydroxylations).[140] Because $Fe^{III}$-OOH, another plausible intermediate, was also unreactive in olefin

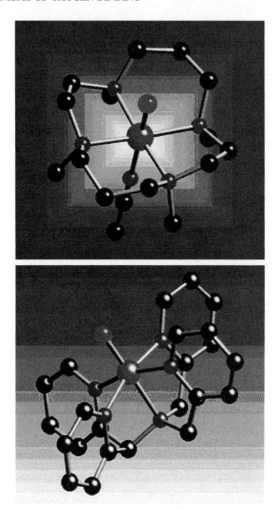

**FIGURE 4.36** X-ray structures of nonheme iron(IV)-oxo complexes. (See the color version of this figure in Color Plates section.)[57]

epoxidation, the still-unobserved $Fe^V=O$ intermediate, which may result from acid-assisted O–O bond heterolysis of the coordinated hydroperoxide, was proposed. Similarly, aqueous $(H_2O)_5Fe^{IV}=O$ transfers an oxygen atom to sulfoxides, forming sulfones, but relatively slow reaction rates exclude ferryl as a possible intermediate in Fenton chemistry, where much faster iron-catalyzed oxidation of sulfoxides with $H_2O_2$ was observed.[141] Hydroxyl radicals must be responsible for Fenton oxidation of sulfoxides. Somewhat limited oxidizing power of nonheme iron(IV)-oxo complexes is surprising and deserves further investigation. It is already clear that the reactivity of these intermediates can be modulated by changing the additional ligand *trans* to Fe=O moiety.[57,58]

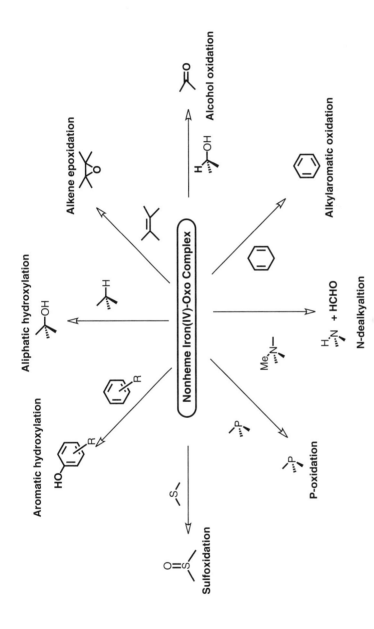

**FIGURE 4.37** Oxidation reactions mediated by mononuclear nonheme iron(IV)-oxo complexes.[58]

177

A possible approach to increasing the reactivity of metal-oxo intermediates is to combine two (or more) such moieties in one dinuclear complex. Indeed, dinuclear or polynuclear metal centers are well known in enzymes, including methane monooxygenase and multicopper oxidases. The best known synthetic complexes that can be classified as high-valent metal-oxos include bis-μ-oxo dicopper(III) complexes and analogous bis-μ-oxo iron(III)iron(IV) complexes, often referred to as "diamond core" structures.[85] Some recent studies indicate that dinuclear complexes bearing terminal high-valent metal-oxo moieties can be even more reactive than their "diamond core" counterparts.[142,143] Exploring the reactivity of di- and polynuclear high-valent metal-oxo complexes will certainly lead to productive strategies for substrate oxidations.

## 4.6   CONCLUDING REMARKS

Mechanistic studies provide the bedrock for understanding chemical reactivity and have direct impact on the design of molecules with tailored properties for new catalytic systems. From the perspective of small molecule activation, it is very important to map out the mechanistic details of multielectron processes, which ultimately result in the E–E bond cleavage. This problem attracted attention of numerous researchers in oxygen activation chemistry, where stepwise $O_2$ reduction appears to be beneficial for peroxide generation and oxidation of thermodynamically less challenging substrates, while four-electron process is desirable for generating highly reactive metal-based oxidants and for energy conversion (in fuel cells, synthetic models of photosynthesis, etc.). Dramatic progress was accomplished in understanding the mechanisms of natural dioxygen carriers and redox enzymes, as well as in designing their structural and functional models. This work will provide a starting point for the mechanism-based design of new metal-catalyzed small-molecule activation and utilization reactions. Rewardingly, similar chemical principles operate in dioxygen activation by metalloenzymes and synthetic transition metal complexes.

Although dioxygen binding to a transition metal center inherently involves an electron transfer, formation of the metal–dioxygen bond tends to be an inner-sphere process controlled by the rates of ligand substitution. Dioxygen coordination to vacant or labile metal sites tends to be an associative, entropically controlled reaction. In the absence of extreme steric hindrance, this process is very rapid for 3d-metal ions, but it can be significantly slower for 4d- and 5d-metals. Dioxygen affinity is controlled by both steric and electronic factors, with the former affecting primarily the $O_2$ binding rates and the latter affecting the $O_2$ dissociation rates. The initial end-on $O_2$ coordination to mononuclear complexes is rate limiting and can be followed, in certain cases, by the formation of the second metal–oxygen bond affording side-on dioxygen adducts. Depending on the relative stability of consecutive oxidation states of the metal, these dioxygen adducts can be formulated as superoxo or peroxo complexes.

Dioxygen binding to dinuclear metal centers also starts with the formation of the first metal–oxygen bond; dioxygen coordination is again controlled by ligand substitution and occurs rapidly at sterically accessible vacant sites. The subsequent step, coordination of the bound $O_2$ to the second metal center, highly depends on the geometry of the dinuclear metal complex and can be retarded or even completely shut down by excessive steric bulk of the multidentate ligands. Bridging $O_2$ in dinuclear complexes may have either side-on or end-on geometry, giving rise to $\mu\text{-}\eta^2{:}\eta^2$, $\mu\text{-}\eta^1{:}\eta^1$, or $\mu\text{-}\eta^1{:}\eta^2$ structures.

O–O bond cleavage usually requires bimetallic activation, although final products may be mononuclear or dinuclear metal-oxo complexes. Avoiding one-electron processes is important for selective substrate oxidations; formally, two-electron processes may result from a single metal ion increasing its oxidation state by 2 or from a pair of metal ions acting in concert, with each metal increasing its oxidation state by 1. Rapid interconversion of dinuclear metal-peroxo and high-valent oxo intermediates is typical for side-on dicopper–dioxygen adducts.

Metal-superoxo and metal-peroxo complexes can participate in various substrate oxidation reactions, such as electron transfer, hydrogen atom or hydride abstraction, proton-coupled electron transfer, and oxygen atom transfer. However, these intermediates, especially when dioxygen is coordinated side-on, tend to be relatively sluggish oxidants. Their reactivity with reducing substrates can be increased by electrophilical activation (for example, by coordinating peroxide ligands to a high-valent metal ion or by inter- or intramolecular protonation). Alternatively, one can take advantage of proximity effects: substrate oxidation is greatly facilitated in the immediate vicinity of the metal–dioxygen moiety. Similar proximity effects may account for the selectivity of enzymatic oxidations.

High-valent metal-oxo intermediates are believed to be rather potent oxidants. Although various metal-oxo intermediates can be prepared in reactions between low-valent metal complexes and strong oxygen-atom donors, very few of these species form directly from synthetic metal complexes and dioxygen. Attempts to apply these complexes in catalysis face additional complication: it is difficult to regenerate the low-valent metal starting material and to close the catalytic cycle. Metalloenzymes overcome these obstacles by using sacrificial reducing agents; O–O bond cleavage is also assisted by controlled delivery of protons. Controlled delivery of protons and electrons remains challenging in synthetic systems. Systematic mechanistic studies will be instrumental in addressing these challenges.

## REFERENCES

1. Backvall, J.-E. *Modern Oxidation Methods*; Wiley-VCH: Weinheim, 2004; p 336.
2. Cornell, C. N.; Sigman, M. S. Molecular oxygen binding and activation: oxidation catalysis. In *Activation of Small Molecules*; Tolman, W. B., Ed.; Wiley-VCH: Weinheim, 2006; pp 159–186.

3. Yamamoto, S. The 50th anniversary of the discovery of oxygenases. *IUBMB Life* **2006**, *58*, 248–250.

4. Ho, R. Y. N.; Liebman, J. F.; Valentine, J. S. Overview of energetics and reactivity of oxygen. In *Active Oxygen in Chemistry*, Foote, C. S.; Valentine, J. S.; Greenberg, A.; Liebman, J. F.; Eds.; Blackie Academic and Professional: London, 1995; pp 1–23.

5. Sawyer, D. T. *Oxygen Chemistry*; Oxford University Press: New York, 1991.

6. Foote, C. S.; Valentine, J. S.; Greenberg, A.; Liebman, J. F. *Active Oxygen in Chemistry*; Blackie Academic & Professional, Chapman & Hall: Glasgow, 1995.

7. Valentine, J. S.; Foote, C. S.; Greenberg, A.; Liebman, J. F. *Active Oxygen in Biochemistry*; Blackie Academic & Professional, Chapman & Hall: Glasgow, 1995.

8. Meunier, B. *Biomimetic Oxidations Catalyzed by Transition Metal Complexes*; Imperial College Press: London, 2000.

9. Simandi, L. I. *Advances in Catalytic Activation of Dioxygen by Metal Complexes*; Kluwer: Dordrecht, 2003.

10. Borovik, A. S.; Zinn, P. J.; Zart, M. K. Dioxygen binding and activation: reactive intermediates. In *Activation of Small Molecules*; Tolman, W. B.; Wiley-VCH: Weinheim, 2006; pp 187–234.

11. Nam, W.; *Acc. Chem. Res.* 2007, *40* (7) (*Special Issue on Dioxygen Activation*).

12. van Eldik, R. *Chem. Rev.* 2005, *105* (6) (*Special Issue Inorganic and Bioinorganic Mechanisms*).

13. Holm, R. H.; Solomon, E. I.; *Chem. Rev.* 2004, *104* (2) (*Special Issue Biomimetic Inorganic Chemistry*).

14. Bakac, A. Mechanistic and kinetic aspects of transition metal oxygen chemistry *Progr. Inorg. Chem.* **1995**, *43*, 267–351.

15. Bakac, A. Dioxygen activation by transition metal complexes. Atom transfer and free radical chemistry in aqueous media. *Adv. Inorg. Chem.* **2004**, *55*, 1–59.

16. Bakac, A. Kinetic and mechanistic studies of the reactions of transition metal-activated oxygen with inorganic substrates. *Coord. Chem. Rev.* **2006**, *250*, 2046–2058.

17. Kryatov, S. V.; Rybak-Akimova, E. V.; Schindler, S. Kinetics and mechanisms of formation and reactivity of non-heme iron oxygen intermediates. *Chem. Rev.* **2005**, *105*, 2175–2226.

18. Greenwood, N. N.; Earnwhaw, A. *Chemistry of the Elements*, 2nd edition; Elsevier: Amsterdam, 1997, p 1341.

19. Bielski, B. H. J.; Cabelli, D. E.; Arudi, R. L.; Ross, A. B. Reactivity of perhydroxyl-superoxide radicals in aqueous solution. *J. Phys. Chem. Ref. Data* **1985**, *14*, 1041–1100.

20. Kurtz, D. M., Jr. *Dioxygen-binding proteins*. In *Comprehensive Coordination Chemistry II*; McCleverty, J. A.; Meyer, T. J.; Elsevier; Oxford, UK, Vol. 8, 2004, pp 229–260.

21. Jameson, G. B.; Ibers, J. A. Dioxygen carrier. In *Biological Inorganic Chemistry: Structure and Reactivity*; Bertini, I.; Gray, H. B.; Stiefel, E. I.; Valentine, J. S.; University Science Books: Sausalito, CA, 2007; pp 354–388.

22. Stenkamp, R. E. Dioxygen and hemerythrin. *Chem. Rev.* **1994**, *94*, 715–726.

23. Solomon, E. I.; Brunold, T. C.; Davis, M. I.; Kemsley, J. N.; Lee, S.-K.; Lehnert, N.; Neese, F.; Skulan, A. J.; Yang, Y.-S.; Zhou, J. Geometric and electronic structure/function correlations in non-heme iron enzymes. *Chem. Rev.* **2000**, *100*, 235–349.

24. de Waal, D. J. A.; Gerber, T. I. A.; Low, W. J.; van Eldik, R. Kinetics and mechanism of the dioxygen uptake of the four-coordinate (X=Cl) and five-coordinate (X=I, SCN, PPh$_3$) complexes Ir(cod)(phen)X. *Inorg. Chem.* **1982**, *21*, 2002–2006.

25. Armstrong, G. D.; Sykes, A. G. Reactions of O$_2$ with hemerythrin, myoglobin, and hemocyan: effects of, D$_2$O on equilibration rate constants evidence for H-bonding. *Inorg. Chem.* **1986**, *25*, 3155–3139.

26. Lloyd, C. R.; Eyring, E. M.; Ellis, W. R., Jr. Uptake and release of O$_2$ by myohemerythrin: evidence for different rate-determining steps and a caveat. *J. Am. Chem. Soc.* **1995**, *117*, 11993–11994.

27. Projahn, H.-D.; Schindler, S.; van Eldik, R.; Fortier, D. G.; Andrew, C. R.; Sykes, A. G. Formation and deoxygenation kinetics of oxyhemerythrin and oxyhemocyanin: a pressure dependence study. *Inorg. Chem.* **1995**, *34*, 5935–5941.

28. Lloyd, C. R.; Raner, G. M.; Moser, A.; Eyring, E. M.; Ellis, W. R., Jr. Oxymyohemerythrin: discriminating between O$_2$ release and autoxidation. *J. Inorg. Biochem.* **2000**, *81*, 293–300.

29. Zahir, K.; Espenson, J. H.; Bakac, A. Reactions of polypyridylchromium(II) ions with oxygen: determination of the self-exchange rate constant of O$_2$/O$_2^-$. *J. Am. Chem. Soc.* **1988**, *110*, 5059–5063.

30. Geletii, Y. V.; Hill, C. L.; Atalla, R. H.; Weinstock, I. A. Reduction of O$_2$ to superoxide anion (O$_2^-$) in water by heteropolytungstate cluster-anions. *J. Am. Chem. Soc.* **2006**, *128*, 17034–17042.

31. Collman, J. P.; Fu, L. Synthetic models for hemoglobin and myoglobin. *Acc. Chem. Res.* **1999**, *32*, 455–463.

32. Momenteau, M.; Reed, C. A. Synthetic heme dioxygen complexes. *Chem. Rev.* **1994**, *94*, 659–698.

33. Jones, R. D.; Summerville, D. A.; Basolo, F. Synthetic oxygen carriers related to biological systems. *Chem. Rev.* **1979**, *79*, 139–179.

34. Niederhoffer, E. C.; Timmons, J. H.; Martell, A. E. Thermodynamics of oxygen binding in natural and synthetic dioxygen complexes. *Chem. Rev.* **1984**, *84*, 137–203.

35. Rybak-Akimova, E. V.; Marek, K.; Masarwa, M.; Busch, D. H. The dynamics of formation of the O$_2$-Co(II) bond in the cobalt(II) cyclidene complexes. *Inorg. Chim. Acta* **1998**, *270*, 151–161.

36. Zhang, M.; van Eldik, R.; Espenson, J. H.; Bakac, A. Volume profiles for the reversible binding of dioxygen to cobalt(II) complexes: evidence for a substitution-controlled process. *Inorg. Chem.* **1994**, *33*, 130–133.

37. Busch, D. H.; Alcock, N. W. Iron and cobalt "lacunar" complexes as dioxygen carriers. *Chem. Rev.* **1994**, *94*, 585–623.

38. Korendovych, I. V.; Roesner, R. R.; Rybak-Akimova, E. V. Molecular recognition of neutral and charged guests using metallomacrocyclic hosts. In *Advances in Inorganic Chemistry*; van Eldik, R.; B.-J. Kristen, Eds.; Academic Press: Amsterdam, **2007**; Vol. 59, pp 109–173.

39. Itoh, S. Mononuclear copper active–oxygen complexes. *Curr. Opin. Chem. Biol.* **2006**, *10*, 115–122.

40. Lewis, E. A.; Tolman, W. B. Reactivity of dioxygen–copper-systems. *Chem. Rev.* **2004**, *104*, 1047–1076.

41. Hatcher, L. Q.; Karlin, K. D. Ligand influences in copper–dioxygen complex-formation and substrate oxidations. *Adv. Inorg. Chem.* **2006**, *58*, 131–184.

42. Wuertele, C.; Gaoutchenova, E.; Harms, K.; Holthausen, M. C.; Sundermeyer, J.; Schindler, S. Crystallographic characterization of a synthetic 1: 1 end-on copper–dioxygen adduct complex. *Angew. Chem., Int. Ed.* **2006**, *45*, 3867–3869.

43. Fry, H. C.; Scaltrito, D. V.; Karlin, K. D.; Meyer, G. J. The rate of $O_2$ and CO binding to a copper complex, determined by a "flash-and-trap" technique, exceeds that for hemes. *J. Am. Chem. Soc.* **2003**, *125*, 11866–11871.

44. Vaska, L. Oxygen-carrying properties of a simple synthetic system. *Science* **1963**, *140*, 809–810.

45. Cotton, F. A.; Wilkinson, G.; Murillo, C. A.; Bochmann, M. *Advanced Inorganic Chemistry*; Wiley: New York, 1999, p 1355.

46. Valentine, J. S. The dioxygen ligand in mononuclear group VIII transition metal complexes. *Chem. Rev.* **1973**, *73*, 235–245.

47. Atwood, J. D. Organoiridium complexes as models for homogeneously catalyzed reactions. *Coord. Chem. Rev.* **1988**, *83*, 93–114.

48. Vaska, L.; Chen, L. S.; Miller, W. V. Oxygenation and related addition reactions of isostructural $d^8$ complexes of cobalt, rhodium, and iridium. A quantitative assessment of the role of the metal. *J. Am. Chem. Soc.* **1971**, *93*, 6671–6673.

49. Vaska, L.; Chen, L. S.; Senoff, C. V. Oxygen-carrying iridium complexes: kinetics, mechanism, and thermodynamics. *Science* **1971**, *174*, 587–589.

50. Theopold, K. Dioxygen activation by organometallics of early transition metals. *Topics in Organometallic Chemistry* **2007**, *26*, 17–37.

51. Kieber-Emmons, M. T.; Riordan, C. G. Dioxygen activation at monovalent nickel. *Acc. Chem. Res.* **2007**, *40*, 609–617.

52. Seo, M. S.; Kim, J. Y.; Annaraj, J.; Kim, Y.; Lee, Y.-M.; Kim, S.-J.; Kim, J.; Nam, W. [Mn (tmc)($O_2$)]$^+$: a side-on peroxide manganese(III) complex bearing a non-heme ligand. *Angew. Chem., Int. Ed.* **2007**, *46*, 377–380.

53. Cramer, C. J.; Tolman, W. B. Mononuclear Cu–$O_2$ complexes: geometries, spectroscopic properties, electronic structure, and reactivity. *Acc. Chem. Res.* **2007**, *40*, 601–608.

54. Aboelella, N. W.; Kryatov, S. V.; Gherman, B. F.; Brennessel, W. W.; Young, V. G., Jr.; Sarangi, R.; Rybak-Akimova, E. V.; Hodgson, K. O.; Hedman, B.; Solomon, E. I.; Cramer, C. J.; Tolman, W. B. Dioxygen activation at a single copper site: structure, bonding, and mechanism of formation of 1: 1 Cu–$O_2$ adducts. *J. Am. Chem. Soc.* **2004**, *126*, 16896–16911.

55. Balch, A. L.; Chan, Y.-W.; Cheng, R.-J.; La Mar, G. N.; Latos Grazynski, L.; Renner, M. W. Oxygenation patterns for iron(II) porphyrins. Peroxo and ferryl ($Fe^{IV}O$) intermediates detected by 1H nuclear magnetic resonance spectroscopy during the oxygenation of (tetramesitylporphyrin)iron(II). *J. Am. Chem. Soc.* **1984**, *106*, 7779–7785.

56. Ghiladi, R. A.; Kretzer, R. M.; Guzei, I.; Rheingold, A. L.; Neuhold, Y.-M.; Hatwell, K. R.; Zuberbuhler, A. D.; Karlin, K. D. ($F_8$TPP)$Fe^{II}$/$O_2$ reactivity studies {$F_8$TPP = tetrakis (2,6-difluorophenyl)porphyrinate(2-)}: spectroscopic (UV-visible and NMR) and kinetic study of solvent-dependent (Fe/$O_2$ = 1: 1 or 2: 1) reversible $O_2$ reduction and ferryl formation. *Inorg. Chem.* **2001**, *40*, 5754–5767.

57. Que, L., Jr. The road to non-heme oxoferryls and beyond. *Acc. Chem. Res. 40* (7), **2007**, 493–500.

58. Nam, W. High-valent iron(IV)-oxo complexes of heme and non-heme ligands in oxygenation reactions. *Acc. Chem. Res. 40* (7), **2007**, 522–531.

59. Kim, S. O.; Sastri, C. V.; Seo, M. S.; Kim, J.; Nam, W. Dioxygen activation and catalytic aerobic oxidation by a mononuclear nonheme iron(II) complex *J. Am. Chem. Soc.* **2005**, *127*, 4178–4179.

60. Trnka, T.; Parkin, G. A survey of terminal chalcogenido complexes of the transition metals: trends in their distribution and the variation of their M=E bond lengths *Polyhedron* **1997**, *16*, 1031–1045.

61. Nugent, W. A.; Mayer, J. M. *Metal-Ligand Multiple Bonds*; Wiley: New York, 1988.

62. Holm, R. H. Metal-centered oxygen atom transfer reactions. *Chem. Rev.* **1987**, *87*, 1401–1449.

63. Hill, C. L. Confirmation of the improbable. *Nature* **2008**, *455*, 1045–1047.

64. Goldberg, D. P. Corrolazines: new frontiers in high-valent metalloporphyrinoid stability and reactivity. *Acc. Chem. Res.* **2007**, *40*, 626–634.

65. Liston, D. J.; West, B. O. Oxochromium compounds. 2. Reaction of oxygen with chromium(II) and chromium(III) porphyrins and synthesis of a μ-oxo chromium porphyrin derivative. *Inorg. Chem.* **1985**, *24*, 1568–1576.

66. Nemes, A.; Bakac, A. Disproportionation of aquachromyl(IV) ion by hydrogen abstraction from coordinated water. *Inorg. Chem.* **2001**, *40*, 2720–2724.

67. Collman, J. P.; Brauman, J. I.; Fitzgerald, J. P.; Sparapany, J. W.; Ibers, J. A. Reversible binding of dinitrogen and dioxygen to ruthenium picnic-basket porphyrins. *J. Am. Chem. Soc.* **1988**, *110*, 3486–3495.

68. Groves, J. T.; Quinn, R. Aerobic oxidation of olefins with ruthenium porphyrin catalysts. *J. Am. Chem. Soc.* **1985**, *107*, 5790–5792.

69. Lai, T.-S.; Kwong, H.-L.; Zhang, R.; Che, C.-M. Stoichiometric enantioselective alkene epoxidation with a chiral dioxoruthenium(VI) D4-porphyrinato complex. *Dalton Trans.* **1998**, 3559–3564.

70. Zierkiewicz, W.; Privalov, T. A computational study of oxidation of ruthenium porphyrins via ORu$^{IV}$ and ORu$^{VI}$O species. *Dalton Trans.* **2006**, 1867–1874.

71. Pestovsky, O.; Bakac, A. Identification and characterization of aqueous ferryl(IV) ion. *ACS Symp. Ser. (Ferrates)* **2008**, *985*, 167–176.

72. Warburton, P. R.; Busch, D. H. Dynamics of iron(II) and cobalt(II) dioxygen carriers. In *Perspectives on Bioinorganic Chemistry*; JAI Press: London, 1993; Vol. 2, pp 1–79.

73. Korendovych, I. V.; Kryatov, S. V.; Rybak-Akimova, E. V. Dioxygen activation at non-heme iron: insights from rapid kinetic studies. *Acc. Chem. Res.* **2007**, *40*, 510–521.

74. Seibig, S.; van Eldik, R. Kinetics of [Fe$^{II}$(edta)] oxidation by molecular oxygen revisited. New evidence for a multistep mechanism. *Inorg. Chem.* **1997**, *36*, 4115–4120.

75. Bianchini, C.; Zoellner, R. W. Activation of dioxygen by cobalt group metal complexes *Adv. Inorg. Chem.* **1997**, *44*, 263–339.

76. Lee, D.; Lippard, S. J. Nonheme di-iron enzymes. In *Comprehensive Coordination Chemistry II*; McCleverty, J. A.; Meyer, T. J.; Elsevier: Oxford, UK, 2004; Vol. 8, pp 309–342.

77. Tshuva, E. Y.; Lippard, S. J. Synthetic models for non-heme carboxylate-bridged diiron metalloproteins: strategies and tactics. *Chem. Rev.* **2004**, *104*, 987–1012.

78. Shan, X.; Que, L., Jr. Intermediates in the oxygenation of a nonheme diiron(II) complex, including the first evidence for a bound superoxo species. *Proc. Natl. Acad. USA* **2005**, *102*, 5340–5345.

79. Zhang, X.; Furutachi, H.; Fujinami, S.; Nagamoto, S.; Maeda, Y.; Watanabe, Y.; Kitagawa, T.; Suzuki, M. Structural and spectroscopic characterization of (μ-OH or μ-O)(μ-peroxo)diiron(III) Complexes. *J. Am. Chem. Soc.* **2005**, *127*, 826–827.

80. Dong, Y.; Yan, S.; Young, V. G., Jr.; Que, L., Jr. Crystal structure analysis of a synthetic non-heme diiron-$O_2$ adduct: insight into the mechanism of oxygen activation. *Angew. Chem., Int. Ed. Engl.* **1996**, *35*, 618–620.

81. Kim, K.; Lippard, S. J. Structure and Mössbauer spectrum of a (μ-1,2-peroxo)bis(μ-carboxylato)diiron(III) model for the peroxo intermediate in the methane monooxygenase hydroxylase reaction cycle. *J. Am. Chem. Soc.* **1996**, *118*, 4914–4915.

82. Ookubo, T.; Sugimoto, H.; Nagayama, T.; Masuda, H.; Sato, T.; Tanaka, K.; Maeda, Y.; Okawa, H.; Hayashi, Y.; Uehara, A.; Suzuki, M. Cis-μ-1,2-peroxo diiron complex: structure and reversible oxygenation. *J. Am. Chem. Soc.* **1996**, *118*, 701–702.

83. Feig, A. L.; Becker, M.; Schindler, S.; van Eldik, R.; Lippard, S. J. Mechanistic studies of the formation and decay of diiron(III) peroxo complexes in the reaction of diiron(II) precursors with dioxygen. *Inorg. Chem.* **1996**, *35* (9), 2590–2601.

84. Ghosh, A.; Tiago de Oliveira, F.; Yano, T.; Nishioka, T.; Beach, E. S.; Kinoshita, I.; Münck, E.; Ryabov, A. D.; Horwitz, C. P.; Collins, T. J. Catalytically active μ-oxodiiron (IV) oxidants from iron(III) and dioxygen. *J. Am. Chem. Soc.* **2005**, *127* (8), 2505–2513.

85. Que, L., Jr.; Tolman, W. B. Bis(μ-oxo)dimetal "Diamond" cores in copper and iron complexes relevant to biocatalysis. *Angew. Chem., Int. Ed.* **2002**, *41*, 1114–1137.

86. LeCloux, D. D.; Barrios, A. M.; Lippard, S. J. The reactivity of well defined diiron(III) peroxo complexes toward substrates: addition to electrophiles and hydrocarbon oxidation. *Bioorg. Med. Chem.* **1999**, *7*, 763–772.

87. Feig, A. L.; Masschelein, A.; Bakac, A.; Lippard, S. J. Kinetic studies of reactions of dioxygen with carboxylate-bridged diiron(II) complexes leading to the formation of (μ-oxo)diiron(III) complexes. *J. Am. Chem. Soc.* **1997**, *119* (2), 334–342.

88. Kryatov, S. V.; Taktak, S.; Korendovych, I. V.; Rybak-Akimova, E. V.; Kaizer, J.; Stéphane Torelli, S.; Shan, X.; Mandal, S.; MacMurdo, V.; Mairata i Payeras, A.; Que, L., Jr. Dioxygen binding to complexes with $Fe^{II}_2$(μ-OH)$_2$ cores: steric control of activation barriers and $O_2$-adduct formation. *Inorg. Chem.* **2005**, *44* (1), 85–99.

89. Costas, M.; Cady, C. W.; Kryatov, S. V.; Ray, M.; Ryan, M. J.; Rybak-Akimova, E. V.; Que, L., Jr. Role of carboxylate bridges in modulating nonheme diiron(II)/$O_2$ reactivity. *Inorg. Chem.* **2003**, *42*, 7519–7530.

90. Kryatov, S. V.; Chavez, F. A.; Reynolds, A. M.; Rybak-Akimova, E. V.; Que, L., Jr.; Tolman, W. B. Mechanistic studies on the formation and reactivity of dioxygen adducts of diiron complexes supported by sterically hindered carboxylates. *Inorg. Chem.* **2004**, *43*, 2141–2150.

91. Kryatov, S. V.; Rybak-Akimova, E. V.; MacMurdo, V. L.; Que, L., Jr. A mechanistic study of the reaction between a diiron(II) complex[$Fe^{II}_2$(μ-OH)$_2$(6-Me$_3$-TPA)$_2$]$^{2+}$ and $O_2$ to form a diiron(III) peroxo complex. *Inorg. Chem.* **2001**, *40*, 2220–2228.

92. LeCloux, D. D.; Barrios, A. M.; Mizoguchi, T. J.; Lippard, S. J. Modeling the diiron centers of non-heme iron enzymes. Preparation of sterically hindered diiron(II)

tetracarboxylate complexes and their reactions with dioxygen. *J. Am. Chem. Soc.* **1998**, *120*, 9001–9014.

93. Tolman, W. B.; Que, L., Jr. Sterically hindered benzoates: a synthetic strategy for modeling dioxygen activation at diiron active sites in proteins. *J. Chem. Soc., Dalton Trans.* **2002**, 653–660.

94. Lee, D.; Pierce, B.; Krebs, C.; Hendrich, M. P.; Huynh, B. H.; Lippard, S. J. Functional mimic of dioxygen-activating centers in non-heme diiron enzymes: mechanistic implications of paramagnetic intermediates in the reactions between diiron(ii) complexes and dioxygen. *J. Am. Chem. Soc.* **2002**, *124*, 3993–4007.

95. Zhao, M.; Song, D.; Lippard, S. J. Water induces a structural conversion and accelerates the oxygenation of carboxylate-bridged non-heme diiron enzyme synthetic analogues. *Inorg. Chem.* **2006**, *45*, 6323–6330.

96. Zhao, M.; Helms, B.; Slonkina, E.; Friedle, S.; Lee, D.; DuBois, J.; Hedman, B.; Hodgson, K. O.; Frechet, J. M. J.; Lippard, S. J. Iron complexes of dendrimer-appended carboxylates for activating dioxygen and oxidizing hydrocarbons. *J. Am. Chem. Soc.* **2008**, *130*, 4352–4363.

97. Mirica, L. M.; Ottenwaelder, X.; Stack, T. D. P. Structure and spectroscopy of copper–dioxygen complexes. *Chem. Rev.* **2004**, *104*, 1013–1045.

98. Schindler, S. Reactivity of copper(I) complexes towards dioxygen. *Eur. J. Inorg. Chem.* **2000**, 2311–2326.

99. Karlin, K. D.; Kaderli, S.; Zuberbuhler, A. D. Kinetics and thermodynamics of copper(I)/dioxygen interaction. *Acc. Chem. Res.* **1997**, *30*, 139–147.

100. Suzuki, M. Ligand effects on dioxygen activation by copper and nickel complexes: reactivity and intermediates. *Acc. Chem. Res.* **2007**, *40*, 609–617.

101. Tueklar, Z.; Jacobson, R. R.; Wei, N.; Murthy, N. N.; Zubieta, J.; Karlin, K. D. Reversible reaction of $O_2$ (and CO) with a copper(I) complex. X-ray structures of relevant mononuclear Cu(I) precursor adducts and the *trans*-($\mu$-1,2-peroxo)dicopper(II) product. *J. Am. Chem. Soc.* **1993**, *115*, 2677–2689.

102. Jacobson, R. R.; Tueklar, Z.; Farooq, A.; Karlin, K. D.; Liu, S.; Zubieta, J. A copper–oxygen ($Cu_2O_2$) complex. Crystal structure and characterization of a reversible dioxygen binding system. *J. Am. Chem. Soc.* **1988**, *110*, 3690–3692.

103. Kitajima, N.; Fujisawa, K.; Fujimoto, C.; Morooka, Y.; Hashimoto, S.; Kitagawa, T.; Toriumi, K.; Tatsumi, K.; Nakamura, A. A new model for dioxygen binding in hemocyanin. Synthesis, characterization, and molecular structure of the $\mu$-$\eta^2$:$\eta^2$ peroxo dinuclear copper(II) complex, $[Cu(HB(3,5-R_2pz)_3]_2(O_2)$ (R = isopropyl and Ph). *J. Am. Chem. Soc.* **1992**, *114*, 1277–1291.

104. Kitajima, N.; Fujisawa, K.; Morooka, Y.; Toriumi, K. $\mu$-$\eta^2$:$\eta^2$-Peroxo binuclear copper complex, $[Cu(HB(3,5-(Me_2CH)_2pz)_3]_2(O_2)$. *J. Am. Chem. Soc.* **1989**, *111*, 8975–8976.

105. Funahashi, Y.; Nishikawa, T.; Wasada-Tsutsui, Y.; Kajita, Y.; Yamaguchi, S.; Arii, H.; Ozawa, T.; Jitsukawa, K.; Tosha, T.; Hirota, S.; Kitagawa, T.; Masuda, H. Formation of a bridged butterfly-type $\mu$-$\eta^2$:$\eta^2$-peroxo dicopper core structure with a carboxylate group. *J. Am. Chem. Soc.* **2008**, *130*, 16444–16445.

106. Mahapatra, S.; Halfen, J. A.; Wilkinson, E. C.; Pan, G.; Wang, X.; Young, V. G., Jr.; Cramer, C. J.; Que, L., Jr.; Tolman, W. B. Structural, spectroscopic, and theoretical characterization of bis-($\mu$-oxo)dicopper complexes, novel intermediates in copper-mediated dioxygen activation. *J. Am. Chem. Soc.* **1996**, *118*, 11555–11574.

107. Zhang, C. X.; Kaderli, S.; Costas, M.; Kim, E.-i.; Neuhold, Y.-M.; Karlin, K. D.; Zuberbuhler, A. D. Copper(I)–dioxygen reactivity of $[(L)Cu^I]^+$ (L-tris(2-pyridylmethyl)amine: kinetic/thermodynamic and spectroscopic studies concerning the formation of $Cu–O_2$ and $Cu_2.O_2$ adducts as a function of solvent medium and 4-pyridyl ligand substituent variations. *Inorg. Chem.* **2003**, *42*, 1807–1824.

108. Weitzer, M.; Schindler, S.; Brehm, G.; Hormann, E.; Jung, B.; Kaderli, S.; Zuberbuhler, A. D. Reversible binding of dioxygen by the copper(I) complex with tris(2-dimethylaminoethyl) amine ($Me_6$tren) ligand. *Inorg. Chem.* **2003**, *42*, 1800–1806.

109. Jensen, M. P.; Que, E. L.; Shan, X.; Rybak-Akimova, E. V.; Que, L., Jr. Spectroscopic and kinetic studies of the reaction of $[Cu^I(6-PhTPA)]^+$ with $O_2$. *Dalton Trans.* **2006**, 3523–3527.

110. Cahoy, J.; Holland, P. L.; Tolman, W. B. Experimental studies of the interconversion of $\mu$-$\eta^2{:}\eta^2$-peroxo- and bis($\mu$-oxo)dicopper complexes. *Inorg. Chem.* **1999**, *38*, 2161–2168.

111. Itoh, S.; Taki, M.; Nakao, H.; Holland, P. L.; Tolman, W. B.; Que, L., Jr.; Fukuzumi, S. Aliphatic hydroxylation by a bis($\mu$-oxo)dicopper(III) complex. *Angew. Chem., Int. Ed.* **2000**, *39*, 398–400.

112. Horwitz, C. P.; Winslow, P. J.; Warden, J. T.; Lisek, C. A. Reaction of the manganese(II) Schiff-base complexes (X-SALPRN)$Mn^{II}$ [X = H, 5-Cl, 5-$CH_3O$; SALPRN = 1,3-bis (salicylideneamino)propane] with dioxygen and reactivity of the oxygenated products. *Inorg. Chem.* **1993**, *32*, 82–88.

113. Lee, D.-H.; Lucchese, B.; Karlin, K. D. Multimetal oxidases. In *Comprehensive Coordination Chemistry II*; McCleverty, J. A.; Meyer, T. J., Eds.; Elsevier: Amsterdam, 2004, Vol. 8, pp 437–457.

114. Solomon, E. I.; Augustine, A. J.; Yoon, J. $O_2$ reduction to $H_2O$ by multicopper oxidases. *Dalton Trans.* **2008**, 3921–3932.

115. Solomon, E. I.; Sarangi, R.; Woertink, J. S.; Augustine, A. J.; Yoon, J.; Ghosh, S. $O_2$ and $N_2O$ activation by bi-, tri-, and tetranuclear Cu clusters in biology. *Acc. Chem. Res.* **2007**, *40*, 581–591.

116. Chufan, E. E.; Puiu, S. C.; Karlin, K. D. Heme-copper/dioxygen adduct formation, properties, and reactivity. *Acc. Chem. Res.* **2007**, *40*, 563–572.

117. Collman, J. P.; Decreau, R. A. Functional biomimetic models for the active site in the respiratory enzyme cytochrome c oxidase *Chem. Commun.* **2008**, 5065–5076.

118. Collman, J. P.; Boulatov, R.; Sunderland, C. J.; Fu, L. Functional analogues of cytochrome c oxidase, myoglobin, and hemoglobin. *Chem. Rev.* **2004**, *104*, 561–588.

119. Collman, J. P.; Decreau, R. A.; Yan, Y.; Yoon, J.; Solomon, E. I. Intramolecular single-turnover reaction in a cytochrome c oxidase model bearing a Tyr244 mimic. *J. Am. Chem. Soc.* **2007**, *129*, 5794–5795.

120. Collman, J. P.; Devaraj, N. K.; Decreau, R. A.; Yang, Y.; Yan, Y.-L.; Ebina, W.; Eberspacher, T. A.; Chidsey, C. E. D. A cytochrome c oxidase model catalyzes oxygen to water reduction under rate-limiting electron flux. *Science* **2007**, *315*, 1565–1568.

121. Groves, J. T. High-valent iron in chemical and biological oxidations *J. Inorg. Biochem.* **2006**, *100*, 434–447.

122. Sheng, X.; Horner, J. H.; Newcomb, M. Spectra and kinetic studies of the compound I derivative of cytochrome P450 119 *J. Am. Chem. Soc.* **2008**, *130*, 13310–13320.

123. Newcomb, M.; Hollenberg, P. F.; Coon, M. J. Multiple mechanisms and multiple oxidants in P450-catalyzed hydroxylations *Arch. Biochem. Biophys.* **2003**, *409* (1), 72–79.

124. Nam, W.; Cytochrome P450. In *Comprehensive Coordination Chemistry II*; McCleverty, J. A.; Meyer, T. J., Eds.; Elsevier: Amsterdam, 2004, Vol. 8, pp 281–307.

125. Chen, J.; Stubbe, J. Bleomycins: new methods will allow reinvestigation of old issues. *Curr. Opin. Chem. Biol.* **2004**, *8*, 175–181.

126. Burger, R. M. Cleavage of nucleic acids by bleomycin. *Chem. Rev.* **1998**, *98*, 1153–1169.

127. Chow, M. S.; Liu, L. V.; Solomon, E. I. Further insights into the mechanism of the reaction of activated bleomycin with DNA. *Proc. Natl. Acad. Sci. USA* **2008**, *105*, 13241–13245.

128. Costas, M.; Mehn, M. P.; Jensen, M. P.; Que, L., Jr. Dioxygen activation at mononuclear nonheme iron active sites: enzymes, models, and intermediates. *Chem. Rev.* **2004**, *104*, 939–986.

129. Maiti, D.; Lee, D.-H.; Gaoutchenova, K.; Wuertele, C.; Holthausen, M. C.; Narducci Sarjeant, A. A.; Sundermeyer, J.; Schindler, S.; Karlin, K. D. Reactions of a copper(II) superoxo complex lead to C–H and O–H substrate oxygenation: modeling copper-monooxygenase C–H hydroxylation. *Angew. Chem., Int. Ed.* **2008**, *47*, 82–85.

130. Wertz, D. L.; Valentine, J. S. Nucleophilicity of iron-peroxo porphyrin complexes. *Struct. Bond.* **2000**, *97*, 38–60.

131. Osako, T.; Ohkubo, K.; Taki, M.; Tachi, Y.; Fukuzumi, S.; Itoh, S. Oxidation mechanism of phenols by dicopper–dioxygen ($Cu_2/O_2$) complexes. *J. Am. Chem. Soc.* **2003**, *125*, 11027–11033.

132. Park, M. J.; Lee, J.; Suh, Y.; Kim, J.; Nam, W. Reactivities of mononuclear non-heme iron intermediates including evidence that iron(III)-hydroperoxo species is a sluggish oxidant. *J. Am. Chem. Soc.* **2006**, *128*, 2630.

133. Yin, G.; Buchalova, M.; Danby, A. M.; Perkins, C. M.; Kitko, D.; Carter, J. D.; Scheper, W. M.; Busch, D. H. Olefin epoxidation by hydrogen peroxide adduct of a novel non-heme manganese(IV) complex: demonstration of oxygen transfer by multiple mechanisms. *Inorg. Chem.* **2006**, *45*, 3467–3474.

134. Mayer, J. M. Hydrogen atom abstraction by metal-oxo complexes: understanding the analogy with organic radical reactions. *Acc. Chem. Res.* **1998**, *31*, 441–450.

135. Roth, J. P.; Yoder, J. C.; Won, T.-J.; Mayer, J. M. Application of the Marcus cross relation to hydrogen atom transfer reactions. *Science* **2001**, *294*, 2524–2526.

136. Yin, G.; Danby, A. M.; Kitko, D.; Carter, J. D.; Scheper, W. M.; Busch, D. H. Oxidative reactivity difference among the metal oxo and metal hydroxo moieties: pH dependent hydrogen abstraction by a manganese(IV) complex having two hydroxide ligands. *J. Am. Chem. Soc.* **2008**, *130*, 16245–16253.

137. Rosenthal, J.; Nocera, D. G. The role of proton-coupled electron transfer in O–O bond activation. *Acc. Chem. Res.* **2007**, *40*, 543–553.

138. Mayer, J. M. Proton-coupled electron transfer: a reaction chemist's view. *Annu. Rev. Phys. Chem.* **2004**, *55*, 363–390.

139. Rohde, J.-U.; In, J.-H.; Lim, M. H.; Brennessel, W. W.; Bukowski, M. R.; Stubna, A.; Münck, E.; Nam, W.; Que, L., Jr. Crystallographic and spectroscopic characterization of a non-heme Fe(IV)=O complex. *Science* **2003**, *299*, 1037–1039.

140. Mas-Balleste, R.; Que, L. Jr. Iron-catalyzed olefin epoxidation in the presence of acetic acid: insights into the nature of the metal-based oxidant. *J. Am. Chem. Soc.* **2007**, *129*, 15964–15972.

141. Pestovsky, O.; Stoian, S.; Bominaar, E. L.; Shan, X.; Münck, E.; Que, L., Jr.; Bakac, A. Aqueous $Fe^{IV}=O$: spectroscopic identification and oxo-group exchange. *Angew. Chem., Int. Ed.* **2005**, *44*, 6871–6874.

142. Xue, G.; Fiedler, A. T.; Martinho, M.; Münck, E.; Que, L. Jr. Insights into P-to-Q conversion in the catalytic cycle of methane monooxygenase from a synthetic model point of view. *Proc. Natl. Acad. Sci. USA* **2008**, *105*, 20615–20620.

143. Rowe, G.; Rybak-Akimova, E. V.; Caradonna, J. P. Unraveling the reactive species of a functional non-heme iron monooxygenase model using stopped-flow UV-vis spectroscopy. *Inorg. Chem.* **2007**, *46*, 10594–10606.

# 5 Activation of Molecular Hydrogen

GREGORY J. KUBAS and DENNIS MICHAEL HEINEKEY

## 5.1 INTRODUCTION

Dihydrogen ($H_2$) has long been vital in catalytic processes such as hydrogenation and conversions of organic compounds. It is also being considered as a future energy storage medium, and hydrogen production and storage are at the forefronts of research. The $H_2$ molecule is held together by a very strong two-electron H–H bond and is only useful chemically when the two H atoms are split apart in a controlled fashion. However, the molecular level mechanism by which the $H_2$ molecule binds and the H–H bond cleaves was established only relatively recently. The electronically saturated dihydrogen molecule had not previously been observed to chemically bind to a metal center, often the first step in breaking a strong bond. The discovery by Kubas and coworkers in 1984 of coordination of a nearly intact $H_2$ molecule to a metal complex ($L_nM$; L = ligand) caught this in intimate detail and led to a new paradigm in chemistry (Scheme 5.1).[1–4]

The $H_2$ binds side-on (referred to as $\eta^2$) to the metal primarily via donation of its two σ-bonding electrons to a vacant metal orbital to form a *stable* $H_2$ complex. It is remarkable that this already strongly bonded electron pair can be donated to a metal to form a nonclassical two-electron, three-center bond, as in other "electron-deficient" molecules such as diborane ($B_2H_6$).

The fact that the existence of dihydrogen complexes remained hidden for so long is remarkable. Metal dihydrides formed by oxidative addition of the H–H bond to a metal center had early on been known to be a part of catalytic cycles.[5,6] Although some type of metal–$H_2$ interaction was assumed to be a forerunner in dihydride formation, it was thought to be an unobservable intermediate. As previously detailed by Kubas, the complex $W(CO)_3(PR_3)_2(H_2)$ was the first molecular compound synthesized and isolated entirely under ambient conditions that contained the $H_2$ molecule (albeit "stretched") other than elemental $H_2$ itself! The H–H bond length in $W(CO)_3$ $(P^iPr_3)_2(H_2)$ (0.89 Å) is elongated ~20% over that in free $H_2$ (0.74 Å), showing that $H_2$ is not physisorbed but rather chemisorbed, where the bond is "activated" toward rupture. The latter process and all its structure/bonding/dynamical ramifications are

*Physical Inorganic Chemistry: Reactions, Processes, and Applications* Edited by Andreja Bakac
Copyright © 2010 by John Wiley & Sons, Inc.

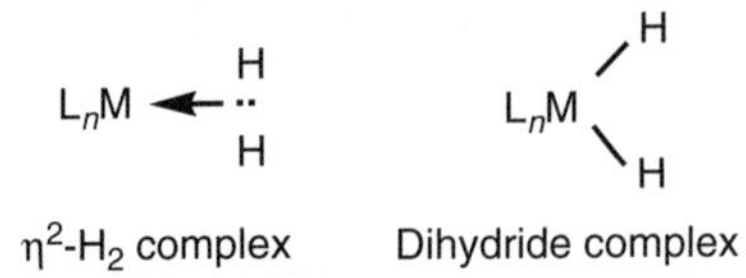

$$L_nM \longleftarrow \overset{H}{\underset{H}{\cdots}} \qquad L_nM \overset{\nearrow H}{\underset{\searrow H}{}}$$

$\eta^2$-H$_2$ complex          Dihydride complex

**SCHEME 5.1**

the reasons why the finding of a H$_2$ complex was important: it is the prototype for activation of *all* σ-bonds, including C–H bonds in hydrocarbons. This chapter will focus primarily on the physical chemistry and spectroscopic aspects of dihydrogen complexes. For example, IR and NMR spectroscopies will be highlighted because they were crucial to both the original discovery and subsequent understanding of H$_2$ binding and activation on metal centers. In this and other aspects of H$_2$ coordination chemistry, isotope effects are of utmost importance and will also be discussed in some detail. Irrefutable evidence for H$_2$ binding in W(CO)$_3$(P$^i$Pr$_3$)$_2$(H$_2$) was difficult to obtain because pinpointing H positions crystallographically is difficult in the presence of heavy atoms. As will be shown, IR spectroscopy gave the first clue that H$_2$ was bound molecularly and not atomically. Here, the large isotopic shifts observed in the spectra of H$_2$ versus D$_2$ complexes aided identification of the vibrations due to bound dihydrogen, which turned out to be much different than those in metal hydrides.

Ultimately, $^1$H NMR spectroscopy turned skeptics into believers: the HD complex W(CO)$_3$(P$^i$Pr$_3$)$_2$(HD) was synthesized and showed a large HD coupling constant that proved that the coordinated H–D bond was mostly intact and not split to give a hydride–deuteride such as W(H)(D)(CO)$_3$(P$^i$Pr$_3$)$_2$. The NMR showed a 1:1:1 triplet (deuterium spin = 1) with $J_{HD} = 33.5$ Hz, nearly as high as in HD gas, 43.2 Hz. Observation of $J_{HD}$ higher than that for a dihydride complex (>2 Hz) became the premier criterion for a H$_2$ complex, as will be described.

The appreciation of H$_2$ complexes was delayed by the notion that such complexes could not be stable relative to classical dihydrides. Even the theoretical basis for interaction of H$_2$ and other σ-bonds with a metal was still undeveloped at the time of the initial discovery. Ironically, a computational paper by Saillard and Hoffmann[7] in 1984 on the bonding of H$_2$ and CH$_4$ to metal fragments such as Cr(CO)$_5$ was published only shortly after our publication of the W–H$_2$ complex (without mutual knowledge). Such interplay between theory and experiment has continued to be an extremely valuable synergistic relationship.[3,4,8] The innate simplicity of H$_2$ was attractive computationally, but the structure/bonding/dynamics of H$_2$ complexes turned out to be extremely complex and led to extensive study (>300 computational papers).

Initially, H$_2$ binding in M(CO)$_3$(PR$_3$)$_2$(H$_2$) seemed unique because the bulky phosphines sterically inhibited formation of a seven-coordinate dihydride via oxidative addition. However, complexes with very simple ligands such as amine, CO, and even water have now been identified, indicating electronic considerations are also important. The hundreds of H$_2$ complexes synthesized after the initial discovery could not initially have been imagined, and it was difficult at first to know where to search for new ones. It would take over a year before others were identified, notably by Morris,

Crabtree, Chaudret, and Heinekey. These researchers have since performed synthetic, reactivity, and NMR studies on $H_2$ complexes[9–16,114] and were eventually joined by more than 100 investigators worldwide. Remarkably, several complexes initially believed to be hydrides were revealed to be $H_2$ complexes by Crabtree[9,17] in 1986 using as criteria the short proton NMR relaxation times of $H_2$ ligands. Particularly interesting was $RuH_2(H_2)(PPh_3)_3$, first reported in 1968.[18] This complex exhibits unusual $H_2$ lability, which led Singleton in 1976 to describe it as a characteristic of "$H_2$-like bonding."[19] However, attempts to prove $H_2$ binding here were problematic, even long after $H_2$ binding was established.[20]

Over 600 $H_2$ complexes are known (most of them stable) for nearly every transition metal and type of coligand and are the focus of 1500 publications, dozens of reviews, and three monographs.[2–4,8–16,114,21–29] The view on $H_2$ complexes has shifted from significance in basic science to more practical energy-relevant aspects, for example, $H_2$ production and storage. An important question after their discovery was are $H_2$ complexes relevant in catalysis, that is, does direct transfer of hydrogen from a $H_2$ ligand to a substrate occur? The answer is yes, and $H_2$ complexes are now believed to be crucial intermediates in hydrogen activation, particularly in *heterolytic* splitting processes.

This chapter will focus on spectroscopic characterization and other methodologies based on physical chemistry for studying interactions of $H_2$ with metal centers. Often the only evidence for formation of certain types of molecular $H_2$ complexes is spectroscopic, particularly solution- and solid-state NMR spectroscopy. NMR spectroscopy is by far the most commonly used and reliable diagnostic for identifying molecular versus atomic hydrogen coordination and the degree of activation of the $H_2$ ligand in metal–$H_2$ complexes. This physical chemistry method along with vibrational and neutron spectroscopy will thus be the primary focal points of this chapter. Isotope effects and thermodynamic and kinetic measurements on $H_2$ complexes will also be covered.

## 5.2   SYNTHESIS OF $H_2$ COMPLEXES

Several synthetic routes to $H_2$ complexes are available. Simplest is addition of $H_2$ gas to a coordinatively unsaturated complex such as $W(CO)_3(PR_3)_2$. Displacement by $H_2$ of weakly bound "solvento" ligands such as $CH_2Cl_2{}^{30}$ in $[Mn(CO)_3\{P(OCH_2)_3CMe\}_2(CH_2Cl_2)]^+$ or $H_2O$ from $[Ru(H_2O)_6]^{2+}$ to form $[Ru(H_2O)_5(H_2)]^{2+}$ (Scheme 5.2) under high $H_2$ pressure is also effective.

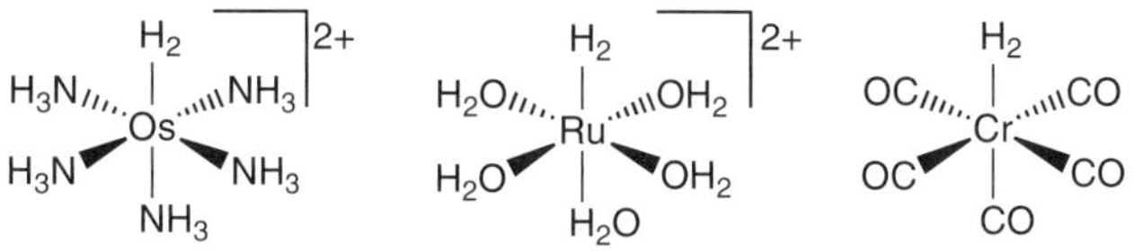

**SCHEME 5.2**

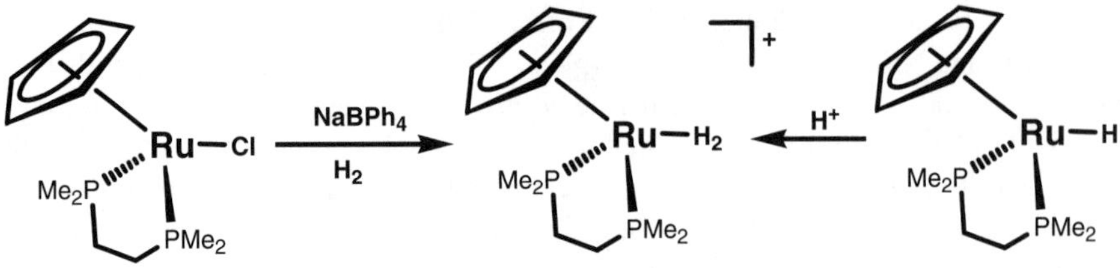

**SCHEME 5.3**

Although most metal–dihydrogen chemistry is carried out in organic solvents, aqueous systems have been found that have potential applications in biomimetic and green chemistry, for example, chemoselective hydrogenation catalysis.[31] Protonation of a hydride complex by acids is often used[13,24] and is widely applicable because it does not require an unsaturated precursor that may not be available. An example is shown in Scheme 5.3.

Kinetic studies of the protonation of metal hydrides with acids to form $H_2$ complexes of the type $[FeH(H_2)P_4]^+$ ($P_4 = 2$dppe or $P(C_2H_4PPh_2)_3$) show first-order dependence on concentration of both complex and acid.[32] This reaction in THF involves direct attack of acid at one hydride in $FeH_2P_4$, apparently by the mechanism in Scheme 5.4, where a type of hydrogen bonding interaction ("dihydrogen bonding") is likely to be the first step.

Importantly, this is the *microscopic reverse* of *heterolytic cleavage* of hydrogen, a crucial step in stoichiometric and catalytic reactions of $H_2$ as will be discussed below.

Most $H_2$ complexes contain low-valent metals with $d^6$ electronic configurations. Reversibility of $H_2$ binding is often a key feature; that is, $H_2$ can be removed simply upon exposure to vacuum and readded many times at ambient temperature/pressure. Virtually all $H_2$ (and other σ) complexes are diamagnetic, and dihydrogen ligands have not been definitively shown to bridge metals. Surprisingly, the coligands on $H_2$ complexes can be simple classical nitrogen-donor ancillary ligands such as ammonia, as in $[Os(NH_3)_5(H_2)]^{2+}$ (Scheme 5.2),[33] which has a very long H–H distance ($d_{HH}$), ~1.34 Å,[34] more characteristic of a dihydride, which it was initially believed to be.[35] Complexes containing only $H_2O^{36}$ or $CO^{37,38}$ coligands are also known, although they are marginally stable (Scheme 5.2). Heinekey showed that the pentacarbonyl complexes were stable enough to obtain low-temperature NMR spectra that proved that the $H_2$ was bound in similar fashion to that in $M(CO)_3(PR_3)_2(H_2)$. Importantly, all these findings demonstrated that sterically demanding coligands such as bulky

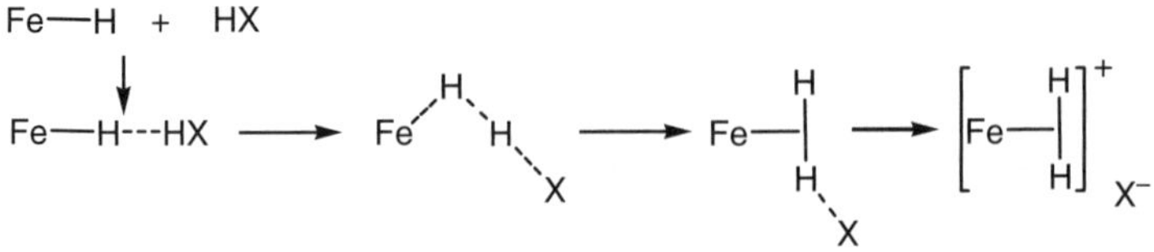

**SCHEME 5.4**

phosphines were unnecessary for $H_2$ coordination per se, although they could impart thermal stability electronically by increasing the electron richness of the metal center.

Determining the presence of a $H_2$ ligand and its $d_{HH}$ is nontrivial since even neutron diffraction has limited applicability and can give foreshortened $d_{HH}$ due to rapid $H_2$ rotation/libration.[39] As will be discussed below, $^1J_{HD}$ is the best criterion and values determined in solution correlate well with $d_{HH}$ in the solid state. Vibrational spectroscopy can also be useful, although the bands due to $H_2$ ligands can be weak and/or obscured, as will be discussed in Section 5.8. Inelastic neutron scattering is a powerful spectroscopic tool developed by Juergen Eckert and coworkers at Los Alamos. Rapid $H_2$ rotation in solid $H_2$ complexes provides unequivocal evidence for molecular $H_2$ binding and also the presence of $M \rightarrow H_2$ backdonation (BD).[40]

## 5.3   STRUCTURE, BONDING, AND DYNAMICS OF $H_2$ COMPLEXES

The three-center metal–$H_2$ interaction complements classical Werner-type coordination complexes where a ligand donates electron density through its nonbonding electron pair(s) and $\pi$-complexes such as olefin complexes in which electrons are donated from bonding $\pi$-electrons (Scheme 5.5).

It is remarkable that the *bonding* electron pair in $H_2$ can further interact with a metal center as strongly as a nonbonding pair in some cases. The resulting side-on bonding in M–$H_2$ and other $\sigma$-complexes is analogous to the 3c–2e bonding in carbocations and diborane. $H_2$ is thus a weak Lewis base that can bind to strong electrophiles, but transition metals are unique in stabilizing $H_2$ and other $\sigma$-bond complexes by *backdonation* of electrons from a filled metal d-orbital to the $\sigma^*$ antibonding orbital of $H_2$ (Scheme 5.5), a critical interaction unavailable to main group atoms.[3,4,8] The backdonation is analogous to that in the Dewar–Chatt–Duncanson model[41,42] for $\pi$-complexes such as ethylene complexes.

Backdonation of electrons from the metal center to $H_2$ is crucial not only in stabilizing M–$H_2$ bonding but also in activating the bond toward homolytic cleavage.[3,4,8] If it becomes too strong, for example, by increasing electron-donor strength of coligands on the metal center, the H–H bond cleaves to form a dihydride because of overpopulation of the $H_2 \sigma^*$ orbital. There is often a fine line between $H_2$ and dihydride coordination, and in some cases *equilibria* exist in solution for $W(CO)_3(PR_3)_2(H_2)$ (Scheme 5.6), showing that side-on coordination of $H_2$ is the first step in H–H cleavage.

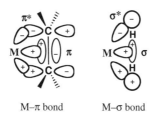

M–$\pi$ bond          M–$\sigma$ bond

**SCHEME 5.5**

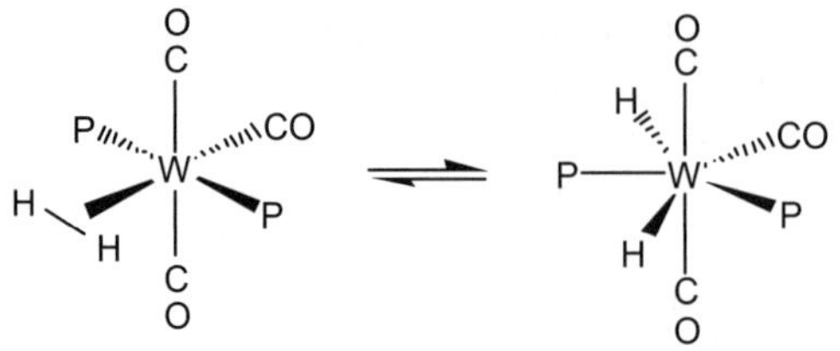

**SCHEME 5.6**

Although electronic factors for oxidative addition of $H_2$ are well established, the role of steric factors is less clear. Bulky phosphines can inhibit $H_2$ splitting: for less bulky R = Me, the equilibrium lies completely to the right; that is, the complex is a *dihydride*.[43] However, as shown above, $H_2$ complexes are also stable with only small coligands L such as $NH_3$ (Scheme 5.2), in some cases with greatly elongated $d_{HH}$, two further paradigm shifts. This led to extensive efforts to vary M, L, and other factors to study stretching of the H–H bond. Within the large regime of hundreds of $L_nM$–$H_2$ complexes, the reaction coordinate for the activation of $H_2$ on a metal (Scheme 5.3) shows $d_{HH}$ varying enormously, from 0.82 to 1.5 Å (Scheme 5.7).[3,10–15,17,21–24,30,34–40,43–50]

Although the $d_{HH}$ ranges shown are arbitrary, each category of complexes has distinct properties. The $d_{HH}$ is relatively short (0.8–1.0 Å), and $H_2$ is reversibly bound, in "true" $H_2$ complexes best exemplified by $W(CO)_3(PR_3)_2(H_2)$, much as in physisorbed $H_2$ where $d_{HH}$ is <0.8 Å. Elongated $H_2$ complexes ($d_{HH} = 1–1.3$ Å)[14,34,48,49–51] were first clearly identified in 1991 in $ReH_5(H_2)(PR_3)_2$ where neutron diffraction showed a $d_{HH}$ of 1.357(7) Å.[49] Complexes with $d_{HH} > 1.3$ Å are now viewed as "compressed hydrides," with NMR features differing from elongated $H_2$ complexes; for example, $J_{HD}$ increases with temperature for the former and decreases for the latter.[51] These are relative terms since a near *continuum* of $d_{HH}$ has been observed.

Activation of $H_2$ is very sensitive to metal, ligand, and charge; for example, changing R from phenyl to alkyl in $Mo(CO)(H_2)(R_2PC_2H_4PR_2)_2$ leads to splitting of $H_2$.[39] Strongly donating L, third-row M, and neutral charge favor elongation or splitting of H–H, while first-row M, electron-withdrawing L, and positive charge (cationic complex) favor $H_2$ binding and shorten $d_{HH}$. The ligand *trans* to $H_2$ has a

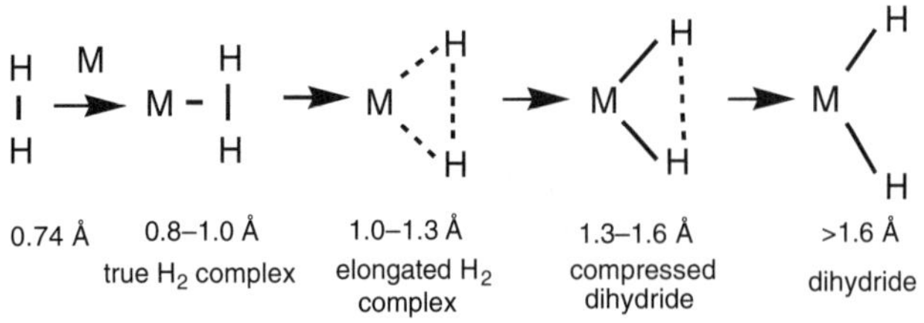

| 0.74 Å | 0.8–1.0 Å | 1.0–1.3 Å | 1.3–1.6 Å | >1.6 Å |
|--------|-----------|-----------|-----------|--------|
|        | true $H_2$ complex | elongated $H_2$ complex | compressed dihydride | dihydride |

**SCHEME 5.7**

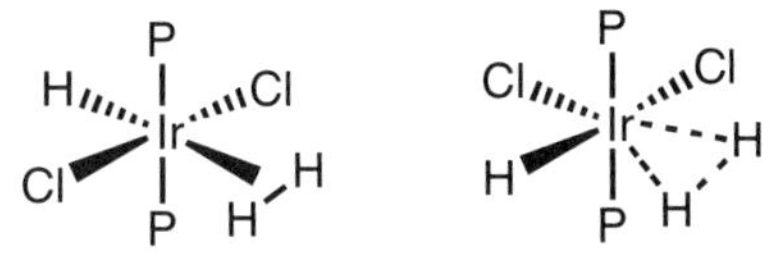

H–H = <0.9 Å        H–H = 1.11 Å

**SCHEME 5.8**

powerful influence: strong $\pi$-acceptors such as CO (and also strong $\sigma$-donors such as H) greatly reduce backdonation and normally keep $d_{HH} < 0.9$ Å. Thus, one can favor a $H_2$ complex by placing the $H_2$ *trans* to a strong $\pi$-acceptor. Conversely, mild $\sigma$-donors such as $H_2O$ or $\pi$-donors such as Cl *trans* to $H_2$ elongate $d_{HH}$ (0.96–1.34 Å), as dramatically demonstrated by the isomers in Scheme 5.8.[52] The *cis*-dichloro complex is actually a "compressed trihydride" ($d_{HH} \sim 1.5$ Å) in solution, but in the solid state it is an elongated $H_2$ complex ($d_{HH} = 1.11$ Å) due to Ir–Cl $\cdots$ H–Ir hydrogen bonding, illustrating the hypersensitivity of $d_{HH}$ to both intra- and intermolecular effects.[53]

Exceptions exist: the isomers of $Cr(CO)_4(PMe_3)(H_2)$ have similar $J_{HD}$ ($\sim 34$ Hz, hence $d_{HH} \sim 0.86$ Å) whether $H_2$ is *trans* to CO or $PMe_3$[38] most likely because the metal center is very *electron-poor* and backdonation is very weak.

The question can be asked: at what point is the H–H bond actually "broken"? Computational analyses suggest this occurs at a H–H distance near 1.48 Å, that is, twice the normal length,[54] but in reality little H–H bonding interaction remains for $d_{HH} > 1.1$ Å.[14] In certain "elongated" $H_2$ complexes, for example, $[OsCl(H_2)(dppe)_2]^+$, the potential energy surface for bond cleavage is extremely flat and the energy barrier for stretching the H–H bond from 0.85 Å all the way to 1.6 Å is calculated[14,51] to be astonishingly low, ca. 1 kcal/mol! The bonding of the $H_2$ ligand can thus be viewed as highly delocalized here: the H atoms undergo large amplitude vibrational motion along the reaction coordinate for H–H breaking. Remarkably, $d_{HH}$ is both temperature and isotope dependent in certain complexes such as $[CpM(diphosphine)(H_2)]^{n+}$ (M = Ru, Ir; $n = 1$, 2).[55,56] These phenomena illustrate the highly dynamic behavior of coordinated $H_2$, which can even exhibit unusual quantum mechanical phenomena such as rotational tunneling[40] and NMR exchange coupling.[57] M–H₂ interactions are among the most dynamic, complex, and enigmatic chemical topologies known. The $H_2$ ligand can bind/dissociate, reversibly split to dihydride, rapidly rotate, and exchange with *cis* hydrides, all on the same metal. In some cases, these dynamics cannot be frozen out on the NMR timescale even at low temperatures.

## 5.4 REACTIVITY OF H₂ COMPLEXES: ACIDITY AND HETEROLYSIS OF H–H BONDS

Aside from loss of $H_2$, reactions of M–$H_2$ complexes are dominated by homolytic cleavage of $H_2$ (oxidative addition) and heterolytic cleavage, essentially deprotonation of bound $H_2$ on electrophilic metal centers (Scheme 5.9).[10]

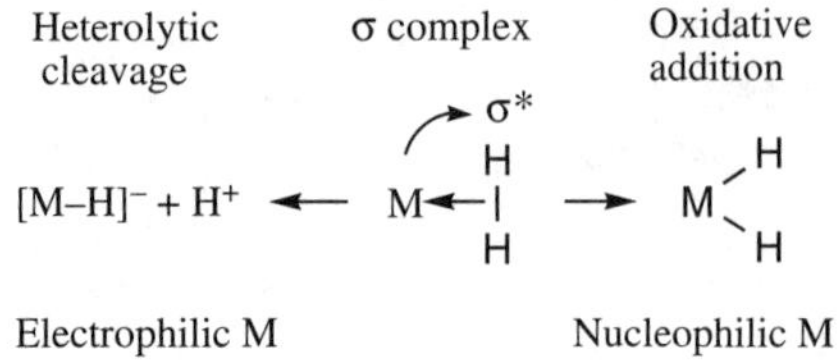

Heterolytic       σ complex       Oxidative
cleavage                            addition

$$[M-H]^- + H^+ \longleftarrow M{\leftarrow}\begin{array}{c} H \\ | \\ H \end{array} \longrightarrow M{<}\begin{array}{c} H \\ \\ H \end{array}$$

Electrophilic M                        Nucleophilic M

**SCHEME 5.9**

Dihydrogen complexes have several advantages in catalytic and other reactions. Foremost is that the formal *oxidation state of the metal does not change* on binding of $H_2$, whereas formation of a dihydride formally increases the metal oxidation state by two. $H_2$ ligands can also have far greater thermodynamic *and* kinetic acidity than hydrides, which is important in the ability of acidic $H_2$ ligands to protonate substrates such as olefins and $N_2$. In heterolytic cleavage,[10,26,58,59] the $H_2$ ligand is deprotonated and the remaining hydrogen ligates to the metal as a hydride. Both pathways have been identified in catalytic hydrogenation and may also be available for other σ-bond activations, for example, C–H cleavage. Heterolysis of H–H bonds via proton transfer to a basic site on a *cis* ligand or to an external base is a crucial step in many industrial and biological processes that involve direct reaction of $H_2$ ligands. $H_2$ complexes can undergo heterolysis via two primary pathways (Scheme 5.10).

*Intramolecular* heterolysis is extremely facile for proton transfer to a ligand L such as H or Cl positioned *cis* to the $H_2$ ligand. The proton can also end up transferring to a *trans* ligand[60] or to the counteranion X of a cationic complex (thus releasing an acid, HX). *Intermolecular* heterolysis involves protonation of an external base B, for example, an ether solvent, to give a metal hydride ($H^-$ fragment) and the conjugate acid of the base, $HB^+$. This is the reverse of protonation reactions used to synthesize $H_2$ complexes (all reactions in Scheme 5.10 can be reversible), and the $[HB]^+$ formed can relay the proton to internal or external sites (base-assisted heterolysis).

Crabtree first demonstrated heterolysis of $H_2$ by showing that the $H_2$ in $[Ir^IH(H_2)(LL)(PPh_3)_2]^+$ is deprotonated by LiR in preference to the hydride ligand.[61] A milder base, $NEt_3$, was shown by Heinekey[62] to more rapidly deprotonate the $\eta^2$-$H_2$ tautomer

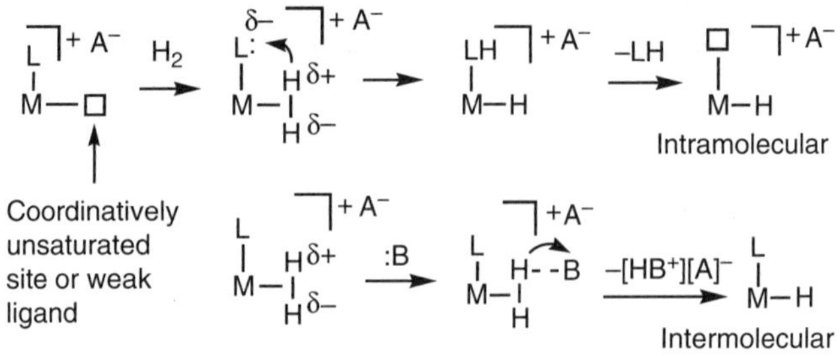

**SCHEME 5.10**

in an equilibrium mixture of $[CpRuH_2(dmpe)]^+$ and $[CpRu(H_2)(dmpe)]^+$. The $H_2$ ligand has greater kinetic acidity because deprotonation of a $H_2$ complex involves *no change in coordination number or oxidation state*. Thus, $H_2$ gas can be turned into a strong acid: free $H_2$ is an extremely weak acid ($pK_a \sim 35$ in THF),[63] but binding it to an electrophilic cationic metal increases the acidity spectacularly, *up to 40 orders of magnitude*. The $pK_a$ can become as low as $-6$; that is, $\eta^2$-$H_2$ *can become more acidic than sulfuric acid* as shown by Morris[10,11,59] and later Jia.[22] Electron-deficient cationic $H_2$ complexes with electron-withdrawing ligands such as CO and short H–H bonds (<0.9 Å), for example, $[Re(H_2)(CO)_4(PR_3)]^+$,[64] are among the most acidic. Positive charge increases acidity: $W(CO)_3(PCy_3)_2(H_2)$ is deprotonated only by strong bases,[65] but on oxidation to $[W(CO)_3(PCy_3)_2(H_2)]^+$ it becomes acidic enough to protonate weakly basic ethers.[66] As will be shown below, such ability is relevant to processes such as ionic hydrogenation and the function of metalloenzymes such as hydrogenases.

Complexes with $H_2$ ligands are often highly dynamic. Extremely rapid exchange between the hydrogens in an $\eta^2$-$H_2$ ligand and a hydride ligand located *cis* to it commonly occurs in solution. This process is facilitated by a *cis interaction*, that is, a weak hydrogen bonding like interaction between the $H_2$ and H ligands observable in the solid state.[3,8,67] The intermediate can be considered to be a "trihydrogen" ($H_3$) complex,[68,69] as exemplified by studies carried out on bis(cyclopentadienyl)Mo-type complexes. The *ansa*-bridged complexes $[Me_2X(C_5R_4)_2MoH(H_2)]^+$ (X = C, R = H; X = Si, R = Me) have been independently determined by both Heinekey[70] and Parkin[71] to be thermally labile *dihydrogen/hydride* complexes, the first complexes with $d^2$ electronic configurations to have *cis* hydride–dihydrogen ligands (Scheme 5.11).

Rapid dynamic processes interchange the hydride and dihydrogen moieties in these complexes. The bound $H_2$ ligand in **1** exhibits hindered rotation with $\Delta G^{\ddagger}_{150} = 7.4\,kcal/mol$. However, H-atom exchange is still rapid at temperatures down to 130 K. The dynamic process envisaged is depicted in Scheme 5.12, with the central Mo–trihydrogen structure representing a transition state for atom transfer from one side of the molecule to the other.

Many similar systems are known; for example, evidence exists for intermediacy of a $M(H_3)$-like complex in facile H-atom exchange in $ReH_2(H_2)(CO)(PR_3)_3$,[72] which can be exceedingly fast even at $-140°C$.[71,73–77] Remarkably, the barrier for hydrogen exchange in $IrClH_2(H_2)(P^iPr_3)_2$ is only 1.5 kcal/mol even in the *solid state*.[74,75]

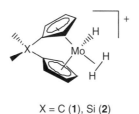

X = C (**1**), Si (**2**)

**SCHEME 5.11**

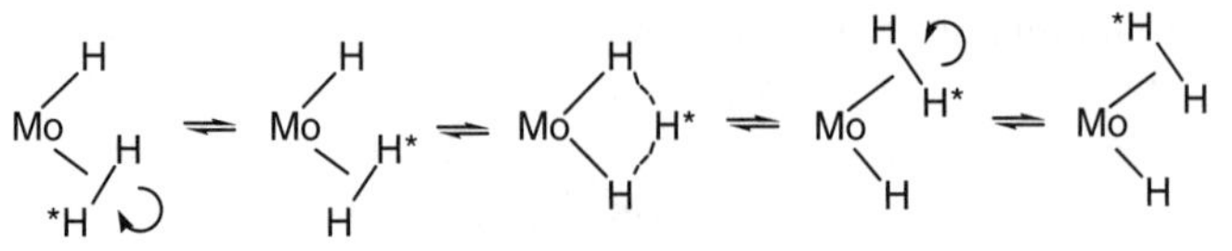

**SCHEME 5.12**

*Direct transfer* of hydrogens from a $H_2$ ligand can occur in catalytic hydrogenation. While difficult to prove conclusively, there is evidence in *ionic hydrogenation* processes where an organometallic hydride, for example, $CpMoH(CO)_3$, plus a strong acid, for example, $HO_3SCF_3$, reduce ketones.[78,79] An acidic $H_2$ complex is believed to be involved in proton transfer to the organic substrate. An elegant example of catalysis employing heterolysis of $H_2$ in a commercially valuable industrial process is the asymmetric hydrogenation of ketones to alcohols catalyzed by the ruthenium system of Noyori (Scheme 5.13).[80,81]

This conversion is catalyzed by *trans*-$RuCl_2[(S)$-binap]$[(S,S)$-dpen] (binap = [1,1'-binaphthalene-2,2'-diylbis(diphenylphosphane)]; dpen = diphenylethylenediamine) and is remarkable in several respects. The reaction is quantitative within hours, gives enantiomeric excesses (ee) up to 99%, shows high chemoselectivity for carbonyl over olefin reduction, and the substrate-to-catalyst ratio is >100,000. The nonclassical metal–ligand bifunctional catalytic cycle is mechanistically novel compared to that of the structurally similar classical ruthenium hydrogenation catalysts. Solvent (alcohol) assists the heterolysis of $H_2$ here, and many new "NH effect" bifunctional mechanisms involving amine, amido, and imido ligands are emerging wherein assistance by solvent, water, protons, and coligands aids heterolysis.[82–85] Once thought to be a mature field, catalytic hydrogenation centered on heterolytic splitting is greatly

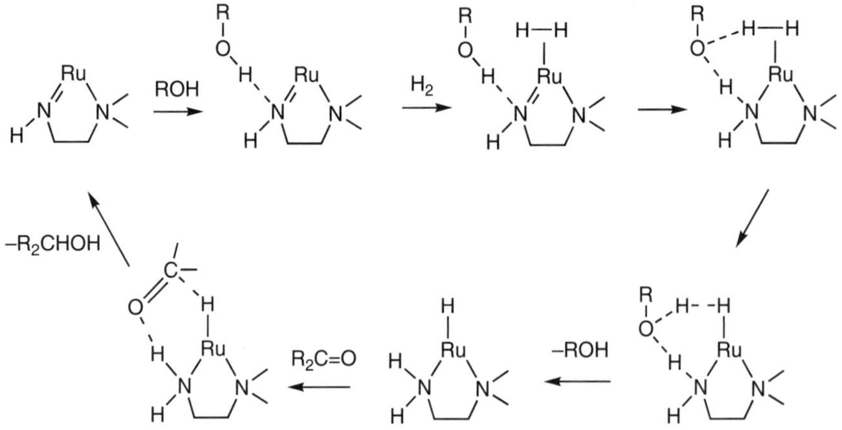

**SCHEME 5.13**

expanding in scope to include even purely main group systems (see below and Scheme 5.15).

## 5.5   ACTIVATION OF $H_2$ IN BIOLOGICAL AND NONMETAL SYSTEMS

Hydrogenases ($H_2$ases) are redox enzymes in microorganisms that catalyze $H_2 \rightleftharpoons 2H^+ + 2e^-$ to either utilize $H_2$ as an energy source or dispose of excess electrons as $H_2$.[86–89] Biologically unprecedented CO and CN ligands were found crystallographically to be present in the dinuclear active sites of $H_2$ases (e.g., Scheme 5.14),[90] which are remarkably organometallic-like and have been extensively modeled for biomimetic $H_2$ production.[88,89,91–96]

These enzymes evolved billions of years ago to utilize the CO ligand, whose strong *trans* influence favors reversible $H_2$ binding and heterolysis.[3,26] This diiron center (and similar centers in Ni–Fe $H_2$ases) presumably transiently binds and heterolytically splits $H_2$, most likely at a binding site *trans* to the bridging CO, where a proton transfers to a thiolate ligand or other Lewis basic site.[89] Such heterolysis has been shown to occur on a mononuclear Fe complex with a pendant nitrogen base that acts as a proton relay[94] as well as on nickel centers.[97] A $H_2$ complex of a H-ase model, $[Ru_2(\mu\text{-}H)(\mu\text{-}S_2C_3H_6)_2(H_2)(CO)_3(PCy_3)_2]^+$, is known, albeit with Ru instead of Fe.[98] A *mononuclear* iron hydrogenase has now been characterized that contains two *cis*-CO ligands where calculations indicate hydrogen is split via an unexpected dual-pathway mechanism.[99,100]

$H_2$ can also be activated at nonmetals, for example, the bridging sulfides in $Cp_2Mo_2S_4$ that react with $H_2$ to form SH ligands perhaps via a four-center $S_2H_2$ transition state.[101] Metal-free hydrogenation of ketones on strong bases such as *t*-BuOK occurs under harsh conditions, apparently via base-assisted heterolysis of $H_2$.[102,103] Thus, $H_2$ is a very weak acceptor (Lewis acid) via electron donation to its $\sigma^*$ orbital and can interact with the O in alkoxide or metal oxides and undergo heterolysis.[3] Significantly, the first example of *reversible* splitting of $H_2$ on a *nonmetal center* has been found.[104] The phosphine borane in Scheme 5.15 has a strong Lewis acidic center (boron) linked to a Lewis basic site (phosphorus).

As suggested by theoretical calculations,[105] it is likely that $H_2$ heterolysis takes place at boron where proton transfer from a $H_2$-like complex to the basic phosphorus site occurs to form a phosphenium borate. Simple combinations of Lewis acids and

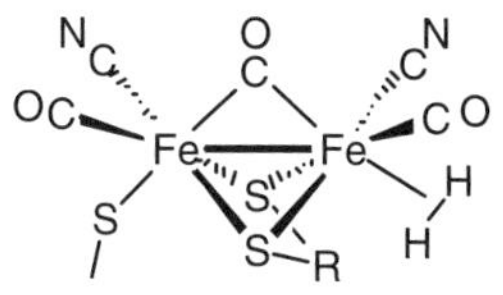

**SCHEME 5.14**

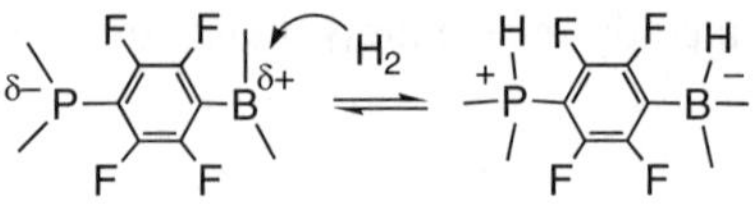

**SCHEME 5.15**

bases that are sterically precluded from forming dative bonds with each other ("frustrated Lewis pairs") can also effect heterolysis of $H_2$ and can even act as metal-free hydrogenation catalysts.[106] New examples include combinations of amines and boranes such as $B(C_6F_5)_3$.[107]

## 5.6   $H_2$ STORAGE AND PRODUCTION

$H_2$ is currently being touted as a fuel of the future, but difficult challenges exist. Materials for $H_2$ storage are difficult to design because, although $H_2$ can be readily extruded from a variety of compounds, it can be difficult to readd to give a regenerable system. The materials must also be light and contain >6% by weight $H_2$, reducing prospects for known facile reversible systems such as metal–$H_2$ or hydride complexes. Amine borane, $H_3NBH_3$, is a popular candidate and also combines both Lewis acidic (B) and basic (N) centers. Here, however, these centers are directly bonded, whereas the acidic and basic sites are separated by linkers in the phosphine–borane in Scheme 5.15. The metal-free aspect is relevant because precious metals such as platinum are often used in catalysis and can be environmentally unfriendly as well as costly or in short supply. Materials such as metal-organic frameworks (MOFs)[108–110] are now being examined for $H_2$ storage and have huge surface area capable of binding a large number of $H_2$ molecules. As will be discussed in Section 5.8.7, neutron scattering spectroscopy by Eckert is critical in determining whether $H_2$ binds to unsaturated metal centers as in organometallics and/or is physisorbed in the framework. An excellent example here of $H_2$ directly binding to the metal center is $H_2$ absorbed on a Cu-exchanged ZSM-5 zeolite.[111] Calculations indicate complexes with multiple $H_2$ ligands, that is, $Cr(H_2)_6$, may be stable,[112] and metal ion species such as $[M(H_2)_n]^+$ ($n = 6$–$10$) have a fleeting gas-phase existence,[113] but isolation in condensed phases will be problematic. The highest number of $H_2$ ligands coordinated to a single metal center in stable fashion is only two, the best example of which is $RuH_2(H_2)_2(PR_3)_2$.[15,114]

Production of $H_2$ fuel from water via solar energy is of exceedingly high interest.[115] Catalysis may involve $H_2$ complexes at least as intermediates, and $H_2$ complexes had previously been implicated in solar energy conversion schemes based on photoreduction of water.[116] Industrially important water-gas shift and related $H_2$-producing reactions undoubtedly proceed via transient $H_2$ complexes.[117] Biomimetic $H_2$ production, particularly solar driven (via photocatalysis), is also a challenge and may take a cue from models of the active site of hydrogenase coupled with models of nature's photosystems.[91–93] Here, the formation of H–H bonds from protons and electrons, the microscopic reverse of $H_2$ heterolytic splitting, will be crucial in leading to formation of $H_2$. This is very rapid at the Fe sites in $H_2$ases, and catalysts based on inexpensive first-row transition metals rather than precious metals will be critical. Coupling $H_2$ production catalysts with photochemical water splitting will require fine-tuning of electrochemical potentials for tandem catalysis schemes. Homogeneous *molecular* catalyst systems should have an advantage over heterogeneous catalysts because such fine-tuning can be facilitated merely by ligand variation. Clearly, the very large body of knowledge developed over the past 25 years in the realm of molecular $H_2$ coordination and activation will be useful in this endeavor.

This chapter will focus on spectroscopic characterization and other physical chemistry-based methodologies for studying interactions of $H_2$ with metal centers. Often the only evidence for formation of molecular $H_2$ complexes is spectroscopic, particularly solution- and solid-state NMR methodologies. NMR is by far the most commonly used and reliable diagnostic for identifying the degree of activation of the $H_2$ ligand in metal–$H_2$ complexes. This method will thus be the primary focus of this chapter, along with vibrational and neutron spectroscopy. Thermodynamic and kinetic measurements on $H_2$ complexes will also be covered.

## 5.7   STRUCTURE DETERMINATION FOR DIHYDROGEN COMPLEXES

### 5.7.1   Diffraction Methods

X-ray diffraction is a standard method employed by chemists to determine the structure of new complexes. In the case of transition metal hydride and dihydrogen complexes, precise location of metal-bound hydrogen atoms by X-ray diffraction is very difficult. Superior structural information is provided by neutron diffraction, but the requirement for large well-formed single crystals has limited this method to a small subset of the known complexes.[118]

Neutron diffraction data were crucial in the original characterization by Kubas of the first dihydrogen complexes, although the quality of the room-temperature neutron diffraction data set collected for $W(CO)_3(P^iPr_3)_2(H_2)$ was not optimal due to some disorder of the phosphine ligands (Figure 5.1). Nevertheless, the presence of a bound dihydrogen ligand was suggested and later verified by a high-quality low-temperature X-ray diffraction study.[1]

Subsequent diffraction studies revealed that even at low temperature, thermal motion of bound dihydrogen is significant, which can have important consequences for attempts to determine the HH distance, which is the most important structural parameter. For example, a careful low-temperature X-ray diffraction study of $Cr(CO)_3(P^iPr_3)_2(H_2)$ gave $d_{HH} = 0.67(5)$ Å, which is shorter than the $d_{HH} = 0.74$ Å found in $H_2$ gas.[119] A similar result was reported from a neutron diffraction study of $Mo(CO)(dppe)_2(H_2)$, where $d_{HH} = 0.736(10)$ Å was reported.[39] These anomalous results arise from correlated motion of the two H atoms in the bound dihydrogen

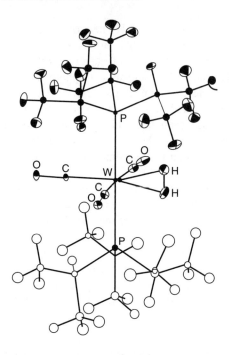

**FIGURE 5.1** Neutron structure of $W(CO)_3(P^iPr_3)_2(H_2)$. Disorder is present in lower phosphine ligand.

ligand. Under standard refinement procedures, rotation or librational motions have the effect of shortening the apparent $d_{HH}$ (Scheme 5.16).

The $d_{HH}$ can be corrected for thermal motion,[120] but difficulties in accounting for the internal vibrational motion of the H atoms can lead to an overestimation of the correction factor and large uncertainties in $d_{HH}$. For example, the corrected $d_{HH}$ for $Mo(CO)(dppe)_2(H_2)$ is estimated to be 0.85–0.88 Å. A summary of neutron diffraction data for dihydrogen complexes, including some that have been corrected for thermal motion, is provided by Koetzle.[121]

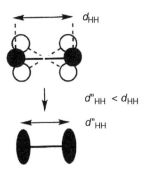

**SCHEME 5.16**

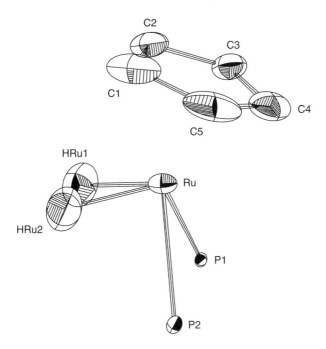

**FIGURE 5.2**    ORTEP diagram for $[Cp^*Ru(dppm)(H_2)]^+$, with methyl and phenyl groups omitted. (Reprinted with permission from Ref. 122. Copyright 1994 American Chemical Society.)

As noted above, a continuum of values for $d_{HH}$ is now available, ranging from very short as in the original Kubas complexes to conventional dihydride complexes with $d_{HH} \geq 1.5$ Å. A relatively small number of complexes with intermediate values of $d_{HH}$ known as elongated dihydrogen complexes have been reported.[14] An example of this type of complex is provided by $[Cp^*Ru(dppm)(H_2)]^+$. A value of $d_{HH} = 1.10 \pm .03$ Å was reported using neutron diffraction by Morris, Koetzle, and coworkers in 1994 (Figure 5.2).[122]

This Ru complex has subsequently been studied computationally, with very interesting results suggesting new interpretations for vibrational spectra (see Section 5.8.5) and predicting novel isotope effects (Section 5.9.5). The NMR spectroscopy of isotopically substituted variants of this complex is very interesting (Section 5.7.3).

### 5.7.2  Solid-State NMR Spectroscopy

The direct measurement of dipolar couplings in solid-state $^1H$ NMR is a potentially general approach to this problem that requires modest quantities of solid sample. Since dipolar coupling between the bound hydrogen atoms is proportional to $(d_{HH})^{-3}$, this method gives very precise values for $d_{HH}$. The requirement for deuterium substitution of nonhydride hydrogen atoms can be avoided by using selective pulse sequences.[123] The $d_{HH}$ measured by this method for W(CO)$_3$(P-Cy$_3$)$_2$(H$_2$) is 0.88 Å. Longer values, up to 1.02 Å, are found for Ru complexes, as shown in Figure 5.3.

The temperature dependence of the Pake pattern can be used to deduce that the bound dihydrogen ligand undergoes a torsional or hindered rotation motion around an axis perpendicular to the metal–dihydrogen axis. The bound hydrogen is characterized as a rigid planar rotator. In some cases, the potential surface for this rotation can be characterized by these measurements.

Related studies have made use of $^2$H NMR spectroscopy in the solid state. The theoretical foundation for the study of dynamics in dihydrogen complexes was developed by Buntkowsky and coworkers.[124] Solid-state $^2$H NMR spectroscopy has been used to study the dynamics of bound D$_2$ in *trans*-[Ru(D$_2$)Cl(dppe)$_2$]PF$_6$, where evidence was found for coherent rotational tunneling, with a barrier of 6.2 kcal/mol.[125] In more recent work, solid-state $^2$H NMR spectroscopy has been used to determine the structure and dynamics of several dihydrogen Ru complexes.[126]

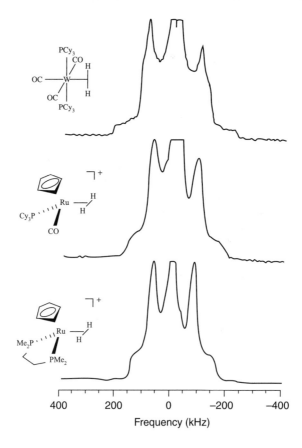

**FIGURE 5.3**  Measurement of dipolar couplings (Pake patterns) in solid-state $^1$H NMR spectra of H$_2$ complexes. (Reprinted with permission from Ref. 123. Copyright 1990 Elsevier.)

### 5.7.3  Solution NMR Spectroscopy

A solution $^1H$ NMR technique that uses the measurement of dipole–dipole relaxation rates was originally developed to measure $d_{HH}$ by Crabtree[127] and coworkers, and later refined by Halpern and coworkers.[128] This method requires the measurement at various temperatures of the spin lattice relaxation time $(T_1)$ of the hydride resonance. If the temperature corresponding to the maximum rate of relaxation (minimum $T_1$) can be reached, a value for $d_{HH}$ can be extracted. One complicating factor in interpretation of such data arises from rapid reorientation of the bound $H_2$ ligand, which is observed in a subset of hydrogen complexes. This issue was first addressed by Morris[129] and subsequently analyzed in depth by Caulton, Gusev, and coworkers.[130]

An example of the application of this technique is shown below for the simple complex $Cr(CO)_5(H_2)$.[37,38] For this hydrogen complex, minimum relaxation times are 22 ms (500 MHz) or 31 ms (750 MHz) at 190–200 K. Analysis of the $T_{1min}$ data for $Cr(CO)_5(H_2)$ leads to $d_{HH} = 0.87\,\text{Å}$ (500 MHz) and 0.86 Å (750 MHz) with the assumption of rapid $H_2$ rotation. The assumption of slow $H_2$ rotation gives values of $d_{HH} = 1.09\,\text{Å}$ (500 MHz) and 1.08 Å (750 MHz) (Figure 5.4).

These data, combined with HD coupling data (see below), suggest that the dihydrogen ligand in $Cr(CO)_5(H_2)$ is reorienting very rapidly. This observation is consistent with vibrational spectroscopy in liquid xenon, where Poliakoff and coworkers have suggested that the $H_2$ unit in $Cr(CO)_5(H_2)$ is essentially spinning freely on the basis of the large linewidths for the $\nu_{HH}$ absorptions (see Section 5.8.2).[131]

A drawback to this method (as well as neutron diffraction data) is that the derived values of $d_{HH}$ can be affected by the rapid rotational motion of the bound dihydrogen ligand observed in some complexes. If the hydrogen rotation is relatively slow or very rapid compared to the rate of molecular tumbling, the analysis is straightforward.

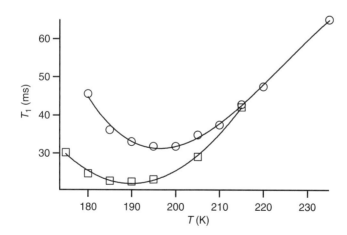

**FIGURE 5.4** $T_{1min}$ data for $Cr(CO)_5(H_2)$. (Reprinted with permission from Ref. 38. Copyright 2006 American Chemical Society.)

If rotation is intermediate between these extremes, a complete analysis is very complex.

A very useful and general solution NMR method that has been widely employed relies upon the measurement of H–D couplings in the bound dihydrogen ligand, where one H atom has been replaced by D. A variety of approaches have been employed to introduce a single deuteron (nuclear spin $I = 1$), the simplest of which is to employ HD gas in the synthesis reaction. In free HD gas, the coupling between H and D ($^1J_{HD}$) is 43 Hz. When HD gas reacts with a transition metal precursor to form a dihydride complex, the resulting two-bond coupling between H and D ($^2J_{HD}$) is typically very small, ca. 2–3 Hz. Dihydrogen complexes have $^1J_{HD}$ values between these two limits, and the value of $^1J_{HD}$ is inversely related to the internuclear distance $d_{HH}$. This empirical correlation is anchored by data from solid-state NMR and neutron diffraction determinations as outlined above.[132] Quantitatively,

$$d_{HH}(\text{Å}) = 1.44 - 0.0168 J_{HD} \qquad (5.1)$$

An essentially identical linear relationship between $J_{HD}$ and $d_{HH}$ was predicted by quantum chemical calculations.[133]

The measurement of HD coupling has been used to derive reliable values for $d_{HH}$ in a large number of hydrogen complexes. This is particularly useful for cases where the complexes lack sufficient stability to be isolated at room temperature. For example, $(CO)_5Cr(H_2)$ was characterized as a dihydrogen complex by this method, although this complex cannot be isolated at room temperature. In the monodeuterated species $(CO)_5Cr(HD)$, the observed $J_{HD} = 35.8$ Hz, which leads to $d_{HH} = 0.84$ Å (Figure 5.5). This observation is helpful in interpreting the relaxation time ($T_1$) data as noted above, where two different values for $d_{HH}$ can be derived from the relaxation data. The HD

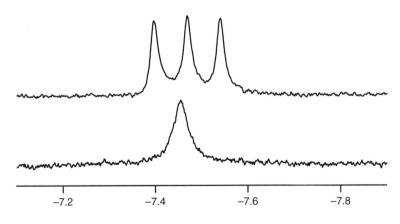

**FIGURE 5.5**  *Bottom*: $^1H$ NMR spectrum of $(CO)_5Cr(H_2)$. *Top*: $^1H$ NMR spectrum of $(CO)_5Cr(HD)$, showing $J_{HD} = 35.8$ Hz. (Reprinted with permission from Ref. 38. Copyright 2006 American Chemical Society.)

coupling data are only consistent with the shorter value for $d_{HH}$ resulting from the assumption of rapid rotation of the bound dihydrogen ligand.

The simple linear relationship between HD coupling and $d_{HH}$ breaks down at longer HH distances, such as those found in elongated dihydrogen species. A more sophisticated analysis of the relationship between HD coupling and $d_{HH}$ has recently been reported by Chaudret, Limbach, and coworkers, which allows for reliable values for $d_{HH}$ to be extracted from HD coupling.[134] A slightly different nonlinear relationship was reported by Heinekey and coworkers (Figure 5.6).[135]

Implicit in the extraction of $d_{HH}$ from values of $J_{HD}$ is the assumption that molecular structure is independent of mass. For some dihydrogen complexes, this may not be correct. As noted above, neutron diffraction gives $d_{HH} = 1.10 \pm 0.03$ Å for $[Cp^*Ru(dppm)(H_2)]^+$. The corresponding value of $J_{HD}$ is *temperature dependent*, ranging from 22.3 Hz at 213 K to 21.1 Hz at 295 K, suggesting that the HD distance increases slightly when the temperature is increased. This unusual behavior was initially attributed to thermal population of vibrational excited states, leading to a longer average distance at higher temperature. Subsequent computational studies of this complex by Lledós, Lluch, and coworkers revealed that the potential energy surface describing the Ru–H and HH interactions is very flat. Thus, large excursions from the equilibrium internuclear distances are possible at modest energies. Lledós and Lluch predicted that this anharmonic potential surface would lead to significant *structural isotope effects*. It was predicted that the deuterium complex would have a DD distance ca. 3% shorter than the distance observed in the protio species.[136]

These predictions were subsequently verified experimentally for $[Cp^*Ru(dppm)(H_2)]^+$ by study of the HD, HT, and DT complexes using $^1H$ and $^3H$ NMR spectroscopy.[55] A significant isotope effect was observed on the values of the HT

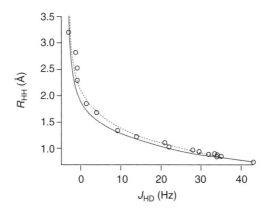

**FIGURE 5.6**  $R_{HH}$ as a function of $J_{HD}$. The solid line represents the relationship reported by Chaudret and coworkers (Ref. 134). The dotted line is derived using slightly different parameters, as reported by Heinekey and coworkers (Ref. 135). (Reprinted with permission from Ref. 135. Copyright 2004 American Chemical Society.)

and DT coupling, consistent with a shortening of the bond, as predicted computationally. This observation represents experimental confirmation of an unprecedented isotope effect upon the structure of a stable molecule. This is another manifestation of the unique structural attributes of dihydrogen and hydride complexes, where the H atoms are often highly delocalized and several alternative *structures* may have very similar energies. Closely related Ru complexes with slightly different ligands allow for the study of complexes with $d_{HH}$ varying from 1 to 1.2 Å. Interestingly, temperature-dependent HD couplings were only observed for molecules with $d_{HH}$ near to 1.10 Å, suggesting that the requirements for a soft potential energy surface are quite exacting.

It is reasonable to ask whether this unusual behavior is common for dihydrogen complexes of other metals. When dicationic complexes such as [Cp*Ir(dmpm)(H$_2$)]$^{2+}$ were prepared, $J_{HD} = 7$–$9$ Hz was observed, which is suggestive of a highly elongated dihydrogen ligand, which could also be described as a compressed dihydride structure. For example, the observed value of $J_{HD}$ for [Cp*Ir(dmpm)(HD)]$^{2+}$ is 8.1 Hz at 240 K, which corresponds to $d_{HH} = 1.45$ Å, in good agreement with the value determined from $T_{1min}$ of 1.49 Å. Interestingly, the HD coupling is temperature dependent, with *higher* values of $J_{HD}$ observed at higher temperatures. This is the opposite trend to the temperature dependence observed for the Ru complexes discussed above. Through computational studies, these observations were interpreted in terms of a potential energy surface with two distinct minima, with different HH distances. The structure with the longer HH distance is slightly more stable, so it is favored at low temperature. Raising the observation temperature increases the fraction of the less stable structure with the shorter HH distance. The barrier for interconversion of the two structures is very low.[135]

## 5.8   VIBRATIONAL SPECTROSCOPIC STUDIES OF H$_2$ BINDING

### 5.8.1   Introduction

Vibrational spectroscopy is not only diagnostic for dihydrogen coordination but as will be described below provided the first clue leading to the discovery of the H$_2$ ligand in W(CO)$_3$(PR$_3$)$_2$(H$_2$). Although IR and (especially) Raman studies of H$_2$ complexes are much more infrequent than NMR characterization, they along with neutron spectroscopy have given valuable information on the structure, bonding, and dynamics of dihydrogen complexes. Isotope effects involving deuterium substitution are strong examples here, as will be detailed below in Section 5.9. Low-temperature vibrational spectroscopy has been invaluable in characterizing thermally unstable H$_2$ complexes formed, for example, in matrix isolation studies. IR studies have also been crucial in the discovery that hydrogenase enzymes that activate H$_2$ contain biologically unprecedented CO and cyanide ligands in their organometallic-like active sites.[137] Other non-NMR spectroscopic techniques have been useful, particularly for *weakly* bound H$_2$, and include inelastic neutron scattering (INS) and mass spectrometry, as will be discussed below.

## 5.8.2   Vibrational Modes for $\eta^2$-$H_2$: Clue to Discovery of $H_2$ Complexes

The complexes formed by addition of $H_2$ to $W(CO)_3(PR_3)_2$ were initially intriguing because of their unusually facile $H_2$ loss *in vacuo* or under argon. However, the anomalous IR frequencies of the isolated solid $H_2$ adducts were the first key indicators to the discovery that the hydrogen was bound *molecularly rather than atomically* (i. e., dihydride). Instead of the expected $\nu_{WH}$ at 1700–2300 cm$^{-1}$ and $\delta_{WH}$ at 700–900 cm$^{-1}$ that would be characteristic of a tungsten hydride complex,[138] *three* widely spaced bands at 1567, 953, and 465 cm$^{-1}$ were observed for $W(CO)_3(P^iPr_3)_2(H_2)$ (Figure 5.7).[2,139]

Importantly, these bands shifted dramatically to 1132, 704, and 312 cm$^{-1}$ when deuterium was substituted for hydrogen, a valuable technique for locating M–H vibrational modes. The isotopic mass shifts ranged from 1.38 to 1.45, close to that expected (1.414). Good fortune also played a role in the discovery: CsBr salt plates were routinely used along with an IR spectrometer capable of recording bands down to 250 cm$^{-1}$. Otherwise, the very weak telltale band at 465 cm$^{-1}$ shifting to 312 cm$^{-1}$ would have escaped notice because most IR spectrometers then in use did not record below 400 cm$^{-1}$. The anomalous low-frequency mode along with the high lability of the $H_2$ suggested that the bonding of the hydrogen to these metal complexes was novel. Subsequent detailed vibrational studies including Raman and neutron

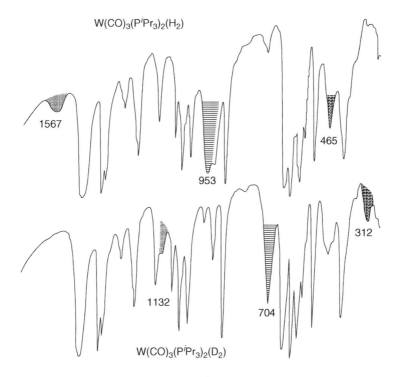

**FIGURE 5.7**   Infrared spectra of $W(CO)_3(P^iPr_3)_2(H_2)$ and $D_2$ analogue in Nujol mulls.

spectroscopic investigations gave unequivocal evidence for side-on bonding of $H_2$ to the metal center, as will be described below.

When diatomic $H_2$ combines with a M–L fragment to form a $\eta^2$-$H_2$, six "new" vibrational modes are created that are related to the "lost" translational and rotational degrees of freedom for $H_2$ (Scheme 5.17).[139]

The H–H stretch, $\nu_{HH}$, is still present, but shifted to much lower frequency and, as will be shown, becomes highly coupled with a $MH_2$ mode. Thus, six fundamental vibrational modes are expected to be formally isotope sensitive: three stretches, $\nu(HH)$, $\nu_{as}(MH_2)$, and $\nu_s(MH_2)$; two deformations, $\delta(MH_2)_{in\text{-}plane}$ and $\delta(MH_2)_{out\text{-}of\text{-}plane}$; and a torsion ($H_2$ rotation), $\tau(H_2)$. These bands shift hundreds of wavenumbers on isotopic substitution with $D_2$ or HD, which greatly facilitates their assignment (Figure 5.7). It is critical to note that the frequencies of these bands for the $\eta^2$-HD

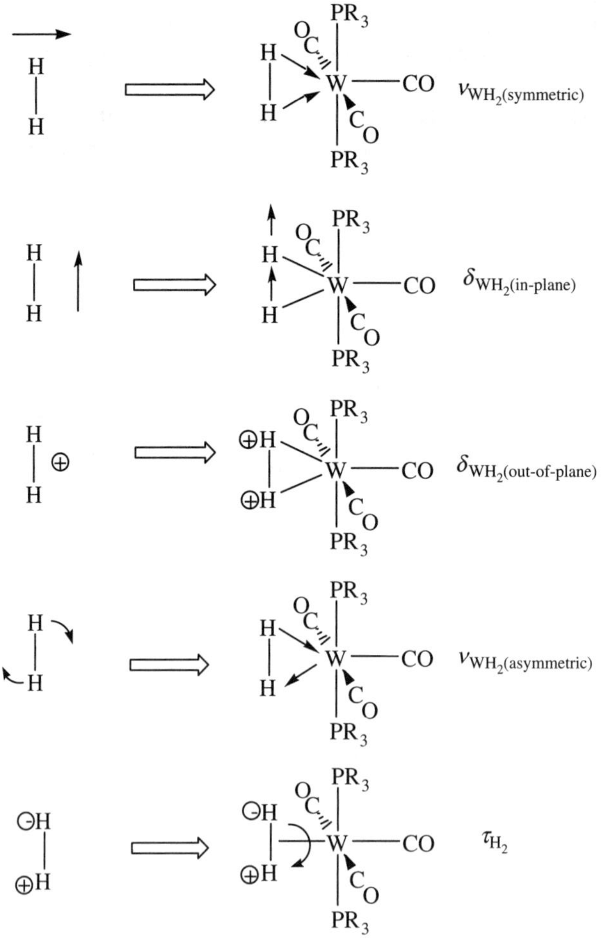

**SCHEME 5.17**

complexes were found to be intermediate to those for the $\eta^2$-HH and $\eta^2$-DD species and *not a superimposition of MH₂ and MD₂ bands* as seen for classical hydrides. This is another valuable diagnostic for distinguishing H₂ versus dihydride coordination, although these vibrational modes are often difficult to observe: the entire set of bands has been identified only in the first H₂ complex, $W(CO)_3(PR_3)_2(H_2)$ (R = Cy, $^iPr$). All but $\nu_s(MH_2)$, observed in both the IR and Raman, are weak, and most of the bands tend to be obscured by other ligand modes. The modes for H₂ rotation about the M–H₂ axis, $\tau(H_2)$, and also $\delta(MH_2)_{out-of-plane}$ near 640 cm$^{-1}$ are observable only by INS methods that will be described in Section 5.8.7 (Figure 5.8).

The frequency of most interest $\nu_{HH}$ is not formally forbidden in the IR of H₂ complexes, but is polarized along the direction of the M–H₂ bond in highly symmetric complexes. Therefore, band intensity arises only from coupling of $\nu_{HH}$ with other modes of the same symmetry such as $\nu_s(MH_2)$ or $\nu_{CO}$ if CO is present, and $\nu_{HH}$ is generally weak. Furthermore, its frequency varies greatly and is often within or near the $\nu_{CH}$ region, an unfortunate position because most ancillary ligands such as phosphines have strong $\nu_{CH}$. Use of perdeuterated phosphine ligands to eliminate such interference enabled location of $\nu_{HH}$ in $W(CO)_3[P(C_6D_{11})_3]_2(H_2)$ as a broad, weak band at 2690 cm$^{-1}$ (Figure 5.9).

The large breadth of the absorption is ascribed to rapid hindered rotation of the H₂ ligand about the W–H₂ axis and a flat rotational potential, as will be discussed below. The HD and DD isotopomers did give increasingly narrower $\nu_{HD}$ and $\nu_{DD}$. As shown

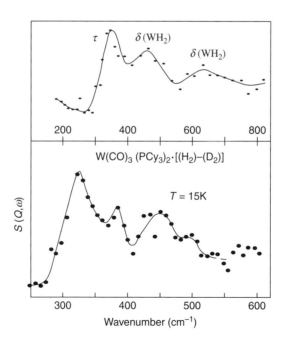

**FIGURE 5.8**    INS spectra of $W(CO)_3(PCy_3)_2(H_2)$ at 15 K obtained as difference spectra by subtracting the spectrum of the D₂ complex from that of the H₂ complex. The lower spectrum with higher resolution shows splitting in the torsional mode (385/325 cm$^{-1}$).

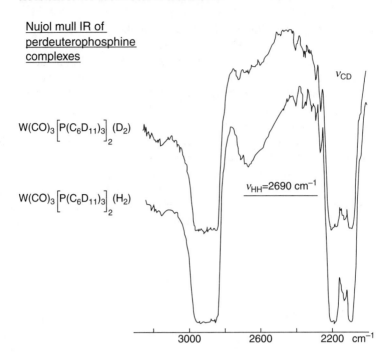

Nujol mull IR of
perdeuterophosphine
complexes

$W(CO)_3\left[P(C_6D_{11})_3\right]_2 (D_2)$

$W(CO)_3\left[P(C_6D_{11})_3\right]_2 (H_2)$

$\nu_{CD}$

$\nu_{HH}=2690$ cm$^{-1}$

3000    2600    2200  cm$^{-1}$

**FIGURE 5.9**    IR spectrum of $W(CO)_3[P(C_6D_{11})_3]_2(H_2)$ and $D_2$ analogue showing $\nu_{HH}$.

in Table 5.1, several other compounds, including surface and cluster species, exhibit $\nu_{HH}$ in the range 2080–3200 cm$^{-1}$, which is considerably lower than that for free $H_2$ gas (4300 cm$^{-1}$).

$\nu_{HD}$ was observed in the IR as a weak broad band at ca. 2360 cm$^{-1}$ in **1**-$d_1$ but $\nu_{DD}$ was obscured by $\nu(CO)$ in **1**-$d_2$. However, Raman spectra of **1**-$d_2$ showed a very broad, barely visible feature centered near 1900 cm$^{-1}$ that apparently mixes with $\nu(CO)_{ax}$ (Figure 5.10)

This results in a shift of 10 cm$^{-1}$ to lower energy on going from the $H_2$ complex to the $D_2$ complex. The asymmetry in the $\nu(CO)_{ax}$ peak for **1**-$d_2$ also suggests the presence of $\nu(DD)$ as an underlying feature, the intensity of which is enhanced by the mixing ($\nu(HH)$ was not observed in Raman spectra, apparently because it is not so enhanced and is thus too weak to be seen in these experiments). The W–$H_2$ stretches were observed clearly in the IR at 1575 cm$^{-1}$ ($\nu_{as}(WH_2)$) and 953 cm$^{-1}$ ($\nu_s(WH_2)$), but in the Raman only $\nu_s(WH_2)$ was seen. $\nu_{as}(WD_2)$ and $\nu_{as}(WHD)$ were partially obscured, and their frequencies could only be estimated.

The torsional mode, $\tau(WH_2)$, was located in the INS difference spectrum (Figure 5.8) as a split mode at 385 and 325 cm$^{-1}$ due to transitions to two split excited librational states ($J = 1, 2$), as will be discussed below. INS spectroscopy is a powerful technique to locate such large amplitude vibrations involving hydrogen. Because deuterium does not scatter, as well as protium, modes not due to $\eta^2$-$H_2$ can be effectively removed by subtracting the spectrum of the $D_2$ complex or a suitable

**TABLE 5.1    IR Frequencies (cm$^{-1}$) for Nominal $\nu_{HH}$ and MH$_2$ Modes in H$_2$ Complexes Compared to $d_{HH}$ (Å)**

| Complex | $\nu$(HH) | $\nu_{as}$(MH$_2$) | $\nu_s$(MH$_2$) | $\delta$(MH$_2$)$^a$ | $d_{HH}$ | Source |
|---|---|---|---|---|---|---|
| CpV(CO)$_3$(H$_2$) | 2642 | | | | | Ref. 160b |
| CpNb(CO)$_3$(H$_2$) | 2600 | | | | | Ref. 160b |
| Cr(CO)$_5$(H$_2$) | 3030 | 1380 | 869, 878 | | | Ref. 131 |
| Cr(CO)$_3$(PCy$_3$)$_2$(H$_2$) | | 1540 | 950 | 563$^b$ | 0.85 | Ref. 119 |
| Mo(CO)$_5$(H$_2$) | 3080 | | | | | Ref. 131 |
| Mo(CO)$_3$(PCy$_3$)$_2$(H$_2$) | ~2950$^c$ | ~1420$^c$ | 885 | 471 | 0.87 | Ref. 1b |
| Mo(CO)(dppe)2(H$_2$) | 2650 | | 875 | | 0.88 | Ref. 39 |
| W(CO)$_5$(H$_2$) | 2711 | | 919 | | | Ref. 131 |
| W(CO)$_3$(P$^i$Pr$_3$)$_2$(H$_2$) | 2695 | 1567 | 953 | 465 | 0.89 | Ref. 1b |
| W(CO)$_3$(PCy$_3$)$_2$(H$_2$) | 2690 | 1575 | 953 | 462 | 0.89 | Ref. 1b |
| WH$_4$(H$_2$)$_4$ | 2500 | 1782 | | 437 | | Ref. 166 |
| Fe(CO)(NO)$_2$(H$_2$) | 2973 | 1374 | ~870 | | | Ref.160c |
| Co(CO)$_2$(NO)(H$_2$) | {3100, 2976}$^d$ | 1345 | 868 | | | Ref. 160c |
| FeH$_2$(H$_2$)(PEtPh$_2$)$_3$ | 2380 | | 850 | 500, 405$^e$ | 0.82 | Ref. 67 |
| RuH$_2$(H$_2$)$_2$(P$^i$Pr$_3$)$_2$ | 2568 | 1673 | 822$^b$ | | 0.92 | Ref. 205 |
| Tp$^*$RuH(H$_2$)$_2$ | 2361 | | | | 0.90 | Ref. 140 |
| Tp$^*$RuH(H$_2$)(THT) | 2250 | | | | 0.89 | Ref. 140 |
| [Os(NH$_3$)$_5$(H$_2$)]$^{2+}$ | 2231$^b$ | | | | [1.34]$^f$ | Ref. 33 |
| [CpRu(dppm)(H$_2$)]$^+$ | 2082$^b$ | 1358$^b$ | 679$^b$ | 486, 397$^b$ | [1.10]$^g$ | Ref. 141 |
| Tp$^*$RhH$_2$(H$_2$) | 2238 | | | | 0.94$^h$ | Ref. 160 |
| Pd(H$_2$) (matrix) | 2971 | 1507 | 950 | | 0.85$^h$ | Ref. 207 |
| Ni(510)–(H$_2$)$^i$ | 3205 | 1185 | 670 | | | Ref. 165 |
| Cu$_3$(H$_2$) (matrix) | 3351, 3232 | | | | | Ref. 208 |

$^a$Presumably in-plane deformation, although assignment has not been made in most cases.
$^b$Assignments unclear; in the case of the elongated Ru and Os complexes, these are highly mixed modes that could involve M–H modes (if present).
$^c$Estimated from observed D$_2$ isotopomer bands.
$^d$Split possibly by Fermi resonance.
$^e$Assignment unclear (data from INS).
$^f$For [Os(ethylenediamine)$_2$(H$_2$)(acetate)]$^+$ (Ref. 34).
$^g$For the Cp$^*$ analogue (Ref. 122).
$^h$Calculated from inelastic neutron scattering data or DFT.
$^i$Data from EELS spectroscopy.

"blank," for example, a similar complex with a non-hydrogen containing ligand in place of H$_2$. Excellent spectra of W(CO)$_3$(PR$_3$)$_2$(H$_2$) (**1**, Figure 5.8) and related complexes are obtained in the range 200–1000 cm$^{-1}$ using neutron spectrometers at Los Alamos National Laboratory. The two WH$_2$ deformational modes at 640 and 462 cm$^{-1}$ are also seen in the INS as broad features. The WH$_2$ deformation around 640 cm$^{-1}$ is obscured in the IR, but the D$_2$ isotopomer gives a band at 442 cm$^{-1}$ for the corresponding WD$_2$ deformation (Figure 5.7). The two WH$_2$ deformational modes at

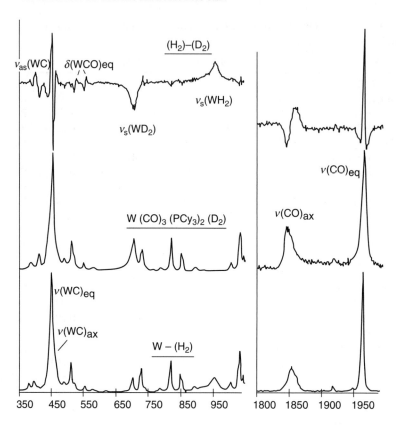

**FIGURE 5.10**    Raman spectra of $W(CO)_3[P(C_6H_{11})_3]_2(H_2)$ and $D_2$ analogue. Samples were powdered complex sealed in melting point capillaries using the 6471 Å line of a Spectra Physics krypton laser and a SPEX double monochromator. Despite the use of low power (ca. 1 mW) and cooling of the sample to 77 K, partial decomposition slowly took place when the sample was illuminated by the laser beam during the course of the experiments.

640 and 462 cm$^{-1}$ were also seen in the INS, as broad features. Several other metal–ligand modes also show small shifts to higher or lower wavenumber upon deuterium substitution due to vibrational coupling between modes that belong to the same symmetry block and that are close in energy. This is especially obvious in the IR spectra of **1** and **1**-$d_2$ near 625 cm$^{-1}$ and Raman spectra in the region below 650 cm$^{-1}$ (Figure 5.10). For example, a band at 400 cm$^{-1}$ due mainly to $\nu_{as}(WC)$ shifts 13 cm$^{-1}$ to *higher* frequency for the $D_2$ complex, presumably because of mixing with $\delta(WD_2)_{\text{out-of-plane}}$. Such shifts have also been observed in other complexes containing both $H_2$ and CO ligands.[131]

Low-temperature stable species with only CO ligands, such as group 6 M $(CO)_5(H_2)$ generated photochemically in liquid rare-gas solvents below $-70°$C, are favorable for study because of little solvent or coligand interference.[131] The $\nu_{HH}$ bands are generally quite broad (<40 cm$^{-1}$ FWHM) even at low temperatures. The

HD and DD isotopomers again give increasingly narrower $\nu_{HD}$ and $\nu_{DD}$. The nature of the metal greatly influences the band position, and third-row transition metals generally give lower frequencies because the H–H bond here is more weakened by higher metal $d_\pi \rightarrow$ H$_2$ $\sigma^*$ backdonation. As will be shown, however, $\nu_{HH}$ is highly coupled with M–H$_2$ vibrational modes, and the degree of bond activation is difficult to correlate with vibrational frequencies.

Modes other than $\nu_{HH}$ have only occasionally been observed in room-temperature stable complexes, and the low-frequency deformations and torsions have been the least observed modes. The lack of these data is partly due to interfering bands from coligands or difficulty in assignment, especially if hydride ligands are also present. This is, for example, the case for Tp*RuH(H$_2$)$_2$, which shows four difficult to assign bands at 458–834 cm$^{-1}$ (Tp* = hydrotris[3,5-dimethylpyrazo-1-yl)borate).[140] Such frequencies are also seen by Raman spectroscopy for [CpRu(dppm)(H$_2$)]BF$_4$, which has an elongated H–H bond (1.10 Å) and the lowest reported value for $\nu_{HH}$, 2082 cm$^{-1}$.[141] As will be shown below, however, the high degree of mode mixing in such elongated H$_2$ complexes renders $\nu_{HH}$ and the other mode representations in Scheme 5.17 for true (unstretched) $\eta^2$-H$_2$ complexes meaningless and new vibrational modes must be defined. The photolytically generated pentacarbonyl and nitrosyl species and the compounds containing H$_2$ bound to a nickel 510 surface or to Pd atoms deposited on rare-gas matrices at 7–12 K are among the few to display MH$_2$ modes aside from the group 6 complexes (Table 5.1). H$_2$ is believed to be bound in $\eta^2$-fashion on the stepped edges of the Ni surface and electron energy loss spectroscopy (EELS) at 100K shows several bands comparable to those for true H$_2$ complexes such as W(CO)$_3$(PCy$_3$)$_2$(H$_2$).

As expected, there is generally a large dependence of $\nu_{HH}$ and MH$_2$ modes on both metal and ligand sets. One might anticipate a good correlation of $\nu_{HH}$ with $d_{HH}$ and the backdonating ability (electron richness) of the metal, as found for $\nu_{NN}$ and $\nu_{CO}$ in similar $\pi$-acceptor N$_2$ and CO ligands. However, this is not the case because of complexity of the bonding and mixing of $\nu$(HH) and $\nu$(MH$_2$) modes, as will be discussed below. In certain cases, H–H and M–H vibrational frequencies have been calculated in order to assess the weakening of the H–H bond on coordination.[142,143] The reduction in frequency of the $\nu_{HH}$ over that in free H$_2$ generally agrees with experimental values, although these data are rather limited.

### 5.8.3 Normal Coordinate Analysis of W(H$_2$)(CO)$_3$(PCy$_3$)$_2$

A normal coordinate analysis of W(H$_2$)(CO)$_3$(PCy$_3$)$_2$ (**1**) provided valuable understanding of the side-on nonclassical M–H$_2$ interaction and the relation of H–H and MH$_2$ frequencies to bond strengths.[139] In addition to the isotopic shifts of the six modes discussed above, several other metal–ligand modes also showed small shifts to higher or lower wavenumber upon deuterium substitution that result from vibrational coupling between modes that belong to the same symmetry block and are close in energy.[144] This was especially obvious in the Raman spectra of **1** and **1**-$d_2$ below 650 cm$^{-1}$; for example, a band at 400 cm$^{-1}$ due mainly to $\nu_{as}$(WC) shifted 13 cm$^{-1}$ to *higher* frequency for the D$_2$ complex, presumably because of mixing

**TABLE 5.2  Observed$^a$ and Calculated Vibrational Frequencies (cm$^{-1}$) and Mode Assignments for 1, 1-d$_2$, and 1-d$_1$**

| Intensity | 1<br>obsd (calcd) | 1-d$_2$<br>obsd (calcd) | 1-d$_1$<br>obsd (calcd) | Assignment |
|---|---|---|---|---|
| IR, w | 2690 (2692.1) | ~1900 (1909.9) | 2360 (2357.8) | $\nu$(HH) |
| IR, R, m | 953 (949.0) | 703 (703.1) | 791 (799.8) | $\nu_s$(WH$_2$) |
| IR, w | 1575 (1574.7) | ~1144 (1136.1) | ~1360 (1357.9) | $\nu_{as}$(WH$_2$) |
| IR, w | 462$^b$ (456.2) | 319 (326.0) | 360 (368.7) | $\delta$(WH$_2$)$_{in\text{-}plane}$ $^c$ |
| INS,$^d$ IR, w | 640$^e$ (640.0) | 442$^f$ (442.0) | | $\delta$(WH$_2$)$_{out\text{-}of\text{-}plane}$ |
| INS, m | 385 (325) | | | $\tau$(WH$_2$) |

$^a$Resolution: 2 cm$^{-1}$.
$^b$Also observed in INS.
$^c$This mode shows a greater observed isotope shift than calculated. It apparently arises from the $\delta$(WH$_2$)$_{in\text{-}plane}$ rocking coordinate coupled strongly with other coordinates.
$^d$Inelastic neutron scattering.
$^e$Observed in INS only.
$^f$Observed in IR only.

with $\delta$(WH$_2$)$_{out\text{-}of\text{-}plane}$. Clearly, the cyclohexyl groups on the phosphines should not show HH, HD, and DD isotope shifts and are not significantly coupled to vibrations other than Cy and P–Cy modes. Therefore, only the "W(H$_2$)(CO)$_3$P$_2$" core fragment was treated in the force field calculations.

The normal coordinate analysis included 13 observed IR and Raman frequencies (W–P modes were not located). Obviously, a large number of force constants had to be fixed in the calculations; for example, force constants of 2–3 were assumed for W–P and for CO-related modes values were assumed based on other systems, such as W(CO)$_6$.[145] The solution is given in Table 5.2, which summarizes only the data for H$_2$-related modes.

The agreement of observed and calculated frequencies was not strong, being off by several wavenumbers in several cases, but overall the analysis sufficed to assign the observed vibrational modes for **1** and its isotopomers.

In order to understand the M–H$_2$ bonding interactions, it was necessary to investigate the reasonableness and meaning of the force constants in Table 5.3 for the hydrogen-related modes.

**TABLE 5.3  Force Constants (mdyn/Å) for Hydrogen-Related Modes in W(CO)$_3$P$_2$(H$_2$) and a Triatomic Model Complex**

| | W(CO)$_3$P$_2$(H$_2$) | W(H$_2$) |
|---|---|---|
| $F_{HH}$ | 1.32 | 1.46 |
| $F_{WH}$(s) | 1.46 | – |
| $F_{WH}$(as) | 1.42 | – |
| $F_{WH}$ | 1.44 | 1.43 |
| $F_{WH,WH'}$ | 0.02 | −0.05 |
| $F_{HH,WH}$ | 0.67 | 0.62 |

The most important mode is $\nu_{HH}$, and the value of 1.3 mdyn/Å for $F_{HH}$ is much smaller than the value for free $H_2$ (5.7 mdyn/Å).[146] This is not surprising, but a standard analysis obtained by multiplying the latter by the ratio of the square of frequencies for bound and free $H_2$ indicates that $F_{HH}$ should have only been lowered to 2.1 mdyn/Å (Equation 5.2).

$$5.7 \times (2690/4395)^2 = 5.7(0.37) = 2.1 \qquad (5.2)$$

The lower than calculated value of $F_{HH}$ would suggest a longer $d_{HH}$ than the value of either 0.82 Å observed by neutron diffraction in $W(CO)_3(P^iPr_3)_2(H_2)$ or 0.89 Å determined by solid-state NMR for both the R = $^i$Pr and Cy species. An empirical correlation between bond length and force constant, known as Badger's rule, $k_e = b/(r_e - a)^3$, in fact predicts $d_{HH}$ to be ~0.94 Å.[147] However, it was clear from the normal coordinate analysis that $\nu_{HH}$ has considerable $\nu_{MH}$ character, so it cannot be treated as an isolated HH mode. In complexes with more activated $H_2$ ligands with longer $d_{HH}$, this mixing of $\nu$(HH) and $\nu$(WH$_2$) is much more extensive and increases to the point at which these modes themselves no longer have meaning and must be redefined as will be shown below. Thus, the values of $\nu_{HH}$ in Table 5.1 are not a reliable predictor of $d_{HH}$ for $H_2$ complexes, especially for elongated complexes where the values listed for $\nu_{HH}$ and $\nu_{WH}$ effectively should be reversed. Similar situations arise for other $\pi$-acceptor ligands with a large degree of M → L back-donation such as metal–ethylene complexes: $\nu_{CC}$ is not a reliable measure of $d_{CC}$ because of normal mode coupling to same-symmetry $C_2H_4$ wagging/deformational modes.[148]

The WH stretching constant is surprisingly as large as that for the HH stretch, and the WH–WH' interaction is negligible. The HH–WH interaction force constant, $F_{HH, WH}$, is very large (0.67 mdyn/Å), indicating that stretching the HH bond leads to strengthening of the HW bonds, and vice versa. Stretching force constants for an *isolated* WH$_2$ group (i.e., neglecting the CO and phosphine ligands) using the frequencies observed for the HH and WH stretches can also be calculated. The results are shown in Table 5.3 for comparison with the more complete treatment. Surprisingly, the agreement is quite good, indicating that such a simplified treatment can yield useful information.

### 5.8.4  Nature of M–$H_2$ Bonding Delineated by Vibrational Analysis

As expected, there is generally a large dependence of the modes for bound $H_2$ on both the metal center and coligands. Vibrational analysis of a M-$\eta^2$-$H_2$ system is further complicated by the three-center, two-electron bonding. The bonding is of the Dewar–Chatt–Duncanson type present in metal–olefin complexes, where there is a strong M → $H_2$ $\sigma^*$ component, $E_{BD}$, to the bonding in addition to electron donation, $E_D$, to the empty metal d-orbital from the $H_2$ electron pair (Scheme 5.18).

Calculations have shown that the $E_{BD}$ component can be energetically as strong as or stronger than $E$.[149] In room-temperature stable complexes such as W

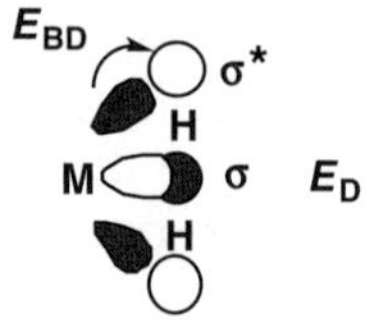

**SCHEME 5.18**

$(CO)_3(PR_3)_2(H_2)$, the $E_{BD}$ bonding component represents about 50% of the overall W–H$_2$ bonding energy. It is weaker in complexes with primarily π-acceptor coligands such as CO, although the calculations show that $E_{BD}$ in Mo(CO)$_5$(H$_2$) still is about one-third of the W–H$_2$ bond energy. One might anticipate a correlation of $\nu_{HH}$ with the BD ability (electron richness) of M, as found for $\nu_{NN}$ and $\nu_{CO}$ in similar π-acceptor N$_2$ and CO ligands. The decrease in $\nu_{HH}$ on going from Mo(CO)$_5$(H$_2$) to Mo(CO)$_3$(P-Cy$_3$)$_2$(H$_2$) to Mo(CO)(dppe)$_2$(H$_2$) shown in Table 5.1 at first glance might seem to reflect increased H–H bond weakening by the increasingly more electron-rich Mo as the number of phosphine donor ligands increases (dppe = diphenylphosphinoethane). However, there are inconsistencies in such correlations. For example, $\nu_{HH}$ decreases in the order Mo > Cr > W for M(CO)$_5$(H$_2$), but Cr should be the worst backbonder and give the highest $\nu_{HH}$ value. Even more disturbingly, the value of $\nu_{HH}$ in W (CO)$_5$(H$_2$) (2711 cm$^{-1}$) is far out of line with the much higher values in the Cr and Mo congeners (3030 and 3080 cm$^{-1}$) and also differs very little (20 cm$^{-1}$) from that in the vastly more electron-rich complex, W(CO)$_3$(PCy$_3$)$_2$(H$_2$). The pentacarbonyl complex is unstable at room temperature and almost certainly has a much shorter $d_{HH}$ and presumably a stronger H–H bond.

It is quite clear from the data in Table 5.1 that $\nu_{HH}$ does not correlate well with $d_{HH}$, and this is presumably because of extensive mode mixing. The normal coordinate analysis of W(CO)$_3$(PCy$_3$)$_2$(H$_2$) treats the W–H$_2$ interaction as a triangulo system, that is, where direct BD electronic interactions exist between W and H atoms (Scheme 5.19, left-hand side) rather than as the strictly three-center bonding representation (Scheme 5.19, right-hand side).

This is confirmed by the fact that the WH stretching force constant is as large as that for the HH stretch and that the HH–HW interaction is very large, indicating that stretching the HH bond leads to strengthening of the HW bonds, and vice versa. This extensive mixing along with the reduction of the $\nu_{HH}$ force constant to one-fourth the value in free H$_2$ indicates that weakening of the H–H bond and formation of W–H bonds is already well along the reaction coordinate to OA in **1**. Furthermore, as the H–H bond becomes more activated on a metal fragment, the observed "$\nu_{HH}$" mode

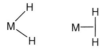

**SCHEME 5.19**

will have increasing M–H character relative to H–H character. Upon H–H cleavage, this mode will then be assimilated into the M–H stretching modes. In elongated $H_2$ complexes with $d_{HH}$ of 1.1 Å or greater, the 2231 and 2082 cm$^{-1}$ frequencies in Table 5.1 for the cationic Os and Ru complexes should be thought more of as $\nu_{OsH}$ or $\nu_{RuH}$ than $\nu_{HH}$, as will be shown below. For a series of complexes, $L_nM(\eta^2\text{-}H_2)$, in which the H–H bond becomes increasingly activated (stretched) and is ultimately broken upon variation of $L_n$, the frequency of the nominal "$\nu_{HH}$" mode would be expected to decrease and eventually "cross over" with that for the increasingly strong $\nu_{MH}$. This essentially represents the reaction coordinate for a bond-breaking process that in effect can be followed by vibrational spectroscopy, a novel situation.

The force constant analysis of **1** indicates that the $H_2$ ligand here is further activated toward oxidative addition to a dihydride than may have been previously thought. It has been a paradox that the $d_{HH}$ in **1** or any of the group 6 complexes in Table 5.1 (0.85–0.89 Å, solid-state NMR) are not as "stretched" as some of those found in later transition element complexes (1.0–1.5 Å), yet the H–H bond in **1** undergoes equilibrium cleavage to the dihydride tautomer in solution (Scheme 5.6). Thus, the observed $d_{HH}$ may not always reflect the degree of "readiness to break"; that is, a very late transition state may exist. At the other end of the spectrum, for $W(CO)_5(H_2)$ and other complexes with very *weakly* bound $H_2$, the T-shaped entity pictured above with one internal coordinate, the H–H stretch, may be a more appropriate model for vibrational analysis. The interaction of $H_2$ with an electrophilic metal has a much stronger $E_D$ than an electron-rich metal, and this offsets the lower $E_{BD}$. For example, a cationic $H_2$ complex such as $[Mn(CO)(dppe)_2(H_2)]^+$ can have remarkably similar properties, for example, $d_{HH}$, to its isoelectronic neutral analogue, $Mo(CO)(dppe)_2(H_2)$.[150] The lack of reliable correlation of vibrational versus other properties brought about by the three-center bonding also extends to the M–$H_2$ modes.

### 5.8.5 Highly Mixed H–H and M–$H_2$ Modes in Elongated $H_2$ Complexes: New Normal Mode Definitions

Complexes containing elongated ("stretched") $H_2$ ligands exhibit even greater mode mixing to the extent that *new normal modes must be defined*. Raman studies of [CpRu(dppm)(H_2)]$^+$ by Chopra et al.[141] show unusually low values for all of the assigned modes in comparison to most of the other complexes. This might be expected for a weakened $\nu(HH)$, but the $\nu(MH_2)$ stretches would have been expected to rise with increasing M–H bond strength. As shown in the calculations by Lluch and Lledós,[14] there is significant anharmonicity in the immediate neighborhood of the potential energy minimum, at least with respect to the H–H and Ru–$H_2$ stretches. This is key to understanding the true delocalized nature of the bonding in these types of complexes as revealed by nuclear motion quantum calculations. These types of calculations, including discrete variable representation (DVR) analysis, reproduce accurate vibrational energy levels and wavefunctions without resorting to the harmonic approximation that is not appropriate for an elongated $H_2$ complex.[151] The computations on a $[CpRu(H_2PCH_2PH_2)_2(H_2)]^+$ model complex determine the nuclear energy levels as a means of obtaining the vibrational energy levels. A contour plot (Figure 5.11) of the

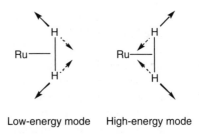

Low-energy mode    High-energy mode

**SCHEME 5.20**

two-dimensional potential energy surface ($d_{HH}$ versus Ru–H$_2$ distance) shows that the valley for the minimum potential energy is oblique with respect to both $\nu_{HH}$ and $\nu(RuH_2)$.

Because only a two-dimensional representation of the PES is calculated, only two normal vibrational modes can be described here that cannot include the asymmetric modes (stretches or deformations). The modes are obviously not a pure H–H stretch plus a pure Ru–H$_2$ stretch but rather are "new" modes that are represented qualitatively in Scheme 5.20.

The first mode essentially parallels the oblique minimum energy path that links the two exits of the potential energy valley in Figure 5.11. Along this normal mode, the changes in energy are very damped, and hence it is labeled as the low-energy mode, which is calculated from the DVR analysis to occur at 555 cm$^{-1}$. This mode essentially corresponds to the reaction coordinate for cleavage of H$_2$ on a metal center, an unprecedented situation. Here, the stretching of the H–H bond leads to shortening of $d_{RuH}$ and thus strengthening of the Ru–H bonds. In the orthogonal high-energy mode, both the H–H bond and the Ru–H bonds stretch or compress simultaneously, which costs much more energy: 2229 cm$^{-1}$ from the DVR calculations. The experimental value of 2082 cm$^{-1}$ in [CpRu(dppm)(H$_2$)]$^+$ would then correspond to the high-energy mode while the 679 cm$^{-1}$ band relates to the 555 cm$^{-1}$ value calculated for the low-energy mode. The high-frequency band can no longer be interpreted as a much weakened H–H stretch because it has a very significant component of Ru–H stretching and in fact must now essentially be regarded to primarily represent the Ru–H mode (mixed with some H–H stretch). Similarly, the low-frequency band should be described as a mode in which the H atoms separate from each other as they approach M, that is, mainly an H–H stretch mixed with some Ru–H stretching. In other words, for elongated H$_2$ complexes *the assignments for the bands as $\nu(HH)$ and $\nu(RuH_2)$ have become in a sense reversed*, although they are still highly mixed.

### 5.8.6   Vibrational Spectroscopy of Unstable Dihydrogen Complexes

Spectroscopic studies of species generated at low temperatures by photodissociation of CO ligands were of key importance to early evidence that "unreactive" small molecules, even saturated molecules containing only σ-bonds such as methane, could

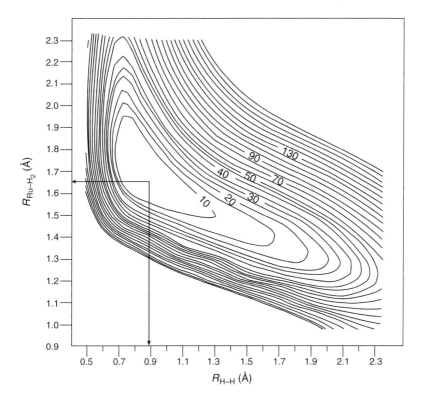

**FIGURE 5.11** Contour plot of the potential energy surface used in the bidimensional calculations. Energy units are kcal/mol relative to the minimum in potential energy. (Reprinted with permission from Ref. 136. Copyright 1997 American Chemical Society.)

interact with metal centers. This was a radical notion at the time, much as that of metal–$H_2$ coordination. A popular early system involved 16e $Cr(CO)_5$ fragments that were highly unstable and generally made by photolysis of $Cr(CO)_6$ in rare-gas matrices at very low temperatures (Equation 5.3).[152–155] These species showed extraordinary binding capabilities toward weak ligands, coordinating $CH_4$ and even rare gases in the vacant site.

$$Cr(CO)_6 \xrightarrow{-CO} Cr(CO)_5 \rightarrow Cr(CO)_5L \quad L = N_2, CH_4, Ar \qquad (5.3)$$

The first spectroscopic evidence for an apparent metal–$H_2$ interaction was obtained in a similar fashion by Sweany just prior to the initial neutron diffraction study of $W(CO)_3(P^iPr_3)_2(H_2)$ in late 1982. Photolysis of $Cr(CO)_6$ in the presence of $H_2$ in a rare-gas matrix was claimed to give $Cr(CO)_5(H_2)$ based on $\nu_{CO}$ frequencies and pattern.[156] However, these results were admitted by the author to be difficult to publish until after (1985) the seminal report of the discovery of stable $H_2$ coordination in W $(CO)_3(PR_3)_2(H_2)$ was published in 1984. At about the same time, related papers also showed that $Cr(CO)_5(H_2)$ could be formed in liquid Xe or cyclohexane but is stable for only seconds at room temperature.[157,158] Much later, Heinekey[37,38] showed that these pentacarbonyl complexes were stable enough to obtain low-temperature NMR spectra that proved that the $H_2$ was bound in similar fashion to that in $M(CO)_3(PR_3)_2(H_2)$, as was discussed previously.

The investigations of low-temperature stable $H_2$ complexes in solid or liquid rare-gas media have continued to be a subdiscipline that has gone hand-in-hand with studies of stable complexes, as shown in reviews by Sweany and Poliakoff.[159,160a] In most cases, the preparations again involve photochemical displacement of CO either in a rare-gas matrix or in liquid Xe solutions at $-70°C$ (a very useful medium), alkane solvents, and even the gas phase (Equation 5.4).

$$L_xM(CO)_n + H_2 \xrightarrow[\text{2–200K}]{hv,\ -CO} L_xM(CO)_{n-1}(H_2) \qquad (5.4)$$

The most intensely studied species continued to be the group 6 pentacarbonyls, $M(CO)_5(H_2)$, and much work was carried out by Poliakoff and coworkers on these and related systems.[131,160–164] In all media, vibrational spectroscopy provided the crucial evidence for $H_2$ rather than dihydride binding. The H–H, H–D, and D–D stretching modes are often observed because of the clear spectroscopic window in rare-gas media, although the experiments are decidedly much more difficult than routine IR or Raman spectroscopy. Perhaps the most novel preparation was Poliakoff's photolysis of the hexacarbonyls impregnated in polyethylene disks under $H_2$ or $N_2$ pressures to give $M(CO)_{6-n}(L)_n$, where $n = 1–2$ for $L = H_2$ and 1–4 for $N_2$.[161] Reactivity followed the order $Mo > Cr > W$, and $H_2$ can displace coordinated $N_2$ in the polyethylene media. $W(CO)_5(H_2)$ was both generated *and dissociated* in this and hypercrosslinked polymer matrices by UV photolysis at 220 and 90 K, respectively.[162] This suggested a potential "UV-activated" mechanism for hydrogen storage and release.

In nearly all cases, these complexes decompose rapidly and irreversibly at or near room temperature because of the weak $H_2$ binding on such CO-rich metals, where less backbonding is present. Their instability is exacerbated because the 16e product of $H_2$ dissociation is extremely reactive since it is not stabilized by internal agostic C–H interactions or solvent binding (hydrocarbon solvents are even more weakly bound than $H_2$). The rate of room-temperature dissociation of $H_2$ from $Cr(CO)_5(H_2)$ in hexane is actually slower than that for many stable species. Thus, this complex and others like it might otherwise be stable under $H_2$. One such complex initially presumed to be unstable, $CpMn(H_2)(CO)_2$, was in fact isolated as a solid from supercritical $CO_2$ (sc$CO_2$) at room temperature in a flow reactor by rapid expansion of the sc$CO_2$ (Scheme 5.21).[163,164]

This complex is more robust than originally thought: displacement of the $\eta^2$-$H_2$ by $N_2$ (500 psi) or ethylene requires 2 h in supercritical $CO_2$. $CpMn(H_2)(CO)_2$ is one of the simplest stable $H_2$ complexes and has the lowest molecular weight (178) and highest percentage of $H_2$ by weight (1.1%) of any isolatable transition metal $H_2$ complex. Although such a complex might appear to have favorable properties for

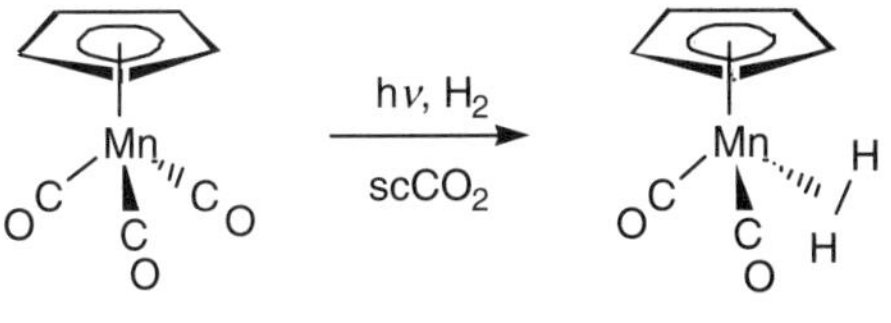

**SCHEME 5.21**

reversible hydrogen storage, the mass percent H$_2$ bound is unfortunately still much too low for practical use.

H$_2$ has been found to molecularly bind to metal surfaces such as Ni(510)[165] where the evidence again is entirely spectroscopic, primarily vibrational. H$_2$ is believed to be bound in $\eta^2$-fashion on the stepped edges of the Ni(510) surface. Although metal surfaces normally cleave H$_2$ to form hydrides, metal atoms on this surface are effectively electronically unsaturated and bind H$_2$ molecularly. Electron energy loss spectroscopy at 100 K shows several bands comparable to those for organometallic H$_2$ complexes (Table 5.1). No such chemisorption is observed on the flat Ni(100) surface that lacks the residual unfilled d states at the step sites that bind the H$_2$. Undoubtedly, H$_2$ coordination is the first step in the dissociation of H$_2$ on metal surfaces to form hydrides and is followed by rapid splitting of H–H analogous to oxidative addition in homogeneous solution activation.

A large series of complexes of the type MH$_x$(H$_2$)$_y$ (M = alkali metal, transition metal, and uranium) has been prepared by Andrews and coworkers by codeposition of laser-ablated metal atoms with neat hydrogen at low temperatures, typically 4 K. An excellent example is WH$_4$(H$_2$)$_4$ (Figure 5.12), which has been studied by matrix isolation IR spectroscopy in solid hydrogen (decomposition occurs above 7 K).[166]

Four bands due to coordinated H$_2$ were located (Table 5.1). A broad mode at 2500 cm$^{-1}$ was assigned to the H–H stretching mode and a mode at 1782 cm$^{-1}$ was assigned to a W–H$_2$ stretch. These can be compared to that for W(CO)$_3$(PR$_3$)$_2$(H$_2$) at 2690 and 1570 cm$^{-1}$. Interestingly, the lower H–H stretch and the higher W–H$_2$ stretch for WH$_4$(H$_2$)$_4$ would seem to suggest that the latter low-temperature stable species has more strongly bound H$_2$ ligands than room-temperature stable W (CO)$_3$(PR$_3$)$_2$(H$_2$) complexes. Calculations showed that the average binding energy per H$_2$ molecule in WH$_4$(H$_2$)$_4$ is 15 kcal/mol, which is somewhat less than the calculated values for W(CO)$_3$(PR$_3$)$_2$(H$_2$), 17–20 kcal/mol. The experimental value for W(CO)$_3$(PCy$_3$)$_2$(H$_2$) is estimated to be near 20 kcal/mol. Thus, the H$_2$ ligand is most likely bound more strongly in the organometallic complexes than in WH$_4$(H$_2$)$_4$, and the position of the IR bands is not a good indicator of M–H$_2$ relative bond energies. This might be expected because of the extensive mode mixing in dihydrogen complexes. A deformational mode was also seen at 437 cm$^{-1}$ in WH$_4$(H$_2$)$_4$, which is comparable to that for W(CO)$_3$(PR$_3$)$_2$(H$_2$) at 450 cm$^{-1}$. This low-frequency mode has been rarely observed in H$_2$ complexes.

Very importantly, molecular hydrogen has been observed to bind to porous materials such as zeolites and MOF complexes at low temperatures, and in some cases even at room temperature. The applications here for reversible hydrogen storage

**FIGURE 5.12**    DFT calculated structure of $WH_4(H_2)_4$.

are approaching the near practical realm. This area will be discussed near the end of this section.

$H_2$ can also weakly interact with *main group* species and is thus amphoteric, behaving as a Lewis base, that is, pure σ donor or, more rarely, as a Lewis acid, that is, an acceptor into the $H_2$ σ* orbital (Scheme 5.22).

Significantly, complexes where $H_2$ can act only as a pure Lewis base are unstable, for example, the triangulo species $H_3{}^+$, which is viewed as $H_2$ binding to highly Lewis acidic $H^+$. This instability attests to the vital role of backdonation from metal d-orbitals in stabilizing σ-ligand binding. Hypervalent main group species such as $CH_5{}^+$ are also rationalized theoretically as highly dynamic $H_2$ complexes of main group cations, that is, $CH_3(H_2)^+$.[167]

### 5.8.7    Inelastic Neutron Scattering Studies of $H_2$ Coordination and Rotation

The $H_2$ ligand undergoes rapid two-dimensional hindered rotation about the M–$H_2$ axis; that is, it spins (librates) in propeller-like fashion with little or no wobbling. This phenomenon has been extensively studied by INS methods by Eckert because it *unequivocally distinguishes molecular $H_2$binding from classical hydride binding*

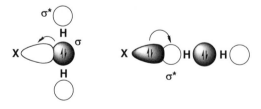

X = M⁺, H⁺, Lewis acids    X = O²⁻, C=C, Lewis bases

**SCHEME 5.22**

where there is no such rotation.[40,168] Furthermore, weak physisorption of H$_2$, for example, van der Waals interaction with main group atoms, can be distinguished from the much stronger binding of H$_2$ to metal centers. This is particularly important in solid-state hydrogen storage materials that cannot easily be studied by NMR or other conventional methods. These discriminating features arise because there is always at least a small to moderate barrier to rotation, $\Delta E$, in metal coordination brought about by M $\rightarrow$ H$_2$ $\sigma^*$ backdonation. The $\sigma$-donation from H$_2$ to M cannot give rise to a rotational barrier since it is completely isotropic about the M–H$_2$ bond. In M (CO)$_3$(PCy$_3$)$_2$(H$_2$), the barrier actually arises from the *disparity* in the BD energies from the d-orbitals when H$_2$ is aligned parallel to P–M–P versus parallel to OC–M–CO, where BD is less (though not zero; Scheme 5.23).

$\Delta E$ varies with M, coligands, and other factors and can be analyzed in terms of the BD and other forces that lead to it, both calculationally and by a series of experiments where metal–ligand (M/L) sets are varied. In most "true" H$_2$ complexes with $d_{HH} < 0.9$ Å, the barrier is only a few kcal/mol and observable only by neutron scattering methods. It can be as low as 0.5 kcal/mol for symmetrical ligand sets, for example, all *cis* L are the same, but has never been measured to be zero because minor geometrical distortions or crystal lattice-related effects are usually present. In the case of complexes with elongated H–H bonds or where rotation is sterically blocked as in [Cp$'_2$M(H$_2$)(L)] + (M = Nb, Ta), much higher barriers of 3–12 kcal/mol are observed by INS or even solution NMR methods.[169] The hindered rotation of $\eta^2$-H$_2$ is thus governed by a variety of forces, which can be divided into bonded (electronic) and nonbonded interactions ("steric" effects). The direct electronic interaction between M and H$_2$ results from overlap of the appropriate molecular orbitals. Nonbonded interactions such as van der Waals forces between the $\eta^2$-H$_2$ atoms and the other atoms on the molecule may vary as $\eta^2$-H$_2$ rotates. The rotational tunneling transition has an approximately exponential dependence on the barrier height, and is therefore extremely sensitive to the latter and thus to even very minor changes in H$_2$ environment (e.g., crystal packing forces). It is this property that is exploited to gain information on the origin of the barrier and to easily distinguish even small variations in H$_2$ binding sites in materials (see below).

Prior to the discovery of H$_2$ complexes, the only systems known containing hydrogen molecules were H$_2$ gas or H$_2$ that was barely affected by its surroundings (as

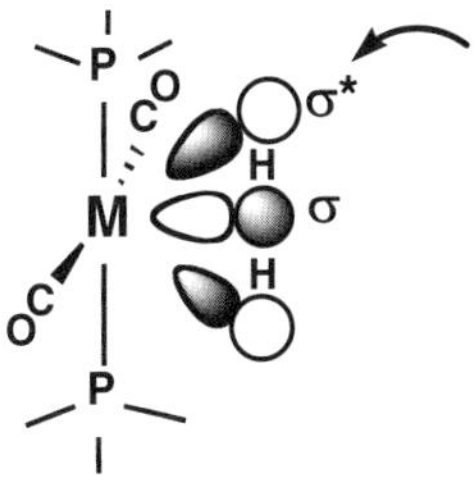

**Evidence for M → H$_2\sigma^*$ backbonding**

For M(CO)$_3$(PCy$_3$)$_2$(H$_2$):

| M | Barrier (kcal) |
|---|---|
| Cr | 1.3 |
| Mo | 1.7 |
| W | 2.2 |

**SCHEME 5.23**

in physisorbed $H_2$). The smallest splittings between the *ortho* and *para* $H_2$ state that had previously been observed were 4.8–10.5 cm$^{-1}$ for $H_2$ in K-intercalated graphite[170] and 30.6 cm$^{-1}$ for $H_2$ in Co ion-exchanged NaA zeolite.[171] In both these cases, $H_2$ is in all likelihood physisorbed as no indication of H–H bond activation could be found. However, for the $M(\eta^2$-$H_2)$ ground librational state, splittings between 17 and 0.6 cm$^{-1}$ are observed at temperatures as high as 200 K. The signals shift to lower energy and broaden but remain visible into the quasielastic scattering region. Observation of rotational tunneling, which is a *quantum mechanical* phenomenon, at such a high temperature is extraordinary.

Considerable molecular level detail on the interaction and binding of $H_2$ with both metal centers and nonmetal substances can be obtained by inelastic neutron scattering from the hindered rotor states of the bound molecule. The transition energies between these quantum mechanical rotational states for an adsorbed hydrogen molecule are very sensitive to the shape and height of the barrier to rotation, which in turn is a reasonably direct measure of the guest–host interactions. For low to medium barrier heights (as in, for example, MOF hydrogen storage materials), the transition between the lowest two states (rotational tunneling transition) decreases approximately exponentially with an increase of the barrier to rotation from the molecule's chemical environment. Moreover, the very large inelastic scattering cross section of $^1$H compared to that of any other atoms present in such systems makes rotational tunneling spectroscopy by INS a highly specific method to characterize the interaction between $H_2$ and its host. In addition to studies of $H_2$ rotational motion, the low- to mid-frequency (200–1000 cm$^{-1}$) region of the neutron vibrational spectrum can be probed to investigate the nature of dihydrogen bonding.[168] For example, deformational modes in $W(CO)_3(PCy_3)_2(H_2)$ have been identified by this technique, as described in Section 5.8.2.

### 5.8.8  Binding of $H_2$ to Highly Porous Solids and INS Studies

Nonmetal highly porous compounds with extremely high surface areas such as carbon-based substances, for example, fullerenes, and MOF materials have been intensely studied as possible lightweight materials for $H_2$ storage. Yaghi and coworkers have, for example, showed that a zinc-based material, MOF-177, has a surface area of more than 5600 m$^2$/g and can store 7.5% by weight of $H_2$ at 77 K.[172] This subject has been reviewed[173] and will not be discussed in detail except for relevance to the structure/bonding principles and methods developed for studying metal–$H_2$ complexes such as inelastic neutron scattering. Such techniques, discussed above, provide a unique tool for investigating the structure, dynamics, and chemical environment of hydrogen in potential hydrogen storage materials. This method has been applied to $H_2$ adsorption at low temperatures (typically 77 K) in porous carbons,[174] zeolites,[171,175] nickel phosphates,[176] and hybrid inorganic–organic compounds,[177,178] and has been described in more detail in the study of hybrid materials.[179] An excellent recent example of the very high value of INS spectroscopy is $H_2$ adsorbed in $NaNi_3(SIPA)_2(OH)(H_2O)_5 \cdot H_2O$, a MOF synthesized by Cheetham shown in Figure 5.13.[180]

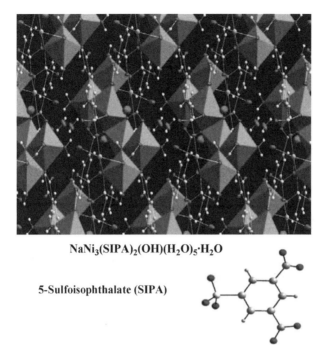

**NaNi₃(SIPA)₂(OH)(H₂O)₅·H₂O**

**5-Sulfoisophthalate (SIPA)**

**FIGURE 5.13**    The crystal structure for hydrated NaNi₃(SIPA)₂(OH)(H₂O)₅·H₂O, viewed in the *ab* plane. NiO₆ octahedra are illustrated as green polygons. Sodium, sulfur, carbon, oxygen, and hydrogen atoms are shown as blue, yellow, gray, red, and white spheres, respectively.(See the color version of this figure in Color Plates section.)

The organic linker here is 5-sulfoisophthalate (SIPA). At the lowest loading of H₂, a strong peak is observed in the rotational tunneling spectra (Figure 5.14) at 4.2 meV along with a weaker peak at 17.3 meV from hindered rotational transitions of the bound H₂ molecule.

The INS spectra of H₂ in NiSIPA appear to strongly suggest that binding of molecular hydrogen first occurs by molecular chemisorption at the unsaturated Ni (II) binding sites created by dehydration, as the series of transitions 31 at 4.1 and 17.3 and 22 meV (not shown) cannot be assigned on the basis of a model for physisorbed H₂ (i.e., double minimum with *two* rotational degrees of freedom), but can be fitted to the model used for coordinated dihydrogen (*planar* rotation in a double-minimum potential). A second site becomes occupied when the H₂ loading is increased to twice the initial loading (Figure 5.14) with a set of transitions at 5.4 meV and about 10 meV that again fit to the model for planar rotation indicative of molecular chemisorption. Two additional binding sites for H₂ become evident at three times the lowest loading, another strong binding site characterized by peaks at 4.8 and 13.8 meV and a second one by a doublet at 8.5 and 9.2 meV. This latter set of transitions, however, corresponds to that for a physisorbed molecule

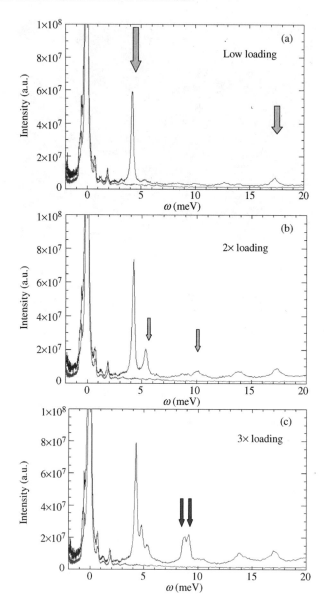

**FIGURE 5.14**    Inelastic neutron spectra of $H_2$ in $NaNi_3(SIPA)_2(OH)(H_2O)_5 \cdot H_2O$ for different $H_2$ loading levels. The upper traces in the spectra are for samples with $H_2$ loading and the lower traces are unloaded samples for comparison. The intensity is expressed in arbitrary units (a.u.). Peaks designated by arrows in spectra (a) and (b) are due to $H_2$ chemisorbed at unsaturated Ni sites (where $H_2$ undergoes planer rotation). New peaks that appear in spectrum (c) on highest ($3\times$) loading (shown by arrows) are due to physisorbed $H_2$ undergoing three-dimensional rotation.

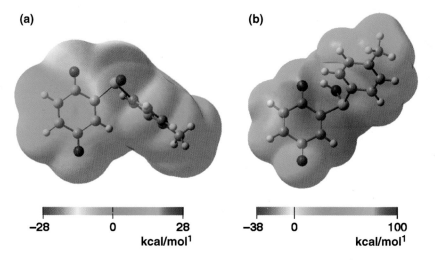

**FIGURE 2.12** Electrostatic potential maps for (a) TolSQ and (b) TolSQH$^+$, calculated with density functional theory at the BLYP/6-31G$^{**}$ level.

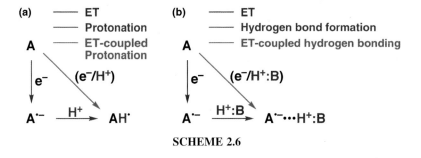

**SCHEME 2.6**

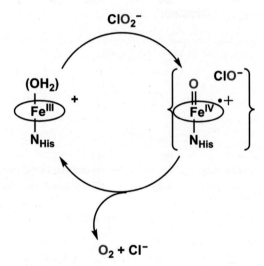

**FIGURE 3.8** Mechanism of chlorite dismutase.

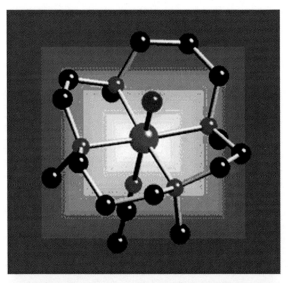

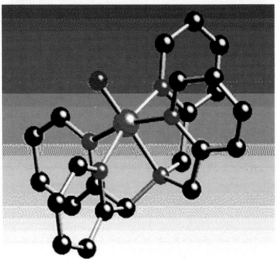

**FIGURE 4.36** X-ray structures of nonheme iron(IV)-oxo complexes.

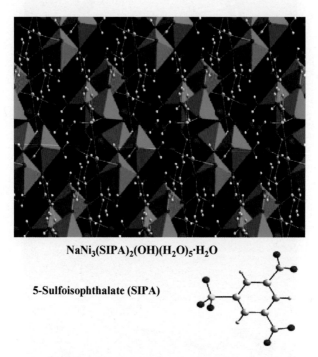

NaNi$_3$(SIPA)$_2$(OH)(H$_2$O)$_5$·H$_2$O

5-Sulfoisophthalate (SIPA)

**FIGURE 5.13** The crystal structure for hydrated NaNi$_3$(SIPA)$_2$(OH)(H$_2$O)$_5$·H$_2$O, viewed in the *ab* plane. NiO$_6$ octahedra are illustrated as green polygons. Sodium, sulfur, carbon, oxygen, and hydrogen atoms are shown as blue, yellow, gray, red, and white spheres, respectively.

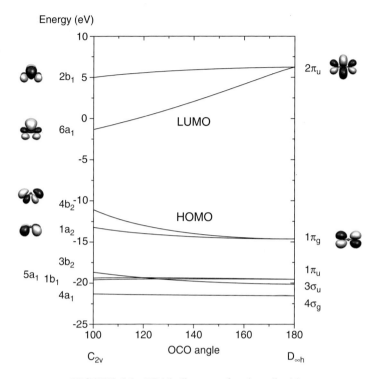

**FIGURE 6.1** Walsh diagram of carbon dioxide.

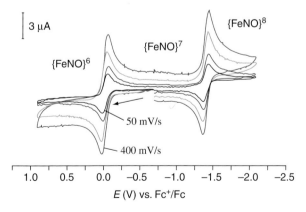

**FIGURE 7.2** Cyclic voltammogram of [Fe(NO)(cyclam-ac)](PF$_6$) in CH$_3$CN at 20°C (0.1 M [N(*n*-Bu)$_4$]PF$_6$ supporting electrolyte, glassy C electrode).

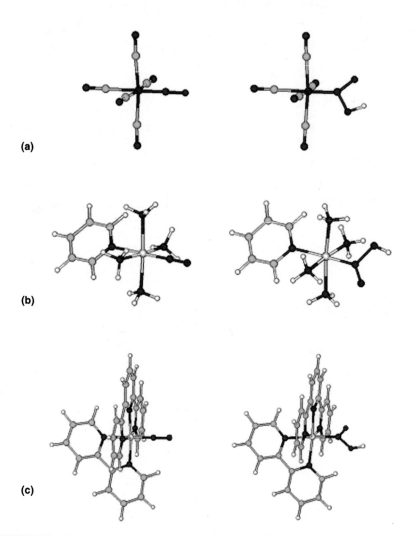

**FIGURE 7.7** Geometries optimized at the B3LYP/6-31G** level (SDD pseudopotentials on the metal centers) for representative members of the set (metallonitrosyls and OH⁻ addition products): (a) $[Fe(CN)_5NO]^{2-}$; (b) *trans*-$[Ru(NH_3)_4NO(py)]^{3+}$; (c) *cis*-$[Ru(bpy)(trpy)NO]^{3+}$.

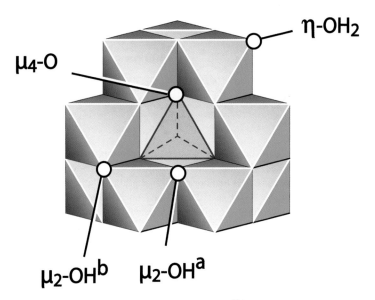

**FIGURE 8.9** Structure of $\varepsilon$-$AlO_4Al_{12}(OH)_{24}(OH_2)_{12}{}^{7+}$ ($\varepsilon$-**Al$_{13}$**, $T_d$ symmetry), viewed along one of the four $C_3$ axes. Reproduced with permission from *Chem. Rev,* **2005**, *106*, 1. Copyright 2006 American Chemical Society.

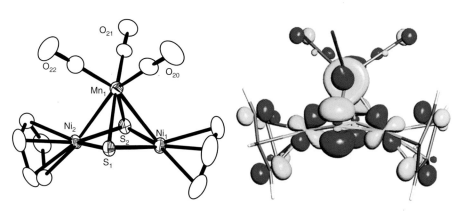

**FIGURE 10.2** An ORTEP diagram of the molecular structure of $Cp_2Ni_2(\mu_3$-$S)_2Mn(CO)_3$ and the computed SOMO. Reproduced with permission from Ref. 30 (chapter 10). Copyright 2004, American Chemical Society.

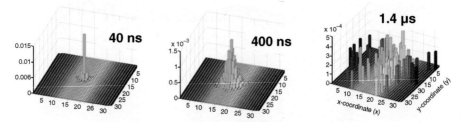

**FIGURE 12.7** Monte Carlo simulations of lateral intermolecular energy transfer across an anatase $TiO_2$ nanocrystallite. A $32 \times 32$ grid of close-packed sensitizer grid with a continuity condition that allowed excited states that hopped off the grid to reappear on the opposite side was used to approximate the $\sim$700 sensitizers found on a $\sim$20 nm spherical crystal. The probability of where a $Ru(dcb)(bpy)_2^*$ excited state would be found at 40 ns, 400 ns, and 1.4 μs that underwent random $(30\,ns)^{-1}$ intermolecular energy hops is shown.

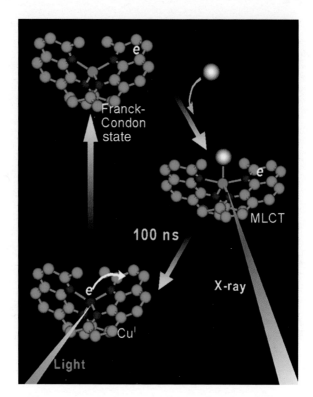

**FIGURE 12.12** X-ray characterization of the equilibrated MLCT excited states of $Cu(dmp)_2^+$ in toluene solution. A pulsed laser was utilized to create the excited state that was subsequently interrogated by a series of X-ray pulses from the advanced light source at Argonne National Laboratory. The results showed that the emissive excited state had five nearest neighbors.

(two-dimensional reorientation) and a barrier of 3.4B. The above data suggest that several accessible, coordinatively unsaturated Ni(II) sites exist in NaNi$_3$(SI-PA)$_2$(OH)(H$_2$O)$_5$·H$_2$O when it is dehydrated at sufficiently high temperature to remove aquo ligands from the Ni octahedra. Remarkably detailed information has also been obtained on the primary binding sites of H$_2$ in a series of metal-organic frameworks composed of Zn$_4$O(O$_2$C-)$_6$ secondary building units with the use of INS.[177,179]

The development of such highly porous solids for reversible molecular H$_2$ binding in the Ni and Zn MOFs is a major challenge in materials science. Several factors besides high surface area influence hydrogen uptake in MOF and other highly porous materials and hence their usefulness for hydrogen storage. A critical property is the heat of adsorption, which is a measure of the strength of interaction between H$_2$ and MOF surfaces. If too weak, as is the case with current MOF compounds, then impractically low temperatures are needed to store the hydrogen. If the interaction is too strong, then energy must be spent to release the hydrogen and there is the possibility for further unfavorable features such as irreversible dissociative binding and slow kinetics. Zhou and coworkers discuss these factors in a recent review article.[173] The MOFs and other highly porous materials containing coordinatively unsaturated metal sites are nevertheless a realistic and promising means of achieving storage for hydrogen and other environmentally relevant small molecules such as methane and carbon dioxide.

A very significant recent finding is unusually strong adsorption of H$_2$ on the Cu$^+$ sites in copper-modified ZSM-5, even at room temperature.[111,181,182] DRIFT and IR transmittance spectra of H$_2$ adsorbed at 77 K or at *room temperature* by the copper-modified ZSM-5 zeolite pre-evacuated or prereduced in CO at 873 K indicated several unusual forms of adsorbed hydrogen.[181] H–H stretching frequencies of adsorbed species at 3075–3300 cm$^{-1}$ are by about 1000 cm$^{-1}$ lower than those in the free hydrogen molecules. This indicates unusually strong perturbation of adsorbed hydrogen by reduced Cu$^+$ ions that has been never before reported either for hydrogen or for adsorption of other molecules by any cationic form of zeolites or oxides. It is clear that the H$_2$ molecules are binding directly to the copper metal sites, and this was confirmed by INS studies by Eckert and coworkers.[111,182] The rotational tunneling spectrum of a Cu-ZSM-5 zeolite after loading with H$_2$ was measured by inelastic incoherent neutron scattering. The peaks observed at ±0.80 and ±1.73 cm$^{-1}$ unambiguously demonstrate the formation of a strongly bound Cu ($\eta^2$-H$_2$) complex.

In summary, in order to reversibly bind molecular H$_2$ for practical applications such as hydrogen storage, it is necessary to design compounds with high surface areas or mimic the nanotube structures of carbon fullerenes, but using much less expensive materials. This is a great opportunity for design of, for example, supramolecular cage-like structures of light main group elements such as boron, oxygen, nitrogen, lithium, and so on with or without metal centers that would help trap molecular hydrogen. As discussed above, H$_2$ molecules have the ability to bind to a large variety of materials as either a Lewis acid or a Lewis base, albeit weakly, and this is the key feature to be explored for new hydrogen storage methods.

## 5.9 ISOTOPE EFFECTS IN $H_2$ LIGAND COORDINATION AND SPLITTING

### 5.9.1 Introduction

An extremely informative general physical chemistry tool involves studies of isotope effects, especially in mechanistic studies. These effects are very valuable in organometallic chemistry, particularly for M(H)(X) systems where X = H, C, Si, and so on. An excellent review of this area is provided by Bullock.[183] Both kinetic and equilibrium (or thermodynamic) effects can provide crucial information about reaction mechanisms that is unavailable from other methodologies. Yet isotope effects often are poorly understood or can even seem to be paradoxical in certain cases. Unlike the situation in organic chemistry, the ability of metal sites, including those in metalloenzymes, to reversibly coordinate substrates prior to rate-determining steps complicates isotope effect "rules" originally formulated by organic chemists. For example, the nature of equilibrium isotope effects (EIEs) for $H_2$ versus $D_2$ addition to metal complexes has been understood only recently. The situation can become even more complex when these ligands undergo homolytic or heterolytic cleavage, either of which can also be reversible. A "normal" isotope effect occurs when the rate of reaction of an unlabeled compound is faster than that for the corresponding labeled species, that is, $k_H/k_D > 1$. It is "inverse" for $k_H/k_D < 1$, and this terminology also applies to EIEs, $K_H/K_D$, which will be discussed first. Deuterium *kinetic* isotope effects (KIEs) are widely used to infer details concerned with reaction mechanisms and the nature of transition states, while EIEs are associated with the site preferences of hydrogen and deuterium and enable one to discern aspects pertaining to molecular structure by using NMR spectroscopy. Primary deuterium isotope effects are often interpreted by using two simple guidelines: the KIE for an elementary reaction is normal ($K_H/K_D > 1$) and the EIE is dictated by deuterium preferring to be located in the site corresponding to the highest frequency oscillation and, as such, may be either normal ($K_H/K_D > 1$) or inverse ($K_H/K_D < 1$). The purpose of this section is to evaluate the applicability of these rules as they apply to the interaction of the H–H bonds with a transition metal center. Importantly, recent experimental and computational studies question the premise that primary EIEs in these systems may be predicted by naively assuming that deuterium prefers to occupy the highest frequency oscillator. Of further significance, the EIEs for formation of $H_2$ σ −complexes by coordination of H–H bonds and also oxidative addition of dihydrogen both exhibit unusual temperature dependences, such that both normal (i.e., $K_H/K_D > 1$) and inverse (i.e., $K_H/K_D < 1$) values may be obtained for the same system. These dependences and other aspects of isotope effects will be discussed in detail below.

### 5.9.2 Equilibrium Isotope Effects for $H_2$ Versus $D_2$ Binding

Deuterium EIEs have been observed for the reversible addition of $H_2$ and $D_2$ to various complexes in solution to form either metal dihydride/dideuteride complexes[184–186] or $H_2/D_2$ complexes (Scheme 5.24).[187–190]

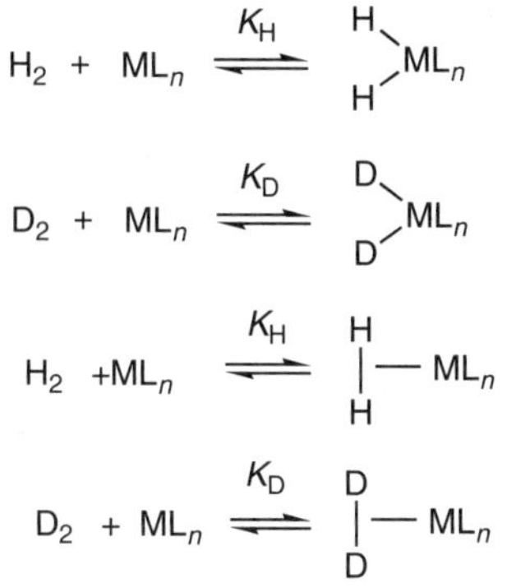

$$H_2 + ML_n \xrightleftharpoons{K_H} \begin{array}{c} H \\ \diagdown \\ \diagup \\ H \end{array} ML_n$$

$$D_2 + ML_n \xrightleftharpoons{K_D} \begin{array}{c} D \\ \diagdown \\ \diagup \\ D \end{array} ML_n$$

$$H_2 + ML_n \xrightleftharpoons{K_H} \begin{array}{c} H \\ | \\ H \end{array}\!\!-ML_n$$

$$D_2 + ML_n \xrightleftharpoons{K_D} \begin{array}{c} D \\ | \\ D \end{array}\!\!-ML_n$$

**SCHEME 5.24**

The EIE values for H$_2$ versus D$_2$ addition are usually inverse over a large temperature range (260–360 K), showing that counterintuitively D$_2$ binds more strongly than H$_2$. The values of $K_H/K_D$ observed thus far are 0.36–0.77 for formation of H$_2$ complexes and 0.47–0.85 for complete OA. An inverse EIE (0.39) also occurs for protonation of metal hydrides to form $\eta^2$-H$_2$ as shown in the reaction Cp$_2$WH$_2$ + H$^+$ → [Cp$_2$WH(H$_2$)]$^+$.[191]

For addition of substrates to the unsaturated precursor complex, Cr(CO)$_3$(PCy$_3$)$_2$, the equilibria shown below are rapidly established under moderate pressures of H$_2$/D$_2$ (1–10 atm) in THF solution (Equations 5.5 and 5.6).

$$Cr(CO)_3(PCy_3)_2 \text{ (soln)} + H_2 \rightleftharpoons Cr(CO)_3(PCy_3)_2(H_2) \text{ (soln)} \qquad (5.5)$$

$$Cr(CO)_3(PCy_3)_2 \text{ (soln)} + D_2 \rightleftharpoons Cr(CO)_3(PCy_3)_2(D_2) \text{ (soln)} \qquad (5.6)$$

Equilibrium constants determined at 13–36°C by IR measurements of $\nu_{CO}$ show that the thermochemical parameters for binding of H$_2$ are $\Delta H^\circ = -6.8 \pm 0.5$ kcal/mol and $\Delta S^\circ = -24.7 \pm 2.0$ eu.[189] For binding of D$_2$, $\Delta H^\circ = -8.6 \pm 0.5$ kcal/mol and $\Delta S^\circ = -30.0 \pm 2.0$ eu; that is, D$_2$ binds better enthalpically ($\Delta\Delta H = 1.8$ kcal/mol), which easily overcomes the disfavored entropy of D$_2$ complexation ($\Delta\Delta S = 5.3$ cal/ (mol deg)). This domination of enthalpy over entropy is general to these primary isotope effects (unlike, for example, H$_2$ versus N$_2$ binding), and thus EIEs are enthalpically driven. It should be kept in mind that EIEs are temperature dependent because of the entropy differences. The EIE for the W analogue cannot be measured directly because of stronger W(H$_2$) bonding, but the equilibrium shown in Equation 5.7 provides a means of determining accurate EIE values:

$$W(CO)_3(PCy_3)_2 \text{ (N}_2\text{)(soln)} + H_2\text{(gas)} \rightleftharpoons W(CO)_3(PCy_3)_2 \text{ (H}_2\text{) (soln)} + N_2\text{(gas)}$$
$$(5.7)$$

Spectroscopic measurements using calibrated $H_2/N_2$ and $D_2/N_2$ gas mixtures give $K_H/K_D = 0.70 \pm 0.15$ in THF solvent at 22 °C.

Data for $H_2$ versus $D_2$ binding are reported for Ir and Os complexes that have both hydride and $H_2$ ligands, that is, $MH_x(H_2)L_n$.[187,188,192] Lower $K_H/K_D$ of 0.36–0.50 occur here, possibly because of a secondary isotope effect due to the hydride ligands (values for $MH_x(D_2)L_n$ cannot be determined because of isotopic exchange, so values for $MD_x(D_2)L_n$ are used). $D_2$ loss from $IrClD_2(D_2)(L)_2$ is energetically $\sim 1$ kcal/mol higher than $H_2$ loss from $IrClH_2(H_2)(L)_2$ ($L = P$-$t$-$Bu_2Me$).[188]

### 5.9.3    Origin of the Inverse EIE for $H_2$ Binding and Temperature Dependency of EIE

Since the free D–D bond is 1.8 kcal/mol stronger than the H–H bond, one might naively expect $H_2$ to bind preferentially over $D_2$. If the $W$–$D_2$ and $W$–$H_2$ bonds were of equal strength in the complexes in Scheme 5.25, the reaction would be predicted to be exothermic by $-1.8$ kcal/mol based on the fact that the D–D bond is that much stronger.

The fact that $\Delta H$ in Scheme 5.25 is actually measured to be *endothermic* by this amount for the Cr complex implies that zero-point and excited-state vibrational energies for the $\eta^2$-$H_2$ species determine the EIE.[189] Furthermore, one must consider more than just the lowering of $\nu_{HH}$ versus $\nu_{DD}$ on coordination. The favoring of the left-hand side of Scheme 5.25 is the inverse of that predicted from simple changes in $\nu_{HH}$ alone, where deuterium should favor the stronger force constant; that is, $D_2$ should prefer to remain unbound compared to $H_2$.

The EIE may be calculated from molecular translational, rotational, and vibrational partition function ratios as described by Bigeleisen and Mayer (Equation 5.8).[193]

$$EIE = MMI \times EXC \times ZPE \qquad (5.8)$$

The calculated EIE is the product of three factors: a rotational and translational factor containing the reduced (classical) rotational and translational partition function ratios of isotopic species (MMI); a factor accounting for contributions from excitations of vibrational energy levels (EXC); and a factor comprising zero-point energy (ZPE) contributions. For $H_2$ complexes, all vibrational spectroscopic data concur that $\nu_{HH}$ and $\nu_{DD}$ (hence bond order) are lowered greatly when $H_2/D_2$ binds to M. This should result in a "normal" equilibrium isotope effect *if changes in the HH(DD) force constant were the major contributor to the EIE*. However, as originally elucidated by Krogh-Jespersen and Goldman[186] and expanded by Bender

$$H_2 + \underset{D}{\overset{D}{|}} \rightarrow W(CO)_3L_2 \;\; \underset{}{\overset{K_H/K_D}{\rightleftharpoons}} \;\; D_2 + \underset{H}{\overset{H}{|}} \rightarrow W(CO)_3L_2$$

**SCHEME 5.25**

**TABLE 5.4   Equilibrium Isotope Effect Contributions From Individual Vibrational Modes (cm$^{-1}$) for H$_2$ (D$_2$) Complexation to W(CO)$_3$(PCy$_3$)$_2$(H$_2$) at T = 300 K**

| Mode (Sym) | H$_2$ (D$_2$) Gas | H$_2$ Complex (D$_2$ Complex) | EXC | ZPE |
|---|---|---|---|---|
| $\nu$(HH) (A$_1$) | 4395 (3118) | 2690 (1900) | 1.000 | 3.215 |
| $\nu$(WH$_2$) (A$_1$) | – | 953 (703) | 0.976 | 0.549 |
| $\delta$(WH$_2$) (B$_1$) | – | 640 (442) | 0.923 | 0.622 |
| $\nu$(WH$_2$) (B$_1$) | – | 1575 (1144) | 0.996 | 0.356 |
| $\delta$(WH$_2$) (B$_2$) | – | 462 (319) | 0.879 | 0.710 |
| $\tau$(WH$_2$) (A$_2$) | – | 355 (251) | 0.856 | 0.780 |

$\Pi_{EXC} = 0.675$; $\Pi_{ZPE} = 0.216$.

and Kubas[189] to the case of H$_2$ complexes, contributions to the ZPE from new vibrational and rotational modes are of critical importance to EIEs when H$_2$ either coordinates or cleaves to dihydride. The change in ZPE for HH(DD) stretching mode contributes a large "normal" factor to the total EIE as expected. The calculated ZPE contribution would predict an EIE of ~3.2 for Equation 5.8 if changes in HH(DD) stretching force constant were the only contributor to the EIE. The five "new" vibrational normal modes in H$_2$ complexes (see the previous section) all contribute modest *inverse* EXC and ZPE factors to the EIE that collectively overcome the strong "normal" ZPE component from $\nu_{HH}$ (Table 5.4). These factors (ZPE = 0.216; EXC = 0.675) multiplied against the MMI term (5.77) predict an overall inverse EIE of 0.78 at 300K for W(CO)$_3$(PCy$_3$)$_2$(H$_2$) (this includes minor contributions from other modes that mix with those in Table 5.4).[189] This value agrees well with the experimental value of $K_H/K_D = 0.70 \pm 0.15$ for Scheme 5.25.

Summarizing, while the [SYM × MMI × EXC] term is not usually invoked when rationalizing EIEs because it is typically a minor component, it can become dominant when either the reactants or products are small so that isotopic substitution has a significant impact on the moment of inertia of one of the molecules. As such, the influence of the [SYM × MMI × EXC] term is particularly relevant to reactions involving H$_2$ and also CH$_4$, and the situation is such that the MMI component favors deuterium residing in the smaller molecule.

### 5.9.4   Temperature Dependence of EIE

The transition between a normal and an inverse EIE reflects the fact that these systems are not characterized by the typical monotonic variation predicted by the van't Hoff relationship. Instead, as discussed in an edifying review by Parkin,[194] the EIEs in these systems are zero at 0K, increase to a value >1, and then decrease to unity at infinite temperature. This unusual behavior is, nevertheless, rationalized by consideration of the individual factors that contribute to the EIE. As discussed above, the EIE may be expressed in the form EIE = SYM × MMI × EXC × ZPE (where SYM is the symmetry factor, MMI is the mass moment of inertia term, EXC is the excitation term, and

ZPE is the zero-point energy term), and the distinctive temperature profile is a consequence of the inverse ZPE (cf. enthalpy) and normal $[SYM \times MMI \times EXC]$ (cf. entropy) components opposing each other and having different temperature dependences. At low temperatures, the ZPE component dominates and the EIE is inverse, while the $[SYM \cdot MMI \cdot EXC]$ component dominates at high temperatures and the EIE is normal (Figure 5.15).

As such, Parkin anticipated that coordination of dihydrogen could also be characterized by a normal EIE at high temperature when the $[SYM \times MMI \times EXC]$ term would dominate. Indeed, this notion was supported by calculations on W $(CO)_5(H_2)$, as illustrated in Figure 5.15. Dihydride and dihydrogen complexes are tautomers and a remaining issue pertains to the site preference of deuterium. In this regard, the EIE for conversion of $W(CO)_5(H_2)$ to the dihydride $W(CO)_5H_2$ was calculated by Parkin to be normal at all temperatures, thereby demonstrating that deuterium favors the nonclassical site in this system (Figure 5.15). This preference is dictated by the ZPE term because substitution of the dihydrogen and hydride ligands by deuterium has relatively little impact on the MMI term due to the large size of the molecules. Furthermore, because $W(CO)_5(H_2)$ and $W(CO)_5H_2$ have the same number of isotopically sensitive vibrations, the normal ZPE term is largely a consequence of the high-energy H–H stretch in $W(CO)_5(H_2)$ becoming a low-energy symmetric bend in $W(CO)_5H_2$.

In summary, the inverse nature of the ZPE term is a consequence of the rotational and translational degrees of freedom of HH becoming low-energy isotopically sensitive vibrations in the product dihydrogen complex, while the normal nature of the $[SYM \times MMI \times EXC]$ component is a consequence of deuterium substitution having a larger impact on the moment of inertia of the smaller molecule. The

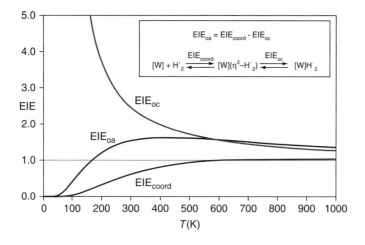

**FIGURE 5.15**    Calculated EIEs as a function of temperature for oxidative addition ($EIE_{oa}$) of $H_2$ and coordination ($EIE_{coord}$) of $H_2$ and $D_2$ to $W(CO)_5$ and for oxidative cleavage of $W(CO)_5(H_2)$ to $W(CO)_5H_2$ ($E_{oc}$).

interactions of H–H (and C–H and presumably other R–H) bonds with transition metal centers are characterized by interesting temperature-dependent deuterium EIEs. It is, therefore, evident that the correct analysis of primary KIEs and EIEs in systems of this type cannot simply be achieved by considering only the ZPEs associated with the high-energy stretching frequencies. As such, it is prudent to consider all isotopically sensitive vibrations before attempting to interpret the significance of an isotope effect. Although the determination of the frequencies of all isotopically sensitive vibrations is experimentally challenging, good estimates may be obtained by using computational methods and the use of the frequencies so obtained provides a useful approach for analyzing isotope effects.

### 5.9.5   EIE for Elongated H$_2$ Complexes and OA Processes

Nuclear motion quantum calculations (DVR methodology, see Section 5.8.5) give somewhat lower EIE values (0.53 at 300K), but unexpectedly show *normal EIE for elongated H$_2$ complex*es such as [Cp*Ru(H $\cdots$ H)(dppm)]$^+$ (EIE = 1.22) and [Os (H $\cdots$ H)Cl(dppe)$_2$]$^+$ (EIE = 1.69). This is proposed to be due to the severe *anharmonicity* in the H$_2$-related vibrational modes in these complexes, which favors addition of H$_2$ over D$_2$. The ZPE factor is affected the most here, and corrections for anharmonicity must be taken into account, especially for complexes with $d_{HH} > 1$ Å, although experimental confirmation is needed.[195]

OA of H$_2$ to WI$_2$(PMe$_3$)$_4$ to give the dihydride WH$_2$I$_2$(PMe$_3$)$_4$ also shows an inverse EIE of 0.63(5) at 60°C because of the large number of isotope-sensitive vibrational modes in the product (two M–H stretching modes and four bending modes) compared to the H$_2$ reactant.[185] The calculated value is 0.73 using a similar approach to that above for formation of H$_2$ complexes and that by Krogh-Jespersen and Goldman for OA of H$_2$ to form IrH$_2$Cl(CO)(PPh$_3$)$_2$ (0.47; experimental, 0.55).[186] Because the MMI factors are similar for dihydride and H$_2$ complex formation, the inverse nature of the EIE may again be traced to the dominant ZPE, which for the dihydride cases is 0.10–0.17.

Because ΔZPE changes for the dihydride and H$_2$ cases are both referenced to free H$_2$(D$_2$), this free energy difference is due to changes in force constants that favor D in the dihydride tautomer. This reasoning is consistent with an overall increase in the force constants (despite the weakening of the bound HH(DD) force constant) when the "loosely bound" H$_2$ ligand in the M(H$_2$) complex fully adds. Indeed, the preference of deuterium as the dideuteride tautomer (where it is more strongly bound to M) might seem to parallel the preference of D in the complex versus that as a free ligand. Related to this is the tendency for D to concentrate in the hydride site in certain hydride(H$_2$) complexes versus that in the $\eta^2$-H$_2$ ligand (Scheme 5.26; Ir = TpIr(PMe$_3$).[196]

The equilibrium constants indicate that the heavier isotope prefers to occupy the hydride site here, but favors $\eta^2$-H$_2$ in [ReH$_2$(H$_2$)(CO)(PR$_3$)$_3$]$^+$. In the latter, the isotope effect was interpreted to be a consequence of a greater vibrational ZPE difference between Re($\eta^2$-HD) and Re($\eta^2$-D$_2$) relative to Re–H and Re–D. The isotopic preferences will be dictated by the changes in *all* the force constants in both

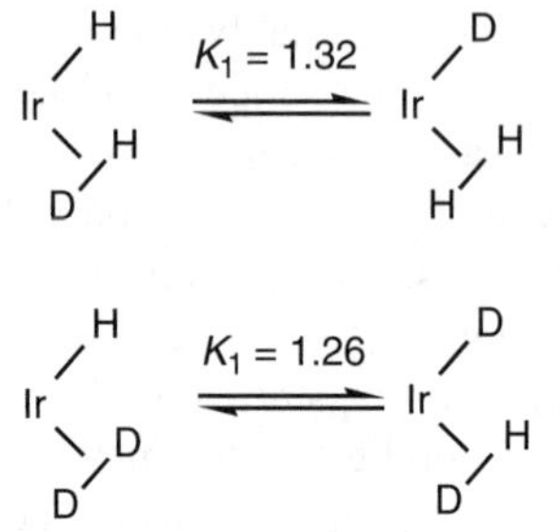

**SCHEME 5.26**

tautomers and depend on relative H–H versus M–H bond strengths. The Ir system has a long $d_{HH}$ of ~1 Å and the Ir–H bonds, particularly to the classical hydride, are undoubtedly stronger than the H–H bond, which would be expected to favor D incorporation. By contrast, the H–H bond in the Re complex is much stronger ($J_{HD} = 34$ Hz); hence, it is a "true" $\sigma$-complex that would bind $D_2$ more strongly than $H_2$. Also in support of this are $K_H/K_D$ values that are 1.5 for tautomerization of $[ReH_2(H_2)(CO)(PMe_3)_3]^+$ to the tetrahydride,[197] and the equilibrium between CpNb $(CO)_3(H_2)$ and $CpNb(CO)_3H_2$ shows similar favoring of D in the nonclassical isomer. The $K_H/K_D = 0.20 \pm 0.05$ for equilibrium isomerization of $[Cp_2WH(H_2)]^+$ to $[Cp_2WH_3]^+$ could reflect a more hydridic nature for this system, that is, a $H_2$ ligand with a weak H–H bond.

### 5.9.6    Kinetic Isotope Effects for $H_2$ Oxidative Addition and Reductive Elimination

There is very limited data on KIEs for $\sigma$-ligand coordination/dissociation or $\sigma$-bond cleavage equilibria as shown in Scheme 5.27.

For $H_2$ loss from the $W(CO)_3(PCy_3)_2$ fragment, $k_{-1} = 469$ s$^{-1}$ for $H_2$ and 267 s$^{-1}$ for $D_2$, giving $k_{-1}^H/k_{-1}^D = 1.7$.[198] Applying the EIE data above and the following expressions, this gives $k_1^H/k_1^D = 1.2$ for $H_2$ binding.

$$K_H/K_D = k_1^H/k_{-1}^H \times k_{-1}^D/k_1^D \tag{5.9}$$

$$k_1^H/k_1^D = K_H/K_D \times k_{-1}^H/k_{-1}^D = 0.7 \times 1.7 = 1.2 \tag{5.10}$$

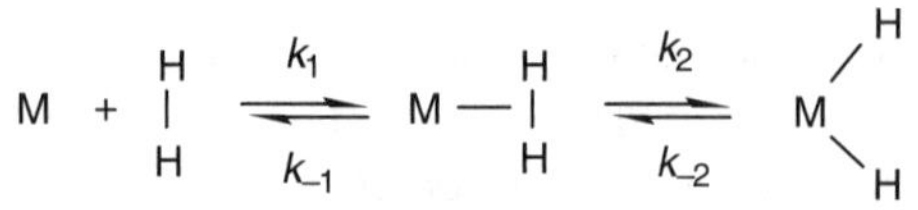

**SCHEME 5.27**

In comparison, the reaction below occurs 1.9 times as fast for $H_2$ as for $D_2$ $(10^4 s^{-1})$.[199]

$$Cr(CO)_5(C_6H_{12}) + H_2 \rightarrow Cr(CO)_5H_2 + C_6H_{12} \qquad (5.11)$$

The subsequent rate of loss of $H_2$ $(2.5 s^{-1})$ is five times as fast as that for $D_2$, consistent with stronger binding of $D_2$ over $H_2$. For *oxidative addition* of $H_2$ to RhCl $(PPh_3)_2$, $k_H/k_D$ is 1.5,[200] and for OA to Vaska's complex, it is smaller, 1.06.[201] For conversion of the dihydride to the $H_2$ complex in Scheme 5.27, $k_{-2} = 37 s^{-1}$ for $H_2$ and 33 $s^{-1}$ for $D_2$, giving $k_H/k_D = 1.08 \pm 0.04$ (the reverse reaction, OA of $H_2$, occurs about 50% slower $(k_2 = 18 s^{-1})$).[198] There is virtually no KIE for H–H bond formation (reductive elimination), which also probably applies to H–H cleavage (OA). The activation energies overlap within experimental error for the $H_2$ and $D_2$ systems in Scheme 5.27 for both the $k_{-1}$ and $k_{-2}$ steps.

In these cases, it is the KIE for binding of $H_2$ in a pre-equilibrium step that determines the overall KIE for OA of $H_2$. In most cases, $H_2$ coordination gives a normal KIE. Since the nature of the transition state is not known and can vary, it is not possible to draw conclusions about the slower rate of binding of $D_2$ versus $H_2$. Dissociation of $H_2$ can give either normal or inverse effects.[183] An overall inverse KIE for reductive elimination of $H_2$ from a dihydride is symptomatic of formation of a $H_2$ complex in an equilibrium step rather than in a single-step process.

Protonation of metal hydrides with HX acids to form $H_2$ ligands (see Scheme 5.3) shows an inverse KIE of 0.21–0.64 for formation of $[FeH(H_2)P_4]^+$ ($P_4$ = 2dppe or P $(C_2H_4PPh_2)_3$). This reaction involves direct attack of HX at one hydride in $FeH_2P_4$, and the inverse effect suggests that protonation occurs through a late transition state with a structure similar to that of the product $H_2$ complex.[32,202–204] Calculated values are similar in magnitude to the observed values, thus supporting the mechanism of protonation of metal hydrides discussed in Section 5.2.

## ACKNOWLEDGMENTS

GJK is grateful for funding from DOE, Basic Energy Sciences, Chemical Sciences that supported the basic research leading to the discovery of $H_2$ complexes and to Los Alamos National Laboratory for LDRD funding. Research by DMH has been supported by the NSF.

## REFERENCES

1. (a) Kubas, G. J.; Ryan, R. R.; Swanson, B. I.; Vergamini, P. J.; Wasserman, H. J. *J. Am. Chem. Soc.* **1984**, *106*, 451; (b) Kubas, G. J.; Unkefer, C. J.; Swanson, B. I.; Fukushima, E. *J. Am. Chem. Soc.* **1986**, *108*, 7000.

2. Kubas, G. J. *Acc. Chem. Res.* **1988**, *21*, 120.

3. Kubas, G. J. *Metal Dihydrogen and σ-Bond Complexes*; Kluwer Academic/Plenum Publishers: New York, 2001.

4. Kubas, G. J. *Chem. Rev.* **2007**, *107*, 4152.

5. James, B. R. *Homogeneous Hydrogenation*; Wiley: New York, 1973.

6. Halpern, J. *J. Organomet. Chem.* **1980**, *200*, 133.

7. Saillard, J.-Y.; Hoffmann, R. *J. Am. Chem. Soc.* **1984**, *106*, 2006.

8. Maseras, F.; Lledós, A.; Clot, E.; Eisenstein, O. *Chem. Rev.* **2000**, *100*, 601.

9. Crabtree, R. H. *Angew. Chem., Int. Ed. Engl.* **1993**, *32*, 789.

10. Jessop, P. G.; Morris, R. H. *Coord. Chem. Rev.* **1992**, *121*, 155.

11. Morris, R. H. *Can. J. Chem.* **1996**, *74*, 1907.

12. Crabtree, R. H. *Acc. Chem. Res.* **1990**, *23*, 95.

13. Heinekey, D. M.; Oldham, W. J., Jr. *Chem. Rev.* **1993**, *93*, 913.

14. Heinekey, D. M.; Lledós, A.; Lluch, J. M. *Chem. Soc. Rev.* **2004**, *33*, 175.

15. Sabo-Etienne, S.; Chaudret, B. *Coord. Chem. Rev.* **1998**, *178–180*, 381.

16. Morris, R. H. *Coord. Chem. Rev.* **2009**, *253*, 1219.

17. Crabtree, R. H.; Hamilton, D. G. *J. Am. Chem. Soc.* **1986**, *108*, 3124.

18. Knoth, W. H. *J. Am. Chem. Soc.* **1968**, *90*, 7172.

19. Ashworth, T. V.; Singleton, E. *J. Chem. Soc., Chem. Commun.* **1976**, 705.

20. Gusev, D. G.; Vymenits, A. B.; Bakhmutov, V. I. *Inorg. Chim. Acta* **1991**, *179*, 195. In 1993, Zilm obtained solid-state $^1$H NMR evidence for $H_2$ coordination ($d_{HH} = 0.93$ Å) on a sample we prepared.

21. Esteruelas, M. A.; Oro, L. A. *Chem. Rev.* **1998**, *98*, 577.

22. Jia, G.; Lau, C.-P. *Coord. Chem. Rev.* **1999**, *190–192*, 83.

23. Esteruelas, M. A.; Oro, L. A. *Adv. Organomet. Chem.* **2001**, *47*, 1.

24. (a) Kuhlman, R. *Coord. Chem. Rev.* **1997**, *167*, 205; (b) Besora, M.; Lledós, A.; Maseras, F. *Chem. Soc. Rev.* **2009**, *38*, 957;

25. McGrady, G. S.; Guilera, G. *Chem. Soc. Rev.* **2003**, *32*, 383.

26. Kubas, G. J. *Adv. Inorg. Chem.* **2004**, *56*, 127.

27. Kubas, G. J. *Catal. Lett.* **2005**, *104*, 79.

28. Dedieu, A., Ed. *Transition Metal Hydrides*; VCH Publishers: New York, 1992.

29. Peruzzini, M.; Poli, R., Ed. *Recent Advances in Hydride Chemistry*; Elsevier: Amsterdam, 2001.

30. Fang, X.; Huhmann-Vincent, J.; Scott, B. L.; Kubas, G. J. *J. Organomet. Chem.* **2000**, *609*, 95.

31. Szymczak, N. K.; Tyler, D. R. *Coord. Chem. Rev.* **2008**, *252*, 212.

32. Basallote, M. G.; Duran, J.; Fernandez-Trujillo, J.; Manez, M. A. *J. Organomet. Chem.* **2000**, *609*, 29, and references therein.

33. Harman, W. D.; Taube, H. *J. Am. Chem. Soc.* **1990**, *112*, 22612.

34. Hasegawa, T.; Li, Z.; Parkin, S.; Hope, H.; McMullan, R. K.; Koetzle, T. F.; Taube, H. *J. Am. Chem. Soc.* **1994**, *116*, 4352.

35. Malin, J.; Taube, H. *Inorg. Chem.* **1971**, *10*, 2403.

36. Aebischer, N.; Frey, U.; Merbach, A. E. *Chem. Commun.* **1998**, 2303.

37. Matthews, S. L.; Pons, V.; Heinekey, D. M. *J. Am. Chem. Soc.* **2005**, *127*, 850.

38. Matthews, S. L.; Heinekey, D. M. *J. Am. Chem. Soc.* **2006**, *128*, 2615.

39. Kubas, G. J.; Burns, C. J.; Eckert, J.; Johnson, S.; Larson, A. C.; Vergamini, P. J.; Unkefer, C. J.; Khalsa, G. R. K.; Jackson, S. A.; Eisenstein, O. *J. Am. Chem. Soc.* **1993**, *115*, 569.

40. Eckert, J.; Kubas, G. J. *J. Chem. Phys.* **1993**, *97*, 2378.

41. Dewar, M. J. S. *Bull. Soc. Chim. Fr.* **1951**, *18*, C79.

42. Chatt, J.; Duncanson, L. A. *J. Chem. Soc.* **1953**, 2939.

43. Heinekey, D. M.; Law, J. K.; Schultz, S. M. *J. Am. Chem. Soc.* **2001**, *123*, 12728.

44. Ingleson, M. J.; Brayshaw, S. K.; Mahon, M. F.; Ruggiero, G. D.; Weller, A. S. *Inorg. Chem.* **2005**, *44*, 3162.

45. Moreno, B.; Sabo-Etienne, S.; Chaudret, B.; Rodriguez, A.; Jalon, F.; Trofimenko, S. *J. Am. Chem. Soc.* **1995**, *117*, 7441.

46. Grellier, M.; Vendier, L.; Chaudret, B.; Albinati, A.; Rizzato, S.; Mason, S.; Sabo-Etienne, S. *J. Am. Chem. Soc.* **2005**, *127*, 17592.

47. Bart, S. C.; Lobkovsky, E.; Chirik, P. J. *J. Am. Chem. Soc.* **2004**, *126*, 13794.

48. Yousufuddin, M.; Wen, T. B.; Mason, S. A.; McIntyre, G. J.; Jia, G.; Bau, R. *Angew. Chem., Int. Ed.* **2005**, *44*, 7227.

49. Brammer, L.; Howard, J. A.; Johnson, O.; Koetzle, T. F.; Spencer, J. L.; Stringer, A. M. *J. Chem. Soc., Chem. Commun.* **1991**, 241.

50. Johnson, T. J.; Albinati, A.; Koetzle, T. F.; Ricci, J.; Eisenstein, O.; Huffman, J. C.; Caulton, K. G. *Inorg. Chem.* **1994**, *33*, 4966.

51. Gelabert, R.; Moreno, M.; Lluch, J. M. *Chem. Eur. J.* **2005**, *11*, 6315.

52. Albinati, A., *et al. J. Am. Chem. Soc.* **1993**, *115*, 7300.

53. Gusev, D. G. *J. Am. Chem. Soc.* **2004**, *126*, 14249.

54. Hush, N. S. *J. Am. Chem. Soc.* **1997**, *119*, 1717.

55. Law, J. K.; Mellows, H.; Heinekey, D. M. *J. Am. Chem. Soc.* **2002**, *124*, 1024.

56. Gelabert, R.; Moreno, M.; Lluch, J. M.; Lledós, A.; Heinekey, D. M. *J. Am. Chem. Soc.* **2005**, *127*, 5632.

57. Sabo-Etienne, S.; Chaudret, B. *Chem. Rev.* **1998**, *98*, 2077.

58. Brothers, P. J. *Prog. Inorg. Chem.* **1981**, *28*, 1.

59. Morris, R. H. In *Recent Advances in Hydride Chemistry*; Peruzzini, M.; Poli, R., Eds.; Elsevier: Amsterdam, 2001, pp 1–38.

60. Schlaf, M.; Lough, A. J.; Morris, R. H. *Organometallics* **1996**, *15*, 4423.

61. Crabtree, R. H.; Lavin, M. *J. Chem. Soc., Chem. Commun.* **1985**, 794.

62. Chinn, M. S.; Heinekey, D. M. *J. Am. Chem. Soc.* **1987**, *109*, 5865.

63. Buncel, E.; Menon, B. *J. Am. Chem. Soc.* **1977**, *99*, 4457.

64. Huhmann-Vincent, J.; Scott, B. L.; Kubas, G. J. *J. Am. Chem. Soc.* **1998**, *120*, 6808.

65. Van Der Sluys, L. S.; Miller, M. M.; Kubas, G. J.; Caulton, K. G. *J. Am. Chem. Soc.* **1991**, *113*, 2513.

66. Bruns, W.; Kaim, W.; Waldhor, E.; Krejcik, M. *Inorg. Chem.* **1995**, *34*, 663.

67. Van Der Sluys, L. S.; Eckert, J.; Eisenstein, O.; Hall, J. H.; Huffman, J. C.; Jackson, S. A.; Koetzle, T. F.; Kubas, G. J.; Vergamini, P. J.; Caulton, K. G. *J. Am. Chem. Soc.* **1990**, *112*, 4831.

68. Brintzinger, H. H. *J. Organomet. Chem.* **1979**, *171*, 337.

69. Burdett, J. K.; Phillips, J. R.; Pourian, M. R.; Poliakoff, M.; Turner, J. J.; Upmacis, R. *Inorg. Chem.* **1987**, *26*, 3054.

70. Pons, V.; Conway, S. L. J.; Green, M. L. H.; Green, J. C.; Herbert, B. J.; Heinekey, D. M. *Inorg. Chem.* **2004**, *43*, 3475.

71. Janak, K. E.; Shin, J. H.; Parkin, G. *J. Am. Chem. Soc.* **2004**, *126*, 13054.

72. Luo, X.-L.; Crabtree, R. H. *J. Am. Chem. Soc.* **1990**, *112*, 6912.

73. Gusev, D. G.; Hubener, R.; Burger, P.; Orama, O.; Berke, H. *J. Am. Chem. Soc.* **1997**, *119*, 3716.

74. Wisniewski, L. L.; Mediati, M.; Jensen, C. M.; Zilm, K. W. *J. Am. Chem. Soc.* **1993**, *115*, 7533.

75. Li, S.; Hall, M. B.; Eckert, J.; Jensen, C. M.; Albinati, A. *J. Am. Chem. Soc.* **2000**, *122*, 2903.

76. Gusev, D. G.; Berke, H. *Chem. Ber.* **1996**, *129*, 1143.

77. Pons, V.; Conway, S. L. J.; Green, M. L. H.; Green, J. C.; Herbert, B. J.; Heinekey, D. M. *Inorg. Chem.* **2004**, *43*, 3475.

78. Bullock, R. M.; Song, J.-S.; Szalda, D. J. *Organometallics* **1996**, *15*, 2504.

79. Guan, H.; Iimura, M.; Magee, M. P.; Norton, J. R.; Zhu, G. *J. Am. Chem. Soc.* **2005**, *127*, 7805.

80. Noyori, R. *Angew. Chem., Int. Ed.* **2002**, *41*, 2008.

81. Ohkuma, T.; Noyori, R. *J. Am. Chem. Soc.* **2003**, *125*, 13490.

82. Heiden, Z. M.; Rauchfuss, T. B. *J. Am. Chem. Soc.* **2009**, *131*, 3593.

83. Friedrich, A.; Drees, M.; Schmedt, J.; Schneider, S. *J. Am. Chem. Soc.* **2009**, *131*, 17552.

84. Ishiwata, K.; Kuwata, S.; Ikariya, T. *J. Am. Chem. Soc.* **2009**, *131*, 5001.

85. Zimmer-De Iuliis, M.; Morris, R. H. *J. Am. Chem. Soc.* **2009**, *131*, 11263.

86. Armstrong, F. A. *Curr. Opin. Chem. Biol.* **2004**, *8*, 133.

87. Volbeda, A.; Fonticella-Camps, J. C. *Coord. Chem. Rev.* **2005**, 1609.

88. Liu, X.; Ibrahim, S. K.; Tard, C.; Pickett, C. J. *Coord. Chem. Rev.* **2005**, 1641.

89. (a) Darensbourg, M. Y.; Lyon, E. J.; Zhao, Z.; Georgakaki, I. P. *Proc. Natl. Acad. Sci. USA* **2003**, *100*, 3683;(b) Tard, C.; Pickett, C. J. *Chem. Rev.* **2009**, *109*, 2245;

90. Peters, J. W.; Lanzilotta, W. N.; Lemon, B. J.; Seefeldt, L. C. *Science* **1998**, *282*, 1853.

91. Capon, J.-F.; Gloagen, F.; Schollhammer, P.; Talarmin, J. *Coord. Chem. Rev.* **2005**, 1664.

92. Sun, L.; Akermark, B.; Ott, S. *Coord. Chem. Rev.* **2005**, 1653.

93. Alper, J. *Science* **2003**, *299*, 1686.

94. Henry, R. M.; Shoemaker, R. K.; Newell, R. H.; Jacobsen, G. M.; DuBois, D. L.; DuBois, M. R. *Organometallics* **2005**, *24*, 2481.

95. Eilers, G.; Schwartz, L.; Stein, M.; Zampella, G.; de Gioia, L.; Ott, S.; Lomoth, R. *Chem. Eur. J.* **2007**, *13*, 7075.

96. Mealli, C.; Rauchfuss, T. B. *Angew. Chem., Int. Ed.* **2007**, *46*, 8942.

97. Wilson, A. D.; Shoemaker, R. K.; Miedaner, A.; Muckerman, J. T.; DuBois, D. L.; Dubois, M. R. *Proc. Natl. Acad. Sci. USA* **2007**, *104*, 6951–6956.

98. Justice, A. K.; Linck, R. C.; Rauchfuss, T. B.; Wilson, S. R. *J. Am. Chem. Soc.* **2004**, *126*, 13214.

99. Shima, S.,*et al. Science* **2008**, *321*, 572.

100. Yang, X.; Hall, M. B. *J. Am. Chem. Soc.* **2009**, *131*, 10901.

101. DuBois, M. R. *Chem. Rev.* **1989**, *89*, 1.

102. Berkessel, A.; Schubert, T. J. S.; Muller, T. N. *J. Am. Chem. Soc.* **2002**, *124*, 8693.

103. Chan, B.; Radom, L. *J. Am. Chem. Soc.* **2005**, *127*, 2443.

104. (a) Welch, G. C.; San Juan, R. R.; Masuda, J. D.; Stephan, D. W. *Science* **2006**, *314*, 1124; (b) Rokob, T. A.; Hamza, A.; Papai, I. *J. Am. Chem. Soc.* **2009**, *131*, 10701; (c) Stephan, D. W. *Dalton. Trans.* **2009**, 3129;

105. Geier, S. J.; Gilbert, T. M.; Stephan, D. W. *J. Am. Chem. Soc.* **2008**, *130*, 12632.

106. Kenward, A. L.; Piers, W. E. *Angew. Chem., Int. Ed.* **2008**, *47*, 38.

107. Sumerin, V.; Schulz, F.; Nieger, M.; Leskela, M.; Repo, T.; Rieger, B. *Angew. Chem., Int. Ed.* **2008**, *47*, 6001.

108. Rosi, N. L.; Eckert, J.; Eddaoudi, M.; Vodak, D. T.; Kim, J.; O'Keeffe, M.; Yaghi, O. M. *Science* 2003, *300*, 1127.

109. Rowsell, J. L. C.; Eckert, J.; Yaghi, O. M. *J. Am. Chem. Soc.* 2005, *127*, 14904.

110. Forster, P. M.; Eckert, J.; Heiken, B. D.; Parise, J. B.; Yon, J. W.; Jhung, S. W.; Chang, J.-S.; Cheetham, A. K. *J. Am. Chem. Soc.* 2006, *128*, 16846.

111. Georgiev, P. A.; Albinati, A.; Mojet, B. L.; Ollivier, J.; Eckert, J. *J. Am. Chem. Soc.* **2007**, *129*, 8086.

112. Gagliardi, L.; Pyykko, P. *J. Am. Chem. Soc.* **2004**, *126*, 15014.

113. Weisshaar, J. C. *Acc. Chem. Res.* **1993**, *26*, 213.

114. Alacaraz, G.; Sabo-Etienne, S. *Acc. Chem. Res.* **2009**, *42*, 1640.

115. (a) Lewis, N. S.; Nocera, D. G. *PNAS* **2006**, *103*, 15729; (b) Nocera, D. G. *Inorg. Chem.* **2009**, *48*, 10001; (c) DuBois, M. R.; DuBois, D. L. *Acc. Chem. Res.* **2009**, *42*, 1974; (d) Dempsey, J. L.; Brunschwig, B. S.; Winkler, J. R.; Gray, H. B. *Acc. Chem. Res.* **2009**, *42*, 1995; (e) Lazarides, T.; McCormick, T.; Du, P.; Luo, G.; Lindley, B.; Eisenberg, R. *J. Am. Chem. Soc.* **2009**, *131*, 9192; (f) Reisner, E.; Powell, D. J.; Cavazza, C.; Fonticella-Camps, J. C.; Armstrong, F. A. *J. Am. Chem. Soc.* **2009**, *131*, 18457; (g) Kohl, S. W.; Weiner, L.; Schwartsburd, L.; Konstantinovski, L.; W. Shimon, L. J.; Ben-David, Y.; Iron, M. A.; Milstein, D. *Science* **2009**, *324*, 74; (h) Esswein, A. J.; Nocera, D. G. *Chem. Rev.* **2007**, *107*, 4022; (i) Le Goff, A.; Artero, V.; Jousselme, B.; Dinh Tran, P.; Guillet, N.; Metaye, R.; Fihri, A.; Palacin, S.; Fontecave, M. *Science* **2009**, *326*, 1384

116. Sutin, N.; Creutz, C.; Fujita, E. *Comments Inorg. Chem.* **1997**, *19*, 67.

117. Torrent, M.; Solà, M.; Frenking, G. *Chem. Rev.* **2000**, *100*, 439.

118. Bau, R.; Drabnis, M. H. *Inorg. Chim. Acta* **1997**, *259*, 27.

119. Kubas, G. J.; Nelson, J. E.; Bryan, C. J.; Eckert, J.; Wisniewski, L.; Zilm, K. *Inorg. Chem.* **1994**, *33*, 2954.

120. Maverick, E. F.; Trueblood, K. N. *THMA11: Program for Thermal Motion Analysis.* UCLA, 1988.

121. Koetzle, T. F. *Trans. Am. Crystallogr. Assoc.* 1997, **31**, 57.

122. Klooster, W. T.; Koetzle, T. F.; Jia, G.; Fong, T. P.; Morris, R. H.; Albinati, A. *J. Am. Chem. Soc.* 1994, **116**, 7677.

123. Zilm, K. W.; Millar, J. M. *Adv. Magn. Opt. Res.* 1990, **15**, 163.

124. Buntkowsky, G.; Limbach, H.-H.; Wehrmann, F.; Sack, I.; Vieth, H.-M.; Morris, R. H. *J. Phys. Chem. A* **1997**, *101*, 4679.

125. Wehrmann, F.; Fong, T. P.; Morris, R. H.; Limbach, H.-H.; Buntkowsky, G. *Phys. Chem. Chem. Phys.* **1999**, *1*, 4033.

126. Walaszek, B.; Adamczyk, A.; Pery, T.; Yeping, X.; Gutmann, T.; de Sousa Amadeu, N.; Ulrich, S.; Breitzke, H.; Vieth, H. M.; Sabo-Etienne, S.; Chaudret, B.; Limbach, H.-H.; Buntkowsky, G. *J. Am. Chem. Soc.* **2008**, *130*, 17502.

127. Hamilton, D. G.; Crabtree, R. H. *J. Am. Chem. Soc.* 1988, **110**, 4126.

128. Desrosiers, P. J.; Cai, L.; Lin, Z.; Richards, R.; Halpern, J. *J. Am. Chem. Soc.* 1991, **113**, 4173.

129. Ricci, J. S.; Koetzle, T. F.; Bautista, M. T.; Hofstede, T. M.; Morris, R. H.; Sawyer, J. F. *J. Am. Chem. Soc.* **1989**, *111*, 8823.

130. Gusev, D. G.; Kuhlman, R.; Renkema, K. B.; Eisenstein, O.; Caulton, K. G. *Inorg. Chem.* **1996**, *35*, 6775.

131. Upmacis, R. K.; Poliakoff, M.; Turner, J. J. *J. Am. Chem. Soc.* **1986**, *108*, 3645.

132. Luther, T. A.; Heinekey, D. M. *Inorg. Chem.* 1998, **37**, 127.

133. Hush, N. S. *J. Am. Chem. Soc.* 1997, **119**, 1717.

134. Gründemann, S.; Limbach, H.-H.; Buntkowsky, G.; Sabo-Etienne, S.; Chaudret, B. *J. Phys. Chem. A* 1999, *103*, 4752.

135. Gelabert, R.; Moreno, M.; Lluch, J. M.; Lledós, A.; Pons, V.; Heinekey, D. M. *J. Am. Chem. Soc.* **2004**, *126*, 8813.

136. Gelabert, R.; Moreno, M.; Lluch, J. M.; Lledós, A. *J. Am. Chem. Soc.* **1997**, *119*, 9840.

137. Bagley, K. A.; Van Garderen, C. J.; Chen, M.; Duin, E. C.; Albracht, S. P. J.; Woodruff, W. *Biochemistry*, **1994**, *33*, 9229.

138. Sweany, R. L.In *Transition Metal Hydrides*; Dedieu, A., Ed.; VCH Publishers: New York, **1992**; pp 65–101.

139. Bender, B. R.; Kubas, G. J.; Jones, L. H.; Swanson, B. I.; Eckert, J.; Capps, K. B.; Hoff, C. D. *J. Am. Chem. Soc.* **1997**, *119*, 9179.

140. Moreno, B.; Sabo-Etienne, S.; Chaudret, B.; Rodriguez, A.; Jalon, F.; Trofimenko, S. *J. Am. Chem. Soc.* **1995**, *117*, 7441.

141. Chopra, M.; Wong, K. F.; Jia, G.; Yu, N.-T. *J. Mol. Struct.* **1996**, *379*, 93.

142. Hay, P. J. *J. Am. Chem. Soc.* **1987**, *109*, 705.

143. Dapprich, S.; Frenking, G. *Angew. Chem., Int. Ed. Engl.* **1995**, *34*, 354.

144. Wilson, E. B., Jr.; Decius, J. C.; Cross, P. C. *Molecular Vibrations*; McGraw-Hill: New York, 1955; pp 197–200.

145. Jones, L. H.; McDowell, R. S.; Goldblatt, M. *Inorg. Chem.* **1969**, *8*, 2349.

146. Levine, I. N. *Molecular Spectroscopy*; Wiley: New York, 1975; p 160.

147. Badger, R. M. *J. Chem. Phys.* **1934**, *2*, 128.

148. Anson, C. E.; Sheppard, N.; Powell, D. B.; Bender, B. R.; Norton, J. R. *J. Chem. Soc., Faraday Trans.* **1994**, *90*, 1449, and references therein.

149. Li, J.; Ziegler, T. *Organometallics* **1996**, *15*, 3844.

150. King, W. A.; Luo, X.-L.; Scott, B. L.; Kubas, G. J.; Zilm, K. W. *J. Am. Chem. Soc.* **1996**, *118*, 6782.

151. (a) Gelabert, R.; Moreno, M.; Lluch, J. M.; Lledós, A. *Chem. Phys.* **1999**, *241*, 155; (b) Torres, L.; Gelabert, R.; Moreno, M.; Lluch, J. M. *J. Phys. Chem. A* **2000**, *104*, 7898;

152. Perutz, R. N.; Turner, J. J. *Inorg. Chem.* **1975**, *14*, 262.

153. Perutz, R. N.; Turner, J. J. *J. Am. Chem. Soc.* **1975**, *97*, 4791.

154. Turner, J. J.; Burdett, J. K.; Perutz, R. N.; Poliakoff, M. *Pure Appl. Chem.* **1977**, *49*, 271.

155. Andrews, L.; Moskovits, M. *The Chemistry and Physics of Matrix Isolated Species*; Elsevier, Amsterdam, 1989.

156. Sweany, R. L. *J. Am. Chem. Soc.* **1985**, *107*, 2374.

157. Upmacis, R. K.; Gadd, G. E.; Poliakoff, M.; Simpson, M. B.; Turner, J. J.; Whyman, R.; Simpson, A. F. *J. Chem. Soc., Chem. Commun.* **1985**, 27.

158. Church, S. P.; Grevels, F.-W.; Hermann, H.; Shaffner, K. *J. Chem. Soc., Chem. Commun.* **1985**, 30.

159. Sweany, R. L.In *Transition Metal Hydrides*; Dedieu, A., Ed.; VCH Publishers: New York, 1992; pp 65–101.

160. (a) Poliakoff, M.; George, M. W. *J. Phys. Org. Chem.* **1998**, *11*, 589; (b) George, M. W.; Haward, M. T.; Hamley, P. A.; Hughes, C.; Johnson, F. P. A.; Popov, V. K.; Poliakoff, M. *J. Am. Chem. Soc.* **1993**, *115*, 2286; (c) Gadd, G. E.; Upmacis, R. K.; Poliakoff, M.; Turner, J. J. *J. Am. Chem. Soc.* **1986**, *108*, 2547.

161. Goff, S. E. J.; Nolan, T. F.; George, M. W.; Poliakoff, M. *Organometallics* **1998**, *17*, 2730.

162. Cooper, A. I.; Poliakoff, M. *Chem. Commun.* **2007**, 2965.

163. Banister, J. A.; Lee, P. D.; Poliakoff, M. *Organometallics* **1995**, *14*, 3876.

164. Lee, P. D.; King, J. L.; Seebald, S.; Poliakoff, M. *Organometallics* **1998**, *17*, 524.

165. Martensson, A.-S.; Nyberg, C.; Andersson, S. *Phys. Rev. Lett.* **1986**, *57*, 2045.

166. Wang, X.; Andrews, L.; Infante, I.; Gagliardi, L. *J. Am. Chem. Soc.* **2008**, *130*, 1972.

167. Marx, D.; Parrinello, M. *Nature* **1995**, *375*, 216.

168. Eckert, J. *Spectrochim. Acta A* **1992**, *48A*, 363.

169. Sabo-Etienne, S.; Rodriguez, V.; Donnadieu, B.; Chaudret, B.; el Makarim, H. A.; Barthelat, J.-C.; Ulrich, S.; Limbach, H.-H.; Moïse, C. *New J. Chem.* **2001**, *25*, 55, and references therein.

170. Beaufils, J. P.; Crowley, T.; Rayment, R. K.; Thomas, R. K.; White, J. W. *Mol. Phys.* **1981**, *44*, 1257.

171. Nicol, J. M.; Eckert, J.; Howard, J. *J. Phys. Chem.* **1988**, *92*, 7117.

172. Yaghi, O. M. *J. Am. Chem. Soc.* **2006**, *128*, 3494.

173. (a) Walker, G., Ed. *Solid-State Hydrogen Storage: Materials and Chemistry*; Woodhead Publishing Limited: Cambridge, UK, 2008. (b) Dinca, M.; Long, J. R. *Angew. Chem., Int. Ed.* **2008**, *47*, 6766; (c) Hoang, T. K. A.; Antonelli, D. M. *Adv. Mater.* **2009**, *21*, 1787; (d) Zhao, D.; Yuan, D.; Zhou, H.-C. *Energy Environ. Sci.* **2008**, *1*, 222; (e) Thomas, K. M. *Dalton Trans.* **2009**, 1487; (f) Murray, L. J.; Dinca, M.; Long, J. R. *Chem. Soc. Rev.* **2009**, *38*, 1294.

174. Brown, C. M.; Yildirim, T.; Neumann, D. A.; Heben, M. J.; Gennett, T.; Dillon, A. C.; Alleman, J. L.; Fischer, J. E. *Chem. Phys. Lett.* **2000**, *329*, 311.

175. (a) MacKinnon, J. A.; Eckert, J.; Coker, D. F.; Bug, A. L. R. *J. Chem. Phys.* **2001**, *114*, 10137; (b) Nouar, F.; Eckert, J.; Eubank, J. F.; Forster, P.; Eddaoudi, M. *J. Am. Chem. Soc.* **2009**, *131*, 2864.

176. Forster, P. M.; Eckert, J.; Chang, J.-S.; Park, S.-E.; Férey, G.; Cheetham, A. K. *J. Am. Chem. Soc.* **2003**, *125*, 1309.

177. (a) Rosi, N. L.; Eckert, J.; Eddaoudi, M.; Vodak, D. T.; Kim, J.; O'Keeffe, M.; Yaghi, O. M. *Science* **2003**, *300*, 1127; (b) Rowsell, J. L. C.; Spencer, E. C.; Eckert, J.; Howard, J. A. K.; Yaghi, O. M. *Science* **2005**, *309*, 1350.

178. Rowsell, J. L. C.; Yaghi, O. M. *Angew. Chem., Int. Ed.* 2005, *44*, 4670.

179. Rowsell, J. L. C.; Eckert, J.; Yaghi, O. M. *J. Am. Chem. Soc.* **2005**, *217*, 14904.

180. Forster, P. M.; Eckert, J.; Heiken, B. D.; Parise, J. B.; Yoon, J. W.; Jhung, S. H.; Chang, J.-S.; Cheetham, A. K. *J. Am. Chem. Soc.* **2006**, *128*, 16846.

181. Serykh, A. I.; Kazansky, V. B. *Phys. Chem. Chem. Phys.* **2004**, *6*, 5250.

182. Georgiev, P. A.; Albinati, A.; Eckert, J. *Chem. Phys. Lett.* **2007**, *449*, 182.

183. Bullock, R. M. *Transition Metal Hydrides*; Dedieu, A., Ed. VCH Publishers: New York, 1992; p 263.

184. Hostetler, M. J.; Bergman, R. G. *J. Am. Chem. Soc.* **1992**, *114*, 7629.

185. (a) Rabinovich, D.; Parkin, G. *J. Am. Chem. Soc.* **1993**, *115*, 353; (b) Hascall, T.; Rabinovich, D.; Murphy, V. J.; Beachy, M. D.; Friesner, R. A.; Parkin, G. *J. Am. Chem. Soc.* **1999**, *121*, 11402;

186. Abu-Hasanayn, F.; Krogh-Jespersen, K.; Goldman, A. *J. Am. Chem. Soc.* **1993**, *115*, 8019.

187. Gusev, D. G.; Vymenits, A. B.; Bakhmutov, V. I. *Inorg. Chem.* **1992**, *31*, 1.

188. Hauger, B. E.; Gusev, D. G.; Caulton, K. G. *J. Am. Chem. Soc.* **1994**, *116*, 208.

189. Bender, B. R.; Kubas, G. J.; Jones, L. H.; Swanson, B. I.; Eckert, J.; Capps, K. B.; Hoff, C. D. *J. Am. Chem. Soc.* **1997**, *119*, 9179.

190. Gusev, D. G.; Bakhmutov, V. I.; Grushin, V. V.; Vol'pin, M. E. *Inorg. Chim. Acta* **1990**, *177*, 115.

191. Henderson, R. A.; Oglieve, K. E. *J. Chem. Soc., Dalton Trans.* **1993**, 3431.

192. Bakhmutov, V. I.; Bertran, J.; Esteruelas, M. A.; Lledós, A.; Maseras, F.; Modrego, J.; Oro, L. A.; Sola, E. *Chem. Eur. J.* **1996**, *2*, 815.

193. Bigeleisen, J.; Mayer, M. G. *J. Chem. Phys.* **1947**, *15*, 261.

194. Parkin, G. *Acc. Chem. Res.* **2009**, *42*, 315.

195. Torres, L.; Gelabert, R.; Moreno, M.; Lluch, J. M. *J. Phys. Chem. A* **2000**, *104*, 7898.

196. (a) Heinekey, D. M.; Oldham, W. J., Jr. *J. Am. Chem. Soc.* **1994**, *116*, 3137; (b) Oldham, W. J., Jr.; Hinkle, A. S.; Heinekey, D. M. *J. Am. Chem. Soc.* **1997**, *119*, 11028.

197. Gusev, D. G.; Nietlispach, D.; Eremenko, I. L.; Berke, H. *Inorg. Chem.* **1993**, *32*, 3628.

198. Zhang, K.; Gonzalez, A. A.; Hoff, C. D. *J. Am. Chem. Soc.* **1989**, *111*, 3627, and references therein.

199. Church, S. P.; Grevels, F.-W.; Hermann, H.; Shaffner, K. *J. Chem. Soc., Chem. Commun.* **1985**, 30.

200. Wink, D. A.; Ford, P. C. *J. Am. Chem. Soc.* **1987**, *109*, 436.

201. Zhou, P.; Vitale, A. A.; San Filippo, J.; Saunders, W. H. *J. Am. Chem. Soc.* **1985**, *107*, 8049.

202. Basallote, M. G.; Duran, J.; Fernandez-Trujillo, J.; Manez, M. A. *J. Chem. Soc., Dalton Trans.* **1998**, 2205.

203. Basallote, M. G.; Duran, J.; Fernandez-Trujillo, J.; Manez, M. A.; Rodriguez de la Torre, J. *J. Chem. Soc., Dalton Trans.* **1998**, 745.

204. Basallote, M. G.; Duran, J.; Fernandez-Trujillo, J.; Manez, M. A. *Organometallics* **2000**, *19*, 5067.

205. Abdur-Rashid, K.; Gusev, D. G.; Lough, A. J.; Morris, R. H. *Organometallics* **2000**, *19*, 1652.

206. Eckert, J.; Albinati, A.; Bucher, U. E.; Venanzi, L. M. *Inorg. Chem.* **1996**, *35*, 1292.

207. Ozin, G. A.; Garcia-Prieto, J. *J. Am. Chem. Soc.* **1986**, *108*, 3099.

208. Hauge, R. H.; Margrave, J. L.; Kafafi, Z. H. *NATO ASI Ser. B* **1987**, *158* (Phys. Chem. Small Clusters), 787.

# 6 Activation of Carbon Dioxide

FERENC JOÓ

## 6.1  INTRODUCTION

Carbon dioxide is a small, simple, and symmetric molecule that can change our lives. It abounds everywhere on Earth mostly as a gas in the atmosphere (approximately $2.5 \times 10^{12}$ tons), in dissolved state in the hydrosphere ($\sim 10^{14}$ tons), and as fixed in carbonate rocks ($\sim 10^{16}$ tons). In the past two centuries, human activity has caused a significant increase in the atmospheric $CO_2$ concentration to above 380 ppm, which has led to an increased greenhouse effect and contributed to the global warming of our planet. It is now generally accepted that strategies for combating global warming must include measures on preventing further increase in the concentration of atmospheric $CO_2$. Chemistry should play an important part in this battle by providing more efficient large-scale procedures with an overall zero $CO_2$ emission. On the other side, chemistry is the central science in making the use of carbon dioxide as an abundant and cheap carbon source possible. Presently, the three major sources of carbon are petroleum, coal, and biomass. Oil reserves are slowly running out while processing coal (another nonrenewable carbon source) into liquid or gaseous fuels and chemicals still has its drawbacks, despite the enormous developments in this field. Biomass is renewable, but its production as a basis for fuel and chemical industry would require huge areas of land; furthermore, the reproduction of biomass is insufficiently slow for this purpose. Eventually, the large-scale capture and reuse of the carbon content of $CO_2$-rich flue gases must be solved. In fact, this is already practiced in isolated cases where the gaseous by-products of a process are comprised almost exclusively of carbon dioxide (synthesis of urea, fermentation technologies such as beer production, etc.). Nevertheless, use of pure $CO_2$ from natural wells is by far the most economical and the carbon dioxide by-product of chemical technologies is most often simply vented to the atmosphere.

Carbon dioxide and carbonates are the most stable forms of carbon and they undergo only a few thermal reactions that can be used for industrial purposes. Large amounts of carbon dioxide are utilized in a reaction with ammonia for the synthesis of

*Physical Inorganic Chemistry: Reactions, Processes, and Applications*  Edited by Andreja Bakac
Copyright © 2010 by John Wiley & Sons, Inc.

urea ($8 \times 10^7$ tons per year) and for producing salicylic acid ($2 \times 10^4$ tons per year) in reactions with phenolates (Kolbe–Schmitt synthesis). Most other reactions of $CO_2$ require activation by metals or metal complexes. Such catalytic reactions include, among others, the synthesis of organic carbonates (important reagents and solvents) and polycarbonates (impact resistant plastics) and the hydrogenation—both homogeneous and heterogeneous—of $CO_2$ to methanol and formic acid (formate salts, amides, and esters).

In this chapter, we discuss the need and the possibilities of catalytic activation of carbon dioxide as well as its most important reactions of synthetic utility. Although important in research and in chemical industry, the reactions that produce expensive waste in an equivalent amount to the desired product (such as Grignard carboxylations and electrochemical reactions with reactive anodes (Al, Mg)[1]) are not discussed. Photosynthesis and enzymatic $CO_2$ activation are not included either, because in their complexity natural processes are so far from the reactions investigated and practiced in synthetic systems that there are only a few points of immediate cross-fertilization of ideas (although such interactions between the fields can be highly rewarding).

Taking the importance of practical carbon dioxide utilization it is no wonder that a large number of books and reviews have been published on this topic. For details, the reader is referred to these—sometimes overlapping—accounts on various aspects of $CO_2$ activation.[2–24]

## 6.2 THE $CO_2$ MOLECULE AND ITS COORDINATION PROPERTIES

Carbon dioxide is a linear, nonpolar molecule. The molecular orbitals relevant to chemical bonding are shown on the right-hand side of the Walsh diagram (Figure 6.1).[25] In the ground state, the highest occupied molecular orbital (HOMO) is $1\pi_g$, which is centered on the oxygen atoms. Consequently, the oxygen atoms of $CO_2$ have nucleophilic character. In contrast, the lowest unoccupied molecular orbital (LUMO; $2\pi_u$) is located mainly on the carbon atom, which, accordingly, is the electrophilic center of the molecule. In most cases (although not exclusively), coordination of $CO_2$ to metal ions involves interactions with both the carbon and oxygen atoms. As can be deduced from the Walsh diagram, backdonation from the filled metal d-orbitals to the LUMO of $CO_2$ should lead to a change of symmetry from $D_{\infty h}$ to $C_{2v}$, which results in lower total energy of the ligand. Indeed, this is what happens, and the preferred configuration of coordinated carbon dioxide in metal complexes has the bent geometry with an average O–C–O angle of 130°.[26] Similar changes of geometry occur on electronic excitation and on placing an extra electron on the LUMO (formation of the radical anion $CO_2^{\bullet-}$, for example, by electrochemical methods).

In the infrared spectrum of free carbon dioxide, the asymmetric stretching frequency is observed at $2349 \, \mathrm{cm}^{-1}$ (gas) and $2342 \, \mathrm{cm}^{-1}$ (solid). The infrared absorption belonging to the bending motion of the molecule is found at $667 \, \mathrm{cm}^{-1}$ (gas). Symmetric stretching of free $CO_2$ can be detected only by Raman spectroscopy

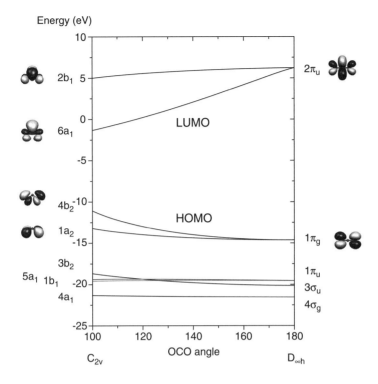

**FIGURE 6.1**    Walsh diagram of carbon dioxide.[25] (See the color version of this figure in Color Plates section.)

and is found in the region of 1285–1388 cm$^{-1}$. On the other hand, in transition metal complexes with a bent $CO_2$ ligand, both $v_{asym}$ and $v_{sym}$ are detected in the infrared region at 1350–1750 and 1140–1300 cm$^{-1}$, respectively.

Metal complexes of carbon dioxide are known in a relatively large number and possess versatile coordination modes (Figure 6.2).[11] In mononuclear complexes, the preferred coordination mode is $\eta^2$-CO and, indeed, the first crystallographically characterized transition metal–$CO_2$ complex, $[Ni(CO_2)(PCy_3)_2]$ (Cy = cyclohexyl), discovered by Aresta et al. belongs to this group (IR: $v = 1740, 1140, 1094$ cm$^{-1}$).[27] The $[MCl(CO_2)(diars)_2]$ (diars = $o$-phenylenebis(dimethylarsine); M = Ir, Rh) complexes of Herskovitz feature the $\eta^1$-C coordination mode (IR: $v = 1550, 1230$ cm$^{-1}$ (Ir) and 1610, 1210 cm$^{-1}$ (Rh)).[28,29] Interestingly, although these compounds were already described in 1977 (Ir) and in 1983 (Rh), since then the number of $CO_2$ complexes with similar $\eta^1$-C coordination has not increased. In contrast, only lately was a uranium complex with a linear, O-bonded carbon dioxide characterized by X-ray diffraction.[30] This interesting complex contains a highly substituted 1,4,7-triazacyclononane ligand that wraps around the U(III) metal center and allows its interaction with $CO_2$ only through a narrow, cylindrical cavity. This way $CO_2$ can bind to uranium only at its oxygen atom and steric constraints also contribute to its

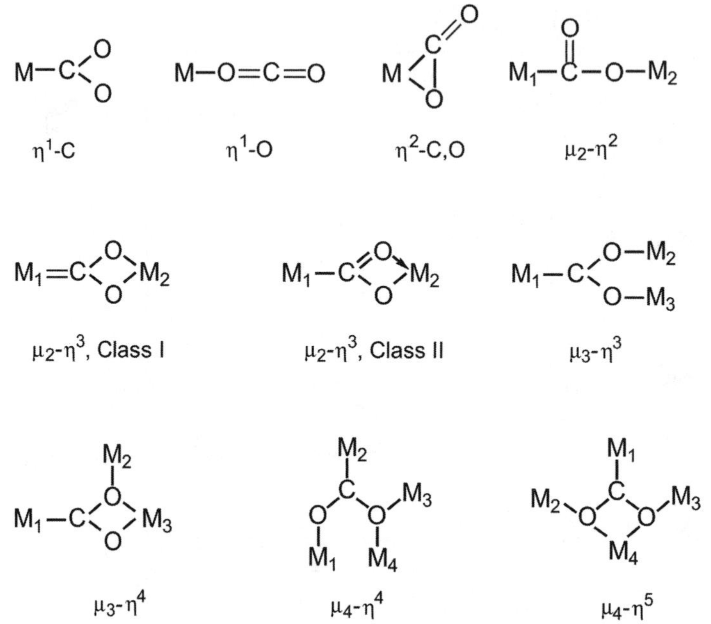

**FIGURE 6.2**    Coordination modes of $CO_2$ in metal complexes.

remaining linear (the OCO angle was found to be 178.0°). Accordingly, there is only a single absorption in the infrared at $2188\ cm^{-1}$ showing a considerable activation of $CO_2$. It was also established by magnetic moment ($\mu_{eff}$) measurements that the U(III) central ion of the complex was oxidized to U(IV) in its reaction with $CO_2$ with a transfer of virtually one electron to the latter, so the resulting compound is better described as a U(IV)–$CO_2^{\cdot-}$ complex.

Other possibilities of coordination involve the simultaneous binding of $CO_2$ to two or more metal centers possessing coordination modes from $\mu_2$-$\eta^2$ (such as in [CpFe (CO)(PPh$_3$)($\mu_2$-C,O-$CO_2$)Re(CO)$_4$(PPh$_3$)]; Cp = cyclopentadienyl, PPh$_3$ = triphenylphosphine))[31] to $\mu_4$-$\eta^5$.[10]

Several complexes containing coordinated carbon dioxide have been characterized by single-crystal X-ray diffraction. The characteristic C–O bond lengths are usually longer than those in the gas-phase free ligand (116.2 pm). In the case of linear $CO_2$ coordination, the C–O (coordinated) distance was found to be 112.2 pm, while the C–O (noncoordinated) distance was 127.7 pm.[30] In complexes with $\mu_2$-$\eta^2$-coordinated $CO_2$, the bond length between the coordinated O and C atoms is generally in the range of 125–129 pm (but can be as high as 137.9 pm), while the one between the C and the noncoordinated O atoms is around 119–126 pm.[26] With increasing hapticity, the C–O bond lengths increase further. Such a bond length increase (signaling the decrease of C–O bond order) can be regarded as a sign of activation. Nevertheless, $CO_2$ complexes do not necessarily show increased reactivity since binding of carbon dioxide to metal ion(s) may result in stable complexes. A nice example is provided by M[Co(salen)($CO_2$)] (M = Li$^+$, Na$^+$, K$^+$; salen = $N,N'$-ethylene bis(salicylaldimine)), where the stability of the complexes is markedly influenced by the counterion: in contrast to the lithium salt, the sodium and potassium salts released $CO_2$ under vacuum or upon sequestering the sodium ion by dicyclohexano-18-crown-6.[32]

Catalytic reactions of carbon dioxide need not necessarily involve coordination complexes of $CO_2$, this seems to be required only for reactions with C–O bond breaking. In other processes (e.g., C–C coupling), the role of the catalyst metal complex may be the activation of the reaction partner (e.g., olefins, alkynes), which then can react with $CO_2$ directly.[33]

## 6.3 PRACTICAL UTILIZATION OF CARBON DIOXIDE

Carbon dioxide is a very stable molecule ($\Delta_f G^\circ_{298} = -394.36\,\text{kJ/mol}$) and its reduction to various products such as formic acid, carbon monoxide, formaldehyde, methanol, or methane (just to name the most important ones) requires considerable amount of energy either in the form of high-energy reducing agents (e.g., $H_2$) or in the form of electrons supplied by electrochemical or photochemical systems. The least energy-demanding routes of $CO_2$ utilization are those where the whole molecule is built into the final products having open-chain (e.g., carboxylates) or cyclic (e.g., lactones) structures. Obviously, there is no sharp distinction between the two categories; for example, production of formic acid by hydrogenation of $CO_2$ belongs to both.

In this section, we treat those synthetic processes that have been practiced on large scale or are close to industrial utilization. In addition, important advances have been made in $CO_2$ activation/utilization that are not yet developed for practical use. Such reactions are treated under "emerging uses" of $CO_2$, although, again, the distinction is not undisputable.

### 6.3.1 Synthesis of Salicylic Acid

Salicylic acid (2-hydroxybenzoic acid) is commercially produced in the reaction of alkali phenolates and carbon dioxide (Scheme 6.1). As originally described by Kolbe in 1859, sodium fenolate saturated by $CO_2$ at 180–200°C yields salicylic acid

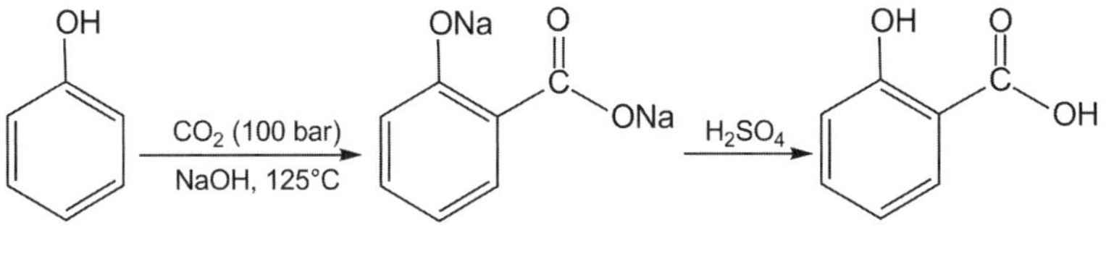

**SCHEME 6.1**  Kolbe–Schmitt synthesis of salicylic acid.

disodium salt and phenol. Later, Schmitt ran the reaction under 5–7 bar $CO_2$ pressure at 120–140°C and obtained complete conversion to sodium salicylate. Potassium phenolates react less readily than the sodium salts. At high temperatures (>200°C), the major product is 4-hydroxybenzoic acid, and accordingly, 2-hydroxybenzoic acid is rearranged to 4-hydroxybenzoic acid on heating. The Kolbe–Schmitt reaction (nowadays practiced commercially at 100 bar $CO_2$ and 125°C) has been the standard procedure for the synthesis of salicylic acid (and other hydroxy acids) for over almost 150 years.[34]

Although the reaction requires no external catalyst, carbon dioxide is activated by the interaction of its electrophilic carbon atom and the negatively polarized carbon in the *ortho* (or *para*) position of the phenolate ring. This mechanism is supported by the fact that even under high $CO_2$ pressure no salicylic acid is formed from phenol. The product is stabilized via an $\alpha \rightarrow \gamma$ proton migration. The free acid is obtained from the sodium salt in reaction with an external proton (usually from sulfuric acid). Formally, this reaction can be regarded as the insertion of $CO_2$ into an aromatic C–H bond; however, the above mechanism disproves this idea.

Salicylic acid is a precursor to aspirin (acetylsalicylic acid), a widely used nonprescription drug for pain and fever relief. Salicylic acid is also a key component of a wide range of cosmetic (mostly skin care) products; in addition, it is also used as food preservative and antiseptic (e.g., in toothpaste).

### 6.3.2  Synthesis of Urea

Urea is one of the largest volume products manufactured directly from carbon dioxide.[35] Reaction of $CO_2$ with $NH_3$ gives ammonium carbamate, which is then thermally dehydrated to urea (Scheme 6.2). The process was originally developed in 1922, and is called the Bosch–Meiser urea process after its inventors. Formation of ammonium carbamate is fast and highly exothermic and under industrial conditions it goes essentially to completion. In contrast, the dehydration process is endothermic, slow, and usually does not reach thermodynamic equilibrium under manufacturing conditions. This does not hinder the practical use of the overall process, since unreacted $NH_3$ and $CO_2$ can be recycled. The conversion of carbon dioxide to urea (a measure of the efficiency of the process) increases with higher temperatures and higher $NH_3/CO_2$ ratios and decreases with increasing $H_2O/CO_2$ ratio. Since the industrial processes work at high pressure and temperature, modeling the technologies requires a great deal of knowledge about the thermodynamics of the process, including the temperature dependence of the equilibrium constants of carbamate

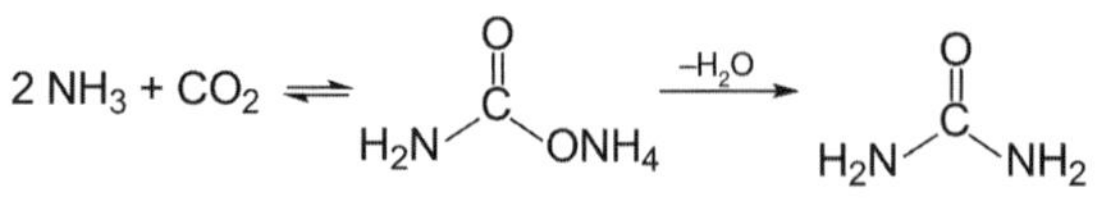

**SCHEME 6.2**   Bosch–Meiser synthesis of urea.

formation and dehydration, the temperature dependence of the gas fugacities, solubilities, and so on.[36]

Presently, two commercial technologies are in use for obtaining approximately 75% of total world urea production. The main differences are in the applied $NH_3/CO_2$ ratios and in the method of separating urea from unreacted starting materials and the recycle of ammonia and carbon dioxide.

In the Stamicarbon process, the $NH_3/CO_2$ molar ratio in the reactor (operated at 140 bar and 180–185°C) is 2.95. In this step, approximately 60% of $CO_2$ and 40% of $NH_3$ are converted, leading to an $NH_3/CO_2$ ratio of about 4.5 in the unreacted gas. The effluent solution from the reactor is stripped at high pressure and temperature with $CO_2$ (reestablishing the proper $NH_3/CO_2$ ratio) and the gases are recycled to the reactor. The urea solution is further purified from residual $NH_3$ and $CO_2$ and finally taken through an evaporation step yielding 99.7% urea melt.

The Snamprogetti technology utilizes a higher $NH_3/CO_2$ molar ratio (3.2–3.4) and a lower $H_2O/CO_2$ ratio (0.4–0.6) at a pressure of 150 bar and at a temperature of 188°C. As a result, 62–64% of the total $CO_2$ entering the reactor is converted to urea. The reactor effluent is taken to a stripper, where it is treated with $NH_3$. The large part of the unconverted carbamate is decomposed here, and the top gases of the stripper ($NH_3$ and $CO_2$) are recycled to the reactor, where the proper $NH_3/CO_2$ molar ratio is reestablished by addition of fresh $CO_2$. The urea solution is purified from residual carbamate, $CO_2$, and $NH_3$ in a medium-pressure (17 bar) and a low-pressure (3.5 bar) step, and finally, the water content is evaporated to yield a 99.8% urea melt, which is then worked up into prills or granules.

It is worth mentioning that both processes employ total recycle of $CO_2$ and $NH_3$ and very efficient methods of heat recovery. As a consequence, discharge of chemicals into the environment is kept at a very low level, characterized by 1 ppm urea and 1 ppm $NH_3$ in discharged process water.

The total world production of urea is about 100 million tons per year. By far the largest part of it is used as a nitrogen fertilizer both in solid form and in solution; this consumes approximately 87% of all urea production. It is also a livestock feed additive (5%) and a raw material for urea-formaldehyde resins (6%) and melamine (1%). Other applications (1%) include its use as deicing agent, raw material for fine chemicals (cyanuric acid, sulfamic acid), formation of crystalline clathrates, and so on.

### 6.3.3 Synthesis of Formic Acid, Formamides, and Formate Esters

Reduction of carbon dioxide to formic acid (and its derivatives) is a very active field of research that has been amply reviewed.[18–20,37,38] While some of the results are close to commercialization, others can be regarded as promising developments. It should also be realized that presently $H_2$ is generated on a fossil fuel basis (natural gas, coal) accompanied by the formation of $CO_2$. Therefore, the hydrogenation of carbon dioxide will not ease the $CO_2$ burden; however, it can be very useful for synthetic purposes.

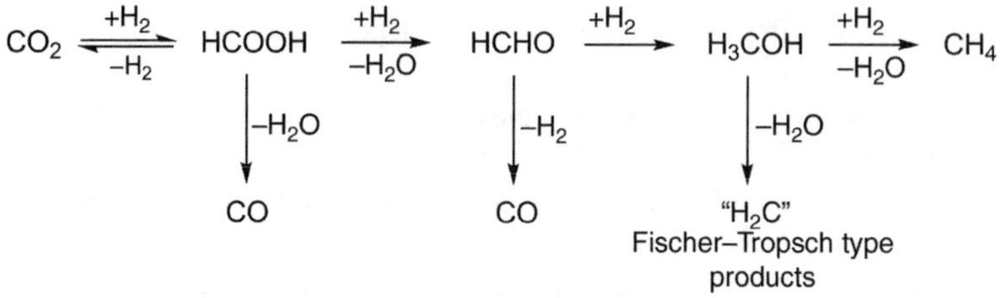

$$CO_2 \underset{-H_2}{\overset{+H_2}{\rightleftharpoons}} HCOOH \xrightarrow[-H_2O]{+H_2} HCHO \xrightarrow{+H_2} H_3COH \xrightarrow[-H_2O]{+H_2} CH_4$$

$$\downarrow{-H_2O} \qquad\qquad \downarrow{-H_2} \qquad\qquad \downarrow{-H_2O}$$

$$CO \qquad\qquad\qquad CO \qquad\qquad\qquad \text{``}H_2C\text{''}$$

Fischer–Tropsch type
products

**SCHEME 6.3**    Products of carbon dioxide hydrogenation.

Stepwise reduction of $CO_2$ with $H_2$ may yield formic acid, formaldehyde, methanol, and finally methane, together with CO or Fischer–Tropsch-type derivatives as shown in Scheme 6.3. The most common product of such a reduction is formic acid. Formation of carbon monoxide, formaldehyde, methanol, and methane has already been reported with molecular catalysts.

Reaction of two gaseous compounds resulting in a liquid product is hindered by a large decrease in entropy, which—depending on the temperature—may make the whole process thermodynamically unfavorable. This is also the case for the hydrogenation of $CO_2$ to HCOOH (Equation 6.1) with $\Delta H_{298}^\circ = -32\,kJ/mol$ but $\Delta G_{298}^\circ = +33\,kJ/mol$. However, in aqueous solution, hydration of the solutes makes the overall entropy difference smaller, and the reaction becomes slightly exergonic with $\Delta G_{298}^\circ = -4\,kJ/mol$ (Equation 6.2).

$$H_2(g) + CO_2(g) \rightleftharpoons HCOOH(l) \tag{6.1}$$

$$H_2(aq) + CO_2(aq) \rightleftharpoons HCOOH(aq) \tag{6.2}$$

According to thermodynamics, therefore, in water this reaction is likely to proceed.[36] Nevertheless, these data refer to standard conditions, and in order to eliminate the kinetic activation barrier at 25°C, highly active catalysts are needed. In addition, the same catalysts can also be active in the reverse process, that is, in the decomposition of formic acid to $H_2$ and $CO_2$; decomposition to CO and $H_2O$ (i.e., the reverse water gas shift) is rarely observed.

Hydrogenation of $CO_2$ can also be shifted to the right by further reactions of HCOOH, which may be simply its neutralization with a base, or reactions with amines or alcohols, yielding formamides or formate esters, respectively. Such an effect of amines or aminoalcohols is especially important since often these compounds are key ingredients of scrubbing mixtures used to recover carbon dioxide from flue gases. Furthermore, in laboratory research, hydrogenation of $CO_2$ is studied in the presence of bases, and while the product is usually mentioned as formic acid, in fact it is a formate salt or an azeotropic mixture, such as that with triethylamine.

### 6.3.3.1  Hydrogenation of Carbon Dioxide Under Subcritical Conditions    The beneficial effect of water was already observed in the early experiments of Inoue et al.[39,40] who discovered that in the presence of a base (NaOH, $NaHCO_3$, $NMe_3$,

$NEt_3$, etc.) transition metal phosphine complexes of 1,2-bis(diphenylphosphino) ethane (dppe) and $PPh_3$, such as $[Pd(dppe)_2]$, $[Ni(dppe)_2]$, $[Pd(PPh_3)_4]$, $[RhCl-(PPh_3)_3]$, $[RuH_2(PPh_3)_4]$, and $[IrH_3(PPh_3)_3]$, catalyzed the formation of HCOOH from $H_2$ and $CO_2$ (25 bar each, room temperature, benzene) with 12–87 turnovers in 20 h. The reaction was substantially accelerated by very small amount of water.

Tsai and Nicholas used $[Rh(NBD)L_3]BF_4$, $L = P(CH_3)_2(C_6H_5)$, as a catalyst precursor for $CO_2$ hydrogenation in THF and also observed acceleration of the reaction in the presence of water.[41] With careful spectroscopic measurements, they could detect the formation of the dihydrides, $[RhH_2(H_2O)L_3]^+$ and $[RhH_2(THF)-L_3]^+$, and also that of the bidentate formato complex, $[RhH(\eta^2-O_2CH)L_3]^+$. It was therefore suggested that the mechanism of the reaction involved the insertion of $CO_2$ into the Rh–H bond of the dihydride yielding the hydridorhodium-formato intermediate, followed by reductive elimination of formic acid, and then oxidative addition of $H_2$ to regenerate the dihydride. In another study by Lau and coworkers,[42] it was suggested that the rate-accelerating effect of water was due to the formation of an intermolecular hydrogen bond between the $H_2O$ ligand in $[RhH_2(H_2O)L_3]^+$ and the incoming $CO_2$ within the insertion transition state, as depicted in Scheme 6.4.

Jessop and coworkers investigated the kinetics of the $CO_2$ hydrogenation in liquid $NEt_3$ and at subcritical $CO_2$ pressures, using the $[RuCl(OAc)(PMe_3)_4]$ precursor.[43] The results showed that the hydrogenation is first order in both $H_2$ and $CO_2$, consistent with a $CO_2$ insertion into the Ru–H bond of a $[RuHX(PMe_3)_3]$ intermediate (X = H or Cl). The same catalyst was also used to study the effect of amines and alcohols on the kinetics of $CO_2$ hydrogenation.

Water-soluble rhodium complexes, such as $[RhCl(mtppts)_3]$ ($mtppts$ = trisulfonated triphenylphosphine) or the ones prepared $in$ $situ$ from $[\{RhCl(COD)\}_2]$ or from $[\{RhH(COD)\}_4]$ and $mtppts$ (P:Rh = 2.6:1), were successfully used by Leitner and coworkers[44] for the hydrogenation of $CO_2$ in aqueous solutions in the presenc of amines or aminoalkanols. There was no formation of HCOOH in the absence of an amine; however, a formic acid concentration of 3.63 M was obtained in an aqueous solution containing 3.97 M $HNEt_2$ (well soluble in water as compared to $NEt_3$). Initial turnover frequencies were substantially higher than any other before, for example, at 81°C and 40 bar total pressure ($CO_2$:$H_2$ = 1:1), a turnover frequency TOF = 7260 $h^{-1}$ was observed (TOF = mol $H_2$/mol catalyst/h). Leitner and coworkers also studied diphosphine complexes of Rh(I), such as $[RhH(dppb)]$ (dppb = 1,4-diphenylphosphinobutane), obtained in the reaction of $[RhH(cod)]_4$ and dppb. The $[Rh(hfacac)(dcpb)]$ complex (hfacac = 1,1,1,5,5,5-hexafluoroacetylacetonate,

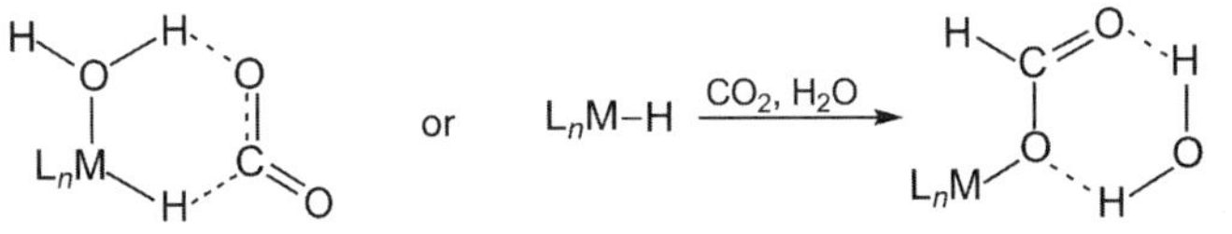

**SCHEME 6.4** Hydrogen bond stabilization of intermediates in carbon dioxide hydrogenation.

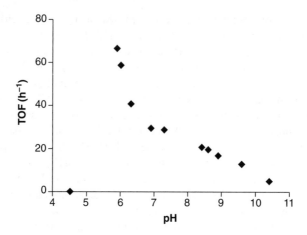

**FIGURE 6.3**   Hydrogenation of bicarbonate catalyzed by [RuCl$_2$(pta)$_4$] as a function of pH. [Ru] = $2.61 \times 10^{-13}$ M, [CO$_2$] + [NaHCO$_3$] = 1 M, 50°C, $p$(H$_2$) = 60 bar. Reprinted with permission from *Inorg. Chem.* **2000**, *39*, 5083. Copyright (2000) American Chemical Society.

dcpb = 1,4-dicyclohexylphosphinobutane) proved particularly active with a TOF = 1335 h$^{-1}$ at only 25°C.

Water-soluble transition metal phosphine complexes catalyze the hydrogenation of bicarbonate with much higher rate than that of CO$_2$.[45–47] For example, with the [RhCl-($m$tppms)$_3$] catalyst ($m$tppms = monosulfonated triphenylphosphine), a TOF = 0.11 h$^{-1}$ was determined under 20 bar CO$_2$ and 60 bar H$_2$ at 24°C. Conversely, the same catalyst hydrogenated NaHCO$_3$ with a TOF = 262 h$^{-1}$. Interestingly, with a [RuCl$_2$(pta)$_4$] catalyst (pta = 1,3,5-triaza-7-phosphaadamantane), the reaction rate of HCO$_3^-$ reduction was substantially increased further by applying CO$_2$ pressure (Figure 6.3).[45] It was concluded that increasing CO$_2$ pressure lowered the pH of the solution and facilitated the formation of [RuHX(pta)$_4$], the presumed catalyst that exists in acidic solutions.

[{RuCl$_2$($m$tppms)$_2$}$_2$] was found to be an excellent catalyst (precursor) for the hydrogenation of bicarbonate to formate especially when an excess of $m$tppms was added. When the reactions were run in the presence of CO$_2$, a very high rate increase was observed: at $p$(H$_2$) = 60 bar, $p$(CO$_2$) = 35 bar at 80°C, a 0.3 M NaHCO$_3$ solution was hydrogenated with a turnover frequency of 9600 h$^{-1}$. Again, it was first suggested that lowering the pH of the aqueous solution by increasing the CO$_2$ pressure facilitated the formation of the true catalytic species. However, DFT calculations revealed that although the prevalent species in solution is [RuH(HCO$_3$)($m$tppms)$_4$] (experimentally observed, isolated, and characterized), the hydrogenation of bicarbonate to formate proceeds through a Ru–CO$_2$ complex. This is obtained by protonation and subsequent dehydration of the bicarbonate ligand in [RuH(HCO$_3$)($m$tppms)$_3$], and this is the step that may become faster with increasing proton concentration under increased CO$_2$ pressure (Scheme 6.5).[48]

Himeda[49] has reported the very active catalyst precursors [Cp*IrCl(DHphen)]Cl and [RuCl(C$_6$Me$_6$)(DHphen)]Cl containing 4,7-dihydroxy-1,10-phenanthroline

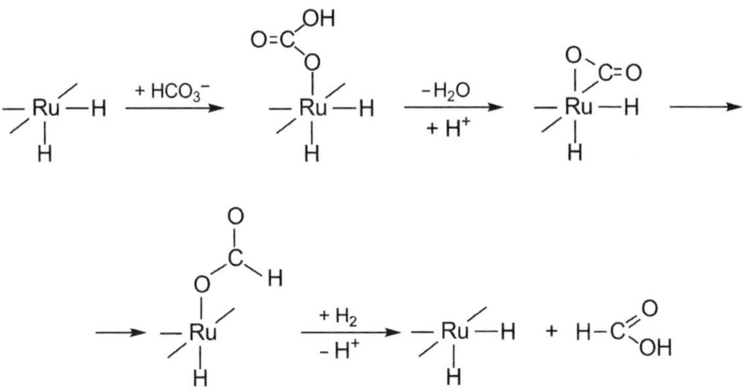

**SCHEME 6.5** Hydrogenation of $HCO_3^-$ via a Ru–$CO_2$ intermediate.

(DHphen) as a ligand ($Cp^* = C_5Me_5$). In basic solution, the DHphen ligand is deprotonated, making it more electron donating. The authors suggested that this factor was responsible for the particularly high activities observed for $CO_2$ hydrogenation in 1 M KOH, which were characterized by a turnover number (TON) 21,000 and an initial TOF 36,400 $h^{-1}$ for the Ir complex at 120°C.

Several patents describe the production of formic acid or formates by hydrogenation of $CO_2$, bicarbonates, or carbonates.[50,51] In water–2-propanol mixtures, the yield of formic acid was a function of the molar composition of the solvent. In the presence of $NEt_3$, at 80°C, and with 27 bar $CO_2$ and 54 bar $H_2$, [{$RuCl_2(CO)_2$}$_n$] catalyzed the hydrogenation of $CO_2$ with a yield of 13% in water, 54.5% in 2-propanol/water 20/80, 60.7% in 2-propanol/water 60/40, and 43.4% in neat 2-propanol.

$K[Ru^{III}Cl(EDTA-H)]$ was found to be a good catalyst (TOF = 375 $h^{-1}$) for the hydrogenation of $CO_2$[52] under mild conditions in aqueous solution (40 °C, with 3 bar $CO_2$ and 17 bar $H_2$). This study is unique in that the primary products of the reaction were formic acid and *formaldehyde*, which later decomposed to give CO, $H_2O$, and $H_2$.

Calcium carbonate was also successfully hydrogenated to calcium formate under $CO_2$/$H_2$ pressure (calcium formate is used in animal nutrition and leather tanning). The catalytic activities of [$RhCl(mtppms)_3$], [{$RuCl_2(mtppms)_2$}$_2$], and [$RuCl_2(pta)_4$] were studied typically at 20 bar $CO_2$ and 20–80 bar $H_2$ in the temperature range of 20–70°C.[53] Under such conditions, the highest TOF (26.6 $h^{-1}$) was observed with [$RhCl(mtppms)_3$]. Importantly, free formic acid was also formed in the reaction. For example, hydrogenation of 1 mmol $CaCO_3$ gave 3.21 mmol $HCO_2^-$ corresponding to 160% of the stoichiometric yield. Of the 3.21 mmol formate, only 1.0 mmol had its carbon source in $CaCO_3$, which means an effective use of $CO_2$ from the gas phase.

The above examples demonstrate that there are several efficient ways to hydrogenate $CO_2$ in the presence of amines or other bases either in aqueous or in nonaqueous solutions. However, the separation of free formic acid from the reaction mixtures is a challenge[4]—ingenious approaches are also found in the patent literature.[50,51] In any case, the catalyst must be removed from the reaction mixtures since it decomposes formic acid at only slightly elevated temperatures. In an attempt to solve the problem of the production of free formic acid, Han and coworkers prepared a task-specific ionic liquid (TSIL) 1-($N$,$N$-dimethylaminoethyl)-2,3-dimethylimidiazolium trifluoromethanesulfonate and used its aqueous solution as solvent for the hydrogenation of $CO_2$ with a heterogenized Ru complex catalyst prepared by the interaction of $RuCl_3$ with thioacetamide-modified silica followed by the addition of $PPh_3$.[54] The reactions were run at 60°C with various $H_2/CO_2$ ratios under 4–18 bar total pressure. Under optimized conditions, 1 equiv of HCOOH per TSIL was formed. After removal of the catalyst by filtration, water was evaporated at 110°C and the remaining liquid was distilled apart at 130°C into free HCOOH and the TSIL. Both the catalysts and the TSIL were recycled.

### 6.3.3.2  *Hydrogenation of CO₂ in Supercritical Carbon Dioxide and the Synthesis of Formamides*

The critical temperature and pressure of carbon dioxide are 31°C and 73 bar, respectively, and scCO$_2$ can dissolve various compounds, large amounts of $H_2$ in particular. In such reaction mixtures, there are much higher concentrations of $CO_2$ and $H_2$ than those available in solutions; furthermore, the reactants have high diffusion rates and the reactivities are not decreased by strong solvation. This gives an excellent opportunity for $CO_2$ hydrogenation in the supercritical state. Indeed, with the scCO$_2$-soluble catalyst precursor, $[RuH_2(PMe_3)_4]$, fast reactions were observed in the presence of $H_2O$ with a TOF $= 1400\,h^{-1}$ at 50°C, 85 bar $H_2$, and 120 bar $CO_2$ with triethylamine as added base.[18–20,55–57] The synthesis of formamides from dialkylamines, $CO_2$, and $H_2$ is slightly exergonic; for example, for the reaction in Equation 6.3,

$$CO_2(aq) + HNMe_2(aq) + H_2(aq) \rightarrow HCONMe_2(l) + H_2O(l) \qquad (6.3)$$

the thermodynamic parameters are the following: $\Delta G° = -0.75\,kJ/mol$, $\Delta H° = -36.3$ kJ/mol, and $\Delta S° = -119\,J/(mol\,K)$. The reaction is catalyzed by complexes similar to those for the hydrogenation of $CO_2$ to formic acid (formates) and can be run in solvents such as hydrocarbons, triethylamine, neat substrates (e.g., $HNMe_2$), or in particular, supercritical $CO_2$. In a few cases, further hydrogenation of the formamide product to methylamine can be detected.

Jessop et al. used $[RuCl_2(PMe_3)_4]$ as catalyst for the synthesis of dimethylformamide (DMF) in scCO$_2$ and could achieve a TON of 420,000 with complete selectivity (Scheme 6.6).[18–20,55–57] It was clearly established that the first step of the reaction is the hydrogenation of $CO_2$ to formic acid (which, indeed, accumulated in the first phase of the reaction) followed by its further reaction with $HNMe_2$ to give DMF. $[RuCl_2(dppe)_2]$ was found even more active by Baiker and coworkers[58] who observed a TON of 740,000. The initial rate of the reaction was also extremely high characterized by a TOF of 360,000 $h^{-1}$. The same group also studied the use of $[RuCl_2(dppe)_2]$ and similar complexes heterogenized by sol–gel method or supported on aerogels; such catalysts also showed high catalytic activity together with good recyclability.[59]

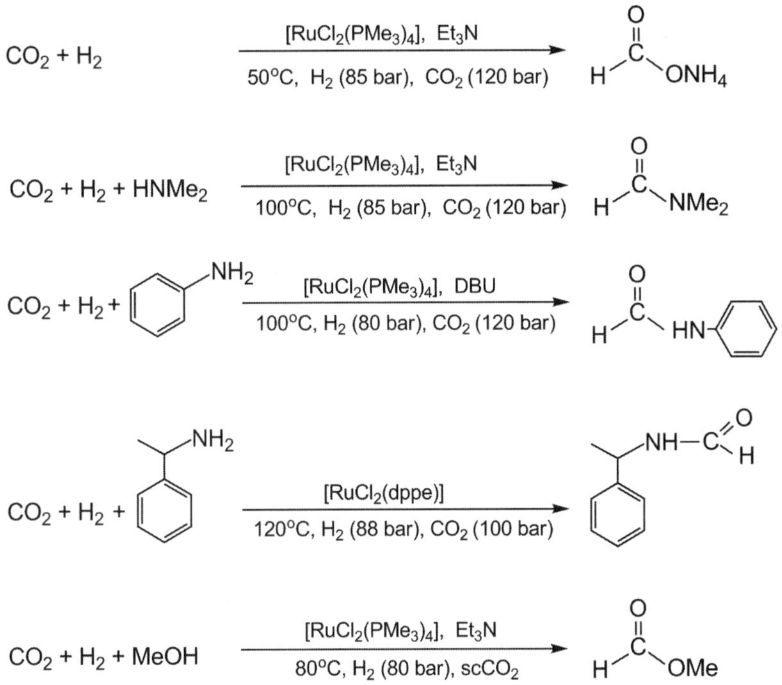

**SCHEME 6.6** $CO_2$-based synthesis of formates, formamides, and formate esters.

Formamides other than DMF are less readily available. A case in the point is the synthesis of formanilide, which proved unsuccessful since HCOOH is not produced in the presence of the weak base aniline. Jessop and coworkers solved this problem by using a stoichiometric quantity of a strong hindered base, DBU (1,8-diazabicyclo [5.4.0]undec-7-ene), which promotes the formation of formic acid without itself giving a formamide. Under such conditions and with [RuCl$_2$(PMe$_3$)$_4$] as catalyst, the reaction of aniline gave formanilide with 72% isolated yield.[43] In principle, both the catalyst and DBU can be recycled; however, this was not investigated. Baiker and coworkers used the [RuCl$_2$(dppe)] catalyst for the high-yield formylation of cyclic secondary amines (pyrrolidine, piperidine, pyrazine, etc.) at 100–120°C and 185–225 bar total pressure. Under such conditions, $R$-$(+)$-1-phenethlyamine was formylated without racemization (Scheme 6.6).[60]

*6.3.3.3 Formate Esters* In the presence of methanol, the hydrogenation of $CO_2$ results in the formation of methyl formate. The overall reaction is exergonic; that is, for the reaction in Equation 6.4,

$$CO_2(aq) + H_2(aq) + CH_3OH(l) \rightarrow HCOOCH_3(l) + H_2O(l) \qquad (6.4)$$

the thermodynamic parameters are the following: $\Delta G° = -5.28$ kJ/mol, $\Delta H° = -15.3$ kJ/mol, and $\Delta S° = -33.6$ J/(mol K). Ethanol and propanol react less readily. One of the most accepted mechanisms consists of the catalytic hydrogenation of $CO_2$ to HCOOH followed by its thermal esterification. Alkyl formate production is generally catalyzed by those complexes that are active in the formation of HCOOH, such as [RhCl(PPh$_3$)$_3$] and [RuCl$_2$(PPh$_3$)$_3$],[38] and by anionic metal carbonyls, for example, [WH(CO)$_5$]$^-$.[61,62] In the presence of Et$_3$N in supercritical $CO_2$, [RuCl$_2$(PMe$_3$)$_4$] catalyzed the formation of methyl formate with a TOF $= 55\,h^{-1}$ and with an overall TON $= 3500$.[57]

Alkyl formates can also be produced in the reaction of $CO_2$, $H_2$, and alkyl halides in the presence of basic salts (such as NaHCO$_3$) to neutralize the hydrohalic acid liberated in the reaction (Equation 6.5):

$$CO_2 + H_2 + RX \rightarrow HCOOR + HX \qquad (6.5)$$

A great deal of mechanistic investigations were done on the reaction with anionic metal carbonyls such as [WCl(CO)$_5$]$^-$ as catalysts.[61,62] Although such studies were extremely helpful in clarifying the way $CO_2$ was transformed to formate esters, the process itself cannot be regarded practical.

Methyl formate is currently produced by the base-catalyzed carbonylation of methanol with CO, and it is unlikely that homogeneous hydrogenation of $CO_2$ in the presence of methanol replaces that process anytime soon. Conversely, one of the major uses of methyl formate, that is, the synthesis of dimethylformamide, may well become obsolete since DMF can be efficiently produced by direct hydrogenation of $CO_2$ in the presence of HNMe$_2$ (see above).

### 6.3.3.4 Hydrogenation of $CO_2$ and Decomposition of HCOOH as Means of Hydrogen Storage and Transport

Transportation and local generation of H$_2$ are important problems of an envisaged "hydrogen economy." Probably the most widely used power generation equipments of that future will be the proton-exchange membrane fuel cells (PEMFCs); however, these require CO-free hydrogen for long-term operation. Various hydrogen storage materials have already been studied or suggested, and formic acid has excellent properties in this respect. As seen earlier in this chapter, HCOOH or HCOOH/Et$_3$N mixtures can be produced by the hydrogenation of $CO_2$, and recently, several catalytic systems have been described for their decomposition.[63,64]

Beller and coworkers investigated the catalytic decomposition of 5HCOOH/2NEt$_3$ mixtures under the action of a large number of soluble and solid catalysts.[65,66] Hydrogen was already generated at room temperature, which can be important for practical applications in portable devices. The best performance was shown by the catalyst obtained *in situ* from RuBr$_3$·xH$_2$O and 3 equiv PPh$_3$. At a reaction temperature of 40°C, the decomposition of formic acid in a 5HCOOH/2NEt$_3$ mixture was catalyzed with an initial turnover frequency of $3630\,h^{-1}$. Under the same conditions, the well-known hydrogenation catalyst [RuCl$_2$(PPh$_3$)$_3$] also showed outstanding activity with a TOF of $2688\,h^{-1}$. Very importantly, no CO was detected in the gas mixture, which makes the generated hydrogen suitable for application in fuel cells. Indeed, this was demonstrated by generating electricity in a H$_2$/O$_2$ PEM fuel cell.

A different approach was followed by Laurenczy and coworkers although the chemistry is similar.[67] They prepared the catalyst by reacting $RuCl_3 \cdot xH_2O$ or $[Ru(H_2O)_6]^{2+}$ with 2 equiv *meta*-trisulfonated triphenylphosphine (*m*tppts) in a 4 M aqueous solution of HCOOH/HCOONa (9:1). Formic acid was slowly decomposed by the resulting Ru(II) species already at 25°C and much faster at elevated temperatures; at 120°C, a TOF $= 460\,h^{-1}$ was determined. No traces of CO could be detected by FTIR spectroscopy. Using a specially designed high-pressure reactor, a given gas release rate (at a given pressure) could be kept constant by automatic control of the feed rate of liquid formic acid. This makes possible the fast delivery of even large quantities of virtually CO-free hydrogen.

Heterogeneous catalytic hydrogenation of bicarbonate in aqueous solution is a well-known process.[68,69] Coupled to a catalytic decomposition of formate back to $H_2$ and $HCO_3^-$, this reaction was also suggested as a method for storage and transport of hydrogen.[70] The best catalysts of bicarbonate hydrogenation consist of metallic palladium, either supported, such as Pd/C, or colloidal, stabilized by β-cyclodextrin. The latter[71] actively catalyzes the photochemically assisted reduction of $HCO_3^-$. The bimetallic Pd–Au/C and Pd–Ag/C catalysts show much higher stability against poisoning than Pd/C. Nevertheless, the activities of such catalysts are well below those of the soluble Ru complexes described above.

### 6.3.4   Synthesis of Organic Carbonates and Polycarbonates

Organic carbonates have several large-scale applications such as solvents and fuel additives.[5,13] Among the open-chain carbonates, dimethyl carbonate (DMC) and diphenyl carbonate (DPC), while of the cyclic carbonates, ethylene carbonate (EC) and propylene carbonate (PC) are the most important chemicals industrially. Furthermore, organic carbonates and isocyanates are important chemicals for the plastics industry as raw materials for polycarbonates and polyurethanes. Presently, organic carbonates are obtained by oxidative carbonylation of alcohols (with CO), while isocyanates are produced in the reaction of amines with noxious phosgene, the replacement of which is highly desirable. Some cyclic carbonates are now industrially produced from carbon dioxide.[72] In case this procedure is developed further, carbonates can play a key role in solving the problem of phosgene-free manufacturing of isocyanates; DMC and DPC are promising alternatives to phosgene in such reactions. In addition, DMC can replace other toxic reagents, such as dimethyl sulfate or methyl chloroformate.

The most straightforward synthesis of acyclic carbonates from $CO_2$ is its dehydrative condensation with alcohols (Scheme 6.7). This reaction is catalyzed by a variety of soluble and solid catalysts, with organometallic Sn derivatives such as $Bu_2Sn(OMe)_2$ playing a significant role. Typical reaction conditions include 140–180°C and up to 300 bar $CO_2$ pressure. A particular problem is that water accumulation affects unfavorably the chemical equilibrium; furthermore, it acts as a catalyst poison and therefore it must be removed from the reaction mixture. Orthoesters (e.g., trimethyl orthoacetate) and acetals (e.g., dimethyl acetal) that work as internal water scavengers can be used as starting materials instead of the

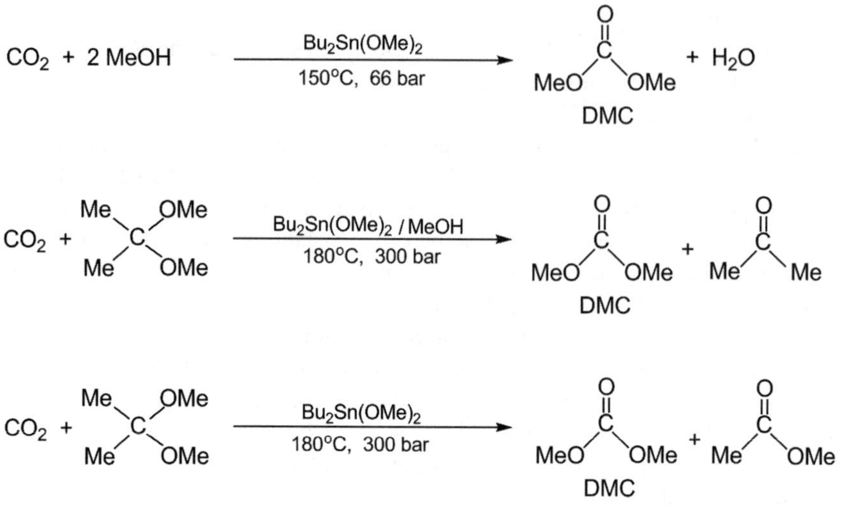

**SCHEME 6.7**  CO$_2$-based syntheses of dimethyl carbonate.

alcohols. Cheap molecular sieves cannot be applied as dehydrating agents at the high reaction temperature but work well when placed into a separate unit kept at room temperature as part of a recycling synthesis loop.[73]

Cyclic carbonates are easily obtained from CO$_2$ and oxiranes (Scheme 6.8).[13,21] The reaction is catalyzed by a large number of catalysts including metal salts and onium derivatives. Typical catalysts are halide salts such as Et$_4$NBr and KI, but salts of Sn, Ni, Zn, Cu, and so on can also be used. Imidazolium- and pyridinium-based ionic liquids are also good catalysts for this transformation, which is not unexpected taking that in general quaternary ammonium (and phosphonium) salts serve as catalysts for the reaction of epoxides and CO$_2$.[74] Task-specific ionic liquids facilitate the absorption and simultaneous reaction of CO$_2$ with oxiranes to give cyclic carbonates in good yields.[75,76] The reactivity of oxiranes follows the order: propylene oxide > styrene

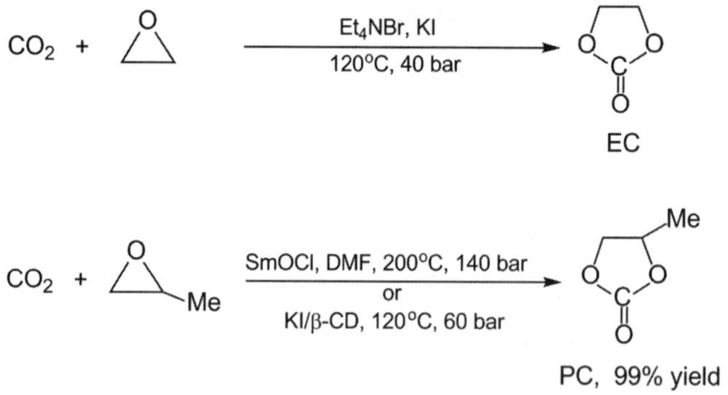

**SCHEME 6.8**  CO$_2$-based synthesis of ethylene carbonate and propylene carbonate.

oxide > ethylene oxide > chloromethyl ethylene oxide (epichlorohydrin). In many of the synthetic procedures, the target carbonates themselves can be used as solvents. Carbonate yields are increased by increasing $CO_2$ pressure, and supercritical $CO_2$ can be used advantageously for such syntheses. In contrast to oxiranes, the product carbonates are insoluble in sc$CO_2$ and this leads to their separation as a distinct phase. However, reactions in sc$CO_2$ require catalysts soluble in that phase. Propylene carbonate was synthesized with high yield (93%) and high selectivity (99%) with polyfluoroalkylphosphonium iodide catalysts, such as $[(C_6F_{13}C_2H_4)_3MeP]I$. The sc$CO_2$-insoluble product was continuously removed from the reactor while the catalyst-containing supercritical fluid phase was maintained by supplying appropriate amounts of propene and $CO_2$.[77] Potassium iodide together with β-cyclodextrin has excellent catalytic properties at 120°C and 60 bar $CO_2$ pressure.[78]

An interesting feature that made the commercialization of the synthesis of linear carbonates possible is that both linear and cyclic carbonates easily undergo transesterification with alcohols (Scheme 6.9). This process is used industrially for the synthesis of diphenylcarbonate from DMC.[72] The coproduct methanol is recycled to DMC. Nevertheless, once a highly efficient direct $CO_2$ process for an organic carbonate is found, transesterification makes possible the production of any other carbonates.

Cyclic carbonates can also be obtained in the reaction of diols and urea, an activated form of $CO_2$ (Scheme 6.10).[79] The ammonia produced in the reaction can be recycled via the Bosch–Meiser process.

Aromatic polycarbonates are transparent and highly impact resistant materials, and these characteristics make them important engineering plastics with an annual production of about 2 million tons; this market is steadily increasing.[22] In addition, the

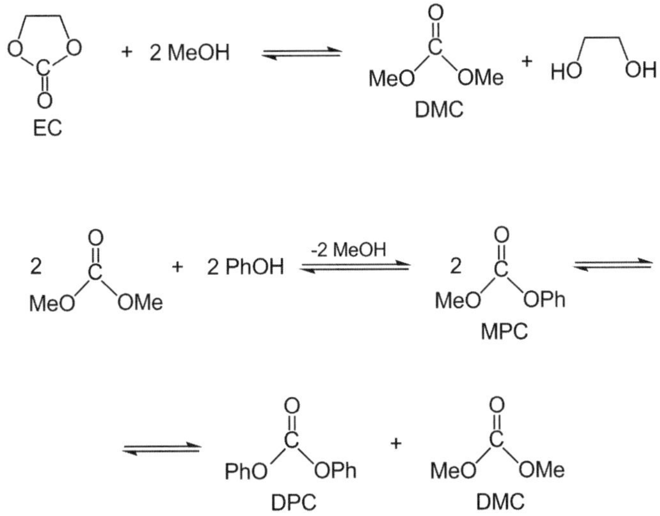

**SCHEME 6.9**    Synthesis of dimethyl carbonate and diphenyl carbonate by transesterification.

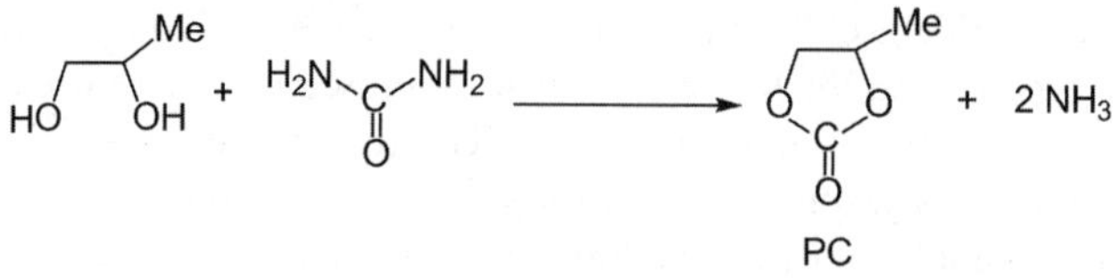

**SCHEME 6.10** The use of urea as masked $CO_2$ for the synthesis of propylene carbonate.

alternating copolymers of oxiranes and $CO_2$ are biodegradable and have high oxygen permeability.

The formation of polycarbonates from epoxides and $CO_2$ catalyzed by $Et_2Zn/H_2O$ (1:1) was first discovered in 1969 by Inoue and coworkers[80,81] Since then, most studies concerned the catalytic effect of Zn derivatives. In general, Zn(II) carboxylates or alkoxides containing bi- or trifunctional groups (such as glutarate or pyrogallol) were found to be most active.[17] Accordingly, the generally accepted mechanism of polycarbonate formation assumes the cooperation of adjacent Zn centers (Scheme 6.11).

Polycarbonates from DPC and bisphenol A (Scheme 6.12) are produced commercially.[21] The technology starts with formation of EC from $CO_2$ and ethylene oxide. EC is then transformed to DMC in a transesterification reaction. DPC is manufactured by the reaction of DMC with phenol in the presence of $[Pb(OPh)_2]$ as catalyst. The final products are the polycarbonate and the ethylene glycol coproduct of EC transesterification. The process does not use phosgene or other halogenated raw materials or solvents.

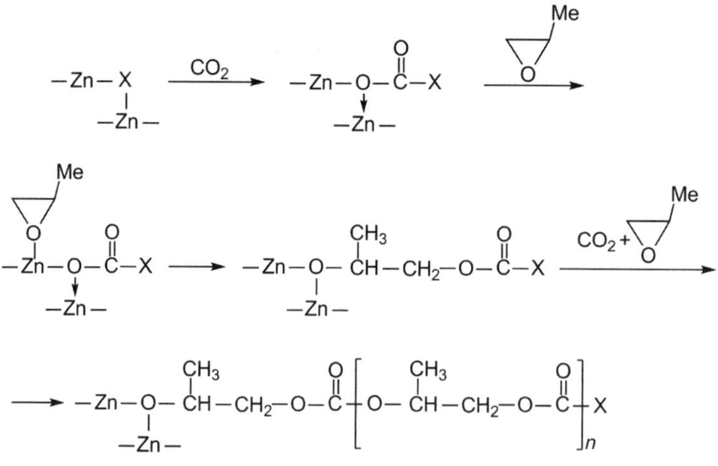

**SCHEME 6.11** Heterogeneous catalysis of the copolymerization of carbon dioxide and propylene carbonate on Zn-based catalysts.

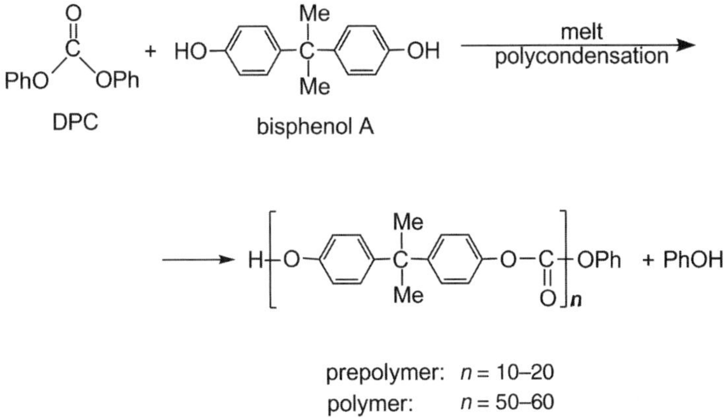

prepolymer: $n = 10–20$
polymer:     $n = 50–60$

**SCHEME 6.12**   Polycarbonates from diphenyl carbonate and bisphenol A.

## 6.4   EMERGING APPLICATIONS OF CARBON DIOXIDE AS C$_1$ FEEDSTOCK

### 6.4.1   Direct Carboxylation of Hydrocarbons

Synthesis of salicylic acid by the Kolbe–Schmitt process (see above) is an example for the direct carboxylation of aromatics.[34] At the same time, it is also an example for the limitations of such reactions. Formal insertion of $CO_2$ into a C–H bond can usually be achieved with compounds containing active hydrogens (such as phenolates), and perhaps more importantly, the product carboxylic acids have to be stabilized by salt formation. Liberation of the free acid from such carboxylates requires a mineral acid and produces 1 equiv of inorganic salt (waste) (Scheme 6.1).

Aromatic carboxylic acids were obtained in good to excellent yield by carboxylation of aromatics with a carbon dioxide–Al$_2$Cl$_6$/Al system at moderate temperatures (20–80°C).[82] In a stoichiometric reaction, the dichloroaluminate salts of carboxylic acids were formed. Experimental results and DFT calculations suggested that the most probable pathway involved activation of $CO_2$ by the superelectrophilic aluminum chloride and its reaction with the aromatics in a typical electrophilic substitution.

Acetic acid is an important commodity chemical, and the direct carboxylation of methane with $CO_2$ is of obvious industrial interest. This reaction has been observed with a [Pd(OAc)$_2$]/[Cu(OAc)$_2$]/O$_2$/CF$_3$COOH system, albeit with low yields.[83]

Alkynes are easily carboxylated by $CO_2$ and usually offer lactones or pyrones as products. However, in several cases carboxylic acids can also be obtained. For example, 2-butyne reacted with $CO_2$ in the presence of a [Ni(CDT)] (CDT = 1,5,9-cyclododecatriene) catalyst and $N,N,N',N'$-tetramethyl-1,2-diamine (TMEDA) as base to yield 2-methylcrotonic acid.[21] Similarly, the reaction of terminal alkynes was catalyzed by [Ni(COD)] in a highly regio- and chemoselective manner in the presence of DBU as base affording $E$-acrylic acids in >85% yield (Scheme 6.13).[84]

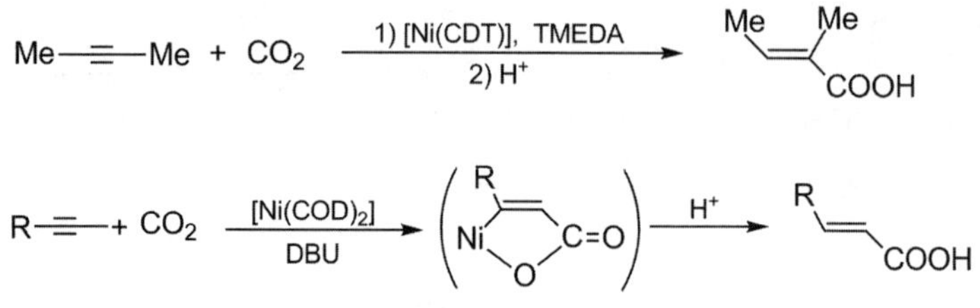

$$Me\text{---}\!\equiv\!\text{---}Me \ + \ CO_2 \quad \xrightarrow[\text{2) } H^+]{\text{1) [Ni(CDT)], TMEDA}} \quad \overset{Me \quad Me}{\diagdown\!\diagup}\overset{}{\underset{COOH}{}}$$

$$R\text{---}\!\equiv\!\text{---} + \ CO_2 \quad \xrightarrow[\text{DBU}]{\text{[Ni(COD)}_2]} \quad \left( \overset{R}{\underset{Ni \diagdown O \diagup}{}} C\!=\!O \right) \quad \xrightarrow{H^+} \quad \overset{R}{\underset{COOH}{}}$$

**SCHEME 6.13** Synthesis of butenoic acids from alkynes and $CO_2$.

Alkenes can react with $CO_2$ to give a variety of products (esters, lactones, and metallacyclic derivatives). Of these, the formation of unsaturated carboxylic acids is outstandingly important. Catalytic synthesis of acrylic or methacrylic acid from ethene and $CO_2$ could provide an entry point to acrylate and methacrylate polymers. However, although acrylate-containing Fe, Mo, W, and Ni complexes could be obtained, these were not able to release acrylic acid and enter a new catalytic cycle.[85] Density functional theory calculations[33] were made on the coupling of ethene and $CO_2$ catalyzed by [Ni(bipy)(CDT)] (bipy = 2,2'-bipyridyl), the product of which was a stable five-membered nickelalactone. It was concluded that the existence of a [Ni(bipy) $(CO_2)$(ethene)] mixed-ligand complex is not a prerequisite of the coupling; in fact, the stability of such a complex toward $CO_2$ decomposition is so low that its presence in the reaction mixture is quite unlikely. Very recently, it has been shown by Aresta, Pápai, and coworkers that the reaction of ethene with the preformed cationic complexes [(L-L)Pd(COOMe)]($OSO_2CF_3$) (L-L = bipyridyl; 2-(2(diphenylphosphino)ethyl)pyridine, or dppe) yielded methyl acrylate and ethyl acrylate in good yields (Scheme 6.14).[86] The highest yield was obtained with dppe as ligand and DMF as solvent. Under $CO_2$/ethene pressure, the reaction was clearly catalytic with an increased yield of ethyl acrylate, albeit the overall TON was low. Propene reacted similarly, but gave methyl methacrylate as the sole product and with lower yields than those observed with ethene. Detailed DFT calculations revealed that the reaction proceeds through the formation of a cyclic metallalactone species with an active participation of the solvent (DMF). The calculations also showed that esterification of the carboxylic moiety facilitates the release of acrylate from the metal coordination sphere. These studies represent a significant step toward the design of active catalytic systems for obtaining important unsaturated esters.

### 6.4.2 Synthesis of Carbamates

Carbamic acid esters (urethanes) and their derivatives are important precursors to agrochemicals (herbicides, fungicides, and pesticides) as well as to pharmaceuticals. Also, they can be transformed to isocyanates, which are major building blocks for polyurethanes, widely used in construction, transportation, and several other fields. Presently, the main technical process for the manufacture of isocyanates is phosgenation of the corresponding amines. Since the global consumption of polyurethane raw

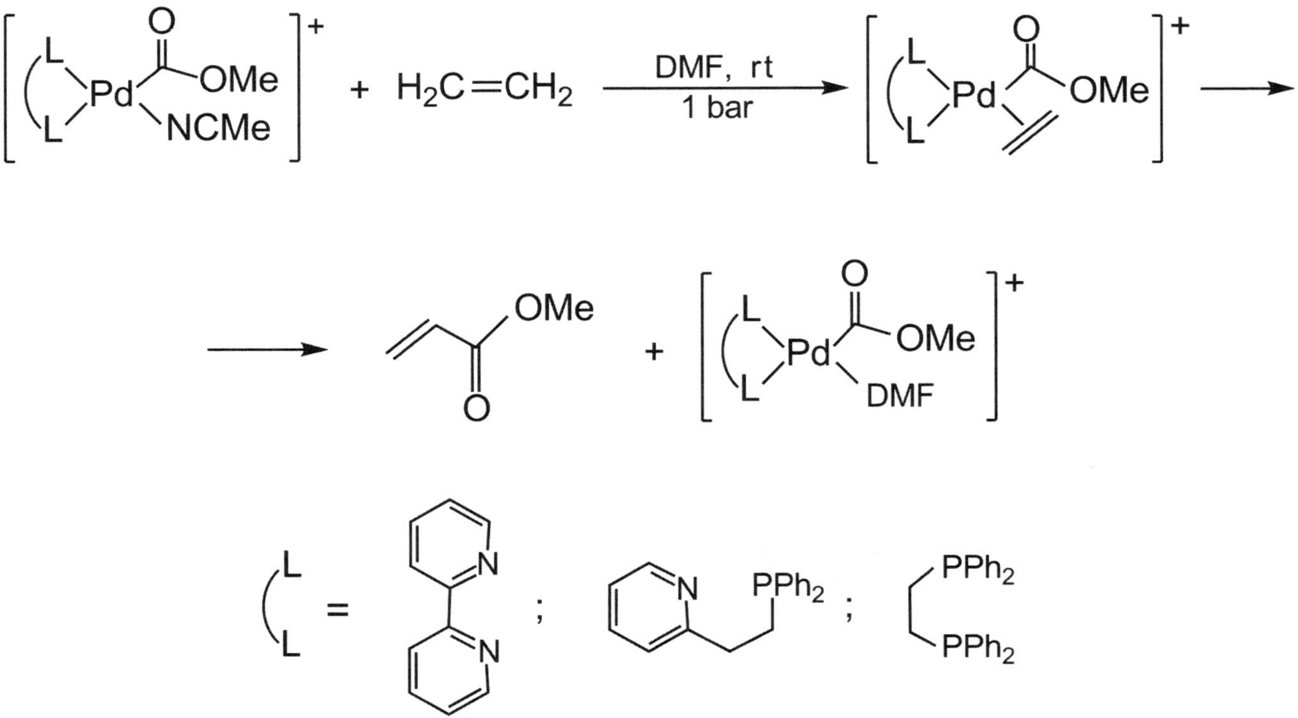

**SCHEME 6.14** Formation of methyl acrylate from ethene and a preformed Pd metallocarboxylate; model reaction for catalytic synthesis of acrylates.

$$CO_2 + RNH_2 + R'OH \underset{Me_2C(OEt)_2}{\overset{cat.}{\rightleftharpoons}} \underset{RNH}{\overset{O}{\underset{}{\parallel}}}\underset{}{C}\underset{OR'}{} + H_2O$$

R' = Me, Et

Cat.: $Bu_2SnO$, $Me_2SnCl_2$, or $Cs_2CO_3$

**SCHEME 6.15**   Direct synthesis of carbamates from $CO_2$, amines, and alcohols.

materials is above 12 million metric tons (2007),[87] there is an immense interest to produce polyurethanes on $CO_2$ basis.

Carbamates are synthesized in the reaction of amines, carbon dioxide, a base, and an alkylating agent, such as organic halides, alcohols, organic carbonates, epoxides, alkynes, and so on (Scheme 6.15). Halide-free routes to urethanes are highly desirable. The reactions can be run advantageously in supercritical $CO_2$. In a reaction reminiscent to the production of linear carbonates, $CO_2$ was shown to react with $t$-butylamine and ethanol applying an acetal, $Me_2C(OMe)_2$ as dehydrating agent, and $Bu_2SnO$ as catalyst.[88] At high $CO_2$ pressure and high temperature, the reaction afforded the corresponding urethane, $t$-BuNH–COOEt, in 84% yield. Carbamic esters have also been obtained under mild conditions (25 bar $CO_2$ pressure) from various amines, alcohols, and $CO_2$ with basic catalysts such as $Cs_2CO_3$ with up to 56% amine conversion and 79% carbamate selectivity.[89]

One important side reaction of urethane production is N-alkylation of the amine leading to loss of reactive primary or secondary amines. Preformed ammonium carbamates were successfully transformed to urethanes in the presence of crown ethers, which directed the reaction into O-alkylation.[90]

As mentioned above, organic carbonates can also be used as alkylating agents for the production of carbamic esters. Since such carbonates (e.g., dimethyl carbonate) can be produced on $CO_2$ basis (see above), procedures can be envisaged for extensive use of $CO_2$ for production of urethanes and polyurethanes.

### 6.4.3   Synthesis of Lactones and Pyrones

Dienes, allenes, and alkynes react with carbon dioxide to yield cyclic lactones—the catalysts include various Ni and Pd complexes.[4] With certain diynes, alternating copolymerization with $CO_2$ results in the formation of poly(2-pyrones) (Scheme 6.16).

Of these reactions perhaps the most important one is the telomerization of 1,3-butadiene with $CO_2$.[91] The reaction is carried out under mild conditions (80°C, 40 bar total pressure) with a combination of $[Pd(acac)_2]$ (acac = acetylacetonate) and tertiary phosphines such as $PPh_3$, $P^iPr_3$, or $PCy_3$ as catalyst. This reaction can yield several products; however, under optimal conditions the main product is the δ-lactone (2-ethyliden-6-hepten-5-olide) (Scheme 6.17).

The yield and selectivity strongly depend on the solvent used, and the best results with respect to the formation of δ-lactone can be achieved with nitriles or organic carbonates (which themselves may be derived from $CO_2$).[92] The importance of these

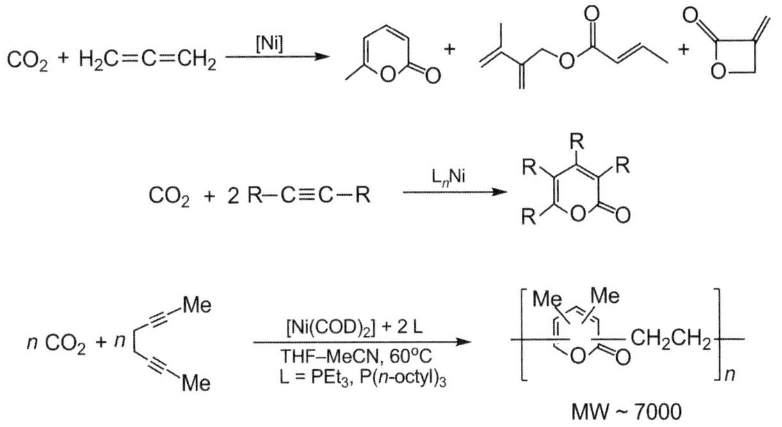

**SCHEME 6.16**  Formation of pyrones and polypyrones in the reaction of $CO_2$ with allenes and alkynes.

results is given by the fact that δ-lactone can be transformed by various procedures (hydrogenation, hydroformylation, hydroamination, hydroaminomethylation, etc.) to a broad variety of useful products.[4] Detailed studies have shown that 2-ethylheptanol is readily available from the δ-lactone (i.e., from $CO_2$ and 1,3-butadiene) (Scheme 6.18) and can be further converted to di(2-ethylheptyl)phthalate, a possible substitute for the toxic plasticizer di(2-ethylhexyl)phthalate.[93,94]

Telomerization of 1,3-butadiene with $CO_2$ has been thoroughly studied under conditions of a miniplant.[91,92] Although acetonitrile and linear carbonates (EC, PC)

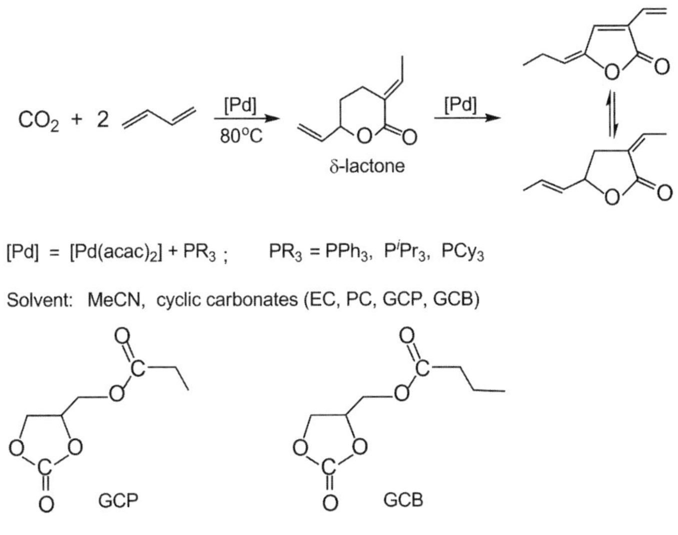

[Pd] = [Pd(acac)$_2$] + PR$_3$ ;    PR$_3$ = PPh$_3$, P$^i$Pr$_3$, PCy$_3$

Solvent: MeCN, cyclic carbonates (EC, PC, GCP, GCB)

**SCHEME 6.17**  Telomerization of $CO_2$ and butadiene.

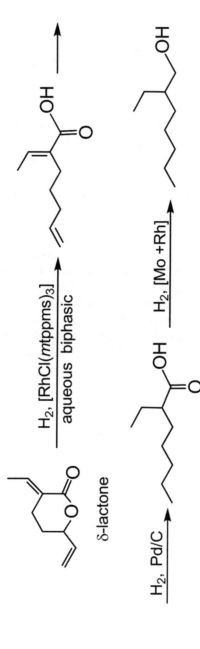

**SCHEME 6.18** Synthesis of 2-ethylheptanol from the δ-lactone obtained in telomerization of $CO_2$ and butadiene.

were suitable solvents from the point of view of the chemical transformation, however, due to their low boiling points, in the separation step, these were the ones removed by distillation from the high boiling δ-lactone. Consequently, the catalyst had to be recycled in solution with part of the product, which was disadvantageous with respect to the space–time yield of the process and formation of by-products. With solvents such as glycerol carbonate propionate (GCP) and glycerol carbonate butyrate (GCB) having higher boiling points (126–128 and 133–135°C at 0.01 mbar) than that of the δ-lactone (76°C at 0.13 mbar), the latter could be separated by distillation and the catalyst could be recycled dissolved in the organic carbonate. Under such conditions, 22% yield of the δ-lactone was obtained at 34% butadiene conversion in 4 h; however, the overall yield can be increased with continuous recycling of unreacted educts.

### 6.4.4  Reforming of Methane to Synthesis Gas and the Production of Methanol

The reaction of CH$_4$ with CO$_2$ (dry reforming of methane) produces CO and H$_2$ in a ratio close to unity (Equation 6.6). The other two ways to get CO + H$_2$ from CO$_2$ and CH$_4$ is the partial oxidation of methane, and its steam reforming (wet reforming) (Equations 6.7 and 6.8). The combination of these three processes is called tri-reforming, which allows synthesis gas production with higher H$_2$/CO ratios, and in which the exothermic partial oxidation of methane in part compensates for the large energy demand of wet and dry reforming. In dry reforming, the main reaction is accompanied by the reverse water gas shift reaction (Equation 6.9) and by coke formation via the Boudouard reaction (CO disproportionation, Equation 6.10) and methane decomposition (Equation 6.11).[6,95,96]

$$CH_4 + CO_2 \rightleftharpoons 2CO + 2H_2 \qquad \Delta H^{\circ}_{298} = 247 \, kJ/mol \qquad (6.6)$$

$$CH_4 + \tfrac{1}{2}O_2 \rightleftharpoons CO + 2H_2 \qquad \Delta H^{\circ}_{298} = -36 \, kJ/mol \qquad (6.7)$$

$$CH_4 + H_2O \rightleftharpoons CO + 3H_2 \qquad \Delta H^{\circ}_{298} = 206 \, kJ/mol \qquad (6.8)$$

$$CO_2 + H_2 \rightleftharpoons CO + H_2O \qquad \Delta H^{\circ}_{298} = 41 \, kJ/mol \qquad (6.9)$$

$$2CO \rightleftharpoons C + CO_2 \qquad \Delta H^{\circ}_{298} = -172 \, kJ/mol \qquad (6.10)$$

$$CH_4 \rightleftharpoons C + 2H_2 \qquad \Delta H^{\circ}_{298} = 75 \, kJ/mol \qquad (6.11)$$

Ni-based heterogeneous catalysts on oxide supports can be used for CO$_2$ reforming of methane. These suffer from rapid deactivation due to carbon deposition and sintering of the metal particles on the surface of the support; however, Ni–La$_2$O$_3$ was found to prevent coke formation. In general, supported Pt catalysts on ZrO$_2$, Al$_2$O$_3$, or MgO show better stability.[96,97] Typically, the reactions are run at

600–800°C, $CH_4/CO_2$ ratio of 1 resulting in around 90% conversion of both gases. The process can also be carried out under cold nonthermal and transitional nonthermal plasma conditions;[98,99] in the latter case, the plasma temperature may be as high as 2000–3000°C.[100]

The $CO_2$ reforming of methane is an attractive process since it consumes two greenhouse gases and produces important feedstock that can be further processed to methanol or to higher hydrocarbons by the Fischer–Tropsch methodology. It is often suggested that the process can have a role in converting $CO_2$-rich natural gas from remote sources into liquid fuels with better transportation properties than those of natural gas itself.[95]

Methanol has numerous important uses in the chemical and energy (fuel) sector and it has even been suggested as the central chemical of a future "methanol economy."[101] It can be obtained by the hydrogenation of both carbon monoxide and carbon dioxide. Interestingly, when $CO_2$ is mixed into the $CO/H_2$ feed, significant improvements in methanol yield and energetics are observed. In fact, in the process applying Cu–ZnO on basic oxide support, up to 30% of $CO_2$ is mixed into the feed. The reaction proceeds at 250–300°C and at 50–100 bar total pressure.[6] The direct hydrogenation of $CO_2$ to methanol can be carried out with Pd-modified or $ZrO_2$-doped Cu–ZnO catalyst at 250°C and 50 bar with high selectivity.[102] Silica-supported Cu–ZnO catalysts showed long-term stability on stream and were studied in a pilot plant. However, large-scale production of methanol by direct $CO_2$ hydrogenation is yet to start.

Methanol formation is rarely observed under homogeneous conditions. One such system applied $[Ru_3(CO)_{12}]$ as catalyst and KI as promoter in $N$-methylpyrrolidone solvent.[103] However, the attainable reaction rate and yield were low (TON for MeOH = 11.5 in 5 h at 200°C and 80 bar $H_2/CO_2 = 3$).

### 6.4.5 Electrochemical and Photochemical Reduction of Carbon Dioxide

The reduction of $CO_2$ requires electron transfer in one-electron or multielectron steps either from reducing agents, for example, $H_2$, or electrochemically. $H_2$ can also be produced by water splitting either electrochemically or photochemically. For efficient electrochemical reduction of dissolved $CO_2$, electron transfer catalysts (electron relays, mediators), usually transition metal complexes, are required while photochemical systems need also a photosensitizer. The two approaches can be combined to photoelectrochemical systems, as well.

In multielectron transfer processes, the reduction of $CO_2$ can yield formic acid, carbon monoxide, formaldehyde, methanol, or methane; that is, the primary electrochemical process supplies $C_1$ compounds. These reactions can proceed at reasonable reduction potentials between $-0.24$ and $-0.61$ V (NHE) (Equations (6.12–6.16); the reduction potentials, $E°$, refer to pH 7 in aqueous solutions versus NHE), while the formation of the $CO_2^{•-}$ radical anion is estimated to take place at $-2.1$ V.[104] Reduction of CO (in the presence of $H^+$) supplies $CH_2^•$ radicals that may yield methane directly or leads to higher hydrocarbons (e.g., ethene or ethane) by recombination.[24,105] Efficient formation of ethene (together with alcohols) was already reported in 1988 by electrolysis of aqueous $CO_2$ at a copper cathode.[106]

$$CO_2 + 2H^+ + 2e^- \rightleftharpoons HCO_2H \qquad E^\circ = -0.61 \text{ V} \qquad (6.12)$$

$$CO_2 + 2H^+ + 2e^- \rightleftharpoons CO + H_2O \qquad E^\circ = -0.52 \text{ V} \qquad (6.13)$$

$$CO_2 + 4H^+ + 4e^- \rightleftharpoons HCHO + H_2O \qquad E^\circ = -0.48 \text{ V} \qquad (6.14)$$

$$CO_2 + 6H^+ + 6e^- \rightleftharpoons CH_3OH + H_2O \qquad E^\circ = -0.38 \text{ V} \qquad (6.15)$$

$$CO_2 + 8H^+ + 8e^- \rightleftharpoons CH_4 + 2H_2O \qquad E^\circ = -0.24 \text{ V} \qquad (6.16)$$

For electrochemical reductions, widely used electron transfer catalysts include $Ni^{2+}$ and $Co^{2+}$ complexes of macrocycles such as tetraaza macrocycles (cyclams, porphyrins), and in many cases the reductions are carried out at Cu or Ni cathodes.[107] In protic systems (water, methanol), an important competing process is the reduction of $H^+$ to $H_2$, which often leads to low selectivity of $CO_2$ reduction. Consequently, the selectivity (and product distribution) in such systems is pH-dependent. Oxygen-containing $C_n$ molecules (ethanol, propanol, acetone, oxalate, pyruvate, glyoxylate, glycolate, acetylacetate, etc.) are formed mostly, but not exclusively in aprotic solutions with solvents such as MeCN, DMF, DMSO, and so on. Interestingly, with homogeneous catalysts in non-electrochemical systems, hydrogenation of $CO_2$ further than formic acid is only rarely observed (and even then homogeneity of those systems was not investigated).

Electrochemistry can also be applied to facilitate the reaction of $CO_2$ with olefins and epoxides to form carboxylic acids and cyclic carbonates, respectively;[108–110] related reactions were discussed in previous paragraphs in this chapter. The aim of the electrochemical approach is achieving higher reaction rates and selectivities under mild conditions.

$Ni^{2+}$ and $Co^{2+}$ complexes mentioned above can serve as photoreduction catalysts also in photochemical systems (Scheme 6.19). For the same purpose,

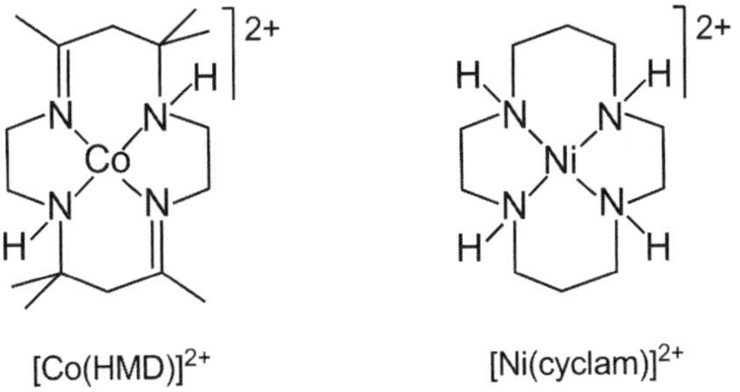

[Co(HMD)]$^{2+}$           [Ni(cyclam)]$^{2+}$

**SCHEME 6.19** Tetraazamacrocyclic complexes for catalytic electrochemical and photochemical reduction of carbon dioxide.

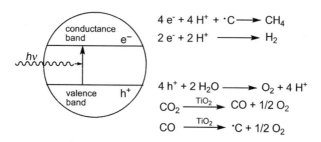

**SCHEME 6.20**   Charge separation and reactions of $CO_2$ reduction on the surface of $TiO_2$ upon illumination.

$[Ru(bipy)_2(CO)_2]^{+}$,[23] Ru colloids,[111] or Pd complexes of dendrimeric phosphines[112] were also found suitable. By far the most widely used soluble photosensitizer is $[Ru(bipy)_3]^{2+}$; however, $[ReCl(bipy)(CO)_3]$ and $[ReBr(bipy)(CO)_3]$ were also applied. The latter two Re complexes served as electron transfer catalysts, too, so the photoreduction of $CO_2$ proceeded without further additives.[113,114] Heterogeneous photosensitizers, most of all $TiO_2$, are subject to extensive studies.

One of the biggest problems of electrochemical and photochemical $CO_2$ reductions is concerned with the source of electrons. The generation of electricity—if not on nuclear or renewable bases—may produce more $CO_2$ than what is reduced in the electrochemical systems. Similarly, most photochemical systems work only in the presence of a sacrificial electron donor ($Et_3N$, triethanolamine, ascorbic acid, etc.), which is more expensive than the products of $CO_2$ photoreduction. Nevertheless, there are examples of formation of $CH_4$ (together with traces of ethene and ethane) and CO solely from the reaction of $CO_2$ and $H_2O$ on the surface of $TiO_2$ illuminated by ultraviolet light (Scheme 6.20).[115] Since $TiO_2$ is cheap, nontoxic, and readily available, these are encouraging results. However, the efficiency of such systems is still very low.

## 6.5   CONCLUSIONS

Chemical use of carbon dioxide requires its activation. As shown in the preceding paragraphs, there are many possibilities for obtaining useful products from $CO_2$—and for doing that efficiently and economically. Eventually, a chemical industry based on $CO_2$ as primary raw material may be able to supply new solvents, fuels, engineering and other plastics, and so on.[116] The transition to this state should be accompanied by decreased emissions of $CO_2$ (or other greenhouse gases) to the atmosphere and by the use of nontoxic chemicals. However, many processes required for such immense changes are only in the stage of development or in their infancy. Besides, since $CO_2$ is the most oxidized state of carbon, its reactions require a net energy input. A supply of this energy from fossil sources may facilitate advances in some fields (e.g., the replacement of phosgene in syntheses) but would further increase atmospheric $CO_2$ concentrations. The key question of winning the race against catastrophic global warming is to move away from the use of fossil sources, ultimately to solar energy. Extensive research on carbon dioxide utilization (among others) is a must to achieve this goal.

## ACKNOWLEDGMENTS

Support of our research into $CO_2$ activation underpinning this chapter was provided by the Hungarian National Research and Technology Office, National Research Fund (NKTH-OTKA K 68482).

## REFERENCES

1. Kamekawa, H.; Senboku, H.; Tokuda, M. *Electrochim. A* **1997**, *42*, 2117–2123.

2. Aresta, M., Ed. *Carbon Dioxide Recovery and Utilization*; Kluwer: Dordrecht, 2003.

3. Behr, A., Ed. *Carbon Dioxide Activation by Metal Complexes*; VCH: Weinheim, 1988.

4. Behr, A. *Angewandte homogene Katalyse*; Wiley-VCH: Weinheim, 2008.

5. Arakawa, H.; Aresta, M.; Armor, J. N.; Barteu, M. A.; Beckman, E. J.; Bell, A. T.; Bercaw, J. E.; Creutz, C.; Dinjus, E.; Dixon, D. A.; Domen, K.; DuBois, D. L.; Eckert, J.; Fujita, E.; Gibson, D. H.; Goddard, W. A.; Goodman, D. W.; Keller, J.; Kubas, G. J.; Kung, H. H.; Lyons, J. E.; Manzer, L. E.; Marks, T. J.; Morokuma, K.; Nicholas, K. M.; Periana, R.; Que, L.; Rostrup-Nielson, J.; Sachtler, W. M. H.; Schmidt, L. D.; Sen, A.; Somorjai, G. A.; Stair, P. C.; Stults, B. R.; Tumas, W. *Chem. Rev.* **2001**, *101*, 953–996.

6. Aresta, M. In *Activation of Small Molecules*; Tolman, W. B., Ed., Wiley-VCH: Weinheim, **2006**; pp 1–41.

7. Aresta, M.; Dibenedetto, A. *Catal. Today* **2004**, *98*, 455–462.

8. Behr, A. *Angew. Chem.* **1988**, *100*, 681–698.

9. Aresta, M.; Quaranta, E.; Tommasi, I.; Giannoccaro, P. *Gaz. Chim. Ital.* **1995**, *125*, 509–537.

10. Gibson, D. H. *Chem. Rev.* **1996**, *96*, 2063–2095.

11. Gibson, D. H. *Coord. Chem. Rev.* **1999**, *185-186*, 335–355.

12. Yin, X.; Moss, J. R. *Coord. Chem. Rev.* **1999**, *181*, 27–59.

13. Sakakura, T.; Choi, J.-C.; Yasuda, H. *Chem. Rev.* **2007**, *107*, 2365–2387.

14. Walther, D.; Ruben, M.; Rau, S. *Coord. Chem. Rev.* **1999**, *182*, 67–100.

15. Leitner, W. *Coord. Chem. Rev.* **1996**, *153*, 257–284.

16. Palmer, D. A.; van Eldik, R. *Chem. Rev.* **1983**, *83*, 651–731.

17. Darensbourg, D. J.; Holtcamp, M. W. *Coord. Chem. Rev.* **1996**, *153*, 155–174.

18. Jessop, P. G.; Ikariya, T.; Noyori, R. *Chem. Rev.* **1995**, *95*, 259–272.

19. Jessop, P. G.; Joó, F.; Tai, C.-C. *Coord. Chem. Rev.* **2004**, *248*, 2425–2442.

20. Jessop, P. G. In *The Handbook of Homogeneous Hydrogenation*; de Vries, J. G.; Elsevier, C. J., Eds.; Wiley-VCH: Weinheim, **2007**; pp 489–511.

21. Omae, I. *Catal. Today* **2006**, *115*, 33–52.

22. Darensbourg, D. J. *Chem. Rev.* **2007**, *107*, 2388–2410.

23. Fujita, E. *Coord. Chem. Rev.* **1999**, *185-186*, 373–384.

24. Tanaka, K. *Bull. Chem. Soc. Jpn.* **1998**, *71*, 17–29.

25. Walsh diagram of $CO_2$ based on DFT calculations. Courtesy of Dr. I. Pápai, Chemical Research Center, Hungarian Academy of Sciences, Budapest, Hungary, 2008.

26. Based on a survey of the Cambridge Structural Database. Courtesy of Dr. A. Bényei, Institute of Physical Chemistry, University of Debrecen, Hungary, 2008.

27. Aresta, M.; Nobile, C. F.; Albano, V. G.; Forni, E.; Manassero, M. *J. Chem. Soc., Chem. Commun.* **1975**, 636–637.

28. Herskovitz, T. *J. Am. Chem. Soc.* **1977**, *99*, 2391–2392.

29. Calabrese, J. C.; Herskovitz, T.; Kinney, J. B. *J. Am. Chem. Soc.* **1983**, *105*, 5914–5915.

30. Castro-Rodriguez, I.; Nakai, H.; Zakharov, L. N.; Rheingold, A. L.; Meyer, K. *Science* **2004**, *305*, 1757–1759.

31. Gibson, D. H.; Ye, M.; Richardson, J. F. *J. Am. Chem. Soc.* **1992**, *114*, 9716–9717.

32. Fachinetti, G.; Floriani, C.; Zanazzi, P. F. *J. Am. Chem. Soc.* **1978**, *100*, 7405–7407.

33. Pápai, I.; Schubert, G.; Mayer, I.; Besenyei, G.; Aresta, M. *Organometallics* **2004**, *23*, 5252–5259.

34. Lindsey, A. S.; Jeskey, H. *Chem. Rev.* **1957**, *57*, 583–620.

35. http://www.the-innovation-group.com/ChemProfiles/Urea.htm.

36. http://.Cheresources.com/ureamodeling.pdf.

37. Joó, F. *Aqueous Organometallic Catalysis*; Kluwer: Dordrecht, 2001.

38. Leitner, W. *Angew. Chem.* **1995**, *107*, 2391–2405; *Angew. Chem., Int. Ed. Engl.***1995**, *34*, 2207–2221.

39. Inoue, Y.; Sasaki, Y.; Hashimoto, H. *J. Chem. Soc., Chem. Commun.* **1975**, 718–719.

40. Inoue, Y.; Izumida, H.; Sasaki, Y. *Chem. Lett.* **1976**, 863–864.

41. Tsai, J.-C.; Nicholas, K. M. *J. Am. Chem. Soc.* **1992**, *114*, 5117–5124.

42. Yin, C.; Xu, Z.; Yang, S.-Y.; Ng, S. M.; Wong, K. Y.; Lin, Z.; Lau, C. P. *Organometallics* **2001**, *20*, 1216–1222.

43. Munshi, P.; Heldebrandt, D.; McKoon, E.; Kelly, P. A.; Tai, C.-C.; Jessop, P. G. *Tetrahedron Lett.* **2003**, *44*, 2725–2727.

44. Gassner, F.; Leitner, W. *J. Chem. Soc., Chem. Commun.* **1993**, 1465–1466.

45. Laurenczy, G.; Joó, F.; Nádasdi, L. *Inorg. Chem.* **2000**, *39*, 5083–5088.

46. Joó, F.; Laurenczy, G.; Nádasdi, L.; Elek, J. *Chem. Commun.* **1999**, 971–972.

47. Kathó, Á.; Opre, Z.; Laurenczy, G.; Joó, F. *J. Mol. Catal. A: Chem.* **2003**, *204-205*, 143–148.

48. Kovács, G.; Schubert, G.; Joó, F.; Pápai, I. *Catal. Today* **2005**, *115*, 53–60.

49. Himeda, Y. *Eur. J. Inorg. Chem.* **2007**, 3927–3941.

50. Drury, D. J.; Hamlin, J. E.US Patent 4,474,959, 1984, to BP Chemicals.

51. Anderson, J. J.; Drury, D. J.; Hamlin, J. E.; Kent, A. G. US Patent 4,855,496, 1989, to BP Chemicals.

52. Taqui Khan, M. M.; Halligudi, S. B.; Shukla, S. *J. Mol. Catal.* **1989**, *57*, 47–60.

53. Jószai, I.; Joó, F. *J. Mol. Catal. A: Chem.* **2004**, *224*, 87–91.

54. Zhang, Z.; Xie, Y.; Li, W.; Hu, S.; Song, J.; Jiang, T.; Han, B. *Angew. Chem.* **2008**, *120*, 1143–1145; *Angew. Chem., Int. Ed.* **2008**, *47*, 1127–1129.

55. Jessop, P. G.; Ikariya, T.; Noyori, R. *Nature* **1994**, *368*, 231–233.

56. Jessop, P. G.; Hsiao, Y.; Ikariya, T.; Noyori, R. *J. Am. Chem. Soc.* **1996**, *118*, 344–355.

57. Ikariya, T.; Jessop, P. G.; Hsiao, Y.; Noyori, R. US Patent 5,763,662, 1998, to Research and Development Corporation of Japan and NKK Corporation, Japan.

58. Kröcher, O.; Köppel, R. A.; Baiker, A. *Chem. Commun.* **1997**, 453–454.

59. Schmid, L.; Rohr, M.; Baiker, A. *Chem. Commun.* **1999**, 2303–2304.

60. Schmid, L.; Canonica, A.; Baiker, A. *Appl. Catal. A: Gen.* **2003**, *255*, 23–33.

61. Darensbourg, D. J.; Ovalles, C. *J. Am. Chem. Soc.* **1987**, *109*, 3330–3336.

62. Darensbourg, D. J.; Kudaroski, R. *Adv. Organomet. Chem.* **1983**, *22*, 129–168.

63. Enthaler, S. *ChemSusChem* **2008**, *1*, 801–804.

64. Joó, F. *ChemSusChem* **2008**, *1*, 805–808.

65. Loges, B.; Boddien, A.; Junge, H.; Beller, M. *Angew. Chem.* *120*, 4026-4029; *Angew. Chem., Int. Ed.* **2008**, *47*, 3962–3965.

66. Boddien, A.; Loges, B.; Junge, H.; Beller, M. *ChemSusChem* **2008**, *1*, 751–758.

67. Fellay, C.; Dyson, P. J.; Laurenczy, G. *Angew. Chem.* **2008**, *120*, 4030–4032; *Angew. Chem., Int. Ed.* **2008**, *47*, 3966–3968.

68. Stalder, C. J.; Chao, S.; Summers, D. P.; Wrighton, M. S. *J. Am. Chem. Soc.* **1983**, *105*, 6318–6320.

69. Wiener, H.; Sasson, Y.; Blum, J. *J. Mol. Catal.* **1986**, *35*, 277–284.

70. Zaidman, B.; Wiener, H.; Sasson, Y. *Int. J. Hydrogen Energy* **1986**, *11*, 341–347.

71. Mandler, D.; Willner, I. *J. Am. Chem. Soc.* **1987**, *109*, 7884–7885.

72. Fukuoka, S.; Kawamura, M.; Komiya, K.; Tojo, M.; Hachiya, H.; Hasegawa, K.; Aminaka, M.; Okamoto, H.; Fukawa, I.; Konno, S. *Green Chem.* **2003**, *5*, 497–507.

73. Choi, J. C.; He, L. N.; Yasuda, H.; Sakakura, T. *Green Chem.* **2002**, *4*, 230–234.

74. Gu, Y.; Shi, F.; Deng, Y. *J. Org. Chem.* **2004**, *69*, 391–394.

75. Wong, W.-L.; Chan, P.-H.; Zhou, Z.-Y.; Lee, K.-H.; Cheung, K.-C.; Wong, K.-Y. *ChemSusChem* **2008**, *1*, 67–70.

76. Zhang, S.; Chen, Y.; Li, F.; Lu, X.; Dai, W.; Mori, R. *Catal. Today* **2006**, *115*, 61–69.

77. He, L.-N.; Yasuda, H.; Sakakura, T. *Green Chem.* **2003**, *5*, 92–94.

78. Song, J.; Zhang, Z.; Han, B.; Hu, S.; Li, W.; Xie, Y. *Green Chem.* **2008**, *10*, 1337–1341.

79. Li, Q.; Zhao, N.; Wei, W.; Sun, Y. *Stud. Surf. Sci. Catal.* **2004**, *153*, 573–576.

80. Inoue, S. *CHEMTECH* **1976**, 588.

81. Sugimoto, H.; Inoue, S. *J. Polym. Sci. A: Polym. Chem.* **2004**, *42*, 5561–5573.

82. Olah, G. A.; Török, B.; Joschek, J. P.; Bucsi, I.; Esteves, P. M.; Rasul, G.; Surya Prakash, G. K. *J. Am. Chem. Soc.* **2002**, *124*, 11379–11391.

83. Kurioka, M.; Nakata, K.; Jinkotu, T.; Taniguchi, Y.; Takaki, K.; Fujiwara, Y. *Chem. Lett.* **1995**, 244.

84. Saito, S.; Nakagawa, S.; Koizumi, T.; Hirayama, K.; Yamamoto, Y. *J. Org. Chem.* **1999**, *64*, 3975–3978.

85. Alvarez, R.; Carmona, E.; Galindo, A.; Gutiérrez, E.; Marin, J. M.; Monge, A.; Poveda, M. L.; Ruiz, C.; Savariaoult, J. M. *Organometallics* **1989**, *8*, 2430–2439.

86. Aresta, M.; Pastore, C.; Giannoccaro, P.; Kovács, G.; Dibenedetto, A.; Pápai, I. *Chem. Eur. J.* **2007**, *13*, 9028–9034.

87. Avar, G. *Kunststoffe Int.* **2008**, 123–127.

88. Abla, M.; Choi, J.-C.; Sakakura, T. *Chem. Commun.* **2001**, 2238–2239.

89. Ion, A.; van Doorslaer, C.; Parvulescu, V.; Jacobs, P.; de Vos, D. *Green Chem.* **2008**, *10*, 111–116.

90. Aresta, M.; Quaranta, E. *Tetrahedron* **1992**, *48*, 1515–1530.

91. Behr, A.; Becker, M. *Dalton Trans.* **2006**, 4607–4613.

92. Behr, A.; Bahke, P.; Klinger, B.; Becker, M. *J. Mol. Catal. A: Chem.* **2007**, *267*, 149–156.

93. Behr, A.; Brehme, V. A. *J. Mol. Catal. A: Chem.* **2002**, *187*, 69–80.

94. Behr, A.; Urschey, M.; Brehme, V. A. *Green Chem.* **2003**, *5*, 198–204.

95. Siddle, A.; Pointon, K. D.; Judd, R. W.; Jones, S. L.; Fuel processing for fuel cells: a status review and assessment of prospects **2003**, Available at http://www.berr.gov.uk/files/file15218.pdf.

96. O'Connor, A. M.; Schuurman, Y.; Ross, J. R. H.; Mirodatos, C. *Catal. Today* **2006**, *115*, 191–198.

97. Yang, M.; Papp, H. *Catal. Today* **2006**, *115*, 199–204.

98. Czernichowski, A.; Wesolowska, K.; Czernichowski, J. US Patent 7,459,594, 2008, to Ceramatec Inc.

99. Cho, W.; Ju, W.-S.; Lee, S.-H.; Baek, Y.-S.; Kim, Y.-C. *Stud. Surf. Sci. Catal.* **2004**, *153*, 205–208.

100. Kalra, C.; Cho, Y.; Gutsol, A.; Fridman, A.; Rufael, T. S. *Prep. Pap. Am. Chem. Soc. Div. Fuel. Chem.* **2004**, *49*, 280–281. Available at http://plasma.mem.drexel.edu/publications/documents/PreprintSyn-gas.pdf.

101. Olah, G. A.; Goeppert, A.; Surya Prakash, G. K. *Beyond Oil and Gas: The Methanol Economy*; Wiley-VCH: Weinheim, 2006.

102. Yang, C.; Ma, Z.; Zhao, N.; Wie, W.; Hu, T.; Sun, Y. *Catal. Today* **2006**, *115*, 222–227.

103. Tominaga, K.; Sasaki, Y.; Watanabe, T.; Saito, M. *Bull. Chem. Soc. Jpn.* **1995**, *68*, 2837–2842.

104. Grant, J. L.; Goswami, K.; Spreer, L. O.; Otvos, J. W.; Calvin, M. *J. Chem. Soc., Dalton Trans.* **1987**, 2105–2109.

105. Kaneco, S.; Iiba, K.; Ohta, K.; Mizuno, T. *Energy Sources* **2000**, *22*, 127–135.

106. Hori, Y.; Murata, A.; Takahashi, R.; Suzuki, S. *J. Chem. Soc., Chem. Commun.* **1998**, 17–19.

107. Tanaka, K.; Ooyama, D. *Coord. Chem. Rev.* **2002**, *226*, 211–218.

108. Chiozza, E.; Desigaud, M.; Greiner, J.; Duñach, E. *Tetrahedron Lett.* **1998**, *39*, 4831–4834.

109. Bringmann, J.; Dinjus, E. *Appl. Organomet. Chem.* **2001**, *15*, 135–140.

110. Tascedda, P.; Weidmann, M.; Dinjus, E.; Duñach, E. *Appl. Organomet. Chem.* **2001**, *15*, 141–144.

111. Willner, I.; Maidan, R.; Mandler, D.; Dürr, H.; Dörr, G.; Zengerle, K. *J. Am. Chem. Soc.* **1987**, *109*, 6080–6086.

112. Miedaner, A.; Curtis, C. J.; Barkley, R. M.; DuBois, D. L. *Inorg. Chem.* **1994**, *33*, 5482–5490.

113. Hawecker, J.; Lehn, J.-M.; Ziessel, R. *J. Chem. Soc., Chem. Commun.* **1983**, 536–538.

114. Kutal, C.; Corbin, A. J.; Ferraudi, G. *Organometallics* **1987**, *6*, 553–557.

115. Tan, S. S.; Zou, L.; Hu, E. *Catal. Today* **2006**, *115*, 269–273.

116. Yu, K. M. K.; Curcic, I.; Gabriel, J.; Tsang, S. C. E. *ChemSusChem* **2008**, *1*, 893–899.

## Note added in proof

A recently published useful review:

Sakakura, T; Kohno, K: The synthesis of organic carbonates from carbon dioxide. *Chem. Commun.* **2009**, 1312–1330.

# 7 Chemistry of Bound Nitrogen Monoxide and Related Redox Species

JOSÉ A. OLABE

## 7.1 INTRODUCTION

Some homo- and heteronuclear small diatomic ($H_2$, $N_2$, $O_2$, CO, and NO) and triatomic ($CO_2$, $N_2O$, and $NO_2$) molecules engage in strong interactions with transition metals, with consequent changes in structure and reactivity. We deal with the chemistry of NO and related redox derivatives, $NO^+$ (nitrosonium) and $NO^-$/ HNO (nitroxyl),[1] as a case study for using state-of-the-art spectroscopic and kinetic methodologies together with modern theoretical calculations in order to visualize the impact of nitrosyl coordination.

The focus on the chemistry of NO evolved from an environmental concern mainly related to gaseous pollutant effects leading to acid rain and ozone layer depletion[2] to the newly raised studies on the biological significance, as evidenced by NO biosynthesis in mammals comprising the oxidation of L-arginine, catalyzed by the heme-based NO synthase (NOS) enzyme. Currently investigated modern topics deal with the mechanisms involved in a variety of roles displayed by NO related to blood pressure control, neuronal communication, cytotoxic effects, and others.[3]

NO is ubiquitously placed as a moderately stable radical molecule. It is an intermediate in the natural redox cycles comprising the interconversion of nitrates to ammonia driven by bacteria containing iron or copper enzymes (namely, $NO_2^-$, NO, or $N_2O$ reductases).[4] Therefore, studies with metalloproteins and with adequately designed model complexes are of permanent interest for disclosing the diverse and complex mechanisms occurring in the biological systems.[5]

The chemistry of free NO, either in gaseous or in solution media, is mainly determined by its radical properties. According to the electronic configuration $(1\sigma_s^b)^2(1\sigma_s^*)^2(2\sigma_s^b)^2(2\sigma_s^*)^2(2\pi_{xy}^b)^4(2\sigma_z^b)^2(2\pi_{xy}^*)^1$, we anticipate a potential redox activity leading to $NO^+$, given the antibonding character of the single upper electron, as well as a reactivity with other radical partners such as $O_2$, $O_2^-$, $NO_2$, or transition

*Physical Inorganic Chemistry: Reactions, Processes, and Applications* Edited by Andreja Bakac
Copyright © 2010 by John Wiley & Sons, Inc.

metals. On the other hand, it also suggests the accessibility of the $^3NO^-$ species, isoelectronic with $^3O_2$. In aqueous solution, both $NO^+$ and $NO^-/HNO$ are very reactive precursors of $NO_2^-$ and $N_2O$, respectively.[1] The differential reactivity of the three nitrosyl diatomic entities has been highlighted in a biorelevant framework.[6] Aqueous NO reductive chemistry was recently reinvestigated by pulse radiolysis and flash photolysis techniques,[7] providing new evidence on the fundamental properties of ground-state (GS) $^3NO^-$ and $^1HNO$, as well as their reactivities with NO to generate the $N_2O_2^-$ radical and the closed-shell $N_3O_3^-$ anion, which unimolecularly decays to the final products, $N_2O$ and $NO_2^-$.[7a] The revised properties of the hyponitrite radicals, $ONNO^-$ and $ONNOH$, the adducts of NO and nitroxyl ($^3NO^-$ and $^1HNO$), are most relevant to the aqueous redox chemistry leading to or originating from NO.[7b] We do not address here these aspects of free NO solution chemistry. Instead, we focus on the changes in structure and reactivity brought out by coordination of NO,[8,9] which (depending on the metal) may conserve its radical character or provide a noticeable stabilization of *formally* considered $NO^+$ and $NO^-/HNO$ ligands.

## 7.2   BONDING IN METAL NITROSYLS: STRUCTURE AND REACTIVITY (THE ENEMARK–FELTHAM FORMALISM)

NO binds covalently to transition metals M forming a diversity of structural types comprising mononuclear, NO bridging, and cluster compounds.[8–10] Within the first class, main coordination numbers are 6, 5, and 4, as sketched in Table 7.1, revealing an important structural result, namely, the adoption of linear or bent geometries of the MNO fragments.

   This situation was tentatively rationalized in the early 1940s by assigning the linear and bent geometries to nitrosyl ligands described as $NO^+$ or $NO^-$, respectively. However, an inspection of the bending angles found in the literature[10,11] shows that *three* rather than two limiting geometries are encountered, within the experimental uncertainties. A widely accepted molecular orbital (MO)-based approach proposed by Enemark and Feltham[11] constitutes an adequate formalism for describing and predicting the main geometrical features of metallonitrosyls. By using the symbolism $\{MNO\}^n$, where $n$ is the electron content comprising the metal d and $\pi_{NO}^*$ orbitals, the overall electronic distribution in the MNO fragment is defined without assumptions on the actual degree of electron density on M and NO. The observation of linear or differentially bent MNO groups can be rationalized in terms of coordination number (CN), number of electrons ($n$), and the nature of the occupied MOs that arise from correlation (Walsh) diagrams based on a one-electron description of the systems.[12] A linear MNO geometry can be predicted for complexes having $n \leq 6$, and increasingly bent structures correspond to the systems with $n = 7$ and 8. Avoided in this formalism is the establishment of extreme oxidation states for the metal and nitrosyl group as well as the consideration of the number and type of coligands L in the considered complex, say $[ML_xNO]$. Indeed, the L ligands *do* influence the electronic density on each of the M, N, and O atoms and exert a control of the reactivity, though this is

**TABLE 7.1    Typical Coordination Geometries of Nitrosyl Complexes**

considered a perturbation on the main features determined by MNO and particularly by the value of $n$.[10–12]

By centering our discussion on pseudo-octahedral NO complexes, both $\sigma$- and $\pi$-interactions between M and NO must be considered. Figure 7.1 shows a simplified diagram of the calculated MO energy levels for $\{MNO\}^n$ complexes, with $n = 6, 7$, and 8.

For the most commonly found linear nitrosyl complexes with $n = 6$, the electrons occupy the predominantly metal orbitals, $e_1$ and $b_2$ (Figure 7.1a). The first strongly bonding orbital results from the mixing of degenerate metal $d_{xz}$, $d_{yz}$, and $\pi^*_{NO}$ orbitals. This "backbonding" interaction depends on M, its charge and electronic configuration, and the donor–acceptor character of the coligands L. Figure 7.1b shows the lower symmetry situation for $n = 7$ and 8, attained with the bent MNO structures, with the bonding $e_1$ and antibonding $e_2$ MOs loosing their degeneracy. We highlight the two mixed antibonding MOs thus generated from $e_2$ of $a'$ symmetry: $(\pi^*_{NO}, d_{z^2})$ and $(\pi^*_{NO}, d_{yz})$. Table 7.2 includes a selection of six-coordinate compounds with $n = 6, 7$, and 8 for which structural information is available.[13]

The main geometrical parameters are included, namely, the M–N and N–O distances and the MNO angle, together with the stretching N–O frequencies $\nu_{NO}$

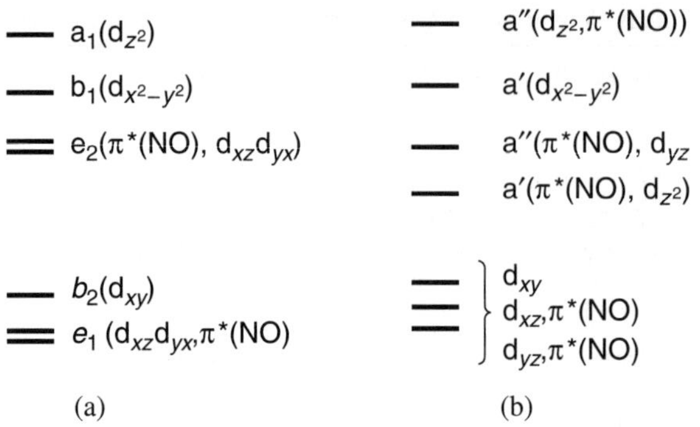

(a)

— $a_1(d_{z^2})$

— $b_1(d_{x^2-y^2})$

= $e_2(\pi^*(NO), d_{xz}d_{yx})$

— $b_2(d_{xy})$

= $e_1(d_{xz}d_{yx},\pi^*(NO))$

(b)

— $a''(d_{z^2},\pi^*(NO))$

— $a'(d_{x^2-y^2})$

— $a''(\pi^*(NO), d_{yz}$

— $a'(\pi^*(NO), d_{z^2})$

= $\left.\begin{array}{l} d_{xy} \\ d_{xz},\pi^*(NO) \\ d_{yz},\pi^*(NO) \end{array}\right\}$

**FIGURE 7.1**  Arrangement of molecular orbital diagram in six-coordinate complexes, in terms of the Enemark–Feltham formalism, $\{MNO\}^n$ (see the text), with M–N–O in (a) linear situation, $n=6$, and (b) angular situation, $n=7$ and 8.

and the total spin state $S$. Thus, a preliminary and comparative insight on the structure and reactivity of nitrosyl complexes can be afforded.

For going a step further in the elucidation of the real electron density in the MNO moieties, complementary physical inorganic techniques are currently employed, namely, the different spectroscopies, such as vibrational (infrared (IR) and Raman), electronic (UV-Vis), electron paramagnetic resonance (EPR), Mössbauer, magnetic circular dichroism (MCD), resonance Raman (RR), X-ray absorption (XAFS), and so on, and theoretical calculations.[12,14] A combined approach is desirable, as the interpretation of the results for a given technique may be controversial or sometimes ambiguous. A simplified view to the electron density problem deals with the use of *limiting* oxidation states for M and NO, based on an extreme ionic model, hiding the covalent nature of the M–N–O interaction. We will critically analyze the validity of these descriptions by considering in an orderly way the situations found for complexes with $n=6$, 7, and 8. Though with an emphasis on six-coordination, we address some specific issues involving five-coordination that are interesting in their own right, or most relevant to bioinorganic chemistry. The quite diverse reactivity modes comprise the coordination (formation and dissociation) of the nitrosyl ligands on the metal ions, electrophilic and nucleophilic additions of different substrates on the MNO fragments, and electron transfer reactions leading to more oxidized or reduced N-containing species.[5,8] Figure 7.2 shows a cyclic voltammetric (CV) experiment with an iron nitrosyl complex able to be reversibly reduced or oxidized, allowing for the electrogeneration of redox states with $n=6$, 7, and 8 for the same $FeL_5$ moiety.[13a]

In addition to the most commonly found end-on M–NO binding mode, we consider the existence of stable linkage isomers generated through reversible photoinduced (and eventually thermal) transformations.[15] Photoreactivity may also lead to the irreversible delivery of NO from $\{MNO\}^n$ complexes.[16]

**TABLE 7.2   Total Spin State ($S$), Nitrosyl Stretching Frequencies $\nu_{NO}$, Relevant Distances $d_{M-N}$ and $d_{N-O}$, and Angles $\angle_{MNO}$ for Selected Nitrosyl Complexes, $[MX_5NO]^x$, Ordered According to the $\{MNO\}^n$ Description ($n = 6$, 7, and 8)$^a$**

| Compound | $S$ | $\nu_{NO}$ (cm$^{-1}$) | $d_{M-N}$ (Å) | $d_{N-O}$ (Å) | $\angle_{MNO}$ (°) | Source |
|---|---|---|---|---|---|---|
| $n = 6$ | | | | | | |
| [Fe(cyclam-ac)NO](PF$_6$)$_2$ | 0 | 1904 | 1.663(4) | 1.132(5) | 175.5(3) | Ref. 13a |
| [Fe(pyS$_4$)NO]PF$_6$ | 0 | 1893 | 1.634(3) | 1.141(3) | 179.5(3) | Ref. 13b |
| [Fe(PaPy$_3$)NO](ClO$_4$)$_2$ | 0 | 1919 | 1.677(2) | 1.139(3) | 173.1(2) | Ref. 13c |
| [Fe(pyN$_4$)NO]Br$_3$$^b$ | | 1926 | 1.67 | 1.15 | 179 | Ref. 13d |
| [Fe(TpivPP)(NO$_2$)NO] | 0 | 1893 | 1.668(2) | 1.132(3) | 180.0 | Ref. 13e |
| | | | | | | |
| Na$_2$[Fe(CN)$_5$NO]·2H$_2$O | 0 | 1945 | 1.6656(7) | 1.1331(10) | 176.03(7) | Ref. 13f |
| Na$_2$[Ru(CN)$_5$NO]·2H$_2$O | 0 | 1926 | 1.776(3) | 1.127(6) | 173.9(5) | Ref. 13g |
| Na$_2$[Os(CN)$_5$NO]·2H$_2$O | 0 | 1897 | 1.774(8) | 1.14(1) | 175.5(7) | Ref. 13h |
| K$_3$[Mn(CN)$_5$NO] | 0 | 1725 | 1.66(1) | 1.21(2) | 174(1) | Ref. 13i |
| (PPh$_4$)$_2$[OsCl$_5$NO] | 0 | 1802 | 1.830(5) | 1.147(4) | 178.5(8) | Ref. 13j |
| K[IrCl$_5$NO] | 0 | 1952 | 1.780(11) | 1.124(17) | 174.3(11) | Ref. 13k |
| | | | | | | |
| $n = 7$ | | | | | | |
| [Fe(cyclam-ac)NO](PF$_6$) | 1/2 | 1615 | 1.722(4) | 1.166(6) | 148.7(4) | Ref. 13a |
| [Fe(pyS$_4$)NO] | 1/2 | 1648 | 1.712(3) | 1.211(7) | 143.8(5) | Ref. 13b |
| [Fe(PaPy$_3$)NO](ClO$_4$) | 1/2 | 1613 | 1.7515(16) | 1.190(2) | 141.29(15) | Ref. 13c |
| [Fe(pyN$_4$)NO]Br$_2$ | 1/2 | 1620 | 1.737(6) | 1.175(8) | 139.4(5) | Ref. 13d |
| K(222)[Fe(TpivPP)(NO$_2$)NO] | 1/2 | 1668 | 1.840(6) | 1.134(8) | 137.4(6) | Ref. 13l |
| Na$_3$[Fe(CN)$_5$NO]·2NH$_3$$^b$ | 1/2 | 1608 | 1.737 | 1.162 | 146.6 | Ref. 13m |
| [Fe(Me$_3$TACN)(N$_3$)$_2$NO] | 3/2 | 1690 | 1.738(5) | 1.142(7) | 155.5(10) | Ref. 13n,o |
| [Fe(L$^{pr}$)(NO)]$^c$ | 1/2,3/2 | 1682 | 1.749(2) | 1.182(3) | 147.0(2) | Ref. 13p |
| | | | | | | |
| $n = 8$ | | | | | | |
| Mb$^{II}$HNO$^d$ | 0 | 1385 | 1.82(2) | 1.24(1) | 131(6) | Ref. 13q |
| [Fe(CN)$_5$HNO]$^{3-}$$^b$ | 0 | 1338 | 1.783 | 1.249 | 137.5 | Ref. 13r |
| [CoCl(en)$_2$NO](ClO$_4$) | 0 | 1611 | 1.820(11) | 1.043(7) | 124.4(11) | Ref. 13s |
| [Ru(py$^{bu}$S$_4$)HNO] | 0 | 1358 | 1.875(7) | 1.242(9) | 130.0(6) | Ref. 13t |
| [OsCl$_2$(CO)(PPh$_3$)$_2$HNO] | 0 | 1410 | 1.915(6) | 1.193(7) | 136.9(6) | Ref. 13u |
| [IrHCl$_2$(PPh$_3$)$_2$HNO] | 0 | 1493 | 1.879(7) | 1.235(11) | 129.8(7) | Ref. 13v |
| K[Pt(NO$_2$)$_4$(H$_2$O)NO]·H$_2$O | 0 | 1655 | 2.10(2) | 1.19 | 129(2) | Ref. 13x |

$^a$Abbreviations used for the ligands: cyclam-ac = monoanion of 1,4,8,11-tetraazacyclotetradecane-1-acetic acid; pyS$_4$ = dianion of 2,6-bis(2-mercaptophenylthiomethyl)pyridine; PaPy$_3$ = monoanion of N,N-bis(2-pyridylmethyl)amine-N-ethyl-2-pyridine-2-carboxamide; pyN$_4$ = 2,6-C$_5$H$_3$N[CMe(CH$_2$NH$_2$)$_2$]$_2$; TpivPP = dianion of $\alpha,\alpha,\alpha,\alpha$-tetrakis(o-pivalamidophenyl)-porphyrin; K(222) = Kryptofix-222; Me$_3$TACN N,N',N''-trimethyl-1,4,7-triazacyclononane; L$^{pr}$ = dianion of 1-isopropyl-4,7-(4-tert-butyl-2-mercaptoben-zyl)-1,4,7-triazacyclononane; Mb = myoglobin; py$^{bu}$S$_4$ = dianion of 2,6-bis(2-mercapto-3,5-di-tert-butyl-phenylthio)dimethylpyridine; PPh$_3$ = triphenylphosphine.
$^b$Distances and angles for the nitrosyl containing moieties are theoretically predicted values.
$^c$The solid contains ca. 50% of each isomer.
$^d$Distances and angles from XAFS spectroscopy.

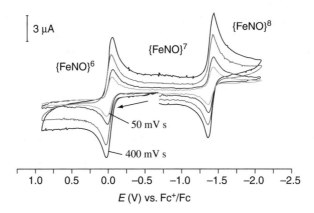

**FIGURE 7.2**    Cyclic voltammogram of [Fe(NO)(cyclam-ac)](PF$_6$) in CH$_3$CN at 20°C (0.1 M [N(n-Bu)$_4$]PF$_6$ supporting electrolyte, glassy C electrode). (See the color version of this figure in Color Plates section.) Ref. 13a

## 7.3  n = 6: LINEAR COMPLEXES WITH DOMINANT ELECTROPHILIC REACTIVITY

### 7.3.1  X-Ray Structure, Magnetism, IR, UV-Vis, and Mössbauer Spectroscopies, and Theoretical Evidence

Complexes of the {MNO}$^6$ class are most commonly found and can be prepared by a direct mixing of NO with Fe(III) precursors[5,8] in a "reductive nitrosylation" reaction (Equation 7.1):

$$[Fe^{III}L_5H_2O]^x + NO \rightleftharpoons [ML_5NO]^x + H_2O \qquad (7.1)$$

For L = CN$^-$, the low-spin Fe(III) complex leads to sodium nitroprusside (SNP) as an *irreversibly* formed product. The proposed mechanism implies a slow associative, reductive step by NO, followed by the fast coordination of NO$^+$ to the [Fe$^{II}$(CN)$_5$H$_2$O]$^{3-}$ intermediate.[17] Faster reversible reactions are onset with high-spin Fe(III) porphyrins, as shown in Section 7.3.2.[5]

Nitrous acid may react with Fe(II) precursors with subsequent proton-assisted dehydration of bound HNO$_2$:

$$[Fe^{II}L_5H_2O]^{(x-1)+} + HNO_2 + H^+ \rightleftharpoons [ML_5NO]^{x+} + 2H_2O \qquad (7.2)$$

Other methods may involve reductive or oxidative processes (namely, using HNO$_3$ and Na$_4$[Fe(CN)$_6$] to obtain SNP, or oxidizing reduced amino ligands to the NO$^+$ state). In nonaqueous media, NOBF$_4$ can be made to react directly with five-coordinate Fe(II) complexes.[8]

The multiple-bonding situation described in Figure 7.1 reflects in comparatively short, strong M–N bonds and a consequent great inertness of the complexes toward nitrosyl dissociation, as evidenced by the lack of kinetic data with *classical* coordination compounds for the release of $NO^+$. The frequent problems caused by disorder limit the accuracy of the relevant lengths and angles, particularly of the N–O distances, which show no significant departure from that seen in NO (1.15 Å) for the different MNO coordination modes.[18] The MNO angles are, as predicted by the Enemark–Feltham formalism,[11] in the range of 170–180°, generally independent of the coligands involved, with the exception of a few complexes with strong N-donor *trans*-L ligands, not included in Table 7.2.[8] A bent structure with an angle of 160° has been found in the model complex $[Fe(OEP)(NO)\{S\text{-}2,6\text{-}(CF_3CONH_2)_2C_6H_3\}]$ (OEP = octaethylporphyrin).[19] The difference between axial N- and S-donor-coordinated ferric heme nitrosyls has been traced to a $\sigma$-*trans* effect of the coordinated S on the bound NO mediated by a NO $\sigma^*$ orbital. Thus, the Fe–N and N–O bonds are also weakened, with $\nu_{NO} = $ ca. $1850 \, cm^{-1}$.[19,20]

IR spectroscopy is a powerful tool for describing the bonding in these complexes, as reflected in the comparatively high $\nu_{NO}$ frequencies usually above $1800 \, cm^{-1}$. Table 7.2 shows that the values of $\nu_{NO}$ for the more reduced complexes with $n = 7$ and 8 are less than $1700 \, cm^{-1}$. Consistent with the generalized EPR-silent behavior, a reasonable limiting structure for the $n = 6$ complexes implies the idealized presence of the $NO^+$ ligand, which means that NO acts as a three-electron donor to the metal. However, the spread of ca. $150 \, cm^{-1}$ in the values of $\nu_{NO}$ must reflect a variable degree of backbonding, which becomes more evident for the stronger $\pi$-donor metals and/or poor electron-acceptor coligands L. The necessary care in interpreting $\nu_{NO}$ can be appreciated with the series of $[Fe(CN)_5NO]^{2-}$, $[Mn(CN)_5NO]^{3-}$, and $[V(CN)_5NO]^{5-}$ complexes, all with $n = 6$, $^1A_1$ ground states, and values of $\nu_{NO}$ at 1939, 1725, and $1575 \, cm^{-1}$, respectively.[21] The large decrease in $\nu_{NO}$ correlates with the increase in backbonding in the ground state, as evidenced by the trends in the population of the $e_1$ level, which acquires more $\pi^*_{NO}$ character in going from Fe (25%) to Mn (42%), to V (73%). Thus, in contrast with the proposed $d^6 M^{II}NO^+$ for the iron complex, a formal electronic configuration $d^2(\pi^*_{NO})^4$ has been suggested for the vanadium complex, with manganese showing an intermediate situation that might be still described as $Mn^INO^+$. Overall, the higher $\nu_{NO}$ values reflect an increasing "nitrosonium" character of the nitrosyl ligand for a series of $d^6$ low-spin $[ML_5NO]^x$ complexes. We could assign some confidence to this approach within the range of $\nu_{NO}$ values of ca. $1800$–$2000 \, cm^{-1}$.

UV-Vis spectroscopy has been a pioneering tool for describing the detailed electronic structure and corresponding optical transitions for $n = 6$ complexes. The $[M(CN)_5NO]^{x-}$ complexes were considered by Manoharan and Gray using semiempirical methods (self-consistent charge and configuration, SCCC), supported by IR (as seen above) and single-crystal polarized absorption UV-Vis measurements.[21] The computed energy levels for SNP include the $\sigma$, $\pi^b$, and $\pi^*$ orbitals for both $CN^-$ and NO. Most relevant is the fact that the $e_2$ MO, derived mainly from $\pi^*_{NO}$, lies mainly between the metal MOs, $b_2(xy)$ and the $b_1(x^2 - y^2)$, $a_1(z^2)$ set (Figure 7.1a). Diamagnetism is consistent with the proposed ground state $(e_1)^4(b_2)^2 = {}^1A_1$. The composition of the highest filled MO, $b_2$, obtained by population analysis, is about 85% $d_{xy}$, 14% $\pi^b_{CN}$, and only 1.6% $\pi^*_{CN}$, revealing a poor backdonation to $CN^-$. In contrast, as anticipated above, the $e_1$ MO is also metal-centered, with ca. 25% $\pi^*_{NO}$, indicating substantial Fe–NO backbonding. The results allow interpreting the available spectral information for SNP. The two lower energy bands correspond to MLCT transitions from $e_1$ and $b_2$ to the vacant $e_2$ MO.[22]

The need of an adequate modeling of the solution environment has been emphasized, related to some controversy in the literature on the work with SNP performed without such a consideration.[13r] Thus, the composition of the HOMO has been alternatively described with a dominant contribution of cyanides, instead of metal. In these studies, the anion has been treated as an isolated species. In fact, even the use of dielectric continuum models may become insufficient, and the need of considering the specific interactions between the bound cyanides and the solvent (or the counter-cation) has been highlighted[23,24] in the context of interpreting the IR and electronic spectral results.[24] Figure 7.3a shows a correlation of $\nu_{NO}$ with the Gutmann's acceptor number (AN) for SNP dissolved in different media, reflecting the strong donor abilities of cyanides, which in turn modulate the electronic density on the $NO^+$ ligand. It can be seen that $\nu_{NO}$ decreases with decreasing AN, reflecting a stronger backbonding from the more electron-rich fragments $Fe^{II}(CN)_5$ to the $NO^+$ ligand.

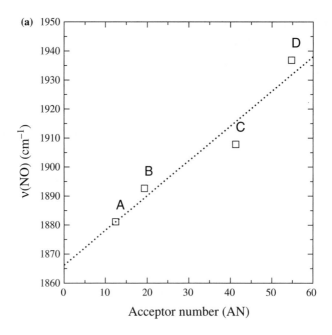

**FIGURE 7.3** Plots of (a) $\nu_{NO}$, IR stretching frequency, and (b) $\nu_{MLCT}$, lowest energy visible band, for $(TBA)_2[Fe(CN)_5NO]$, against solvent acceptor number in the Gutmann's scale. A = acetone; B = acetonitrile; C = methanol; D = water. Ref. 24

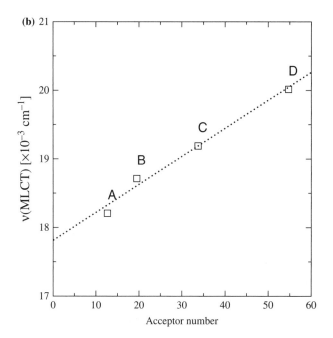

**FIGURE 7.3**   (*Continued*).

Significant energy shifts were also observed for the electronic transitions in the different solvents, as seen in Figure 7.3b.[24]

The modeling of such a type of strong interaction is not a trivial task, and a fair approach for SNP and its reduced derivatives consisted in using a point-charged model mimicking the effect of the solvent. The point charges were located along the axis behind the cyano ligands, with the magnitudes adjusted in order to attain systems that are defined neutral as a whole.[13r,21,22] These donor–acceptor interactions are also established with ammine complexes.[23]

A bonding picture similar to the one in SNP has been described for other iron complexes by using DFT calculations, considering the solvent effects through a continuum model approach (namely, $[Fe(cyclam-ac)NO]^{2+}$, $[Fe(pyN_4)(NO)]^{3+}$, and $[Fe(pyS_4)(NO)]^{+}$, cf. Table 7.2).[13] By working with some nonheme analogues of nitrile hydratase (NHase), a NO-binding Fe(III) enzyme, the chemical bonding at the BLYP-optimized ground state was evaluated by natural bond orbital (NBO) analysis of the appropriate density matrix obtained from B3LYP single point calculations.[25] For three different complexes affording either carboxamido- or thiolate-containing coligands, the $Fe^{II}NO^{+}$ contribution to the ground state was found dominant, with one of them showing some minor contribution of the limiting $Fe^{III}NO^{\bullet}$ structure, based on the calculation of partial spin density on $\pi_{NO}^{*}$. Time-dependent DFT (TDDFT) calculations have been performed with a series of *trans*-$[Ru(NH_3)_4(L)NO]^{x+}$ complexes, where L = $NH_3$, $H_2O$, pyrazine (pz), pyridine (py) ($x = 3$), $Cl^{-}$, and $OH^{-}$ ($x = 2$).[26] The nature of the HOMO depends on the coligands.

With $L = NH_3$, $H_2O$, $Cl^-$, and $OH^-$, a metal-centered situation was found, in contrast with the other complexes showing the filled frontier orbitals mostly located on pz and py, probably because the strong $\pi$-interaction in Ru(II)–NO$^+$ lowers enough the energy of the metal orbitals. Similarly, the [Ru(bpy)(tpm)NO]$^{3+}$ complex shows the HOMO strongly localized on the bpy ligand.[27] In contrast with the above descriptions for the HOMO, the LUMO becomes predominantly $\pi^*_{NO}$ in all of the nitrosyl complexes calculated so far, although with significant metal d participation.

Mössbauer spectroscopy has been used for describing the electron density in complexes of the $n = 6$ series.[13a,c,d,e,l,p,28] Small isomer shifts ($\delta$), ca. 0 mm/s, and large quadrupole splitting ($\Delta E_Q$) parameters were measured for a number of complexes. The values are detailed in Table 7.3 and have been used to support the limiting Fe$^{II}$NO$^+$ description,[13p] rejecting the alternative formulation as Fe(IV) ($S = 1$) containing a NO$^-$ ($S = 1$) ligand.[28]

### 7.3.2    Bonding and Dissociation Reactions in Nitrosyl Porphyrins: How is NO Released from the "Ferri-Hemes"?

The reactions of NO with the high-spin "ferri-hemes" (Equation 7.3), either in model porphyrin complexes or in proteins (with $H_2O$ being eventually replaced by other ligands such as histidine or a thiolate),[5] show some similarities with the classical complexes discussed in Section 7.3.1 (cf. Equation 7.1).

$$[Fe^{III}(por)(H_2O)_2]^x + NO \rightleftharpoons [Fe(por)(H_2O)NO]^x + H_2O \quad k_{on}, k_{off} \quad (7.3)$$

The products are EPR-silent, with essentially linear FeNO arrangements for both five- and six-coordinate nitrosyl compounds and values of $\nu_{NO}$ around 1900 cm$^{-1}$. These features are characteristic of the Fe$^{II}$NO$^+$ limiting description.

However, with all the *classical* complexes studied so far, reaction 7.1 appears to be irreversible, as expected for a strongly bound NO$^+$ ligand. This is in contrast with Equation 7.3 showing *reversibility* for a great variety of model metalloporphyrins and proteins, with values of the corresponding rate constants, $k_{on}$ and $k_{off}$, that have been determined by stopped-flow and flash photolysis techniques.[5] Values of $k_{off}$ are in the range 1–50 s$^{-1}$, which may be considered a striking high reactivity. The facile release of NO from a Fe(III)–heme arrangement is a crucial step in the mechanistic chemistry of the NO$_2^-$ reductase enzymes,[4] and also appears to determine the physiological activity of the nitrophorins (NPs), which are NO-carrying Fe(III)–heme proteins found in the saliva of blood-sucking insects, which favor the dissociation of NO upon injection of the saliva into the tissues of the victim, a dilution process associated with an increase in pH from about 5.6 to 7.4.[29]

We do not consider here the detailed mechanistic aspects of reactions such as 7.3, which have been much considered[5,30] and are currently under revision on the basis of using modern spectroscopic and kinetic methodologies (by the way, pH changes have been found to strongly influence NO reactivity through the potential accessible deprotonation of the aqua ligand).[31] Instead, we focus on some arguments favoring

TABLE 7.3  Zero-Field Mössbauer Parameters of {FeNO}ⁿ Complexes with $n = 6$ and 7

| | $n = 6, S = 0$ | | | | $n = 7, S = 1/2$ | | | Source |
|---|---|---|---|---|---|---|---|---|
| | $\delta$ (mm/s)[a] | $\Delta E$ (mm/s)[b] | $T$ | | $\delta$ (mm/s)[a] | $\Delta E$ (mm/s)[b] | $T$ | |
| [Fe(cyclam-ac)NO]²⁺ | 0.01 | +1.76 | 80 | [Fe(NO)(cyclam-ac)]⁺ | 0.26 | +0.74 | 80 | Ref. 13a |
| [FeCl(cyclam)NO]²⁺ | 0.04 | +2.05 | 4.2 | [Fe(cyclam)(NO)Cl]⁺ | 0.27 | +1.26 | 4.2 | Ref. 28 |
| [Fe(NO)(pyS₄)]⁺ | 0.04 | −1.63 | 4.2 | [Fe(NO)(pyS₄)]⁰ | 0.33 | −0.40 | 4.2 | Ref. 13p |
| [Fe(NO)(pyN₄)]³⁺ | 0.04 | +1.84 | 77 | [Fe(NO)(pyN₄)]²⁺ | 0.31 | +0.84 | 77 | Ref. 13d |
| [Fe(PaPy₃)NO]²⁺ | −0.05 | +0.85 | 80 | [Fe(PaPy₃)(NO)]⁺ | 0.18 | 0.66 | 80 | Ref. 13c |
| [Fe((NO)(TpivPP)(NO₂)] | 0.09 | 1.43 | 4.2 | [Fe(TpivPP)(NO₂)(NO)]⁻ | 0.35 | 1.20 | 4.2 | Ref. 13e, 1 |
| Na₂[Fe(CN)₅NO]·2H₂O[c] | 0.00 | +1.72 | 77 | | | | | Ref. 75 |

[a]Isomer shift versus $\alpha$-Fe at 298 K. See Table 7.2 for the identification of ligands.
[b]Quadrupole splitting. The sign, where explicitly indicated, was established from applied field Mössbauer spectra.
[c]Values of $\delta$ and $\Delta E$ for MS1: 0.18 and +2.75; for MS2: 0.20 and +2.85, respectively.[75]

either one or the other limiting electron configurations, $Fe^{II}NO^+$ or $Fe^{III}NO^\bullet$, for describing the product in reaction 7.3. Although this question has been defined as semantic,[4b] given the accepted covalent nature of MNO, we believe that it merits some comment, as a reasonable explanation for the lack of structure–reactivity correlations is still absent.

A peculiar feature of the NP–NO complexes is the strong ruffling of the heme (with alternate *meso*-carbons shifted significantly above and below the mean plane of the porphyrin ring, to produce a very nonplanar porphyrin macrocycle). This has been analyzed by F. Ann Walker in connection with a probable electron configuration for NPs, namely, $(d_{xz}, d_{yz})^4(d_{xy})^1$.[29] The ruffling would stabilize the latter configuration of low-spin Fe(III), with the $d_{xy}$ orbital unable to overlap with the half-filled $\pi^*$ orbital of NO because of unfavorable symmetry. This predominant $Fe^{III}NO^\bullet$ electron distribution would facilitate reversible dissociation of NO. In this context, the Mössbauer results with these "diamagnetic" species apparently do not allow distinguishing between the truly paired situation, as in $Fe^{II}NO^+$ (suggested by IR, high $\nu_{NO}$), and the strongly antiferromagnetically coupled $Fe^{III}(d_{xy})$–$NO^\bullet$ configuration.

The electronic structure of $[Fe(TPP)(1\text{-MeIm})(NO)]BF_4$ has been studied using nuclear resonance vibrational spectroscopy (NRVS), coupled to normal coordinate analysis and DFT calculations.[32] The results support the $Fe^{II}NO^+$ GS description, with strong Fe–N and N–O bonds. Quite interestingly, a low-spin $Fe^{III}NO^\bullet$ ($S = 0$) state exists as a stable minimum at a surprisingly low energy, only 1–3 kcal/mol above the $Fe^{II}NO^+$ GS. In addition, the $Fe^{II}NO^+$ potential energy surface (PES) crosses the low-spin $Fe^{III}NO^\bullet$ energy surface at a very small elongation (only 0.05–0.1 A) of the Fe–NO bond from the equilibrium distance. This implies that ferric heme nitrosyls with the latter GS might exist upon small steric or electronic perturbations that may occur in protein active sites, as for axial cysteinate coordination, known to stabilize Fe(III). The properties of both GSs are very different; for $Fe^{III}NO^\bullet$, the Fe–NO and N–O bonds are distinctively weaker. The PES calculations further reveal that the thermodynamic weakness of the Fe–NO bond in ferric heme nitrosyls is an intrinsic feature that relates to the properties of an additional *high-spin* $Fe^{III}NO$ ($S = 2$) state that appears at low energy and is dissociative with respect to the Fe–NO bond. Altogether, release of NO from six-coordinated ferric heme nitrosyls requires the system to pass through at least *three* different electronic states, a process that is remarkably complex and unprecedented for transition metal nitrosyls. Certainly, it provides a support to the fast NO release found in related heme nitrosyl systems as NPs.

Alternatively, the NO release mechanism of NP4 has been discussed using state-of-the-art computer techniques, namely, molecular dynamics (MD) simulations to disclose NP4 conformational changes at pH 5.6 and 7.4.[33] The free energy profile for NO release has been computed and hybrid quantum mechanical/molecular mechanics (QM/MM) used to analyze the heme–NO structure and the Fe–NO bond strength in the different NP4 conformations. The results provide the molecular basis to explain that NO escape from NP4 is determined by differential NO migration rates and not by a difference in the Fe–NO bond strength. In contrast to most heme proteins that control ligand affinity by modulating the bond strength to the iron, NP4 has evolved a cage mechanism that traps the NO at low pH and releases it upon cage opening when the pH rises. This model implicitly discards any influence of pH on the Fe–NO dissociation reactivity arising in acid–base changes in the first or near coordination sphere (namely, $OH^-/H_2O$ versus histidine).[31]

### 7.3.3 Nucleophilic Addition Reactions of $NO^+$ Complexes: Kinetic and Computational Studies with $OH^-$

In contrast with the substitution inert character of the M–NO bond, the assignment of bound $NO^+$ as an electrophilic moiety predicts that the delocalized LUMO in MNO may be the site of attack for a variety of nucleophiles, as described by Equation 7.4:

$$[ML_5NO]^x + B \rightleftharpoons [ML_5N(O)B]^x \qquad (7.4)$$

The stoichiometry of reaction 7.4 and the subsequent reactivity of the adduct intermediates have been studied for a vast amount of metals (mainly ruthenium) and L coligands ($CN^-$, $NH_3$, $Cl^-$, polypyridines, EDTA, etc.), with different B nucleophiles such as $OH^-$, S-binding species ($SR^-$, $SH^-$, and $SO_3^{2-}$), N-binding bases ($NH_3$, amines, $NH_2OH$, $HN_3$, and $N_2H_4$), and others.[8,34]

A detailed kinetic and mechanistic study has been done for the reaction of $OH^-$ with SNP[35] (Equation 7.5):

$$[Fe(CN)_5NO]^{2-} + 2OH^- \rightleftharpoons [Fe(CN)_5NO_2]^{4-} + H_2O \qquad (7.5)$$

The studies have been extended to a great variety of $[ML_5NO]^x$ complexes.[27,36] As an example, Figure 7.4 shows the successive spectra for the absorbance increase of the nitro product at 288 nm, starting from $[Ru(bpy)(tpm)(NO)]^{3+}$.[27] Table 7.4 displays the relevant kinetic parameters, together with values of $\nu_{NO}$ and $E_{NO^+/NO}$, for a selected group of nitrosyl complexes all showing the same global stoichiometry.

The dependence of the pseudo-first-order rate constants $k_{obs}$ on the concentration of $OH^-$ can be appreciated in Figure 7.5, and is represented by

$$k_{obs} = a[OH^-] + b/[OH^-] \qquad (7.6)$$

An expression of the same form has been derived for $k_{obs}$ by using a steady-state treatment, assuming the following general mechanistic scheme:

$$[ML_5NO]^x + OH^- \rightleftharpoons \{[ML_5NO]^x \cdot OH^-\} \quad K_{ip} \qquad (7.7)$$

$$\{[ML_5NO]^x \cdot OH^-\} \rightleftharpoons [ML_5NO_2H]^{x-1} \quad K_8, k_8, k_{-8} \qquad (7.8)$$

$$[ML_5NO_2H]^{x-1} + OH^- \rightleftharpoons [ML_5NO_2]^{x-2} + H_2O \quad K_9 \qquad (7.9)$$

Reaction 7.7 comprises a fast association preequilibrium, prior to the relevant nucleophilic addition step (Equation 7.8), which leads to the nitrous acid adduct

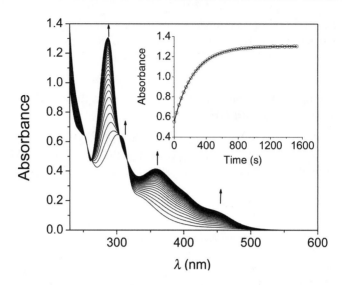

**FIGURE 7.4**  Successive spectra for the reaction of $5.4 \times 10^{-5}$ [Ru(bpy)(tpm)NO]$^{3+}$ with OH$^-$: $I = 1$ M, $T = 25°$C, and [OH$^-$] $= 2.2 \times 10^{-9}$ M. *Inset*: Absorption increase of product with time at 288 nm. Ref. 27

intermediate. The latter may go back to the reactants or react as in Equation 7.9 to form the final product. The values of $a$ and $b$ in Equation 7.6 can be traced to $a = k_{OH}$ and $b = k_{OH}/K_{eq}$, with $k_{OH} = K_{ip}k_9$ and $K_{eq} = K_{ip}k_8k_9$. Thus, values of $k_{OH}$ (M$^{-1}$ s$^{-1}$) and $K_{eq}$ (M$^{-2}$) can be obtained through an adequate fitting of Equation 7.6. Then, the value of $k_8$ (s$^{-1}$) can be calculated using estimated values of $K_{ip}$ (Table 7.4). As the influence of the $k_{OH}/K_{eq}$[OH$^-$] term becomes negligible when [OH$^-$] and $K_{eq}$ are sufficiently high, $k_{OH}$ may be obtained from the linear plot of $k_{obs}$ against [OH$^-$] under these conditions.

Figure 7.6 shows a plot of $\ln k_{OH}$ against $E_{NO^+/NO}$. A very good correlation is obtained ($r^2 = 0.993$) for most of the complexes studied, with the exception of those corresponding to the *trans*-[Ru(py)$_4$(L)NO]$^{x+}$ series, which lie in a parallel line, showing lower rates than expected, probably because of steric hindrance.

The slope of the main line is 20.2 V$^{-1}$. Remarkably, the correlation spans about 10 orders of magnitude in the values of $k_{OH}$, covering a range of around 1 V in the redox potentials. Figure 7.6 is a linear free energy relation (LFER), as frequently found in the correlation of kinetic versus thermodynamic parameters for a set of reactions governed by the same mechanism.[37] The value of the slope is close to the one predicted for LFERs in weakly coupled outer-sphere, one-electron transfer reactions (19.4 V$^{-1}$ or $0.5/RT$), following Marcus' treatment for cross-reactions. Marcus extended the theory to inner-sphere processes, with the prediction that the same slope could also be found in the case of substitution reactions proceeding through an associative mechanism.[38]

Table 7.4 shows that the increase in rate constants and redox potentials goes in parallel with an *increase* in *both* the activation enthalpies and entropies. While the

**TABLE 7.4** Addition Rate Constants, Activation Parameters, and Corresponding $\nu_{NO}$, $E_{NO+/NO}$, and $K_{eq}$ Values for Different $\{MX_5NO\}^n$ Complexes[a]

| Compound | $k_{OH}$ ($M^{-1} s^{-1}$)[b] | $k_8$ ($s^{-1}$)[c] | $\Delta H^{\#}$ (kJ/mol) | $\Delta S^{\#}$ (J/(K mol)) | $E_{NO+/NO}$ (V) | $\nu_{NO}$ ($cm^{-1}$) | $K_{eq}$ ($M^{-2}$)[d] |
|---|---|---|---|---|---|---|---|
| (1) cis-[Ru(AcN)(bpy)$_2$NO]$^{3+}$ | $5.60 \times 10^6$ | $2.31 \times 10^6$ | $89 \pm 1$ | $159 \pm 5$ | 0.35 | 1960 | $2.1 \times 10^{23}$ |
| (2) cis-[Ru(bpy)(trpy)NO]$^{3+}$ | $3.17 \times 10^5$ | $1.31 \times 10^5$ | $83 \pm 7$ | $120 \pm 20$ | 0.25 | 1946 | |
| (3) cis-[Ru(bpy)$_2$(NO$_2$)NO]$^{2+}$ | $5.06 \times 10^4$ | $2.75 \times 10^4$ | $100 \pm 3$ | $164 \pm 8$ | 0.18 | 1942 | |
| (4) cis-[Ru(bpy)$_2$ClNO]$^{2+}$ | $8.5 \times 10^3$ | $4.6 \times 10^3$ | $91 \pm 4$ | $135 \pm 10$ | 0.05 | 1933 | $1.6 \times 10^{15}$ |
| (5) trans-[NCRu(py)$_4$CNRu(py)$_4$NO]$^{3+}$ | $9.2 \times 10^3$ | $3.4 \times 10^3$ | $62 \pm 1$ | $-6 \pm 5$ | 0.22 | 1917 | $3.2 \times 10^{15}$ |
| (6) trans-[RuClNO(py)$_4$]$^{2+}$ | $4.6 \times 10^1$ | $3.1 \times 10^1$ | | | 0.09 | 1910 | |
| (7) trans-[Ru(NCS)NO(py)$_4$]$^{2+}$ | $2.03 \times 10^2$ | $1.36 \times 10^2$ | | | 0.12 | 1902 | |
| (8) trans-[RuNO(OH)(py)$_4$]$^{2+}$ | $2.4 \times 10^{-1}$ | $1.6 \times 10^{-1}$ | | | $-0.22$ | 1866 | |
| (9) trans-[Ru(NH$_3$)$_4$NO(pz)]$^{3+}$ | $1.77 \times 10^2$ | $9.55 \times 10^2$ | $76 \pm 2$ | $54 \pm 6$ | $-0.11$ | 1942 | $6.0 \times 10^8$ |
| (10) trans-[Ru(NH$_3$)$_4$(mic)NO]$^{3+}$ | $3.3 \times 10^1$ | $1.8 \times 10^1$ | $78 \pm 1$ | $44 \pm 4$ | $-0.18$ | 1940 | $5.9 \times 10^7$ |
| (11) trans-[Ru(Clpy)(NH$_3$)$_4$NO]$^{3+}$ | $2.60 \times 10^1$ | $1.40 \times 10^1$ | | | $-0.19$ | 1927 | $6.0 \times 10^6$ |
| (12) trans-[Ru(NH$_3$)$_4$NO(py)]$^{3+}$ | $1.45 \times 10^1$ | $7.8 \times 2 \times 10^0$ | | | $-0.22$ | 1931 | $2.2 \times 10^5$ |
| (13) trans-[Ru(4-Mepy)(NH$_3$)$_4$NO]$^{3+}$ | $9.54 \times 10^0$ | $5.14 \times 10^0$ | $75 \pm 1$ | $26 \pm 4$ | $-0.25$ | 1934 | $7.7 \times 10^5$ |
| (14) trans-[Ru(hist)(NH$_3$)$_4$NO]$^{3+}$ | $7.6 \times 10^{-1}$ | $4.12 \times 10^{-1}$ | | | $-0.39$ | 1921 | $4.6 \times 10^{13}$ |
| (15) [Fe(CN)$_5$NO]$^{2-}$ | $5.5 \times 10^{-1}$ | $3.9 \times 10^0$ | 53 | $-49$ | $-0.29$ | 1945 | $1.5 \times 10^5$ |
| (16) [Ru(CN)$_5$NO]$^{2-}$ | $9.5 \times 10^{-1}$ | $6.4 \times 10^0$ | 57 | $-54$ | $-0.35$ | 1926 | $4.4 \times 10^6$ |
| (17) [Os(CN)$_5$NO]$^{2-}$ | $1.37 \times 10^{-4}$ | $8.63 \times 10^{-4}$ | 80 | $-73$ | $-0.68$ | 1897 | $4.2 \times 10^1$ |

[a] Adapted from Ref. 36.
[b] Derived from the rate law.
[c] Obtained through $k_8 = k_{OH}/K_{ip}$, with $K_{ip}$ estimated according to an electrostatic model.
[d] Values obtained from the literature.

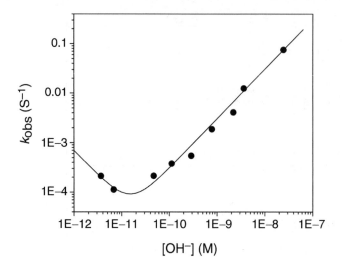

**FIGURE 7.5**  Dependence of $k_{obs}$ on [OH$^-$] for the reaction of $(2.5 - 5.5) \times 10^{-5}$ M [Ru(bpy)(tpm)NO]$^{3+}$ with OH$^-$: $T = 25°$C and $I = 1$ M. Ref. 27

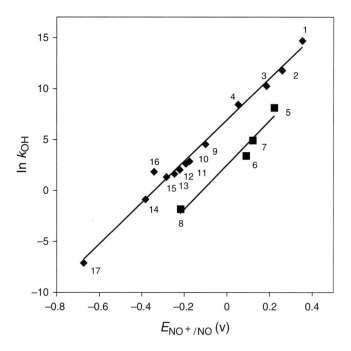

**FIGURE 7.6**  Plot of $\ln k_{OH}$ against $E_{NO^+/NO}$ for the reactions of a series of [MX$_5$NO]$^n$ complexes with OH$^-$. See Table 7.4 for the assignment of numbers.

trends in the entropies can be reasonably explained considering the different changes in solvation, related to the reactions of OH$^-$ with complexes carrying equal or opposite charges, the consideration of enthalpy changes is not so straightforward. It has been proposed that the rate of addition in Equation 7.8 is controlled mainly by the energetically costly steps involving the reorganization of the linear MNO moiety to angular M–NO$_2$H, as seen below.[36]

### 7.3.3.1 Influence of the Coligands on Reactivity

Although the UV-Vis spectral results account for the reversible formation of $[ML_5NO_2]^{(n-2)}$, direct evidence on the proposed $[ML_5NO_2H]^{(n-1)}$ intermediates (Equation 7.8) is not generally available.[8,34] This is a frequently found situation in the studies of electrophilic reactions of bound nitrosyl, where the claimed adduct intermediates react usually very fast. Quantum chemical calculations become helpful for the characterization and stability analysis of the adduct intermediates. By choosing a representative set of $\{ML_5NO\}^x$ complexes, associated with different values of $\ln k_{OH}$, the attention focused on reactants and products of reactions 7.7 and 7.8. Geometry optimization shows that the reactants and products are true minima in the potential hypersurface, with no negative components in the calculated Hessian. The optimized geometries are shown in Figure 7.7 for three representative examples.

Different basis sets have been used throughout the calculations, from the simplest 3-21G to those based on pseudopotentials for the metal centers. The geometrical parameters for the groups more strongly involved in the reaction (MNO, MNO$_2$H) have been provided, and charges on the electrophilic centers have been calculated. The increase in $k_{OH}$ correlates with the decrease in the NO distance and a simultaneous increase in the M–N one. The charge computations deal with the consideration of the N atom, the NO group, or the MNO group for an accurate definition of the electronic characteristics of the electrophilic center. Better correlations have been obtained by defining a group charge $q_{MNO}$, which correlates very well with $\ln k_{OH}$. This is a clear demonstration that the major influence on the rates of the electrophilic reactions is exerted by the ancillary coligands on the delocalized MNO, a result that should be emphasized, despite being a reasonably predictable one. Backdonation to the antibonding MNO orbital determines the N charge and NO interatomic distance. Given that this effect is smaller for the coligands more characterized as acceptors, both the electron density and the NO distance decrease when going from the cyano to the polypyridine coligands.[36]

It is worth mentioning that $\Delta E$ for reaction 7.8 ($\Delta E$ = energy of the products – energy of the reactants) also correlates with the observed trends in $\ln k_{OH}$. An explanation follows from the fact that larger $\ln k_{OH}$ values also correlate with lower energy values of the calculated LUMOs. This indicates that the addition product of the reaction, far from being only a stable intermediate, is in several cases more stable than the reactants. As the addition can be formally described as a nucleophilic attack of OH$^-$ to the LUMO of linear nitrosyl, occupation of the degenerate $e_2$ orbital (mainly $\pi^*_{NO}$) splits the energy, lowering the symmetry from $C_{4v}$ to $C_s$. Bond formation is associated with the interaction of the p orbital of OH$^-$, mainly centered on the O atom, with the $a'$ orbital, stabilized after splitting (Figure 7.1b). Thus, the NO distance

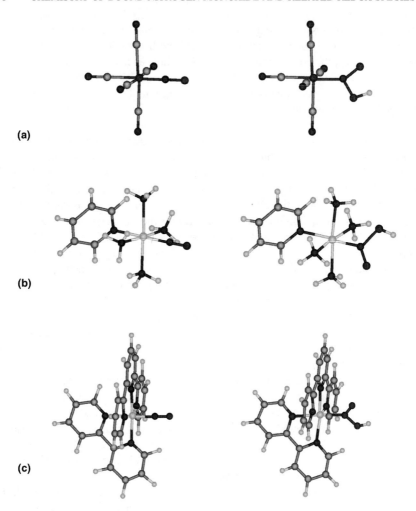

**FIGURE 7.7**    Geometries optimized at the B3LYP/6-31G$^{**}$ level (SDD pseudopotentials on the metal centers) for representative members of the set (metallonitrosyls and OH$^-$ addition products): (a) [Fe(CN)$_5$NO]$^{2-}$; (b) *trans*-[Ru(NH$_3$)$_4$NO(py)]$^{3+}$; (c) *cis*-[Ru(bpy)(trpy)NO]$^{3+}$. (See the color version of this figure in Color Plates section.) Ref. 36

increases for all the members of the series. In this process, a lower energy of the LUMO favors the stabilization of OH$^-$, an effect that is reflected in larger negative $\Delta E$ values. The energy of the LUMO then becomes an important determinant of the energy involved in this process. The variation in the ONO and MNO planar angles along the members of the series shows that geometry closest to that imposed by a sp$^2$ hybridization of the N atom is attained for the stronger stabilization of the OH$^-$ addition product.[36] The linear $\rightarrow$ bent reorganization corresponds to a *formal* reduction, namely, the conversion of an {MNO}$^6$ species into an {MNO}$^8$ one, in the Enemark–Feltham notation.[11]

The comparison of the structural parameters calculated with different basis sets shows that both LANL2DZ and SDD give very similar results. The overestimation of the NO distance has been previously analyzed by Gorelsky and Lever[39a] using a similar calculation level, and also by Boulet et al.[39b] and Wanner et al. using the ADF program.[40] According to previous data, it seems that triple-$\zeta$ basis sets including diffusion and polarization functions are sometimes necessary for a closer approximation to the experimental $d_{NO}$ values.

### 7.3.3.2 Calculated Reaction Mechanism for SNP

The complete reaction profile has been analyzed for SNP including polarization functions in the basis. Structural parameters and spectroscopic data are given in Table 7.5 for the different steps of the reaction. The effect of the basis reflects in the NO distance, which decreases when polarization functions are considered in the calculations. Better values are obtained when pseudopotentials are used for the Fe atom. The overall increase of the bond lengths for $[Fe(CN)_5NO_2H]^{3-}$, compared with SNP, reflects the increase in the electron population in the delocalized antibonding system. The energy of the $OH^-$ addition product is calculated 0.12 au higher than the energy of the reactants. The energy of the TS is also higher, by 0.123 au, and very close to the product of reaction 7.8 (see Figure 7.8).

**TABLE 7.5   Selected Distances ($d$, Å), Angles ($\triangle$, °), and Stretching IR Frequencies ($\nu$, $cm^{-1}$) Calculated for the Different Steps of Reaction 7.5 (Including Intermediate $[XNO_2H]^{3-}$) at the B3LYP/6-31G** Level**

|  | $[XNO]^{2-}(exp)^a$ | $[XNO]^{2-}$ | TS[b] | $[XNO_2H]^{3-}$ | $[XNO_2]^{4-}$ |
|---|---|---|---|---|---|
| $d_{FeC}$ ax | 1.9257(9) | 1.9694 | 1.9878 | 1.9888 | 1.987 |
| $d_{FeC}$ eq[c] | 1.935 | 1.9595 | 1.9890 | 1.9880 | 2.011 |
| $d_{FeN}$ | 1.6656(7) | 1.6155 | 1.8223 | 1.813 | 2.104 |
| $d_{CN}$ ax | 1.1591(12) | 1.1683 | 1.1751 | 1.1755 | 1.184 |
| $d_{CN}$ eq[c] | 1.1613 | 1.1691 | 1.1761 | 1.1782 | 1.1826 |
| $d_{NO}$ | 1.1331(10) | 1.1604 | 1.2255 | 1.2275 | 1.2642 |
| $d_{NO}(H)$ |  |  | 1.4536 | 1.4713 | 1.2642 |
| $d_{OH}$ |  |  | 0.9801 | 0.9784 |  |
| $\triangle FeNO$ | 176.03(7) | 179.96 | 134.69 | 133.11 | 122.48 |
| $\triangle ONO$ |  |  | 109.25 | 107.98 | 115.02 |
| $\triangle CFeC$ eq[d] | 176.63(4) | 180.00 |  | 173.5 | 179.83 |
| $\nu_{CN}$ | 2147–2177 | 2161–2170 |  | 2100–2120 | 2043 |
| $\nu_{NO}$ | 1943 | 1907 |  | 1567, 1266, 789 | 1317, 1351, 802 |
| $\nu_{FeN}$ | 658 | 712 |  | 575 | 574 |

Experimental values are given when available ($X = [Fe(CN)_5]$).
[a] Adapted from Ref. 36.
[b] Transition state.
[c] The fourth digit averaged.
[d] Backward the $NO_2H$ group.

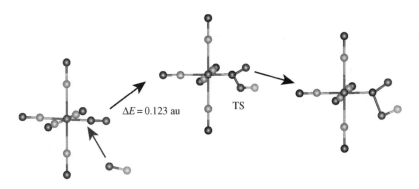

**FIGURE 7.8** Optimized geometries for the initial steps of the reaction of $[Fe(CN)_5NO]^{2-}$ with $OH^-$. More details are available in the text. Ref. 36

The main structural difference between TS and product deals with the value of the FeNOH torsion angle, $164.35°$ in the TS versus $179.99°$ in the addition product [Fe $(CN)_5NO_2H]^{3-}$. Accordingly, the energy involved in the evolution from TS to the intermediate adduct is mainly associated with the change in hybridization of the O atom, from $sp^3$ in the $OH^-$ ion to $sp^2$ in the coordinated adduct. The energy cost when going from reactants to the TS is largely associated with the electronic reorganization of the N atom. Table 7.5 shows some selected frequencies for the calculated species, with values for SNP that agree with those calculated using pseudopotentials for Fe. Although underestimated relative to our experimental data, they show coincidence with previous experiments, and also with previous theoretical results. For $[M(CN)_5NO_2]^{4-}$, the frequency values at 1351, 1317, and $802 \, cm^{-1}$ correspond to the stretchings and deformations of the nitro group in related complexes. The value of $\nu_{CN}$ at $2043 \, cm^{-1}$ is typical of M(II) pentacyano complexes with moderately $\pi$-accepting sixth ligands, in contrast with the high values of $\nu_{CN}$ in SNP, at ca. $2160 \, cm^{-1}$. The $\nu_{Fe-N}$ stretching has a lower value than that in SNP, as expected for a weaker Fe–N bond. An analysis of the frequencies for the $[M(CN)_5NO_2H]^{3-}$ intermediate supports the protonation of the nitro group. A similar frequency pattern has been calculated for the other adducts, using pseudopotentials on the M centers.[39a]

Electronic transitions have been calculated within a TDDFT approach. The results are remarkably good when dealing with excited states of valence type but may be affected by incorrect asymptotic behavior of the potential when excitations involve transitions to unbound orbitals, as is the case of the analyzed anions. However, the results obtained for SNP are in good agreement with the experimental data, increasing the confidence in the results for the other species, which are, at present, a prediction. Calculations of MLCT transitions for SNP are in good agreement with experiment,[22] confirming the assignment of the lowest excitation to a transition between $b_2$, $d_{xy}$, and $e_2$ MOs. Lower energy occupied orbitals are involved in the higher energy transitions. The calculations also support reasonable assignments for the $[M(CN)_5NO_2H]^{3-}$ and $[M(CN)_5NO_2]^{4-}$ ions.[36]

### 7.3.4    Reactions of SNP with N-Binding Nucleophiles: the Case of $N_2H_4$

Small nitrogenated molecules are active nucleophiles toward bound $NO^+$, as shown by the following reaction stoichiometries:[8,34]

$$[Fe(CN)_5NO]^{2-} + NH_3 + OH^- \rightarrow [Fe(CN)_5H_2O]^{3-} + N_2 + H_2O \qquad (7.10)$$

$$[Fe(CN)_5NO]^{2-} + NH_2R + OH^- \rightarrow [Fe(CN)_5H_2O]^{3-} + N_2 + ROH \qquad (7.11)$$

$$[Fe(CN)_5NO]^{2-} + NH_2OH + OH^- \rightarrow [Fe(CN)_5H_2O]^{3-} + N_2O + H_2O \qquad (7.12)$$

$$[Fe(CN)_5NO]^{2-} + HN_3 + OH^- \rightarrow [Fe(CN)_5H_2O]^{3-} + N_2 + N_2O \qquad (7.13)$$

$$[Fe(CN)_5NO]^{2-} + N_2H_4 + OH^- \rightarrow [Fe(CN)_5H_2O]^{3-} + N_2O + NH_3 \qquad (7.14)$$

Reactions 7.10–7.14 may be described as additions of the N atom of the nucleophile on the N atom of the MNO fragment, coupled with deprotonation, as suggested by pH-dependent rate laws with a first-order behavior in complex and nucleophile concentrations, and followed by rapid adduct reorganizations. The negative activation entropies support an associative mechanism in the initial addition step. The adduct reorganizations generate the different gaseous products, $N_2$ and/or $N_2O$. Theoretical (DFT) characterizations of the adduct intermediates have been reported,[41] as previously detailed for the $OH^-$ additions.[36]

We focus on a comprehensive kinetic and mechanistic work reported for reaction 7.14.[42] Scheme 7.1 describes the proposed steps for the addition of $N_2H_4$ ($k_{N_2H_4} = 0.43$ $M^{-1} s^{-1}$, pH 9.4, 25°C), with subsequent deprotonation and N–N cleavage, leading to $NH_3$ and to the side-on $\eta^2$-$N_2O$ and end-on $\eta^1$-$N_2O$ intermediate isomers. The other

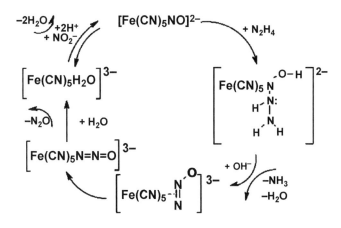

**SCHEME 7.1**

final products are free $N_2O$ and $[Fe(CN)_5H_2O]^{3-}$, which is able to further coordinate more nitrite (as $NO^+$).

In this way, a catalytic reduction of $NO_2^-$ by $N_2H_4$ ensues in the appropriate conditions. Interestingly, the addition of $N_2H_4$ to other nitrosyl complexes leads to azide complexes, not to $N_2O$. The attack of N-binding nucleophiles on bound $NO^+$ is at the heart of the mechanisms of $NO_2^-$ reductions in soils by bacteria and reducing enzymes, evolving to gaseous products, $N_2/N_2O$.[4] As anticipated for reaction 7.2, the coordination of $NO_2^-$ is considered the first step toward further reactivity.

Reaction 7.14 has been studied using labeled SNP ($^{15}NO$), and the gaseous product has been quantitatively identified as $^{14}N^{15}NO$, with no label at $NH_3$. This fact, together with DFT evidence shown in Figure 7.9, supports the proposed catalytic process.

The prediction of the $N_2O$ linkage isomers relates to the results on the NO analogues described in Section 7.6. Direct spectroscopic evidence exists only for the coordination of $\eta^1$-$N_2O$ on some Ru and Os complexes.[8a] The involvement of $\eta^2$-$N_2O$ and $\eta^1$-$N_2O$ in reaction 7.14 is supported by the geometrical and IR parameters derived from the DFT treatment.

In the reactions of Me-substituted derivatives of $N_2H_4$ adding to SNP, closely related stoichiometries leading to $N_2O$ have been found for methylhydrazine and 1,1-dimethylhydrazine, forming methylamine and dimethylamine as products, respectively (cf. reaction 7.14). The related mechanisms support an attack through the $NH_2$ groups, with the rates decreasing by about a factor of 10 for each Me substitution.

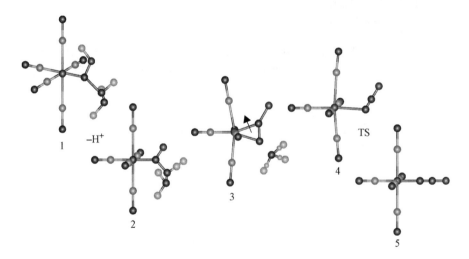

**FIGURE 7.9** DFT-calculated geometries in the initial steps of the reaction of [Fe(CN)$_5$NO]$^{2-}$ with $N_2H_4$, rendering the $N_2O$-bound intermediates. The structures correspond to singular points in the potential hypersurface, calculated at a B3LYP/6-31G** level. Relative energies ($y$-coordinate) are not drawn to scale. From left to right: (1) [Fe(CN)$_5$N(OH)NHNH$_2$]$^{2-}$; (2) [Fe(CN)$_5$N(O)NHNH$_2$]$^{3-}$; (3) [Fe(CN)$_5$-$\eta^2$-$N_2O$]$^{3-}$; (4) TS structure; (5) [Fe(CN)$_5$-$\eta^1$-$N_2O$]$^{3-}$. Ref. 42

A parallel path with a different product distribution has been found for methylhydrazine at pH >8, involving an adduct formation through the N atom vicinal to the Me group. Remarkably, the reaction of SNP with 1,2-dimethylhydrazine follows a route with a very different stoichiometry (Equation 7.15). It comprises a full six-electron reduction of $NO^+$ to $NH_3$, with formation of azomethane.

$$[Fe(CN)_5NO]^{2-} + 3MeHNNHMe \rightarrow [Fe(CN)_5NH_3]^{3-} + H_3O^+ + 3MeNNMe$$

$$(7.15)$$

The mechanism in reaction (7.15) probably involves two-electron reduced intermediates, $[Fe(CN)_5HNO]^{3-}$ and $[Fe(CN)_5NH_2OH]^{3-}$.

## 7.4   $n = 7$: PARTIALLY BENT MNO COMPLEXES—A DIVERSE STRUCTURAL AND REACTIVITY PICTURE

### 7.4.1   Six- and Five-Coordinate Complexes: IR, EPR, and Mössbauer Spectroscopies

Direct mixing of *stoichiometric* NO with Fe(II) complexes leads to {MNO}[7] species:[8]

$$[Fe^{II}L_5H_2O]^x + NO \rightleftharpoons [ML_5NO]^x + H_2O \qquad (7.16)$$

Disproportionation may be induced in excess NO conditions. The NO complexes can also be generated through the chemical or electrochemical reduction/oxidation of adequate precursors (cf. Figure 7.2).[8]

Complexes with $n = 7$ are not as widespread as with $n = 6$.[8,10] All the complexes included in Table 7.2 are bent, with ∠MNO ca. 140–150° and values of $\nu_{NO}$ in the range 1600–1700 cm$^{-1}$, which are significantly lower than the ones described for $NO^+$ species. When the same metal/coligand environment is considered, the elongation of $d_{M-N}$ and $d_{N-O}$ can be clearly observed, as predicted by the partial occupation of the $\pi^*$-type MNO orbital.

By using EPR spectroscopy, complexes with ground states $S = 3/2$ or $S = 1/2$ can be identified. Some five- and six-coordinate complexes with nonheme ferrous centers in metalloproteins react reversibly with NO forming nitrosyl species with $S = 3/2$. X-ray, RR, XAFS, MCD, and Mössbauer spectroscopies and theoretical calculations support a best description as $Fe^{III}NO^-$ (high-spin ferric, $S = 5/2$, antiferromagnetically coupled to $NO^-$, $S = 1$).[12,13o] Similar descriptions have been proposed for the classical complexes $[Fe(Me_3TACN)(N_3)_2NO]$[13n,o] and $[Fe(EDTA)NO]$[13o] and for the "brown-ring" compound, $[Fe(H_2O)_5NO]^{2+}$.[43] An interesting series of trigonal bipyramidal nonheme iron nitrosyl complexes with tripodal ligands derived from tris (N-R-carbamoylmethyl)amine, with R = isopropyl (${}^iPr$), cyclopentyl (cyp), and 3,5-dimethylphenyl (dmp), has been studied through X-ray, EPR, Mössbauer, and magnetic data.[44] The R groups form cavities around the metal ion that influence the structure and particularly the degree of FeNO bending, ranging from 178.2° (R = $^i$Pr) to 160.3° (R = dmp). This is supported by the EPR measurements, $S = 3/2$, showing a significantly more rhombic spectrum for the highly bent ligand. The combined approach supports the $Fe^{III}NO^-$ description. Finally, the $[Fe(L^{Pr})NO]$ complex displays spin equilibrium between the valence tautomers $S = 1/2$ and $S = 3/2$ in the solid state.[13p]

The $S = 1/2$ situation is generally found with relatively strong ligand field systems. An experimental and theoretical EPR study has been done with the series of $[M(CN)_5NO]^{3-}$ anions (M = Fe, Ru, and Os).[40] By reducing the $NO^+$ precursors in acetonitrile frozen solutions at 3.5 K, the *in situ* generated species showed typical axial spectra with one $^{14}N$ hyperfine coupling constant. The results are summarized in Table 7.6, suggesting the identification of the paramagnetic species as $[M^{II}(CN)_5NO^{\bullet}]^{3-}$.

High-level DFT calculations (ADF/BP and G98/B3LYP) for all three systems were performed. The addition of one electron to the $NO^+$ precursors causes the most pronounced changes in the M–NO and N–O lengths and in the M–N–O angles (close to 145°), as described above. The calculations confirm the lowering of symmetry through the removal of degeneracy of the $e_2$ orbitals and the occupation of one singly occupied MO, SOMO, of $a'$ symmetry. The calculated compositions of the SOMOs for the three ions are included in Table 7.6.

For $[Os^{II}(CN)_5NO^{\bullet}]^{3-}$, Figure 7.10 indicates that the spin density is not only confined to the nitrosyl part of the molecule (with about two-thirds share on the nitrogen atom), the SOMO also has a sizable metal contribution.

**TABLE 7.6   Comparison of Experimental and Calculated g Values$^a$ and $^{14}N$ Hyperfine Constants A (mT)$^b$ for [M(CN)$_5$NO]$^{3-}$ Complexes$^c$**

|  | $[Fe(CN)_5NO]^{3-}$ | | $[Ru(CN)_5(NO)]^{3-}$ | | $[Os(CN)_5(NO)]^{3-}$ | |
|---|---|---|---|---|---|---|
|  | Exp.$^d$ | Calc. | Exp.$^e$ | Calc. | Exp.$^e$ | Calc. |
| $g_1$ | 1.99 | 2.015 | 2.004 | 2.000 | 1.959 | 2.002 |
| $g_2$ | 1.99 | 1.995 | 2.002 | 1.991 | 1.931 | 1.940 |
| $g_3$ | 1.92 | 1.893 | 1.870 | 1.803 | 1.634 | 1.583 |
| $g_1 - g_3$ | 0.07 | 0.122 | 0.134 | 0.197 | 0.325 | 0.419 |
| $g_{av}$$^f$ | 1.967 | 1.968 | 1.959 | 1.932 | 1.847 | 1.824 |
| $A_1$ | n.a. | 0.73 | n.o. | 0.51 | n.o. | 0.58 |
| $A_2$ | 2.8$^g$ | 3.16 | 3.8 | 3.26 | 3.5 | 3.28 |
| $A_3$ | n.a. | 0.65 | n.o. | 0.36 | n.o. | 0.44 |
| Spin $\delta$, N(O) | 62 (22) | | 66 (24) | | 65 (25) | |

$^a$Spin-restricted calculations, including spin–orbit coupling.
$^b$Calculations using scalar relativistic UKS-ZORA approach.
$^c$Adapted from Ref. 40.
$^d$From Ref. 45a in aqueous solution. EPR measurements at 77 K.
$^e$From electrolysis in $CH_3CN$/0.1 M $Bu_4NPF_6$. EPR measurements at 3.5 K.
$^f$Calculated from $g_{av} = [g_1{}^2 \pm g_2{}^2 \pm g_3{}^2)/3]^{1/2}$.
$^g$Extracted from Ref. 45a.

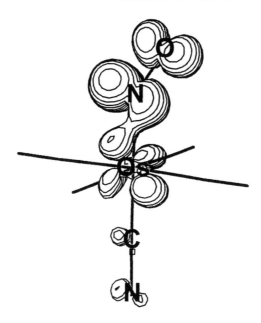

**FIGURE 7.10**  Spin density distribution within $[Os(CN)_5NO]^{3-}$.

The comparisons with EPR data for NO bound to solid supports, copper, or other ruthenium complex fragments confirm the predominantly ligand-centered spin. The effect of the strongly increasing spin–orbit coupling from Fe < Ru < Os is clearly evident from both the experimental and calculated data (Table 7.6). The effect is most pronounced in the Os system, where $g_3$ and the calculated isotropic value $g_{av}$ are lowest and the total $g$ anisotropy $g_1 - g_3$ is largest. This work established a firm basis for assigning the EPR spectra of $n = 7$ systems with $S = 1/2$, contributing to a confirmation of the early work with $[Fe(CN)_5NO^•]^{3-}$,[45a] and allowed for discriminating between the EPR spectra of the latter species and the one for the *trans* labilized $[Fe(CN)_4NO]^{2-}$ (Section 7.4.2),[45b] which at that time was erroneously reported as $[Fe(CN)_5NOH]^{2-}$. The success of the interpretation has been later extended to a great variety of $[M^{II}L_5NO^•]^x$ complexes (M = Ru, Fe), as described in Figure 7.11 with $[Ru(bpy)(tpm)NO]^{2+}$.[13a–c,45c]

EPR studies have been also performed with metallonitrosyl porphyrins affording $S = 1/2$.[18,46] For the six-coordinate $[Fe(TpivPP)(NO_2)NO]$,[131] the unpaired electron has been assigned to a SOMO with a highly predominant $d_{z^2}$ character, suggesting a nearly pure $Fe^INO^+$ distribution. The EPR spectra of some nitrosyl metalloproteins (MbNO and HbNO) were interpreted on similar grounds.[46a] However, recent evidence with $[Fe(TPP)(1-MeIm)NO]$ from MCD, Mössbauer, and IR spectroscopies and DFT calculations shows a nitrosyl-centered SOMO (ca. 20% in Fe), suggesting a best description as $Fe^{II}NO^•$, as stated above for other half-spin systems.[40,45c] This has been confirmed for several $[Ru(TPP)(NO)(X)]$ ions (X = 4-CNpy, py, 4-*N,N*-dimethylaminopy), showing the typically invariant

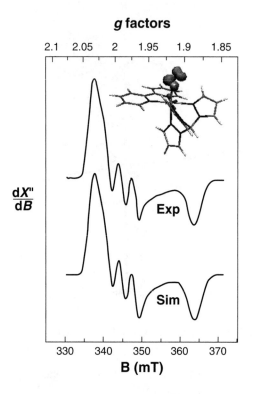

**FIGURE 7.11** EPR spectrum of $[Ru(bpy)(tpm)NO^{\bullet}]^{2+}$. *Top right*: DFT-calculated spin density in vacuum (B3LYP level, LanLDz basis set). *Middle*: Spectrum of the electrogenerated cation in $CH_3CN/0.1$ M $Bu_4NPF_6$ at 110 K. *Bottom*: Computer simulated spectrum. Ref. 27

$g$ characteristics: $g_1 > 2$, $g_2 \approx 2.0$, $g_3 < 2$, $A_2(^{14}N) \sim 32$ G, and a spin density of $\sim 65\%$ on the NO ligand.[46d]

Reasonable assignments of the complex pattern of electronic UV-Vis absorption bands, strongly dependent on the nature of the metal and coligands, have been reported through DFT calculations, which also allow to estimate the mixed compositions of the HOMO and LUMO. The electronic absorption bands are significantly shifted with respect to those observed in the corresponding $\{MNO\}^6$ systems.[27,46]

Finally, Mössbauer spectroscopy proved to be useful for discriminating between analogue nitrosyl complexes with $n = 6$ and 7. Table 7.3 shows significantly greater and smaller values of $\delta$ and $\Delta E_Q$, respectively, for the complexes with $n = 7$ containing the $NO^{\bullet}$ ligand. In both cases, $n = 6$ and 7, the measured values are independent of the coligands. The shifts do not reflect oxidation state changes at the iron site, as proposed earlier,[28] but the different degrees of backbonding with the changing chemical character of the NO ligand. The calculated field gradient tensors show the correct trend with respect to experiment. A rationale for the correlation has been provided.[13a] Thus, the more strongly backbonding the axial ligands are, the more electron density is preferentially taken out of the iron 3d shell, the 3d shielding decreases, and the electron density at the iron nucleus increases. At the same time, the metal–ligand bond strength shrinks, which leads to a distortion of the valence 4s contribution to the bonding orbitals to contract, which further increases the electron density at the nucleus. By comparing with data obtained with carbonyl analogues, the $\pi$-accepting ability has been proposed to follow the trend $NO^+ > CO > NO^{\bullet}$, consistent with the dissociation kinetics evidence discussed below (Section 7.4.3). DFT calculations also support the smaller degree of backbonding for NO (20% mixture of $a'$, $\pi^*_{NO}$ in the bonding MO) compared to $NO^+$ (28%).

### 7.4.2   The *trans* Effect: Keys for a Signaling Role of NO

An important structural fact for $[ML_5NO]^x$ complexes with $n = 7$ is the elongation of the M–L bond in the *trans* ligand to NO, which sometimes may lead to dissociation.[8,12] This has a bioinorganic relevance, because NO is an essential cellular signaling agent, associated with its coordination to the ferroheme center in soluble guanylyl cyclase (sGC) and the subsequent release of the proximal histidine ligand, thus triggering a chain of events leading to vascular response.[3,47a] The mechanistic details on this important bioreaction are still under close scrutiny,[47b] because *two* NO molecules appear to be involved in the enzyme activation process.

*trans* labilizations of the chloride ligand have been observed for $[Ru^{II}Cl(cyclam)NO]^+$,[48a] $[Ru^{II}(NH_3)_4(L)NO^{2+}]$,[48b] and $[Os^{II}Cl_5NO]^{3-}$.[13j] They arise probably in repulsion effects of the unpaired electron in the SOMO (with some $d_{z^2}$ character) with the donor *trans* ligands. Interestingly, the complexes $[cis\text{-}Os^{II}(bpy)_2ClNO]^+$, $[Ru^{II}(bpy)(tpm)NO]^{2+}$, and $[cis\text{-}Ru(L^{py})NO]^{2+}$ behave as robust species, probably because the *trans* position is occupied by the chelating ligand; however, the DFT calculations indicate some relative elongation of the *trans* Ru–N bond for the last two complexes.

The *trans* effect has been quantitatively demonstrated for $[Fe^{II}(CN)_5NO]^{3-}$, as described by Equation 7.17, with $K_{17} = 6.75 \times 10^{-5}\,M$.[49]

$$[Fe(CN)_5NO]^{3-} \rightleftarrows [Fe(CN)_4NO]^{2-} + CN^- \tag{7.17}$$

The pH influences the equilibrium concentrations of $[Fe(CN)_5NO]^{3-}$ and $[Fe(CN)_4NO]^{2-}$, with the latter increasing in acid medium, due to HCN formation. The two anions can be distinguished through the UV-Vis, IR, and EPR spectra and DFT calculations.[13m,45] A salt of the $[Fe(CN)_4NO]^{2-}$ complex has been isolated, and the X-ray structure reveals a square pyramidal cyano fragment with $\nu_{NO}$ at $1755\,cm^{-1}$. This is distinctively high compared to that in $[Fe(CN)_5NO]^{3-}$ at ca. $1600\,cm^{-1}$, suggesting a limiting $Fe^INO^+$ electronic distribution.[50]

Also, a distinctive EPR spectrum for $[Fe(CN)_4NO]^{2-}$ as compared to $[Fe(CN)_5NO]^{3-}$ has been reported, by working with solutions obtained upon reduction of 90% $^{13}C$-labeled SNP with dithionite, leading to a $g$ value of 2.024.[45b] The spectrum could be interpreted in terms of coupling to a single $^{14}N$ nucleus, $A(^{14}N) = 15.2\,G$, and to four $^{13}C$ nuclei. The $A(^{14}N)$ value is very similar to that observed in solution for the square pyramidal $[Fe(NO)(S_2CNMe_2)_2]$.[51] Extended Hückel calculations confirmed the experimental geometry and the value of $A(^{14}N)$. Removal of the axial cyanide from $[Fe^{II}(CN)_5NO]^{3-}$ to give $[Fe(CN)_4NO]^{2-}$ causes the iron $d_{z^2}$ orbital to shift from being an almost pure metal orbital to an admixed character with the iron $4p_z$ and with $\sigma(NO)$ to give a SOMO $\sigma$-orbital bonding over the whole FeNO fragment. Note that $[Fe^{II}(CN)_5NO^\bullet]^{3-}$ has a SOMO of $\pi$-type with respect to the FeNO direction, with $A(^{14}N) = 38$ G.

The five-coordinate *bent* metalloporphyrins were considered under the same framework.[18,46a] Figure 7.12 shows the EPR spectrum of $[Fe(TPP)NO]$.[46e]

For the related $[Fe(OEP)NO]$, the unpaired electron has been assigned to a SOMO with a highly predominant $d_{z^2}$ character, suggesting a nearly pure $Fe^INO^+$ distribution, as for $[Fe(CN)_4NO]^{2-}$. Recent MCD, Mössbauer, and IR spectroscopies and DFT calculations support this conclusion for $[Fe(TPP)NO]$, although with a more mixed electronic distribution, ca. 50% in Fe and NO.[46b] The electronic structure of the latter complex has been compared in detail with the six-coordinate derivative, $[Fe(TPP)(1-MeIm)(NO)]$, by using complementary UV-Vis absorption, IR, Raman, and $^1H$ NMR spectroscopies.[46c] The experimental and theoretical evidence show that binding of the *trans* ligand weakens the Fe–NO bond. Both five- and six-coordinate complexes contain bent FeNO angles of ca. 140°. In the five-coordinate complex, donation from the singly occupied $\pi^*$ orbital of NO into $d_{z^2}$ of Fe(II) leads to the formation of a Fe–NO $\sigma$-bond. In addition, a medium–strong $\pi$-backbond is present in both the complexes. The most important difference is the stronger $\sigma$-bond for the

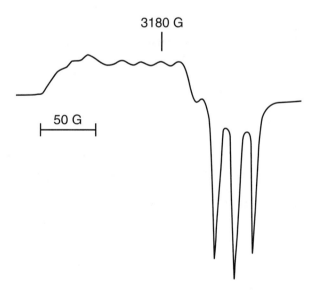

**FIGURE 7.12** X-band EPR spectrum in toluene glass (120K) for the five-coordinate Fe(TPP) (NO) ($g_1 = 2.102$, $A_1(^{14}N) = 12.4$; $g_2 = 2.064$, $A_2(^{14}N) = 16.6$; $g_3 = 2.010$, $A_3(^{14}N) = 16.2$). The units for the $A(^{14}N)$ hyperfine splittings are $10^{-4}$ cm$^{-1}$. Ref. 46e

five-coordinate species, which also leads to a significant transfer of spin density from NO to iron, as also shown by MCD spectra.[46b] Thus, the complex has a noticeable $Fe^I NO^+$ character relative to the six-coordinate analogue, which is a $Fe^{II}NO^•$(radical) complex.

As an important conclusion for the $\{FeNO\}^7$ systems, a versatile picture arises, depending on the CN and type of coligands, allowing for different limiting descriptions such as $Fe^{II}NO^•$ and others involving some variable degree of electron transfer to or from the metal: $Fe^I NO^+$ or $Fe^{III}NO^-$.

Two related mononuclear copper(I) nitrosyl complexes containing pyrazolyl methane and borate derivatives have been studied by X-ray, IR, EPR, and DFT calculations.[52] They pertain to the $\{CuNO\}^{11}$ class, which are relevant as intermediates in the reactions of copper proteins with NO.[4,9] The interpretation of the data allowed assigning the complexes as limiting $Cu^I NO^•$, with end-on coordination of the bent nitrosyl.

### 7.4.3  NO Ligand Interchange: Disproportionation Reactions

Little kinetic work has been carried out with $NO^•$ as a ligand in classical complexes.[5,8] Careful removal of impurities in the gas stream and absence of $NO_2^-$ and/or $NO_2$ in solution must be ensured. NO binds reversibly to *high-spin* $[Fe^{II}L_xH_2O]$ complexes (L = EDTA, NTA, and derivatives), which are potential catalysts for NO removal from gas streams.[31] As described above, the electronic structure of the nitrosylated products ($S = 3/2$) has been described as $Fe^{III}NO^-$ (Equation 7.18):

$$[Fe^{II}L_xH_2O] + NO \rightleftharpoons [Fe^{III}L_x(NO^-)] + H_2O \quad k_{on}, k_{off} \qquad (7.18)$$

By using fast techniques, values of $k_{on}$ were found in the range $10^6$–$10^8 \, M^{-1}s^{-1}$, at 25°C, whereas $k_{off}$ values vary between $10^{-1}$ and $10^3 \, s^{-1}$, depending on L. Similarly, for $[Fe(H_2O)_5NO]^{2+}$, $k_{on} = 1.41 \times 10^6 \, M^{-1}s^{-1}$ and $k_{off} = 3.2 \times 10^3 \, s^{-1}$.[43] Water exchange measurements and activation parameters support a dissociative interchange ($I_d$) mechanism.[31]

The $[Fe^{II}(CN)_5NO]^{3-}$ ion may be considered a typical model for $d^6$, *low-spin* $M^{II}NO^•$ systems. Figure 7.13a shows the spectral changes for the reaction of NO with $[Fe^{II}(CN)_5H_2O]^{3-}$ (Equation 7.19).[53]

$$[Fe^{II}(CN)_5H_2O]^{3-} + NO \rightleftharpoons [Fe^{II}(CN)_5NO^•]^{3-} + H_2O \quad k_{on}, k_{off} \qquad (7.19)$$

The inset of Figure 7.13a shows that the initially attained absorption at 350 nm corresponding to $[Fe^{II}(CN)_5NO]^{3-}$ decays subsequently in excess of NO, suggesting decomposition (see below). Independent of this complication, the faster traces reflect a well-behaved pseudo-first-order process. The values of $k_{obs}$ depend linearly on the concentration of NO, and a value of $k_{on} = 250 \, M^{-1}s^{-1}$ can be derived. This value is similar to others measured for the coordination of several ligands (CO, $NH_3$, py, etc.) into $[Fe(CN)_5H_2O]^{3-}$, reflecting a dissociative mechanism (probably of D type) in

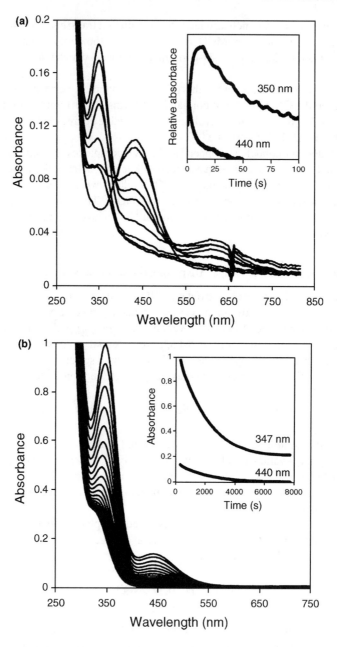

**FIGURE 7.13**    (a) Successive spectra in the reaction of $5.0 \times 10^{-5}$ M $[Fe(CN)_5H_2O]^{3-}$ with $1.8 \times 10^{-3}$ M NO: pH 10, $I = 0.1$ M, and $T = 25.4°C$. Decay of the aqua ion at 440 nm and buildup of $[Fe(CN)_5NO]^{3-}$ at 350 nm. Absorptions at ca. 600 nm reflect the presence of $[Fe(CN)_4NO]^{2-}$. *Inset*: Time dependence of fast reactant decay and product formation. The further decay at 350 nm reflects product decomposition (see the text). (b) Dissociation of NO from $[Fe(CN)_5NO]^{3-}$ in the presence of free cyanide: pH 10.2, $I = 0.1$ M, $T = 50.4°C$, and cycle time 312 s. *Inset*: Kinetic traces at 440 and 347 nm. Ref. 53

**TABLE 7.7   Rate Constants and Activation Parameters for the Complex Formation ($k_{on}$) and Dissociation ($k_{off}$) Reactions with Selected Complexes of the $[Fe^{II}(CN)_5L]^{n-}$ Series (cf. Reaction 7.19)$^a$**

| Ligand L | $k_{on}$ (M$^{-1}$ s$^{-1}$), $k_{off}$ (s$^{-1}$) | $\Delta H^{\neq}$ (kJ/mol) | $\Delta S^{\neq}$ (J/(K mol)) | $\Delta V^{\neq}$ (cm$^3$/mol) |
|---|---|---|---|---|
| NO$^{+b}$ | | | | |
| CO | 310, <10$^{-8}$ | 63, – | 15, – | –, – |
| CN$^-$ | 30, 4 × 10$^{-7}$ | 76.9, – | 42, – | 13.5, – |
| NO | 250, 1.6 × 10$^{-5}$ | 70, 106 | 34, 20 | 17.4, 7.1 |
| DMSO | 240, 7.5 × 10$^{-5}$ | 64.4, 110 | 16.7, 46 | – |
| pz | 380, 4.2 × 10$^{-4}$ | 64.4, 110.5 | 20.9, 58.6 | –, 13.0 |
| his | 315, 5.3 × 10$^{-4}$ | 64.4, 105.4 | 21, 46.0 | 17.0, – |
| NH$_3$ | 365, 1.75 × 10$^{-2}$ | 62, 102 | 10, 68 | 14.4, 16.4 |

$^a$Adapted from Ref. 8b. Reaction conditions: $T$ = ca. 25°C, $I$ = 0.5–1 M.
$^b$Unmeasurable formation and undetectable dissociation reactions.

which the release of water is rate-limiting, as supported by the positive activation enthalpies, entropies, and volumes (Table 7.7). It can be concluded that NO behaves like other Lewis base ligands, at least with the Fe(II) metal centers, without a particular influence of the single unpaired electron on the mechanism of the formation reaction.

Figure 7.13b shows the spectral changes upon dissociation of NO from [Fe(CN)$_5$NO]$^{3-}$ (reverse of Equation 7.19).[53] According to an accepted mechanistic description for the dissociation of X from $[Fe^{II}(CN)_5X]^{3-}$ complexes,[8,31] Equation 7.20 describes the rate-determining decay of the reactant and Equation 7.21 shows the formation of the final product under conditions of an excess of CN$^-$, which acts as a fast scavenger of the aqua complex.

$$[Fe(CN)_5NO]^{3-} + H_2O \rightleftharpoons [Fe(CN)_5H_2O]^{3-} + NO \quad \text{slow} \quad (7.20)$$

$$[Fe(CN)_5H_2O]^{3-} + CN^- \rightleftharpoons [Fe(CN)_6]^{4-} + H_2O \quad \text{fast} \quad (7.21)$$

The value of $k_{off}$ ($k_{-20}$) = 1.6 × 10$^{-5}$ s$^{-1}$ (25.0°C, pH 10), obtained under saturation conditions, is included in Table 7.7 together with a list of related $k_{-X}$ values obtained for other $[Fe^{II}(CN)_5X]^{n-}$ complexes.

The trend is consistent with the magnitude of the $\sigma$–$\pi$ interactions in the Fe$^{II}$–X bond. It can be concluded that NO$^\bullet$ is a moderate-to-strong ligand, weaker than carbonyl or cyanide, and certainly weaker than NO$^+$ (lower $\pi$-acceptor ability of NO$^\bullet$, despite a significant $\sigma$-contribution, cf. the Mössbauer results in Table 7.3). Reliable dissociation rate data for [ML$_5$NO$^\bullet$] systems are scarce. For several members of the [Ru$^{II}$(NH$_3$)$_4$(L)NO]$^{2+}$ series, values of $k_{off}$ ($k_{-NO}$) have been estimated by CV, upon reduction of the corresponding NO$^+$ complexes.[48b] The values appear as strikingly high (range 1–10$^{-4}$ s$^{-1}$), compared to the quoted one for [Fe(CN)$_5$NO]$^{3-}$. Indeed, the electron-donor coligands aid in NO labilization.

The value of $k_{off}$ for $[Fe(CN)_5NO]^{3-}$ is relevant to the rapid vasorelaxation (minute timescale) effected by solutions of SNP after injection in the bodily fluids.[54] The initial reductive event is likely as described in reaction (7.22), followed by further NO release to the medium.

$$[Fe(CN)_5NO]^{2-} + SR^- \rightarrow [Fe(CN)_5NO]^{3-} + 1/2RS\text{-}SR \qquad (7.22)$$

In the above context, the spontaneous thermal decomposition process of equilibrated solutions of $[Fe(CN)_5NO]^{3-}$ and $[Fe(CN)_4NO]^{2-}$ (Equation 7.17) has been studied.[55] Scheme 7.2 summarizes the diverse reactivity picture and the combined use of different spectroscopic and kinetic evidence for the appropriate characterization of the intermediates and final products.

The pH conditions are crucial for analyzing the results. At pH 7, $[Fe(CN)_4NO]^{2-}$ becomes predominant, and also decays slowly with $k_{off} =$ ca. $10^{-5}$ s$^{-1}$ (25°C). A faster decomposition of $[Fe(CN)_4NO]^{2-}$ occurs in the minute timescale at pH 4–5, with successive cyanide and NO release and the generation of Prussian blue-type precipitates (Scheme 7.2, below). We may still share the currently accepted idea that $[Fe(CN)_4NO]^{2-}$ is the necessary precursor for a fast NO release, *if* the cyano ligands are exposed to donor interactions with specific acceptor protein sites, thus promoting decomposition, even under pH 7 physiological conditions.[54]

At pH >8, $[Fe(CN)_5NO]^{3-}$ becomes predominant, and the slow NO$^\bullet$ release is followed by the formation of an EPR-silent intermediate (**I**) with a characteristic UV-Vis spectrum and $\nu_{NO}$ at 1695 cm$^{-1}$. **I** is a precursor of NO disproportionation into

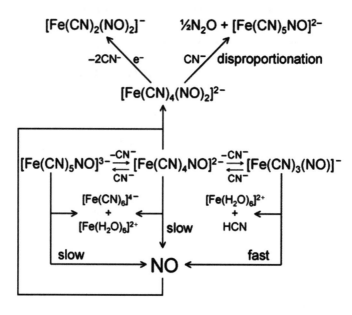

**SCHEME 7.2**   Ref. 97c

$[Fe(CN)_5NO]^{2-}$ and $N_2O$, displaying a rigorous 1:0.5 molar stoichiometry. The overall evidence suggests that **I** is a dinitrosyl species, *trans*-$[Fe(CN)_4(NO)_2]^{2-}$, with preliminary DFT calculations supporting the above formulation, consistent with a recent description of an iron dinitrosyl porphyrin, $[Fe(NO)_2(por)]$.[56] The formation of *trans*-$[Fe(CN)_4(NO)_2]^{2-}$ is supported by independent kinetic data on the reaction of $[Fe(CN)_4NO]^{2-}$ with excess NO, leading to a second-order rate law, $k = 4.3 \times 10^4$ $M^{-1} s^{-1}$. A related dinitrosyl compound, $[Fe(NO)_2(pyS_4)]$, has been obtained by mixing $[Fe(pyS_4)]$ with an excess of NO.[13b] The synthetic, structural, and reactivity aspects of dinitrosyl compounds are quite relevant to the chemistry of NO reductases.[4]

Also remarkable in Scheme 7.2 is the appearance of a final decomposition product associated with a new EPR signal, indicative of the presence of the so-called "$g = 2.03$" dinitrosyls, which are biologically relevant and labile species, active toward vasodilation.[57] Their general formula is $[Fe(L)_2(NO)_2]$, in a pseudotetrahedral arrangement, with L = thiolates, imidazolates, and so on as coligands. In the reported reaction conditions of Scheme 7.2, L should be necessarily cyanide.

### 7.4.4 Electrophilic Addition of Oxygen

In the same way as electrophilic reactivity can be predicted for $NO^+$ complexes, we may anticipate a nucleophilic reactivity for the more electron-rich NO complexes. Although nitrosyl protonation reactions seem not to occur at the $\{MNO\}^7$ moieties, some complexes in this series have been proved to be oxygen-sensitive. Figure 7.14

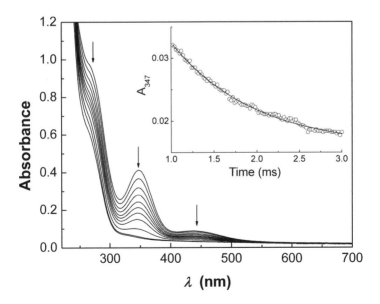

**FIGURE 7.14** Reaction of $O_2$ with the $[Fe(CN)_5NO^\bullet]^{3-}$ ion. Successive UV-Vis spectra for the titration of $10^{-4}$ M $[Fe(CN)_5NO^\bullet]^{3-}$ with $2.6 \times 10^{-4}$ M $[O_2]$: pH 10, $I = 0.1$ M, $5 \times 10^{-4}$ M $CN^-$, and $T = 25°C$. *Inset*: Stopped-flow trace for the decay of $[Fe(CN)_5NO^\bullet]^{3-}$ at 347 nm. Ref. 58

shows the decay of $[Fe(CN)_5NO]^{3-}$ with successive additions of dissolved $O_2$,[58] showing the following stoichiometry (Equation 7.23):

$$4[Fe(CN)_5NO]^{3-} + O_2 + 2H_2O \rightarrow 4[Fe(CN)_5NO]^{2-} + 4OH^- \qquad (7.23)$$

Under excess $O_2$ conditions, $[Fe(CN)_5NO]^{3-}$ decays exponentially in a stopped-flow timescale (Figure 7.14, inset). The experimental pseudo-first-order rate constant $k_{obs}$ correlates linearly with $[O_2]$, leading to a global second-order rate law: $-1/4d[Fe(CN)_5NO^{3-}]/dt = k_{23}[Fe(CN)_5NO^{3-}][O_2]$, with $k_{23} = (3.5 \pm 0.2) \times 10^5\ M^{-1}s^{-1}$ at 25°C, pH 10. The activation parameters are $\Delta H^{\#} = 40\ kJ/mol$ and $\Delta S^{\#} = 12\ J/(K\ mol)$. In all the experiments, an excess of free $CN^-$ had to be used to minimize *trans* labilization of this ligand (Equation 7.17). The rate constant was insensitive to changes in pH (9–11) and ionic strength (0.1–1 M). However, for pH <10 and without extra cyanide, the oxidation rate *decreased* markedly.

The above results cannot be accommodated by an outer-sphere mechanism because of the endergonic character of the first one-electron transfer process. Alternative $O_2$ coordination steps promoted by the dissociation of NO or $CN^-$ have also been discarded. Instead, reaction 7.24 was proposed as the initial step of reaction 7.23.

$$[Fe^{II}(CN)_5NO]^{3-} + O_2 \underset{k_{-ad}}{\overset{k_{ad}}{\rightleftharpoons}} [Fe^{III}(CN)_5N(O)O_2]^{3-} \qquad (7.24)$$

Reaction 7.24 describes the formation of a new covalent bond between bound NO and $O_2$ (Figure 7.15). Rather than proposing the formation of a Fe(II)-nitrosyldioxyl radical, the DFT computations suggest a two-electron reduction for $O_2$, with the binding of a peroxynitrite anion to Fe(III).

In related studies with $Mb^{II}NO$ (Section 7.8), an isomerization of the N-bound to an O-bound peroxynitrite adduct had been proposed, subsequent to a reaction similar to 7.24, in order to explain the formation of $NO_3^-$ as a final product.[59] Instead, reaction 7.25 comprises a fast bimolecular formation of $[Fe(CN)_5NO_2]^{3-}$, which may further react leading to SNP in Equation 7.26.

$$[Fe(CN)_5N(O)O_2]^{3-} + [Fe(CN)_5NO]^{3-} \rightarrow 2[Fe(CN)_5NO_2]^{3-} \qquad (7.25)$$

$$[Fe(CN)_5NO_2]^{3-} + [Fe(CN)_5NO]^{3-} + H_2O \rightarrow 2[Fe(CN)_5NO]^{2-} + 2OH^- \quad (7.26)$$

Both reactions 7.25 and 7.26 probably involve several steps. The oxidation equivalents remain bound to the metal all along the reaction, leading to the experimentally found 4:1 global stoichiometry, without other detectable by-products.

Assuming steady-state conditions for $[Fe(CN)_5N(O)O_2]^{3-}$, we get $-d[Fe(CN)_5NO^{3-}]/dt = 4k_{ad}k_{25}[O_2][Fe(CN)_5NO^{3-}]^2/(k_{-ad} + k_{25}[Fe(CN)_5NO^{3-}])$. With $k_{25}[Fe(CN)_5NO^{3-}] > k_{-ad}$, this expression reduces to the observed rate law, with $k_{23} = k_{ad}$.

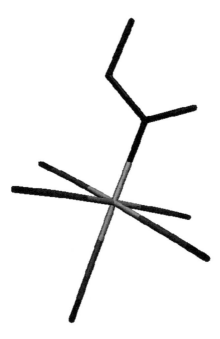

**FIGURE 7.15** DFT-optimized geometry of the Fe(III) peroxynitrite adduct formed in the initial step of the reaction of $[Fe(CN)_5NO^{\bullet}]^{3-}$ with $O_2$.

Second-order rate laws have also been found for the $[Ru(bpy)(tpm)NO]^{2+}$ and $[Ru(NH_3)_5NO]^{2+}$ complexes reacting with $O_2$. As the spin density distribution along the different $\{MNO\}^7$ moieties remains essentially invariable (Section 7.4.1), it is reasonable to expect similar reactivity patterns for the NO complexes. The $[Fe(CN)_5NO]^{3-}$ and $[Ru(NH_3)_5NO]^{2+}$ complexes (affording $E_{NO^+/NO}$ values near to $-0.10$ V) react with very similar addition rate constants. However, the $[Ru(bpy)(tpm)NO]^{2+}$ ion (with $E_{NO^+/NO} = 0.55$ V) showed a much lower value of $k_{ad}$, by five orders of magnitude. If a tentative plot is built up with $\ln k_{ad}$ against $E_{NO^+/NO}$ for the above three complexes, a linear trend can be appreciated, with a *negative slope* of $-18.4 \pm 0.9$ V$^{-1}$. The value is in close agreement with the theoretically predicted Marcus-type behavior for bimolecular reactions occurring with associative character, $19.4$ V$^{-1}$.[38] Not unexpectedly, the plot appears very similar to the one showed in Figure 7.6 for the *electrophilic* addition reactions of $[ML_5(NO^+)]$ complexes with $OH^-$ as a nucleophile, although with a *positive slope*.

It can be concluded that six-coordination is a necessary condition to achieve autoxidation of NO complexes. We stated above on the rate decrease for reaction 7.23 with decreasing pH, suggesting the *unreactivity* of $[Fe(CN)_4NO]^{2-}$. Also, the picket-fence compound $[Fe(TpivPP)NO]$ reacts with $O_2$ in nonaqueous medium *only* in the presence of pyridine, to give $[Fe(TpivPP)(NO_2)(py)]$.[60a] A product with bound $NO_2^-$ is also obtained in the autoxidation of the nonheme $[Fe(PaPy_3)NO]^+$ complex in acetonitrile solution.[60b] In both cases, the two-electron processes (with a 2:1 Fe:$O_2$ stoichiometry) formally involve one-electron oxidations of $NO^{\bullet}$ to $NO_2^-$ and of Fe(II) to Fe(III).

Indeed, that the redox potentials of the $MNO^+/MNO^{\bullet}$ couples could quantitatively predict the NO autoxidation reactivities appears quite significant. More work on additional, well-characterized NO complexes is desirable in order to extend the mechanistic analysis and eventually validate the predictive approach. It must be recognized that one of the routes for the decay of free NO in biologically relevant solutions is through the reaction described by Equation 7.27:

$$4NO + O_2 + 2H_2O \rightarrow 4H^+ + 4NO_2^- \tag{7.27}$$

Reaction 7.27 is termolecular, with $k = 2.88 \times 10^6 \, M^{-2} \, s^{-1}$.[61] Thus, NO is expected to survive a long time under the dilute NO concentrations in the bodily fluids, unless immune response conditions are generated. The previous discussion shows that $NO^{\bullet}$ coordination compounds could react with $O_2$ in order to provide a fast route to $NO^{\bullet}$ consumption. However, the complexes such as $[Fe(CN)_5NO]^{3-}$ could hardly compete with other main sinks for NO reactivity, namely, the very fast processes involving the reactions of free NO with sGC or with $HbO_2$.[1,3]

## 7.5   $n = 8$: STRONGLY BENT $NO^-$/HNO COMPLEXES—PROTONATION, DISSOCIATION, AND OTHER REACTIONS

Well-characterized complexes of the $\{MNO\}^8$ class are scarce (Table 7.2).[62] Five-coordinate species with bound $NO^-$ and square pyramidal geometries are dominant,[10,18] probably related to the great *trans*-effect of $NO^-$. Most common are the cobalt nitrosyl porphyrins.[18] By reacting with monodentate ligands ($NCS^-$, $Cl^-$, $Br^-$, etc.), six-coordinate complexes may be obtained. Reaction 7.28 describes the preparation of the first compound of this type studied by X-ray methods.[13s]

$$[Co(en)_2NO]^{2+} + Cl^- \rightarrow trans\text{-}[CoCl(en)_2NO]^+ \tag{7.28}$$

The reduced form of aquacobalamin (vitamin $B_{12r}$, $Cbl^{II}$) binds NO under physiological conditions yielding a diamagnetic six-coordinate product with a weakly bound $\alpha$-dimethylbenzimidazole and a bent nitrosyl coordinated to cobalt at the $\beta$-site of the corrin ring. Like other cobalt nitrosyl porphyrins, it has been described as $Co^{III}NO^-$, on the basis of UV-Vis, $^1H$, $^{31}P$, and $^{15}N$ NMR data.[63a] Recently, evidence for the $Co^{III}NO^-$ description has been obtained through the X-ray structure of nitroxylcob (III)alamine.[63b] Table 7.2 includes a six-coordinate platinum complex, $K[Pt(H_2O)(NO_2)_4NO]$,[13x] that has been described as a $Pt^{IV}NO^-$ moiety.

The lack of isolated Fe–HNO complexes is a remarkable drawback. Table 7.2 includes data for the two-electron reduction product of SNP, theoretically predicted to be a stable species containing bound HNO,[13r] and for the first hemoprotein nitroxyl derivative, $Mb^{II}HNO$,[13q] which has been prepared in aqueous solution by reduction of $Mb^{II}NO^{64}$ and has been characterized by using different spectroscopies.[13q,64] The $^1H$

NMR spectrum in $D_2O$ shows a signal at 14.8 ppm, which transforms into a doublet by using $^{15}N$ ($J_{NH} = 72\,Hz$). This evidence, together with $\nu_{NO}$ at 1385 cm$^{-1}$ obtained by RR spectroscopy, provides a strong support to the identity of the HNO ligand in Mb$^{II}$HNO, complemented by XANES and XAFS spectroscopies and by a study of the $^1H$ NMR structure at the heme pocket.

For [Fe(cyclam-ac)NO],[13a] electrochemical and IR measurements including $^{15}N$ and $^{18}O$ labeling, as well as DFT calculations, support the presence of NO$^-$ in acetonitrile. Indeed, there is strong evidence for the formation of a complex with $n = 8$, although the lack of structural or $^1H-^{15}N$ NMR information precludes a clear discrimination between NO$^-$ and HNO.

Only three crystalline structures have been reported for six-coordinate HNO complexes (Table 7.2).[13t–v] Representative synthetic approaches are described by reactions 7.29 and 7.30.

$$[Os^{II}Cl(CO)(PPh_3)_2NO] + HCl \rightarrow [Os^{II}(Cl)_2(CO)(PPh_3)_2HNO] \qquad (7.29)$$

$$[Ru^{II}(py^{bu}S_4)NO]^+ + H^- \rightarrow [Ru^{II}(py^{bu}S_4)HNO] \qquad (7.30)$$

Additional M–HNO complexes were more recently well characterized in nonaqueous solutions.[65] Among them, the first porphyrin derivative, [Ru$^{II}$(TTP)(1-MeIm)HNO], was obtained similarly as in reaction 7.30.[65a] The [Re$^I$(CO)$_3$(PPh$_3$)$_2$HNO] complex was prepared by two-electron oxidation of the hydroxylamine precursor with Pb(Ac)$_4$.[65b] A new alternative synthetic route for the latter Re$^I$ complex proceeds by direct insertion of NO$^+$ into the metal–hydride bond in Re$^I$(H)(CO)$_2$(PPh$_3$)$_2$.[65c] Finally, *cis,trans*-Re$^I$Cl(CO)$_2$(PR$_3$)$_2$HNO (R = Ph, Cy) have been obtained through similar protonation reactions as in Equation 7.29.[65d]

All the NO$^-$/HNO complexes display common structural and spectroscopic features in the solid state and in solution. They are all EPR-silent (d$^6$ low-spin metal, singlet NO$^-$/HNO), with an MNO angle near to 120°, reflecting a more pronounced bending than for $n = 7$ systems, as seen in Table 7.2. The combined $^1H/^{15}N$ NMR evidence has been crucial for the conclusive identification of the HNO ligand in solution for most of the above reported complexes. The $\nu_{NO}$ values appear at ca. 1300–1400 cm$^{-1}$ for the M$^{II}$HNO complexes, but values $\geq 1500$ cm$^{-1}$ have been observed with the metals in higher formal oxidation states, namely, Co$^{III}$NO$^-$ or Pt$^{IV}$NO$^-$ complexes. Overall, they reflect a diminished bond order for NO$^-$/HNO in comparison with NO$^\bullet$ and NO$^+$ complexes, consistent with a full population of the a′ antibonding MO (Figure 7.1b). The reasons for the preferential occurrence of bound NO$^-$ or HNO constitute an open issue. M$^{III}$/M$^{IV}$ complexes apparently favor the NO$^-$ situation. The expected lower π-donor ability may explain the comparative high $\nu_{NO}$ values as well as the more facile deprotonation of HNO, in contrast with the M$^{II}$ systems.

The NO$^-$ species are expected to behave as strong nucleophiles, rapidly abstracting protons from an adequate source, namely, water. The recently reported p$K_a$ of free $^1$HNO/$^3$NO$^-$ is ∼11.4, and probably ca. 23 for $^1$HNO/$^1$NO$^-$.[66] No data are available on the p$K_a$ values for metal-bound $^1$HNO/$^1$NO$^-$ in aqueous solutions.

Unfortunately, most of the M–HNO complexes are insoluble in water, in contrast with $Mb^{II}HNO$, which shows to be remarkably inert toward the release of HNO (hour timescale). The latter fact supports the proposal of HNO being a long-lived intermediate in the disproportionation reaction of $NH_2OH$ catalyzed by pentacyanoferrates, leading to bound $NO^+$ as the final oxidized product,[67] or in the six-electron reduction of SNP with 1,4-dimethylhydrazine (Equation 7.15).[42] On the basis of crystallographic observation of some intermediates and DFT calculations, HNO has been proposed as a necessary intermediate in the six-electron reduction of $NO_2^-$ to $NH_3$ catalyzed by cytochrome $c$ nitrite reductase.[68]

The reported HNO complexes are air-sensitive, but the products and mechanisms have not been studied in detail. Some five-coordinate $[Co^{III}L_4NO^-]$ complexes react with $O_2$ in nonaqueous media, *only* in the presence of nitrogen and phosphorus bases to yield the corresponding nitrocompounds, $[CoL_4(NO_2)B]$.[69a] The rates of these autoxidation reactions strongly depend on the nature of the *trans*-ligand to $NO^-$ in the $Co^{III}$ complexes, and this was interpreted as influencing the nucleophilicity of the $\{CoNO\}^8$ moieties. With $[Ir^{III}Cl(CO)(NO)(PPh_3)_2X]$ complexes ($X = I^-$, $Br^-$, $Cl^-$, $NCS^-$, etc.), autoxidation leads to $NO_3^-$.[69b] Other complexes may form mixtures of $NO_2^-$ and $NO_3^-$.

## 7.6   LINKAGE ISOMERS: END-ON, $\eta^1$-ON "ISONITROSYL," AND SIDE-ON, $\eta^2$-NO

In 1977, Hauser et al. reported the production of a long-lived state of SNP by irradiation with blue–green light at 80 K, as part of a Mössbauer spectroscopic study.[70] The new species (called a *metastable* state, MS1) showed a larger quadrupole splitting and a more positive isomer shift than SNP. A second metastable state, MS2, was put into evidence by Güida et al.,[71] through the IR measurements of single-crystal plates of SNP and the barium derivative. MS2 showed a similar, slightly more positive shift as MS1 in the Mössbauer spectrum compared to SNP (Table 7.3).

Both MS1 and MS2 were detected through differential scanning calorimetry (DSC), after irradiation with light of about 350–590 nm.[72] Figure 7.16 shows a DSC run obtained at a constant rate of temperature increase by monitoring the heat flow to the sample. The peaks indicate the heat released upon relaxation of higher energy species to the GS. By measuring the integrated areas, values of the enthalpies could be determined for MS1 (57 kJ/mol) and MS2 (36.3 kJ/mol). Thus, fractional conversion percentages can be estimated; they are also available through the IR or Mössbauer experiments.

The completely reversible deexcitation of the MS states into the GS can be obtained by illumination in the red and near-IR region of 600–900 nm or by heating over the decay temperatures of about 150 K (MS2) and 200 K (MS1), both processes following first-order kinetics. Illumination with near-IR light (900–1200 nm) partly transfers MS1 into MS2. Figure 7.17 describes the spectral regions for the population, depopulation, and transfers among MS1, MS2, and GS.[73]

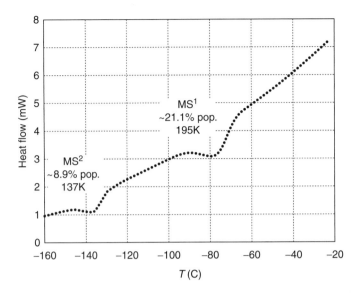

**FIGURE 7.16** Differential scanning calorimetry curve for a laser-irradiated crystal of $Na_2[Fe(CN)_5NO]\cdot2H_2O$. Heating rate, 4°C/min. Ref. 13f

The ES arises by an electronic transition, proposed to be from $b_2$ to the $e_2$ LUMO in SNP (Figure 7.1). Further relaxation of the ES intermediate into MS1, MS2, or back to the GS is described in Figure 7.18, showing potential energy surfaces with the relative energies of each species.

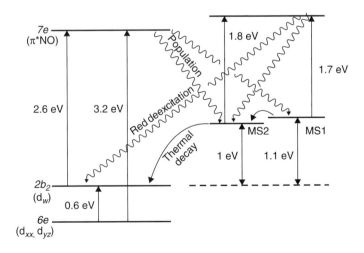

**FIGURE 7.17**    Relative positions of the metastable states MS1 and MS2 of $Na_2[Fe(CN)_5NO]\cdot$ $2H_2O$, their excited states, and excited levels of the ground state. Ref. 13f

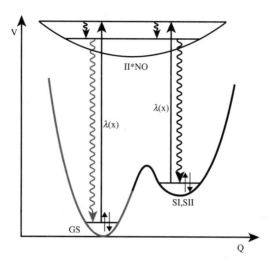

**FIGURE 7.18** Potential scheme of the ground and metastable states of $Na_2[Fe(CN)_5NO] \cdot 2H_2O$, MS1 and MS2. The $\pi^*(NO)$ orbital is the intermediate state for the relaxation into MS1, MS2, or back into the ground state. Ref. 73

In 1985, Yang and Zink performed ES Raman experiments with solutions of SNP, using 9 ns pulses.[74] A new peak at $1835 \, cm^{-1}$ and others in the $500–700 \, cm^{-1}$ region suggested a relaxed ES with an undefined "bent" structure. The $1835 \, cm^{-1}$ frequency is notoriously coincident with the value of $\nu_{NO}$ in MS1 (Table 7.3). Theoretical calculations indicate that the relaxed ES is close in energy and geometry to the GS/MS1 transition state, thus providing an explanation for the considerable efficiency of the isomerization process.[75] At 80K, the lifetime of the ES for SNP is in the range of $20–80 \, \mu s$, long enough for the rearrangement of the rest electron density on Fe. Consequently, the radiationless decay from the ES occurs into the new potential minima of MS1 and MS2, still maintaining the diamagnetic GS configuration on iron, $(d_{xz,yz})^4(d_{xy})^2$. With the latter evidence, each of the MS states is considered a *linkage isomer* of bound nitrosyl, not an ES species. Transient kinetic experiments using absorption spectroscopy with single crystals and *aqueous solutions* of SNP at *room temperature* have been recently reported.[76] By exposing to nanosecond laser pulses, MS2 has been populated with a buildup time faster than 1 ns, via a singlet ES transition, and decays monoexponentially by thermal deexcitation. The lifetime at 302 K is $1.8 \times 10^{-7}$ s in single crystals and $1.1 \times 10^{-7}$ s in aqueous solutions.

The strongest experimental evidence on the detailed structure of MS1 and MS2 relies on the low-temperature X-ray photocrystallographic results, pioneered by the group of P. Coppens in 1994.[13f] The difference Fourier maps of the electronic density allowed obtaining the relevant distances and angles for each species, present together with the GS complex of SNP upon excitation. Figure 7.19 shows a drawing with the geometric changes associated with the GS and MS states.

For MS1, a lengthening of the bond from Fe to the proximal atom of nitrosyl (N for the GS, O for MS1) by 0.053(6) Å is accompanied by a cooperative change in the

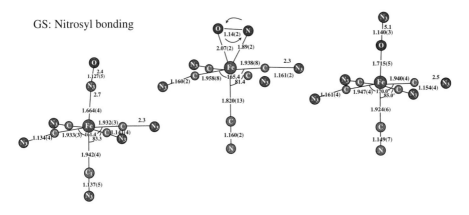

**FIGURE 7.19**  Bond lengths (in Å) and angles for ground state and metastable states (MS1 and MS2) of $Na_2[Fe(CN)_5NO] \cdot 2H_2O$, obtained at 50K. Ref. 77

angles between the *trans*-equatorial ligands, which increase slightly. On the other hand, in MS2 the equatorial ligands are repelled by the *side-on* NO group and the $C_{eq}$–Fe–$C_{eq(trans)}$ angle decreases. The Fe–C(axial) bond is considerably shortened in MS2 (not in MS1) by 0.106 Å. Surprisingly, the N–O bond lengths are little changed from the GS value.

Studies with polarized exciting light on the orthorhombic SNP crystals showed that the saturation population strongly depends on the direction of polarization of light and can be as high as 50% for MS1. As MS2 reaches saturation much quicker than MS1, it is proposed to be an intermediate along the GS → MS1 reaction coordinate. DFT calculations by Delley et al.[75] and Boulet et al.[39b] agree with this proposal, although the quantitative geometry and energy changes of the different states must be considered as crude estimates.

The initial Mössbauer spectra for MS1 and MS2 indicated a lower electron density at the nucleus compared to the GS, though without identifying the real changes occurring in the nitroprusside anion. A combined approach of Mössbauer results and DFT calculations confirmed the isonitrosyl conformation (FeON) of the $NO^+$ group in MS1, also suggesting that MS2 could be best represented by a dynamic conformation in which a bending vibration of FeON and a rotation of the $NO^+$ group around the main molecular axis simultaneously occur.[77] Nuclear inelastic scattering (NIS) of synchrotron radiation was used with guanidinium nitroprusside, $(CN_3H_6)_2[Fe(CN)_5NO]$, at 77 K. Comparisons with DFT-simulated NIS spectra provided strong evidence only for the isonitrosyl structure of MS1.[78]

IR[79] and Raman[80] spectroscopies have been most successful for contributing to proper assignments. Table 7.8 shows how some internal vibration bands of the anion (N–O and Fe–N stretchings, $\delta_{FeNO}$ bending) are shifted to lower energies, together

**TABLE 7.8    Infrared Spectral Frequencies (77 K) of Na$_2$[Fe(CN)$_5$NO]·2H$_2$O at GS, MS1, and MS2 States$^a$**

|  | GS | MS1 | MS2 |
|---|---|---|---|
| $\nu_{CN}$ (cm$^{-1}$) | 2177 | 2168 | 2180 |
|  | 2168 | 2159 | 2165 |
|  | 2163 | 2153 sh | 2149 |
|  | 2146 | 2138 | 2133 |
| $\nu_{NO}$ (cm$^{-1}$) | 1960 | 1835 | 1664 |
| $\delta_{FeNO}$ (cm$^{-1}$) | 667 | 582 | 596 |
| $\nu_{Fe-N}$ (cm$^{-1}$) | 657 | 565 | 547 |
| $\nu_{FeC}$ (cm$^{-1}$) | 414 | 405 |  |
| $\nu_{FeC}$ (cm$^{-1}$) | 410 | 399 |  |

$^a$From Ref. 79.

with a minor decrease in the C–N and Fe–C stretchings. This confirms that the most significant changes occur at the FeNO moiety.

A low-temperature IR experimental and theoretical study with MS1 and MS2 derived from SNP, isotopically normal and enriched with $^{15}$NO and N$^{18}$O, was performed by irradiating with 488 nm light and subsequently with 1064 nm light, thus reaching a high population of MS2.[79] In this way, support has been provided in favor of the end-on and side-on geometries. Additional evidence on the more questioned characterization of MS2 is based on the $^{16/18}$O isotope shift of $\delta_{FeNO}$. Besides, the Raman study[80] showed a downshift from 669 to 655 cm$^{-1}$ of the $\delta_{FeNO}$ band of the GS upon $^{15}$NO substitution, while the corresponding band for MS1 at 582 cm$^{-1}$ was nearly unshifted, as expected if the NO ligand in MS1 is bound through the oxygen atom, and in agreement with calculated downshifts for the GS and the isonitrosyl MS1 structures.

The linkage isomers have also been characterized for other [ML$_5$NO]$^x$ complexes (L = mixed-type ligands such as NH$_3$, py, bpy, NO$_2^-$, halides, porphyrins, etc.).[15] Although most of the complexes afford the {MNO}$^6$ configuration, linkage isomers have been found in other, more reduced systems, namely, {MNO}$^7$ (five-coordinate iron nitrosyl porphyrins),[81] {MNO}$^8$ ([Pt(NH$_3$)$_4$Cl(NO)]Cl$_2$]),[82] and {MNO}$^{10}$ ([Ni(NO)$\eta^5$-Cp]).[83,84] The recently reported Pt complex describes the generation of a unique MS isomer by irradiating in the red spectral range. In contrast with the generalized situation for {MNO}$^6$ systems, the proposed Pt–N–O to Pt–O–N conversion implies strongly bent species in both cases ($\sim$120°, theoretically calculated values), with an *increase* in $\nu_{NO}$ from 1673 to 1793 cm$^{-1}$, in good agreement with DFT calculations. The GS of the complex has been described as limiting Pt$^{II}$NO$^+$, though the alternative Pt$^{IV}$NO$^-$ description could be possible (cf. Ref. 13x).

The characteristic decay temperatures, $T_d$, for the recovery of the GS by heating the MS states strongly depend on the ML$_5$ fragments.[15] For the MS1 isomers, they tend to correlate with $\nu_{NO}$, and with the suggested order of increasing π-donor ability of the *trans*-ligand to NO, with a minor influence of the equatorial ligands. As the NO bond gets weaker, $\nu_{NO}$ and $T_d$ decrease. The weak π-donor in the *trans*-position with respect to NO seems to stabilize the L–M–ON fragment with the MS1 conformation. These correlations must be analyzed with caution up to the present stage of results.

Linkage isomerizations have also been observed for other small molecules such as $N_2$, $NO_2^-$, $NCS^-$, $SO_2$, and dimethylsulfoxide (DMSO).[15] The isomers are likely participants as intermediates in biorelevant ligand interchange or addition processes, in the NO recombination after photolysis, or even in the route to NO release from the ES in aqueous, room-temperature photochemistry (Section 7.7). Indeed, the photo-switchable nitrosyl compounds are an attractive class of materials with favorable photochromic and photorefractive properties, potentially useful for optical and biomedical applications.

## 7.7  PHOTOCHEMICAL REACTIVITY

Another manifestation of the electrophilicity of bound $NO^+$ is given by photochemical activation, achieved by exciting electronic transitions that usually appear in the near UV-Vis region, finally leading to released NO.[16] Equation 7.31 describes this process for SNP, which has been studied earlier.[85]

$$[Fe(CN)_5NO]^{2-} + h\nu \rightarrow [Fe^{III}(CN)_5H_2O]^{2-} + NO \qquad (7.31)$$

The initial step in reaction 7.31 has been described as an electron promotion from the $e_1$ MO to the $e_2$ ES (Figure 7.1a). This implies the corresponding decrease and increase in the electronic populations, respectively, and both effects should weaken the Fe–NO bond. The assignment is supported by irradiation experiments at different wavelengths, with a maximum quantum yield, $\Phi = 0.35$–0.37, being reached at 366–313 nm, where ESs of charge transfer character are populated. Irradiation at longer wavelengths leads to a decrease in $\Phi$ (0.18 at 435 nm). Remarkably, the irradiation at 480 nm evidenced no reaction, suggesting that the photoactive transition does not involve the nearly nonbonding $b_2$ MO. The same quantum yields were obtained at a fixed irradiation wavelength by quantifying either of the products, $[Fe^{III}(CN)_5H_2O]^{2-}$ or NO.[86] Pulsed laser, flash photolysis experiments indicate that the NO ejection from the ES occurs faster than the microsecond timescale. NO release has also been observed by irradiating the ruthenium and osmium pentacyano complexes, with decreasing quantum yields, interpreted as due to the increasing M–NO strength in the ES, M = Fe < Ru < Os, as well as to the greater competitive photophysical deactivations of the ES occurring for ruthenium and osmium, traced to the spin–orbit coupling influence.[86] Related ruthenium nitrosyl complexes with a formal $Ru^{II}NO^+$ configuration (namely, [*trans*-Ru[NH$_3$]$_4$L(NO)]$^{3+}$, *cis*-[Ru (bpy)$_2$L(NO)]$^{3+}$, *trans*-[RuCl([15]aneN$_4$)NO]$^{2+}$, with L = py, 4-Mepy, 4-Acpy, etc.; [15]aneN$_4$ = 1,4,8,12-tetraazacyclopentadecane) are also photoactive toward NO release, yielding the corresponding Ru(III) aqua complexes.[87] Interestingly, the NHase enzyme, which is thermally robust toward NO dissociation, releases NO under irradiation.[88] Though the majority of photoactive systems afford the $n = 6$ distribution, NO release has also been found for some complexes with $n = 7$.[87] There is a current interest in fundamental and applied studies in this field in order to achieve the controlled delivery of NO through a judicious selection of the complexes. With the model nonheme analogues of NHase, the questions are placed on the identification of the more efficient electronic transitions, eventually related to the nature of the coligands that bind in a *trans*-position with respect to $NO^+$. One of the goals deals with the ability of achieving the release of NO through the absorption of low energy, even infrared light, able to use the potential NO precursor drugs for a controlled delivery of NO in photodynamic therapy, to destroy cancer cells inducing cellular apoptosis.[88]

## 7.8   O-BOUND AND N-BOUND PEROXYNITRITE COMPLEXES

The small radical molecules $O_2$, $O_2^-$, NO, and $NO_2$ all react with biological targets, mostly through metal-mediated processes.[89] Besides, their mutual interactions may result in a variety of products and responses in the cell, including gene regulation and transcription. Indeed, reactive oxygen and nitrogen species are central to the life and death cellular decisions. One such species is the peroxynitrite ion, ONOO (oxoperoxonitrate(1−)).[90]

Solutions of $ONOO^-$ are conveniently prepared by the reaction of $NO_2^-$ with $H_2O_2$,[90a] and can also be generated at a nearly diffusion controlled rate either by mixing NO with $O_2^-$ [89–91] or by reacting $^3NO^-$ with $O_2$ following photolysis of $N_2O_3^{2-}$, derived from Angeli's salt, in alkaline medium.[66] The $NO/O_2^-$ reaction is supposed to occur *in vivo*, provided the production of the reactants is physiologically coordinated in a spatiotemporal sense.[89] $ONOO^-$ is stable at pH >7, but its protonated form peroxynitrous acid (HOONO, $pK_a$ 6.8) isomerizes to $NO_3^-$ in about 1 s at pH 7 and 25°C. Peroxynitrite is considered a likely mediator of NO biochemistry and oxidative/nitrative stress injury. It is a potent oxidant with the potential to destroy critical cellular components. It is capable of damaging DNA, initiate lipid peroxidation, or modify aromatic or sulfur-containing amino acid residues. It also nitrates aromatic compounds through catalytic processes with metal complexes or metalloproteins. Moreover, it reacts with $CO_2$ to yield $ONOOCO_2^-$, a stronger nitrating agent than peroxynitrite.[89–91] Figure 7.20 includes an overview of the relevant interactions of NO, $O_2^-$, and $ONOO^-$,[90a] showing that $ONOO^-$ can react with biological targets, can decompose to $NO_3^-$, or undergo metal-catalyzed reactions. A detailed analysis of the complex and variable pattern of $ONOO^-$ reactivity is out of scope of this chapter.

Most relevant is the recognized ability of $ONOO^-$ toward metal coordination, particularly with heme proteins (metMb and metHb) and metalloporphyrins. Dioxygen iron complexes of myoglobin ($MbO_2$) and hemoglobin ($HbO_2$) react rapidly with NO, $k = 4$–$9 \times 10^7\,M^{-1}\,s^{-1}$, at pH 7 and 20°C (Equation 7.32), and these reactions constitute a main sink for NO in the biological fluids.[1,3,5]

$$Hb^{II}O_2 + NO \rightarrow Hb^{III} + NO_3^-  \qquad (7.32)$$

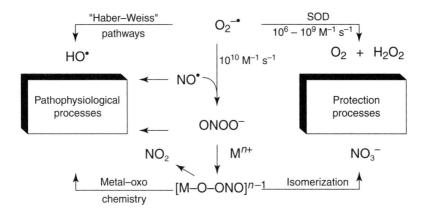

**FIGURE 7.20**   Interactions of superoxide, NO, and peroxynitrite.

Diffusion of NO into the distal pocket is considered to be the rate-limiting step. An intermediate, [Hb$^{III}$OONO], in reaction 7.32 has been proposed on the basis of rapid-scan UV-Vis spectroscopy and EPR evidence.[92] The mechanism of decay of the peroxynitrite intermediates is still a matter of debate.[89–91,93] An alternative route to isomerization implies the homolysis of the O–O bond in [Hb$^{III}$OONO], leading to $NO_2$ and Fe$^{IV}$O–protein intermediates (Figure 7.20).

The very fast reactions of ONOO$^-$ with transition metals afford an opportunity to develop new catalytic pharmaceuticals for the interception of ONOO$^-$.[90a] Metalloporphyrin complexes of Fe(III) and Mn(III) efficiently catalyze its isomerization to $NO_3^-$, and also show a high biological activity *in vivo*.[90] Several nonporphyrin peroxynitritometal complexes have been studied, and a general picture for their formation (starting with dioxygen complexes plus NO) and further reactivity is displayed in Scheme 7.3, encompassing the possibility of generation of $NO_3^-$ or $NO_2^-$ as final products.[93,94]

Although discrete peroxynitritometal complexes are rare, a solid has been isolated, formulated as [(Et)$_4$N]$_3$[Co$^{III}$(CN)$_5$OONO]$^{3-}$, through the reaction of [Co(CN)$_5$O$_2$]$^{3-}$ with NO.[95a] The complex is stable in the dark at pH 6, and slowly

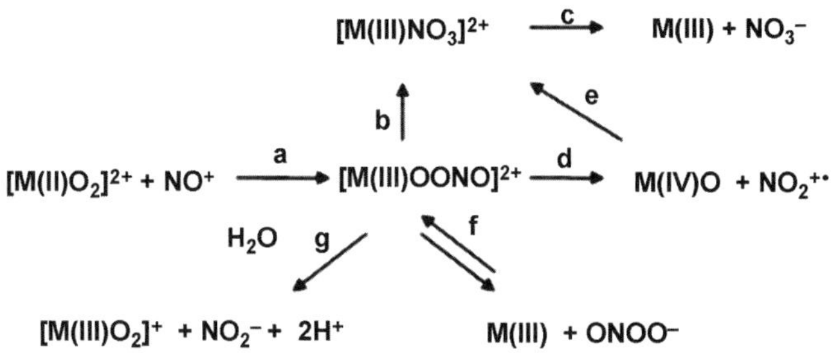

$$[M(III)NO_3]^{2+} \xrightarrow{\text{c}} M(III) + NO_3^-$$

$$\uparrow b \qquad \qquad \nwarrow e$$

$$[M(II)O_2]^{2+} + NO^+ \xrightarrow{\text{a}} [M(III)OONO]^{2+} \xrightarrow{\text{d}} M(IV)O + NO_2^{+\bullet}$$

$$H_2O \quad g \swarrow \qquad \qquad \searrow f$$

$$[M(III)O_2]^+ + NO_2^- + 2H^+ \qquad \qquad M(III) + ONOO^-$$

**SCHEME 7.3**

generates the $NO_3^-$ complex at pH 2 (Scheme 7.3: a, b). It is destroyed by photolysis leading to $[Co^{III}(CN)_5H_2O]^{2-}$. A similar intermediate was postulated in the related reaction of NO with a cyclam derivative of Rh(II)–$O_2$, leading to $NO_3^-$, seemingly through a combination of routes a–c and a, d, e, c in Scheme 7.3, the latter one involving homolysis.[94,95b] In the reactions of $Ti^{IV}O_2^+$ and peroxynitrous acid, reaction paths such as f and g in Scheme 7.3 have been proposed.[93]

The reaction of NO with a $Cu^I/O_2$ adduct containing a tetradentate, substituted tris (2-aminoethyl)amine (tren) ligand, $[(TMG_3tren)Cu^{II}(O_2^{\bullet-})]^+$, has been reported.[96] The proposed $[(TMG_3tren)Cu^{II}(OONO)]^+$ complex decomposes thermally to $[(TMG_3tren)Cu^{II}(ONO)]^+$ plus $O_2$. As no solid is available, the mechanistic evidence on the formation and reactivity comes from electrospray ionization mass spectrometry (ESI-MS) using $^{16}O$ and $^{18}O$, EPR spectra, and DFT calculations. The crystal structure of the O-bound $ONO^-$ product has been established, and strong differences appear with the EPR spectra of the $ONO^-$ and $OONO^-$ complexes. There is no formation of $NO_3^-$, discarding an isomerization process of O-bound peroxynitrite. This is the first discrete $Cu^{II}OONO$ complex showing a peculiar reactive chemistry of peroxynitrite, with likely evidence that O–O cleavage of the latter species has occurred.

Skibsted and Bohle afforded kinetic and mechanistic studies on the autoxidation process of $Mb^{II}NO$, the pigment found in cured meat, whose products are $Mb^{III}$ and $NO_3^-$.[97] A detailed revision of this reaction was performed by using spectrophotometric measurements with singular value decomposition and global analysis of the absorption spectra.[97c] Two consecutive pseudo-first-order reactions were detected, with $k_{obs}$ values of about $10^{-4} s^{-1}$ at 20°C, indicating the growth and decay of a reaction intermediate, as in reaction 7.33:

$$[Mb^{II}NO] + O_2 \rightleftharpoons \text{intermediate} \rightarrow Mb^{III} + NO_3^- \qquad (7.33)$$

In the first step, the rate depends linearly on the oxygen pressure at low pressures. The spectrum of the intermediate is very similar to the one for $Hb^{III}OONO$, previously identified for reaction 7.32. On this basis, and considering the measured activation parameters (including volumes) and making a comparative analysis with previous work, a unified approach has been proposed for reactions 7.32 and 7.33, comprising two different situations related to the $NO/O_2$ supply.

Reaction 7.33, at the upper part of Scheme 7.4, implies that NO, initially bound to Fe(II) in Mb, is displaced by $O_2$ in a reversible ligand exchange reaction prior to an irreversible electron transfer. The ligand exchange process is dissociative in nature and depends on bond breaking, with NO being trapped in a protein cavity. Both ligands are in proximity to the Fe(II) center prior to the second step, an intramolecular rearrangement to $NO_3^-$.

The mechanistic analysis discards the previously considered alternatives for an initial N-bound peroxynitrite intermediate[97a,b] arising in an associative step, followed by isomerization to the O-bound species and $NO_3^-$ release. Also considered is the intrinsic ambiguity with respect to the operability of a closely related limiting dissociative mechanism for NO release, given that the rate of formation of the

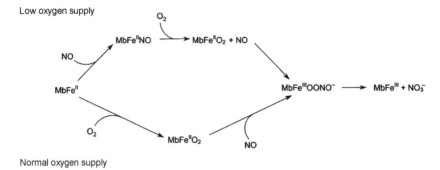

**SCHEME 7.4**

intermediate is similar to the rate of NO dissociation from $[Mb^{II}NO] = 1.2 \times 10^{-4}\,s^{-1}$, pH 7, 22°C.

The reaction of $[CrO_2]^{2+}$ with NO has been studied ensuing the photolysis of a Cr (II) nitrosyl complex (Equation 7.34):[94,98]

$$[CrNO]^{2+} + O_2 + h\nu \rightarrow [CrO_2]^{2+} + NO \rightarrow [CrOONO]^{2+} \qquad (7.34)$$

The reactive route forming the transient O-bound peroxynitrite appears similar to the thermal reaction described above for $Mb^{II}NO$. The final products are the chromium (III) nitrato complex and free $NO_3^-$. The intermediacy of $NO_2$, formed by homolysis of $[CrOONO]^{2+}$, suggests that route a, d in Scheme 7.3 is operative.

Summing up, we may conclude that insufficient evidence exists in the literature on the real participation of N-bound peroxynitrite intermediates, although with the recognition that a recent proposal provides experimental and theoretical support for the intermediacy of such a species in the oxidation of triphenylphosphine and cyclohexene by $O_2$, catalyzed by Naflon-bound six-coordinate (nitro) cobalt porphyrin complexes.[99] In this context, the results and mechanistic interpretations on the autoxidation of $[Fe(CN)_5NO]^{3-}$ (Equation 7.23)[58] may be highlighted (Section 7.4.4). In fact, the above commented ambiguity on the possible rate-limiting NO dissociation is absent for this reaction, because $k_{-NO}$ is 10-fold slower, and the measured oxidation rate is comparatively very fast. The stoichiometry and the DFT-calculated results on the structure of the N-bound peroxynitrite intermediate are valuable points in the proposed mechanism. It should be remarked that the $NO^+$-bound product is equivalent to $NO_2^-$; that is, no isomerization to $NO_3^-$ has been possible because of the higher competitive reactivity of $[Fe^{III}(CN)_5N(O)OO]^{3-}$ with the initial reactant, $[Fe(CN)_5NO]^{3-}$ (Equation 7.25).

From the studies discussed above, it can be seen that the $NO_2$ radical becomes a crucial intermediate, probably responsible in part for the biological damage attributed to peroxynitrite, as a result of O–O bond homolysis. $NO_2$ is well known as cause of environmental pollution[100] and has received less attention than NO in solution kinetic and mechanistic studies.[101] This is probably because $NO_2$ has a very short lifetime, with rapid disproportionation to $NO_2^-$ and $NO_3^-$. The reaction of the superoxochromium(III) complex, $Cr_{aq}OO^{2+}$, with $NO_2$ could be studied by generating $NO_2$ by laser flash photolysis of $(NH_3)_5CoNO_2^{2+}$ in the presence of excess $Cr_{aq}OO^{2+}$.[102] The equilibrium reaction 7.35 was studied through competition experiments with scavengers for $NO_2$.

$$Cr_{aq}OONO_2^{2+} \rightleftharpoons Cr_{aq}OO^{2+} + NO_2 \qquad (7.35)$$

It was concluded that the reverse reaction in Equation 7.35 involves a radical coupling. The rate constant for the forward homolytic reaction was estimated as $k_H = 197\,s^{-1}$ in 40% acetonitrile; the rapid homolysis, traced to the weak N–O bond in the peroxynitrate complex, leaves little time for it to engage in bimolecular reactions with added substrates.

## 7.9 SIMULTANEOUS METAL COORDINATION OF NO AND OTHER NON-INNOCENT LIGANDS: WHERE THE ADDED ELECTRONS GO?

Nitrosyl is the archetypal "non-innocent" ligand because of its facile interconversion between three oxidation states, including a spin-bearing radical form, as has been emphasized in this chapter. This has been critically considered through a detailed evaluation of the different spectroscopic indicators in order to describe the electron density for the $n = 6$, 7, and 8 distributions.

A more complicated situation arises in some delocalized molecules in which the coligand may also be redox active, as with metallonitrosyl porphyrins.[46d] We have selected a series of five-coordinate NO adducts of bis(dithiolene)–iron complexes, which may constitute a maximally five-membered electron transfer series [Fe(NO) (dithiolene)$_2$]$^z$ ($z = +1$, 0, $-1$, $-2$, $-3$).[103]

We describe the series in Chart 7.1. In principle, the redox activity may arise from the dithiolene, coordinated either as a closed-shell dianion ($S_L = 0$) or as a $\pi$-radical monoanion ($S_L = 1/2$). Nitrosyl might be present as $NO^+$ ($S = 0$), NO ($S = 1/2$), or $NO^-$ ($S = 0$ or 1). The central iron ion can possess a $d^6$ ($S_{Fe} = 0$, 1, 2) or $d^5$ ($S_{Fe} = 1/2$, 3/2, 5/2) configuration.

We restrict our analysis to one of the sets of complexes discussed in the original work, namely, that with R = $p$-tolyl, named as [1a]$^z$ in Chart 7.1. Two members of the series have been isolated as crystalline solids, and the relevant distances and angles obtained by X-ray diffraction are displayed in Table 7.9, together with the results

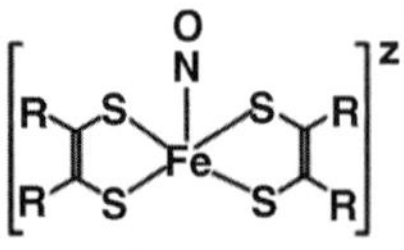

$z = 1+, 0,1–, 2–, (3–)$

**CHART 7.1**

**TABLE 7.9  Spectroscopic (IR, UV-Vis, EPR, Mössbauer, and X-Ray) and Electrochemical Data for the $[Fe(NO)(L)_2]^z$ Complex Series (L = Dithiolene Derivative; z = +1, 0, −1, −2)[a]**

| | $[1a]^+$ | $\xrightarrow{e^-}$ | $[1a]^0$ | $\xrightarrow{e^-}$ | $[1a]^-$ | $\xrightarrow{e^-}$ | $[1a]^{2-}$ |
|---|---|---|---|---|---|---|---|
| $\nu_{NO}$ (cm$^{-1}$)[b] | 1833 | | 1800, 1783 | | 1758 | | 1575, 1530 |
| $E_{1/2}$ (V)[c] | | 0.17 | | −0.44 | | −1.22 | |
| $d_{S-C}$ (av) (Å) | | | 1.71 | | 1.75 | | |
| $d_{C-C}$ (av) (Å) | | | 1.39 | | 1.36 | | |
| $\lambda_{max}$ (nm) | 460 (1.0) | | 384 (1.5) | | 310 (3.7) | | 316 sh (3.0) |
| ($10^4\varepsilon$, M$^{-1}$cm$^{-1}$)[d] | 523 (1.0) | | 490 (0.7) | | 410 sh (1.0) | | 420 (1.0) |
| | 858 (2.7) | | 607 (0.5) | | 678 (0.12) | | 710 (0.12) |
| | | | 855 (1.1) | | 1560 (0.6) | | |
| $g_x$[e] | — | | 2.0224 | | — | | 2.0496 |
| $g_y$[e] | — | | 2.0127 | | — | | 2.0297 |
| $g_z$[e] | — | | 1.999 | | — | | 2.0097 |
| $A_{xx}(^{14}N)$[f] | — | | n.o. | | — | | 0.0 |
| $A_{yy}(^{14}N)$[f] | — | | n.o. | | — | | 14.4 |
| $A_{zz}(^{14}N)$[f] | — | | n.o. | | — | | 15.6 |
| $\delta$ (mm/s)[g,h] | 0.07 | | 0.06 | | 0.04 | | 0.20 |
| $\Delta E_Q$ (mm/s)[g,i] | 1.40 | | 1.70 | | 1.88 | | 1.16 |
| Proposed e$^-$ distribution | $[Fe^{II}(NO^+)(L^\bullet)_2]^+$ | | $[Fe^{II}(NO^+)(L^\bullet)(L)]^0$ | | $[Fe^{II}(NO^+)(L)_2]^-$ | | $[Fe^{II}(NO^\bullet)(L)_2]^{2-}$ |

[a] Data from Ref. 103.
[b] In KBr disk.
[c] In CH$_2$Cl$_2$ versus Fc$^+$/Fc, 20°C.
[d] In CH$_2$Cl$_2$, 25°C.
[e] X-band EPR data, S = 1/2, in frozen CH$_2$Cl$_2$, 30 K.
[f] Nitrogen hyperfine coupling constants ($\times 10^4$, cm$^{-1}$). n.o.: not observed.
[g] Zero-field Mössbauer parameters, 80 K.
[h] Isomer shifts versus $\alpha$-Fe at 298 K.
[i] Quadrupole splitting.

obtained with X-band EPR, Mössbauer, UV-Vis, and IR spectroscopies. The onset of the different members of the redox series could be detected through CV and spectroelectrochemical measurements with controlled potential electrolysis in $CH_2Cl_2$. Magnetic susceptibilities in the solid state were also complementarily obtained.

The first column in Table 7.9 deals with the most oxidized member of the series, [1a]$^+$, which has been isolated as a diamagnetic salt with [BF$_4$]$^-$ and was described as an {FeNO}$^6$ species ($S = 0$). The value of $\nu_{NO}$ at 1833 cm$^{-1}$ supports a Fe$^{II}$NO$^+$ assignment, with two bound dithiolene radical monoanions: [Fe$^{II}$(NO$^+$)(L$^{1\bullet}$)$_2$]$^+$. The latter description for the dithiolene ligands is confirmed by the UV-Vis spectral data. In particular, the intense absorption maximum at ~850 nm was assigned as a ligand-to-ligand charge transfer (LLCT) band of a nearly planar (L$^\bullet$)Fe$^{II}$(L$^\bullet$) unit. Similar absorptions found with related iron–dithiolene complexes not containing the Fe–NO moiety provide strong evidence for the LLCT transition as an excellent marker for the Fe$^{II}$(L$^{1\bullet}$)$_2$ unit.

Through a one-electron reduction (cf. the CV data), a neutral solid paramagnetic complex [1a]$^0$ can be obtained ($S = 1/2$). The IR spectrum reveals two values of $\nu_{NO}$, at 1800 and 1783 cm$^{-1}$, still indicative of an {FeNO}$^6$ moiety. This suggests that reduction of [1a]$^+$ is a ligand-based process: (L$^\bullet$)$^-$ + e$^-$ → (L$^1$)$^{2-}$. Crystallographic results show that the NO group is bound in the apical position of a square-based pyramidal FeS$_4$N polyhedron. The FeNO group is linear, with the central iron ~0.50 Å above the best plane of the sulfur atoms. The two basal S$_4$ planes roughly face each other, giving rise to four weak S ⋯ S interactions, which may provide a pathway for the observed very weak antiferromagnetic spin exchange coupling ($J = -1.1$ cm$^{-1}$) in solid [1a]$^0$. The geometrical features of the FeNO moiety in [1a]$^0$ also correspond with {FeNO}$^6$, which has been described in the literature as diamagnetic. Thus, the latter unit carries a +3 charge, which requires two dithiolene ligands to adopt two different oxidation levels: one closed-shell dianion and a $\pi$-radical monoanion. The average C–S distances are comparatively short (1.70 Å) and the average "olefinic" C–C distance is long, at 1.39 Å, with respect to the next one-electron reduced member of the series, [1a]$^-$ (see Chart 7.2).

In principle, as a preliminary conclusion, the following three formal electron distributions may be envisaged for [1a]$^0$: [{FeNO}$^7$(L$^\bullet$)$_2$], [{FeNO}$^6$(L$^\bullet$)(L)], and [{FeNO}$^5$(L)$_2$]. The latter option was discarded, since species with this distribution have not been characterized to date. The X-ray data suggest an electronic structure with $n = 6$, [Fe$^{II}$(NO$^+$)(L$^\bullet$)(L)]$^0$ ↔ [Fe$^{II}$(NO$^+$)(L)(L$^\bullet$)]$^0$, with a class III (delocalized) chelating ligand mixed valency. The EPR spectrum confirms this assignment,

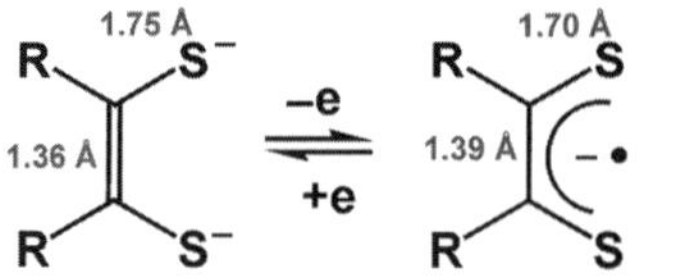

**CHART 7.2**

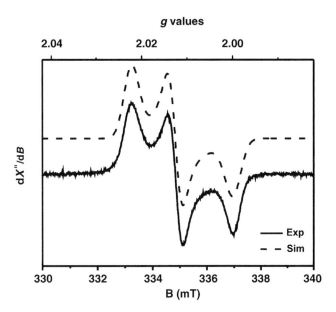

**FIGURE 7.21**  X-band EPR spectrum of $[Fe(NO)(L^{1\bullet})(L^1)]$, $[1a]^0$, in frozen $CH_2Cl_2$ at 20 K. $g$ values are given in Table 7.9. $L^1 = S_2C_2R_2$, $R = p$-tolyl. Ref. 103

allowing to discard the $n = 7$ arrangement. Thus, Figure 7.21 shows the EPR spectrum of $[1a]^0$, consisting of a rhombic signal with small $g$ anisotropy ($g = 2.02$–1.999) and no N-hyperfine splitting. The spectrum is typical of an S-centered radical, consistent with the presence of one $\pi$-radical monoanionic ligand $(L^\bullet)^-$.

Additional one-electron reduction leads to diamagnetic $[1a]^-$, isolated as a solid with $[Co(Cp)_2]^+$. The value of $\nu_{NO}$ is $1758\,cm^{-1}$, again suggesting that the $\{FeNO\}^6$ unit is conserved. The X-ray results show that the FeNO geometry in $[1a]^-$ is very similar to that from $[1a]^0$, although with a significant difference: the average C–S bond length at 1.75 Å and the average C–C bond at 1.36 Å have increased and decreased, respectively, in comparison to the same bonds in $[1a]^0$. This is a clear indication that reduction of $[1a]$ involves only the dithiolene ligands to form $[1a]^-$. Thus, the metrical details in $[1a]^-$ suggest an electronic structure as in $[Fe^{II}(NO^+)(L^1)_2]^-$.

Further reduction of $[1a]^-$ to $[1a]^{2-}$ generates a paramagnetic species ($S = \frac{1}{2}$), with $\nu_{NO}$ at $1575\,cm^{-1}$. The large downshift of $\nu_{NO}$ is indicative of reduction at the FeNO moiety, typical of a $\{FeNO\}^7$ distribution for the product. The EPR spectrum in Figure 7.22 confirms this prediction.

By comparing with Figure 7.21, it can be seen that $[1a]^{2-}$ displays a first-order hyperfine interaction with the nitrogen nucleus ($^{14}N$, $I = 1$) of the NO group, $A$ ($^{14}N$) $= \sim 15\,G$. The spectrum resembles those reported for other five- and six-coordinate $\{MNO\}^7$ ($S = 1/2$, M = Fe, Ru, Os) species, as previously shown in Section 7.4, and is best described as low-spin Fe(II) bound to a neutral $NO^\bullet$ radical.

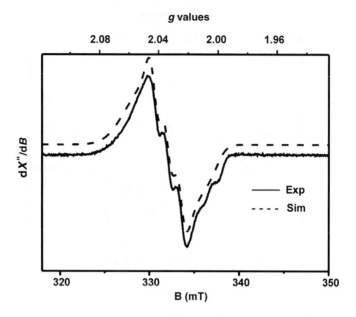

**FIGURE 7.22**   X-band spectrum of $[Fe(NO^{\bullet})(L^2)_2]^{2-}$, $[2a]^{2-}$, in frozen $CH_2Cl_2$ at 30 K. $g$ values are given in Table 7.9. $L^2 = S_2C_2R_2$, $R =$ bipyridyl (a similar spectrum was found for $[1a]^{2-}$). Ref. 103

Unfortunately, no crystal structure is available in order to put in evidence the predictable bending of FeNO in $[1a]^{2-}$.

The Mössbauer data presented in Table 7.9 are confirmative of the above assignments. For $[1a]^+$, $[1a]$, and $[1a]^-$, isomer shifts are all observed in the narrow range of 0.08–0.01 mm/s and quadrupole splittings at 1.4–2.0 mm/s, a clear indication that the $\{MNO\}^6$ moieties are retained irrespective of the charge of the complexes or dithiolene oxidation level. In contrast, the $[1a]^{2-}$ complex shows an increase and decrease of the above parameters, respectively, suggesting an $\{MNO\}^7$ moiety, in agreement with previous reports (cf. Table 7.3).

By performing the CV experiments, only three reversible waves are found, with half-wave potentials as detailed in Table 7.9. The fourth expected reduction eventually leading to $[1a]^{3-}$ has not been observed, although it has been recorded for a related complex with the maleonitrile-1,2-dithiolate ligand, at −1.83 V. The $[1a]^{3-}$ complex should contain the $\{MNO\}^8$ moiety. As discussed in Section 7.5, iron complexes of this configuration are scarce, and very elusive to a proper characterization.

## ACKNOWLEDGMENTS

To the University of Buenos Aires and to ANPCYT and CONICET, for academic and financial support.

# REFERENCES

1. Feelisch, M.; Stamler, J. S., Eds. *Methods in Nitric Oxide Research*; Wiley: Chichester, UK, 1996.

2. (a) Armor, J. N., Ed. *Environmental Catalysis, ACS Symp. Ser.* 552, 1993; (b) Richter-Addo, G. B.; Legdzins, P. *Metal Nitrosyls*; Oxford University Press: New York, 1992.

3. Ignarro, L. J., Ed. *Nitric Oxide, Biology and Pathobiology*; Academic Press: San Diego, CA, 2000.

4. (a) Wasser, I. M.; de Vries, S.; Moënne-Loccoz, P.; Schröder, I.; Karlin, K. D. *Chem. Rev.* **2002**, *102*, 1201;(b) Averill, B. A. *Chem. Rev.* **1996**, *96*, 2951.

5. Ford, P. C.; Lorkovic, I. M. *Chem. Rev.* **2002**, *102*, 993.

6. Stamler, J. S.; Singel, D. J.; Loscalzo, J. *Science* **1992**, *258*, 1898.

7. (a) Lymar, S. V.; Shafirovich, V.; Poskrebyshev, G. A. *Inorg. Chem.* **2005**, *44*, 5212; (b) Poskrebyshev, G. A.; Shafirovich, V.; Lymar, S. V. *J. Am. Chem. Soc.* **2004**, *126*, 891.

8. (a) Olabe, J. A.; Slep, L. D.In *Comprehensive Coordination Chemistry II: from Biology to Nanotechnology*; McCleverty, J. A.; Meyer, T. J., Eds.; Elsevier: Oxford, 2004, Vol. 1, p 603; (b) Roncaroli, F.; Videla, M.; Slep, L. D.; Olabe, J. A. *Coord. Chem. Rev.* **2007**, *251*, 1903.

9. (a) McCleverty, J. A. *Chem. Rev.* **2004**, *104*, 403; (b) Lee, D. H.; Mondal, B.; Karlin, K. D. In *Activation of Small Molecules*; Tolman, W. B., Ed.; Wiley-VCH: Weinheim, Germany, 2006; Chapter 2.

10. Feltham, R. D.; Enemark, J. H. *Top. Inorg. Organomet. Stereochem.* **1981**, *12*, 155.

11. Enemark, J. H.; Feltham, R. D. *Coord. Chem. Rev.* **1974**, *13*, 339.

12. Westcott, B. L.; Enemark, J. H.In *Inorganic Electronic Structure and Spectroscopy*; Solomon, E. I.; Lever, A. B. P., Eds.; Wiley: New York, 1999; Vol. 2, p 403.

13. (a) Garcia Serres, R.; Grapperhaus, C. A.; Bothe, E.; Bill, E.; Weyhermuller, T.; Neese, F.; Wieghardt, K. *J. Am. Chem. Soc.* **2004**, *126*, 5138; (b) Sellmann, D.; Blum, N.; Heinemann, F. W.; Hess, B. A. *Chem. Eur. J.* **2001**, *7*, 1874; (c) Afshar, R. K.; Patra, A. K.; Bill, E.; Olmstead, M. M.; Mascharak, P. K. *Inorg. Chem.* **2006**, *45*, 3774; (d) Pitarch López, J.; Heinemann, F. W.; Prakash, R.; Hess, B. A.; Horner, O.; Jeandey, C.; Oddou, J. J.; Latour, J. M.; Grohmann, A. *Chem. Eur. J.* **2002**, *8*, 5709; (e) Ellison, M. K.; Schultz, C. K.; Scheidt, W. R. *Inorg. Chem.* **1999**, *38*, 100; (f) Carducci, M. D.; Pressprich, M. R.; Coppens, P. *J. Am. Chem. Soc.* **1997**, *119*, 2669; (g) Olabe, J. A.; Gentil, L. A.; Rigotti, G. E.; Navaza, A. *Inorg. Chem.* **1984**, *23*, 4297; (h) Baraldo, L. M.; Bessega, M. S.; Rigotti, G. E.; Olabe, J. A. *Inorg. Chem.* **1994**, *33*, 5890; (i) Gans, P.; Sabatini, A.; Sacconi, L. *Inorg. Chem.* **1966**, *5*, 1877; (j) Singh, P.; Sarkar, B.; Sieger, M.; Niemeyer, M.; Fiedler, J.; Zalis, S.; Kaim, W. *Inorg. Chem.* **2006**, *45*, 4602; (k) Bottomley, F.; Clarkson, S. G.; Tong, S. B. *J. Chem. Soc., Dalton Trans.* **1974**, 2344; (l) Nasri, H.; Ellison, M. K.; Chen, S.; Huynh, B. H.; Scheidt, W. R. *J. Am. Chem. Soc.* **1997**, *119*, 6274; (m) Nast, R.; Schmidt, J. *Angew. Chem., Int. Ed. Engl.* **1969**, *8*, 383; (n) Pohl, K.; Wieghardt, K.; Nuber, B.; Weiss, J. *J. Chem. Soc., Dalton Trans.* **1987**, 187; (o) Brown, C. A.; Pavlovsky, M. A.; Westre, T. E.; Zhang, Y.; Hedman, B.; Hodgson, K. O.; Solomon, E. I. *J. Am. Chem. Soc.* **1995**, *117*, 715; (p) Li, M.; Bonnet, D.; Bill, E.; Neese, F.; Weyhermuller, T.; Blum, N.; Sellmann, D.; Wieghardt, K. *Inorg. Chem.* **2002**, *41*,

3444; (q) Immoos, C. E.; Sulc, F.; Farmer, P. J.; Czarnecki, K.; Bocian, D.; Levina, A.; Aitken, J. B.; Armstrong, R.; Lay, P. A. *J. Am. Chem. Soc.* **2005**, *127*, 814; (r) González Lebrero, M.; Scherlis, D. A.; Estiú, G. L.; Olabe, J. A.; Estrin, D. A. *Inorg. Chem.* **2001**, *40*, 4127; (s) Snyder, D. A.; Weaver, D. L. *Inorg. Chem.* **1970**, *9*, 2760; (t) Sellmann, D.; Gottschalk-Gaudig, T.; Haussinger, D.; Heinemann, F. W.; Hess, B. A. *Chem. Eur. J.* **2001**, *7*, 2099; (u) Wilson, R. D.; Ibers, J. A. *Inorg. Chem.* **1979**, *18*, 336; (v) Melenkivitz, R.; Hillhouse, G. L. *Chem. Commun.* **2002**, 660; (x) Peterson, E. S.; Larsen, R. D.; Abbott, E. H. *Inorg. Chem.* **1988**, *27*, 3514.

14. Solomon, E. I.; Lever, A. B. P., Eds. *Inorganic Electronic Structure and Spectroscopy*; Wiley: New York, 1999; Vols. 1 and 2.

15. (a) Coppens, P.; Novozhilova, I.; Kovalevsky, A. *Chem. Rev.* **2002**, *102*, 861; (b) Coppens, P.; Fomitchev, D. V.; Carducci, M. D.; Culp, K. *J. Chem. Soc., Dalton Trans.* **1998**, 865.

16. Ford, P. C.; Wecksler, S. *Coord. Chem. Rev.* **2005**, *249*, 1382.

17. Roncaroli, F.; Olabe, J. A.; van Eldik, R. *Inorg. Chem.* **2002**, *41*, 5417.

18. (a) Scheidt, W. R.; Ellison, M. K. *Acc. Chem. Res.* **1999**, *32*, 350; (b) Wyllie, G. R. A.; Scheidt, W. R. *Chem. Rev.* **2002**, *102*, 1067.

19. Xu, N.; Powell, D. R.; Cheng, L.; Richter-Addo, G. B. *Chem. Commun.* **2006**, 2030.

20. Paulat, F.; Lehnert, N. *Inorg. Chem.* **2007**, *46*, 1547.

21. Manoharan, P. T.; Gray, H. B. *Inorg. Chem.* **1966**, *5*, 823.

22. Manoharan, P. T.; Gray, H. B. *J. Am. Chem. Soc.* **1965**, *87*, 3340.

23. Chen, P.; Meyer, T. J. *Chem. Rev.* **1998**, *98*, 1439.

24. Estrin, D. A.; Baraldo, L. M.; Slep, L. D.; Barja, B. C.; Olabe, J. A.; Paglieri, L.; Corongiu, G. *Inorg. Chem.* **1996**, *35*, 6327.

25. Greene, S. N.; Richards, N. G. *Inorg. Chem.* **2004**, *43*, 7030.

26. Gorelsky, S. I.; da Silva, S.; Lever, A. B. P.; Franco, D. W. *Inorg. Chim. Acta*, **2000**, *300–302*, 698.

27. Videla, M.; Jacinto, J. S.; Baggio, R.; Garland, M. T.; Singh, P.; Kaim, W.; Slep, L. D.; Olabe, J. A. *Inorg. Chem.* **2006**, *45*, 8608.

28. Hauser, C.; Glaser, T.; Bill, E.; Weyhermuller, T.; Wieghardt, K. *J. Am. Chem. Soc.* **2000**, *122*, 4353.

29. Walker, F. A. *J. Inorg. Biochem.* **2005**, *99*, 216.

30. Ford, P. C.; Fernandez, B. O.; Lim, M. D. *Chem. Rev.* **2005**, *105*, 2439.

31. van Eldik, R. *Coord. Chem. Rev.* **2007**, *251*, 1649 and references therein.

32. Praneeth, V. K. K.; Paulat, F.; Berto, T. C.; DeBeer, G. S.; Nather, C.; Sulok, C. D.; Lehnert, N. *J. Am. Chem. Soc.* **2008**, *130*, 15288.

33. Marti, M. A.; Gonzalez Lebrero, M. C.; Roitberg, A.; Estrin, D. *J. Am. Chem. Soc.* **2008**, *130*, 1611.

34. Bottomley, F.In *Reactions of Coordinated Ligands*; Braterman, P. S., Ed.; Plenum Publishing: New York, 1989; Vol. 2.

35. (a) Swinehart, J. H.; Rock, P. A. *Inorg. Chem.* **1966**, *5*, 573; (b) Masek, J.; Wendt, H. *Inorg. Chim. Acta* **1969**, *3*, 455.

36. Roncaroli, F.; Ruggiero, M. E.; Franco, D. W.; Estiu, G. L.; Olabe, J. A. *Inorg. Chem.* **2002**, *41*, 5760.

37. Wilkins, R. G. *Kinetics and Mechanism of Reactions of Transition Metal Complexes*, 2nd edition; VCH Publishers: New York, 1991.

38. Marcus, R. A. *J. Phys. Chem.* **1968**, *72*, 891.

39. (a) Gorelsky, S. I.; Lever, A. B. P. *Int. J. Quant. Chem.* **2000**, *80*, 636; (b) Boulet, P.; Buchs, M.; Chermette, H.; Daul, C.; Gilardoni, F.; Rogemond, F.; Schlapfer, C. W.; Weber, J. *J. Phys. Chem. A* **2001**, *105*, 8991;**2001**, *105*, 8999.

40. Wanner, M.; Scheiring, T.; Kaim, W.; Slep, L. D.; Baraldo, L. M.; Olabe, J. A.; Zalis, S.; Baerends, E. J. *Inorg. Chem.* **2001**, *40*, 5704.

41. Olabe, J. A.; Estiú, G. L. *Inorg. Chem.* **2003**, *42*, 4873.

42. Gutiérrez, M. M.; Amorebieta, V. T.; Estiú, G. L.; Olabe, J. A. *J. Am. Chem. Soc.* **2002**, *124*, 10307.

43. Wanat, A.; Schneppensieper, T.; Stochel, G.; van Eldik, R.; Bill, E.; Wieghardt, K. *Inorg. Chem.* **2002**, *41*, 4.

44. Ray, M.; Golombek, A.; Hendrich, M. P.; Yap, G. P. A.; Liable-Sands, L. M.; Rheingold, A. L.; Borovik, A. S. *Inorg. Chem.* **1999**, *38*, 3110.

45. (a) van Voorst, J. D. W.; Hemmerich, P. *J. Chem. Phys.* **1966**, *45*, 3914; (b) Glidewell, C.; Johnson, I. L. *Inorg. Chim. Acta* **1987**, *132*, 145; (c) Frantz, S.; Sarkar, B.; Sieger, M.; Kaim, W.; Roncaroli, F.; Olabe, J. A.; Zalis, S. *Eur. J. Inorg. Chem.* **2004**, *2902*.

46. (a) Salerno, J. C.In *Nitric Oxide, Principles and Actions*; Lancaster, J. Jr., Ed.; Academic Press: 1996; (b) Praneeth, V. K. K.; Neese, F.; Lehnert, N. *Inorg. Chem.* **2005**, *44*, 2570; (c) Praneeth, V. K. K.; Näther, C.; Peters, G.; Lehnert, N. *Inorg. Chem.* **2006**, *45*, 2795; (d) Singh, P.; Das, A. K.; Sarkar, B.; Niemeyer, M.; Roncaroli, F.; Olabe, J. A.; Fiedler, J.; Zális, S.; Kaim, W. *Inorg. Chem.* **2008**, *47*, 7106; (e) Wayland, B. B.; Olson, L. W. *J. Am. Chem. Soc.* **1974**, *96*, 6037.

47. (a) Ballou, D. P.; Zhao, Y.; Brandish, P. E.; Marletta, M. A. *Proc. Natl. Acad. Sci. USA* **2002**, *99*, 12097; (b) Martí, M. A.; Capece, L.; Crespo, A.; Doctorovich, F.; Estrin, D. A. *J. Am. Chem. Soc.* **2005**, *127*, 7721.

48. (a) Lang, D. R.; Davis, J. A.; Lopes, L. G. F.; Ferro, A. A.; Vasconcellos, L. C. G.; Franco, D. W.; Tfouni, E.; Wieraszko, A.; Clarke, M. J. *Inorg. Chem.* **2000**, *39*, 2294; (b) Toledo, J. C.; dos Santos Lima Neto, B.; Franco, D. F. *Coord. Chem. Rev.* **2005**, *249*, 419.

49. Cheney, R. P.; Simic, M. J.; Hoffman, M. Z.; Taub, I. A.; Asmus, K. D. *Inorg. Chem.* **1977**, *16*, 2187.

50. Schmidt, J.; Kühr, H.; Dorn, W. J.; Kopf, W. L. *Inorg. Nucl. Chem. Lett.* **1974**, *10*, 55.

51. Davies, G. R.; Jarvis, J. A. J.; Kilbourn, B. T.; Mais, R. H. B.; Owston, P. G. *J. Chem. Soc.* **1979**, 1275.

52. Fujisawa, K.; Tateda, A.; Miyashita, Y.; Okamoto, K.; Paulat, F.; Praneeth, V. K. K.; Merkle, A.; Lehnert, N. *J. Am. Chem. Soc.* **2008**, *130*, 1205.

53. (a) Roncaroli, F.; Olabe, J. A.; van Eldik, R. *Inorg. Chem.* **2003**, *42*, 4179; (b) Olabe, J. A. *Dalton Trans.* **2008**, 3633.

54. Butler, A. R.; Megson, I. L. *Chem. Rev.* **2002**, *102*, 1155.

55. Roncaroli, F.; van Eldik, R.; Olabe, J. A. *Inorg. Chem.* **2005**, *44*, 2781.

56. (a) Patterson, J. C.; Lorkovic, I. M.; Ford, P. C. *Inorg. Chem.* **2003**, *42*, 4902; (b) Conradie, J.; Wondimagegn, T.; Ghosh, A. *J. Am. Chem. Soc.* **2003**, *125*, 4968.

57. (a) Tsai, F. T.; Chiou, S. J.; Tsai, M. C.; Tsai, M. L.; Huang, H. W.; Chiang, M. H.; Liaw, W. F. *Inorg. Chem.* **2005**, *44*, 5872–5881; (b) Lee, C. M.; Chen, C. H.; Chen, H. W.; Hsu, J. L.; Lee, G. H.; Liaw, W. F. *Inorg. Chem.* **2005**, *44*, 6670–6679.

58. Videla, M.; Roncaroli, F.; Slep, L. D.; Olabe, J. A. *J. Am. Chem. Soc.* **2007**, *129*, 278.

59. Moller, J. K. S.; Skibsted, L. H. *Chem. Rev.* **2002**, *102*, 1167.

60. (a) Cheng, L.; Powell, D. R.; Khan, M. A.; Richter-Addo, G. B. *Chem. Commun.* **2000**, 2301; (b) Patra, A. K.; Rowland, J. M.; Marlin, D. S.; Bill, E.; Olmstead, M. M.; Mascharak, P. K. *Inorg. Chem.* **2003**, *42*, 6812.

61. Goldstein, S.; Czapski, G. *J. Am. Chem. Soc.* **1995**, *117*, 12078 and references therein.

62. (a) Farmer, P. J.; Sulc, F. *J. Inorg. Biochem.* **2005**, *99*, 166; (b) Miranda, K. M. *Coord. Chem. Rev.* **2005**, *249*, 433.

63. (a) Wolak, M.; Zahl, A.; Schneppensieper, T.; Stochel, G.; van Eldik, R. *J. Am. Chem. Soc.* **2001**, *123*, 9780; (b) Hannibal, L.; Smith, C. A.; Jacobsen, D. W.; Brasch, N. E. *Angew. Chem., Int. Ed.* **2007**, *46*, 5140.

64. Lin, R.; Farmer, P. J. *J. Am. Chem. Soc.* **2000**, *122*, 2393.

65. (a) Lee, J.; Richter-Addo, G. B. *J. Inorg. Biochem.* **2004**, *98*, 1247; (b) Southern, J. S.; Hillhouse, G. L.; Rheingold, A. L. *J. Am. Chem. Soc.* **1997**, *119*, 12406; (c) Melenkevitz, R.; Southern, J. S.; Hillhouse, G. L.; Concolino, T. E.; Liable-Sands, L. M.; Rheingold, A. L. *J. Am. Chem. Soc.* **2002**, *124*, 12068; (d) Southern, J. S.; Green, M. T.; Hillhouse, G. L.; Guzei, I. A.; Rheingold, A. L. *Inorg. Chem.* **2001**, *40*, 6039.

66. Shafirovich, V.; Lymar, S. V. *Proc. Natl. Acad. Sci. USA* **2002**, *99*, 7340.

67. Alluisetti, G.; Almaraz, A. E.; Amorebieta, V. T.; Doctorovich, F.; Olabe, J. A. *J. Am. Chem. Soc.* **2004**, *126*, 13432.

68. Einsle, O.; Messerschmidt, A.; Huber, R.; Kroneck, P. M. H.; Neese, F. *J. Am. Chem. Soc.* **2002**, *124*, 11737.

69. (a) Clarkson, S. G.; Basolo, F. *Inorg. Chem.* **1973**, *12*, 1528; (b) Kubota, M.; Phillips, D. A. *J. Am. Chem. Soc.* **1975**, *97*, 5637.

70. Hauser, U.; Oestreich, V.; Rohrweck, H. D. *Z. Phys. A* **1977**, *280*, 17.

71. Güida, J. A.; Piro, O. E.; Shaiquevich, P. S.; Aymonino, P. J. *Solid State Commun.* **1986**, *57*, 175; **1988**, *66*, 1007.

72. Zöllner, H.; Woike, Th.; Krasser, W.; Haussühl, S. *Z. Kristallogr.* **1989**, *188*, 139.

73. Gutlich, P.; Garcia, Y.; Woike, Th. *Coord. Chem. Rev.* **2001**, *219–221*, 839.

74. Yang, Y. Y.; Zink, J. I. *J. Am. Chem. Soc.* **1985**, *107*, 4799.

75. Delley, B.; Schefer, J.; Woike, Th. *J. Chem. Phys.* **1997**, *107*, 10067.

76. Schaniel, D.; Woike, Th.; Merschjann, C.; Imlau, M. *Phys. Rev. B* **2005**, *72*, 195119.

77. Rusanov, V.; Stankov, Sv.; Trautwein, A. H. *Hyperfine Interact.* **2002**, *144/145*, 307.

78. Paulsen, H.; Rusanov, V.; Benda, R.; Herta, C.; Schünemann, V.; Janiak, C.; Dorn, T.; Chumakov, A. I.; Winkler, H.; Trautwein, A. X. *J. Am. Chem. Soc.* **2002**, *124*, 3077.

79. Chacón Villalba, M. E.; Güida, J. A.; Varetti, E. L.; Aymonino, P. J. *Inorg. Chem.* **2003**, *42*, 2622.

80. Morioka, Y.; Takeda, S.; Tomizawa, H.; Miki, E. *Chem. Phys. Lett.* **1998**, *292*, 625.

81. (a) Cheng, L.; Novozhilova, I.; Kim, C.; Kovalevsky, A.; Bagley, K. A.; Coppens, P.; Richter-Addo, G. E. *J. Am. Chem. Soc.* **2000**, *122*, 7142; (b) Lee, J.; Kovalevsky, A. Y.;

Novozhilova, I. V.; Bagley, K. A.; Coppens, P.; Richter-Addo, G. B. *J. Am. Chem. Soc.* **2004**, *126*, 7180.

82. Schaniel, D.; Woike, Th.; Delley, B.; Biner, D.; Krämer, K. W.; Güdel, H. U. *Phys. Chem. Chem. Phys.* **2007**, *9*, 5149.

83. Crichton, O.; Rest, J. *J. Chem. Soc., Dalton Trans.* **1977**, 986.

84. Schaiquevich, P. S.; Güida, J. A.; Aymonino, P. J. *Inorg. Chim. Acta* **2000**, *277*, 277.

85. Wolfe, S. K.; Swinehart, J. H. *Inorg. Chem.* **1975**, *14*, 1049.

86. Videla, M.; Braslavsky, S. E.; Olabe, J. A. *Photochem. Photobiol. Sci.* **2005**, *4*, 75.

87. (a) Prakash, R.; Czaja, A. U.; Heinemann, F. H.; Sellmann, D. *J. Am. Chem. Soc.* **2005**, *127*, 13758; (b) Sauaia, M. G.; de Souza Oliveira, F.; Tedesco, A. C.; Santana da Silva, R. *Inorg. Chim. Acta* **2003**, *355*, 191; (c) de Souza Oliveira, F.; Togniolo, V.; Tedesco, A. C.; Santana da Silva, R. *Inorg. Chem. Commun.* **2004**, *7*, 160; (d) Tfouni, E.; Krieger, M.; McGarvey, B. R.; Franco, D. F. *Coord. Chem. Rev.* **2003**, *236*, 57.

88. Rose, M. J.; Mascharak, P. K. *Curr. Opin. Chem. Biol.* **2008**, *12*, 238.

89. Bohle, D. S. *Curr. Opin. Chem. Biol.* **1998**, *2*, 194.

90. (a) Groves, J. T. *Curr. Opin. Chem. Biol.* **1999**, *3*, 226; (b) Koppenol, W. H.In *Metals in Biology*; Siegel, H., Ed.; **1999**; Vol. 36, p. 597 Marcel Dekker, New York; (c) Szabo, C.; Ischiropoulos, H.; Radi, R. *Nat. Rev.* **2007**, *6*, 662.

91. Goldstein, S.; Lind, J.; Merénhi, G. *Chem. Rev.* **2005**, *105*, 2457.

92. (a) Herold, S. *FEBS Lett.* **1998**, *439*, 85; (b) Olson, J. S.; Foley, E. W.; Rogge, C.; Tsai, A. L.; Doyle, M. L.; Lemon, D. D. *Free Radic. Biol. Med.* **2004**, *36*, 685.

93. Herold, S.; Koppenol, W. H. *Coord. Chem. Rev.* **2005**, *249*, 499.

94. Bakac, A. *Adv. Inorg. Chem.* **2004**, *55*, 1.

95. (a) Wick, P.; Kissner, R.; Koppenol, W. H. *Helv. Chim. Acta* **2000**, *83*, 748; (b) Pestovsky, O.; Bakac, A. *J. Am. Chem. Soc.* **2002**, *124*, 1698.

96. Maiti, D.; Lee, D. H.; Narducci Sarjeant, A. A.; Pau, M. Y. M.; Solomon, E. I.; Gaoutchenova, K.; Sundermeyer, J.; Karlin, K. D. *J. Am. Chem. Soc.* **2008**, *130*, 6700.

97. (a) Andersen, H. J.; Skibsted, L. H. *J. Agric. Food Chem.* **1992**, *40*, 1741; (b) Arnold, E. V.; Bohle, D. S. *Methods Enzymol.* **1996**, *269*, 41; (c) Moller, J. K. S.; Skibsted, L. H. *Chem. Eur. J.* **2004**, *10*, 2291.

98. Nemes, A.; Pestovsky, O.; Bakac, A. *J. Am. Chem. Soc.* **2002**, *124*, 421.

99. Goodwin, J. A.; Coor, J. L.; Kavanagh, D. F.; Sabbagh, M.; Howard, J. W.; Adamec, J. R.; Parmley, D. J.; Tarsis, E. M.; Kurtikyan, T. S.; Hovhannisyan, A. A.; Desrochers, P. J.; Standard, J. M. *Inorg. Chem.* **2008**, *47*, 7852.

100. Lerdau, M. T.; Munger, J. W.; Jacob, D. J. *Science* **2000**, *289*, 2291.

101. Huie, R. E.; Neta, P. *J. Phys. Chem.* **1986**, *90*, 1193.

102. Pestovsky, O.; Bakac, A. *Inorg. Chem.* **2003**, *42*, 1744.

103. Ghosh, P.; Stobie, K.; Bill, E.; Bothe, E.; Weyhermuller, T.; Ward, M. D.; McCleverty, J. A.; Wieghardt, K. *Inorg. Chem.* **2007**, *46*, 522.

# 8 Ligand Substitution Dynamics in Metal Complexes

THOMAS W. SWADDLE

## 8.1 INTRODUCTION

With the passing of Nobel Laureate Henry Taube in 2005 and Priestley Medalist Fred Basolo in 2007, the era of pioneering studies in inorganic reaction mechanisms drew to a close. In particular, the field of the kinetics and the mechanism of ligand substitution in metal complexes in solution, which was launched by the seminal reviews of Taube in 1952[1] and Basolo in 1953[2] and gained much momentum from Basolo and Pearson's classic monograph of 1958[3] (revised 1967[4]), may now be considered to be a mature discipline in which relatively few fundamental developments are anticipated. It is nevertheless of increasing importance in disciplines such as aquatic geochemistry, organometallics, biomedical sciences, catalysis, and computational chemistry.

A vast amount of information is available on ligand substitution dynamics and has been amply summarized and discussed in several monographs,[3–15] conference proceedings,[16,17] specialized series[18,19] and reviews[1,2,20–22] that may be consulted for detailed information. A comprehensive review here would be redundant. This chapter seeks instead to give an overview of some basic features of the field, concentrating on the factors that control the rates of substitution reactions of simple metal ion complexes (Werner complexes) in solution and providing a historical perspective on particular issues that have attracted the author's attention over the past five decades. Special consideration will be given to solvent exchange reactions, which continue to attract interest as simple processes that expose the intrinsic substitutional reactivity of metal ions, provide tractable systems for computational studies, and, in current studies,[23] serve as models for geochemical processes. In the spirit of the title of this book, the focus will be on inorganic systems; reactions of biological molecules[24] and organometallic mechanisms[15] will not be considered. Photochemical reactions of transition metal complexes are reviewed in Chapter 6 in Volume 1.

*Physical Inorganic Chemistry: Reactions, Processes, and Applications*   Edited by Andreja Bakac
Copyright © 2010 by John Wiley & Sons, Inc.

## 8.2    THERMODYNAMICS, KINETICS, AND MECHANISM

A chemical reaction will proceed to products in significant yield only if the standard Gibbs free energy change of reaction $\Delta G^0$ is negative or at worst slightly positive, since the equilibrium constant $K^0$ for the reaction under standard conditions is given by

$$\ln K^0 = -\Delta G^0/RT \qquad (8.1)$$

Neither $\Delta G^0$ nor $K^0$ depends upon the mechanism of the reaction. But even if $\Delta G^0$ is strongly negative, the yield of thermodynamically favored products may be negligible if the reaction proceeds too slowly on the human timescale or if the slowness of a critical step in the reaction sequence relative to some alternative steers the reaction to other (metastable) products. Thus, as Taube[1] emphasized, we need to understand what makes some ligand substitution reactions fast and others slow. The *mechanism* of reaction is a simplified hypothetical model, an approximation to reality that purports to trace the progress of the system from reactants to products, and is significant only insofar as it helps us understand the kinetics and stereochemistry of the reaction (rather than vice versa as some workers tend to believe).

For example, $Co(NH_3)_6{}^{3+}$ is thermodynamically unstable in aqueous acidic solution with respect to decomposition to $Co^{3+}(aq) + 6NH_4{}^+$, for which $\Delta G^0 = -186$ kJ/mol under standard conditions;[25] in fact, the instability is greater than this quantity because $Co^{3+}(aq)$ itself undergoes spontaneous reduction to $Co(H_2O)_6{}^{2+}$. Yet such is the *kinetic* inertness of cobalt(III) ammine and amine complexes that much of the early work on the kinetics and mechanism of reactions of metal complexes[1-7,26] focused on cobalt(III) complexes because they provide relatively unreactive frameworks against which processes occurring at an active site can be investigated and products can be unambiguously isolated and identified (e.g., in the context of stereochemical change). As a measure of this kinetic inertness, the initial step in the hydrolysis of $Co(NH_3)_6{}^{3+}$ in aqueous $HClO_4$

$$Co(NH_3)_6{}^{3+} + H^+ + H_2O \rightarrow Co(NH_3)_5OH_2{}^{3+} + NH_4{}^+ \qquad (8.2)$$

has a $[H^+]$-independent first-order rate constant $k_{8.2}$ that is conveniently measurable only at temperatures well above the normal boiling point ($k_{8.2} = 8 \times 10^{-5}\,s^{-1}$ at 141°C and the saturated vapor pressure) and is negligible at near-ambient temperatures because of a high enthalpy of activation ($\Delta H_{8.2}^{\ddagger} = 153$ kJ/mol). For the second step

$$Co(NH_3)_5OH_2{}^{3+} + H^+ + H_2O \rightarrow Co(NH_3)_4(OH_2)_2{}^{3+} + NH_4{}^+ \qquad (8.3)$$

there is a similar $[H^+]$-independent hydrolysis pathway with $k_{8.3} = 1.3 \times 10^{-4}\,s^{-1}$ (141°C) and $\Delta H_{8.3}^{\ddagger} = 175$ kJ/mol, but this is accompanied by a pathway with dependence on $1/[H^+]$ ($k_{8.3}' = 6 \times 10^{-5}$ mol/(L s), $\Delta H_{8.3}'^{\ddagger} = 182$ kJ/mol). In subsequent steps, paths with inverse dependence on $[H^+]$ become dominant and the incursion of relatively rapid redox processes leading to $Co^{2+}(aq)$, $NH_4^+$, $N_2$, $N_2O$, and some $O_2$ obscures simple ligand substitution.[27]

Several points emerge from this example. First, what is described above is the *stoichiometric mechanism*—that is, the progress from reactants to products via identifiable chemical intermediates—that evidently proceeds preferentially by paths facilitated by dissociation of $Co-OH_2^{3+}$ to give $Co-OH^{2+} + H^+$ (whence the inverse $[H^+]$ dependence) once bound aqua ligands become available. Such acid dissociation of aqua ligands is markedly endothermic and consequently becomes very important in aqueous systems at high temperatures.[28] Second, $H^+$ does not catalyze ammine displacement, despite its presence in Equation 8.2; in other words, although a chemically balanced equation dictates the equilibrium expression, it does not necessarily tell us what mechanism the reaction will follow (although the forward rate expression divided by that for the reverse reaction must give the equilibrium equation). Third, because the concentration of water, as the solvent, remains effectively constant, the kinetics give no information on the *intimate mechanism* of hydrolysis—that is, whether or not an incoming water molecule begins to bind to the $Co^{III}$ center before the bond to the departing ammine ligand is completely broken (associative and dissociative activation, respectively)—and this is seen to be a minor issue in this context. Finally, as Taube recognized,[1] the relative inertness of cobalt(III) ammines toward substitution can be traced to the electronic configuration of the central metal ion (spin-paired $3d^6$, or $t_{2g}^6 e_g^0$ in $O_h$ symmetry), as discussed in Section 8.4.1.

Other examples of kinetic inertness of specific metal complexes thwarting thermodynamic exigency abound. Thus, hexacyanoferrates(II) and –(III) (ferro- and ferricyanide ions) have low short-term toxicity in acid solution such as in the stomach, but in time hydrolyze to liberate highly toxic HCN.

## 8.3  MECHANISTIC CLASSIFICATIONS

Substitution mechanisms of transition metal complexes differ from those of familiar organic reactions in several respects:

(a) Transition metal complexes can often expand or reduce their coordination numbers to give intermediates of significant stability. By contrast, in aliphatic substitution, three- or five-coordination at carbon is transient, often occurring on the timescale of a single molecular vibration.

(b) Metal complexes are frequently charged, resulting in important electrostatic interactions such as ion pairing with ionic or dipolar incoming ligands.

(c) The wide variety of coordination geometries of metal complexes contrasts with the tetrahedral geometry of aliphatic carbon, leading to difficulties in dealing with the stereochemistry of substitution at metal centers (for example, it is not obvious whether there is an analogue of the Walden inversion in octahedral substitution).

(d) Transition metal complexes often engage readily in redox processes, so that an initial redox step may facilitate ligand substitution by generating a more labile intermediate.

(e) Nucleophilicity toward metal centers is not necessarily related to nucleophilicity toward carbon; moreover, Brønsted bases may induce substitution in metal complexes by initial abstraction of an acidic proton from a ligand rather than by nucleophilic attack at the metal center (a conjugate base (CB) reaction mechanism)—this is especially important in reactions of simple aquated metal ions, in which the bound aqua groups can be more acidic than, say, hydrofluoric or acetic acid, and in ammine complexes of $Co^{III}$ and $Ru^{III}$, which are highly susceptible to hydrolysis by the conjugate base route.

Consequently, the enormous body of mechanistic information on substitution at carbon has been of only limited value in understanding substitution in metal complexes.

### 8.3.1 First Steps

Early attempts to establish mechanistic models for substitution at metal centers used the labels $S_N1$ and $S_N2$ (substitution, nucleophilic, unimolecular or bimolecular) inherited from Ingold's attempts[26] to extend his classic studies in organic reaction mechanisms to substitution at elements other than carbon. Unfortunately, Ingold's attribution of the discontinuity in reaction rates in the replacement of one $Cl^-$ in $cis$-$Co(en)_2Cl_2^+$ by various anions $A^-$ in methanol in the sequence

$$NO_3^- = Cl^- = NCS^- < NO_2^- < N_3^- \ll CH_3O^- \qquad (8.4)$$

to a changeover from $S_N1$ to $S_N2$ due to increasing nucleophilic power of $A^-$ turned out to be wrong. As successfully argued by Basolo and Pearson,[3,4] the faster rates for $NO_2^-$, $N_3^-$, and especially $CH_3O^-$ arise because of proton abstraction from the $-NH_2$ groups of en (ethylenediamine (1,2-diaminoethane)) by these Brønsted bases, generating a conjugate base complex that is much more labile than the parent $cis$-$Co(en)_2Cl_2^+$ ($NO_3^-$, $Cl^-$, and $NCS^-$ have negligible Brønsted basicity). Since other evidence (Sections 8.4.1 and 8.5) indicates that ligand substitution at $Co^{III}$ centers is generally dissociatively activated, Basolo and Pearson labeled the mechanism for $Cl^-$ substitution by $A^-$ in $cis$-$Co(en)_2Cl_2^+$ by the first three $A^-$ in sequence 8.4 as $S_N1$, as did Ingold, and the latter three as $S_N1CB$. Convincing evidence for the $S_N1CB$ interpretation over $S_N2$ in $Co^{III}$ complexes for the simple replacement of a unidentate ligand comes from the absence of a dependence of the hydrolysis rates of complexes on the concentration of Brønsted bases when there are no abstractable protons present on the "spectator" (i.e., nonreacting) ligand atoms, as in $trans$-$Co(py)_4Cl_2^+$ (py = pyridine), $cis$-$Co(bpy)_2(OAc)_2^+$ (bpy = 2,2'-bipyridyl), $trans$-$Co(bpy)_2(NO_2)_2^+$, $trans$-$Co(tep)_2Cl_2^+$ (tep = $(C_2H_5)_2PCH_2CH_2P(C_2H_5)_2$), $Co(diars)_2Cl_2^+$ (in methanol) (diars = $o$-phenylenebis(dimethylarsine)), $Co(CN)_5Br^{3-}$, and $Co(CN)_5I^{3-}$. The last two cases show that this effect is unrelated to the overall charge on the complex.[4]

Most water-soluble metal complexes, unlike typical organic reactants, are cations or anions and are therefore subject to Coulombic ion–ion interactions in solution. In essence, these are of two kinds: the Debye–Hückel or ionic atmosphere type, which affects the activity coefficients of the complex ion and hence the kinetics of its reactions, and ion association—usually considered simply as anion–cation pair formation.[29] For cationic substrates in particular, pairing with an anionic incoming ligand may give an illusion of bimolecularity (an $S_N2$ mechanism) when in fact the reaction may be dissociatively activated within the ion pair or "encounter complex".

Furthermore, for reactions in water that involve a change in charge of the complex ion, or ones in which it is desired to vary the $H^+$ concentration substantially to investigate conjugate base pathways, it is customary to use a swamping concentration of an electrolyte with negligible complexing tendency—usually an alkali metal perchlorate or, to avoid risk of explosion, trifluoromethanesulfonate (triflate)—to maintain constant ionic strength $I$ with the expectation that activity coefficients will then remain constant over the course of the reaction. The risk is then that interactions between the reactant and the swamping electrolyte, or diminution of the activity of the solvent if very high concentrations are used, may mask the very factors that one is trying to expose.

Yet another complication in mechanistic assignment in donor solvents is that a solvent complex intermediate may be the initial product, so that formation of the final product with the expected incoming ligand may involve displacement of a solvent ligand rather than the ligand originally present.

In summary, the $S_N1/S_N2$ dichotomy, imported from organic chemistry in the 1950s and based on molecularity, is not well suited to describing substitution reaction mechanisms of complex ions in solution. Besides, not all ligands in reactions of metal complexes (e.g., $NO^+$) are appropriately described as nucleophiles.

### 8.3.2    The Langford–Gray Classification

In light of the foregoing, Langford and Gray[5] proposed a nomenclature comprising three *operationally distinguishable* mechanistic classes without reference to nucleophilicity:

1. *Associative* (**A**) The incoming ligand binds to the metal center in the first step to form an intermediate of expanded coordination number, "detectable by departure from strict second-order kinetics when the reaction is followed in the direction for which the transition state lies after the intermediate."[5]

2. *Dissociative* (**D**) The departing ligand leaves completely in the first step, forming an intermediate of reduced coordination number that is detectable by its selective reactivity.

3. *Interchange* (**I**) A concerted process (i.e., one without detectable intermediates) in which the departure of the leaving group and the entry of the incoming ligand occur on a timescale that is short relative to the lifetime of the encounter complex or second coordination sphere (solvation sheath, etc.). If the reaction

rate is insensitive to the nature of the entering ligand, the reaction may be said to be dissociatively activated (dissociative interchange, $\mathbf{I_d}$); conversely, if the rate shows comparable dependences on the nature of both entering and leaving groups, the reaction may be described as associatively activated (associative interchange, $\mathbf{I_a}$).

The interchange mechanism accommodates not only reactions within ion-paired precursor assemblages but also those in which the essential steps take place within a relatively immobile "cage" of solvent or other molecules.

The Langford–Gray system has gained wide acceptance among inorganic chemists, but, as may be expected with any system of scientific models, it has some limitations. In particular, with regard to the $\mathbf{I_a}/\mathbf{I_d}$ distinction, just how much sensitivity to the nature of the incoming ligand is required to warrant the $\mathbf{I_a}$ label? Furthermore, the notion of *operational* definitions (i.e., the identification of a scientific concept with a *unique set* of operations that defines it), due originally to Bridgman,[30] may lead to the assignment of an $\mathbf{I_a}$ mechanism by one type of measurement and $\mathbf{I_d}$ by another. Similarly, kinetic evidence for the existence of intermediates or lack thereof may conflict with spectroscopic or other information, and in any event departure from apparent second-order kinetics is also characteristic of $\mathbf{I}$ mechanisms that occur within ion pairs as noted below. Ironically, Bridgman, who pioneered the use of high-pressure experimentation that has proved so valuable in mechanistic chemistry, had a profound distaste for mechanistic models.[30,31]

### 8.3.3 Mechanistic Criteria

**8.3.3.1 The Associative Mechanism**    The $\mathbf{A}$ mechanism can be expected to predominate in complexes that can expand their coordination number relatively easily. The obvious candidates are the square planar complexes of $Pt^{II}$, $Pd^{II}$, and $Au^{III}$, for which many comparable five-coordinate complexes are also known. The salient feature of ligand substitution kinetics in these square planar complexes, however, is not the detectability of five-coordinate intermediates—the reaction rates are usually accurately second order overall—but rather the wide-ranging dependence of the second-order rate constants on the nature of the incoming ligand. This implies that in the scheme

$$ML_3X + A \underset{k_{-1}}{\overset{k_1}{\rightleftharpoons}} ML_3XA \overset{k_2}{\longrightarrow} ML_3A + X \tag{8.5}$$

the $k_1$ step is rate determining, such that $[ML_3XA]$ is always small and $d[ML_3XA]/dt \approx 0$ (the "steady-state approximation"), in which case the observed reaction rate is

$$d[ML_3A]/dt = \{k_1k_2/(k_{-1} + k_2)\}[ML_3X][A] = k_{obsd}[ML_3X][A] \tag{8.6}$$

Thus, if $[A] \gg [ML_3X]$ so that $[A]$ does not vary significantly over the course of the reaction, a pseudo-first-order rate constant $k'_{obsd} = k_{obsd}[A]$ measured at each of

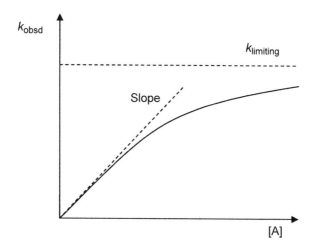

**FIGURE 8.1** Dependence of observed rate constant $k_{obsd}$ on the concentration of the incoming ligand A, with A in large excess.

several [A] should be a linear function of [A]; any falloff from linearity at the higher [A] as in Figure 8.1 would imply (if ionic strength effects can be neglected) that $[ML_3XA]$ is no longer small and $k_1$ is no longer exclusively rate determining. In the extreme case, the rate-determining step shifts to $k_2$—that is, the intermediate $ML_3XA$ forms as a substantial fraction of the total $[ML_3X]$ in a relatively rapid preequilibrium with a formation constant $K_{int} = k_1/k_{-1}$, and the rate equation for A in excess, when depletion of $ML_3X$ is allowed for, becomes

$$d[ML_3A]/dt = \{k_2 K_{int}[A]/(1 + K_{int}[A])\}[ML_3X]_{total} \qquad (8.7)$$

in which the term in braces becomes independent of [A] at sufficiently large [A] (i.e., if $K_{int}[A] \gg 1$), so that $k_{obsd}$ tends to a limit as in Figure 8.1. Such cases are not normally encountered in practice, but the point is made that a departure of a plot of $k'_{obsd}$ versus [A] from linearity is not necessarily inconsistent with an **A** mechanism—in fact, the definition of this mechanism by Langford and Gray (Section 8.3.2) specifically accommodates this. We return to Equation 8.7 below in consideration of interchange mechanisms.

The key characteristic of associative activation, then, is not necessarily conformity to Equation 8.6 but rather a marked dependence of the reaction rate on the nature of the incoming ligand A. In the case of ligand exchange reactions (i.e., where bound X and free X exchange), this criterion is inapplicable, and the defining features are strongly negative entropies $\Delta S^{\ddagger}$ and volumes $\Delta V^{\ddagger}$ of activation, arising mainly from loss of translational degrees of freedom and from compaction, respectively, on bond formation between the incoming ligand and the metal center.

The possibility exists, however, that associatively and dissociatively activated substitution pathways could coexist. Early work[5] on the replacement kinetics of X by A in $ML_3X$ (mostly for $M = Pt^{II}$) suggested rate equations of the type

$$d[ML_3A]/dt = k_{obsd}[ML_3X] = (k_1 + k_2[A])[ML_3X] \tag{8.8}$$

in which the very small contribution from the [A]-independent $k_1$ term implied the presence of a minor dissociative pathway operating in parallel with the greatly predominant **A** pathway represented by the $k_2$ term. For example, Gray and Olcott[32] studied the replacement of the aqua ligand in $Pt(dien)OH_2^{2+}$ (dien = diethylene-triamine) by the relatively weak nucleophiles $Cl^-$ and $NO_2^-$ in aqueous solution, and reported a very small intercept in plots of $k_{obsd}$ versus [A], corresponding to a barely detectable contribution (first-order rate constant $\sim 3 \times 10^{-4} \, s^{-1}$ at 25°C) from a dissociatively activated pathway. Kotowski et al.,[33] however, repeated this study with an extended set of A and found the intercepts on the $k_{obsd}$ axis to be zero within the experimental uncertainty in all cases. Unambiguous cases of dissociatively activated substitution in square planar complexes of $Pt^{II}$ seem to be confined to *cis* organo-metallic complexes such as *cis*-$Pt(Ph)_2(Me_2SO)_2$[34] (Ph = phenyl; Me = methyl), *cis*-$Pt(Ph)_2(Me_2S)_2$,[35] and $Pt(bph)(SR_2)_2$[36] in nonaqueous solvents; characteristically, these pathways show no dependence of rate on the nature of the entering ligand and, for ligand exchange reactions, strongly positive entropies of activation.

### 8.3.3.2 The Dissociative Mechanism

In the **D** mechanism for substitution of X in $ML_nX$, the M–X bond is completely broken and the $ML_n$ intermediate has "lost all memory" of X before bonding to the incoming ligand A begins. Thus, applying the steady-state approximation $(d[ML_n]/dt \approx 0)$ to the mechanism 8.9

$$ML_nX \underset{k_r}{\overset{k_f}{\rightleftharpoons}} ML_n + X \xrightarrow{+A,k_A} ML_nA + X \tag{8.9}$$

with the simplifying conditions that X is the solvent and A is in large excess (so that [A] and [X] remain effectively constant and [X] can be subsumed into $k_r$), we have

$$d[ML_nA]/dt = k_{obsd}[ML_nX], \text{ where } k_{obsd} = k_f k_A[A]/(k_r + k_A[A]) \tag{8.10}$$

This has the same form with respect to [A] as Equation 8.7—thus, the rate expression is not necessarily definitive of mechanism, but the physical meanings of the constituent rate constants will be different for **A** and **D** processes. The question then becomes whether these constituent rate constants can be identified from an independent type of experiment—in particular, when X is the solvent, whether $k_r$ derived from Equation 8.10 is equal to the rate constant $k_{ex}$ measured for solvent exchange.

For many years, the studies of Haim, Grassi, and Wilmarth[37–39] in the 1960s on the substitution of water in $Co^{III}(CN)_5OH_2^{2-}$ by various anions $A^-$, as well as the reverse processes, were generally considered to provide the archetypal demonstration of the **D** mechanism in an aqueous medium. These anion–anion reactions are good test cases in that there should be no tendency for a long-lived $Co(CN)_5^{2-}$ intermediate to hinder the departure (i.e., retain any "memory") of $A^-$. Indeed, initial findings were that the kinetic data appeared to fit Equation 8.10, as the **D** mechanism requires, but attempts to obtain $k_{ex}$ values for the $Co(CN)_5{}^{18}OH_2{}^{2-}/H_2O$ exchange reaction by precipitation of the Co complex were unsuccessful. In the 1980s, the system was revisited by Burnett et al.[40–42] and by Haim,[43] and doubts arose concerning the case for a common $Co(CN)_5{}^{2-}$ intermediate in all these reactions. For example, analysis of the kinetic data for the $Co(CN)_5OH_2{}^{2-}/A^-$ reactions according to Equation 8.10 requires that a departure from first-order dependence on $[A^-]$ be observed (cf. Figure 8.1), as was the case for $A^- =$ thiocyanate, but reexamination of the case $A^- =$ azide showed no significant deviation from first-order behavior. Furthermore, Burnett et al.[42] found a marked leaving group effect on the relative efficiencies of both anionic and neutral nucleophiles in competition for the putative $Co(CN)_5{}^{2-}$ intermediate in aquation of $Co(CN)_5A^{3-}$—in other words, the intermediate does *not* completely lose its "memory" of the departed ligand. Finally, in 1988, Bradley et al.[44] succeeded in measuring $k_{ex}$ for the $Co(CN)_5{}^{18}OH_2{}^{2-}/H_2O$ exchange directly by examining the $^{18}O$ content of the solvent and found a value moderately but distinctly higher $(3.5 \times 10^{-3}\,s^{-1}$ at 40°C) than any $k_r$ calculated from Equation 8.10 for the net substitution reactions $(6.1 \times 10^{-4}\,s^{-1}$[41] to $2.0 \times 10^{-3}\,s^{-1}$[43] at 40°C—the range in $k_r$ with different A is itself a hint that no common long-lived $Co(CN)_5{}^{2-}$ intermediate is involved). Thus, rigorous tests for a **D** mechanism fail in detail for this system, and the mechanism is better described as $\mathbf{I_d}$—that is, with a $Co(CN)_5{}^{2-}$ intermediate that is short lived relative to the relaxation of its molecular environment (including the departure of $A^-$). Nevertheless, the positive volume of activation $\Delta V_{ex}^{\ddagger}$ for the $Co(CN)_5OH_2{}^{2-}/H_2O$ exchange $(+7\,cm^3/mol^{-1}$; see Section 8.4)[44] together with similar values for the $Co(CN)_5OH_2{}^{2-}/A^-$ reactions (A = Br, I, NCS)[45] does support a strongly dissociative mode of activation for water displacement in $Co(CN)_5OH_2{}^{2-}$ at least.

For substitution in octahedral complexes in aqueous media, a true **D** mechanism is improbable, since the small molecular size of the abundant solvent water and its high affinity for metal ion centers will keep the lifetime of an intermediate of reduced coordination number short,[46] although arguments for a **D** mechanism can be made in special cases.[15] Certainly, complexes of reduced coordination number can coexist in solution in equilibrium with the parent complex, as in the case of $Co(H_2O)_4{}^{2+}$ that becomes detectable along with the majority species $Co(H_2O)_6{}^{2+}$ in visible spectra of $Co^{2+}$(aq) at elevated temperatures,[47–49] but these do not necessarily provide an effective pathway for ligand exchange (which, for water exchange on $Co(H_2O)_6{}^{2+}$, evidently occurs via an $\mathbf{I_d}$ mechanism involving transient pentacoordination[21]).

*8.3.3.3 Interchange Mechanisms*    As we proceed through the rest of this chapter, the following points will be made: (i) the **I** mechanism is best regarded as a continuum with varying degrees of $\mathbf{I_a}$ and $\mathbf{I_d}$ character, so that the widespread practice of insisting upon classifying a particular reaction as *either* $\mathbf{I_a}$ *or* $\mathbf{I_d}$ can be counterproductive; (ii) the mechanism of reaction for complexes of a particular metal ion $M^{z+}$ can be moved to and fro along the $\mathbf{I_a}$–$\mathbf{I_d}$ continuum by changing the ligand set; and (iii) a practical distinction between $\mathbf{I_a}$ and **A** mechanisms is the energetic inaccessibility of dissociative pathways relative to associative ones. At this point, however, we consider only the characteristics of the rate equation that typifies an **I** mechanism.

In an interchange mechanism, any intermediate of expanded or reduced coordination number ($\mathbf{I_a}$ and $\mathbf{I_d}$, respectively) is kinetically undetectable,[5] meaning in effect that it is short lived on the nanosecond timescale of relaxation of the surroundings of the reacting metal complex (solvation sheath, ion pair, etc.)—the act of substitution of X by A in $ML_nX$, if it occurs at all, is completed within the lifetime of the $\{ML_nX, A\}$ encounter complex. This does not mean that the observed substitution rate is necessarily fast; rather, the final ascent of the activation energy barrier will occur only very rarely in the course of picosecond-scale jostling within in the encounter complex, but, when it is successful, it can progress to products before the local environment has time to relax.[46] In other words, A is locally present as X leaves, and may ($\mathbf{I_a}$) or may not ($\mathbf{I_d}$) begin bonding to $ML_n$ before the M–X bond is completely broken, but in any event the "memory" of X is not lost before M–A bond formation is completed. In practice, the rates of $\mathbf{I_a}$ processes will continue to show a significant sensitivity to the identity of A after the thermodynamics of formation of the encounter complex has been taken into account, whereas those of $\mathbf{I_d}$ processes will not.

The encounter complex $\{ML_nX, A\}$ may be taken to form in a rapid (time frame 1 ns or shorter) preequilibrium with formation constant $K_{ec}$ and replacement of X by A to occur internally within it with a first-order rate constant $k_i$. For A in large excess over $ML_nX$ (i.e., [A] effectively constant over the course of the reaction), the observed pseudo-first-order rate constant $k_{obsd}$ at a given [A] will be given by

$$k_{obsd} = k_i K_{ec}[A]/(1 + K_{ec}[A]) \qquad (8.11)$$

which has the same overall form as Equations 8.7 and 8.10 but with different meanings for the parameters. Thus, the mathematical form of the dependence of substitution rates on [A] cannot distinguish between **A**, **D**, and $\mathbf{I_a/I_d}$ mechanisms without independent information to identify the constants—in all cases, a plot of $k_{obsd}$ versus [A] could in principle take the form of Figure 8.1. For the **I** mechanisms, in the limit of low [A] ($K_{ec}[A] \ll 1$), $k_{obsd}$ becomes proportional to [A] with slope $k_i K_{ec}$, whereas at high [A] $k_{obsd}$ will tend to a maximum $k_{limiting} = k_i$; if this limit is not closely approached, the internal substitution rate constant $k_i$ may still be extracted if $K_{ec}$ is independently known—for example, where the encounter complex is an ion pair (formation constant $K_{IP}$).

Independent measurements of ion pair formation, however, may give values of $K_{IP}$ that are technique dependent (i.e., $K_{IP}$, determined for charge transfer in an ion pair by UV spectroscopy, may not represent the same physical phenomenon as ion pair effects on electrode potentials, conductance, diffusion, or substitution kinetics). Theoretical approaches to estimating $K_{IP}$ by Bjerrum, Fuoss, Barthel, Justice and Justice, and others are useful as first approximations and serve to emphasize that low solvent dielectric constants (relative permittivity, $\varepsilon$) as well as high ionic charges are key factors in giving high $K_{IP}$ values, but all suffer from limitations—not least in defining what constitutes an ion pair.[29] For substitution of X = $H_2O$ in a cationic complex $ML_nX^{z+}$ in aqueous solution ($\varepsilon = 78.3$ at 25°C) by a univalent anion $A^-$ via an ion pair

$\{ML_nX^{z+}, A^-\}$, $K_{IP}$ is often not large enough for limiting behavior as in Figure 8.1 to be observed, particularly where L is also $H_2O$, and may not be very different from one $A^-$ to another. In such cases, any substantial variation in $k_{obsd}$ with $A^-$ for a given substrate (e.g., $Cr(H_2O)_6^{3+}$) may be taken as evidence for some degree of selectivity toward $A^-$—that is, to associative activation.

It has long been recognized[50] that substitution of X in $ML_nX$ by A in a donor solvent such as water may proceed by initial replacement of X by solvent, followed by replacement of the coordinated solvent by A. Thus, the replacement of coordinated solvent, either by another solvent molecule (solvent exchange, rate constant $k_{ex}$) or by a new ligand A, is of key importance in the study of ligand substitution processes. For the replacement of solvent in $ML_n$(solvent) by A, we need to know how $k_i$ compares with the rate constant $k_{ex(ec)}$ for solvent exchange within the $\{ML_n$(solvent), A$\}$ encounter complex, since $k_{ex(ec)}$ sets an upper limit for $k_i$ in $I_d$ processes (solvent being always in large excess). Conversely, if $k_i$ varies widely with the nature of A, and especially if it exceeds $k_{ex(ec)}$ in some cases, A is displaying its nucleophilicity and the mechanism is $I_a$. If, however, A is a Brønsted base and either the substrate or the solvent has ionizable protons, the conjugate base effects noted in Section 8.3.1 will obscure nucleophilic power.

It follows that kinetic evidence for the $I_a$ mechanism in solvent substitution reactions is best sought in nonaqueous donor solvents without readily ionizable protons, such as DMF (DMF = $N,N$-dimethylformamide), in which $\varepsilon$ is low enough to allow ion pairing to approach saturation (for DMF, $\varepsilon = 36.7$ at 25°C). Studies[51–53] of the kinetics of substitution of coordinated DMF on $Cr(DMF)_6^{3+}$ in DMF solvent by A = $Cl^-$, $Br^-$, $NCS^-$, or $N_3^-$ (pseudo-first-order rate constants $k_{obsd}$ at a given [A]) in conjunction with the rate of solvent exchange in both free $Cr(DMF)_6^{3+}$ (rate constant $k_{ex}^0$, observable with bulky tetraphenylborate as the counterion; perchlorate reduced $k_{ex}$ slightly through ion pairing) and ion pairs $\{Cr(DMF)_6^{3+}, A^-\}$ ($k_{exec}$) have provided a fairly complete picture of the dynamics of an $I_a$ process. Figure 8.2 summarizes a reevaluation of data from the literature[51–53] for the system $Cr(DMF)_6^{3+} + Cl^-$ in DMF at 71.1°C and shows the saturation in both $k_{obsd}$ (cf. Figure 8.1) and $k_{ex}$ brought about by ion pairing of $Cr(DMF)_6^{3+}$ with $A^- = Cl^-$ in DMF. The curves in Figure 8.2 are nonlinear least-squares fits of the data to Equations 8.11 and 8.12

$$k_{ex} = (k_{ex}^{lim}K_{IP}[A^-] + k_{ex}^0)/(1 + K_{IP}[A^-]) \tag{8.12}$$

which give the limiting rate constants $k_i = (9.1 \pm 0.4) \times 10^{-5}\,s^{-1}$ for substitution of $Cl^-$ for coordinated DMF and $k_{ex}^{lim} = (4.1 \pm 0.2) \times 10^{-5}\,s^{-1}$ for DMF exchange within the ion pair. From the chloride substitution data, $K_{IP} = (2.8 \pm 0.3) \times 10^2$ L/mol, while the less marked effect of $Cl^-$ on the DMF exchange data gives $K_{IP} = (2.2 \pm 0.6) \times 10^2$ L/mol, in agreement within the error limits, confirming that the same ion pairing is affecting both the substitution and solvent exchange processes. Thus, chloride substitution is slightly but distinctly faster than DMF exchange within the $\{Cr(DMF)_6^{3+}, Cl^-\}$ ion pair, consistent with an $I_a$ mechanism rather than $I_d$ in view of the predominance of DMF molecules over chloride.

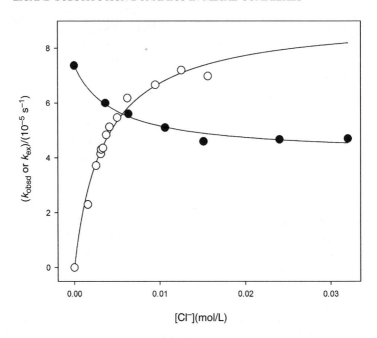

**FIGURE 8.2**  Dependence of rate constants $k_{obsd}$ (open symbols) for substitution of $Cl^-$ for one DMF, and $k_{ex}$ (filled symbols) for exchange of coordinated DMF with solvent, on $Cr(DMF)_6^{3+}$ in DMF solvent at 71.1°C.

Unequivocal evidence for associative activation within $\{Cr(DMF)_6^{3+}, A^-\}$ ion pairs in DMF is given in Table 8.1, which shows that azide ion attacks the Cr center some 100 times faster than solvent exchange and 650 times faster than bromide. This is of particular interest in that no comparable result can be obtained for the reaction of $N_3^-$ with $Cr(H_2O)_6^{3+}$ in water because of the Brønsted basicity of azide that causes the reaction to proceed via $HN_3$ and $Cr(H_2O)_5OH^{2+}$, but in anhydrous DMF the relative nucleophilic power of anionic reagents toward $Cr^{III}$ is clearly displayed: $N_3^- \gg NCS^- > Cl^- > Br^- \gg ClO_4^-, BPh_4^-$.

Several other significant results emerge from Table 8.1. Although perchlorate ion does not complex significantly with $Cr^{III}$ under these conditions, it is nevertheless seen to be involved in ion pairing, albeit with a relatively low $K_{IP}$, and this reduces the solvent exchange rate measurably. Ion pairing in general is seen to have only a limited effect on the solvent exchange rate constant. For net substitution of $A^-$ into $Cr(DMF)_6^{3+}$, $k_i$ varies 650-fold over a range of $A^-$ for which $K_{IP}$ varies only fourfold (Table 8.1), allowing a clear distinction to be made between ion association effects and genuine nucleophilicity of $A^-$. It can be seen from this and also from the solvent exchange data that ion pairing does not greatly influence the intrinsic substitutional reactivity of metal complexes. For aqueous reactions such as that of $Cr(H_2O)_6^{3+}$ with $A^-$, for which $K_{IP}$ is small enough at practical ionic strengths $I$ (e.g., for $\{Cr(H_2O)_6^{3+}, Cl\}$, $K_{IP} = 4.0$ L/mol at $I = 1.0$ mol/L and 25°C, falling to 0.3 L/mol for

**TABLE 8.1    Ion Pair Formation and Limiting Rate Constants Associated with Substitution of DMF by Anions A$^-$ and Solvent Exchange in Cr(DMF)$_6$$^{3+}$ in DMF Solution**[a]

| A$^-$ | $k_i$ ($10^{-5}$ s$^{-1}$) | $k_{ex}^{lim}$ ($10^{-5}$ s$^{-1}$) | $K_{IP}$ ($10^2$ L/mol) |
|---|---|---|---|
| None[b] | – | 7.4 | – |
| ClO$_4$$^-$ | – | 5.7 | 0.3 |
| Br$^-$ | 0.65 | 4.4 | 1.4[c] |
| Cl$^-$ | 9.1 | 4.1 | 2.8 |
| NCS$^-$ | 17.3 | ~5.5 | 0.73 |
| N$_3$$^-$ | 424.0 | | 3.3[d] |

[a]Data from Refs 51–53 recalculated for the case A $=$ Cl; 71.1°C except as stated.
[b]Counterion tetraphenylborate (nonpairing, noncomplexing).
[c]90°C.
[d]50°C.

$I = 4.4$ mol/L[54]) that curvature of plots such as Figure 8.1 is barely detectable, the implication is that the relative specific rate constants $k_{obsd}/[A^-]$ give at least an approximate measure of the nucleophilic power of A$^-$ (other than Brønsted bases) toward the metal center.

Finally, the utility of solvent exchange reactions in assessing the reactivities of metal centers will be evident, since solvent exchange is a symmetrical process with zero free energy change (except for small isotope effects when labeled solvent is used to follow the exchange process—e.g., DMF-$h_7$ replacing DMF-$d_7$ in the foregoing example). Furthermore, the complicating effects of solvational changes that occur when anionic reagents attack or are released from cationic metal centers are avoided in solvent exchange reactions. Thus, the influence of metal ion properties such as charge, size, and electronic configuration are most clearly exposed in solvent exchange reactions. While it is true that one cannot obtain mechanistic information on solvent exchange reactions by varying the reagent (solvent) concentration (except possibly in mixed solvents—Section 8.4.2), a mechanistic interpretation of the effect of pressure $P$ on $k_{ex}$, usually expressed as a volume of activation $\Delta V_{ex}^{\ddagger}$

$$\Delta V^{\ddagger} = -RT(\partial \ln k/\partial P)_T \qquad (8.13)$$

is relatively straightforward for solvent exchange because of the absence of solvational change and of differences between the incoming and outgoing ligands. For associatively activated solvent exchange processes, one expects $\Delta V_{ex}^{\ddagger}$ to be negative because of bond formation en route to the transition state, and conversely for dissociatively activated solvent exchange, $\Delta V_{ex}^{\ddagger}$ is expected to be positive because of the predominance of bond breaking.

This mechanistic use of $\Delta V_{ex}^{\ddagger}$ was first proposed in 1958 by Taube,[55] who found $\Delta V_{ex}^{\ddagger} = +1.2$ cm$^3$/mol for the exchange of solvent water on Co(NH$_3$)$_5$OH$_2$$^{3+}$ consistent with an I$_d$ mechanism as expected from other information. Taube, in conversation with the present author in 1966, conceded that the smallness of this pressure effect had discouraged him from pursuing the idea further, but agreed that it might be worth revisiting for other solvent exchange reactions. Indeed, in 1971 a more substantial pressure effect on a solvent exchange rate was found for the exchange of water on $Cr(H_2O)_6^{3+}$:[56] $\Delta V_{ex}^{\ddagger} = -9.3 \, cm^3/mol$ (subsequently revised to $-9.6 \, cm^3/mol$ following correction for the concurrent conjugate base pathway[57]), indicating an $\mathbf{I_a}$ mechanism as suggested by other information[58] (Section 8.5). Since then, $\Delta V_{ex}^{\ddagger}$ measurements have become a routine tool in the elucidation of inorganic substitution mechanisms.[21,59] For the $Cr(DMF)_6^{3+}$/DMF exchange discussed above, $\Delta V_{ex}^{\ddagger} = -6.3 \, cm^3/mol$, confirming that this cation reacts by an associatively activated mechanism.[52]

## 8.4   THE SIMPLEST LIGAND SUBSTITUTION REACTIONS: SOLVENT EXCHANGE

Slow solvent exchange rates at metal centers are generally measured either mass spectrometrically by use of isotopically labeled solvent tracers such as $H_2^{18}O$ if the exchange can be efficiently quenched during sampling (often followed by chemical conversion, for example, of $H_2O$ to $CO_2$ for mass spectrometry),[60] or by *in situ* measurement of NMR line intensity evolution if the tracer contains a NMR-active nucleus. For many faster reactions, NMR relaxation methods can give residence times of an NMR-active nucleus in the solvent (e.g., $H_2^{17}O$) within the first coordination sphere of the metal from line-broadening measurements and indeed most data on the solvent exchange phenomenon have come from this technique. The rates of some solvent exchange reactions of particular interest, such as $Mo(H_2O)_6^{3+}/H_2O$ (which also presents difficulties due to high sensitivity to $O_2$), fall between the effective time frames of the available techniques, while others such as $In^{3+}(aq)/H_2O$ and $Zn^{2+}(aq)/H_2O$ are too fast for the dynamic NMR method.[61] The very fast rates (subnanosecond timescales) for water exchange on most metal ions of periodic groups 1 and 2 lie outside the NMR time frame and have been inferred indirectly from other methods such as ultrasound absorption.

The experimental limitations of all these methods, however, should be borne in mind when solvent exchange rates are considered. The mass spectrometric method may be invalidated by incomplete quenching or by "induced exchange" or isotopic fractionation resulting from chemical reaction during the sample isolation procedure. NMR line broadening can be caused by several factors other than chemical exchange—measurements over a range of temperatures are necessary to distinguish chemical kinetics from physical factors, but the separation may be incomplete so that the temperature dependences of the derived $k_{ex}$ may themselves be subject to serious errors. Thus, Helm and Merbach[59] note that the 11 $k_{ex}$ values reported between 1967 and 1989 for the exchange of $Ni(CH_3CN)_6^{2+}$ with $CH_3CN$ solvent, as measured by $^1H$ or $^{14}N$ NMR line broadening, ranged from $2.0 \times 10^3$ to $14.5 \times 10^3 \, s^{-1}$ at 25°C. Worse still, the enthalpies $\Delta H_{ex}^{\ddagger}$ and the corresponding entropies $\Delta S_{ex}^{\ddagger}$ of activation extracted from the temperature dependence of the line broadening ranged from 40 to 68 kJ/mol and from $-33$ to $+50$ J/(K mol), respectively. At least part of this excessive scatter is due to enthalpy–entropy error correlation—that is, an error $\delta H$ in $\Delta H_{ex}^{\ddagger}$ implies a compensatory error $T\delta S$ in $\Delta S_{ex}^{\ddagger}$ such that $k_{ex}$ may be a reliable parameter even if $\Delta H_{ex}^{\ddagger}$ and $\Delta S_{ex}^{\ddagger}$ are not. The latter parameters from NMR line broadening are therefore not tabulated in this article (values are available in Refs [15,21,22], and [59]). Furthermore, the extraction of $\Delta S^{\ddagger}$ from an Eyring plot involves in effect an extrapolation to infinite temperature. In any event, the common supposition that the sign of $\Delta S_{ex}^{\ddagger}$ might be used as a criterion of mechanism (negative for $\mathbf{I_a}$, positive for $\mathbf{I_d}$) is unreliable for solvent exchange data derived from NMR relaxation.

Volumes of activation $\Delta V_{ex}^{\ddagger}$ for solvent exchange on metal ions are less susceptible to compensatory errors than are $\Delta H_{ex}^{\ddagger}$ and $\Delta S_{ex}^{\ddagger}$, since in the integrated form of Equation 8.13 assuming constant $\Delta V_{ex}^{\ddagger}$

$$\ln k = \ln k^0 - P\Delta V^{\ddagger}/RT \tag{8.14}$$

there is no long extrapolation to evaluate the other parameter, the logarithm of the rate constant $k^0$ at zero applied pressure—which, unlike $\Delta S_{ex}^{\ddagger}$, is directly measurable. If, however, $k_{ex}$ is obtained by NMR line broadening, the variable pressure experiment must be run under conditions at which $k_{ex}$ is the only significant contributor to the line width (corrected for field inhomogeneity, etc.). In practice, the simple linear Equation 8.14 fits most solvent exchange rate data within the experimental uncertainty—that is, $\Delta V_{ex}^{\ddagger}$ is indeed sensibly constant over the usual experimental applied pressure range (0–200 MPa). This is because significant pressure dependences of $\Delta V^{\ddagger}$ in ligand substitution reactions are typically due to solvational change (e.g., in aquation of $Co(NH_3)_5SO_4^+$ to $Co(NH_3)_5OH_2^{3+} + SO_4^{2-}$, where increased solvation stabilizes the developing charges and the $\ln k$ versus $P$ plot is strongly curved[62]), whereas there is normally no net solvational change in solvent exchange reactions. Nevertheless, the three $\Delta V_{ex}^{\ddagger}$ values reported independently for the $Ni(CH_3CN)_6^{2+}/CH_3CN$ exchange[59] are disappointingly inconsistent (+6, +10, and +12 cm$^3$/mol), despite their lesser susceptibility to errors, but at least are all clearly positive, consistent with an $\mathbf{I_d}$ mechanism.

Given that $\Delta V_{ex}^{\ddagger}$ can be mechanistically informative (indeed, it is the *only* credible mechanistic criterion for solvent exchange reactions unless the solvent concentration can be varied with an inert diluent[63,64]—which, however, may introduce uncertainties due to preferential solvation), the question arises as to what the numerical magnitude of $\Delta V_{ex}^{\ddagger}$ should be for the limiting $\mathbf{D}$ and $\mathbf{A}$ cases at least. An empirical approach that involves an empirical relation of the absolute partial molar volume $V_{abs}^0$ of a metal ion in solution to the number $n$ of solvent molecules in its first coordination sphere is available for aqueous systems;[65] it is restricted, however, to homoleptic aqua complexes and to date has not been extended to nonaqueous systems for want of sufficient experimental values of $V_{abs}^0$ and $n$ in those solvents. For aqueous systems, the partial molar volumes $V_{abs}^0$ of metal ions $M^{z+}(aq)$ in solution at 25°C, 0.1 MPa, and infinite dilution, relative to $V_{abs}^0 = -5.4$ cm$^3$/mol for $H^+(aq)$, are given by Equation 8.15[65,66]

$$V_{abs}^0 (\text{cm}^3/\text{mol}) = 2.523 \times 10^{-6}(r + \Delta r)^3 - 18.07n - 417.5z^2/(r + \Delta r) \tag{8.15}$$

in which $n$ is the number of water molecules in the first coordination sphere of $M^{z+}(aq)$ and $r$ is the effective ionic radius of $M^{z+}$ in pm appropriate to $n$ ($r$ increases by about 6 pm for each increase of 1 in $n$);[67] $\Delta r$, which represents the effective thickness of the first coordination sphere of aqua ligands, was determined empirically as 238.7 pm. From this, one can estimate that the limiting volume change for the formation of an intermediate $M(H_2O)_5{}^{z+} + H_2O$ from $M(H_2O)_6{}^{z+}$ (and thus for $\Delta V_{ex}^{\ddagger}$ for water exchange on $M(H_2O)_6{}^{z+}$ by a simple **D** mechanism) should range from about $+14$ cm$^3$/mol for $r = 40$ pm to $+12$ cm$^3$/mol for $r = 120$ pm, with very little dependence on $z$. For a simple **A** mechanism, the corresponding volume changes are $-14$ to $-12$ cm$^3$/mol. Thus, for an $I_d$ mechanism for water exchange on typical $M(H_2O)_6{}^{z+}$, we may expect $+13 > \Delta V_{ex}^{\ddagger} > 0$ cm$^3$/mol and for $I_a$, $-13 < \Delta V_{ex}^{\ddagger} < 0$ cm$^3$/mol.

It should be borne in mind that the uncertainty in the calculated limiting $\Delta V_{ex}^{\ddagger}$ values is on the order of 1–2 cm$^3$/mol, arising largely from the uncertainty in $r$ as it changes along with the change in $n$. On the other hand, in response to the criticism of Rotzinger,[68] we may note that the incorporation of the dependence of $r$ on $n$ into this empirically based treatment means that volumetric effects due to increases or decreases of the M–O bond lengths of the spectator aqua ligands during the respective **A** or **D** activation processes are adequately taken into account. In addition, the objection[68] that the limiting $\left| \Delta V_{ex}^{\ddagger} \right|$ values calculated above are overestimates because the transition state for an **A** mechanism may be reached with an incomplete formation of the bond of the incoming water to M (and correspondingly for a **D** mechanism with incomplete breaking of the M-OH$_2$ bond) misses the point that these $\left| \Delta V_{ex}^{\ddagger} \right|$ values are maxima calculated for limiting cases.

### 8.4.1 Water Exchange Reactions

Water exchange on metal ions in solution is of particular importance in that net ligand substitution reactions in aqueous solution generally occur on a timescale similar to water exchange. Thus, the exchange reactions provide an important measure of the intrinsic reactivity of the metal aqua ion without the distractions that accompany net chemical change. Furthermore, as discussed in Section 8.8, water exchange reactions have recently attracted interest among geochemists as benchmarks for chemical processes that occur in the weathering and deposition of minerals in aqueous environments.[23] Lincoln[69] has recently reviewed mechanistic studies of metal aqua complexes.

With regard to reaction mechanism, the $\Delta V_{ex}^{\ddagger}$ data of Table 8.2 may be interpreted on the basis of Equation 8.15—that is, to indicate that most water exchange reactions of simple aqua complexes are best described as **I** processes. This is expressed diagrammatically in Figure 8.3, the concept of which[70] was derived from a potential energy plot devised by More O'Ferrall[71] to express the gradation in organic reaction mechanisms between limiting $S_N1$ and $S_N2$ cases. In Figure 8.3, the quantities plotted are $\Delta V_{ex}^{\ddagger}$ for the symmetrical water exchange reaction that range between limits set by Equation 8.15 of $-13.5$ cm$^3$/mol for an **A** mechanism (top right corner) and $+13.5$ cm$^3$/mol (bottom left) for **D**; the reaction volume trajectories from top left to

**TABLE 8.2    Rate Constants $k_{ex}$ and Volumes of Activation $\Delta V_{ex}^{\ddagger}$ for Water Exchange on Selected Metal Aqua Complexes and Their Conjugate Bases**

| Metal Ion | d Configuration in $O_h$ Ligand Field | $k_{ex}$ $(s^{-1})^a$ | $\Delta V_{ex}^{\ddagger}$ $(cm^3/mol)^a$ | Radius$^b$ of $M^{z+}$ (pm) |
|---|---|---|---|---|
| $Be(H_2O)_4^{2+}$ | | $7.3 \times 10^2$ | $-13.6$ | 27 |
| $Mg(H_2O)_6^{2+}$ | | $6.7 \times 10^5$ | $+6.7$ | 72 |
| $V(H_2O)_6^{2+}$ | $t_{2g}^3 e_g^0$ | 87 | $-4.1$ | 79 |
| $Cr(H_2O)_6^{2+}$ | $(t_{2g}^3 e_g^1)^c$ | $>10^8$ | | 80 |
| $Mn(H_2O)_6^{2+}$ | $t_{2g}^3 e_g^2$ | $2.1 \times 10^7$ | $-5.4$ | 83 |
| $Fe(H_2O)_6^{2+}$ | $t_{2g}^4 e_g^2$ | $4.4 \times 10^6$ | $+3.8$ | 78 |
| $Co(H_2O)_6^{2+}$ | $t_{2g}^5 e_g^2$ | $3.2 \times 10^6$ | $+6.1$ | 74.5 |
| $Ni(H_2O)_6^{2+}$ | $t_{2g}^6 e_g^2$ | $3.2 \times 10^4$ | $+7.2$ | 69 |
| $Cu(H_2O)_6^{2+d}$ | $(t_{2g}^6 e_g^3)^c$ | $4.4 \times 10^9$ | $+2.0$ | 73 |
| $Zn(H_2O)_6^{2+}$ | $t_{2g}^6 e_g^4$ | $\geq 5 \times 10^7$ | | 74 |
| $Ru(H_2O)_6^{2+}$ | $t_{2g}^6 e_g^0$ | $1.8 \times 10^{-2}$ | $-0.4$ | |
| $Pd(aq)^{2+d}$ | | $5.6 \times 10^2$ | $-2.2$ | 64 |
| $Pt(aq)^{2+d}$ | | $3.9 \times 10^{-4}$ | $-4.6$ | 60 |
| $Eu(H_2O)_7^{2+}$ | | $4.4 \times 10^9$ | $-11.3$ | 120 |
| $Al(H_2O)_6^{3+}$ | | 1.29 | $+5.7$ | 53.5 |
| $AlOH(aq)^{2+d}$ | | $3.1 \times 10^4$ | | 53.5 |
| $Ga(H_2O)_6^{3+}$ | | $4.0 \times 10^2$ | $+5.0$ | 62 |
| $Ti(H_2O)_6^{3+}$ | $t_{2g}^1 e_g^0$ | $1.8 \times 10^5$ | $-12.1$ | 67 |
| $V(H_2O)_6^{3+}$ | $t_{2g}^2 e_g^0$ | $5.0 \times 10^2$ | $-8.9$ | 64 |
| $Cr(H_2O)_6^{3+}$ | $t_{2g}^3 e_g^0$ | $2.4 \times 10^{-6}$ | $-9.6$ | 61.5 |
| $Cr(H_2O)_5OH^{2+}$ | $t_{2g}^3 e_g^0$ | $1.8 \times 10^{-4}$ | $+2.7$ | 61.5 |
| $Fe(H_2O)_6^{3+}$ | $t_{2g}^3 e_g^2$ | $1.6 \times 10^2$ | $-5.4$ | 64.5 |
| $Fe(H_2O)_5OH^{2+}$ | $t_{2g}^3 e_g^2$ | $1.2 \times 10^5$ | $+7.0$ | 64.5 |
| $Ru(H_2O)_6^{3+}$ | $t_{2g}^5 e_g^0$ | $3.5 \times 10^{-6}$ | $-8.3$ | 68 |
| $Ru(H_2O)_5OH^{2+}$ | $t_{2g}^5 e_g^0$ | $5.9 \times 10^{-4}$ | $+0.9$ | 68 |
| $Rh(H_2O)_6^{3+}$ | $t_{2g}^6 e_g^0$ | $2.2 \times 10^{-9}$ | $-4.2$ | 66.5 |
| $Rh(H_2O)_5OH^{2+}$ | $t_{2g}^6 e_g^0$ | $4.2 \times 10^{-5}$ | $+1.5$ | 66.5 |
| $Ir(H_2O)_6^{3+}$ | $t_{2g}^6 e_g^0$ | $1.1 \times 10^{-10}$ | $-5.7$ | 68 |
| $Ir(H_2O)_5OH^{2+}$ | $t_{2g}^6 e_g^0$ | $5.6 \times 10^{-7}$ | $-0.2$ | 68 |
| $Gd(H_2O)_8^{3+}$ | | 83.0 | $-3.3$ | 105.3 |
| $Tb(H_2O)_8^{3+}$ | | 55.8 | $-5.7$ | 104.0 |
| $Dy(H_2O)_8^{3+}$ | | 43.4 | $-6.0$ | 102.7 |
| $Ho(H_2O)_8^{3+}$ | | 21.4 | $-6.6$ | 101.5 |
| $Er(H_2O)_8^{3+}$ | | 13.3 | $-6.9$ | 100.4 |
| $Tm(H_2O)_8^{3+}$ | | 9.1 | $-6.0$ | 99.4 |
| $Yb(H_2O)_8^{3+}$ | | 4.7 | | 98.5 |

$^a$At 25°C; for $\Delta H_{ex}^{\ddagger}$ and $\Delta S_{ex}^{\ddagger}$ values and data sources, see Refs 21, 59, and 15.
$^b$Ionic radii for the appropriate coordination geometry and spin state.[67]
$^c$Nominal—this hypothetical configuration forces Jahn–Teller distortion to approximately $D_{4h}$ symmetry for six-coordination.
$^d$See text.

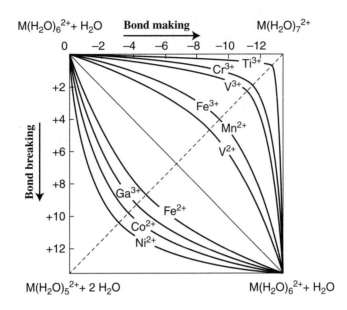

**FIGURE 8.3** Schematic representation of volume changes ($cm^3/mol$) accompanying water exchange on octahedral hexaaqua complexes. The instantaneous volume of the system is the algebraic sum of the ordinate and abscissa values, with the sum on the dashed diagonal line representing the transition state for each trajectory. Reproduced with permission from *Chem. Rev,* **2005,** *105*, 1923. Copyright 2005 American Chemical Society.

bottom right are arbitrarily sketched but are constrained to cross the **A–D** diagonal where the algebraic sum of their coordinates equals the experimental $\Delta V_{ex}^{\ddagger}$. A simple progression from reactants to the (identical) products is assumed. The essential point is that all the trajectories lying within the square represent **I** mechanisms with various degrees of bond making and bond breaking—that is, of $\mathbf{I_a}$ and $\mathbf{I_d}$ *character*; there is no delineation between $\mathbf{I_a}$ and $\mathbf{I_d}$ *mechanisms.* Kowall et al.,[61] however, note that nonlimiting values of $\Delta V_{ex}^{\ddagger}$ could be consistent with **A** or **D** mechanisms if the transition state for the rate-determining step is not essentially the same as the seven- or five-coordinate intermediates described above—for example, if the foregoing simple picture is complicated by second coordination sphere interactions as suggested by computer modeling of water exchange on $Al^{3+}$(aq), $Ga^{3+}$(aq), and $In^{3+}$(aq). Nonetheless, the most negative $\Delta V_{ex}^{\ddagger}$ values observed to date, $-13.6$ and $-12.1\ cm^3/mol$ for water exchange on $Be(H_2O)_4^{2+}$[64] and $Ti(H_2O)_6^{3+}$,[72] respectively, may be compared with $-12.9$ and $-13.4\ cm^3/mol$ calculated for a simple **A** mechanism for those particular ions according to Equation 8.15, implying that the predictions of limiting $\Delta V_{ex}^{\ddagger}$ values based on Equation 8.15 are indeed meaningful within the uncertainty of the calculation ($\pm 1$–$2\ cm^3/mol$). To date, no $\Delta V_{ex}^{\ddagger}$ values close to the upper limit have been found for water exchange on homoleptic metal aqua complexes, consistent with the view expressed in Section 8.3.3.2 and elsewhere[46] that a true **D** mechanism is unlikely in such cases.

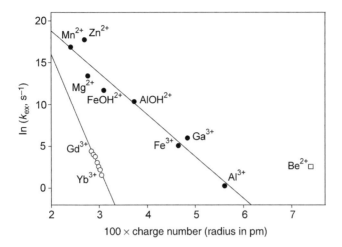

**FIGURE 8.4**  Dependence of water exchange rate constants for ions with no ligand field stabilization energy on the surface potential of the central metal ion, using traditional ionic radii from Shannon.[67]

One might anticipate naïvely that values of $\ln k_{ex}$ (which is related to the free energy of activation, $\Delta G_{ex}^{\ddagger}$, through the Eyring equation $\ln k_{ex} = \ln(k_B T/h) - \Delta G_{ex}^{\ddagger}/RT$) should correlate inversely with the ionic charge-to-radius ratio of the central metal ion $M^{z+}$, since the higher the electrostatic potential experienced by the ligands, the more firmly they should be held and the slower the expected exchange rate, regardless of the mechanism. Figure 8.4 shows that this has some validity for spherically symmetrical $M^{z+}$—that is, those with $d^0$, high-spin $d^5$ ($t_{2g}{}^3 e_g{}^2$), or $d^{10}$ outer electronic configurations—but the typical aqua complexes (filled symbols in Figure 8.4) follow a different linear correlation from the eight-coordinate later lanthanides, and four-coordinate $Be^{2+}$(aq) is more labile than either trend would predict. The artificiality of "ionic radii", which are derived from an arbitrary splitting of M–O or M–F distances in crystals, may contribute to the differences in these correlations—for example, use of Shannon's preferred "crystal radii" (= conventional ionic radii + 14 pm)[67] brings $Be(H_2O)_4{}^{2+}$ closer to the nonlanthanide plot, although the lanthanides remain separate. Nevertheless, Figure 8.4 implies that the lower limit of $k_{ex} \sim 5 \times 10^7 \text{ s}^{-1}$ set[61] for water exchange on $Zn^{2+}$(aq) is in fact close to the actual value, and furthermore that the strong acceleration of water exchange on forming the conjugate bases of $Fe(H_2O)_6{}^{3+}$ and $Al(H_2O)_6{}^{3+}$ is primarily due to the reduction in charge on deprotonation.

More important, the failure of many transition metal aqua ions to fit the correlations of Figure 8.4 highlights the influence of d electron configuration on the reactivity of metal aqua ions in substitution reactions. The importance of d electron configuration was first noted by Taube in 1952[1] and explained qualitatively in terms of valence bond theory. Taube, with his predilection for simple test tube demonstrations, distinguished "labile" metal complexes (ones which underwent substitution within the time of mixing) from "inert" ones, the latter being typically octahedral complexes with $d^3$ or low-spin $d^5$ and $d^6$ configurations and square planar complexes of $Pd^{II}$, $Pt^{II}$, and $Au^{III}$. Taube attributed the fact that the $V(H_2O)_6^{3+}/H_2O$ exchange is faster than the $V(H_2O)_6^{2+}/H_2O$ (Table 8.2), despite the higher charge, to the availability of a vacant interaxial d orbital ($d_{xy}$, $d_{xz}$, or $d_{yz}$; $t_{2g}$ symmetry in an $O_h$ environment) on $V^{3+}$ to form a bond with an electron pair from the incoming ligand, so facilitating associative activation. Conversely, the absence of vacant $t_{2g}$ orbitals confers relative inertness unless there are electrons in the axial orbitals ($d_{x^2-y^2}$ or $d_{z^2}$; $e_g$ symmetry) that will weaken the metal–ligand bonds and favor dissociative processes.

Basolo and Pearson[3,4] developed a semiquantitative approach using crystal field theory, which treats the ligands as point charges or point dipoles interacting electrostatically with the d electrons of $M^{z+}$. Subsequently, most discussion has used molecular orbital (MO) concepts, recognizing both $\sigma$ and $\pi$ interactions between the ligands and the d electrons of $M^{z+}$ (ligand field theory); for aqua complexes, only $\sigma$ interactions need be considered and these are adequately approximated by the crystal field approach. When $M^{z+}$ is surrounded by a regular octahedral ($O_h$) field of six anionic or dipolar ligands on the Cartesian $x$, $y$, and $z$ axes, the potential energy of an electron in an axial d orbital ($e_g$ orbitals, which are antibonding in the MO $\sigma$ sense) is elevated through interactions with the ligands by an amount conventionally defined as $6Dq$ whereas that of an electron in an interaxial d orbital ($t_{2g}$; nonbonding in $\sigma$ interactions) is stabilized by an amount $4Dq$, both relative to the hypothetical energy of a d electron in a field equivalent to that of six ligands smeared out in spherical symmetry around $M^{z+}$ (Figure 8.5). The total splitting of the two sets of d orbitals

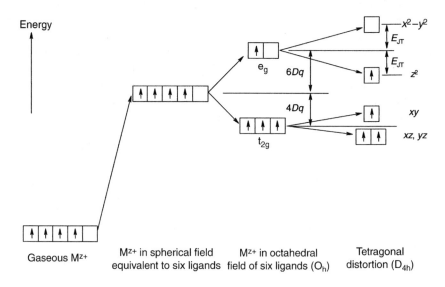

Gaseous $M^{z+}$    $M^{z+}$ in spherical field    $M^{z+}$ in octahedral    Tetragonal
equivalent to six ligands    field of six ligands ($O_h$)    distortion ($D_{4h}$)

**FIGURE 8.5**    Splitting of d orbitals in octahedral and tetragonally distorted octahedral ligand fields. For a high-spin $d^4$ ion, the presence of an odd electron forces splitting of the $e_g$ set because the energy of the system is thereby reduced by an amount $E_{JT}$ (Jahn–Teller effect), but the presence of the odd electron in the $d_{z^2}$ orbital labilizes the ligands on the $z$ axis.

(more generally, the separation between the $d_{xy}$ and the $d_{x^2-y^2}$ orbitals) is the quantity $10Dq$, which is measurable by UV-visible–near IR spectroscopy. For example, a $Cr^{3+}$ ion in an octahedral complex (d electron configuration $t_{2g}^{3}e_g^{0}$) is stabilized by a total ligand field stabilization energy (LFSE) of $12Dq$. One can argue that removal of one of the six ligands in a **D** process to leave a square pyramidal $Cr(H_2O)_5^{3+}$ ion will reduce the LFSE by $2Dq$ that, for $Cr(H_2O)_6^{3+}$, amounts to about 42 kJ/mol. The observed $\Delta H_{ex}^{\ddagger}$ for the $Cr(H_2O)_6^{3+}/H_2O$ exchange is 109 kJ/mol,[57] whereas that for the analogous $Fe(H_2O)_6^{3+}/H_2O$ reaction, in which there is no LFSE to be lost, is 64 kJ/mol;[73] thus, the fact that the $Cr^{III}$ water exchange is some $7 \times 10^7$ times slower than the $Fe^{III}$ can be ascribed to the loss of LFSE in the activation process. Put another way, the $Cr(H_2O)_6^{3+}/H_2O$ exchange deviates from the correlation of Figure 8.4 by an amount corresponding to $2Dq$ in energy.

The apparent near-quantitative success of this crude argument is, of course, fortuitous—the $\Delta V_{ex}^{\ddagger}$ value ($-9.6\,cm^3/mol$[57]) shows the reaction mechanism to be not **D** but **I$_a$**—but it does show that ligand field effects in transition metal aqua complexes can produce strong deviations from expectations based on Figure 8.4. Basolo and Pearson[3,4] attempted to place the concept of ligand field effects on substitution kinetics on a quantitative footing by calculating the ligand field contributions to the activation energy (LFAE) for both associatively and dissociatively activated substitution reactions of octahedral complexes; in principle, this could predict the reaction mechanism on the basis of whichever pathway had the lower calculated activation energy. Assumptions, however, were required regarding the geometry of the transition states—for dissociative processes, the intermediate of reduced coordination number might ideally be square pyramidal ($C_{4v}$) or trigonal bipyramidal ($D_{3h}$), and for associative activation the seven-coordinate intermediates considered were an octahedral wedge ($C_{2v}$, from flanking attack of the incoming ligand on the outgoing) or pentagonal bipyramid ($D_{5h}$)—and also concerning the bond lengths in the transition state, particularly in the octahedral wedge model. LFAEs for exchange reactions proceeding by hypothetical square pyramidal and octahedral wedge transition states are given in Table 8.3, in which "low spin" refers to those cases in which the ligand field is strong enough to enforce spin pairing. In principle, the lowest LFAE for different assumed transition state geometries for a given metal ion should give a clue as to the reaction mechanism, but consideration of the $\Delta V_{ex}^{\ddagger}$ values of Table 8.2 and the LFAEs of Table 8.3 shows that this succeeds for $Cr(H_2O)_6^{3+}$ (**I$_a$**) but fails for $Ni(H_2O)_6^{2+}$ (**I$_d$**)—this may be attributed to the lower charge on the $Ni^{2+}$ aqua ion, which will facilitate ligand dissociation. As Basolo and Pearson emphasized,[4] LFAEs are to be applied in addition to the more general centrosymmetric contributions that are crudely represented in Figure 8.4. Moreover, there is no explicit consideration of covalency in the metal–ligand interactions, or of the effects of the surrounding solvent.

Many attempts have been made to refine the ligand field treatment of substitution kinetics, and computer modeling of solvent exchange in particular (Section 8.7) has lately become almost a discipline in its own right, but there is as yet no panacea. The basic LFAE concept, however, does provide readily comprehensible qualitative

**TABLE 8.3    Ligand (Crystal) Field Contributions to the Activation Energy for Exchange Reactions of Transition Metal Ions Assuming Square Pyramid and Octahedral Wedge Transition States to Represent Dissociative and Associative Activation, Respectively**[a]

| Number of d Electrons | LFAE/$Dq$ | |
|---|---|---|
| | Square Pyramid | Octahedral Wedge |
| 1 or 6 | −0.57 | −2.08 |
| 2 or 7 | −1.14 | −0.68 |
| 3 or 8 | 2.00 | 1.80 |
| 4 or 9 | −3.14 | −2.79 |
| 0, 5, or 10 | 0 | 0 |
| 4 low spin | 1.43 | −0.26 |
| 5 low spin | 0.86 | 1.14 |
| 6 low spin | 4.00 | 3.63 |
| 7 low spin | −1.14 | −0.98 |

[a]From Ref. 4.

explanations for the relative reactivity of many transition metal aqua complexes in the suitably simple case of water exchange. Some examples follow.

1. The greater lability of $V(H_2O)_6^{3+}/H_2O$ relative to $V(H_2O)_6^{2+}/H_2O$ can be attributed to a negative LFAE ($-0.68Dq$) in the former case and a positive one ($+1.80Dq$) in the latter for an assumed octahedral wedge transition state, given that $\Delta V_{ex}^{\ddagger}$ is consistent with an $\mathbf{I_a}$ mechanism in either case. It should be borne in mind that $Dq$ for a +3 ion is typically about twice that for a +2 ion. On the same basis, the very high lability of $Ti(H_2O)_6^{3+}$ (the fastest of the +3 ions) may be explained in terms of a calculated LFAE of $-2.08Dq$ that greatly reduces the activation energy for an octahedral wedge transition state, consistent with the near-limiting $\mathbf{A}$ mechanism implied by the $\Delta V_{ex}^{\ddagger}$ value.

2. In general, spin-paired $d^5$ ($t_{2g}^5 e_g^0$; $Ru^{III}$) and $d^6$ ($t_{2g}^6 e_g^0$; $Rh^{III}$, $Ir^{III}$) configurations (Table 8.2) confer kinetic inertness. Water exchange on both $Rh(H_2O)_6^{3+}$ and $Ir(H_2O)_6^{3+}$ occurs by an $\mathbf{I_a}$ mechanism, as indicated by $\Delta V_{ex}^{\ddagger}$ and supported by modeling.[74] Data for water exchange on $Co(H_2O)_6^{3+}$, however, are not available; for this 3$d^6$ ion, $Dq$ is just large enough to enforce spin pairing in the ground state, but it has been conjectured that its substitution reactions may proceed through an accessible labile high-spin state, and besides it is slowly reduced by water to labile $Co^{2+}(aq)$ that could labilize $Co(H_2O)_6^{3+}$ through bimolecular electron transfer. On descending a periodic group, $Dq$ and hence LFAEs increase markedly for a given oxidation state and ligand set, so that the $Ir(H_2O)_6^{3+}/H_2O$ exchange is extremely slow with $\Delta H_{ex}^{\ddagger}$ as high as 131 kJ/mol.[75]

3. For comparisons that include $Co^{III}$ complexes, we may consider the aqua exchange data for $M(NH_3)_5OH_2^{3+}$ in Table 8.4.[55,76,77] The volumes of activation, as well as other information discussed in Section 8.5, indicate dominant

**TABLE 8.4  Water Exchange Kinetics for Some Pentaam(m)ine Aqua Complexes**[a]

| Complex | d Electron Configuration | LFAE[b] (kJ/mol) | $k_{ex}^{298}$ $(10^{-5}\,s^{-1})$ | $\Delta H_{ex}^{\ddagger}$ (kJ/mol) | $\Delta S_{ex}^{\ddagger}$ (J/(K mol)) | $\Delta V_{ex}^{\ddagger}$ (cm$^3$/mol) |
|---|---|---|---|---|---|---|
| $Ru(NH_3)_5OH_2^{3+}$ | 4d: $t_{2g}^5 e_g^0$ | 1.14Dq = 46 | 23.0 | 91.5 | −8 | −4.0 |
| $Cr(NH_3)_5OH_2^{3+}$ | 3d: $t_{2g}^3 e_g^0$ | 1.8Dq = 45 | 5.2 | 97.1 | 0 | −5.8 |
| $Cr(MeNH_2)_5OH_2^{3+}$ | | 1.8Dq = 44 | 0.41 | 98.5 | −18 | −3.8 |
| $Co(NH_3)_5OH_2^{3+}$ | 3d: $t_{2g}^6 e_g^0$ | 4.0Dq = 102$^c$ | 0.57 | 111.3 | +28 | +1.2 |
| $Co(MeNH_2)_5OH_2^{3+}$ | | 4.0Dq = 98$^c$ | 70.0 | 99.0 | +27 | +5.7 |
| $Rh(NH_3)_5OH_2^{3+}$ | 4d: $t_{2g}^6 e_g^0$ | 3.6Dq = 142 | 0.87 | 102.9 | +3 | −4.1 |
| $Rh(MeNH_2)_5OH_2^{3+}$ | | 3.6Dq = 139 | 1.06 | 112.7 | +38 | +1.2 |
| $Ir(NH_3)_5OH_2^{3+}$ | 5d: $t_{2g}^6 e_g^0$ | 3.6Dq = 173 | 0.0061 | 117.6 | +11 | −3.2 |

[a] Data from Refs 76,77, and 140.
[b] For an octahedral wedge transition state, except as indicated.
[c] For a square pyramidal transition state.

$\mathbf{I_d}$ character for $Co(NH_3)_5OH_2^{3+}$ but $\mathbf{I_a}$ for the other $M(NH_3)_5OH_2^{3+}$ listed, and LFAEs in Table 8.4 are calculated accordingly;[4] in practice, the choice of assigned mechanism does not significantly affect the following observations. For $Rh(NH_3)_5OH_2^{3+}$ and $Ir(NH_3)_5OH_2^{3+}$, the absolute values of the LFAEs substantially exceed the experimental $\Delta H_{ex}^{\ddagger}$ even before any centrosymmetric or covalent interactions are allowed for, and clearly should only be used as a qualitative guide to ligand field effects on substitution reaction rates. Moreover, although the LFAEs do indeed increase because of increasing $Dq$ from Co to Rh to Ir, regardless of the assigned mechanism, the experimental $\Delta H_{ex}^{\ddagger}$ and $\Delta S_{ex}^{\ddagger}$ for $Co(NH_3)_5OH_2^{3+}$ are anomalously high relative to the trend set by the Rh$^{III}$ and Ir$^{III}$ analogues, resulting in $k_{ex}$ being surprisingly similar for $Co(NH_3)_5OH_2^{3+}$ and $Rh(NH_3)_5OH_2^{3+}$. This and other evidence (Section 8.5) implies that Co$^{III}$ complexes behave to some degree atypically in ligand substitution reactions, which is ironic inasmuch as the majority of mechanistic studies intended to elucidate octahedral substitution have been conducted on Co$^{III}$ complexes. These anomalies may reflect the unusually small ionic radius (54.5 pm, cf. Table 8.2)[67] of the Co$^{3+}$ ion in octahedral environments, resulting in steric congestion in its complexes and a predisposition to dissociative activation in their substitution reactions, as indicated by the positive $\Delta V_{ex}^{\ddagger}$ for the $Co(NH_3)_5OH_2^{3+}/H_2O$ exchange reaction. Put differently, ligand substitution in +3 transition metal complexes other than Co$^{III}$, including the later lanthanides, normally occurs by processes with dominant $\mathbf{I_a}$ character, as evidenced in Tables 8.2 and 8.4 by negative $\Delta V_{ex}^{\ddagger}$ values.

4. Ruthenium(III) complexes are seen in Tables 8.2 and 8.3 to be kinetically comparable to the corresponding chromium(III) species, despite a larger $Dq$ for Ru$^{III}$, because the LFAE factor is lower for low-spin d$^5$ than for d$^3$. For $Ru(H_2O)_6^{3+}$, $\Delta V_{ex}^{\ddagger}$ indicates a much more associative mode of activation than for $Ru(H_2O)_6^{2+}$, which seems typical of octahedral transition metal complexes

and may be attributed to the direct electrostatic effect of the higher oxidation state as well as its indirect effect that increases $Dq$ and hence the LFAE.

5. All the high-spin octahedral metal +2 aqua complexes of Table 8.2 are labile in the Taube classification, and with the exceptions of $Cr^{II}$ and $Cu^{II}$, the sequence of lability $V^{2+} \ll Cr^{2+} > Mn^{II} > Fe^{II} > Co^{II} > Ni^{II} \ll Cu^{II} > Zn^{2+}$ follows predictions based on LFAE. In particular, the configurations $t_{2g}^3 e_g^0$ ($V^{II}$) and $t_{2g}^6 e_g^2$ ($Ni^{II}$) should confer the greatest LFAEs ($2Dq$ in either case, assuming a dissociative mechanism) and hence the slowest reaction rates. The volumes of activation are consistent with mainly $\mathbf{I_a}$ character for $V^{2+}$ and $Mn^{2+}$, with progressively dominant $\mathbf{I_d}$ character on going to $Fe^{2+}$, $Co^{2+}$, and $Ni^{2+}$. The assignment of an $\mathbf{I_a}$ mechanism to the $Mn(H_2O)_6^{2+}$ water exchange might seem to be inconsistent with the apparent lack of a marked sensitivity of $Mn^{2+}$ to the nature of the incoming ligand in its net substitution reactions. This may simply be a matter of a lack of appropriate data for these very fast reactions, but in any event it has long been recognized by physical organic chemists[78] that the faster an associative process is, the less selectivity it will show—and the $Mn^{II}$ substitution reactions are very fast by any standards. For the $M(H_2O)_6^{2+}/H_2O$ exchanges (M = Mn, Fe, Co, and Ni), combination of $\Delta V_{ex}^{\ddagger}$ with the conventional molar volumes $V_{conv}^0$ ($V_{conv}^0$ for $H^+(aq) = 0$) gives the respective conventional molar volumes $V_{conv}^{0\ddagger}$ of the transition states as $-22.8$, $-21.5$, $-19.3$, and $-21.2$ cm$^3$/mol, which are effectively constant (given the additive experimental uncertainties) relative to the $>12$ cm$^3$/mol spread in $\Delta V_{ex}^{\ddagger}$.[79] Thus, the cause of the variation in $\Delta V_{ex}^{\ddagger}$, and hence in the relative importance of associative versus dissociative activation within a common $\mathbf{I}$ mechanism, resides mainly in the initial states of the aqua ions—the large $Mn^{2+}$ ion can easily expand its coordination number, and the small $Ni^{2+}$ ion cannot. A similar conclusion may be drawn for $M(NH_3)_5OH_2^{3+}/H_2O$ exchanges.[79]

6. For $Ru(H_2O)_6^{2+}$, which is low-spin $4d^6$ ($t_{2g}^6 e_g^0$), the LFAE of $4Dq$ (for an assumed dissociative mechanism, since $\Delta V_{ex}^{\ddagger}$ indicates only very weak associative character) puts it on the border of Taube's inert category whereas most octahedral 2+ ions are labile. At the other extreme, the very high $k_{ex}$ for $Cr^{2+}$ and $Cu^{2+}$ can be ascribed to the major Jahn–Teller distortions that result when metal ions with a single electron in an axial d orbital ($Cr^{2+}$), or two electrons in one axial d orbital and only one in the other ($Cu^{2+}$), are placed in a quasi-octahedral environment; the geometry must spontaneously distort, normally with tetragonal elongation (four equatorial ligands close in and two remote axial ligands; $D_{4h}$). This is illustrated for a high-spin $d^4$ ($Cr^{2+}$) complex in Figure 8.5, in which the driving force of the Jahn–Teller effect is seen to be the energy $E_{JT}$. Indeed, the $Cr(H_2O)_6^{2+}$ unit in the high-spin solid Tutton salt $(NH_4)_2[Cr(H_2O)_6](SO_4)_2$[80] is tetragonally distorted with a 16% elongation of the axial Cr–O bond lengths relative to the equatorial,[81] and it is reasonable to assume that this structure persists in solution. Clearly, the two remote ligands could be easily exchanged with solvent by either an $\mathbf{I_a}$ or an $\mathbf{I_d}$ mechanism, and vibrations of the complex should result in rapid inversion of the axial and

equatorial positions on a femtosecond to picosecond timescale.[82] Unfortunately, $\Delta V_{ex}^{\ddagger}$ for the easily oxidized $Cr^{2+}$(aq) has not been measured.

7. The very high $k_{ex}$ ($4.4 \times 10^9\,s^{-1}$) and low $\Delta H_{ex}^{\ddagger}$ ($11.5\,kJ/mol$)[83] for water exchange on $Cu^{2+}$(aq) can obviously be attributed to the Jahn–Teller effect, but it is not entirely clear how this extreme lability relates to the structure of the aqua ion. Conventional wisdom, drawing on early X-ray diffraction measurements on aqueous solutions of $Cu^{2+}$(aq) (e.g., by Magini[84]) and analogies with the structures of various solid-state $Cu^{II}$ compounds (e.g., $[Cu(H_2O)_6](ClO_4)_2$[85]), holds that $Cu^{II}$ exists in aqueous solution as a tetragonally distorted six-coordinate species (approximate $D_{4h}$ symmetry). EPR data on frozen $Cu^{2+}$(aq) solutions suggest that the aqua complex may indeed be six-coordinate but not centrosymmetric—some tetrahedral distortion is also present, reducing the symmetry to approximately $D_{2d}$.[86] In 2001, however, Pasquarello et al.[87] interpreted neutron diffraction data and molecular dynamics simulations on aqueous $Cu(ClO_4)_2$ in terms of a five-coordinate $Cu^{2+}$(aq). Furthermore, studies of aqueous $Cu^{2+}$ solutions using X-ray absorption spectroscopy in conjunction with density functional theory (DFT)[88,89] have suggested a square pyramidal $Cu(H_2O)_5^{2+}$ with an elongated axial Cu–O bond and a distortion toward $D_{2d}$ geometry. On the other hand, in 2002, Persson et al.[90] used EXAFS and large angle X-ray diffraction of $Cu^{2+}$ aqua ion salts in both solid phase and aqueous solution to reaffirm presence of elongated octahedral $Cu(H_2O)_6^{2+}$ ions with four equatorial Cu–O bonds of length 195 pm and two axial Cu–O of 229 pm. The molecular dynamics simulations of Schwenk and Rode[91] support the assignment of the elongated octahedral structure to $Cu^{2+}$(aq) in solution, with fluxional rearrangement of the Cu–O bond lengths on a timescale probably even shorter[82] than the 5 ps estimated from NMR measurements by Powell et al.[83] The basic problem is that the fitting of data from the various X-ray methods requires assumptions about the structure and consequently is not unambiguously definitive of the coordination number. As for the various computer modeling methods, Rotzinger[92] cautions that these tend to be biased toward either higher or lower coordination numbers, depending on the method used (Section 8.7). Thus, the question of the geometric consequences of the Jahn–Teller effect in aqueous $Cu^{2+}$(aq) remains open, although the weight of the evidence favors an elongated octahedral hexaaqua complex. Besides, as Rotzinger notes,[92] a coordinatively unsaturated $Cu(H_2O)_5^{2+}$ complex would be expected to undergo water exchange through an **A** or **$I_a$** mechanism with a substantially negative volume of activation, but the observed $\Delta V_{ex}^{\ddagger}$ ($+2.0 \pm 1.5$ $cm^3/mol$),[93] though approximate, is clearly positive, implying predominant **$I_d$** from a six-coordinate ground state. Finally, the empirical Equation 8.15 gives $V_{abs}^0 = -37\,cm^3/mol$ for $Cu^{2+}$(aq) with $n = 6$ and $-19\,cm^3/mol$ for $n = 5$, whereas the experimental value (adapted from Millero[94]) is $-33\,cm^3/mol^{-1}$. Whatever the geometric consequences of the Jahn–Teller distortion in aqueous $Cu^{2+}$(aq), the kinetic result is extraordinarily fast water exchange, with even faster scrambling of the aqua ligand positions (lifetimes in the femtosecond–

picosecond range[82]) within the first coordination sphere. The rate constant given in Table 8.2 is calculated on the basis of six-coordination.

8. The exchange of water on $Pt^{2+}$(aq) is six orders of magnitude slower than on $Pd^{2+}$(aq), and this might be qualitatively attributed, on a variant of LFAE as calculated for square planar complexes, to the increase in $Dq$ on descending a periodic group, but once again it transpires that the structure of the ground-state aqua complexes in solution, long assumed to be square planar $M(H_2O)_4^{2+}$, needs to be reconsidered. Given the prevalence of **A** mechanisms in square planar $Pt^{II}$ complexes (Sections 8.3.3.1 and 8.6), the observation[95] of $\Delta V_{ex}^{\ddagger} = -2.2 \pm 0.2\,cm^3/mol$ for $Pt^{2+}$(aq) rather than a much more negative number has been puzzling, although there is no equivalent of Equation 8.15 to indicate what the limiting $\Delta V_{ex}^{\ddagger}$ value might be for an **A** reaction of a square planar aqua complex. In 2005, Lincoln[69] conjectured that a weakly bonded water molecule might be present on one side or the other of the square planar $Pt(H_2O)_4^{2+}$ unit in the ground state in solution, so minimizing the volume change in forming a five-coordinate transition state. Indeed, in 2008, Jalilehvand and Laffin[96] used EXAFS to show that $Pt^{2+}$(aq) has four equatorial aqua ligands with Pt–O distances of 201 pm and either one or two axial water(s) at 239 pm; the *cis*-diaquadiammineplatinum(II) ion, which is relevant to the dynamics of action of the cancer drug cisplatin (Section 8.6), has a similar structure in solution.

9. The reactivities of the conjugate bases $MOH^{2+}$(aq) relative to the parent $M(H_2O)_6^{3+}$ in water exchange (Table 8.2) show no particular pattern traceable to LFAE effects except that the largest conjugate base labilization occurs for the very small central $Al^{3+}$ ion for which there are no d electrons and hence no LFAE. Normally, reversible proton loss from $M(H_2O)_6^{3+}$ to give $M(H_2O)_5OH^{2+}$ is fast relative to the subsequent water exchange on the latter, which, according to $\Delta V_{ex}^{\ddagger}$ (Table 8.2), is always more dissociatively activated than water exchange on the parent hexaaqua complex. The $Al^{3+}$(aq) hydrolysis, however, is unique among those tabulated in that dissociation of a water molecule is actually synchronous with proton loss from $Al(H_2O)_6^{3+}$:[97]

$$Al(H_2O)_6^{3+} \rightleftharpoons Al(H_2O)_4OH^{2+} + H_2O + H^+ \qquad (8.16)$$

Equation 8.16 implies that three protons are exchanged for every one oxygen, and indeed the first-order exchange rate constants are $(9 \pm 1) \times 10^4\,s^{-1}$ for $H$[98,99] compared to $3.1 \times 10^4\,s^{-1}$ for $^{17}O$ for the $H^+$-dependent water exchange pathway. Equation 8.16 also explains the observed overall $\Delta V_{ex}^{\ddagger}$ of $-0.7\,cm^3/mol$ for the $H^+$-dependent exchange pathway, which at first sight would seem to imply that the conjugate base reacts more associatively than the parent hexaaqua ion ($\Delta V_{ex}^{\ddagger} = +5.7\,cm^3/mol$), against both reasonable expectations[59] and all evidence for other conjugate base reactions as in Table 8.2; in the mechanism proposed in Equation 8.16, however, the large positive volume change on forming the five-coordinate $AlOH^{2+}$(aq) is effectively canceled by a negative one of associative water attack to effect water exchange on this coordinatively unsaturated ion. Computer modeling supports this mechanism, showing that deprotonation of $Al(H_2O)_6^{3+}$ is accompanied in less than 1 ps by loss of one aqua ligand to form $Al(H_2O)_4OH^{2+}$, the most stable form of the conjugate base.[97,100] Such a reduction in coordination number on hydrolysis of $Al(H_2O)_6^{3+}$ had been suggested in 1991 by Martin,[101] who noted that no such aqua ligand loss seemed to occur on hydrolysis of $Fe(H_2O)_6^{3+}$—this is consistent with $\Delta V_{ex}^{\ddagger} = -5.4$ and $+7.0\,cm^3/mol$ for water exchange on $Fe(H_2O)_6^{3+}$ and $FeOH^{2+}(aq)$, respectively. The anomalous hydrolytic behavior of $Al(H_2O)_6^{3+}$ can be ascribed to steric consequences of the unusually small radius of $Al^{3+}$ in six-coordination (53.5 pm, cf. 57 pm calculated for the radius of the cavity defined by six closest packed water molecules)—reduction of the charge from +3 to +2 on removal of a proton diminishes the electrostatic hold on the aqua ligands, and one of these leaves promptly to reduce steric congestion.

10. As one traverses the lanthanide (Ln) period, the partial molar volumes of $Ln^{3+}$ ions in water from La to Nd follow Equation 8.15 with coordination number $n = 9$, and from Tb to Lu with $n = 8$, with Eu and to some extent Gd falling in between.[66] Indeed, spectroscopic data show $Eu^{3+}(aq)$ to be a mixture of $Eu(H_2O)_8^{3+}$ and $Eu(H_2O)_9^{3+}$. Definitive rate constants for water exchange on the early lanthanide(III) ions are not currently available, but $k_{ex}$ for $Ln(H_2O)_8^{3+}$ falls progressively from Ln = Gd to Yb, with $\Delta V_{ex}^{\ddagger}$ hovering around $-6\,cm^3/mol$ for all except Gd ($-3\,cm^3/mol$). These data are as expected for large, spherically symmetrical ions for which expansion of the coordination number from 8 to 9 is fairly easy, favoring an $I_a$ mechanism, but becomes more difficult as the radius of the central ion decreases and hence steric congestion increases. For the lanthanide series, there are no partially occupied d orbitals, and ligand field effects on the deep-seated 4f electrons are of negligible importance in the context of LFAEs. Of the bivalent lanthanide ions, only $Eu^{2+}$ is of importance (though easily oxidized) as an aqueous ion, which interestingly exists as an equilibrium mixture of a seven-coordinate and a minor eight-coordinate aqua complex[102] and consequently undergoes facile water exchange through associative attack on $Eu(H_2O)_7^{2+}$. The ion also has a very low charge-number to metal-ion-radius ratio. These factors result in a very high $k_{ex}$.[103]

11. For actinide ions, 6d electrons are generally absent and the 5f electrons are not sensitive to ligand field effects. Accordingly, water exchange on these large ions is generally fast: for example, for $M(H_2O)_{10}^{4+}$ (M = Th or U), $k_{ex}$ is on the order of $10^7\,s^{-1}$,[59] while for $UO_2(H_2O)_5^{2+}$, $k_{ex}^{298} = 1.3 \times 10^6\,s^{-1}$.[104,105]

Finally, for this section, it is appropriate to reiterate that prediction of limits for $\Delta V_{ex}^{\ddagger}$ from Equation 8.15 as in Figure 8.3 is applicable only to water exchange reactions of homoleptic metal aqua ions or their conjugate bases, although the predictions may serve as a rough guide for other water exchange processes. For example, $\Delta V_{ex}^{\ddagger}$ for water exchange on the water-soluble iron(III) porphyrin $Fe(TMPS)(H_2O)_2{}^{3-}$ (TMPS = *meso*-tetra(sulfonatomesityl)porphine) is $+12\,cm^3/mol$,[106] which certainly indicates a strongly dissociative mechanism but not necessarily a limiting **D** process.

### 8.4.2 Nonaqueous Solvent Exchange Reactions

One obvious difference between solvent exchange reactions in water and in nonaqueous solvents is that the molecular bulk of the latter solvents can be much greater, so that associative activation may be hindered. Indeed, for solvent exchange on $Be(solvent)_4{}^{2+}$, use of an inert diluent (nitromethane) to vary the nonaqueous solvent concentration showed that the observed rate constant $k_{obsd}$ comprised the sum of contributions $k_1$ from a dissociative pathway independent of [solvent] and $k_2$[solvent] from an associative pathway that was eliminated on going to more bulky solvents. Both paths contributed significantly to solvent exchange in DMF. Moreover, whereas $\Delta V_{ex}^{\ddagger}$ for the $Be(H_2O)_4{}^{2+}/H_2O$ exchange ($-13.6\,cm^3/mol$) stands at the limit predicted for an **A** mechanism, $\Delta V_{ex}^{\ddagger}$ for $Be(solvent)_4{}^{2+}/solvent$ exchanges in organic media becomes increasingly positive as the bulk of the solvent molecule increases ($-4.1$, $-3.1$, $-2.5$, $+10.3$, and $+10.5\,cm^3/mol$ for TMP (trimethylphosphate), DMF, DMSO (dimethylsulfoxide), DMPU (DMPU = dimethylpropyleneurea), and TMU (tetramethylurea), respectively), consistent with a changeover to **I** and possibly **D** mechanisms because of steric hindrance to associative activation at the very small $Be^{2+}$ center ($r = 27\,pm$).[64] These results provide confirmation that $\Delta V_{ex}^{\ddagger}$ is a valid criterion of mechanism in solvent exchange reactions.

Figure 8.6 shows that nonaqueous solvent exchange rates on bivalent metal ions of the first transition series follow the same general pattern as for water as solvent, albeit somewhat slower as is generally the case for organic solvents, and similar rationales apply. The progression toward increasing dissociative activation noted for water for $Mn^{2+}$ through $Ni^{2+}$ is evident in $\Delta V_{ex}^{\ddagger}$ (Table 8.5) for all the solvents listed. The $\Delta V_{ex}^{\ddagger}$ values in Table 8.5 show that associative activation is important for exchanges on the $+3$ ions listed (though less so for $Fe^{3+}$), and for $Mn^{2+}$ except in DMF—it appears that, of the solvents listed, the relatively bulky DMF most favors dissociative processes. For lanthanide(III) ions in DMF, the nine–eight coordination equilibrium noted for aqueous solutions in Section 8.4.1 is evident as early as $Ln = Ce$, Pr and Nd. For the very rapid DMF exchange on $Ln(DMF)_8{}^{3+}$, $\Delta V_{ex}^{\ddagger}$ rises from about $+5$ to $+12\,cm^3/mol$ on going from Tb to Yb as an $I_d$ mechanism evolves into **D** as the $Ln^{3+}$ radius shrinks; a spectrophotometric measurement of the equilibrium volume change for $Nd(DMF)_9{}^{3+} \rightarrow Nd(DMF)_8{}^{3+} + DMF$ gave $+10\,cm^3/mol$. These mechanisms were confirmed by solvent dilution with $CD_3NO_2$: for Tb, the exchange rate constant depended on [DMF] whereas for Yb it did not.[63] Consistent with this, $k_{ex}$ as a function of decreasing ionic radius goes through a minimum at $Ho(DMF)_8{}^{3+}$ as the minor contribution of associative activation to an $I_d$ process is sterically squeezed out and dissociative activation becomes increasingly favored (Table 8.6). Evolution of the relatively low $\Delta H_{ex}^{\ddagger}$ and negative $\Delta S_{ex}^{\ddagger}$ values to high $\Delta H_{ex}^{\ddagger}$ and positive $\Delta S_{ex}^{\ddagger}$ on going from $Tb^{3+}$ to $Yb^{3+}$ supports this interpretation.

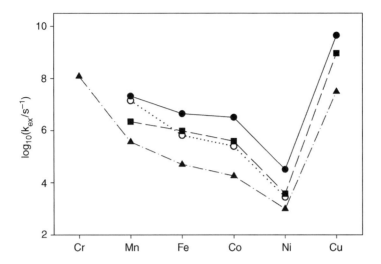

**FIGURE 8.6**   Rate constants for solvent exchange at 25°C on bivalent metal ions. Filled circles, solid line: water; squares, dashed line: DMF; open circles, dotted line: acetonitrile; triangles, dots and dashes: methanol.

As noted above, individual values of $\Delta H_{ex}^{\ddagger}$ and $\Delta S_{ex}^{\ddagger}$ obtained from NMR measurements are not always reliable, but the suite of related measurements in Table 8.6 on the $Ln(DMF)_8^{3+}/DMF$ reaction series shows systematic trends and provides a paradigm of the relation between these parameters and reaction mechanism. Associative activation achieves low $\Delta H_{ex}^{\ddagger}$ through the energy return of partial bond formation, but at the cost of reduced degrees of freedom including the localization of the incoming ligand (low $\Delta S_{ex}^{\ddagger}$), whereas the larger enthalpy input

**TABLE 8.5   Volumes of Activation $\Delta V_{ex}^{\ddagger}$ ($cm^3$/mol) for Solvent Exchange on First-Row Transition Metal Ions**[a]

| $M^{z+}$ in $M(solvent)_6^{z+}$ | Solvent | | | | |
|---|---|---|---|---|---|
| | $H_2O$ | $CH_3OH$ | $CH_3CN$ | DMF | $DMSO$[b] |
| $Mn^{2+}$ | −5.4 | −5.0 | −7.0 | 2.4 | |
| $Fe^{2+}$ | 3.8 | 0.4 | 3.0 | 8.5 | |
| $Co^{2+}$ | 6.1 | 8.9 | 7.7 | 6.7 | |
| $Ni^{2+}$ | 7.2 | 11.4 | 9.6 | 9.1 | |
| $Cu^{2+}$ | 2.0 | 8.3 | | 8.4 | |
| $Ti^{3+}$ | −12.1 | | | −5.7 | |
| $V^{3+}$ | −8.9 | | | | −10.1 |
| $Cr^{3+}$ | −9.6 | | | −6.3 | −11.3 |
| $Fe^{3+}$ | −5.4 | | | −0.9 | −3.1 |

[a] Data from Ref. 59.
[b] Here and elsewhere in this chapter, $\Delta V_{ex}^{\ddagger}$ data on DMSO solutions are limited because DMSO freezes under pressures of a few MPa at near-ambient temperatures.

**TABLE 8.6  Exchange of $N,N$-Dimethylformamide Solvent on Lanthanide(III) Ions[a]**

| Ln in Ln(DMF)$_8^{3+}$ | $k_{ex}^{298}$ ($10^7$ s$^{-1}$) | $\Delta H_{ex}^{\ddagger}$ (kJ/mol) | $\Delta S_{ex}^{\ddagger}$ (J/(K mol)) | $\Delta V_{ex}^{\ddagger}$ (cm$^3$/mol)[b] |
|---|---|---|---|---|
| Tb | 1.9 | 14 | −58 | +5.2 |
| Dy | 0.63 | 14 | −69 | +6.1 |
| Ho | 0.36 | 15 | −68 | +5.2 |
| Er | 1.3 | 24 | −30 | +5.4 |
| Tm | 3.1 | 33 | +10 | +7.4 |
| Yb | 9.9 | 39 | +40 | +11.8 |

[a] Data from Ref. 63.
[b] At 235–255 K.

for independent bond breaking in dissociative activation is compensated by gains in configurational degrees of freedom on going to the transition state. By way of a geographical analogy, one may cross a mountain range at low altitude (low $\Delta H_{ex}^{\ddagger}$) but only by a sharply limited number of choices of passes (low $\Delta S_{ex}^{\ddagger}$), or cross with many more choices of routes (high $\Delta S_{ex}^{\ddagger}$) by going to high altitude (high $\Delta H_{ex}^{\ddagger}$). The compensation between $\Delta H_{ex}^{\ddagger}$ and $T\Delta S_{ex}^{\ddagger}$ means that the impact of mechanistic variation on the rate constant is often less than one might expect.

For nonaqueous solvent exchange reactions, assignments of mechanism from $\Delta V_{ex}^{\ddagger}$ alone are somewhat arbitrary because analogues of Equation 8.15 have not been developed for nonaqueous media. Although some large values of $\left|\Delta V_{ex}^{\ddagger}\right|$ have been recorded for nonaqueous solvent exchange (e.g., $-20$ cm$^3$/mol for Sc(TMP)$_6^{3+}$/ TMP[59]), it has been noted[15] that $\Delta V_{ex}^{\ddagger}$ values for nonaqueous solvent exchange, such as those listed in Table 8.5, are often not very different from those for the corresponding water exchange reactions, despite much larger molar volumes for organic solvents relative to water, which may seem surprising because solvent molar volume appears in Equation 8.15 (18.07 cm$^3$/mol for water at 25°C). For example, $\Delta V_{ex}^{\ddagger}$ values for aceto-, propio-, butyro-, isobutyro-, valero-, and benzonitrile solvent exchange on Ni (solvent)$_6^{2+}$ all lie within the narrow range 12.4–14.4 cm$^3$/mol, consistent with a common dissociatively activated mechanism, despite the wide variation in the molecular bulk of the solvents.[107] For the analogous exchange reactions on Mn(solvent)$_6^{2+}$, $\Delta V_{ex}^{\ddagger}$ ranges from $-4.2$ cm$^3$/mol for acetonitrile to $+2.5$ cm$^3$/mol for isobutyronitrile, which in the light of the Ni$^{2+}$ results is indicative of a progression from $\mathbf{I_a}$ to $\mathbf{I_d}$ as the molecular bulk of the nitriles increases (cf. Table 8.5).[107] The near constancy of $\Delta V_{ex}^{\ddagger}$ for Ni$^{2+}$ despite a large variation in solvent bulk, however, raises the question of whether limits for $\Delta V_{ex}^{\ddagger}$ for **A** and **D** mechanisms in organic solvents could ever be estimated from an equation analogous to Equation 8.15. That equation is an empirical correlation based on an abundance of experimental values of $V_{abs}^0$ and $n$ for aqueous ions, in which $\Delta r$ is an adjustable parameter representing the effective thickness of the first coordination sphere of aqua ligands and 18.07 cm$^3$/mol is the molar volume of solvent water. There are no sufficiently large sets of experimental $V_{abs}^0$ and $n$ values available to derive a correlation similar to Equation 8.15 for any one organic solvent, but if there were, then large values of the solvent molar volume would be accompanied by large values of $V_{abs}^0$ and hence of the fitted parameter $\Delta r$, and it seems unlikely that a predictively useful correlation would emerge. To paraphrase Jordan,[15] the lack of any obvious dependence of $\Delta V_{ex}^{\ddagger}$ on the molar volume of organic solvents may mean that only the "business end" of a ligand contributes significantly to volume effects when the coordination number of the central metal ion changes; the very small water molecule, which consists of only a single ligand atom (O) with two H atoms attached, is "all business", and hence Equation 8.15 and the derived limiting $\Delta V_{ex}^{\ddagger}$ values are realistic for aqueous systems.

One advantage of solvent exchange studies using organic solvents is that the conjugate base pathways that complicate water exchange reactions are absent. The opposite is true for exchange of the very basic solvent, anhydrous liquid ammonia, with metal ammine complexes: for the $Cr(NH_3)_6^{3+}/NH_3$ exchange, added $NH_4^+$ salts strongly suppress the rate of $^{15}NH_3$ exchange (in inverse proportion to $[NH_4^+]$ above $[NH_4NO_3] = 1$ mmol/L) while added $NH_2^-$ accelerates exchange sharply.[108,109] An added complication is that, because of the low dielectric constant of liquid $NH_3$ ($\varepsilon \approx 15$ at 20°C), there is extensive ion pairing between $Cr(NH_3)_6^{3+}$ and the anions present that also reduces the exchange rate, nitrate more so than perchlorate. For 0.010 mol/L $[Cr(NH_3)_6](ClO_4)_3$ in liquid $NH_3$ under saturated vapor pressure without added salts, the extrapolated $^{15}NH_3$ exchange rate constant is $4.2 \times 10^{-4}$ s$^{-1}$ at 25°C, or two orders of magnitude faster than $Cr(H_2O)_6^{3+}/H_2O$ but similar to $Cr(H_2O)_5OH^{2+}/H_2O$ (Table 8.2), and the entire observable exchange rate appears to be carried by the conjugate base ion pair $\{Cr(NH_3)_5NH_2^{2+}, ClO_4^-\}$. For the much less acidic ion $Ni(NH_3)_6^{2+}$, neither $NH_4^+$ nor other added ions had any effect on the rate ($k_{ex}^{298} = 4.7 \times 10^4$ s$^{-1}$, or similar to that for the $Ni(H_2O)_6^{2+}/H_2O$ exchange).[110]

## 8.5   SUBSTITUTION IN OCTAHEDRAL COMPLEXES

The rate constant for water exchange on a metal aqua complex provides an index of the reactivity of that complex in net substitution reactions if the reaction mechanism is **D** or **I$_d$**, in which case $k_{ex}$ will determine the rate constant $k_{obsd}$ for substitution by incoming ligand A once ion pairing is taken into account (Section 8.3.3.3). Where associative activation is significant, substantial variation of $k_{obsd}$ with A should be evident even after ion pairing is allowed. For a series of reactions $ML_5OH_2^{3+} + A^-$, ion pairing in water is not extensive and, to a fair approximation, can be regarded as roughly constant throughout the series. For the aquation of complexes $Co^{III}(NH_3)_5X^{2+}$,

$$Co(NH_3)_5X^{2+} + H_2O \underset{k_{an}}{\overset{k_{aq}}{\rightleftharpoons}} Co(NH_3)_5OH_2^{3+} + X^- \qquad (8.17)$$

Langford[111] noted that $\log k_{aq}$ was a linear function of $-\log Q$, where $Q (= k_{an}/k_{aq})$ is the stability constant of $Co(NH_3)_5X^{2+}$, with slope 1.0, and took this to indicate that the transition state closely resembled the products, implying an **I$_d$** or **D** mechanism. A more direct interpretation of the slope of this linear free energy relationship (LFER)[112] is that the rate constant $k_{an}$ for the *reverse* of reaction 8.17 ("anation") must be essentially the same for all the anions $X^-$, which is indicative of an $I_d$ or $D$ process. The volumes of activation $\Delta V_{aq}^{\neq 0}$ at zero applied pressure for the aquation reaction 8.17 with $X^{z-} = H_2O$, $NO_3^-$, $Br^-$, $Cl^-$, and $SO_4^{2-}$ are $+1.2$,[55] $-6.3$, $-9.2$, $-10.6$, and $-18.5$ cm³/mol,[62] respectively, and the corresponding equilibrium volumes of reaction $\Delta V$ for the aquations are 0, $-6.0$, $-9.6$, $-10.4$, and $-15.2$ cm³/mol,[113] which are consistent with Langford's interpretation. The wide range in both $\Delta V_{aq}^{\neq 0}$ and $\Delta V$ values reflects primarily the hydration of the liberated $X^{z-}$; as seen for $X = H_2O$, the contribution of bond breaking in this series to $\Delta V_{aq}^{\neq 0}$ and $\Delta V$ is minimal. The marked pressure dependence of $\Delta V_{aq}^{\neq}$ for anionic X can be related to the hydration of $X^{z-}$ as it is released.[62]

For the aquation of $Cr^{III}(H_2O)_5X^{2+}$, however, Swaddle and Guastalla[114] found a similar LFER but with slope 0.59, which, following Langford's argument, implies a transition state just over half-way between reactants and products—in other words, an $I_a$ mechanism. A simpler interpretation[115] of the slope is that the rate constants $k_{an}$ for the reaction of $Cr(H_2O)_6^{3+}$ with $X^-$ vary markedly with the identity of $X^-$, which is the hallmark of associative activation. The propensity of $Cr(H_2O)_6^{3+}$ to $I_a$ substitution was subsequently verified through its markedly negative $\Delta V_{ex}^{\neq}$ for water exchange[56,57] and competition studies.[58] Volumes of activation for aquation in the $Cr^{III}(NH_3)_5X^{2+}$ series plot linearly against the reaction volume, as noted above for $Co(NH_3)_5X^{2+}$, but the slope is only about 0.6, implying an $I_a$ mechanism for the $Cr^{III}$ ammines.[116]

The Langford–Gray distinction between $I_a$ and $I_d$ mechanisms is arbitrary—the former is said to apply when "the reaction rate is approximately as sensitive (or more sensitive) to variation of the entering group as to variation of the leaving group," and the latter when "the reaction rate is much more sensitive to variation of the leaving group than to variation of the entering group"[5]—but there is clearly a continuum of degrees of sensitivity to the nature of the entering group (selectivity)[46,70,117] that needs to be described in a quantitative manner. Sasaki and Sykes[118] defined a parameter $R$ as the ratio of the rate constant for thiocyanate anation to that for chloride for a particular metal aqua complex; this proposal has the virtue that the necessary data are available for a useful range of metal aqua substrates, although the concept is limited because $NCS^-$ may attach to "hard"[119,120] metal centers such as $Cr^{III}$ through its N atom and "soft" centers such as the heavier metals of second and third transition series through the S end. An alternative parameter $S$ defines selectivity for some particular metal aqua complex $M$ relative to a reference aqua complex $M_{ref}$ for as wide a range as possible of rate constants $k_{an}$ for incoming ligands A displacing an aqua ligand from a metal complex:

$$S = (\delta \ln k_{an}^M)/(\delta \ln k_{an}^{Mref}) \tag{8.18}$$

where $\delta \ln k_{an}^M$ is the change in ln(rate constant) as A is varied.[46,70] This allows the use of a range of A, not necessarily the same for all $M$. A suitable universal reference substrate $M_{ref}$ is $Cr(H_2O)_6^{3+}$ for which $S$ is defined as 1.0. Values of $R$ and $S$ are given for some tervalent metal ions in Table 8.7. These show that high selectivity parallels the associative character inferred from the $\Delta V_{ex}^{\neq}$ data of Table 8.2—for example, the conjugate base species $Cr(H_2O)_5OH^{2+}$ and $Fe(H_2O)_5OH^{2+}$ show less selectivity, and

**TABLE 8.7    Selectivity Parameters $R$ and $S$ at 25°C**

| Substrate | $R$ | $S$ |
|---|---|---|
| $Mo(H_2O)_6^{3+}$ | 62 | 1.1 |
| $Cr(H_2O)_6^{3+}$ | 55 | 1.0 |
| $Co(H_2O)_6^{3+}$ | $\geq 43$ | |
| $V(H_2O)_6^{3+}$ | $\geq 36$ | |
| $Fe(H_2O)_6^{3+}$ | 14 | 0.7 |
| $Cr(NH_3)_5OH_2^{3+}$ | 6 | 0.3 |
| $Co(H_2O)_5OH^{2+}$ | 2 | |
| $Cr(H_2O)_5OH^{2+}$ | 1 | 0.4 |
| $Rh(NH_3)_5OH_2^{3+}$ | 0.6 | |
| $Fe(H_2O)_5OH^{2+}$ | 0.6 | 0.1 |
| $Co(NH_3)_5OH_2^{3+}$ | 0.5 | $-0.1$ |

less associative character in solvent exchange, than do their parent hexaaqua complexes.

A severe limitation in setting up LFERs and selectivity scales for metal aqua or ammine complexes is that incoming ligands $A^-$ that have appreciable Brønsted basicity, such as $F^-$ and $N_3^-$, will tend to abstract a proton from an aqua ligand and react as HA with the labilized conjugate base $M(H_2O)_{n-1}OH^{(z-1)+}$. The microscopic reverse of this is the preferential separation of HA from $M(H_2O)_{n-1}OH^{(z-1)+}$ in an aquation reaction rather than of $A^-$ from $M(H_2O)_n^{z+}$—the so-called proton ambiguity. The same problem arises with ammine complexes, for example, in the formation or aquation of $Co(NH_3)_5N_3^{2+}$, since amido complexes of $Co^{III}$, $Ru^{III}$, and so on undergo substitution much more rapidly than the parent ammines. These alternative pathways are indistinguishable through their rate equations, but may be revealed through anomalous activation parameters;[112] for example, for aquation of $Co(NH_3)_5N_3^{2+}$, $\Delta V_{aq}^{\ddagger} = +17\,cm^3/mol$ rather than the negative value expected if $N_3^-$ were to separate from $Co(NH_3)_5^{3+}$.[62]

### 8.5.1    Spectator Ligands and Stereochemical Change

Spectator ligands—those not directly involved in the ligand substitution process—can affect the reaction rate electronically (*cis* and *trans* effects), through steric hindrance or acceleration, or through influencing solvation.

*trans* directed labilization is generally less effective in octahedral than in $d^8$ square planar complexes, but nevertheless can have important consequences in transition metal chemistry. Comparison of water exchange rates for $M(NH_3)_5OH_2^{3+}$ and $M(H_2O)_6^{3+}$ (Tables 8.2 and 8.4) for M = Cr, Ru, Rh, and Ir indicates a modest labilizing effect of $NH_3$ relative to $H_2O$, and for $Cr(NH_3)_n(H_2O)_{6-n}^{3+}$ this has been shown to be a *trans* directed effect.[121] In the complexes $Cr^{III}(H_2O)_5X^{2+}$, $H_2^{18}O$ exchange rate data show that the aqua ligand *trans* to X is labilized in the sequence $I^- > Br^- > Cl^- > NCS^- > H_2O$.[122] For loss of X from $Rh^{III}(en)_2LX$ systems, Poë and Vuik[123] found an intrinsic kinetic *trans* effect series $L = I^- \gg Br^- > NH_3 > OH^- \geq Cl^-$. More impressive are the strong *trans* labilization effects of ligands such as $S_2O_3^{2-}$ and especially $SO_3^{2-}$ in $Co^{III}$ am(m)ine complexes; the S-bonded $SO_3^{2-}$ ligand in $Co^{III}L_4(SO_3)X$ (L = amine or 1/2(en)) labilizes *trans* X ligands some $10^8$ times more effectively than does $NH_3$.[124] In $Co^{III}$ complexes at least, powerful *trans* activators are said to work by favoring a limiting **D** mechanism.[125] Other powerful *trans* labilizing ligands include $CN^-$, isonitriles, CO, NO, and alkyl groups. One consequence of the high labilizing power of $CN^-$ is that aquation of $Cr(CN)_6^{3-}$ is highly stereospecific, proceeding via *fac*-$Cr(H_2O)_3(CN)_3$ and *cis*-$Cr(H_2O)_4(CN)_2^+$ to $Cr(H_2O)_5CN^{2+}$; conversely, the direct synthesis of this last complex from $Cr(H_2O)_6^{3+}$ is impossible because $CN^-$ substitution will promptly proceed beyond the first step.[126] In general, the kinetic *trans* effect is associated with a structural *trans* effect involving *trans* bond lengthening and a minor *cis* bond lengthening—there appears to be no case in which the *trans* bond lengthening is accompanied by *cis* bond shortening (in contrast to tetragonal Jahn–Teller distortions).[127]

Although any *cis* labilization accompanying a substantial kinetic *trans* effect is small,[127] some spectator ligands do exert an important labilizing effect on ligands *cis* to them but with little static (structural) effect. Jordan[15] gives the order of *cis* activating power in octahedral metal complexes (primarily organometallics) as $NO_3^- > RCO_2^- > Cl^- \approx Br^- >$ pyridine $> I^- > PR_3 >$ CO, $H^-$, which is roughly opposite to the order of *trans* effects. For example, the thermal aquation of $Cr(NH_3)_5ONO_2^{2+}$ in acidic solution gives only 33% $Cr(NH_3)_5OH_2^{3+}$, the balance of the products being *cis*-$Cr(NH_3)_4(OH_2)_2^{3+}$ and more highly aquated species, plus $NH_4^+$.[128] Even dry, solid $[Cr(NH_3)_5ONO_2](NO_3)_2$ decomposes in the dark over a few days liberating $NH_3$; no solid halopentaamminechromium(III) nitrate or perchlorate salts decompose detectably on this timescale. The most likely explanation is that nitrate ion can form a transient chelate complex $Cr(NH_3)_4(O_2NO)^{2+}$ following expulsion of one $NH_3$ in an associative *cis* attack by the $NO_3^-$ ligand; in aqueous solution, the chelate then aquates to *cis*-$Cr(NH_3)_4(ONO_2)OH_2^{3+}$, which can either aquate to *cis*-$Cr(NH_3)_4(OH_2)_2^{3+}$ or else go on to form $Cr(NH_3)_3(O_2NO)OH_2^{2+}$ with expulsion of another $NH_3$ followed by dechelation and uptake of another water molecule and so on until the nitrato ligand is itself aquated off. That this sequence occurs for nitrato complexes of $Cr^{III}$ but apparently not $Co^{III}$ can be ascribed to the proclivity of $Cr^{III}$ centers to associative activation; significantly, base hydrolysis of $Cr(NH_3)_5ONO_2^{2+}$, which proceeds relatively rapidly by a dissociatively activated conjugate base pathway (see below), goes cleanly to $Cr(NH_3)_5OH^{2+}$.

Bakac and coworkers[129] recently provided a striking example of the synthetic utility of the *cis* labilizing effect of the nitrato ligand in $Cr^{III}$ complexes. Reaction of $(H_2O)_5CrONO_2^{2+}$ with dilute (10–100 mmol/L) $H_2O_2$ in acidic aqueous solution gave the hydroperoxochromium(III) complex $CrOOH^{2+}$(aq) in high (>50%) non-equilibrium yield, together with minor amounts of the superoxochromium(III) analogue $CrOO^{2+}$(aq) and some $Cr(H_2O)_6^{3+}$ aquation product. Direct reaction of $H_2O_2$ with $Cr(H_2O)_6^{3+}$ at low pH does not produce significant amounts of $CrOOH^{2+}$(aq), which was formerly prepared by reduction of $CrOO^{2+}$(aq) recovered in small yield from the oxidation of $Cr^{2+}$(aq) by $O_2$. Since both $CrOOH^{2+}$(aq) and $CrOO^{2+}$(aq) can act as precursors of genotoxic $Cr^V$, these results may have implications for the toxicology of ostensibly benign $Cr^{III}$ in the presence of phys-iological peroxide and environmental nitrate or other potentially bidentate ligands.[129]

A corollary of the proposed nitrato *cis* activation mechanism is that other potentially bidentate ligands such as $RCO_2^-$, $SO_4^{2-}$, $SO_3^{2-}$ (if O-bonded), and $CO_3^{2-}$ should activate *cis* ligands in $Cr^{III}$ complexes, which are predisposed to associative processes, though probably not in $Co^{III}$ analogues, which are not. Indeed, attempts to make $Cr^{III}$ analogues of the familiar $Co(NH_3)_5OCO_2^+$ and $Co(NH_3)_4(O_2CO)^+$ salts yield only $Cr(OH)_3$, and efforts in the writer's laboratory to make $Cr(NH_3)_5SO_4^+$ have been unsuccessful. Similarly, $Cr(H_2O)_5OSO_2^+$ is much more labile in substitution even than $Cr(H_2O)_5OH^{2+}$, but unlike the powerfully *trans* activating S-bonded $SO_3^{2-}$ ligand in $Co^{III}$ complexes, the sulfito ligand bonds through O to $Cr^{III}$ and is known to be at least partially a *cis* activator.[130] In the aquation of $Cr(NH_3)_5O_2CR^{2+}$, carboxylate aquation prevails for $R = CF_3$ or $CCl_3$, but *cis* ammonia loss is the primary result for $R = CH_3$, $CH_2Cl$, and $CHCl_2$, the rate increasing with the increasing basicity of the carboxylate ligand.[131]

The strong conjugate base activation resulting from deprotonation of an am(m)ine ligand in base hydrolysis of cobalt(III) complexes (Sections 8.3.1 and 8.4.1) is almost always a *cis* effect.[132] Base hydrolysis of $Co^{III}$ complexes evidently involves a five-coordinate intermediate that may survive long enough to undergo substantial isomer-ization before resuming six-coordination (in the limit, a dissociative conjugate base or $D_{CB}$ mechanism), in contrast to simple aquation reactions of $Co^{III}$ ammine complexes that show more limited stereochemical change (an $I_d$ mechanism). For example, in their classic study of base hydrolysis of *trans*-$(^{15}NH_3)Co^{III}(NH_3)_4X^{2+}$ ($X = Cl$, Br, $NO_3$), Buckingham et al.[133] were able to show that the products were 50% *cis*-$(^{15}NH_3)Co(NH_3)_4OH^{2+}$ and 50% *trans*-$(^{15}NH_3)Co(NH_3)_4OH^{2+}$ regardless of the identity of X, giving evidence for a common intermediate of sufficient lifetime to have largely "lost its memory" of X. The percentage of *trans* product, however, is greater than the maximum 33% expected if an initially formed square pyramidal five-coordinate intermediate relaxes *fully* to a trigonal bipyramid that subsequently accepts the incoming ligand in the equatorial plane. Since a square pyramidal intermediate would accept an incoming ligand at the vacant position to form exclusively a *trans* product, the 50% yield of the *trans* isomer suggests that at least some of this intermediate acquires a sixth ligand before rearranging to a trigonal bipyramid. This implies that the time frame for stereochemical relaxation is not very different from that for resumption of six-coordination. Indeed, it is not clear whether the five-coordinate intermediate lives long enough to discriminate between potential new ligands or whether it quickly scavenges a molecule from its immediate environment.[132] In the latter case, the mechanism might be better labeled $I_{dCB}$.

The striking labilization of $Co^{III}$ am(m)ine complexes following abstraction of a N proton seems more than can be attributed to overall charge reduction (in MO terms, a $\sigma$ effect) and was explained by Pearson and Basolo[134] as arising from the interaction of the $\pi$ (2p) electrons of the newly created amido N atom with a Co 3d orbital, so destabilizing the six-coordinate parent complex but stabilizing a trigonal bipy-ramidal five-coordinate intermediate. This explanation is consistent with the general observation that substitution reactions of *trans*-$Co^{III}(en)_2LX$ typically result in stereochemical change when L is a potential $\pi$ donor[4] and has gained widespread acceptance. It is not, however, entirely satisfactory—base hydrolyses of $Cr^{III}$, $Rh^{III}$, $Ir^{III}$, and especially $Ru^{III}$ ammines all proceed rapidly by conjugate base mechanisms but are stereoretentive, and for $Co^{III}$ there is evidence, as noted above, that at least the initial form of the five-coordinate intermediate is square pyramidal. One alternative suggestion is that the labilization of the conjugate base could derive from a single electron transfer to the central $M^{III}$ from the amido N to give labile $M^{II}$ transiently; this could explain why the relatively easily reducible $Co^{III}$ and $Ru^{III}$ ammines are much more susceptible to base hydrolysis than those of $Cr^{III}$, $Rh^{III}$, and $Ir^{III}$.[132] It may suffice to say that amido ligands in $Co^{III}$ complexes exert a powerful labilizing effect, probably mainly through their powerful $\pi$ donor capability, and act most effectively from a position *cis* to the leaving ligand.

According to Vanquickenborne and Pierloot,[135] stereomobility in low-spin $d^6$ ($t_{2g}^6$ $e_g^0$) complexes is possible only if a spin-state change, such as to a low-lying quintet, occurs somewhere along the reaction coordinate, which means that stereochemical change will become less likely as $Dq$ increases. Consequently, although significant stereochemical change accompanies aquation and especially base hydrolysis of low-spin $Co^{III}$ complexes, substitution in $Rh^{III}$ and $Ir^{III}$ complexes is stereoretentive. More precisely,[68] if a trigonal bipyramidal $Co^{III}$ or $Fe^{II}$ species exists with a quintet state lying close to the singlet ground state of the square pyramidal intermediate, stereochemical change is possible; no such quintet state is accessible for $Rh^{III}$ or $Ir^{III}$. Low-spin $Fe^{III}$ complexes ($3d^5$) will substitute with stereochemical change if there is a trigonal bipyramidal intermediate with a sextet state lying close to the doublet state of the initially formed square pyramidal intermediate. Ruthenium(III) am(m)ines (low-spin $4d^5$), however, although at least as highly labilized as their $Co^{III}$ analogues in base hydrolysis, are stereotetentive in substitution reactions, presumably because of their higher $Dq$. Chromium(III) am(m)ines ($3d^3$) are less susceptible to base hydrolysis (relative to nonconjugate base substitution) than their $Co^{III}$ counterparts and are stereoretentive in substitution reactions, probably because of a somewhat greater disposition toward $I_a$ behavior in substitution (Table 8.7)—either an octahedral wedge seven-coordinate intermediate or a square pyramidal five-coordinate intermediate will lead to stereochemical retention unless further isomerization processes occur—but in any event the spin-change mechanism proposed for $d^6$ complexes is inapplicable to $d^3$.

### 8.5.2 Steric Effects

Following the traditional lore of physical organic chemistry, it may be expected that addition of bulky substituents on spectator ligands should hinder the approach of an incoming ligand (associative activation) but accelerate the departure of the replaced ligand through relief of steric congestion (dissociative activation). For the replacement of chloride by water in $M(RNH_2)_5Cl^{2+}$, $k_{R=Me}/k_{R=H} = 22$, 0.030 (25°C), and 0.50 (85°C) for M = Co, Cr, and Rh, respectively; for the base hydrolysis, the corresponding ratios are $1.5 \times 10^4$, 225, and 29 and are much larger than can be accounted for by changes in the acidities of the $RNH_2$ ligands.[136–138] These results are consistent with expectations for steric effects in $\mathbf{I_d}$, $\mathbf{I_a}$, and mid-range $\mathbf{I}$ mechanisms for Co, Cr, and Rh, respectively, in the aquation reactions, and for $\mathbf{D_{CB}}$, $\mathbf{I_{Dcb}}$, and mild $\mathbf{I_{dCB}}$ in the corresponding base hydrolyses. For aquation and base hydrolysis of $Cr(RNH_2)_5Cl^{2+}$, progression from R = methyl through ethyl and $n$-propyl to $n$-butyl results in a less dramatic increase in rate, presumably because increases in either congestion or inductive electron release from the amine ligands favors the dissociative component, and/or because of increasing inhibition of solvation of the complex.[136] Surprisingly, however, the Cr–Cl bond is some 3 pm shorter in $Cr(MeNH_2)_5Cl^{2+}$ than in $Cr(NH_3)_5Cl^{2+}$, opening the possibility that the slower aquation rate in $Cr(MeNH_2)_5Cl^{2+}$ arises from a stronger Cr–Cl bond rather than steric hindrance to associative activation; at the same time, the Cr–N bonds all lengthen by 3 pm.[139] A lesser M–Cl bond shortening on N-methylation of $M(NH_3)_5Cl^{2+}$ occurs for M = Rh and Ir, whereas for M = Co there is no change in the Co–Cl bond length. In none of these complexes is there structural evidence of substantial steric strain in the ground state.

Lay[139] has attributed the reactivity differences to M–Cl $\pi$ interactions in the ground state and developed from this a case for describing Cl substitution in all these cases, including Cr, as $\mathbf{I_d}$. For water exchange on $M(RNH_2)_5OH_2^{3+}$, however, there is no possibility of M-OH$_2$ $\pi$ bonding, yet the same pattern of retardation on going from R = H to R = Me for M = Cr, and acceleration for M = Co, is observed (Table 8.4); for M = Rh, there is almost no change. More significantly, $\Delta V_{ex}^{\ddagger}$ shows unambiguously that in all cases going from M = H to M = Me in $M(RNH_2)_5OH_2^{3+}$ results in an increase in the dissociative character in the $\mathbf{I_a}$–$\mathbf{I_d}$ continuum, that for M = Cr the $\mathbf{I_a}$ label remains appropriate, and that water exchange on $Co(NH_3)_5OH_2^{3+}$ is fairly described as $\mathbf{I_d}$ and not at all close to the limiting $\mathbf{D}$ case. González and Martínez[140] have confirmed this through studies on the rates of substitution of the aqua ligand in $M(RNH_2)_5OH_2^{3+}$ with a range of anions ($H_2PO_4^-$, $H_2PO_3^-$, $CF_3CO_2^-$, $Br^-$, $Cl^-$, and $NCS^-$)—for M = Cr, selectivity toward the entering ligand decreases on going from R = H to R = Me, and so the span of rate constants narrows, as expected for predominantly $\mathbf{I_a}$ behavior, whereas for M = Co no such narrowing of the span of rates occurs.

A possible further factor in the origin of the kinetic effects of N-methylation of $Co^{III}$ ammines emerges from the observation[141] that in the aquation of $Co(NH_3)_5DMF^{3+}$ and cis- and trans-$MeNH_2Co(NH_3)_4DMF^{3+}$, the rate constants $k^{298}/10^{-5}\,s^{-1}$ are 0.2, 0.3, and 0.8, respectively, indicating that the trans-MeNH$_2$ is exerting a modest *trans* labilization electronically. Since steric congestion and solvation effects due to the single methyl group in the first two complexes should be minimal, it may be inferred that the acceleration of $\mathbf{I_d}$ $Co^{III}$ substitutions on going from $Co(NH_3)_5$- to $Co(MeNH_2)_5$-, and the retardation of the corresponding $\mathbf{I_a}$ $Cr^{III}$ reactions, is due in part to inductive electronic release from the methyl substituents to the metal center. This effect, however, is not large, so it can be concluded that contributions from steric as well as solvational[136] and LFAE (Table 8.4) effects are also present. In summary, it may be stated that going from $M(NH_3)_5$- to $M(MeNH_2)_5$- does diminish the $\mathbf{I_a}$ and augment the $\mathbf{I_d}$ characters, respectively, in octahedral ligand substitution, and may be used with caution as a criterion for the dominance of one or the other activation mode within the continuum of the interchange mechanism.

### 8.5.3 Chelate Complexes

Chemists can intuitively appreciate that a metal chelate complex of a multidentate ligand will ordinarily be much more stable than a complex of the same metal ion with the corresponding number of comparable unidentate ligands, but a quantitative understanding of this "chelate effect" has proved elusive. For example, direct comparison of the stability constants for the multidentate and unidentate complexes is not possible because the units will be different, and the same problem carries over into comparison of naïvely calculated reaction entropies. Analysis of the kinetics of the multistep formation and solvolysis of a chelate complex such as $(H_2O)_4Ni(en)^{2+}$ (Figure 8.7; stability constant $4.0 \times 10^7$ L/mol at 25°C) can give a clearer picture. Early attempts,[4] however, suffered from incomplete information and from the preconception that the first coordination step ($k_{12}$ or $k_{34}$) will be rate determining because ring closure ($k_{35}$) must be fast—an idea that seems to have originated with Schwarzenbach's observation[142] that once one donor atom of a multidentate ligand is attached to the metal ion, the concentration of other donor atoms becomes locally very high. Jordan[15] points out that the local concentration of donor atoms should not affect the rate if the reaction is dissociatively activated, but this is an oversimplification because in an $I_d$ process such as is typical of $Ni^{II}$ complexes, the transient intermediate of reduced coordination number has to scavenge a replacement ligand from whatever is within easy reach. That said, Jordan's careful analysis[15,143] of the $Ni(H_2O)_6^{2+}$-en reaction gives $k_{43} \approx 900$ L/(mol s), $k_{34} \approx 15$ s$^{-1}$, $k_{35} \approx 1.2 \times 10^5$ s$^{-1}$, and $k_{53} \approx 0.14$ s$^{-1}$ (Figure 8.7). These data clearly show that the high stability of $(H_2O)_4Ni(en)^{2+}$ originates mainly in the surprisingly slow, rate-determining, ring-opening step ($k_{53}$); the ring closure rate constant $k_{35}$ is only about four times faster than water exchange on $Ni(H_2O)_6^{2+}$ (Table 8.2; comparison with $k_{43}$ is not directly possible because of the different units).

The $Ni(H_2O)_6^{2+}$-en example serves as a paradigm for chelation reactions of metal aqua ions. The role of $H^+$(aq) may be noted as favoring the aquation of typical metal

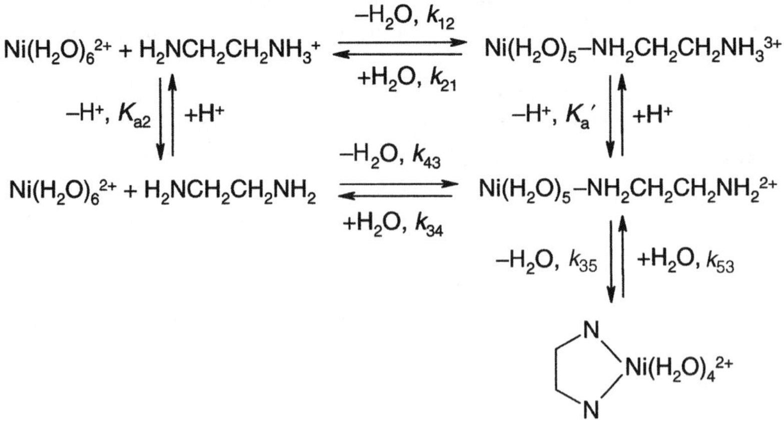

**FIGURE 8.7**    Formation and aquation of $Ni(en)(H_2O)_4^{2+}$; notation as per Jordan.[15]

chelates—it is not involved in the initial rate-determining ring-opening step because there is no free N electron pair to protonate, but once the free end of unidentate en is available, protonation can assist ligand loss in the same manner as it accelerates loss of basic ligands such as azide, fluoride, acetate, or carbonate.[112] We note in passing that the loss of protonated carbonate involves $MO–CO_2$ rather than $M–OCO_2$ bond breaking[144] and that this is a common mechanism of aquation of basic oxoanionic ligands such as O-bonded sulfite.[145] For the carbonato ligand as a bidentate chelate in $(NH_3)_4CoO_2CO^+$, $H^+$-induced ring opening involves Co–O bond breaking but separation of $CO_2$ proceeds by C–O fission.[146]

## 8.6  SUBSTITUTION IN SQUARE PLANAR COMPLEXES

Interest in the dynamics of substitution in square planar $Pt^{II}$ complexes[147] has a long history, largely because the reaction rates involved are slow enough to be readily accessible by conventional techniques and show marked dependences on the natures of incoming, outgoing, and spectator ligands—in short, they provide a paradigm of associative activation in inorganic chemistry. This interest continues today largely because of the application of $Pt^{II}$ complexes related to $cis$-$Pt(NH_3)_2Cl_2$ (cisplatin) in cancer therapy; cisplatin binds to purine bases on DNA in a cancer cell, and the resulting intrastrand cross-linking destroys the cell.[148–150] Some 3000 $Pt^{II}$ complexes have been investigated to date in this context, largely with the objective of finding drugs with equal or better anticancer activity but less toxic side effects. The kinetics of aquation and competitive complex formation of the $Pt^{II}$ complexes with components of biological media as well as sites on the target DNA are essential information for drug development; for example, the interaction of aquated cisplatin with phosphate and especially carbonate[151] can diminish the pool of the active drug. The efficacy of cisplatin is due in part to the favorable magnitudes of aquation and subsequent complex formation rates—the chloride ligands must aquate slowly enough to allow drug to reach its target, but fast enough to present itself to the DNA as $cis$-$Pt(NH_3)_2(OH_2)_2^{2+}$; the $Pt(NH_3)_2^{2+}$ cross-linking unit must then remain attached to the DNA long enough to destroy the cell. In this respect, the analogous square planar $Pd^{II}$ complexes substitute too rapidly.

Consideration of Figure 8.5 shows that, for $d^8$ metal ions in weak to moderate homoleptic ligand fields, the high-spin $t_{2g}^6e_g^2$ configuration will favor six-coordination and a regular $O_h$ geometry, but that if the ligand field strength represented by $Dq$ becomes large enough, the complex will stabilize itself by an amount $2E_{JT}$ (minus the electron pairing energy) by undergoing tetragonal ($D_{4h}$) distortion strong enough to pair the two higher energy electrons in the $d_{z^2}$ orbital. This distortion typically results in loss of the two ligands on the $z$ axis. Thus, while most $Ni^{2+}$ complexes are octahedral (or, more rarely, tetrahedral as in $NiCl_4^{2-}$) and paramagnetic, the strong ligand $CN^-$ forces spin pairing to give the diamagnetic square planar $Ni(CN)_4^{2-}$— this "either/or" imposition of alternative geometries is not to be confused with the continuously variable Jahn–Teller distortion that arises from an orbital degeneracy. Almost as effective at energy minimization in the $Ni^{2+}$ cyanide case is the formation

**TABLE 8.8  Kinetics of Bimolecular Cyanide Exchange on Aqueous Tetracyanometalates at pH $\geq 6$**

| Complex | $k_2^{298}$ (L/(mol s)) | $\Delta H_2^{\ddagger}$ (kJ/mol) | $\Delta S_2^{\ddagger}$ (J/(K mol)) | $\Delta V_2^{\ddagger}$ (cm$^3$/mol) |
|---|---|---|---|---|
| Ni(CN)$_4^{2-}$ $^a$ | $2.3 \times 10^6$ | 21.6 | $-51$ | $-19$ |
| Pd(CN)$_4^{2-}$ $^a$ | 82 | 23.5 | $-129$ | $-22$ |
| Pt(CN)$_4^{2-}$ $^a$ | 11 | 25.1 | $-142$ | $-27$ |
| Au(CN)$_4^{-}$ $^b$ | $6.2 \times 10^3$ | 40.0 | $-38$ | $+2$ |

$^a$Ref. 152.
$^b$Ref. 153.

of the spin-paired, five-coordinate ($D_{3h}$) Ni(CN)$_5^{3-}$ ion with a full complement of 18 electrons in its valence shell. Consequently, the second-order exchange reaction between the 16-electron, coordinatively unsaturated Ni(CN)$_4^{2-}$ and CN$^-$ in solution proceeds very rapidly through the intermediate Ni(CN)$_5^{2-}$ that exists in detectable amounts in solution in equilibrium with Ni(CN)$_4^{2-}$; this is a clear example of an **A** mechanism. Cross[147] points out that such $16 \rightarrow 18 \rightarrow$ 16-electron pathways are common to many catalytic processes. As we descend the periodic table, however, $Dq$ increases markedly, further stabilizing the square planar species, with the result that Pd(CN)$_4^{2-}$ and Pt(CN)$_4^{2-}$ have much increased stability relative to five-coordinate intermediates and undergo cyanide exchange much more slowly (Table 8.8).[152,153]

For M(H$_2$O)$_4^{2+}$, substitution of an aqua ligand by A$^{z-}$ takes place some $10^5$-fold faster for M = Pd than for M = Pt; typical rate constants are given in Table 8.9, which also shows that the charge on A is not important—the fastest reactions listed are with neutral thiourea—and hence ion pairing is not a prerequisite for reaction, in contrast to **I**$_a$ processes. Recent evidence[96] that at least one (and probably two) extra water molecule(s) is (are) located at relatively long M-OH$_2$ distances above (and below) the MO$_4$ plane in aqueous solution implies that the intimate mechanism may not be the simple **A** process that is commonly invoked.

If, for simplicity, we disregard the likely presence of additional weakly bound solvent molecules, the availability of space above and below the plane of a square planar complex means that associative attack is favored, and consequently dissociative

**TABLE 8.9  Bimolecular Rate Constants $k_2$ (L/(mol s)) for the Formation of M(H$_2$O)$_3$A$^{(2-z)+}$ from M(H$_2$O)$_4^{2+}$ and A$^{z-}$ in Aqueous Solution$^a$**

| A$^{z-}$ | M = Pd | M = Pt |
|---|---|---|
| Cl$^-$ | $1.8 \times 10^4$ | $2.7 \times 10^{-2}$ |
| Br$^-$ | $9.2 \times 10^4$ | 0.21 |
| I$^-$ | $1.1 \times 10^6$ | 7.7 |
| SCN$^-$ | $4.4 \times 10^5$ | 1.3 |
| DMSO | 2.5 | $8.4 \times 10^{-5}$ |
| Me$_2$S | $1.5 \times 10^5$ | 3.6 |
| (NH$_2$)$_2$CS | $9.6 \times 10^5$ | 14 |

$^a$Data from Ref. 22.

ligand loss in square planar $Pd^{II}$, $Pt^{II}$, and $Au^{III}$ complexes is rare except in complexes such as $cis$-$Ph_2Pt^{II}(Me_2S)_2$[35] in which organic spectator ligands can stabilize a 14-electron, three-coordinate intermediate (Section 8.3.3.1). Steric hindrance by bulky ligands does retard the associative processes markedly, as expected, but not sufficiently to allow dissociative pathways to emerge.[154] An important factor inviting associative attack is the "soft" character[119] of the central metal ions, which means that attacking ligands A with "soft" characteristics such as $I^-$, thiolates, $CN^-$, or organophosphines will be highly effective. For these, the reaction rates will be characterized by relatively low $\Delta H^{\ddagger}$ because of the strength of the forming M–A bond, and this will help offset the negative $\Delta S^{\ddagger}$ typical of associative activation. Representative activation parameters are given in Table 8.8.

As may be expected from the predominance of associative character, selectivity is much more evident in square planar than in octahedral substitution. The nucleophilicity of incoming ligands A toward $Pt^{II}$, relative to the solvent, can be represented by a parameter $n^0_{Pt}$ defined as log $(k_A/k_{solvent}) = n^0_{Pt}$, where the standard substrate is $trans$-$Pt^{II}(py)_2Cl_2$ at 30°C and the solvent is methanol ($k_{solvent}$ is divided by the molar concentration of the solvent to give a dimensionless ratio). This can then be generalized to the Equation 8.19

$$\log k_A = s n^0_{Pt} + \log k_{solvent} \tag{8.19}$$

for other substrates and other solvents; the constant $s$, the slope of the linear plot of $\log k_A$ against $n^0_{Pt}$, is a nucleophilic discrimination (selectivity) factor for the particular substrate and the solvent—the larger the $s$, the more selective the substrate is. Values of $n^0_{Pt}$ for numerous nucleophiles have been tabulated by Basolo and Pearson[4] and define the sequence of increasing nucleophilicity toward $Pt^{II}$ as $H_2O \sim$ MeOH $\ll Cl^- < NH_3 < py < NO_2^- < PhSH < Br^- < Et_2S < Me_2S < I^- < Me_2Se <$ $SCN^- < SO_3^{2-} < Ph_3As < CN^- < (MeO)_3P < SeCN^- < (NH_2)_2CS < S_2O_3^{2-} <$ $Et_3As < Ph_3P < Et_3P$. Leaving group effects on the reaction rate follow essentially the reverse of this sequence. For $Pd^{II}$, a similar $n^0_{Pd}$ scale could in principle be established, but the availability of data is limited by the much faster rates of substitution at $Pd^{II}$. For $Au^{III}$, many ligands of interest are susceptible to oxidation by the metal center,[155] but generally substitution rates are orders of magnitude faster than for corresponding reactions of $Pt^{II}$ and are even more sensitive to the identity of the incoming ligand.[156]

The kinetic effect of spectator ligands $trans$ to the leaving group is much more prominent in square planar than in octahedral substitution and is an important factor in the design of syntheses of $Pt^{II}$ compounds. Almost always ligand substitution in square planar complexes is stereoretentive,[147] so the $trans$ effect is easy to trace. As a time-honored example dating from 1844, Rieset found that heating solid $[Pt(NH_3)_4]Cl_2$ to 250°C gave a pale yellow compound, later shown to be $trans$-$Pt(NH_3)_2Cl_2$, whereas Peyrone showed that reaction of $K_2[PtCl_4]$ with aqueous ammonia gave an orange isomer of $Pt(NH_3)_2Cl_2$ that turned out to be the $cis$ form because the $trans$ labilizing effect of $Cl^-$ through $Pt^{II}$ is stronger than that of $NH_3$. The general sequence of increasing $trans$ labilizing power at $Pt^{II}$ is $H_2O \sim OH^- \sim F^-$ $< NH_3 < py < Cl^- < Br^- < I^- < R_2S < R_3P, H^- < CO, \eta^2$-$C_2H_4 < CN^-$. The similar low ranking of $H_2O$ and $OH^-$ in this series serves to make the point that conjugate base mechanisms are of little importance in $Pt^{II}$ substitution, which is understandable because they operate by promoting dissociative and not associative activation.

An example of a strong *trans* effect is provided by $PtCl_3(\eta^2\text{-}C_2H_4)^-$ (Zeise's anion), in which the ethene is bonded sideways on to Pt with its molecular axis perpendicular to the square plane. Ethene exerts a *trans* labilizing effect so powerful that Otto and Elding[157] had to resort to cryo-stopped-flow diode array spectrophotometry at 223 K to obtain rate constants $k_{obsd}$ for replacement of the *trans*-Cl by various ligands A in methanol. The rate equation is $k_{obsd} = k_1 + k_2[A]$, where $k_1^{223} = k_{MeOH}^{223}[MeOH] = 1.2 \times 10^2 \, s^{-1}$ and $k_2^{223} = 5.1 \times 10^2, 3.5 \times 10^3, 1.2 \times 10^4$, and $5.6 \times 10^4 \, L/(mol \, s)$ for $A = Br^-, I^-, N_3^-$, and $SCN^-$, respectively. As a benchmark, the rate constant for replacement of one $Cl^-$ in $PtCl_4^{2-}$ by $Br^-$ in water at 298K is $4.8 \times 10^{-5} \, L/(mol \, s)$.[158] Because of the strong *trans* labilizing effect of $\eta^2\text{-}C_2H_4$, Zeise's anion exchanges bound and free ethene rapidly in methanol ($k_2 = 2.1 \times 10^3 \, L/(mol \, s)$) at 298 K).[159] The static effect of ethene on the *trans*-Pt–Cl bond length is very small, presumably because $\sigma$ electron release and $\pi^*$ electron acceptance by $\eta^2\text{-}C_2H_4$ counteract each other, indicating that the origin of the labilization lies in stabilization of the associative transition state rather than any weakening of the *trans*-Pt–Cl bond in the ground state.

The kinetic *trans*-effect of a ligand L in *trans*-$LPt^{II}L_2'X$ complexes can operate by $\sigma$ electron release from a highly polarizable ("soft") L with fairly obvious consequences for the departure of X, or by $\pi$ electron withdrawal from the filled $d_{xz}$ Pt orbital into a vacant $\pi$ orbital of L ("back-donation"), above and below the plane of the complex ($xy$), so facilitating the entry of an incoming nucleophile A and stabilizing a five-coordinate intermediate. The $\pi$ acceptor orbital on L may be a vacant $p_z, d_{xz}$ or (as in the case of $\eta^2\text{-}C_2H_4$) an antibonding $\pi^*$ MO of L, and the $\pi$ acceptance of electron density from Pt on to L can stimulate $\sigma$ release from L to Pt—a synergic or "push–pull" effect. The *trans* labilizing effect of $\eta^2\text{-}C_2H_4$ and CO is primarily due to $\pi^*$ acceptance, whereas $H^-$ and $CH_3^-$ are essentially $\sigma$ donors, and $CN^-$, thiourea, $I^-$, and triorganophosphines act through both $\sigma$ donation and $\pi^*$ acceptance of electron density. This Dewar–Chatt–Duncanson model[160,161] has served well for over half a century, and has received support from more recent theoretical treatments.[162]

Increasing the steric bulk of spectator ligands can reasonably be expected to retard an associatively activated process in square planar complexes. Indeed, as we go from $Pd(dien)OH_2^{2+}$ to $Pd(Me_5dien)OH_2^{2+}$ to $Pd(Et_5dien)OH_2^{2+}$, the rate constant $k_{ex}^{298}$ for water exchange falls from 5100 to 187 to $2.9 \, s^{-1}$, and a similar pattern is found for the corresponding ligation reactions, but values of $\Delta V_{ex}^{\ddagger}$ for the water exchange reactions are $-3, -7$, and $-8 \, cm^3/mol$, respectively.[163] Thus, the water exchange reaction remains associatively activated as steric hindrance increases, evidently because for $Pd^{II}$, like $Pt^{II}$, dissociatively activated pathways are energetically inaccessible in the absence of electronic effects involving strong $\sigma$ donor ligands such as alkyl or aryl ligands (Section 8.3.3.1). This stands in sharp contrast to the question of steric hindrance in interchange reactions of octahedral complexes $M^{III}(RNH_2)_5X$, in which there is a move toward more dissociative character in the $I_a/I_d$ continuum as we go from H to Me and beyond (Section 8.5.2). In light of this absence of an accessible dissociative contribution, the mechanism of ligand substitution in $Pd^{II}$ and $Pt^{II}$ complexes may be truly described as **A** rather than $\mathbf{I_a}$.

## 8.7  COMPUTER MODELING OF LIGAND SUBSTITUTION PROCESSES

An early attempt by Connick and Alder[164] (C&A) to simulate "rare" solvent exchange in metal aqua complexes by computer modeling was limited to a two-dimensional model with 90 particles including a central metal ion with four close-packed coordinated solvent molecules in the plane of the model. Exchange was "rare" in the sense that even the fastest solvent exchange events were very infrequent on the timescale of the multitude of solvent molecule collisions that provided the activation. In order to achieve particle (solvent) exchange within the limitations of computer time, the events were constrained to proceed in the forward direction along the reaction coordinate—that is, toward a successful ascent of the activation barrier. Although relatively primitive by today's standards, this study produced informative insights into the exchange process. As the two exchanging particles moved to cross the activation barrier, they were seen to occupy space between the inner coordination and second coordination spheres—one moving out from the inner sphere, and one coming in from the second—in the manner envisaged in the definition of an interchange mechanism. The three spectator ligands underwent significant angular, but not radial, displacements during the exchange process. Most important, although the particles present generally moved distances less than their own radii in the many hundreds of collisions in the simulation, it was clear that the exchange pathway involved a highly correlated collective movement of *all* the particles present. Interestingly, the crossing of the activation barrier resembled diffusion, with possible reversals and recrossings (cf. solvent dynamical control of reaction rates,[165] more commonly encountered in electron transfer processes), rather than the unidirectional crossing assumed in transition state theory.

Since the publication of the C&A simulation in 1983, there have been immense improvements in both computer hardware and the relevant software. Nevertheless, there continue to be limitations as to what can be achieved in modeling even such a "simple" ligand substitution process as solvent exchange because, to be realistic, a quantum mechanical (QM) treatment of properties of the metal complex and the intimate substitution process needs to be combined with a molecular mechanics (MM) treatment of the collective motions of some 500 surrounding solvent molecules in a classical force field in three dimensions (a QM/MM simulation).[82] The latter requirement was made clear by the C&A article, and a recent DFT study[166] of bond lengths in complexes of the type $MX_xL_{n-x}$ has emphasized the need to include a realistic treatment of the surrounding condensed phase in order to account even for static phenomena such as the structural *trans* effect. There is also the difficulty, noted in the C&A article, that successful ligand substitution events are very rare on the picosecond timescale of MM simulations. Consequently, many modelers have chosen instead to calculate the energies of various postulated transition states for ligand substitution, all too often in the "gas phase" (more correctly *in vacuo*) for computational convenience.

Gas-phase computations, then, may be of limited relevance to ligand exchange processes in solution and may give quite incorrect results. For example, in 1994, Åkesson et al.[167] presented large-basis-set self-consistent-field (SCF) gas-phase calculations that strongly supported an $I_d$ mechanism for water exchange on bivalent first transition period metal aqua ions $M(H_2O)_6^{2+}$ for $M = V$ through Zn, whereas soon thereafter Rotzinger reported Hartree–Fock or complete active space (CAS) SCF calculations indicating agreement with experimentalists' assignment of an $I_a$ mechanism for $M = V$[168] and Mn[169] and dissociative activation for the other $M^{2+}$. Tsutsui, Wasada and Funahashi[170] used *ab initio* MO calculations to argue that associative processes were possible for water exchange on any $M^{2+}$(aq) having less than seven 3d electrons. Rotzinger[168] acknowledged the need to include bulk solvent in the calculations and subsequently[171] reported that treatment of the solvent even as a simple dielectric continuum (i.e., a medium with time-averaged solvent molecular properties) gave activation energies in better agreement with experiment than did the gas-phase models; such a model, however, cannot give entropies or free energies of activation, and so prediction of rate constants remained out of reach. For water exchange on $Ni(H_2O)_6^{2+}$, classical and quantum mechanical MM simulations by Inada et al.[172] indicated that water exchange proceeds through a five-coordinate intermediate with a lifetime of about 2.5 ps, so the mechanism can be classified as $I_d$ on the basis of this timescale, in agreement with experimental mechanistic assignments.

For water exchange on tervalent transition metal aqua ions, computations are generally in agreement that $I_a$ mechanisms are the norm except in their conjugate bases, in which dissociative activation becomes dominant.[167–169,171,173] For the group 13 cations, an $I_d$ mechanism is indicated by both Hartree–Fock computations and $\Delta V_{ex}^{\ddagger}$ measurements.[61] Molecular dynamics simulations of coordination numbers, volume properties, and reaction mechanisms among the tervalent lanthanides in water have been carried out by Merbach's group and the results agreed reasonably well with deductions from experiment.[127,174–177]

In Section 8.5.2, the controversy over the mechanistic consequences of N-methylation of the ammine ligands in $M^{III}(NH_3)_5X$ was summarized. Rotzinger[178] has carried out *ab initio* quantum mechanical calculations that indicated that water exchange on $Cr(NH_2Me)_5OH_2^{3+}$ would proceed with dissociative activation because of the bulkiness of the methylamine ligands—in other words, that the $I_a$ mechanism proposed for water exchange on $Cr(NH_3)_5OH_2^{3+}$ is suppressed in favor of $I_d$ by steric hindrance, which is the traditional view. The calculations also showed that the energy barrier to rearrangement of the putative square pyramidal intermediates $Cr(NH_3)_5^{3+}$ and $Cr(NH_2Me)_5^{3+}$ to trigonal bipyramidal isomers was considerably higher than that for water exchange, so that the dissociative substitution pathway should be stereo-retentive—as experiment has indicated for $Cr^{III}$ substitution reactions in general.

In 2005, Rotzinger[92] gave an assessment of the various MO and DFT methods used to calculate geometries and energies of metal aqua ions in the gas phase and noted that different computational methods may disagree on such basic matters as the coordination number, which tends to be overestimated by Hartree–Fock but underestimated by DFT calculations with obvious consequences for predicting the preferred mechanism of water exchange. The problem is frequently compounded by the use of inadequate basis sets for the metal ion and for the ligands—particularly for more general ligand substitution reactions in which the latter are anionic or $\pi$-bonding.[68] Further common failings are the inadequate treatment of electron correlation and/or hydrogen bonding.[68]

Erras-Hanauer et al.,[179] Rotzinger,[68] and Rode et al.[82] have recently published reviews of the methodology and results of modeling of solvent exchange on metal ions, and these articles should be consulted for a detailed assessment. It is clear that, even with present-day hardware and software, compromises have to be made in order to complete a QM/MM simulation within an acceptable time. Given the enormous success of DFT in transition metal chemistry, the Car–Parrinello methodology[180] that unifies DFT with molecular dynamics would seem to be the most promising approach to the modeling of ligand field substitution dynamics, at least for the present time.

Computer modeling may fail to satisfy chemists of the older school inasmuch as readily identifiable influences on kinetic behavior such as ligand field effects (LFAE) and bulk solvent properties (dielectric constant, viscosity, etc.) are replaced by the totality of a simulation that may well give a useful result but at the same time obscure the major underlying factors that make one compound behave differently from others. There may also be unease over the fact that choices are made between several alternative computational methods to suit the particular chemical system, which suggests that an investigator might choose a procedure that would give results that met with his or her prejudices. Whichever procedure is chosen, the consequences of the inevitable approximations involved must be carefully assessed. Evans et al.[181] explored the exchange of aqua ligands in $Al(H_2O)_6^{3+}$ with up to 12 surrounding water molecules by both *ab initio* and molecular dynamics methods, and obtained contradictory predictions of the stereochemical pathways; the *ab initio* calculations were strongly affected by the details of the surrounding cluster of water molecules, whereas molecular dynamics simulations gave generally more convincing results, particularly when guided by prior knowledge from experiment of what those results should be. Evans et al. urge caution in the application of *ab initio* simulation to complicated processes such as ligand exchange, recommending that such static computations be accompanied by molecular dynamic simulations and that the results be regarded as hypotheses to be tested experimentally. These recommendations complement Rotzinger's conclusion[68] that contemporary modeling techniques, properly applied, serve to confirm the mechanistic assignments made for solvent exchange by experimentalists, and in particular vindicate the use of $\Delta V_{ex}^{\ddagger}$ in this regard.

## 8.8   LIGAND SUBSTITUTION KINETICS AND COMPUTER MODELING IN AQUATIC GEOCHEMISTRY

At the time of writing, much of the current experimental and computer modeling activity in the fields described in this chapter is being conducted by geochemists, notably W. H. Casey and his coworkers.[23,182–196] The kinetics and mechanisms of dissolution (weathering) of minerals in aqueous environments bear striking resemblances to those of ligand substitution in metal complexes.[185] For example, the rates of dissolution per unit mineral surface area of end-member orthosilicate minerals $(M^{II}_2SiO_4)$ at pH 2 show a linear log–log relationship to $k_{ex}$ for water exchange on $M^{2+}(aq)$ (Figure 8.8).[186] Such relationships offer geochemists the power to predict geochemical reaction rates from the now large database of $k_{ex}$ values.[21,59,69,177] Furthermore, it has long been recognized that biologically produced carboxylic acids, amino acids, phenols, humic acids, and other potential complexing (particularly chelating) agents greatly accelerate the dissolution of metallic minerals such as bunsenite (NiO)[187,188] and are involved in the transport of the released metal ions as complexes in the aquatic environment; clearly, knowledge of the ligand substitution kinetics of model metal complexes is essential in this connection. Because of the complexity of the interactions of extended mineral surfaces with the aqueous environment and the difficulty of carrying out relevant experiments *in situ* in heterogeneous systems, geochemists have come to rely heavily—perhaps too heavily—on computer modeling, and information from ligand substitution dynamics at the molecular level provides an important factual underpinning for such studies. Water exchange reactions are especially useful in this regard. On the other hand, as noted in Section 8.7, the emerging recognition of the limitations of computer models (particularly *ab initio* calculations) of even such simple systems as water exchange on

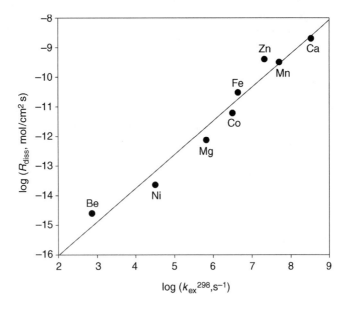

**FIGURE 8.8**    Correlation of rates of dissolution $R_{diss}$ of orthosilicates $M_2SiO_4$ in water at pH 2 and 25°C with water exchange rate constants $k_{ex}$ for $M^{2+}(aq)$. Data from Lincoln and Merbach[21] and Casey and Westrich.[186]

$M(H_2O)_n{}^{z+}$ serves as a guide for geochemists as to how modeling should be done if the results are to be truly meaningful.[68,181]

Given the difficulty of carrying out experiments on mineral–water interactions on extended surfaces, experimental investigations of water exchange on polyoxoions with dimensions in the range of 1–5 nm are appealing as providing approximations to the behavior of not only extended surfaces but also particulate geological material. Such studies also fit in with the current preoccupation of the inorganic chemistry community with nanomaterials and nanotechnology.[189] Aluminum is the third most common element in the Earth's crust (after O and Si), and consequently hydrolytic oligomers of $Al^{III}$ such as $AlO_4Al_{12}(OH)_{24}(OH_2)_{12}{}^{7+}$ (**Al$_{13}$**) and $Al_2O_8Al_{28}(OH)_{56}(OH_2)_{26}{}^{18+}$ (**Al$_{30}$**) are of special geochemical interest because they are well characterized, with structures accurately known from X-ray diffraction measurements, and can serve as experimentally tractable models for reactions occurring at the structurally similar surfaces of such Al-containing minerals as bayerite ($\gamma$-Al(OH)$_3$) and $\gamma$-alumina.[189–191]

The common form of **Al$_{13}$** is the $\varepsilon$-isomer of the five possible Keggin structures, which consist of a central AlO$_4$ unit surrounded by 12 AlO$_6$ octahedra in four planar blocks of three. The Keggin isomers are interconvertible by rotation of the four Al$_3$ blocks relative to each other; the $\varepsilon$-**Al$_{13}$** structure has T$_d$ symmetry (Figure 8.9). The central AlO$_4$ tetrahedron shares each of its oxygens with three AlO$_6$ octahedra ($\mu_4$-O). There are two kinds of bridging ($\mu_2$) hydroxide: those forming the edge-sharing links between separate Al$_3$ units ($\mu_2$-OH$^a$), and those edge-shared between octahedra

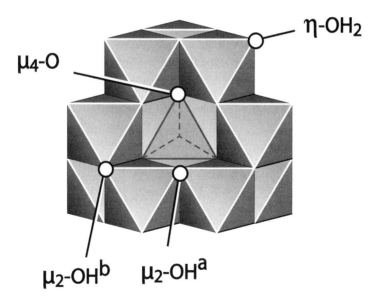

**FIGURE 8.9**    Structure of $\varepsilon$-AlO$_4$Al$_{12}$(OH)$_{24}$(OH$_2$)$_{12}{}^{7+}$ ($\varepsilon$-**Al$_{13}$**, T$_d$ symmetry), viewed along one of the four C$_3$ axes. Reproduced with permission from *Chem. Rev.* **2006**, *106*, 1. Copyright 2006 American Chemical Society. (See the color version of this figure in Color Plates section.)

within the $Al_3$ units ($\mu_2$-$OH^b$). All 12 terminal $H_2O$ ligands ($\eta$-$H_2O$) are equivalent. Oxygen-17 NMR relaxation measurements[191,192] showed that the apical water ligands ($\eta$-$H_2O$) on $\varepsilon$-$\mathbf{Al_{13}}$ exchange with solvent water at pH 5.1–5.3 with $k_{ex}^{298} = 1.1 \times 10^3 \, s^{-1}$, which is about 1000 times faster than water exchange on $Al(H_2O)_6^{3+}$ (Table 8.2). The much slower exchange of the $\mu_2$-$OH$ sites with solvent water was followed by $^{17}O$ NMR line intensity measurements at $+35$ ppm, where the resonances happen to coincide, giving a biphasic decay pattern corresponding to a fast exchange and a concurrent slow one. These measurements, however, had to be restricted to a very narrow temperature range (282–292 K), and the very high $\Delta H_{ex(1)}^{\ddagger}$ ($\sim$200 kJ/mol) and $\Delta S_{ex(1)}^{\ddagger}$ ($\sim$400 J/(K mol)) values obtained for the faster exchange are therefore suspect because of error correlation (Section 8.4); however, the short extrapolation to 298 K is reasonable and gave $k_{ex(1)}^{298} = 1.6 \times 10^{-2} \, s^{-1}$. For the slower $\mu_2$-$OH$ exchange, $k_{ex(2)}^{298} = 1.6 \times 10^{-5} \, s^{-1}$. These values appeared to be independent of pH, but the experiments were restricted to the narrow pH range over which $\varepsilon$-$\mathbf{Al_{13}}$ is sufficiently stable in solution. Unfortunately, the $^{17}O$-NMR resonances of the two $\mu_2$-$OH$ sites coincide and this precludes an unambiguous assignment of the two rate constants to particular sites. It is believed,[190] however, that the faster rate constant $k_{ex(1)}$ refers to the $\mu_2$-$OH^a$ sites, which is reasonable as it implies that the $Al_3$ blocks are more persistent than the links between them. The $\mu_4$-$O$ sites are so kinetically inert that no solvent $^{17}O$ exchanged with them over several days.

If the unique central $Al^{III}$ in $\varepsilon$-$\mathbf{Al_{13}}$ is replaced by $Ga^{III}$ ($\varepsilon$-$\mathbf{GaAl_{12}}$), $k_{ex(1)}^{298}$ and $k_{ex(2)}^{298}$ become 890 and 39 times smaller, respectively, than in $\varepsilon$-$\mathbf{Al_{13}}$.[193] If the central $Al^{III}$ is replaced by $Ge^{IV}$,[194] major changes to the labilities of the $\mu_2$-$OH$ sites again result, with $k_{ex(1)}^{298}$ becoming too fast to measure by the NMR line intensity technique, although direct comparison with $\varepsilon$-$\mathbf{Al_{13}}$ is inappropriate because the overall charge in $\varepsilon$-$\mathbf{GeAl_{12}}$ is $+8$ rather than $+7$ and moreover the rates of $^{17}O$ exchange at the $\mu_2$ sites do increase significantly with increasing $[H^+]$. Nevertheless, it is clear that the central metal ion M in $\varepsilon$-$\mathbf{MAl_{12}}$, though itself inert with respect to ligand substitution, exerts a surprisingly large influence on O site lability three bonds away, and molecular dynamics computations confirm that the whole polyoxoanion responds to a substitution event at any one site.[195] It follows from this that exchange mechanisms in aluminum oxide/hydroxide minerals in aquatic environments may be much more complicated than the simple surface phenomenon that had been naïvely envisaged; on the other hand, the observed influence of highly coordinated oxygens, such as the $\mu_4$-$O$ in boehmite ($\gamma$-$AlO(OH)$), on the reactivities of $Al^{III}$ minerals is now understandable.[195]

The $\varepsilon$-$\mathbf{Al_{13}}$ example gives an indication of the difficulties awaiting the experimentalist in the study of hydrolytic oligomers. The Casey group has therefore sought out "well-behaved" polyoxoions in the hope of gaining insights into the general principles of reaction dynamics in such nanospecies and has settled upon the decaniobate ion $H_xNb_{10}O_{28}^{(6-x)-}$ as a suitable candidate (though not one of intrinsic geochemical importance) because it does not protonate strongly when dissolved in water, is stable at near-neutral pH, and shows distinct $^{17}O$ resonances for all seven different O sites in this ion.[196] Their results are not easily summarized, but a major finding of this study is, once again, that even the simplest bond breaking process in a polyoxoion of nanometer dimensions cannot be considered in isolation from all the other atoms in the molecule. This places restrictions on how computer modeling of nanospecies should be conducted and echoes the concerns expressed in Section 8.7.

## 8.9    CLOSING REMARKS

The assignment of intimate mechanisms to ligand substitution reactions may be less important than is usually supposed. For a wide enough range of concentrations of the incoming ligand A, the rate equations for the **D**, **A**, and **I** mechanisms are all of the same gross mathematical form as represented by Figure 8.1; they differ only in the meaning attached to the parameters in those equations. Limiting **D** mechanisms may not be relevant to most aqueous systems because the ubiquitous, very small, strong nucleophile $H_2O$ will normally scavenge intermediates of reduced coordination number efficiently. The **A** mechanism, rather than $I_a$, is paramount in reactions of square planar complexes of $Pd^{II}$, $Pt^{II}$, and $Au^{III}$ because of the coordinative unsaturation and "softness" of the metal centers, and the lack of energetically accessible dissociatively activated pathways except in a few organometallic cases. Substitution in octahedral complexes, especially in water, generally takes place by an interchange mechanism, and although it is convenient to label reactions of some complexes (e.g., $Co^{III}$ am(m)ines) as $I_d$ and others (e.g., $Cr^{III}$ aqua complexes) as $I_a$, there is no sharp demarcation between these categories. Rather, the **I** mechanism should be recognized as a continuum stretching from near-**A** to near-**D** mechanisms with varying admixtures of **a** and **d** characters (cf. the $S_N1/S_N2$ continuum now recognized by physical organic chemists[71,197]). Thus, changes in the spectator ligands can move the mechanistic proclivity of a particular metal ion along the $I_a$–$I_d$ continuum, as in the case of the replacement of various X in $Cr^{III}(NH_2R)_5X$ where going from R = H to R = $CH_3$ increases the $I_d$ character at the expense of $I_a$. Such mechanistic variability will influence the selectivity shown by a metal ion toward incoming ligands, but because an increase in dissociative character usually results in increases in both $\Delta H^{\ddagger}$ and $T\Delta S^{\ddagger}$ that tend to cancel each other, the net effect of such mechanistic shifts on the important quantity, the rate constant, may be less than anticipated.

Computer modeling has become enormously popular since 1980, the cost and capabilities of computer hardware and software having decreased greatly while the costs of, and environmental and safety constraints on, experimental work have risen sharply. Modeling can certainly be a very valuable adjunct to experimental work, but taken in isolation it is vulnerable to serious errors arising from the inevitable approximations. As noted in Sections 8.7 and 8.8, it is important to include the whole of a reacting system, including the surrounding solvent, if a computation is to be at all realistic—gas phase *ab initio* or DFT calculations taken in isolation can be seriously misleading.

Ligand substitution studies continue apace in organometallic and bioinorganic chemistry, where the ligand systems are much more complicated than those considered in this chapter. It is in geochemical contexts, however, that the kinetics and mechanisms of simple substitution reactions at metal centers have recently found new relevance, and conversely it is most gratifying to see that valuable feedback into the basic physical–inorganic foundations of ligand substitution kinetics is now coming from geochemists.

## ACKNOWLEDGMENTS

I thank the National Research Council of Canada and latterly the Natural Sciences and Engineering Research Council of Canada for continuous support to my work on inorganic reaction mechanisms from 1964 to 2009, my students and research associates for their outstanding contributions throughout this period, and Prof. W. H. Casey (University of California, Davis) for providing Figure 8.9.

## REFERENCES

1. Taube, H. *Chem. Rev.* **1952**, *50*, 69.
2. Basolo, F. *Chem. Rev.* **1953**, *52*, 459.
3. Basolo, F.; Pearson, R. G. *Mechanisms of Inorganic Reactions,* 1st edition, Wiley: New York, 1958.
4. Basolo, F.; Pearson, R. G. *Mechanisms of Inorganic Reactions,* 2nd edition, Wiley: New York, 1967.
5. Langford, C. H.; Gray, H. B. *Ligand Substitution Processes*; W. A. Benjamin, Inc.: Reading, MA, 1966.
6. Sykes, A. G. *Kinetics of Inorganic Reactions*; Pergamon: Oxford, 1966.
7. Tobe, M. L. *Inorganic Reaction Mechanisms*; Nelson: London, 1972.
8. Burgess, J. *Metal Ions in Solution*; Ellis Horwood: Chichester, UK, 1978.
9. Wilkins, R. G. *Kinetics and Mechanism of Reactions of Transition Metal Complexes*, 2nd edition; VCH: Weinheim, Germany, 1991.
10. Henderson, R. A. *The Mechanisms of Reactions at Transition Metal Sites*; Oxford University Press: Oxford, 1993.
11. Atwood, J. D. *Inorganic and Organometallic Reaction Mechanisms,* 2nd edition; VCH: New York, 1997.
12. Richens, D. T. *The Chemistry of Aqua Ions*; Wiley: Chichester, UK, 1997.
13. Tobe, M. L.; Burgess, J. *Inorganic Reaction Mechanisms*; Addison Wesley Longman: Harlow, Essex, UK, 1999.
14. Hay, R. W. *Reaction Mechanisms of Metal Complexes*; Horwood: Chichester, UK, 2000.
15. Jordan, R. B. *Reaction Mechanisms of Inorganic and Organometallic Systems,* 3rd edition; Oxford University Press: Oxford, 2007.
16. Murmann, R. K., Ed. *Mechanisms of Inorganic Reactions*; American Chemical Society: Washington, DC, 1965.
17. Rorabacher, D. B.; Endicott, J.F., Eds. *Mechanistic Aspects of Inorganic Reactions*; American Chemical Society: Washington, DC, 1982.
18. Sykes, A. G., Ed. *Advances in Inorganic and Bioinorganic Mechanisms*; Academic Press: New York, 1982–1986; Vols. 1–4.

19. Twigg, M. V., Ed. *Mechanisms of Inorganic and Organometallic Reactions*; Plenum Press: New York, 1983–1994; Vols. 1–8.

20. Margerum, D. W.; Cayley, G. R.; Weatherburn, D. C.; Pagenkopf, G. K. In *Coordination Chemistry. ACS Monograph 174*; Martell, A. E., Ed.; American Chemical Society: Washington, DC, 1978; Vol. 2.

21. Lincoln, S. F.; Merbach, A. E. *Adv. Inorg. Chem.* **1995**, *42*, 1.

22. Richens, D. T., *Chem. Rev.* **2005**, *105*, 1961.

23. Casey, W. H.; Swaddle, T. W. *Rev. Geophys.* **2003**, *41* (2), 4–1.

24. Williams, R. J. P.; Fraústo da Silva, J. J. R. *The Biological Chemistry of the Elements: The Inorganic Chemistry of Life*; 2nd edition; Oxford University Press: Oxford, 2001.

25. Wagman, D. D.; Evans, W. H.; Parker, V. B.; Schumm, R. H.; Halow, I.; Bailey, S. M.; Churney, K. L.; Nuttall, R. L. *The NBS Tables of Chemical Thermodynamic Properties*; National Bureau of Standards: Washington, DC, 1982.

26. Ingold, C. K. *Substitution at Elements Other Than Carbon*; Weizmann Science Press of Israel: Jerusalem, 1959.

27. Newton, A. M.; Swaddle, T. W. *Can. J. Chem.* **1974**, *52*, 2751.

28. Baes, C. F. Jr.; Mesmer, R. E. *The Hydrolysis of Cations*; Wiley–Interscience: New York, 1976.

29. Marcus, Y.; Hefter, G. *Chem. Rev.* **2006**, *106*, 4585.

30. Bridgman, P. W. *The Logic of Modern Physics*; The Macmillan Co.: New York, 1928.

31. Swaddle, T. W. *Rev. Phys. Chem. Jpn.* **1980**, *50*, 232.

32. Gray, H. B.; Olcott, R. J. *Inorg. Chem.* **1962**, *1*, 481.

33. Kotowski, M.; Palmer, D. A.; Kelm, H. *Inorg. Chim. Acta* **1980**, *44*, L113.

34. Lanza, S.; Minnitti, D.; Romeo, R.; Moore, P.; Sachinidis, J.; Tobe, M. L. *J. Chem. Soc., Chem. Commun.* **1984**, 542.

35. Alibrandi, G.; Bruno, G.; Lanza, S.; Minnitti, D.; Romeo, R.; Tobe, M. L. *Inorg. Chem.* **1987**, *26*, 185.

36. Plutino, M. R.; Scolaro, L. M.; Romeo, R.; Grassi, A. *Inorg. Chem.* **2000**, *39*, 2712.

37. Haim, A.; Wilmarth, W. K. *Inorg. Chem.* **1962**, *1*, *573*, 583.

38. Haim, A.; Grassi, R. J.; Wilmarth, W. K.; Mechanisms of Inorganic Reactions. *ACS Adv. Chem. Ser.* **1967**, *49*, 31.

39. Haim, A.; Grassi, R. J.; Wilmarth, W. K. *Inorg. Chem.* **1967**, *6*, 237.

40. Burnett, M. G.; Gilfillan, W. M. *J. Chem. Soc. Dalton Trans.* **1981**, 1578.

41. Abou-El-Wafa, M. H. M.; Burnett, M. G.; McCullagh, J. F. *J. Chem. Soc., Dalton Trans.* **1986**, 2083.

42. Abou-El-Wafa, M. H. M.; Burnett, M. G.; McCullagh, J. F. *J. Chem. Soc., Dalton Trans.* **1987**, 1059, 2311.

43. Haim, A. *Inorg. Chem.* **1982**, *21*, 2887.

44. Bradley, S. M.; Doine, H.; Krouse, H. R.; Sisley, M. J.; Swaddle, T. W. *Aust. J. Chem.* **1988**, *41*, 1323.

45. Palmer, D. A.; Kelm, H. *Z. Anorg. Allg. Chem.* **1979**, *450*, 50.

46. Swaddle, T. W. *Comments Inorg. Chem.* **1991**, *12*, 237.

47. Swaddle, T. W.; Fabes, L. *Can. J. Chem.* **1980**, *58*, 1418.

48. Fedorchuk, C.; Swaddle, T. W. *J. Phys. Chem. A* **2000**, *104*, 5651.
49. Faherty, K. P.; Thompson, C. J.; Aguirre, F.; Michne, J.; Metz, R. B. *J. Phys. Chem. A* **2001**, *105*, 10054.
50. Pearson, R. G.; Moore, J. W. *Inorg. Chem.* **1964**, *3*, 1332.
51. Lo, S. T. D.; Watts, D. W. *Aust. J. Chem.* **1975**, *28*, 491.
52. Lo, S. T. D.; Swaddle, T. W. *Inorg. Chem.* **1975**, *14*, 1878.
53. Lo, S. T. D.; Swaddle, T. W. *Inorg. Chem.* **1976**, *15*, 1881.
54. Wrona, P. K. *Inorg. Chem.* **1984**, *23*, 1558.
55. Hunt, H. R.; Taube, H. *J. Am. Chem. Soc.* **1958**, *80*, 2642.
56. Stranks, D. R.; Swaddle, T. W. *J. Am. Chem. Soc.* **1971**, *93*, 2783.
57. Xu, F.-C.; Krouse, H. R.; Swaddle, T. W. *Inorg. Chem.* **1985**, *24*, 267.
58. Carey, L. R.; Jones, W. E.; Swaddle, T. W. *Inorg. Chem.* **1971**, *8*, 1566.
59. Helm, L.; Merbach, A. E. *Chem. Rev.* **2005**, *105*, 1923.
60. Hunt, J. P.; Plane, R. A. *J. Am. Chem. Soc.* **1954**, *76*, 5960.
61. Kowall, T.; Caravan, P.; Bourgeois, H.; Helm, L.; Merbach, A. E. *J. Am. Chem. Soc.* **1998**, *120*, 6569.
62. Jones, W. E.; Carey, L. R.; Swaddle, T. W. *Can. J. Chem.* **1972**, *50*, 2739.
63. Pisaniello, D. L.; Helm, L.; Meier, P. F.; Merbach, A. E. *J. Am. Chem. Soc.* **1983**, *105*, 4528.
64. Pittet, P.-A.; Elbaze, G.; Helm, L.; Merbach, A. E. *Inorg. Chem.* **1990**, *29*, 1936.
65. Swaddle, T. W. *Inorg. Chem.* **1983**, *22*, 2663.
66. Swaddle, T. W.; Mak, M. K. S. *Can. J. Chem.* **1983**, *61*, 473.
67. Shannon, R. D. *Acta Crystallogr., Sect. A* **1976**, *32*, 751.
68. Rotzinger, F. P. *Chem. Rev.* **2005**, *105*, 2003.
69. Lincoln, S. F. *Helv. Chim. Acta* **2005**, *88*, 523.
70. Swaddle, T. W. *Adv. Inorg. Bioinorg. Mech.* **1983**, *2*, 95.
71. More O'Ferrall, R. A. *J. Chem. Soc. B* **1970**, 274.
72. Hugi, A. D.; Helm, L.; Merbach, A. E. *Inorg. Chem.* **1987**, *26*, 1763.
73. Grant, M.; Jordan, R. B. *Inorg. Chem.* **1981**, *20*, 55.
74. De Vito, D.; Weber, J.; Merbach, A. E. *Inorg. Chem.* **2004**, *43*, 858.
75. Cusanelli, A.; Frey, U.; Richens, D. T.; Merbach, A. E. *J. Am. Chem. Soc.* **1996**, *118*, 5265.
76. Doine, H.; Ishihara, K.; Krouse, H. R.; Swaddle, T. W. *Inorg. Chem.* **1987**, *26*, 3240.
77. González, G.; Moullet, B.; Martínez, M.; Merbach, A. E. *Inorg. Chem.* **1994**, *33*, 2330.
78. Leffler, J. E.; Grunwald, E. *Rates and Equilibria of Organic Reactions*; Wiley: New York, 1963.
79. Swaddle, T. W. *J. Chem. Soc., Chem. Commun.* **1982**, 832.
80. Earnshaw, A.; Larkworthy, L. F.; Patel, K. C.; Beech, G. *J. Chem. Soc. A* **1969**, 1334.
81. Figgis, B. N.; Kucharski, E. S.; Reynolds, P. A. *Acta Crystallogr., Sect. B* **1990**, *46*, 577.
82. Rode, B. M.; Schwenk, C. F.; Hofer, T. S.; Randolf, B. R. *Coord. Chem. Rev.* **2005**, *249*, 2993.
83. Powell, D. H.; Helm, L.; Merbach, A. E. *J. Chem. Phys.* **1991**, *95*, 9258.
84. Magini, M. *Inorg. Chem.* **1982**, *21*, 1535.

85. Gallucci, J.; Gerkin, R. E. *Acta Crystallogr., Sect. C* **1989**, *45*, 1279.

86. Peisach, J.; Mims, W. B. *Chem. Phys. Lett.* **1976**, *37*, 307.

87. Pasquarello, A.; Petri, I.; Salmon, P. S.; Parisel, O.; Car, R.; Tóth, É.; Powell, D. H.; Fischer, H. E.; Helm, L.; Merbach, A. E. *Science* **2001**, *291*, 856.

88. Benfatto, M.; D'Angelo, P.; Della Longa, S.; Pavel, N. V. *Phys. Rev. B* **2002**, *65*, 174205.

89. Frank, P.; Benfatto, M.; Szilyagi, R. K.; D'Angelo, P.; Della Longa, S.; Hodgson, K. O. *Inorg. Chem.* **2005**, *44*, 1922.

90. Persson, I.; Persson, P.; Sandström, M.; Ullström, A.-S. *J. Chem. Soc., Dalton Trans.* **2002**, 1256.

91. Schwenk, C. F.; Rode, B. M. *J. Am. Chem. Soc.* **2004**, *126*, 12786.

92. Rotzinger, F. P. *J. Phys. Chem. B* **2005**, *109*, 1510.

93. Powell, D. H.; Furrer, P.; Pittet, P.-A.; Merbach, A. E. *J. Phys. Chem.* **1995**, *99*, 16622.

94. Millero, F. J. In *Water and Aqueous Solutions: Structure, Thermodynamics, and Transport Processes*; Horne, R. A., Ed; Wiley–Interscience: 1972; Chapter 13.

95. Helm, L.; Elding, L. I.; Merbach, A. E. *Inorg. Chem.* **1985**, *24*, 1719.

96. Jalilehvand, F.; Laffin, L. J. *Inorg. Chem.* **2008**, *47*, 3248.

97. Swaddle, T. W.; Rosenqvist, J.; Yu, P.; Bylaska, E.; Phillips, B. L.; Casey, W. H. *Science* **2005**, *308*, 1450.

98. Fong, D. W.; Grunwald, E. *J. Am. Chem. Soc.* **1969**, *91*, 2413.

99. Holmes, L. P.; Cole, D. L.; Eyring, E. M. *J. Phys. Chem.* **1968**, *72*, 301.

100. Hanauer, H.; Puchta, R.; Clark, T.; van Eldik, R. *Inorg. Chem.* **2007**, *46*, 1112.

101. Martin, R. B. *J. Inorg. Biochem.* **1991**, *44*, 141.

102. Moreau, G.; Helm, L.; Purans, J.; Merbach, A. E. *J. Phys. Chem. A* **2002**, *106*, 3034.

103. Caravan, P.; Tóth, É.; Rockenbauer, R.; Merbach, A. E. *J. Am. Chem. Soc.* **1999**, *121*, 10403.

104. Farkas, I.; Bányai, I.; Szabó, Z.; Wahlgren, U.; Grenthe, I. *Inorg. Chem.* **2000**, *39*, 799.

105. Szabó, Z.; Toraishi, T.; Vallet, V.; Grenthe, I. *Coord. Chem. Rev.* **2006**, *250*, 784.

106. Schneppensieper, T.; Zahl, A.; van Eldik, R. *Angew. Chem., Int. Ed.* **2001**, *40*, 1678.

107. Funahashi, S.; Inada, Y. *Bull. Chem. Soc. Japan* **2002**, *75*, 1901.

108. Swaddle, T. W.; Coleman, L. F.; Hunt, J. P. *Inorg. Chem.* **1963**, *2*, 950.

109. Glaeser, H. H.; Hunt, J. P. *Inorg. Chem.* **1964**, *3*, 1245.

110. Hunt, J. P.; Dodgen, H. W.; Klanberg, F. *Inorg. Chem.* **1964**, *2*, 478.

111. Langford, C. H. *Inorg. Chem.* **1965**, *4*, 265.

112. Swaddle, T. W. *Coord. Chem. Rev.* **1974**, *14*, 217.

113. Spiro, T. G.; Revesz, A.; Lee, J. *J. Am. Chem. Soc.* **1968**, *90*, 4000.

114. Swaddle, T. W.; Guastalla, G. *Inorg. Chem.* **1968**, *7*, 1915.

115. Espenson, J. H. *Inorg. Chem.* **1969**, *8*, 1554.

116. Guastalla, G.; Swaddle, T. W. *Can. J. Chem.* **1973**, *51*, 821.

117. Lay, P. A. *Comments Inorg. Chem.* **1991**, *9*, 235.

118. Sasaki, Y.; Sykes, A. G. *J. Chem. Soc., Dalton Trans.* **1975**, 1048.

119. Pearson, R. G. *J. Chem. Educ.* **1968**, *45*, 64.

120. Pearson, R. G. *Inorg. Chem.* **1988**, *27*, 734.

121. Mønsted, L.; Mønsted, O. *Acta Chem. Scand.* **1982**, *36*, 365, 555.

122. Bracken, D. E.; Baldwin, H. W. *Inorg. Chem.* **1974**, *13*, 1325.

123. Poë, A.; Vuik, C. *Can. J. Chem.* **1975**, *53*, 1842.

124. Jackson, W. G. *Inorg. Chem.* **1988**, *27*, 777.

125. Byrd, J. E.; Wilmarth, W. K. *Inorg. Chim. Acta Rev.* **1971**, *5*, 7.

126. Schilt, A. A.; Schaap, W. B. *Inorg. Chem.* **1973**, *12*, 1424.

127. Anderson, K. M.; Orpen, A. G. *Chem. Commun.* **2001**, 2682.

128. Guastalla, G.; Swaddle, T. W. *Can. J. Chem.* **1974**, *52*, 527.

129. Cheng, M.; Song, W.; Bakac, A. *Eur. J. Inorg. Chem.* **2008**, 4687.

130. Carlyle, D. W.; King, E. L. *Inorg. Chem.* **1970**, *9*, 2333.

131. Zinato, E.; Furlani, C.; Lanna, G.; Riccieri, P. *Inorg. Chem.* **1972**, *11*, 1746.

132. Tobe, M. L. *Adv. Inorg. Bioinorg. Mech.* **1983**, *2*, 1.

133. Buckingham, D. A.; Olsen, I. I.; Sargeson, A. M. *J. Am. Chem. Soc.* **1967**, *89*, 5129.

134. Pearson, R. G.; Basolo, F. *J. Am. Chem. Soc.* **1956**, *78*, 4878.

135. Vanquickenborne, L. G.; Pierloot, K. *Inorg. Chem.* **1981**, *20*, 3673.

136. Parris, M.; Wallace, W. J. *Can. J. Chem.* **1969**, *47*, 2257.

137. Buckingham, D. A.; Foxman, B. M.; Sargeson, A. M. *Inorg. Chem.* **1970**, *9*, 1790.

138. Swaddle, T. W. *Can. J. Chem.* **1977**, *55*, 3166.

139. Lay, P. A. *Comments Inorg. Chem.* **1991**, *11*, 235.

140. González, G.; Martínez, M. *Inorg. Chim. Acta* **1995**, *230*, 67.

141. Benzo, F.; Bernhardt, P. V.; González, G.; Martínez, M.; Sienra, B. *J. Chem. Soc., Dalton Trans.* **1999**, 3973.

142. Schwarzenbach, G. *Helv. Chim. Acta* **1952**, *35*, 2344.

143. Letter, J. E. Jr.; Jordan, R. B. *J. Am. Chem. Soc.* **1975**, *97*, 2381.

144. Hunt, J. P.; Rutenberg, A. C.; Taube, H. *J. Am. Chem. Soc.* **1952**, *74*, 268.

145. van Eldik, R.; Harris, G. M. *Inorg. Chem.* **1980**, *19*, 880.

146. Posey, F. A.; Taube, H. *J. Am. Chem. Soc.* **1953**, *75*, 4099.

147. Cross, R. J. *Adv. Inorg. Chem.* **1989**, *34*, 219.

148. Jamieson, E. R.; Lippard, S. J. *Chem. Rev.* **1999**, *99*, 2467.

149. Jung, Y.; Lippard, S. J. *Chem. Rev.* **2007**, *107*, 1387.

150. Reedijk, J. *Chem. Commun.* **1996**, 801.

151. Todd, R. C.; Lovejoy, K. S.; Lippard, S. J. *J. Am. Chem. Soc.* **2007**, *129*, 6370.

152. Monlien, F. J.; Helm, L.; Abou-Hamdan, A.; Merbach, A. E. *Inorg. Chem.* **2002**, *41*, 1717.

153. Monlien, F. J.; Helm, L.; Abou-Hamdan, A.; Merbach, A. E. *Inorg. Chim. Acta* **2002**, *331*, 257.

154. Romeo, R. *Comments Inorg. Chem.* **1990**, *11*, 21.

155. Elmroth, S. K. C.; Elding, L. I. *Inorg. Chem.* **1996**, *35*, 2337.

156. Skibsted, L. H. *Adv. Inorg. Bioinorg. Mech.* **1986**, *4*, 137.

157. Otto, S.; Elding, L. I. *J. Chem. Soc., Dalton Trans.* **2002**, 2354.

158. Elding, L. I.; Gröning, A. B. *Chem. Scripta* **1977**, *11*, 8.

159. Plutino, M. R.; Otto, S.; Roodt, A.; Elding, L. I. *Inorg. Chem.* **1999**, *38*, 1233.

160. Dewar, M. J. S. *Bull. Soc. Chim. Fr.* **1951**, *18*, C79.

161. Chatt, J.; Duncanson, L. A. *J. Chem. Soc.* **1953**, 2939.
162. Lin, Z.; Hall, M. B. *Inorg. Chem.* **1991**, *30*, 646.
163. Berger, J.; Kotowski, M.; van Eldik, R.; Frey, U.; Helm, L.; Merbach, A. E. *Inorg. Chem.* **1989**, *28*, 3759.
164. Connick, R. E.; Alder, B. J. *J. Phys. Chem.* **1983**, *87*, 2764.
165. Hynes, J. T. *Annu. Rev. Phys. Chem.* **1985**, *36*, 573.
166. Hocking, R. K.; Deeth, R. J.; Hambley, T. W. *Inorg. Chem.* **2007**, *46*, 8238.
167. Åkesson, R.; Pettersson, L. G. M.; Sandström, M.; Wahlgren, U. *J. Am. Chem. Soc.* **1994**, *116*, 8705.
168. Rotzinger, F. P. *J. Am. Chem. Soc.* **1996**, *118*, 6760.
169. Rotzinger, F. P. *J. Am. Chem. Soc.* **1997**, *119*, 5230.
170. Tsutsui, Y.; Wasada, H.; Funahashi, S. *Bull. Chem. Soc. Japan* **1998**, *71*, 1771.
171. Rotzinger, F. P. *Helv. Chim. Acta* **2000**, *83*, 3006.
172. Inada, Y.; Mohammed, A. M.; Loeffler, H. H.; Rode, B. M. *J. Phys. Chem. A* **2002**, *106*, 6783.
173. Hartmann, M.; Clark, T.; van Eldik, R. *J. Phys. Chem. A* **1999**, *103*, 9899.
174. Kowall, T.; Foglia, F.; Helm, L.; Merbach, A. E. *J. Am. Chem. Soc.* **1995**, *117*, 3790.
175. Kowall, T.; Foglia, F.; Helm, L.; Merbach, A. E. *J. Phys. Chem.* **1995**, *99*, 13078.
176. Kowall, T.; Foglia, F.; Helm, L.; Merbach, A. E. *Chem. Eur. J.* **1996**, *2*, 285.
177. Helm, L.; Merbach, A. E. *Coord. Chem. Rev.* **1999**, *187*, 151.
178. Rotzinger, F. P. *J. Phys. Chem. A* **2000**, *104*, 8787.
179. Erras-Hanauer, H.; Clark, T.; van Eldik, R. *Coord. Chem. Rev.* **2003**, *238–239*, 233.
180. Car, R.; Parrinello, M. *Phys. Rev. Lett.* **1985**, *55*, 2471.
181. Evans, R. J.; Rustad, J. R.; Casey, W. H. *J. Phys. Chem. A* **2008**, *112*, 4125.
182. Casey, W. H.; Rustad, J. R. *Annu. Rev. Earth Planet. Sci.* **2007**, *35*, 21.
183. Spiccia, L.; Casey, W. H. *Geochim. Cosmochim. Acta* **2007**, *71*, 5590.
184. Balogh, E.; Casey, W. H. *Prog. Nucl. Magn. Reson. Spectrosc.* **2008**, *53*, 193.
185. Casey, W. H.; Ludwig, C. *Rev. Mineral.* **1995**, *31*, 87.
186. Casey, W. H.; Westrich, H. R. *Nature* **1992**, *355*, 157.
187. Ludwig, C.; Casey, W. H.; Rock, P. A. *Nature* **1995**, *375*, 44.
188. Ludwig, C.; Devidal, J.-L.; Casey, W. H. *Geochim. Cosmochim. Acta* **1996**, *60*, 213.
189. Casey, W. H.; Rustad, J. R.; Banerjee, D.; Furrer, G. *J. Nanopart. Res.* **2005**, *7*, 377.
190. Casey, W. H. *Chem. Rev.* **2006**, *106*, 1.
191. Phillips, B. L.; Casey, W. H.; Karlsson, M. *Nature* **2000**, *404*, 379.
192. Casey, W. H.; Phillips, B. L.; Karlsson, M.; Nordin, S.; Nordin, J. P.; Sullivan, D. J.; Neugebauer-Crawford, S. *Geochim. Cosmochim. Acta* **2000**, *64*, 2951.
193. Casey, W. H.; Phillips, B. L. *Geochim. Cosmochim. Acta* **2001**, *65*, 705.
194. Lee, A. P.; Phillips, B. L.; Casey, W. H. *Geochim. Cosmochim. Acta* **2002**, *66*, 577.
195. Rustad, J. R.; Loring, J. S.; Casey, W. H. *Geochim. Cosmochim. Acta* **2004**, *68*, 3011.
196. Villa, E. M.; Ohlin, C. A.; Balogh, E.; Anderson, T. M.; Nyman, M. D.; Casey, W. H. *Angew. Chem., Int. Ed.* **2008**, *47*, 4844.
197. Jencks, W. P. *Chem. Soc. Rev.* **1981**, *10*, 345.

# 9    Reactivity of Inorganic Radicals in Aqueous Solution

DAVID M. STANBURY

## 9.1   INTRODUCTION

The objective of this chapter is to describe the various types of inorganic reactions of inorganic radicals in aqueous solution, to develop a comprehensive system of reaction classes, to review the current status of thinking about the reaction mechanisms, and to survey trends in equilibrium and rate constants. It is left as an exercise for the reader to identify features of the reactivity of inorganic radicals that differ qualitatively from those of nonradical species.

In this chapter, we define inorganic radicals as main group molecules or ions that bear one unpaired electron in their ground states and that lack C–C and C–H bonds. By this definition we include stable species such as $ClO_2$ and NO, reactive species such as $I_2^-$ and $NO_2$, and carbon-containing species such as SCN and $CO_2^-$. Reactive species such as $^3O_2$, $^1O_2$, and HNO are excluded. The scope is limited to reactivity in homogeneous aqueous solution in order to place coherent bounds on the subject. A further constraint is that the reaction partners must also be inorganic species, including transition metal complexes. Reactivity discussions are also limited to the behavior at 25°C.

Much of the information related to the reactivity of inorganic radicals is derived from transient techniques such as flash photolysis and pulse radiolysis, and as a consequence, it is largely the subject of elementary reaction steps rather than overall or net reactions. Accordingly, this chapter is organized in terms of specific elementary reaction types rather than the periodic table. It turns out that virtually all of the known reactions can be classified into a limited set of 15 types. The reactivity of the aqueous electron is rather unusual, but it is still accommodated in this classification. The body of this chapter thus consists of 15 sections, each of which describes a specific reaction type, gives a few examples, probes the scope of reactivity within that type, and looks for reactivity patterns and other unifying concepts that can be used to predict the reaction rates. These sections are arranged sequentially roughly by increasing complexity. Thus, we first deal with reactions intrinsic to the individual radicals,

*Physical Inorganic Chemistry: Reactions, Processes, and Applications*   Edited by Andreja Bakac
Copyright © 2010 by John Wiley & Sons, Inc.

and later with reactions between radicals and other solutes. One of the most important properties determining the rates of elementary steps is the standard free energy change; this and allied thermochemical data such as reduction potentials, $pK_a$ values, and other equilibrium constants are the subject of an ongoing IUPAC project "Reduction Potentials of Radicals," the results of which should be published shortly. In the interim, the reader is directed to the author's 1989 review "Reduction Potentials Involving Inorganic Free Radicals in Aqueous Solution."[1] A series of books on "The Chemistry of Free Radicals" has appeared relatively recently, and it contains several sections on the topic of this chapter.[2–5] A major goal of this chapter is to identify meaningful trends in rate constants, and in such an endeavor it is essential to know what the reaction products are. Thus, a multitude of reactions for which rate constants are known but the products are unknown receive little consideration here.

No attempt has been made in this chapter to provide historical context or to assign authorship priority to the chemistry described. The aim is rather to provide the reader access to the most current literature on the subject, with a special emphasis on reviews and compendia.

## 9.2  DIMERIZATION

Dimerization is one of the most common modes of reaction for inorganic free radicals when other reactants are absent, and it is always a potential reaction pathway in any reaction system involving these radicals. These reactions are listed in Table 9.1.

In most cases the dimerization reactions are effectively irreversible, but $NO_2$, $SO_2^-$, and a few other species are exceptions where the dimerization equilibrium constants have been determined.[11] Not all radicals dimerize. For example, NO and $ClO_2$ are quite stable as monomers and exhibit no significant tendency to dimerize; apparently, the dimerization equilibrium constants are unfavorable. In other cases, dimerization has not been demonstrated because the radicals disproportionate instead, as discussed below. However, for those radicals where dimerization has been demonstrated, the rate constants are generally close to the diffusion limit; these rate constants show systematic decreases attributable to repulsion of ionic charges, as demonstrated by the series $H_2PO_4$, $HPO_4^-$, and $PO_4^{2-}$.

The proposed dimerization of $e^-$ (aq) requires special comment: In early reports it was observed that $e^-$ (aq) decays with second-order kinetics; the net reaction is

$$2e^- \text{ (aq)} + 2H_2O \rightarrow H_2 + 2OH^- \tag{9.1}$$

It was suggested that $e_2^{2-}$ (aq) was formed as a reaction intermediate. However, this dielectron species has never been detected directly. The most recent mechanistic proposal for the second-order decay of the hydrated electron suggests that the rate-limiting step corresponds to proton transfer from water to a solvated electron, which is electrostatically induced by the presence of a nearby second $e^-$ (aq). The $H^\bullet$, $e^-$ pair so generated would then combine to form $H^-$, and then $H^-$ would react with water to form $H_2$.[12]

**TABLE 9.1    Rate Constants for Radical Dimerization Reactions**

| Radical | $2k$ ($M^{-1}s^{-1}$) | Source |
|---|---|---|
| $e^-$ (aq) | $5.5 \times 10^9$ | Ref. 6 |
| $O^-$ | $2 \times 10^9$ | Ref. 7 |
| OH | $5.5 \times 10^9$ | Ref. 6 |
| H | $7.8 \times 10^9$ | Ref. 6 |
| I | $8 \times 10^9$ | Ref. 8 |
| ClO | $7 \times 10^9$ | Ref. 8 |
| $BrO_2$ | $3.0 \times 10^9$ | Ref. 8 |
| $SO_2^-$ | $1.1 \times 10^9$ | Ref. 8 |
| $SO_3^-$ | $5.3 \times 10^8$ | Ref. 8 |
| $SO_4^-$ | $8.1 \times 10^8$ | Ref. 8 |
| HS | $6.5 \times 10^9$ | Ref. 8 |
| $S_2O_3^-$ | $8.6 \times 10^8$ | Ref. 8 |
| $N_3$ | $4.4 \times 10^9$ | Ref. 8 |
| $NH_2$ | $2.2 \times 10^9$ | Ref. 8 |
| $NO_2$ | $4.6 \times 10^8$ | Ref. 8 |
| $N_2H_3$ | $2 \times 10^9$ | Ref. 9 |
| $PHO_2^-$ | $3.3 \times 10^9$ | Ref. 10 |
| $H_2PO_4$ | $1 \times 10^9$ | Ref. 8 |
| $HPO_4^-$ | $1.5 \times 10^8$ | Ref. 8 |
| $PO_4^{2-}$ | $3.9 \times 10^7$ | Ref. 8 |
| $CO_2^-$ | $5.0 \times 10^8$ | Ref. 8 |

The oxide radical anion, $O^-$, decays rapidly with second-order kinetics, and the reaction is commonly described as dimerization to form $O_2^{2-}$. This description, however, poses a puzzle, because $O_2^{2-}$ is unknown in aqueous solution. Perhaps, $O_2^{2-}$ is formed fleetingly and then protonated to $HO_2^-$; alternatively, proton transfer from water might occur in a concerted way.

Surprisingly, little information is available on the dimerization of halogen atoms. Dimerization of iodine atoms has been reported only once,[8] no data are available for bromine atom dimerization, and the data for chlorine atom dimerization yield results that are not internally consistent.[13]

ClO is believed to dimerize, but the dimer, $Cl_2O_2$, is unstable and disproportionates to $Cl + ClO_3^-$.[14] In the case of $N_2H_3$, the dimer is inferred kinetically from the delayed production of the disproportionation products.[9]

$N_3$ radicals decompose with second-order kinetics as in $2N_3 \rightarrow 3N_2$. It has been suggested that the dimer $N_6$ is a reaction intermediate, but this species has been detected directly only in a low-temperature matrix.

Dimerization of $SO_3^-$ is demonstrated unambiguously by the yield of the highly inert species $S_2O_6^{2-}$; however, there are two channels for the self-reaction of $SO_3^-$, the other being disproportionation to yield $SO_4^{2-} + SO_2$.[15,16]

$CO_2^-$ is generally regarded to dimerize to oxalate, but in a gamma-radiolysis study the yield of oxalate was reported to decrease at lower pH.[17] It was suggested that in a pH-dependent post-rate-limiting step disproportionation to $CO_2$ + formate could become the major path.

$NO_3$ radicals apparently decompose through first- and second-order paths. The first-order path is the oxidation of water to the hydroxyl radical (discussed below), and the second-order path is suggested to yield $N_2O_6$.[18] The latter pathway, however, is quite slow ($k = 8 \times 10^5\,M^{-1}\,s^{-1}$) and may be unreliable.

## 9.3   DISPROPORTIONATION

The second-order decay of radicals may occur through disproportionation rather than dimerization. Distinction between the two is achieved through detection of the products. Table 9.2 is a rather complete list of these reactions. It shows that the disproportionations of the halogen and pseudohalogen dimer radical anions ($X_2^-$) are quite fast, approaching the diffusion limit. On the other hand, the disproportionations of $NO_2$ and $HO_2$ are distinctly slower. The reactions of $X_2^-$ likely occur through X-atom transfer from one $X_2^-$ to another, yielding $X^-$ and $X_3^-$ directly; $X_3^-$ could then dissociate rapidly to yield $X_2$ and $X^-$. The lower rate constant for decay of $NO_2$ is easily attributed to its more complex mechanism; the details are not yet fully understood, but clearly a molecule of water must be involved in a very specific way. $CO_3^-$ and $SO_3^-$ disproportionate by $O^-$ transfer, a process that has a significant kinetic barrier. It seems likely that the disproportionation of $SO_5^-$ proceeds through an intermediate tetroxide, as is widely reported for the analogous reactions of organoperoxyl radicals. The disproportionation of $HO_2/O_2^-$ is slow and requires special consideration.

$ClO_2$ disproportionates, but only with the assistance of nucleophiles.[23–25] Three pathways are found in the mechanism. The first path has a rate law that is first order in $[ClO_2]$ and $[OH^-]$ and yields $ClO_2^- + ClO_3^-$; it is inferred to have addition of $OH^-$ to $ClO_2$ as the rate-limiting step with disproportionation occurring in a rapid reaction

**TABLE 9.2   Rate Constants for Radical Disproportionation Reactions**

| Reaction | $2k\,(M^{-1}\,s^{-1})$ | Source |
|---|---|---|
| $2HO_2 \rightarrow O_2 + H_2O_2$ | $8.3 \times 10^5$ | Ref. 19 |
| $2Cl_2^- \rightarrow 2Cl^- + Cl_2$ | $2.2 \times 10^9$ | Ref. 8 |
| $2Br_2^- \rightarrow 2Br^- + Br_2$ | $2.2 \times 10^9$ | Ref. 8 |
| $2I_2^- \rightarrow 2I^- + I_2^-$ | $3.2 \times 10^9$ | Ref. 8 |
| $2SO_3^- \rightarrow SO_2 + SO_4^{2-}$ | $4 \times 10^8$ | Refs. 15,16 |
| $2SO_5^- \rightarrow 2SO_4^- + O_2$ | $1 \times 10^8$ | Ref. 20 |
| $2NO_2 + H_2O \rightarrow NO_2^- + NO_3^- + 2H^+$ | $1.0 \times 10^8$ | Ref. 8 |
| $2N_2O_2^- \rightarrow N_2O_2^{2-} + 2NO$ | $8.2 \times 10^7$ | Ref. 21 |
| $2(SCN)_2^- \rightarrow 2SCN^- + (SCN)_2$ | $1.3 \times 10^9$ | Ref. 8 |
| $2CO_3^- \rightarrow CO_2 + CO_4^{2-}$ | $6 \times 10^6$ | Ref. 22 |

of this adduct with $ClO_2$. This addition of $OH^-$ to $ClO_2$ differs from the rapid association reactions described in Section 9.6 because it is quite slow. In the second path, the same $OH^-/ClO_2$ adduct is formed, but it reacts with $ClO_2$ to yield $HO_2^-$ instead of $ClO_3^-$. The third path is second order in $[ClO_2]$, first order in [Nu], and corresponds to the nucleophile-assisted electron transfer mechanism discussed in Section 9.13.

Disproportionation of NO to $N_2O$ and $NO_2^-$ is thermodynamically favorable ($\Delta G^\circ = -544$ kJ per mole of $N_2O$), but it does not occur measurably without catalysis. There are indications that transition metal complexes can catalyze the disproportionation,[26] but the reported rates are quite slow relative to the rates for catalyzed superoxide disproportionation discussed below.

$HO_2/O_2^-$ disproportionation occurs directly and through metal ion catalyzed paths. The direct disproportionation has the rate law

$$\text{rate} = \frac{2(k_{HO_2} + k_{mix}(K_a/[H^+]))[HO_2]_{tot}^2}{(1 + K_a/[H^+])^2} \tag{9.2}$$

where $[HO_2]_{tot}$ is the sum of $[HO_2]$ and $[O_2^-]$, $K_a$ is the acid dissociation constant of $HO_2$ ($pK_a = 4.8$), $k_{HO_2}$ ($= 8 \times 10^5\,M^{-1}\,s^{-1}$) is the bimolecular rate constant for disproportionation of two $HO_2$ molecules, and $k_{mix}$ ($= 9.7 \times 10^7\,M^{-1}\,s^{-1}$) is the bimolecular rate constant for disproportionation of a molecule of $HO_2$ with $O_2^-$.[19] An important consequence of this rate law is that the rates become increasingly slow as the pH rises beyond pH 6. Accurate measurements of the rates at high pH are difficult because catalysis by metal ion impurities becomes highly competitive. Foti et al. have suggested that the $k_{HO_2}$ process has a head-to-tail H-atom transfer mechanism that requires the accepting $HO_2$ radical to be stabilized through hydrogen bonding to the solvent.[27] The rate constant is low because water is not a very good H-bond acceptor.

The catalyzed disproportionation (dismutation) of superoxide has been studied extensively, particularly in regard to its enzymatic catalysis by the superoxide dismutases (SODs). The superoxide dismutases are metalloenzymes and can have Cu–Zn, Fe, Mn, or Ni active sites. Simple inorganic species can also catalyze the reaction.[28] For example, catalysis by $Fe^{2+}/Fe(III)$ at pH 7.2 has the rate law

$$\text{rate} = k_{cat}[O_2^-][Fe]_{tot} \quad k_{cat} = 1.3 \times 10^7\,M^{-1}\,s^{-1} \tag{9.3}$$

Note that under these conditions Fe(III) will precipitate, so special techniques were required to obtain the rate data.[29] Catalysis by some synthetic Mn(II) complexes has an analogous rate law, with values for $k_{cat}$ as large as $1.6 \times 10^9\,M^{-1}\,s^{-1}$.[30] Ni-SOD is extremely active, with $k_{cat} = 1.3 \times 10^9\,M^{-1}\,s^{-1}$.[31] Most of these catalytic reactions are usually considered to have similar ping-pong mechanisms in which superoxide alternately reduces and oxidizes the metal center, thus invoking $Cu^I/Cu^{II}$, $Fe^{II}/Fe^{III}$, $Mn^{II}/Mn^{III}$, and $Ni^{II}/Ni^{III}$ redox cycles:

$$O_2^- + M_{ox} \rightarrow O_2 + M_{red} \quad k_{ox} \tag{9.4}$$

$$O_2^- + M_{red} \, (+2H^+) \rightarrow H_2O_2 + M_{ox} \quad k_{red} \tag{9.5}$$

Under steady-state catalytic conditions, the rates of these two processes are equal, so the overall rate of loss of superoxide is

$$-d[O_2^-]/dt = 2k_{ox}k_{red}[O_2^-][M]_{tot}/(k_{ox} + k_{red}) \tag{9.6}$$

$$k_{cat} = 2k_{ox}k_{red}/(k_{ox} + k_{red}) \tag{9.7}$$

The above equations are written for reactions in terms of $O_2^-$ only, but they apply for reactions involving $O_2^-$ and $HO_2$ variously, so long as the rate constants all pertain to the same pH, and $[O_2^-]$ is understood to refer to the sum of the two species. Good agreement between the measured value of $k_{cat}$ and that calculated from the individually measured values of $k_{ox}$ and $k_{red}$ was obtained with $Fe^{2+}$.[29] Similar agreement has been obtained for catalysis by $Cu^{2+}$,[32] Cu-SOD,[33] Fe-EDTA,[34] and various Mn (II) complexes.[28,35] Interestingly, $Ni^{2+}$(aq) is not a catalyst for superoxide disproportionation, and the only synthetic Ni catalysts are peptide complexes derived from the Ni-SOD enzyme.[36] Effective catalysts for superoxide disproportionation through the ping–pong mechanism must have redox potentials high enough to oxidize superoxide rapidly ($E° > 0\,V$) and low enough to reduce hydroperoxyl rapidly ($E° < 0.8\,V$).

An alternative metal ion-catalyzed mechanism for superoxide disproportionation involves the Lewis acidity of the metal center rather than its redox properties. In this mechanism, the metal center activates coordinated superoxide toward reaction with another superoxide:

$$LM(II) + O_2^- \rightarrow LM^{II}(O_2^-) \tag{9.8}$$

$$LM^{II}(O_2^-) + O_2^- \, (+2H^+) \rightarrow LM(II) + O_2 + H_2O_2 \tag{9.9}$$

Manganese(II) complexes seem to be the best examples of this mechanism,[37] although Th(IV) and U(VI) should also be catalytic.[38]

## 9.4  PROTON TRANSFER REACTIONS

The participation of radicals in acid/base reactions with the solvent occurs for several species, and it often has profound effects on the radical reactivity. For example, deprotonation of $HO_2$ generates $O_2^-$, which is stable with respect to disproportionation. Table 9.3 is a sampling of radicals having known $pK_a$ values.

Most of these reactions reach equilibrium very rapidly, so their rates usually do not influence the overall reactions in which they participate. As a general rule, the $pK_a$

**TABLE 9.3  A Sampling of Radical $pK_a$ Values**

| Reaction | $pK_a$ | Source |
|---|---|---|
| $HO_2 \rightleftharpoons H^+ + O_2^-$ | 4.8 | Ref. 19 |
| $OH \rightleftharpoons H^+ + O^-$ | 11.9 | Ref. 6 |
| $H \rightleftharpoons H^+ + e^-(aq)$ | 9.6 | Ref. 6 |
| $HPO_3^- \rightleftharpoons H^+ + PO_3^{2-}$ | 5.8 | Ref. 8 |
| $H_2PO_4 \rightleftharpoons H^+ + HPO_4^-$ | 5.7 | Ref. 8 |
| $HPO_4^- \rightleftharpoons H^+ + PO_4^{2-}$ | 8.9 | Ref. 8 |
| $Cl\,(+ H_2O) \rightleftharpoons H^+ + ClOH^-$ | 5.3 | Ref. 39 |
| $I\,(+ H_2O) \rightleftharpoons H^+ + IOH^-$ | 13.1 | Ref. 40 |
| $SCN\,(+ H_2O) \rightleftharpoons SCNOH^- + H^+$ | 12.5 | Ref. 41 |

value of radical HX is much lower than the $pK_a$ value of the corresponding nonradical $HX^-$. Another general rule is that successive $pK_a$ values for inorganic radical oxyanions should differ by 4–5 units.[42–44]

The conversion of hydrogen atoms to hydrated electrons is an exception to the rule of rapid reactivity; in this case, the reaction is relatively slow, first order in $[OH^-]$, and has a second-order rate constant of $3 \times 10^7 M^{-1} s^{-1}$.[45] In less alkaline solutions, the direct reaction of the hydrogen atom with water is quite slow and yields $H_2 + OH$. As a result of these low rate constants, the interconversion between hydrogen atoms and hydrated electrons can be catalyzed by weak bases such as $F^-$ and $NH_3$.

The acidic character of the halogen atoms and SCN (X) is not as straightforward as might be suggested by the reactions shown in Table 9.3, because the conjugate bases, $XOH^-$, can dissociate reversibly to form $X^- + OH$. This dissociation causes the degree of ionization of X to depend on the concentration of $X^-$. A further complication is that $X^-$ can associate reversibly with X to form $X_2^-$.

## 9.5  HYDROXYL RADICAL PRODUCTION REACTIONS

There are a number of reactions in which radicals react with water or its components to form hydroxyl radicals. These can be of great importance because they are intrinsic properties of the radicals in aqueous media and because they can interconvert species having highly different properties. Table 9.4 presents a sampling of these radical transformation reactions, selected in several cases because of their reversibility.

As Table 9.4 shows, most of these hydroxyl radical generating reactions are rather slow, so it can be challenging to make accurate measurements on them. In particular, the $NO_3$ reaction has yielded results that vary quite dramatically from one lab to the other.[51] It seems rather unlikely that a uniform method can be taken to explain the rates of these reactions, given that a variety of mechanisms are involved.

**TABLE 9.4   Some Important Reactions of Radicals with Water**

| Reaction | $k_f$ (s$^{-1}$) | $K_{eq}$ (M$^2$) | Source |
|---|---|---|---|
| $Cl + H_2O \rightleftharpoons OH + Cl^- + H^+$ | $1.8 \times 10^5$ | $1.1 \times 10^{-5}$ | Ref. 46 |
| $NO_3 + H_2O \rightleftharpoons OH + NO_3^- + H^+$ | $1.6 \times 10^4$(?) | | Ref. 47 |
| $SO_4^- + H_2O \rightleftharpoons OH + SO_4^{2-} + H^+$ | 360 | $1.0 \times 10^{-3}$ | Ref. 48 |
| $H_2PO_4 + H_2O \rightleftharpoons OH + H_3PO_4$ | $1.4 \times 10^5$ | $3.3^a$ | Ref. 49 |
| $O_3^- + H^+ \rightarrow OH + O_2$ | $9 \times 10^{10b}$ | | Ref. 8 |
| $BrOH^- \rightleftharpoons OH + Br^-$ | $4.2 \times 10^6$ | $3.1 \times 10^{-3a}$ | Ref. 50 |

$^a$Dimensions of M.
$^b$Dimensions of M$^{-1}$s$^{-1}$.

## 9.6   ASSOCIATION WITH NONRADICAL MOLECULES

One of the most common bimolecular reactions of radicals is their association with other nonradical molecules. We have used the term "hemicolligation" to describe this reaction type elsewhere.[11] This mode of reaction is particularly important because many radical precursors can react in this way. For example, bromine atoms are often generated by oxidizing bromide ions, so the reaction $Br + Br^- \rightleftharpoons Br_2^-$ is an unavoidable component in such systems. Association of radicals with $O_2$ is another common process that can be important when atmospheric oxygen is not completely excluded from the reaction mixture. When the radical is the hydrated electron, the association reaction is simply a reduction and is treated separately (Table 9.5).

These reactions are very important because of their profound effect on the reactivity of the radicals involved. For example, iodine atoms are good oxidants and have negligible reducing character, while the $I_2^-$ radical is readily oxidized to $I_2$.

**TABLE 9.5   Some Association Reactions of Radicals**

| Reaction | $k_f$ (M$^{-1}$s$^{-1}$) | Source | $K$ (M$^{-1}$) | Source |
|---|---|---|---|---|
| $H + O_2 \rightarrow HO_2$ | $2.1 \times 10^{10}$ | Ref. 6 | | |
| $H + I^- \rightleftharpoons HI^-$ | $2.8 \times 10^8$ | Ref. 52 | ~300 | Ref. 52 |
| $O^- + O_2 \rightleftharpoons O_3^-$ | $3.6 \times 10^9$ | Ref. 6 | $9 \times 10^5$ | Ref. 53 |
| $OH + Cl^- \rightleftharpoons ClOH^-$ | $4.3 \times 10^9$ | Ref. 6 | 0.7 | Ref. 54 |
| $Cl + Cl^- \rightleftharpoons Cl_2^-$ | $6.5 \times 10^9$ | Ref. 8 | $1.4 \times 10^5$ | Ref. 55 |
| $Br + Br^- \rightleftharpoons Br_2^-$ | $9 \times 10^9$ | Ref. 8 | $3.9 \times 10^5$ | Ref. 56 |
| $I + I^- \rightleftharpoons I_2^-$ | $1.2 \times 10^{10}$ | Ref. 8 | $1.3 \times 10^5$ | Ref. 57 |
| $HS + HS^- \rightarrow H_2S_2^-$ | $5.4 \times 10^9$ | Ref. 8 | | |
| $SO_3^- + O_2 \rightarrow SO_5^-$ | $1.5 \times 10^9$ | Ref. 8 | | |
| $NH_2 + O_2 \rightarrow H_2NO_2$ | $2 \times 10^9$ | Ref. 58 | | |
| $HPO_3^{2-} + O_2 \rightarrow HPO_5^{2-}$ | $1.9 \times 10^9$ | Ref. 8 | | |
| $CCl_3 + O_2 \rightarrow CCl_3O_2$ | $3.3 \times 10^9$ | Ref. 59 | | |
| $SCN + SCN^- \rightleftharpoons (SCN)_2^-$ | $7 \times 10^9$ | Ref. 60 | $2.0 \times 10^5$ | Ref. 61 |

The reversibility of many of these reactions is another important feature, as it can complicate the interpretation of reactivity trends. One notable example is in the chemistry of the $Cl/Cl_2^-$ equilibrium where several reactions that were originally attributed to $Cl_2^-$ were later reassigned to reactions of $Cl$.[62]

Most of these association reactions have rate constants in the range of $2 \times 10^9$ $M^{-1}s^{-1}$, with the hydrogen atom reactions being more variable. The factors leading to these rate variations are not yet understood. A related question is what governs the magnitude of the association equilibrium constants. Why doesn't $Br_2^-$ associate with $O_2$? Why doesn't NO associate with $Cl^-$? The answers to these questions must lie in the nature of the bonding in the adducts, but we have no general guidelines for making predictions.

The hydrated electron adds to many species, and these reactions are discussed in Section 9.9.

## 9.7   COLLIGATION WITH OTHER RADICALS

Asymmetric radical–radical association (colligation) reactions are most easily investigated when the radicals are present at high concentrations or when other potential reactants are absent as in the radiolysis of water. High radical concentrations can readily be achieved when one is a stable species such as NO or $ClO_2$. On the other hand, in any homogeneous reaction system where metal ions of variable valence are not present, the ultimate fate of any unstable radicals must be through dimerization, disproportionation, or asymmetric colligation. A sampling of asymmetric colligation reactions is presented in Table 9.6.

From the rate constants presented in Table 9.6 it can be seen that these reactions tend to be extremely rapid and near the diffusion limit. The similarities to radical dimerization reactions are self-evident.

**TABLE 9.6   Some Asymmetric Colligation Reactions**

| Reaction | $k$ $(M^{-1}s^{-1})$ | Source |
|---|---|---|
| $H + e^-(aq) \, (+ \, H_2O) \rightarrow H_2 + OH^-$ | $2.5 \times 10^{10}$ | Ref. 6 |
| $OH + e^-(aq) \rightarrow OH^-$ | $3.0 \times 10^{10}$ | Ref. 6 |
| $H + OH \rightarrow H_2O$ | $7 \times 10^9$ | Ref. 6 |
| $NO + O_2^- \rightarrow OONO^-$ | $5 \times 10^9$ | Ref. 63 |
| $HO_2 + OH \rightarrow H_2O_3$ | $2.8 \times 10^{10}$ | Ref. 64 |
| $OH + NH_2 \rightarrow NH_2OH$ | $9.5 \times 10^9$ | Ref. 8 |
| $NO + NO_2 \rightleftharpoons N_2O_3 \, (\rightarrow 2HNO_2)$ | $1.1 \times 10^9$ | Ref. 8 |
| $ClO_2 + O^- \rightarrow ClO_3^-$ | $2.7 \times 10^9$ | Ref. 8 |
| $HO_2 + NO_2 \rightleftharpoons HOONO_2$ | $4 \times 10^9$ | Ref. 63 |
| $CO_2^- + NO \rightarrow NOCO_2^-$ | $3.5 \times 10^9$ | Ref. 65 |

**TABLE 9.7   Some Nucleophilic Displacement Reactions**

| Reaction | $K_{eq}$ | Source |
|---|---|---|
| $BrOH^- + Br^- \rightleftharpoons Br_2^- + OH^-$ | 70 | Ref. 50 |
| $IOH^- + I^- \rightleftharpoons I_2^- + OH^-$ | $2.5 \times 10^{-4}$ | Ref. 11 |
| $ClSCN^- + SCN^- \rightleftharpoons (SCN)_2^- + Cl^-$ | $3 \times 10^4$ | Ref. 11 |
| $BrSCN^- + Br^- \rightleftharpoons Br_2^- + SCN^-$ | $1 \times 10^{-3}$ | Ref. 11 |
| $ISCN^- + SCN^- \rightleftharpoons (SCN)_2^- + I^-$ | $2.5 \times 10^{-3}$ | Ref. 11 |
| $(SCN)_2^- + S_2O_3^{2-} \rightleftharpoons SCNS_2O_3^{2-} + SCN^-$ | $1.6 \times 10^2$ | Ref. 11 |

## 9.8   NUCLEOPHILIC DISPLACEMENT REACTIONS

These reactions are defined as the association of one nucleophile with a radical that leads to the displacement of another nucleophile. Several examples of these reactions are shown in Table 9.7. It is generally possible to rewrite these reactions as two component reactions of the radical association with nonradical type. For example, the reaction

$$BrSCN^- + Br^- \rightleftharpoons Br_2^- + SCN^- \tag{9.10}$$

is thermochemically equivalent to the sum of

$$BrSCN^- \rightleftharpoons Br + SCN^- \tag{9.11}$$

and

$$Br + Br^- \rightleftharpoons Br_2^- \tag{9.12}$$

All three reaction types tend to be very fast, so often it is only the overall position of equilibrium that is chemically important.

## 9.9   ELECTRON TRANSFER

Simple bimolecular electron transfer reactions can occur between radicals and nonradical species, and they can also occur between radicals and transition metal complexes. The number of reactions in this category is quite large, and they are notable for the wide range of reported rate constants. A sampling of these reactions, selected for their reversibility, is shown in Table 9.8.

Many electron transfer reactions of inorganic radicals conform to the outer-sphere model and hence can be modeled with the Marcus theory of electron transfer.[71] This model relies, in part, on the concept of self-exchange reactions, and the inference that self-exchange reactions can be defined for radicals. For many years, it was simply

**TABLE 9.8    Some Simple Reversible Electron Transfer Reactions of Radicals**

| Reaction | $k_f$ ($M^{-1} s^{-1}$) | $k_r$ ($M^{-1} s^{-1}$) | Source |
|---|---|---|---|
| $O_3$ (aq) + $ClO_2^- \rightleftharpoons O_3^- + ClO_2$ | $4 \times 10^6$ | $1.8 \times 10^5$ | Ref. 53 |
| $N_3^- + [IrCl_6]^{2-} \rightleftharpoons N_3 + [IrCl_6]^{3-}$ | $1.6 \times 10^2$ | $5.5 \times 10^8$ | Ref. 66 |
| $CO_2^- + Tl^+ \rightleftharpoons CO_2 + Tl$ (aq) | $3.0 \times 10^6$ | $3.5 \times 10^7$ | Ref. 67 |
| $SO_3^{2-} + [Ru(phen)(NH_3)_4]^{3+} \rightleftharpoons SO_3^-$ <br> + $[Ru(phen)(NH_3)_4]^{2+}$ | $3.7 \times 10^4$ | $1.0 \times 10^8$ | Ref. 68 |
| $NO_2^- + [Fe(TMP)_3]^{3+} \rightleftharpoons NO_2$ <br> + $[Fe(TMP)_3]^{2+}$ | $3.9 \times 10^3$ | $1.0 \times 10^7$ | Ref. 69 |
| $O_2 + [Ru(NH_3)_5isn]^{2+} \rightleftharpoons O_2^-$ <br> + $[Ru(NH_3)_5isn]^{3+}$ | $0.11$ | $2.2 \times 10^8$ | Ref. 70 |

phen = 1,10-phenanthroline; TMP = 3,4,5,6-tetramethylphenanthroline; isn = isonicotinamide.

assumed that such reactions could occur, and Marcus theory was used to calculate the self-exchange rate constants. Now, however, there are some examples where the real self-exchange rate constants have been measured, notably for the $NO_2/NO_2^-$ and $O_2/O_2^-$ systems.[72,73] Factors recognized to influence the reaction rate constants include the structural rearrangements of the radical, the reaction driving force, the self-exchange rate constant of the reaction partner, the reaction electrostatics, and the size disparity between the reactants. On the other hand, some of these reactions are too fast to comply with Marcus theory and are inferred to have reduced kinetic barriers arising from an inner-sphere type of mechanism in which the radical orbitals overlap with those of the reaction partners (strong overlap). An example of the latter is the oxidation of NO by $[IrCl_6]^{2-}$.[69] For a third group of reactions, the estimated intrinsic barriers are so small that even within the Marcus model the cross-exchange rate constants are always effectively diffusion controlled; this description is notably applicable to reactions where iodine atoms are reduced to iodide.

A number of ostensibly outer-sphere reactions involving the $O_2^-/O_2$ redox couple were considered to have anomalous rates that were inconsistent with the Marcus model, but many of these deviations are now recognized to arise from size disparities between the small $O_2^-$ radical and the large reaction partners; these size disparities affect the solvent reorganization energies in ways that are not accounted for in the simple Marcus cross relationship.[74]

Nuclear tunneling is potentially a significant consideration in outer-sphere radical electron transfer reactions. The case of reduction of $NO_2$ to $NO_2^-$ is notable in that nuclear tunneling is predicted to increase the self-exchange rate constant by a factor of 79 relative to the classical value.[75] Kinetic isotope effect measurements could provide experimental evidence for nuclear tunneling. $^{18}O/^{16}O$ KIE measurements have indeed provided evidence for nuclear tunneling in reactions involving the $O_2/O_2^-$ redox couple.[76]

Addition of the hydrated electron to another molecule or ion can be considered as an electron transfer reaction, perhaps the ultimate electron transfer reaction, and many such reactions have been observed.[6] When the reaction occurs with no bond cleavage,

**TABLE 9.9    Some Reversible Dissociative/Associative Radical Electron Transfer Reactions**

| Reaction | $k_f$ ($M^{-2}s^{-1}$) | $k_r$ ($M^{-1}s^{-1}$) | Source |
|---|---|---|---|
| $2I^- + [Os(bpy)_3]^{3+} \rightleftharpoons I_2^- + [Os(bpy)_3]^{2+}$ | $3.3 \times 10^4$ | $1.1 \times 10^8$ | Ref. 77 |
| $2Br^- + N_3 \rightleftharpoons Br_2^- + N_3^-$ | $4 \times 10^8$ | $7.3 \times 10^3$ | Ref. 78 |
| $2SCN^- + [Ru(bpy)_3]^{3+} \rightleftharpoons (SCN)_2^- + [Ru(bpy)_3]^{2+}$ | $2 \times 10^7$ | $3.5 \times 10^7$ | [a] |

[a] Rate constants calculated from the data in Figure 4 of Ref. 79.

the rate constants are typically at the diffusion limit as adjusted for electrostatic effects. However, some f-element species break this rule, such as the reactions of $e^-$ (aq) with $Am^{3+}$ ($k = 1.6 \times 10^8 \, M^{-1} \, s^{-1}$), $Er^{3+}$ ($k = 1.0 \times 10^7 \, M^{-1} \, s^{-1}$), and $Tm^{3+}$ ($k = 3.3 \times 10^8 \, M^{-1} \, s^{-1}$).

Somewhat more complex are electron transfer reactions that are coupled to bond cleavage or bond formation (dissociative/associative electron transfer). Some of these appear to be of the outer-sphere type. In Table 9.9 are shown a few examples of such reactions where the radicals undergo reductive cleavage and the rates have been measured in both directions. The roster of such reactions where the rates have been measured in only one direction is vastly greater. Radicals known to undergo reductive cleavage include $I_2^-$, $Br_2^-$, $Cl_2^-$, and $(SCN)_2^-$. Oxidative cleavage can also occur, as discussed below.

When reductive cleavage reactions are considered in the direction in which cleavage occurs, the rates are simply bimolecular and raise no dynamical issues. For the other direction, where the rate laws are third order overall, some chemists may question whether they can have elementary steps that are truly third order, arguing that such a process is statistically improbable. According to this view, some preassociation between two of the reactants must occur. The principle of microscopic reversibility then requires that the reaction in the cleavage direction (bimolecular) must also be a two-step process, the first one producing the same (pre)association complex that is generated in the other direction. As an example, the reduction of $I_2^-$ to two solvent-separated iodide ions must produce first a geminal pair of iodide ions. Simple ion pairing calculations reveal that such an association between two iodide ions is not very favorable, but apparently it is sufficient to explain the rates observed. A further advantage of this view is that it enables the cross relationship of Marcus theory to be adapted to these reactions. In its usual form, the cross relationship is dimensionally incompatible with third-order rate constants. The workaround is to define the iodide self-exchange reaction as occurring in two steps:

$$2I^- \rightleftharpoons (I^-)_2 \tag{9.13}$$

$$(I^-)_2 + {}^*I_2^- \rightarrow I_2^- + ({}^*I^-)_2 \tag{9.14}$$

The other issue is to formulate a model for the potential energy surfaces. In the usual Marcus model, it is assumed that the bond rearrangements are relatively small and can be approximated as distortions along a harmonic oscillator. For bond cleavage reactions, the harmonic oscillator picture is inappropriate since the potential energy surface for two iodide ions is completely repulsive. Expressions have been derived for the reorganizational energy that are based on a Morse potential for this repulsive energy surface. The outcome of these considerations is an adequate interpretation of the rates of outer-sphere cross-exchange reactions involving reductive cleavage of $I_2^-$.[80]

Oxidative cleavage can occur with some radicals, such as when $HO_2$ is oxidized to $O_2 + H^+$. Oxidative cleavage of $HO_2$ is actually a form of proton-coupled electron transfer (PCET), discussed below. It occurs in the reaction of $HO_2$ with $Cu^{2+}$,[32] $Ce^{4+}$, $Am^{4+}$, various Ni(III) complexes, and $[Ru(bpy)_3]^{3+}$.[19]

Hydroxyl radicals can react as net one-electron oxidants as in the reactions

$$OH + [Fe(CN)_6]^{4-} \rightarrow OH^- + [Fe(CN)_6]^{3-} \quad k = 9 \times 10^9 \ M^{-1} \ s^{-1} \tag{9.15}$$

$$OH + [IrCl_6]^{3-} \rightarrow OH^- + [IrCl_6]^{2-} \quad k = 1.2 \times 10^{10} \ M^{-1} \ s^{-1} \tag{9.16}$$

These reactions tend to be very fast, typically at or near the diffusion limit. They might appear to be candidates for a paradigm of outer-sphere electron transfer, but in fact it seems as though all such reactions are really inner-sphere in one way or another.[81]

Hydrogen atoms can act as one-electron reductants as in

$$H + MnO_4^- \rightarrow H^+ + MnO_4^{2-} \quad k = 2.4 \times 10^{10} \ M^{-1} \ s^{-1} \tag{9.17}$$

The extreme differences in solvation between H and $H^+$ are expected to lead to an enormous barrier to outer-sphere electron transfer, so it is inferred that H atoms must react through inner-sphere mechanisms.[82] This logic was supported by an analysis of the low rate constant for the reaction of $Fe^{3+}$(aq). In the case of the actinyl ions, $MO_2^{2+}$ (M = U, Pu, Np, and Am), the rate constants range from $4.5 \times 10^7$ to $1.6 \times 10^9 \ M^{-1} s^{-1}$ and increase uniformly as the metal ion becomes a stronger oxidant.[83] Presumably, the oxo ligands in these actinyl ions provide binding sites for inner-sphere mechanisms. Similarly, the rapid reduction of $[Fe(CN)_6]^{3-}$ by H atoms can be rationalized through bonding between a bound $CN^-$ ligand and a H atom.

Hydrogen atoms can also act as net one-electron oxidants generating $H_2$, as described in Section 9.10.

Radicals can react with transition metal complexes through classical inner-sphere electron transfer mechanisms in which the radical binds directly to the metal center. For example, in the oxidation of $Fe^{2+}$(aq) by $ClO_2$ a $Fe(ClO_2)^{2+}$ intermediate is detected, and its rate of formation is governed by the rate of substitution at the metal center.[84] $(SCN)_2^-$, $I_2^-$, $Br_2^-$, and $O_2^-$ are reported to oxidize $Fe^{II}$, $Co^{II}$, and $Mn^{II}$ complexes of EDTA and $NTA^-$ through inner-sphere mechanisms.[85] Likewise, the oxidation of several tetraazamacrocyclic Ni(II) complexes by $(SCN)_2^-$, $Cl_2^-$, and $Br_2^-$ proceeds through an inner-sphere mechanism.[86] Oxidation of $U^{3+}$ by radicals proceeds likewise.[87] For further discussion of inner-sphere redox reactions, see Section 9.12.

An interesting aspect of several of the inner-sphere reactions mentioned above is that they entail reductive cleavage of radicals, for example,

$$Cl_2^- + Co^{2+} \rightarrow CoCl^{2+} + Cl^- \tag{9.18}$$

The reactions of $Cl_2^-$ and $Br_2^-$ with $Fe^{2+}$ and $Mn^{2+}$ are analogous to the reaction of $Cl_2^-$ with $Co^{2+}$.[88-90] The reactions of $Cl_2^-$, $Br_2^-$, and $I_2^-$ with $Cr^{2+}$ are also similar.[91] The same radicals react with $V^{2+}$ through an outer-sphere mechanism because of the relatively slow rate of ligand exchange.[91] On the other hand, the oxidation of $Cu^{II}(gly_4)$ by $Br_2^-$ yields $Cu^{III}(Gly_4)Br$ through an inner-sphere mechanism.[92] Inner-sphere reductive cleavage also occurs with $(SCN)_2^-$, for example, in its reaction with $Mn^{II}(NTA)$.[8]

## 9.10   HYDROGEN ATOM TRANSFER/PROTON-COUPLED ELECTRON TRANSFER

A common net mode of reactivity is for radicals to act as hydrogen atom acceptors. These reactions can, in theory, proceed through a direct H-atom abstraction process, or they can occur through a sequential process of electron transfer and proton transfer. Experimental distinction between these two mechanisms can be quite difficult. Nevertheless, it is clear that many radicals are thermodynamically powerful hydrogen atom acceptors; a "pecking order" of hydrogen atom abstraction power is established by the thermodynamic data assembled in Table 9.10. This table gives standard reduction potentials for the radicals expressed as reactions in which a hydrogen atom is acquired by the radical, and it also presents the R–H bond enthalpies; these two measures of hydrogen atom acceptance power show some parallels and also some interesting reversals.

According to its calculated reduction potential, the $PO_4^{2-}$ radical is a very strong hydrogen atom acceptor; something might be in error here, because the calculated potential implies that $PO_4^{2-}$ would oxidize water. Nevertheless, $PO_4^{2-}$ is believed to oxidize $NH_3$ through a hydrogen atom transfer (HAT) mechanism.[97]

As mentioned above, hydroxyl radicals are believed to never engage in outer-sphere electron transfer. An alternative mechanism of hydrogen atom abstraction is frequently mentioned, and as Table 9.10 shows, the hydroxyl radical is extremely potent in this regard. Reactivity via a HAT mechanism is well established with organic substrates, as the significant H/D kinetic isotope ratios for reaction with alcohols demonstrate.[98] The argument in favor of a HAT mechanism for the reactions of OH with $Fe^{2+}$ $(k = 4 \times 10^8\,M^{-1}\,s^{-1})$, $Mn^{2+}$ $(k = 3 \times 10^7\,M^{-1}\,s^{-1})$, $Ce^{3+}$ $(k = 3 \times 10^8\,M^{-1}\,s^{-1})$, and $Cr^{3+}$ $(k = 3 \times 10^8\,M^{-1}\,s^{-1})$ proceeds along the lines that the rates do not show the proper trends for outer-sphere electron transfer, exceed the rates of substitution of coordinated water, and are consistent with considerations based on the Polanyi–Semenov rule for HAT.[99] These conclusions are now questionable in view of

**TABLE 9.10   A Thermodynamic Hydrogen Atom Abstraction Pecking Order of Some Radicals**

| Half Reaction | $E°$ (V)$^a$ | $D$(R–H) (kJ)$^b$ |
|---|---|---|
| $PO_4^{2-} + H^+ + e^- = HPO_4^{2-}$ | 2.96 (?)$^c$ | |
| $OH + H^+ + e^- = H_2O$ | 2.72 | 497 |
| $SO_4^- + H^+ + e^- = HSO_4^-$ | 2.55 | |
| $NH_2 + H^+ + e^- = NH_3$ | 2.26 | 453 |
| $H + H^+ + e^- = H_2$ | 2.13 | 436 |
| $ClO + H^+ + e^- = HOCl$ | 1.95 | 393 |
| $O^- + H^+ + e^- = OH^-$ | 1.77 | |
| $N_3 + H^+ + e^- = HN_3$ | 1.61 | 338 |
| $BrO_2 + H^+ + e^- = HBrO_2$ | 1.5 | |
| $HS + H^+ + e^- = H_2S$ | 1.49 | 382 |
| $HO_2 + H^+ + e^- = H_2O_2$ | 1.44 | 369 |
| $NO_2 + H^+ + e^- = HNO_2$ | 1.18 | 328 |
| $ClO_2 + H^+ + e^- = HClO_2$ | 1.17 | |
| $N_2H_3 + H^+ + e^- = N_2H_4$ | 1.11 | 366 |
| $O_2^- + H^+ + e^- = HO_2^-$ | 1.03 | |
| $NF_2 + H^+ + e^- = HNF_2$ | 0.91 | 317 |
| $NO + H^+ + e^- = HNO^d$ | −0.14 | 213 |

$^a$Potentials versus NHE. From Ref. 1 and calculated from $\Delta_f G°$ data in Refs 1, 93, and 94.
$^b$Bond enthalpies from Kerr's tables[95] and data in the NIST WebBook.
$^c$Calculated from data in Ref. 49.
$^d$Data from Ref. 96.

the fivefold lower rate constant currently accepted for the reaction of $Mn^{2+}$ (shown above) and uncertainties regarding the rate constant of the $Cr^{3+}$ reaction.[100] This mechanistic proposal has also been disputed on other grounds.[81] The reaction of OH with $H_2O_2$ to produce $HO_2$ has a H/D KIE of about 3, which provides good support for a HAT mechanism in this reaction.[101] The reactions of OH with $NH_3$,[97] $NH_2Cl$,[102] $H_2PO_2^-$, $H_2PO_3^-$, and $HPO_3^{2-}$ are also considered to have hydrogen atom abstraction mechanisms.[10]

The sulfate radical was long considered as a virtually obligatory outer-sphere oxidant, but the data in Table 9.10 show that it is also a very powerful hydrogen atom abstractor. Arguments in favor of a HAT mechanism for $SO_4^-$ reactions have been made for a number of reactions with organic substrates.[103,104] On the basis of a parallel trend in the rates of oxidation by OH and $SO_4^-$, it has been suggested that the $SO_4^-$ oxidations of $NH_3OH^+$, $H_2PO_3^-$, $HPO_3^{2-}$, $H_2O_2$, $H_2PO_2^-$, $N_2H_5^+$, $N_2H_4$, and $NH_2OH$ all occur through a HAT mechanism.[105] A HAT mechanism has also been inferred for the $SO_4^-$ oxidation of $NH_3$.[97]

Table 9.10 reveals that the amino radical ($NH_2$) and ClO are thermodynamically powerful hydrogen atom abstractors. However, there is scant evidence for their reactivity in this way with inorganic substrates.

Hydrogen atoms are usually thought of as strong reducing agents, but they can react as oxidants through net hydrogen atom abstraction reactions that generate $H_2$. This type of behavior is widely documented for reactions with organic substrates, and it has also been reported for their reactions with $[Co(NH_3)_5O_2CH]^{2+}$, $N_2H_5^+$, $H_3PO_3$, and $H_3PO_4$.[6] The reactions of H atoms with $NH_2OH$ and $NH_3OH^+$ are also believed to have a HAT mechanism.[106] The reaction of H with $N_2H_4$ is considered to be HAT, but the reaction of H with $N_2H_5^+$ is believed to entail another fragmentation process.[107] The mechanism of these $H_2$-generating reactions could be HAT or sequential PCET. In the latter case, the aqueous hydride ion would be an intermediate, which then protonates to form $H_2$:

$$H + N_2H_4 \rightarrow H^- + N_2H_4^+ \tag{9.19}$$

$$H^- + H^+ \rightarrow H_2 \tag{9.20}$$

The evidence for $H^-$ (aq) as a real entity is extremely limited, and estimates of the standard reduction potential of H atoms are highly uncertain; the most recent estimate, however, implies that the hydrogen atom should be a fairly strong simple one-electron oxidant: $E°(H/H^-) = 0.83$ V.[108] This implies that it may be rather difficult to distinguish clearly between the HAT and PCET mechanisms for some H-atom reactions.

The oxide radical ion ($O^-$) is quite unlikely to react as a simple outer-sphere one-electron oxidant because the corresponding $O^{2-}$ product is extremely basic. On the other hand, it is a rather strong oxidant when the net reaction yields $OH^-$. Experimental data on the rates of reaction of $O^-$ are limited, mostly because the species exists only at high pH. One clear example is the reaction of $O^- + H_2$.[109] $O^-$ reacts with $H_2O$ to produce $OH + OH^-$, but it is unclear whether to describe this as the transfer of a H atom or a proton. The body of evidence is considerably greater for its reactivity as a hydrogen atom acceptor with organic substrates. $O^-$ oxidizes species lacking protons such as $I^-$ and $Br^-$; apparently, these reactions occur through addition of $O^-$ to the substrate (coupled somehow with proton addition from the solvent) to yield $XOH^-$, which can then dissociate to yield $X + OH^-$.[40]

Superoxide is believed to react as an effective hydrogen atom acceptor. Thermodynamically, it is one of the weakest oxidants in Table 9.10, but the electron transfer alternative is highly disfavored by the inaccessibility of $O_2^{2-}$ in aqueous solution: Taube estimates that $HO_2^-$ has a $pK_a$ value of 21.[110] Thus, a HAT mechanism is inferred for the oxidations of $[Co^{II}(sep)]^{2+}$, $[Fe^{II}(tacn)_2]^{2+}$, $[Ru^{II}(tacn)_2]^{2+}$, and $[Ru^{II}(sar)]^{2+}$;[111] in these reactions the ligands are the hydrogen atom donors, but it is the metal centers that are oxidized. The bimolecular disproportionation of $HO_2$ is another likely example of hydrogen atom transfer; this mechanism may also apply to the disproportionation via the reaction of $HO_2$ with $O_2^-$.[27] Curiously, the reaction of $O_2^-$ with the hydrated electron is fast and seems not to be pH dependent. This latter reaction may yield genuine $O_2^{2-}$ as an intermediate, or it may be a form of PCET where the solvent donates a proton during the addition of the electron.

One-electron reduction of NO to HNO is unique because of the spin issues involved. Sequential electron–proton transfer would generate $^3NO^-$ as an intermediate, and protonation of this species to give $^1HNO$ is slow.[112] These kinetic considerations add to the overall thermodynamic effects favoring H-atom transfer mechanisms. Despite these expectations, evidence in favor of H-atom transfer to NO is quite sparse. NO has been argued to act as a H-atom acceptor in its reaction with hydroxylamine.[113] This reaction has a rate law first order in [NO] and first order in [$H_2NO^-$], and HNO is inferred as an intermediate. NO also abstracts a hydrogen atom from hyponitrous acid, HONNOH.[114] As Table 9.10 shows, NO is now considered to be a very weak hydrogen atom acceptor, so the reports of its reactivity in this mode may bear reexamination.

Surprisingly, it has been argued that bromine atoms can act as hydrogen atom acceptors, although the supporting experimental data were based only on organic donors.[115] In these reactions, the net product is the bromide ion, but HBr is inferred as an intermediate that releases protons immediately into solution.

## 9.11   OXYGEN ATOM/ANION ABSTRACTION

The formal transfer of a neutral oxygen atom has been reported for some organic reactions of peroxyl radicals;[116] it was also suggested that $CCl_3O_2$ transfers an oxygen atom to iodide, but recent evidence supports an alternative process.[117]

A more common process seems to be the formal abstraction of an $O^-$ radical by another radical. Examples are known where the $O^-$ donor is also radical as in

$$NO + CO_3^- \rightarrow NO_2^- + CO_2 \tag{9.21}$$

Other examples exist where the $O^-$ donor is a nonradical:

$$H + NO_3^- \rightarrow OH^- + NO_2 \tag{9.22}$$

Examples of the first type include the reaction of NO with $CO_3^-$,[65] the reactions of $CO_3^-$, $NO_2$, and $SO_3^-$ with $CO_3^-$,[118] and the disproportionation of $SO_3^-$.[15,16] Examples of the second type include the reactions of H atoms with $NO_2^-$, $HNO_2$, and $NO_3^-$.[119] It has been shown that the reaction of H atoms with $NO_2^-$ is a two-step process in which H atoms first add to $NO_2^-$ to produce the relatively long-lived $HNO_2^-$ radical, and then the $HNO_2^-$ cleaves to yield $OH^- + NO$.[44] Most of these reactions are significantly slower than the diffusion limit, as expected from the barriers arising from making and breaking O–R bonds.

## 9.12   SUBSTITUTION AT METAL CENTERS

Radicals can add at metal centers in coordination complexes. These reactions can occur through replacement of an existing ligand or through addition at a vacant coordination site, although the distinction between these two mechanisms can be vague. The roster of radicals demonstrated to add to metal centers presently consists of H, $O_2^-/HO_2$, NO, $CO_2^-$, and $SO_3^-$. As discussed below, although OH adds formally to many metal centers, there is considerable doubt about describing the reactions mechanistically as ligand substitutions.

One feature of these reactions is the challenge in distinguishing between simple substitution and redox reactions. A radical can add to become a radical ligand, or a change of oxidation state of the metal center can accompany the reaction. For example, when superoxide adds to a Co(II) center there can be ambiguity as to whether the product is a Co(II)-superoxo complex or a Co(III)-peroxo complex. Moreover, there is the possibility that a Co(II)-superoxo complex might form initially, and then intramolecular electron transfer might occur to yield a Co(III)-peroxo complex. Reactions of NO raise similar issues.

Hydrogen atoms can oxidize metal ions through an inner-sphere mechanism as in the net reaction

$$H + Fe^{2+} (+H^+) \rightarrow H_2 + Fe^{3+} \tag{9.23}$$

In several cases hydrido complexes are detected as intermediates, the general mechanism being

$$H + M^{n+} \rightarrow HM^{n+} \tag{9.24}$$

or

$$HM^{n+} + H^+ \rightarrow H_2 + M^{(n+1)+}$$

$$HM^{n+} + H_2O \rightarrow H_2 + M^{(n+1)+} + OH^- \tag{9.25}$$

Such hydrido intermediates have been detected in the reactions of $Cr^{2+}$,[120,121] $Cu^+$,[122] $Fe^{2+}$,[123] and $Ti^{3+}$.[124] The rates of formation of the hydrido species conform to the usual reaction patterns for $I_d$ substitution kinetics. An example of H-atom addition to metal complexes is provided by the reaction of $[Ni(cyclam)]^{2+}$ and a Co (II) tetraazamacrocycle.[82,125]

The hydroxyl radical reacts with many metal centers through an apparent substitution mechanism:

$$OH + M(H_2O)^{n+} \rightarrow M(OH)^{n+} + H_2O \tag{9.26}$$

These are, of course, redox reactions in which the metal center undergoes one-electron oxidation. The mechanism could be displacement of coordinated water accompanied by electron transfer (substitution), hydrogen atom transfer (as discussed in Section 9.10), or outer-sphere electron transfer followed by hydrolysis of the oxidized metal ion. At present, there seems to be no clear experimental basis for resolving this ambiguity.

Superoxide ($O_2^-$) and hydroperoxyl ($HO_2$) are well known to bind to metal ions, and this form of reactivity is of great importance in metal-catalyzed disproportionation

**TABLE 9.11   Stability Constants for Binding of $O_2^-$ by Metal Ions**

| Metal Ion | $K_{eq}$ (M$^{-1}$) | Source |
|---|---|---|
| $Co^{III}(papd)^a$ | $10^{12}$ | Ref. 126 |
| $Co^{III}([14]aneN_4)(H_2O)_2$ | $1.3 \times 10^{15}$ | Refs. 110, 127, 128' |
| $Co^{III}([15]aneN_4)(H_2O)_2$ | $>10^{18}$ | Refs. 110, 128 |
| $Co^{III}(Me_6\text{-}[14]aneN_4)(H_2O)_2$ | $1.1 \times 10^{15}$ | Ref. 127 |
| $Co^{III}(Me_6\text{-}[14]aneN_4)(H_2O)(OH)$ | $2.5 \times 10^{11}$ | Ref. 127 |
| $UO_2^{2+}$ | $1.4 \times 10^7$ | Ref. 110 |
| Th(IV) | $0.9 \times 10^9$ | Ref. 110 |
| $Fe^{2+}$ ($+ HO_2$) | 240 | Ref. 129 |
| $Cr^{3+}$ | $3 \times 10^7$ | Ref. 130 |
| $Cu^{2+}$ ($+ HO_2$) | $5 \times 10^7$ | Ref. 131 |
| $Mn^{2+}$ ($+ O_2^-$) | $2 \times 10^4$ | Ref. 132 |

$^a$ papd = 1,5,8,11,15-pentaazapentadecane.

of superoxide and the understanding of many aspects of metal ion activation of dioxygen and hydrogen peroxide. Despite this obvious importance, direct measurements of the affinity of metal ions for these radicals are quite rare; the data have been supplemented by indirect determinations from equilibrium constants for $O_2$ binding. Table 9.11 summarizes the published equilibrium constants.

The stability constants for $O_2^-$ binding by Co(III) complexes are notably large. As Taube and Bakac have noted, they presumably reflect a degree of $Co^{II}\text{-}O_2$ character in the products.[110,127] Bakac has also noted unusually strong $O_2^-$ binding by Cr (III).[127,133]

$HO_2$ and $O_2^-$ react with a large number of metal ions and complexes, but in most cases it is unclear whether the reactions involve substitution. The few examples where metal binding has been demonstrated include the reactions of $HO_2$ with $Fe^{2+}$,[129,134] $O_2^-$ with $Fe^{III}TMPyP(H_2O)_2$,[135] $O_2^-$ with some macrocyclic $Co^{II}$ complexes and [Co (bpy)$_2$]$^{2+}$,[136] $O_2^-$ with $Co^{II}EDTA^{2-}$ and $Co^{II}NTA^-$,[85] and $HO_2$ with $Th^{IV}$ and $U^{VI}$;[38] $Fe^{II}EDTA$ forms a complex with $O_2^-$.[34] $HO_2$ and $O_2^-$ add to Cu(II)–arginine complexes.[137] $HO_2/O_2^-$ adds to [Mn(EDTA)]$^{2-}$ and [Mn(nta)]$^-$,[85] various other Mn(II) complexes,[138,139] a seven-coordinate Mn(II) complex,[140] and Mn-SOD.[141] It is believed that superoxide is reduced by superoxide reductase by binding $O_2^-$ at a Fe (II) site of the enzyme.[142] The rate of binding of $HO_2$ to $Fe^{2+}$ seems limited by the rate of solvent exchange,[129] but in most of the other reactions mentioned the nature of the rate-limiting process has not been defined. In all cases the adduct is unstable, generally leading to the net oxidation or reduction of $HO_2/O_2^-$; many of these reactions are thus components of a SOD catalytic cycle as discussed in Section 9.3.

Nitric oxide adds to many metal centers, and the data are particularly abundant because of the stability and ease of handling of NO solutions. Several review articles have been published on aspects of this subject recently.[143–149] In many cases the reactions are reversible, so binding equilibrium constants and rate constants for

**TABLE 9.12    Addition of NO to Metal Complexes**

| Metal Reactant | $K$ ($M^{-1}$) | $k_{on}$ ($M^{-1}s^{-1}$) | $k_{off}$ ($s^{-1}$) | Source |
|---|---|---|---|---|
| $(P^{8+})Fe^{III}(H_2O)_2$ | 577 | $1.5 \times 10^4$ | 26 | Ref. 150 |
| $(P^{8+})Fe^{III}(OH)(H_2O)$ | 258 | $1.6 \times 10^3$ | 6.2 | Ref. 150 |
| $(TMPyP^{4+})Fe^{III}(H_2O)_2$ | 491 | $2.9 \times 10^4$ | 59 | Ref. 150 |
| $Fe^{2+}$ | 470 | $1.4 \times 10^6$ | $3.2 \times 10^3$ | Ref. 151 |
| $Fe^{II}$citrate | 670 | $4.4 \times 10^5$ | $6.6 \times 10^2$ | Ref. 152 |
| $Fe^{II}(acac)_2$ | 17 | $4.0 \times 10^2$ | 24 | Ref. 152 |
| $Fe^{II}$EDTA | $2.1 \times 10^6$ | $2.4 \times 10^8$ | 91 | Ref. 153 |
| $Fe^{II}$hedtra | $1.1 \times 10^7$ | $6.1 \times 10^7$ | 4.2 | Ref. 153 |
| $Fe^{II}$nta | $1.8 \times 10^6$ | $2.1 \times 10^7$ | 9.3 | Ref. 153 |
| $Fe^{II}$mida | $2.1 \times 10^4$ | $1.9 \times 10^6$ | 57 | Ref. 153 |
| $Fe^{II}(mida)_2$ | $3.0 \times 10^4$ | $1.8 \times 10^6$ | 62 | Ref. 153 |
| $[Fe^{II}(CN)_5(H_2O)]^{3-}$ | $1.6 \times 10^7$ | 250 | $1.6 \times 10^{-5}$ | Ref. 154 |
| $Fe^{II}$TPPS | $2.4 \times 10^{12}$ | $1.5 \times 10^9$ | $6.3 \times 10^{-4}$ | Ref. 155 |
| $Co^{II}$TPPS | $1.3 \times 10^{13}$ | $1.9 \times 10^9$ | $6.3 \times 10^{-4}$ | Ref. 155 |
| $Cbl(II)^a$ | $3.1 \times 10^7$ | $6 \times 10^8$ | 3.4 | Ref. 156 |
| $Cu^{II}(Ppidtc)_2{}^b$ | $1.4 \times 10^3$ | | | Ref. 157 |
| $Cu^{II}(Deadtc)_2{}^c$ | $2 \times 10^{10}$ | | | Ref. 157 |
| $Cu^{II}(Pdidtc)_2(NO)$ | 407 | | | Ref. 157 |

$^a$Cbl(II) = cobalamin(II).
$^b$Ppidtc = (2-piperidinecarboxy)dithiocarbamate.
$^c$Deadtc = (dihydroxyethyl)dithiocarbamate.

binding and release ($k_{on}$ and $k_{off}$) have been measured. A selection of these data is presented in Table 9.12.

The binding of NO to $Fe^{III}$ is dominated by studies of heme proteins and porphyrin complexes and has been studied intensively. These reactions convert high- or intermediate-spin $Fe^{III}$(Por) into diamagnetic $[Fe-NO]^6$, $Fe^{II}$(Por)(NO$^+$), and the considerable electronic rearrangement involved is believed to contribute to the large equilibrium binding constants. The binding rate constants ($k_{on}$) for the diaquo complexes tend to be quite large and are determined by a dissociative substitution mechanism. $Cyt^{III}$ displays considerably lower rate constants, apparently because the axial sites are occupied by relatively inert imidazole and methionine ligands. Subsequent to most of the review articles mentioned above, further studies have confirmed these general features of NO binding at $Fe^{III}$ porphyrins, and they have shown that at higher pH the $Fe^{III}$(Por)(OH) complexes bind and release NO considerably more slowly.[150,158–160] This lowered reactivity occurs despite the increased rate of water exchange that occurs. Evidently, the mechanism becomes associative, and a high barrier for NO substitution arises because of the larger spin changes involved. There is little information on binding of NO to nonheme $Fe^{III}$ complexes. One notable exception is the net reaction of NO with $[Fe(CN)_5(L)]^{2-}$ to produce $[Fe(CN)_5NO]^{2-}$; in this case, the reaction is catalyzed by substitution at $[Fe(CN)_5(H_2O)]^{3-,147,161}$ which is a more labile species.[154]

Studies of NO binding to Fe(II) are more diverse than for Fe(III), although Fe(II) hemes still attract much attention. NO adds to Fe(II) centers to generate $[Fe-NO]^7$

species of high stability that can have doublet, quartet, or spin crossover ground spin states. The stability constants of about 40 nonheme NO–Fe(II) complexes span a range of four orders of magnitude and have been shown to correlate with the energy of the MLCT band of the Fe–NO complex and the rates of autoxidation.[162] Where comparisons can be made, NO substitution at Fe(II) is generally several orders of magnitude faster than that at the corresponding Fe(III) complex, one exception being cytochrome $c$. A fascinating observation is that the rate constant for binding of NO to $Fe^{II}(TPPS)$ and $Fe^{II}(TMPS)$ is about 100-fold greater than that for the corresponding reactions of CO;[155] comparable results are obtained for binding to myoglobin.[163,164] These rate differences between CO and NO binding arise from the associative character of the reactions and have been attributed to differing barriers arising from the spin-forbidden characteristics of the reactions.[165] The substitution reactions at other Fe(II) complexes usually have $I_d$ mechanisms, an exception being the reaction with $[Fe(nta)(H_2O)_2]^-$, which has an $I_a$ mechanism.[153] Acetate enhances the stability constant for the $Fe^{II}(aq)$ complex, and the stability constant for the dinitrosyl complex has been measured.[166] Reduction of NO to $N_2O$ is catalyzed by nitric oxide reductase, which contains dinuclear heme/nonheme Fe active site; the reduction mechanism is believed to entail binding of two NO molecules, one at the $Fe^{II}$-heme and the other at the nonheme $Fe^{II}$ site.[167]

NO substitution at Ru(III) is generally coupled with internal electron transfer to yield $Ru(II)–(NO^+)$. These reactions differ from substitution reactions at Fe(III) in that the Ru(III) reactants are always low spin. One fascinating reaction is that of NO with $[Ru(NH_3)_6]^{3+}$ to produce $[Ru(NH_3)_5NO]^{3+}$; in acidic media, the reaction is pH independent and first order in both reactants.[168–170] An associative displacement of coordinated $NH_3$ by NO is consistent with these observations, and it is in agreement with more general reports of associative substitution at Ru(III).[171] A similar description applies to the reactions of NO with $[Ru(NH_3)_5Cl]^{2+}$ and $[Ru(NH_3)_5(H_2O)]^{3+}$.[169] The reactions of NO with $cis$- and $trans$-$[Ru^{III}(terpy)(NH_3)_2Cl]^{2+}$ are multistep processes, and the first step is the reversible displacement of $Cl^-$ by NO; surprisingly, it is argued that the second step is the conversion of $Ru^{III}$–NO to $Ru^{II}$–$NO^+$.[172] In further steps, the Ru(terpy) complexes release nitrite and ammonia. NO binds rapidly and strongly to various $Ru^{III}$ polyaminocarboxylates such as Ru(EDTA).[173,174]

NO adds reversibly to reduced cobalamin, Cbl(II).[156] It does not react directly with aquacobalamin(III), $(Cbl^{III}(H_2O))$, but it does add to $Cbl^{III}(NO_2^-)$ and $Cbl^{III}(NO)$.[175] Acid hydrolysis of the dinitroso species releases nitrite, and binding of nitrite to $Cbl^{III}(H_2O)$ generates $Cbl^{III}(NO_2^-)$. This sequence thus affords a nitrite-catalyzed mechanism for NO substitution at $Cbl^{III}(H_2O)$. The reaction of NO with $Co^{III}$ porphyrins is quite complex.[176] In the first step, NO displaces an axial water ligand to form a weakly bound mono NO complex; this mono NO complex reacts with a second molecule of NO to form nitrite and a reduced Co–NO complex. This latter process is called reductive nitrosylation. Manganese(II) porphyrins bind NO very rapidly.[177] Stability constants have been measured for the formation of mono and bis NO complexes of $Cu^{II}(dithiocarbamate)_2$.[157]

The inner-sphere reactions of $SO_3^-$ with metal ions are limited to a study of the reaction with $Fe^{2+}$.[178] Unusual concentration dependences in the kinetics of this reaction have been interpreted with a mechanism involving rapid equilibrium binding of $SO_3^-$ to form a Fe(II) radical complex with a stability constant of $278\,M^{-1}$; rate-limiting internal electron transfer ($k = 3 \times 10^4\,s^{-1}$) then leads to $Fe^{III}(SO_3^{2-})$.

The $CO_2^-$ radical usually reacts as a simple outer-sphere reductant, but there are examples where it adds to metal centers, reducing them in an inner-sphere mechanism. Two such examples are the reactions with $[Ni(cyclam)]^{2+}$ and a Co(II) tetraazamacrocycle.[82,125]

## 9.13   NUCLEOPHILE-ASSISTED ELECTRON TRANSFER REACTIONS

A new class of electron transfer reactions has been identified recently. It is characterized by third-order rate laws that are first order in the concentration of various nucleophiles and second order in radical concentration. In these reactions, the nucleophiles act as catalysts. The phenomenon was first reported for the disproportionation of $ClO_2$ and the reaction of $ClO_2$ with $BrO_2$.[23] In a subsequent report, it was shown to occur in the reaction of $ClO_2$ with $NO_2$ also.[179] We use the latter reaction to illustrate the mechanism: the net reaction of $NO_2$ with $ClO_2$ yields $NO_3^-$ and $ClO_2^-$, and its rates are a strong function of the concentrations of various nucleophiles such as $NO_2^-$, $Br^-$, $CO_3^{2-}$, and so on.[179]

The rate law is thus

$$-d[NO_2]/dt = k_{nu}\,[ClO_2]\,[NO_2][Nu] \tag{9.27}$$

and the rate constant varies from $4.4 \times 10^6\,M^{-2}\,s^{-1}$ for $NO_2^-$ to $2.0 \times 10^3\,M^{-2}\,s^{-1}$ when $H_2O$ is the nucleophile. The proposed mechanism is

$$ClO_2 + NO_2 + Nu \rightarrow NuNO_2^+ + ClO_2^-    k_{nu} \tag{9.28}$$

$$NuNO_2^+ + H_2O \rightarrow Nu + HNO_3 + H^+ \quad \text{(fast)} \tag{9.29}$$

Similar mechanisms are assigned to the reaction of $ClO_2$ with $BrO_2$ and the disproportionation of $ClO_2$. The reverse of the rate-limiting step (Equation 9.28) is formally analogous to the reductive cleavage mechanism discussed below in Section 9.15.

## 9.14   OTHER THIRD-ORDER REACTIONS

The reaction of nitric oxide with oxygen ($NO + O_2$) has an unusual third-order rate law in the gas phase and also in aqueous solution:[180,181]

$$-d[NO]/dt = k[NO]^2[O_2] \tag{9.30}$$

In the gas phase the reaction produces $NO_2$ in a single discernible step, while in aqueous solution the reaction is

$$O_2 + 4NO + 2H_2O \rightarrow 4NO_2^- + 4H^+ \tag{9.31}$$

It is believed that the solution-phase reaction has $NO_2$ production as the rate-limiting step, which is followed by association of $NO_2$ with NO; the $N_2O_3$ then hydrolyzes to yield nitrite. The third-order rate constants for the solution- and gas-phase reactions are essentially equivalent after correction for solvation effects, so it is considered that the transition state is quite similar in the two phases.

The reactions of NO with $Cl_2$ and $P(C_6H_5)_3$ in organic solvents show analogous third-order rate laws,[182,183] and likely will have these rate laws in aqueous solution. Similar remarks pertain to the reactions of NO with amines.[184] The aqueous reaction of NO with sulfite to produce $ON(NO)SO_3^{2-}$ is first order in [NO];[185] apparently, the rate-limiting step is production of $ONSO_3^{2-}$, which adds a second molecule of NO rapidly.

## 9.15   REDUCTIVE CLEAVAGE BY RADICALS

Radicals can induce the reductive cleavage of other species. For example, the O–O bond in $H_2O_2$ is cleaved in its reaction with hydrogen atoms:[186]

$$H + H_2O_2 \rightarrow OH + H_2O \tag{9.32}$$

Hydrogen atoms also cleave O–O bonds in their reactions with $S_2O_8^{2-}$, $H_2SO_5$, and $O_3$, and they cleave the N–O bond in $N_2O$.[6] Other reducing radicals can also perform reductive cleavage, including the solvated electron, $CO_2^-$, and $O_2^-$. Thus, the solvated electron cleaves $NH_2Cl$,[102] $H_2O_2$, $N_2O$, $S_2O_3^{2-}$, $S_2O_8^{2-}$, $H_2NOH$, and many other species.[6] $CO_2^-$ cleaves $N_2O$ and $H_2O_2$.[8] $O_2^-$ cleaves the O–Cl bond in HOCl.[187]

The species undergoing reductive cleavage can itself be a radical, as in the reaction of solvated electrons with $Br_2^-$:

$$e^-(aq) + Br_2^- \rightarrow 2Br^- \tag{9.33}$$

The rates for reductive cleavage reactions are often significantly less than diffusion controlled, as might be anticipated for any reaction where a relatively strong bond is cleaved.

## 9.16   LIGAND REACTIONS

We use the term "ligand reactions" to encompass reactions of radicals with the ligands of transition metal complexes. In some cases, the transition metals merely perturb chemistry that is characteristic of the free ligand, and in other cases, the transition metal centers can have a profound effect. Frequently, the transition metal center undergoes a change in oxidation state, and this can occur while the ligand is reacting with the radical or in a stepwise way. Overall, the diversity of ligand reactions is quite large, and only selected examples can be presented here. The selection of reactions is organized according to the radical reactant.

Hydrated electrons react with certain water-soluble metalloporphyrin complexes, reducing the porphyrin ligands to pi-radical species. When the metal centers are Zn(II), Pd(II), Ag(II), Cd(II), Cu(II), Sn(IV), and Pb(II), the radical complexes are produced at diffusion-controlled rates and decay with second-order kinetics.[188] Fe(III) porphyrins, on the other hand, yield Fe(II) porphyrins.[189] Rather different behavior is seen in the reaction of $e^-(aq)$ with $[Ru(bpy)_3]^{3+}$; here, parallel paths generate the well-known luminescent excited-state $[^*Ru(bpy)_3]^{2+}$ and another reduced intermediate, both of which decay to the ground-state $[Ru(bpy)_3]^{2+}$.[190] In a direct demonstration of the "chemical" mechanism of inner-sphere electron transfer, $[Co^{III}(NH_3)_5L]^{2+}$ complexes where L = nitrobenzoate and dinitrobenzoate react with $e^-(aq)$ to form Co(III)–ligand radical intermediates, which then undergo intramolecular electron transfer to yield Co(II) and L.[191]

Coordinated ligands can also undergo one-electron reduction when $CO_2^-$ is the reducing agent. This occurs for the coordinated bpz ligand in $[Ru(bpz)_3]^{2+}$,[192] and also for an extensive series of related complexes.[193] The rates of these reactions depend on driving force in conformity with the Marcus cross relationship.

Hydrogen atoms add to coordinated benzoate ligands such as in $[Co(NH_3)_5(O_2CPhNO_2)]^{2+}$;[191] these reactions are very rapid, having rates approaching the diffusion limit. Hydrogen atoms also add to the bpy ligand of $[Ru(bpy)_3]^{2+}$.[194]

Hydroxyl radicals add to the ligands of many complexes, and they abstract hydrogen atoms from the ligands of many other complexes. The net effect can include either the oxidation or reduction of the metal center. As an example of radical addition, OH adds to the pyridine ligand of $[Co^{III}(NH_3)_5(py)]^{3+}$; the adduct decays by several pathways, one of which is intramolecular electron transfer to form Co(II).[195] Thus, the oxidation of the ligand by OH induces the reduction of the metal center. When OH reacts with $[Ru^{III}(NH_3)_5(isn)]^{3+}$, the ligand is oxidized and Ru(II) is produced at a diffusion-controlled rate, significantly faster than the reaction of OH with the free isn ligand.[70] The difference between the Ru(III) and Co(III) reactions is thought to arise from the differing symmetries of the redox-active d-orbitals, which allows the Ru(III) reaction to occur in concert with radical attack on the ligand pi system. As an example of hydrogen atom abstraction, the reaction of OH with $[Co(en)_3]^{3+}$ leads to the production of Co(II) and the en is oxidized to the imine; it is assumed that OH abstracts a hydrogen atom from a coordinated en and then Co(III) provides the second oxidizing equivalent.[196]

Simple one-electron oxidation of the ligand can be achieved with several other oxidizing radicals. For example, $(SCN)_2^-$, $Br_2^-$, $Cl_2^-$, $N_3$, and $I_2^-$ oxidize the ligand in various metalloporphyrins such as ZnTMpyP.[197]

As discussed in Section 9.3, superoxide can disproportionate by reacting with superoxo complexes. A U(VI) example is as follows:[38]

$$U^{VI}\text{-}HO_2 + HO_2 \rightarrow U(VI) + O_2 + H_2O_2 \qquad (9.34)$$

Analogous chemistry occurs with Th(IV) and certain Mn(II) complexes.[37,38]

Nitric oxide displays a wonderful diversity of ligand reactions. It is important to distinguish two classes of these reactions: reductive nitrosation and reductive nitrosylation. Reductive nitrosation refers to the addition of NO to a bound amide ligand with concomitant reduction of the metal center as in

$$LM^{III}\text{-}(NHR^-) + NO \rightarrow LM^{II}\text{-}(NHR(NO)) \qquad (9.35)$$

Reductive nitrosylation, on the other hand, can refer to the addition of NO to a metal center $M_{ox}$ with formal reduction of the metal center to yield $M_{red}(NO^+)$, but in the context of ligand reactions reductive nitrosylation refers to the net reactions of NO with metal-bound NO and the ensuing events. Reductive nitrosation of coordinated amines to form nitrosamines occurs through the conjugate base of the amine, and this process has been reported for reactions of NO with $[Ni(tacn)_2]^{3+}$,[198] with methylamine coordinated to a macrocyclic Ni(III) complex,[199] with triglycyl complexes of Fe(III), Ni(III), and Cu(III),[200] and with Cu(II) macrocyclic complexes.[201] Reductive nitrosation of $[Ru(NH_3)_6]^{3+}$ produces $[Ru(NH_3)_5N_2]^{2+}$ with base-catalyzed kinetics; the coordinated $N_2$ is produced by hydrolysis after the nitrosation step.[170]

Reductive nitrosylation occurs for several $Fe^{III}$ porphyrins,[145,147,159] where the net reaction is

$$LFe^{III}(H_2O) + 2NO \rightarrow [LFe(NO)]^- + NO_2^- + 2H^+ \qquad (9.36)$$

These reactions are catalyzed by nucleophiles, including water, hydroxide, and nitrite, and are assigned the following mechanism:

$$LFe^{III}(H_2O) + NO \rightleftharpoons [LFe(NO)] + H_2O \qquad (9.37)$$

$$[LFe(NO)] + Nu \rightleftharpoons [LFe(NO-Nu)] \qquad (9.38)$$

$$[LFe(NO-Nu)] + H_2O \rightarrow [LFe^{II}(H_2O)]^- + NO-Nu^+ \qquad (9.39)$$

$$[LFe^{II}(H_2O)]^- + NO \rightarrow [LFe(NO)]^- + H_2O \qquad (9.40)$$

$$NO-Nu^+ + H_2O \rightarrow NO_2^- + 2H^+ + Nu \qquad (9.41)$$

The production of nitrite in reaction 9.41 is important, because nitrite can also serve as the nucleophile in reaction 9.38. When azide is the nucleophile, the products are $N_2$ and $N_2O$ rather than nitrite, a result of the differing hydrolysis pathway for

$N_3NO$. Fe(II) and other nitrogenous products can be obtained when the nucleophile is $NH_3$, $H_2NOH$, or $N_2H_4$.[202] The mechanism depicted above conveys that the reactions are not strictly described as ligand reactions, but they are included in this section for convenience. Co(III) porphyrins react with NO in excellent analogy with the stoichiometry of Equation 9.36, but the rates are not catalyzed by nucleophiles; the proposed mechanism is quite different from that for the Fe(III) reactions and involves substitution at both axial sites.[147,176]

In another class of reactions, NO is oxidized to $NO_3^-$ or its equivalent by metal–$(O_2/O_2^-)$ complexes. In these reactions, NO first adds to the $O_2^-$ moiety to form a peroxynitrito intermediate that can decompose in various ways.[203] The analogy with the direct reaction of NO with free superoxide to form peroxynitrite and its isomerization to nitrate is clear. An important example is the oxidation of NO to $NO_3^-$ by oxyhemoglobins, where the $Fe–O_2$ reactants have considerable $Fe^{III}–(O_2^-)$ character.[204] In other examples, NO reacts with superoxo complexes of Rh(III) and Cr(III) to produce peroxonitrito intermediates that decompose in complicated ways.[205,206]

NO adds to oxo complexes, including $CrO^{2+}$ and $MbFe^{IV}O$, to form nitrito complexes.[203] Similar reactions of the ferryl(V) moiety in various metalloproteins are likely to become an important physiological consideration.

$NO_2$ adds to oxo complexes such as $CrO^{2+}$,[205] $[(TMPS)Fe^{IV}O]$,[207] and $MbFe^{IV}O$ to form nitrato complexes.[203] It adds reversibly to $CrOO^{2+}$ to form a rather unreactive peroxynitrato complex.[208] These reactions lead to the reduction of the metal center and thus can be considered as good examples of the inner-sphere electron transfer mechanism.

## ACKNOWLEDGMENT

The NSF is thanked for support through grant #CHE-0509889. during the preparation of this chapter.

## REFERENCES

1. Stanbury, D. M. *Adv. Inorg. Chem.* **1989**, *33*, 69–138.

2. Alfassi, Z. B., Ed. *Peroxyl Radicals*; Wiley: New York, 1997; pp 1–535.

3. Alfassi, Z. B., Ed. *N-Centered Radicals*; Wiley: New York, 1998; pp 1–715.

4. Alfassi, Z. B., Ed. *General Aspects of the Chemistry of Radicals*; Wiley: New York, 1999; pp 1–563.

5. Alfassi, Z. B., Ed. *S-Centered Radicals*; Wiley: New York, 1999; pp 1–371.

6. Buxton, G. V.; Greenstock, C. L.; Helman, W. P.; Ross, A. B. *J. Phys. Chem. Ref. Data* **1988**, *17*, 513–886.

7. Alam, M. S.; Janata, E. *Chem. Phys. Lett.* **2006**, *417*, 363–366.

8. Neta, P.; Huie, R. E.; Ross, A. B. *J. Phys. Chem. Ref. Data* **1988**, *17*, 1027–1284.

9. Buxton, G. V.; Stuart, C. R. *J. Chem. Soc., Faraday Trans.* **1996**, *92*, 1519–1525.

10. Shastri, L. V.; Huie, R. E.; Neta, P. *J. Phys. Chem.* **1990**, *94*, 1895–1899.

11. Stanbury, D. M. In *General Aspects of the Chemistry of Radicals*; Alfassi, Z. B., Ed.; Wiley: New York, 1999; pp 347–384.

12. Marin, T. W.; Takahashi, K.; Jonah, C. D.; Chemerisov, S. D.; Bartels, D. M. *J. Phys. Chem. A* **2007**, *111*, 11540–11551.

13. Wagner, I.; Karthauser, J.; Strehlow, H. *Ber. Bunsen-Ges. Phys. Chem.* **1986**, *90*, 861–867.

14. Jia, Z. J.; Margerum, D. W.; Francisco, J. S. *Inorg. Chem.* **2000**, *39*, 2614–2620.

15. Fischer, M.; Warneck, P. *J. Phys. Chem.* **1996**, *100*, 15111–15117.

16. Waygood, S. J.; McElroy, W. J. *J. Chem. Soc., Faraday Trans.* **1992**, *88*, 1525–1530.

17. Flyunt, R.; Schuchmann, M. N.; von Sonntag, C. *Chem. Eur. J.* **2001**, *7*, 796–799.

18. Glass, R. W.; Martin, T. W. *J. Am. Chem. Soc.* **1970**, *92*, 5084–5093.

19. Bielski, B. H. J.; Cabelli, D. E.; Arudi, R. L. *J. Phys. Chem. Ref. Data* **1985**, *14*, 1041–1100.

20. Yermakov, A. N.; Zhitomirsky, B. M.; Poskrebyshev, G. A.; Sozurakov, D. M. *J. Phys. Chem.* **1993**, *97*, 10712–10714.

21. Poskrebyshev, G. A.; Shafirovich, V.; Lymar, S. V. *J. Phys. Chem. A* **2008**, *112*, 8295–8302.

22. Alfassi, Z. B.; Dhanasekaran, T.; Huie, R. E.; Neta, P. *Radiat. Phys. Chem.* **1999**, *56*, 475–482.

23. Wang, L.; Nicoson, J. S.; Hartz, K. E. H.; Francisco, J. S.; Margerum, D. W. *Inorg. Chem.* **2002**, *41*, 108–113.

24. Wang, L.; Margerum, D. W. *Inorg. Chem.* **2002**, *41*, 6099–6105.

25. Odeh, I. N.; Francisco, J. S.; Margerum, D. W. *Inorg. Chem.* **2002**, *41*, 6500–6506.

26. Roncaroli, F.; van Eldik, R.; Olabe, J. A. *Inorg. Chem.* **2005**, *44*, 2781–2790.

27. Foti, M. C.; Sortino, S.; Ingold, K. U. *Chem. Eur. J.* **2005**, *11*, 1942–1948.

28. Cabelli, D. E.; Riley, D.; Rodriguez, J. A.; Valentine, J. S.; Zhu, H. In *Biomimetic Oxidations Catalyzed by Metal Complexes*; Meunier, B., Ed.; Imperial College Press: London, 2000; pp 461–508.

29. Rush, J. D.; Bielski, B. H. J. *J. Phys. Chem.* **1985**, *89*, 5062–5066.

30. Aston, K.; Rath, N.; Naik, A.; Slomczynska, U.; Schall, O. F.; Riley, D. P. *Inorg. Chem.* **2001**, *40*, 1779–1789.

31. Choudhury, S. B.; Lee, J. -W.; Davidson, G.; Yim, Y. -I.; Bose, K.; Sharma, M. L.; Kang, S. -O.; Cabelli, D. E.; Maroney, M. J. *Biochemistry* **1999**, *38*, 3744–3752.

32. Rabani, J.; Klug-Roth, D.; Lilie, J. *J. Phys. Chem.* **1973**, *77*, 1169–1175.

33. Klug-Roth, D.; Fridovich, I.; Rabani, J. *J. Am. Chem. Soc.* **1973**, *95*, 2786–2790.

34. Bull, C.; McClune, G. J.; Fee, J. A. *J. Am. Chem. Soc.* **1983**, *105*, 5290–5300.

35. Bielski, B. H. J.; Cabelli, D. E. In *Active Oxygen in Chemistry*; Foote, C. S.; Valentine, J. S.; Greenberg, A.; Liebman, J. F., Eds.; Blakie Academic and Professional: New York, 1995; 66–104.

36. Shearer, J.; Long, L. M. *Inorg. Chem.* **2006**, *45*, 2358–2360.

37. Cabelli, D. E.; Bielski, B. H. J. *J. Phys. Chem.* **1984**, *88*, 6291–6294.

38. Meisel, D.; Ilan, Y. A.; Czapski, G. *J. Phys. Chem.* **1974**, *78*, 2330–2334.

39. Buxton, G. V.; Bydder, M.; Salmon, G. A.; Williams, J. E. *Phys. Chem. Chem. Phys.* **2000**, *2*, 237–245.

40. Mulazzani, Q. G.; Buxton, G. V. *Chem. Phys. Lett.* **2006**, *421*, 261–265.

41. Behar, D.; Bevan, P. L. T.; Scholes, G. *J. Phys. Chem.* **1972**, *76*, 1537–1542.

42. Czapski, G.; Lymar, S. V.; Schwarz, H. A. *J. Phys. Chem. A* **1999**, *103*, 3447–3450.

43. Lymar, S. V.; Schwarz, H. A.; Czapski, G. *Radiat. Phys. Chem.* **2000**, *59*, 387–392.

44. Lymar, S. V.; Schwarz, H. A.; Czapski, G. *J. Phys. Chem. A* **2002**, *106*, 7245–7259.

45. Han, P.; Bartels, D. M. *J. Phys. Chem.* **1990**, *94*, 7294–7299.

46. Yu, X.-Y. *J. Phys. Chem. Ref. Data* **2004**, *33*, 747–763.

47. Poskrebyshev, G. A.; Neta, P.; Huie, R. E. *J. Geophys. Res.* **2001**, *106*, 4995–5004.

48. Tang, Y.; Thorn, R. P.; Maudlin, R. L.; Wine, P. H. *J. Photochem. Photobiol. A* **1988**, *44*, 243–258.

49. Jiang, P.-Y.; Katsumura, Y.; Domae, M.; Ishikawa, K.; Nagaishi, R.; Ishigure, K.; Yoshida, Y. *J. Chem. Soc., Faraday Trans.* **1992**, *88*, 3319–3322.

50. Mamou, A.; Rabani, J.; Behar, D. *J. Phys. Chem.* **1977**, *81*, 1447–1448.

51. Herrmann, H. *Chem. Rev.* **2003**, *103*, 4691–4716.

52. Bartels, D. M.; Mezyk, S. P. *J. Phys. Chem.* **1993**, *97*, 4101–4105.

53. Kläning, U. K.; Sehested, K.; Holcman, J. *J. Phys. Chem.* **1985**, *89*, 760–763.

54. Jayson, G. G.; Parsons, B. J.; Swallow, A. J. *J. Chem. Soc., Faraday Trans. I* **1973**, *69*, 1597–1607.

55. Yu, X.-Y.; Barker, J. R. *J. Phys. Chem. A* **2003**, *107*, 1313–1324.

56. Liu, Y.; Pimentel, A. S.; Antoku, Y.; Giles, B. J.; Barker, J. R. *J. Phys. Chem. A* **2002**, *106*, 11075–11082.

57. Liu, Y.; Sheaffer, R. L.; Barker, J. R. *J. Phys. Chem. A* **2003**, *107*, 10296–10302.

58. Laszlo, B.; Alfassi, Z. B.; Neta, P.; Huie, R. E. *J. Phys. Chem. A* **1998**, *102*, 8498–8504.

59. Neta, P.; Huie, R. E.; Ross, A. B. *J. Phys. Chem. Ref. Data* **1990**, *19*, 413–505.

60. Baxendale, J. H.; Bevan, P. L. T.; Stott, D. A. *Trans. Faraday Soc.* **1968**, *64*, 2389–2397.

61. Milosavljevic, B. H.; LaVerne, J. A. *J. Phys. Chem. A.* **2005**, *109*, 165–168.

62. Buxton, G. V.; Salmon, G. A. *Prog. React. Kinet. Mech.* **2003**, *28*, 257–297.

63. Goldstein, S.; Lind, J.; Merényi, G. *Chem. Rev.* **2005**, *105*, 2457–2470.

64. Elliot, A. J.; Buxton, G. V. *J. Chem. Soc., Faraday Trans.* **1992**, *88*, 2465–2470.

65. Czapski, G.; Holcman, J.; Bielski, B. H. J. *J. Am. Chem. Soc.* **1994**, *116*, 11465–11469.

66. Ram, M. S.; Stanbury, D. M. *Inorg. Chem.* **1985**, *24*, 4233–4234.

67. Schwarz, H. A.; Dodson, R. W. *J. Phys. Chem.* **1989**, *93*, 409–414.

68. Sarala, R.; Islam, M. S.; Rabin, S. B.; Stanbury, D. M. *Inorg. Chem.* **1990**, *29*, 1133–1142.

69. Ram, M. S.; Stanbury, D. M. *Inorg. Chem.* **1985**, *24*, 2954–2962.

70. Stanbury, D. M.; Mulac, W. A.; Sullivan, J. C.; Taube, H. *Inorg. Chem.* **1980**, *19*, 3735–3740.

71. Stanbury, D. M. In *Electron Transfer Reactions*; Isied, S., Ed.; American Chemical Society: Washington, DC, 1997; pp 165–182.

72. Lind, J.; Shen, X.; Merényi, G.; Jonsson, B. Ö. *J. Am. Chem. Soc.* **1989**, *111*, 7654–7655.

73. Stanbury, D. M.; deMaine, M. M.; Goodloe, G. *J. Am. Chem. Soc.* **1989**, *111*, 5496–5498.

74. Weinstock, I. A. *Inorg. Chem.* **2008**, *47*, 404–406.

75. Stanbury, D. M.; Lednicky, L. A. *J. Am. Chem. Soc.* **1984**, *106*, 2847–2853.

76. Roth, J. P.; Wincek, R.; Nodet, G.; Edmonson, D. E.; McIntire, W. S.; Klinman, J. P. *J. Am. Chem. Soc.* **2004**, *126*, 15120–15131.

77. Nord, G.; Pedersen, B.; Floryan-Løvborg, E.; Pagsberg, P. *Inorg. Chem.* **1982**, *21*, 2327–2330.

78. Alfassi, Z. B.; Harriman, A.; Huie, R. E.; Mosseri, S.; Neta, P. *J. Phys. Chem.* **1987**, *91*, 2120–2122.

79. DeFelippis, M. R.; Faraggi, M.; Klapper, M. H. *J. Phys. Chem.* **1990**, *94*, 2420–2424.

80. Stanbury, D. M. *Inorg. Chem.* **1984**, *23*, 2914–2916.

81. Meyerstein, D. *Acc. Chem. Res.* **1978**, *11*, 43–48.

82. Kelly, C. A.; Mulazzani, Q. G.; Venturi, M.; Blinn, E. L.; Rodgers, M. A. J. *J. Am. Chem. Soc.* **1995**, *117*, 4911–4919.

83. Gogolev, A. V.; Shilov, V. P.; Fedoseev, A. M.; Pikaev, A. K. *Radiat. Phys. Chem.* **1991**, *37*, 531–535.

84. Wang, L.; Odeh, I. N.; Margerum, D. W. *Inorg. Chem.* **2004**, *43*, 7545–7551.

85. Lati, J.; Meyerstein, D. *J. Chem. Soc., Dalton Trans.* **1978**, 1105–1118.

86. Maruthamuthu, P.; Patterson, L. K.; Ferraudi, G. *Inorg. Chem.* **1978**, *17*, 3157–3163.

87. Golub, D.; Cohen, H.; Meyerstein, D. *J. Chem. Soc., Dalton Trans.* **1985**, 641–644.

88. Laurence, G. S.; Thornton, A. T. *J. Chem. Soc., Dalton Trans.* **1973**, 1637–1644.

89. Thornton, A. T.; Laurence, G. S. *J. Chem. Soc., Dalton Trans.* **1973**, 1632–1636.

90. Thornton, A. T.; Laurence, G. S. *J. Chem. Soc., Dalton Trans.* **1973**, 804–813.

91. Laurence, G. S.; Thornton, A. T. *J. Chem. Soc., Dalton Trans.* **1974**, 1142–1148.

92. Kirschenbaum, L. J.; Meyerstein, D. *Inorg. Chem.* **1980**, *19*, 1371–1379.

93. Stanbury, D. M. *Prog. Inorg. Chem.* **1998**, *47*, 511–561.

94. Wagman, D. D.; Evans, W. H.; Parker, V. B.; Schumm, R. H.; Halow, I.; Bailey, S. M.; Churney, K. L.; Nuttall, R. L. *J. Phys. Chem. Ref. Data* **1982**, *11*, (Suppl. 2).

95. Kerr, J. A.; Stocker, D. W. In *CRC Handbook of Chemistry and Physics*; Lide, D. R., Ed.; CRC Press: Boca Raton, FL, 2002; pp 9-52–9-75.

96. Shafirovich, V.; Lymar, S. V. *Proc. Natl. Acad. Sci. USA* **2002**, *99*, 7340–7345.

97. Neta, P.; Maruthamuthu, P.; Carton, P. M.; Fessendon, R. W. *J. Phys. Chem.* **1978**, *82*, 1875–1878.

98. Lossack, A. M.; Roduner, E.; Bartels, D. M. *J. Phys. Chem. A* **1998**, *102*, 7462–7469.

99. Berdnikov, V. M. *Russ. J. Phys. Chem.* **1973**, *47*, 1547–1552.

100. Zhang, H.; Bartlett, R. J. *Environ. Sci. Technol.* **1999**, *33*, 588–594.

101. Stuart, C. R.; Ouellette, D. C.; Elliot, A. J. *AECL Rep.* **2002**, 12107.

102. Poskrebyshev, G. A.; Huie, R. E.; Neta, P. *J. Phys. Chem. A* **2003**, *107*, 7423–7428.

103. Gilbert, B. C.; Smith, J. R. L.; Taylor, P.; Ward, S.; Whitwood, A. C. *J. Chem. Soc., Perkin Trans.* **1999**, *2*, 1631–1637.

104. Khursan, S. L.; Semes'kov, D. G.; Teregulova, A. N.; Safiullin, R. L. *Kinet. Catal.* **2008**, *49*, 202–211.

105. Maruthamuthu, P.; Neta, P. *J. Phys. Chem.* **1978**, *82*, 710–713.

106. Johnson, H. D.; Cooper, W. J.; Mezyk, S. P.; Bartels, D. M. *Radiat. Phys. Chem.* **2002**, *65*, 317–326.

107. Mezyk, S. P.; Tateishi, M.; MacFarlane, R.; Bartels, D. M. *J. Chem. Soc., Faraday Trans.* **1996**, *92*, 2541–2545.

108. Kelly, C. A.; Rosseinsky, D. R. *Phys. Chem. Chem. Phys.* **2001**, *3*, 2086–2090.

109. Hickel, B.; Sehested, K. *J. Phys. Chem.* **1991**, *95*, 744–747.

110. Taube, H. *Prog. Inorg. Chem.* **1986**, *34*, 607–625.

111. Bernhard, P.; Anson, F. C. *Inorg. Chem.* **1988**, *27*, 4574–4577.

112. Shafirovich, V.; Lymar, S. V. *J. Am. Chem. Soc.* **2003**, *125*, 6547–6552.

113. Bonner, F. T.; Wang, N.-Y. *Inorg. Chem.* **1986**, *25*, 1858–1862.

114. Akhtar, M. J.; Bonner, F. T.; Hughes, M. N. *Inorg. Chem.* **1985**, *24*, 1934–1935.

115. Merényi, G.; Lind, J. *J. Am. Chem. Soc.* **1994**, *116*, 7872–7876.

116. Engman, L.; Persson, J.; Merényi, G.; Lind, J. *Organometallics* **1995**, *14*, 3641–3648.

117. Stefanic, I.; Asmus, K.-D.; Bonifacic, M. *J. Phys. Org. Chem.* **2005**, *18*, 408–416.

118. Lilie, J.; Hanrahan, R. J.; Henglein, A. *Radiat. Phys. Chem.* **1978**, *11*, 225–227.

119. Mezyk, S. P.; Bartels, D. M. *J. Phys. Chem. A* **1997**, *101*, 6233–6237.

120. Cohen, H.; Meyerstein, D. *J. Chem. Soc., Dalton Trans.* **1974**, 2559–2564.

121. Ryan, D. A.; Espenson, J. H. *Inorg. Chem.* **1981**, *20*, 4401–4404.

122. Mulac, W. A.; Meyerstein, D. *Inorg. Chem.* **1982**, *21*, 1782–1784.

123. Jayson, G. G.; Keene, J. P.; Stirling, D. A.; Swallow, A. J. *Trans. Faraday Soc.* **1969**, *65*, 2453–2464.

124. Micic, O. I.; Nenadovic, M. T. *J. Chem. Soc., Dalton Trans.* **1979**, 2011–2014.

125. Creutz, C.; Schwarz, H. A.; Wishart, J. F.; Fujita, E.; Sutin, N. *J. Am. Chem. Soc.* **1991**, *113*, 3361–3371.

126. Maeder, M.; Mäcke, H. R. *Inorg. Chem.* **1994**, *33*, 3135–3140.

127. Bakac, A. *Prog. Inorg. Chem.* **1995**, *43*, 267–351.

128. Wong, C. -L.; Switzer, J. A.; Balakrishnan, K. P.; Endicott, J. F. *J. Am. Chem. Soc.* **1980**, *102*, 5511–5518.

129. Mansano-Weiss, C.; Cohen, H.; Meyerstein, D. *J. Inorg. Biochem.* **2002**, *91*, 199–204.

130. Espenson, J. H.; Bakac, A.; Janni, J. *J. Am. Chem. Soc.* **1994**, *116*, 3436–3438.

131. Meisel, D.; Levanon, H.; Czapski, G. *J. Phys. Chem.* **1974**, *78*, 778–782.

132. Jacobsen, F.; Holcman, J.; Sehested, K. *J. Phys. Chem. A* **1997**, *101*, 1324–1328.

133. Bakac, A.; Scott, S. L.; Espenson, J. H.; Rodgers, K. R. *J. Am. Chem. Soc.* **1995**, *117*, 6483–6488.

134. Jayson, G. G.; Parsons, B. J.; Swallow, A. J. *J. Chem. Soc., Faraday Trans. I* **1973**, *69*, 236–242.

135. Solomon, D.; Peretz, P.; Faraggi, M. *J. Phys. Chem.* **1982**, *86*, 1843–1849.

136. Simic, M. G.; Hoffman, M. Z. *J. Am. Chem. Soc.* **1977**, *99*, 2370–2371.

137. Cabelli, D. E.; Bielski, B. H. J.; Holcman, J. *J. Am. Chem. Soc.* **1987**, *109*, 3665–3669.

138. Cabelli, D. E.; Bielski, B. H. J. *J. Phys. Chem.* **1984**, *88*, 3111–3115.

139. Barnese, K.; Gralla, E. B.; Cabelli, D. E.; Valentine, J. S. *J. Am. Chem. Soc.* **2008**, *130*, 4604–4606.

140. Deroche, A.; Morgenstern-Badarau, I.; Cesario, M.; Guilhem, J.; Keita, B.; Nadjo, L.; Houée-Levin, C. *J. Am. Chem. Soc.* **1996**, *118*, 4567–4573.

141. Bull, C.; Niederhoffer, E. C.; Yoshida, T.; Fee, J. A. *J. Am. Chem. Soc.* **1991**, *113*, 4069–4076.

142. Niviére, V.; Fontecave, M. *J. Biol. Inorg. Chem.* **2004**, *9*, 119–123.

143. Ford, P. C. *Pure Appl. Chem.* **2004**, *76*, 335–350.

144. Ford, P. C.; Laverman, L. E.; Lorkovic, I. M. *Adv. Inorg. Chem.* **2003**, *54*, 203–257.

145. Ford, P. C.; Fernandez, B. O.; Lim, M. D. *Chem. Rev.* **2005**, *105*, 2439–2455.

146. Ford, P. C.; Lorkovic, I. M. *Chem. Rev.* **2002**, *102*, 993–1018.

147. Franke, A.; Roncaroli, F.; van Eldik, R. *Eur. J. Inorg. Chem.* **2007**, 773–798.

148. Hoshino, M.; Laverman, L. E.; Ford, P. C. *Coord. Chem. Rev.* **1999**, *187*, 75–102.

149. Wolak, M.; van Eldik, R. *Coord. Chem. Rev.* **2002**, *230*, 263–282.

150. Jee, J.-E.; Wolak, M.; Balbinot, D.; Jux, N.; Zahl, A.; van Eldik, R. *Inorg. Chem.* **2006**, *45*, 1326–1337.

151. Wanat, A.; Schneppensieper, T.; Stochel, G.; van Eldik, R.; Bill, E.; Wieghardt, K. *Inorg. Chem.* **2002**, *41*, 4–10.

152. Littlejohn, D.; Chang, S. G. *J. Phys. Chem.* **1982**, *86*, 537–540.

153. Schneppensieper, T.; Wanat, A.; Stochel, G.; van Eldik, R. *Inorg. Chem.* **2002**, *41*, 2565–2573.

154. Roncaroli, F.; Olabe, J. A.; van Eldik, R. *Inorg. Chem.* **2003**, *42*, 4179–4189.

155. Laverman, L. E.; Ford, P. C. *J. Am. Chem. Soc.* **2001**, *123*, 11614–11622.

156. Wolak, M.; Zahl, A.; Schneppensieper, T.; Stochel, G.; van Eldik, R. *J. Am. Chem. Soc.* **2001**, *123*, 9780–9791.

157. Cachapa, A.; Mederos, A.; Gili, P.; Hernandez-Molina, R.; Dominguez, S.; Chinea, E.; Rodriguiz, M. L.; Feliz, M.; Llusar, R.; Brito, F.; Ruiz de Galaretta, C. M.; Tarbraue, C.; Gallardo, G. *Polyhedron* **2006**, *25*, 3366–3378.

158. Imai, H.; Yamashita, Y.; Nakagawa, S.; Munakata, H.; Uemori, Y. *Inorg. Chim. Acta* **2004**, *357*, 2503–2509.

159. Jee, J.-E.; Eigler, S.; Jux, N.; Zahl, A.; van Eldik, R. *Inorg. Chem.* **2007**, *46*, 3336–3352.

160. Wolak, M.; van Eldik, R. *J. Am. Chem. Soc.* **2005**, *127*, 13312–13315.

161. Roncaroli, F.; Olabe, J. A.; van Eldik, R. *Inorg. Chem.* **2002**, *41*, 5417–5425.

162. Schneppensieper, T.; Finkler, S.; Czap, A.; van Eldik, R.; Heus, M.; Nieuwenhuizen, P.; Wreesmann, C.; Abma, W. *Eur. J. Inorg. Chem.* **2001**, 491–501.

163. Quillin, M. L.; Li, T.; Olson, J. S.; Phillips, G. N.; Dou, Y.; Ikeda-Saito, M.; Regan, R.; Carlson, M.; Gibson, Q. H.; Li, H.-R.; Elber, R. *J. Mol. Biol.* **1995**, *245*, 416–436.

164. Olson, J. S.; Phillips, G. N. *J. Biol. Chem.* **1996**, *271*, 17593–17596.

165. Strickland, N.; Harvey, J. N. *J. Phys. Chem. B* **2007**, *111*, 841–852.

166. Pearsall, K. A.; Bonner, F. T. *Inorg. Chem.* **1982**, *21*, 1978–1985.

167. (a) Collman, J. P.; Yang, Y.; Dey, A.; Decréau, R. A.; Ghosh, S.; Ohta, T.; Solomon, E. I. *Proc. Natl. Acad. Sci. USA* **2008**, *105*, 15660–15665. (b) There are different types of nitric oxide reductases with a great diversity of active sites: see Daiber, A.; Shoun, H.; Ullrich, V. *J. Inorg. Biochem.* **2005**, *99*, 185–193. and Zumft, W. G. *J. Inorg. Biochem.* **2005**, *99*, 194–215.

168. Armor, J. N.; Scheidegger, H. A.; Taube, H. *J. Am. Chem. Soc.* **1968**, *90*, 5928–5929.

169. Czap, A.; van Eldik, R. *Dalton Trans.* **2003**, 665–671.

170. Pell, S. D.; Armor, J. N. *J. Am. Chem. Soc.* **1973**, *95*, 7625–7633.

171. Fairhurst, M. T.; Swaddle, T. W. *Inorg. Chem.* **1979**, *18*, 3241–3244.

172. Czap, A.; Heinemann, F. W.; van Eldik, R. *Inorg. Chem.* **2004**, *43*, 7832–7843.

173. Davies, N. A.; Wilson, M. T.; Slade, E.; Fricker, S. P.; Murrer, B. A.; Powell, N. A.; Henderson, G. R. *Chem. Commun.* **1997**, 47–48.

174. Storr, T.; Cameron, B. R.; Gossage, R. A.; Yee, H.; Skerlj, R. T.; Darkes, M. C.; Fricker, S. P.; Bridger, G. J.; Davies, N. A.; Wilson, N. T.; Maresca, K. P.; Zubieta, J. *Eur. J. Inorg. Chem.* **2005**, 2685–2697.

175. Roncaroli, F.; Shubina, T. E.; Clark, T.; van Eldik, R. *Inorg. Chem.* **2006**, *45*, 7869–7876.

176. Roncaroli, F.; van Eldik, R. *J. Am. Chem. Soc.* **2006**, *128*, 8024–8153.

177. Spasojevic, I.; Batinic-Haberle, I.; Fridovich, I. *Nitric Oxide: Biol Chem.* **2000**, *4*, 526–533.

178. Buxton, G. V.; Barlow, S.; McGowan, S.; Salmon, G. A.; Williams, J. E. *Phys. Chem. Chem. Phys.* **1999**, *1*, 3111–3115.

179. Becker, R. H.; Nicoson, J. S.; Margerum, D. W. *Inorg. Chem.* **2003**, *42*, 7938–7944.

180. Awad, H. H.; Stanbury, D. M. *Int. J. Chem. Kinet.* **1993**, *25*, 375–381.

181. Ford, P. C.; Wink, D. A.; Stanbury, D. M. *FEBS Lett.* **1993**, *326*, 1–3.

182. Lim, M. D.; Lorkovic, I. M.; Ford, P. C. *Inorg. Chem.* **2002**, *41*, 1026–1028.

183. Nottingham, W. C.; Sutter, J. R. *Int. J. Chem. Kinet.* **1986**, *18*, 1289–1302.

184. Bohle, D. S.; Smith, K. N. *Inorg. Chem.* **2008**, *47*, 3925–3927.

185. Littlejohn, D.; Hu, K. Y.; Chang, S. G. *Inorg. Chem.* **1986**, *25*, 3131–3135.

186. Mezyk, S. P.; Bartels, D. M. *J. Chem. Soc., Faraday Trans.* **1995**, *91*, 3127–3132.

187. Candeias, L. P.; Patel, K. B.; Stratford, M. R. L.; Wardman, P. *FEBS Lett.* **1993**, *333*, 151–153.

188. Harriman, A.; Richoux, M. C.; Neta, P. *J. Phys. Chem.* **1983**, *87*, 4957–4965.

189. Wilkins, P. C.; Wilkins, R. G. *Inorg. Chem.* **1986**, *25*, 1908–1910.

190. Jonah, C. D.; Matheson, M. S.; Meisel, D. *J. Am. Chem. Soc.* **1978**, *100*, 1449–1465.

191. Simic, M. G.; Hoffman, M. Z.; Brezniak, N. V. *J. Am. Chem. Soc.* **1977**, *99*, 2166–2172.

192. Venturi, M.; Mulazzani, Q. G.; Ciano, M.; Hoffman, M. Z. *Inorg. Chem.* **1986**, *25*, 4493–4498.

193. D'Angelantonio, M.; Mulazzani, Q. G.; Venturi, M.; Ciano, M.; Hoffman, M. Z. *J. Phys. Chem.* **1991**, *95*, 5121–5129.

194. Baxendale, J. H.; Fiti, M. *J. Chem. Soc., Dalton Trans.* **1972**, 1995–1998.

195. Hoffman, M. Z.; Kimmel, D. W.; Simic, M. G. *Inorg. Chem.* **1979**, *18*, 2479–2485.

196. Shinohara, N.; Lilie, J. *Inorg. Chem.* **1979**, *18*, 434–438.

197. Neta, P.; Harriman, A. *J. Chem. Soc., Faraday Trans.* **1985**, *81*, 123–138.

198. deMaine, M. M.; Stanbury, D. M. *Inorg. Chem.* **1991**, *30*, 2104–2109.

199. Shamir, D.; Zilberman, I.; Maimon, E.; Cohen, H.; Meyerstein, D. *Inorg. Chem. Commun.* **2007**, *10*, 57–60.

200. Shamir, D.; Zilbermann, I.; Maimon, E.; Gellerman, G.; Cohen, H.; Meyerstein, D. *Eur. J. Inorg. Chem.* **2007**, 5029–5031.

201. Khin, C.; Lim, M. D.; Tsuge, K.; Iretskii, A.; Wu, G.; Ford, P. C. *Inorg. Chem.* **2007**, *46*, 9323–9331.

202. Olabe, J. A. *Dalton Trans.* **2008**, 3633–3648.

203. (a) Bakac, A. *Adv. Inorg. Chem.* **2004**, *55*, 1–59. (b) Herold, S.; Koppenol, W. H. *Coord. Chem. Rev.* **2005**, *249*, 499–506.

204. Gardner, P. R.; Gardner, A. M.; Brashear, W. T.; Suzuki, T.; Hvitved, A. N.; Setchell, K. D. R.; Olson, J. S. *J. Inorg. Biochem.* **2006**, *100*, 542–550.

205. Nemes, A.; Pestovsky, O.; Bakac, A. *J. Am. Chem. Soc.* **2002**, *124*, 421–427.

206. Pestovsky, O.; Bakac, A. *J. Am. Chem. Soc.* **2002**, *124*, 1698–1703.

207. Shimanovich, R.; Groves, J. T. *Arch. Biochem. Biophys.* **2001**, *387*, 307–317.

208. Pestovsky, O.; Bakac, A. *Inorg. Chem.* **2003**, *42*, 1744–1750.

# 10 Organometallic Radicals: Thermodynamics, Kinetics, and Reaction Mechanisms

TAMÁS KÉGL, GEORGE C. FORTMAN, MANUEL TEMPRADO, and CARL D. HOFF

## 10.1 INTRODUCTION

Though more difficult to characterize than the "classical" reactions of 16 e⁻ and 18 e⁻ complexes, several reactivity patterns of 17 e⁻ and 19 e⁻ species have emerged and are illustrated generally in Scheme 10.1 where $^{\bullet}ML_n$ is a 17 e⁻ radical species.

Interaction of A–B with the 17 e⁻ radical $^{\bullet}ML_n$ leads to formation of the 19 e⁻ complex $^{\bullet}ML_n(A–B)$. According to the simple molecular orbital theory, such 19 e⁻ complexes are stabilized by the formation of a half bond between the metal and A–B.[1] Much of the reactivity of 17 e⁻ radicals keys off these elusive 19 e⁻ adducts that can form in rapid associative reactions. The 19 e⁻ adduct $^{\bullet}ML_n(A–B)$ can eject a ligand to form a substituted radical complex via the lower reaction in Scheme 10.1. If the A–B bond is particularly weak, homolysis to $A-ML_n$ and a free $^{\bullet}B$ radical can occur. The $^{\bullet}B$ radical generated as such may then combine with $^{\bullet}ML_n(A–B)$ to form $B-ML_n$ or enter different reaction channels. If the A–B bond is strong, a second mole of radical may be required to break the A–B bond. In that case, an intermediate adduct $L_nM(A–B)ML_n$ forms and then fragments to $A-ML_n$ and $B-ML_n$. These are the same products obtained in homolytic (single metal) cleavage, but the important difference is that free $^{\bullet}B$ radicals are not formed in the reaction.

In addition to these atom transfer reactions, electron transfer reactions can occur. Reduction of 17 e⁻ $^{\bullet}ML_n$ to 18 e⁻ $ML_n^-$ anions is a common reaction. Oxidation of 19 e⁻ $^{\bullet}ML_n(A–B)$ adducts to 18 e⁻ $ML_n(A–B)^+$ cations is also frequently observed. Thus, an additional reaction pathway of the intermediate $L_nM(A–B)ML_n$ is disproportionation to an ionic compound $[L_nM(A–B)]^+[ML_n]^-$. The detailed mechanisms of these reactions are more complex than what is shown in Scheme 10.1. Many of the reactions are reversible and, in addition, they can couple to other reactions.

*Physical Inorganic Chemistry: Reactions, Processes, and Applications*   Edited by Andreja Bakac
Copyright © 2010 by John Wiley & Sons, Inc.

**429**

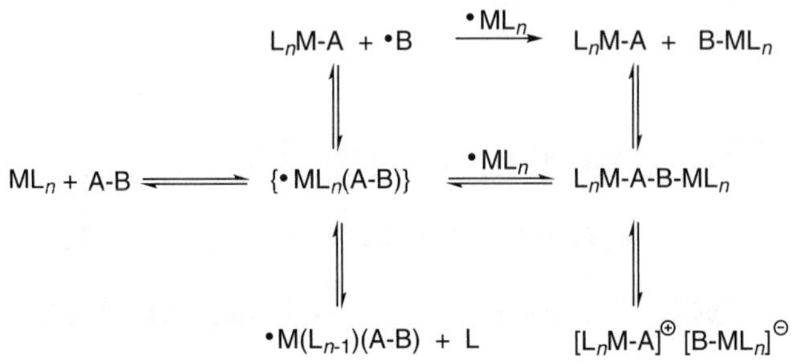

$$L_nM\text{-}A + {}^{\bullet}B \xrightarrow{\;{}^{\bullet}ML_n\;} L_nM\text{-}A + B\text{-}ML_n$$

$$ML_n + A\text{-}B \rightleftharpoons \{{}^{\bullet}ML_n(A\text{-}B)\} \xrightleftharpoons{\;{}^{\bullet}ML_n\;} L_nM\text{-}A\text{-}B\text{-}ML_n$$

$$^{\bullet}M(L_{n\text{-}1})(A\text{-}B) + L \qquad [L_nM\text{-}A]^{\oplus} \; [B\text{-}ML_n]^{\ominus}$$

**SCHEME 10.1**    Typical reactions of 17 e$^-$ radicals $^{\bullet}ML_n$.

Electrochemical and photochemical reactions also play a prominent role in metal radical chemistry.

The goal of this chapter is to provide an introduction to metal radical reactivity. Following brief sections on historical development and methods of characterization of metal radicals, four separate sections describe the chemistry of representative radicals in increasing oxidation number of the metal. Application of the general reactivity pattern shown in Scheme 10.1 varies greatly depending upon the metal complex. For that reason these selected complexes have been dealt with in detail. Low-valent metal carbonyls in the (0) oxidation state are represented by the $^{\bullet}Co(CO)_4$ radical, which also plays a prominent role in hydroformylation catalysis. The $(+1)$ oxidation state is represented in discussion of photochemically and thermally generated Cr, Mo, and W radicals of general formula $^{\bullet}M(CO)_3C_5R_5$. The $(+2)$ oxidation state centers on the discussion of $^{\bullet}Rh(II)(porphyrin)$ and related complexes that bridge the gap between organometallic and bioinorganic chemistry. The $(+3)$ oxidation state is represented by the high-spin $(S = 3/2)$ radical complex $Mo(NRAr)_3$. Then, the ligand-centered reactivity and spin change effects in the reactivity will be briefly summarized. A final section discusses how theoretical advances have played a prominent role in understanding metal radical chemistry. The treatment is meant to illustrate reaction patterns and is not comprehensive.

## 10.2  HISTORICAL

The advent of organic radical chemistry is generally attributed to the discovery of the triphenyl methyl (trityl) radical by Moses Gomberg in 1900.[2] This discovery was aided in part by its accidental exposure to air and subsequent trapping by oxygen to form the peroxide, as shown in Equation 10.1:

$$2Ph_3C\text{-}Cl + Zn \rightarrow ZnCl_2 + 2Ph_3C^{\bullet} \; + O_2 \rightarrow Ph_3C\text{-}O\text{-}O\text{-}CPh_3 \qquad (10.1)$$

The discovery of the simpler but more reactive methyl radical dates back to experiments by Fritz Paneth in the late 1920s showing *reversible* thermal decomposition of tetralkyl leads:[3]

$$Pb(CH_3)_4 \rightleftharpoons {}^{\bullet}CH_3 + {}^{\bullet}Pb(CH_3)_3 \tag{10.2a}$$

$${}^{\bullet}Pb(CH_3)_3 \rightleftharpoons {}^{\bullet}CH_3 + Pb(CH_3)_2 \tag{10.2b}$$

$$Pb(CH_3)_2 \rightleftharpoons {}^{\bullet}CH_3 + {}^{\bullet}Pb(CH_3) \tag{10.2c}$$

$${}^{\bullet}Pb(CH_3) \rightleftharpoons {}^{\bullet}CH_3 + Pb \tag{10.2d}$$

The principal focus of this review is on the chemistry of low-valent transition metal radicals. The metal-based radical chemistry cannot be said to be as advanced as the now sophisticated use of organic radicals. However, rapid progress has been made in this area that can now be said to be an emerging area of inorganic research.[4] There is still considerable room for growth until inorganic radical chemistry can achieve the degree of development in organic radical chemistry described below:

> In recent time the myth of free radicals as highly reactive intermediates has been exploded by the advent of rationally designed, highly efficient free radical chain sequences permitting both the high yielding interconversion of functional groups and the formation of carbon–carbon bonds under mild neutral conditions. The extent of this change in perception is such that free radical chain reactions can now be said to occupy an equal place in the armory of the synthetic organic chemist with long standing two electron concerted processes.[5]

The roots of organometallic radical chemistry are more difficult to trace. In the conclusion of an early review[6] of the chemistry of Vitamin $B_{12}$ model compounds (cobaloximes), it is stated that "the first and simplest cobaloximes" were known since Tschugaeff's time.[7] Calvin et al. reported in 1946 on the reversible binding of oxygen to $Co(salen)_2$ complexes.[8] The 1946 paper refers to work by Pfeiffer in 1933 in which cobaltous bis-salicylaldeheethylenediimine complexes reversibly turn from red to black upon exposure to air—reversible binding of oxygen reminiscent of what prompted Gomberg's discovery as shown in reaction 10.1.

The chemistry of metal carbonyl radicals dates back to the discovery of ${}^{\bullet}V(CO)_6$ in 1959.[9] The electronic structure and reactions of this complex continue to be investigated.[10] The rate of ligand substitution of ${}^{\bullet}V(CO)_6$ would be shown 15 years later to occur at a rate $10^{10}$ times faster than that in $Cr(CO)_6$ and was ascribed to associative rather than dissociative pathways occurring for the radical complex:

$$\underset{18e^-}{Cr(CO)_6} \xrightarrow[\text{slow}]{\text{very}} CO + \underset{16e^-}{Cr(CO)_5} + L \xrightarrow{\text{fast}} \underset{18e^-}{Cr(CO)_5(L)} + CO \tag{10.3}$$

$$\underset{17e^-}{{}^{\bullet}V(CO)_6} + L \underset{}{\overset{\text{rapid}}{\rightleftharpoons}} \underset{19e^-}{\{{}^{\bullet}V(CO)_6(L)\}} \xrightarrow{\text{variable}} \underset{17e^-}{{}^{\bullet}V(CO)_5(L)} + CO \tag{10.4}$$

Reaction 10.4 is one of the prototypical reactions shown in Scheme 10.1 (ligand substitution) and this occurs much more rapidly in radical species. The much greater rate of ligand substitution for radical complexes compared to coordinatively saturated complexes can lead to radical chain substitution reactions. One of the earliest reports is that of Byers and Brown in 1975,[11] as shown in the reaction sequence below:

$$^\bullet X + H\text{-}Re(CO)_5 \rightarrow {}^\bullet Re(CO)_5 + H\text{-}X \tag{10.5}$$

$$L + {}^\bullet Re(CO)_5 \rightarrow {}^\bullet Re(CO)_4 L + CO \tag{10.6}$$

$$^\bullet Re(CO)_4 L + H\text{-}Re(CO)_5 \rightarrow {}^\bullet Re(CO)_5 + H\text{-}Re(CO)_4 L \tag{10.7}$$

$$H\text{-}Re(CO)_5 + L \rightarrow H\text{-}Re(CO)_4 L + CO \tag{10.8}$$

Reaction 10.5 initiates the chain reaction via H-atom transfer to generate $^\bullet Re(CO)_5$. Rapid associative ligand substitution occurs in reaction 10.6. H-atom transfer between the substituted metal radical and the starting hydride completes the chain, as shown in Equation 10.7. All of these reactions occur much faster in radical systems than their saturated counterparts. The result is that ligand substitution in the net reaction 10.8 (reactions 10.6 and 10.7) occurs most rapidly by a pathway involving largely unseen radical intermediates. The paper by Byers and Brown also contains two statements of relevance to those new to the area of organometallic radical chemistry:

> We found it exceedingly difficult to obtain reproducible kinetic results. Under the most rigorous attainable conditions of solvent and reagent purity, with exclusion of light, the thermal reaction in hexane under $N_2$, of $10^{-3}$ M $HRe(CO)_5$ with $\approx 10^{-2}$ M tributyl phosphine, $P(n\text{-}C_4H_9)_3$, exhibited no reaction after 60 days at 25°C.
> Contrary to an earlier report, $HRe(CO)_5$ . . . does not react with dissolved oxygen at room temperature.

The first quotation is of use to those studying kinetics—one of the first signs that a radical process is going on is a rate variance from one day to the next, when background levels of trace radical initiators are at different levels. The second stresses the importance of removing trace radical impurities in synthesis. Very pure metal hydrides (or other complexes) may show a surprising stability provided they are free of all radical contaminants.

The first metallocene radical isolated was cobaltocene—the importance of which is highlighted by its inclusion in an early preparative organometallic text.[12] This 19 e$^-$ complex remains a reagent of choice for electron transfer reactions in nonaqueous solvents.[13] A key question in delocalized organometallic radical complexes is "where is the unpaired electron." An early example of this is $RhCp_2$ ($Cp = \eta^5\text{-}C_5H_5$), which was proposed as early as 1965 by Fishcer and Wawersik[14] to form a dimer through the C–C coupling reaction shown in Equation 10.9.

$$2 \; RhCp_2 \longrightarrow \; Cp\text{–}Rh \cdots \cdots Rh\text{–}Cp \qquad (10.9)$$

This type of coupling reaction through coordinated organic ligands remains a major reaction pathway in organometallic radical chemistry. It implies that a fair amount of electron density in these radicals is located on the organic ligand, and not on the metal. A more detailed discussion of this topic is presented in a subsequent section.

The question of where an electron is located as well as uncertainties in the oxidation state of metals is illustrated by the preparation of the bisdithiolene complex $(C_5H_5)W[(S_2C_2(CF_3)_2]_2$, as shown in Equation 10.10.

$$(10.10)$$

As originally proposed by King and Bisnette[15] in 1967, two limiting tautomeric forms could be proposed. The first would be a formal W(V) bis-thiolate complex, and the second a formal W(I) complex containing coordinated bis-thione donor ligands. The resolution of which tautomer is preferred, or the assignment of intermediate more complex bonding motifs in complexes of this type is a challenge requiring modern theoretical and spectroscopic techniques. The incorporation of non-innocent organic ligands, that is, ligands in which the oxidation state is unclear, changes the bonding landscape for a metal radical—so does the incorporation of "soft"[16] ligands such as thiolates.

Pioneering research into radical reaction mechanisms was performed by the Halpern group in the late 1960s.[17] The $d^7$ radical ion $^\bullet Co(CN)_5^{3-}$ was shown to react with methyl iodide to generate a methyl radical, as shown in Equation 10.11.

$$^\bullet Co(CN)_5^{3-} + I\text{-}CH_3 \rightarrow I\text{-}Co(CN)_5^{3-} + {}^\bullet CH_3 \qquad (10.11)$$

Oxidative addition of $H_2$, as seen in reaction 10.12, was shown to follow the termolecular rate law $-d[H_2]/dt = k[^\bullet Co(CN)_5^{3-}]^2[H_2]^1$.

$$2\,^\bullet Co(CN)_5^{3-} + H_2 \rightarrow 2H\text{-}Co(CN)_5^{2-} \qquad (10.12)$$

This is one of the first specific examples of a ternary transition state for oxidative addition, one of the characteristic reactions of metal radicals as shown generally in Scheme 10.1. In addition, the transition metal–hydrogen bond strength (58 kcal/mol) was determined for $H\text{-}Co(CN)_5^{2-}$—one of the first accurate determinations. A landmark review on organometallic radicals by Lappert and Lednor[18] provides solid introduction to early work in the metalloradical area.

In spite of the significance of the research described above, it is safe to say that radical chemistry was viewed by most investigators at that time as a "curious" area outside the main stream of organometallic chemistry. That was due to the rapid and logical progress in understanding and using the reactions of 16 e$^-$ and 18 e$^-$ complexes that were more readily characterized and studied. Since radicals were known to cause decomposition reactions, great care was generally taken by many investigators to avoid them. A goal of this chapter is to show some of the remarkable advances that have been made in organometallic radical chemistry since these early days. The future should hold even more promise: the ability to design and implement radical reactions similar to that present in organic radical reactions and implicit in the quotation below:

> The catalysis of stannane-mediated radical chain reactions by benzeneselenol once again demonstrates that a single, inefficient propagation step may be advantageously replaced by two well-matched steps. The application of this opens numerous doors previously closed to the synthetic chemist.[19]

## 10.3 SPECTROSCOPIC TECHNIQUES USED TO CHARACTERIZE METAL RADICALS

NMR is perhaps one of the most widely used techniques in the characterization of diamagnetic organometallic species. However, the presence of unpaired electrons in the molecules results in large isotropic shifts and broadened signals. As a result, the detection of coupling and integration of signals is difficult. Moreover, the NMR spectra of paramagnetic species do not provide information either on the chemical environment or on the structure of the compounds. In addition, the resonances attributed to hydrides and H-atoms attached to carbon atoms directly bonded to the metals are generally unobservable. As summarized by Pariya and Theopold[20]:

> It has come to the point where some chemists, faced with a compound that does not exhibit a "normal" $^1$H NMR spectrum, discard it and move on to a more promising project. In effect, our addiction to NMR spectroscopy has allowed the 18-electron rule to become a set of blinders limiting our view.

These authors also advocate increased use of $^2$H NMR for characterization of paramagnetic complexes:

> When the line broadening becomes a serious obstacle, one can take advantage of a phenomenon that results in narrower lines. Specifically, substitution of $^1$H atoms with the $^2$H isotope coupled with $^2$H NMR spectroscopy results in spectral lines that are up to 40 times narrower than those of corresponding $^1$H NMR, due to slower nuclear relaxation of the $^2$H nuclei in paramagnetic compounds.

The definitive characterization method for radical complex is its three-dimensional structure determined by crystallography. Some introduction to the literature is provided here for the interested reader who wishes to look up the data on representative structures. The essentially planar structures of the perchlorinated trityl radical[21] as well as several Si, Ge, and Sn radicals[22] have been determined. The structures of the 18 e$^-$ complex Mn(CO)(dppe)($\eta^5$-C$_6$H$_6$Ph) and the 17 e$^-$ cation radical Mn(CO) (dppe)($\eta^5$-C$_6$H$_6$Ph)$^+$ were determined and found to be quite similar.[23] The most pronounced effect of oxidation on the radical was found to be lengthening of the Mn–P bond distance from 2.221 to 2.338 Å. The structures of Cr(CO)$_2$(PPh$_3$)(Cp)[24] and ($\eta^5$-C$_5$Ph$_5$)Cr(CO)$_3$[25] have been determined. The crystal structure of the 17 e$^-$ radical complex Re(PCy$_3$)$_2$(CO)$_3$ (related to the Mn(CO)$_5$ radical) has been determined and it resembles that of the 18 e$^-$ Kubas[26] complex W(PCy$_3$)$_2$(CO)$_3$. An important difference is that Re(PCy$_3$)$_2$(CO)$_3$ shows no signs in its structure or spectra of the agostic bond that is present in W(PCy$_3$)$_2$(CO)$_3$.

The crystal structure of the paramagnetic Cr(III) complex Cp*Cr(Me)$_2$(pyridine) was discussed by Pariya and Theopold[20] who concluded as follows:

> Two generalizations can be made on our growing structural database. First, paramagnetic compounds do not differ significantly in any structural aspect from comparable 18-electron compounds. For example, the Cp*-ring is bound to the chromium in an entirely normal $\eta^5$-mode, and many of the compounds adopt the familiar "three legged piano stool" geometry. The bond distances and angles are within reasonable expectations.

To a first approximation it would seem that the structures of metal radical complexes are not greatly different from similar nonradical complexes. However, small differences can be quite important and it can be expected that as more data on metal radical structures exists, a better appreciation will emerge.

Many organometallic radical complexes are not stable enough to be isolated and characterized by single-crystal X-ray or neutron diffraction. Careful infrared and UV-Vis spectra provide a wealth of information—comparable to what is available for nonradical complexes. Electrochemical studies, often used to generate radicals, also provide valuable information on oxidation and reduction potentials for these intermediates. Magnetic susceptibility measurements can provide information but generally require relatively high radical concentration for detection. Chemically induced dynamic nuclear polarization (CIDNP) studies,[27] though difficult to quantify, can provide powerful qualitative signs that radical processes are involved in a reaction.

The dominant technique for radical characterization is ESR. It should be kept in mind that a number of free radicals can be ESR silent in solution. For example, $^\bullet$Cr(CO)$_3$Cp* does not display an ESR spectrum in solution, only as a frozen solid. Proving the existence of radicals by ESR has led increasingly to direct experimental information about the singly occupied molecular orbital (SOMO) in which the unpaired electron resides. In favorable cases, spectroscopic data can be simulated to get the coupling constants between the free electron and all magnetic nuclei present in the radical. This can be used to compute spin densities and thus provide information

(a)

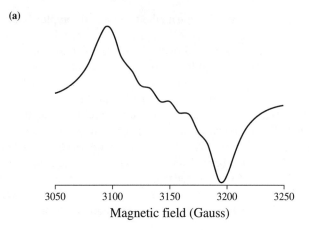

Magnetic field (Gauss)

(b)

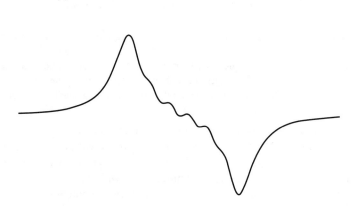

**FIGURE 10.1** Experimental (a) and computed (b) ESR spectral data for the delocalized radical $Cp_2Ni_2(\mu_3\text{-}S)_2Mn(CO)_3$. Reproduced with permission from Ref. 30. Copyright 2004, American Chemical Society.

as to where the electron is located.[28] Illustrative examples will be highlighted below, but it should be pointed out that the relationship between spin densities and observed reactivity remains to be fully developed.

Computational and ESR spectroscopic studies were performed on complexes formally derived from $^{\bullet}Rh(CO)_4$ by ligand substitution. The structures of the complexes were in between square planar and tetrahedral geometry. Definitive data regarding the position of the electron in these complexes (analogues of the important $^{\bullet}Co(CO)_4$ radical discussed in a later section) led to the conclusion that spin density localized on the Rh center was smaller than that on the coordinated ligands, even for CO:

The major part of the spin density is delocalized over the hydrocarbon framework of the trop ligand. Inspection of previously reported monomeric rhodium complexes with

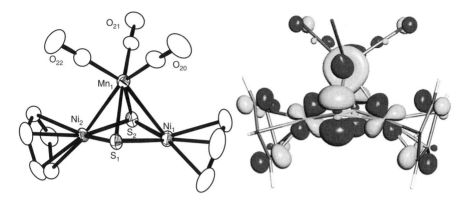

**FIGURE 10.2**    An ORTEP diagram of the molecular structure of $Cp_2Ni_2(\mu_3\text{-}S)_2Mn(CO)_3$ and the computed SOMO. Reproduced with permission from Ref. 30. Copyright 2004, American Chemical Society. (See the color version of this figure in Color Plates section.)

carbonyl, olefin, and/or phosphane ligands show that none of these can be described as true $d^9$-Rh(0) complex; in all of them the odd electron is highly delocalized. This is likely due to the fact that all ligands applied so far have energetically low lying $\pi$-type acceptor orbitals which interact strongly with the d-orbitals at the metal. Evidently, this enhances stability of the formally zero valent rhodium complexes and lowers the reduction potential of the cationic precursor complexes, but a trityl rhodium centered metallo-radical in which the majority of the spin density is localized on the metal center remains to be synthesized and its reactivity studied.[29]

Extension of ESR techniques to radical cluster systems can also be performed successfully. For the $Cp_2Ni_2(\mu_3\text{-}S)_2Mn(CO)_3$ radical,[30] the experimental and computed ESR spectral data are shown in Figure 10.1. The crystal structure and computed SOMO are shown in Figure 10.2. These data represent a relatively complete characterization of a complex multicenter radical.

These data allow computation that the largest spin density, approximately 20%, is located on the Mn atom, but that there is significant electron delocalization to the S and Ni atoms as well as the ligands. The physical properties and chemical reactivities of such delocalized radicals are a matter of current research interest to a number of groups.

## 10.4    REPRESENTATIVE ORGANOMETALLIC RADICALS AND REACTIVITY

### 10.4.1    $Co(CO)_4$ as Representative of Metal Carbonyl Radicals

The oxo process is catalyzed by cobalt and rhodium complexes and is the largest homogeneous process performed industrially:

$$CO + H_2 + \text{olefin} \rightarrow \text{aldehyde} + H_2 \rightarrow \text{alcohol} \qquad (10.13)$$

In the case of cobalt, a key process is hydrogenation of cobalt carbonyl, as shown in Equation 10.14.

$$Co_2(CO)_8 + H_2 \rightarrow 2HCo(CO)_4 \tag{10.14}$$

A key intermediate in the Heck–Breslow mechanism for hydroformylation is proposed to be $HCo(CO)_3$. This $16\,e^-$ coordinatively unsaturated intermediate would be produced by dissociation of CO as shown in Equation 10.15 and rapidly bind olefins as shown in Equation 10.16.

$$HCo(CO)_4 \rightleftharpoons HCo(CO)_3 + CO \tag{10.15}$$

$$HCo(CO)_3 + olefin \rightleftharpoons 2HCo(CO)_3\,(olefin) \tag{10.16}$$

A key experimental fact is that the reaction in Equation 10.13 is inhibited by CO pressure—as would be predicted based on Equations 10.15 and 10.16.

In spite of the prominence of 16/18 $e^-$ complexes in the Heck–Breslow mechanism, a number of observations suggested a possible role, under some conditions of temperature and pressure, for the $^{\bullet}Co(CO)_4$ radical in the hydroformylation process. Photolysis studies in frozen gas matrices gave evidence[31] for the infrared spectrum of the proposed radical species at low temperatures:

$$Co_2(CO)_8 \xrightarrow{\;h\nu\;} 2^{\bullet}Co(CO)_4 \tag{10.17}$$

The structure of $^{\bullet}Co(CO)_4$ is best addressed by computational studies (see later section). Infrared studies under pressure of CO in the temperature range $-165$ to $210°C$ by Bor et al. proposed the presence of $^{\bullet}Co(CO)_4$ as a "third isomer" of $Co_2(CO)_8$.[32]

Some of the most important implications for the role of $^{\bullet}Co(CO)_4$ remain mechanistic in nature.[33] Wegman and Brown[34] proposed a radical pathway for the decomposition of $HCo(CO)_4$ in which in one of the key steps the $^{\bullet}Co(CO)_4$ radical interacts with $HCo(CO)_4$ to form the $^{\bullet}HCo_2(CO)_7$ radical. The role of reaction 10.15 was challenged:

The commonly accepted mechanism for thermal decomposition or substitution of HCo $(CO)_4$ involves a presumed facile loss of CO to form $HCo(CO)_3$. There is, however, little independent evidence that such a thermal process exists.

In spite of years of work on this industrially important system, the role of $HCo(CO)_3$ is not yet fully understood. A recent computational work[35] suggests that loss of CO from $HCo(CO)_4$, as shown in Equation 10.15, is endothermic by 25.7 (loss of equatorial CO) or 36.3 (loss of axial CO) kcal/mol. This would clearly present a large kinetic barrier to any active catalyst and appear to rule out establishment of equilibrium quantities of $HCo(CO)_3$ formed via Equation 10.15.

Additional key experiments revolved around the hydrogenation of arene-substituted olefins[36] by $HCo(CO)_4$ in which CIDNP was observed and a radical pair mechanism proposed, as shown in reactions 10.18–10.20.

$$HCo(CO)_3 + Ph_2C = CH_2 \rightleftharpoons [{}^{\bullet}Co(CO)_4, Ph_2C(\bullet)CH_3] \qquad (10.18)$$

$$[{}^{\bullet}Co(CO)_4, Ph_2C(\bullet)CH_3] \rightarrow {}^{\bullet}Co(CO)_4 + Ph_2C(\bullet)CH_3 \qquad (10.19)$$

$$HCo(CO)_4 + Ph_2C(\bullet)CH_3 \rightarrow {}^{\bullet}Co(CO)_4 + Ph_2C(H)CH_3 \qquad (10.20)$$

It is worth noting that this applies to arene-substituted olefins, and that the organic radical generated may be particularly stable since it resembles a "trityl" radical. The rapid H-atom transfer between trityl radical and $HCo(CO)_4$ was independently studied by Ungvary and Marko (Equation 10.21).

$$HCo(CO)_4 + {}^{\bullet}CPh_3 \rightarrow {}^{\bullet}Co(CO)_4 + HCPh_3 \qquad (10.21)$$

Some caution is indicated before these radical processes are extended to simple olefins where the energetics of the organic radical itself may limit reactions 10.18 and 10.19. Nevertheless, there remains a prominent role of the ${}^{\bullet}Co(CO)_4$ radical in oxo chemistry. This is still being delineated today, most notably in the laboratories of Klingler et al.[37] High-pressure, high-temperature NMR studies in supercritical $CO_2$ and other solvents have led to a number of elegant results in this area, in particular the very rapid exchange of H-atoms between $HCo(CO)_4$ and $Co_2(CO)_8$ that is mediated by the (largely unseen) ${}^{\bullet}Co(CO)_4$ radical:

Thus the exchange of the hydride moiety between the cobalt centers in $HCo(CO)_4$ and $Co_2(CO)_8$ is occurring more than a million times faster than the steady-state transfer for the same hydride moiety to the olefin in the hydroformylation process.[38]

### 10.4.2   M(CO)₃Cp (M = Cr, Mo, W) as Representative of Substituted Metal Carbonyl Radicals

*10.4.2.1   Introduction*   The radical fragments ${}^{\bullet}M(CO)_3(C_5R_5)$ (M = Cr, Mo, W) and simple substituted derivatives have been extensively studied. Ligand substitution of $[Cr(CO)_3Cp]_2$ was found to be extremely facile in work done by Hackett et al.[39] in 1974 where they noted that

Many of the reactions of $[(\pi\text{-}C_5H_5)Cr(CO)_3]_2$ imply that the Cr–Cr bond is weak relative to the metal–metal bonds in its Mo and W counterparts.

Later work would prove this observation correct. Ligand substitution by $PPh_3$ was originally proposed to yield a phosphine-substituted metal–metal-bonded dimer.

$$[Cr(CO)_3Cp]_2 + 2PPh_3 \rightarrow \text{``}[Cr(CO)_2(PPh_3)Cp]_2\text{''} + 2CO \tag{10.22}$$

Baird would later show that the complex prepared by Hackett, O'Neil, and Manning was actually the monomeric Cr(I) radical $^{\bullet}Cr(CO)_2(PPh_3)Cp$.[40] Structural work by Adams et al. confirmed that there was an unusually long Cr–Cr bond in $[Cr(CO)_3Cp]_2$[41] of 3.281 Å. Muetterties and coworkers[42] prepared the trimethyl phosphite-substituted dimeric complex $[Cr(CO)_2(P(OMe)_3)Cp]_2$. This complex was found to have an even longer Cr–Cr bond of 3.343 Å and to largely dissociate in solution to give monomeric radicals, as shown in Equation 10.23.

$$[Cr(CO)_2(P(OMe)_3) Cp]_2 \rightleftharpoons 2^{\bullet}Cr(CO)_2(P(OMe)_3)Cp \tag{10.23}$$

Solutions of these radicals were found to react with $H_2$ in solution at room temperature:

$$2^{\bullet}Cr(CO)_2(P(OMe)_3Cp + H_2 \rightarrow 2H\text{-}Cr(CO)_2(P(OMe)_3Cp \tag{10.24}$$

It was not until careful spectroscopic measurements were made by McLain[43] that the equilibrium in Equation 10.25 was proven unequivocally in solution measurements.

$$[Cr(CO)_3) Cp]_2 \rightleftharpoons 2^{\bullet}Cr(CO)_3) Cp \tag{10.25}$$

In spite of low levels of the $^{\bullet}Cr(CO)_3Cp$ radical present in solution under typical conditions, it was soon realized that the increased reactivity of the Cr–Cr complex could be assigned unequivocally to the presence, albeit in small amounts, of this highly reactive radical. Compared to the cyclopentadienyl system, the pentamethylcyclopentadienyl dimer (which exists in the solid state as a Cr–Cr-bonded dimer) is more extensively dissociated than the unsubstituted cyclopentadienyl system:

$$[Cr(CO)_3Cp^*]_2 \rightleftharpoons 2^{\bullet}Cr(CO)_3Cp^* \tag{10.26}$$

Under typical millimolar concentrations, only a few percent of the $[Cr(CO)_3Cp]_2$ complex dissociates to radicals at room temperature according to Equation 10.25. For $[Cr(CO)_3Cp^*]_2$ under the same conditions, only a few percent remains as Cr–Cr dimer—the complex is almost totally dissociated to radicals. For the pentaphenyl system, even in the solid state there is no Cr–Cr bond complex, only $^{\bullet}Cr(CO)_3C_5Ph_5$ exists.

The concentration of radicals is important in kinetic and mechanistic considerations, particularly when the second-order reactions in metal radical complex are sought. For the Mo and W complexes, due to their stronger M–M bonds, there is no sign of significant radical concentrations present in solution at room temperature for $^{\bullet}M(CO)_2(L)C_5R_5$ (M = Mo, W). An exception is the report by Tyler and coworkers[44] of an equilibrium generating the $^{\bullet}Mo(CO)_3C_5Ph_5$ radical. For the most part, the highly reactive $^{\bullet}M(CO)_3C_5R_5$ radicals are generated by photolysis of the M–M dimeric complexes for M = Mo, W and are thermally accessible for M = Cr. Two separate sections follow: photochemical generation of $^{\bullet}M(CO)_3C_5R_5$ for M = Mo, W and thermal reactions of $^{\bullet}Cr(CO)_3C_5R_5$. The goal of these sections is to give a fairly detailed summary of observed reactivity for these systems. It is important to keep in mind the different radical concentrations and other conditions made in the studies of these reactions. Different reaction channels can be followed under different conditions.

### 10.4.2.2 Photochemical Generation of $^{\bullet}M(CO)_3C_5R_5$ (M = Mo, W)

Photochemical study of the $[M(CO)_3C_5R_5]_2$ (M = Cr, Mo,W) has been extensively studied, and will be discussed in some detail since it is representative of other low-valent metal–metal-bonded complexes and it also allows comparison to reactions of thermally generated radicals. A somewhat historical presentation of developments is made in this section to show that the early insights of a number of talented investigators set the stage for current level of sophistication of knowledge in this system that has been achieved utilizing ultrafast, even femtosecond, techniques.

Haines et al.[45] reported photoreactivity studies of reaction 10.27.

$$[Mo(CO)_3\,Cp]_2 + PPh_3 \xrightarrow{\;h\nu\;} Cp(CO)_2(PPh_3)Mo\text{-}Mo(CO)_3Cp + CO \quad (10.27)$$

In addition to the substituted metal–metal-bonded complex, they observed formation of a disproportionated complex $\{Mo(CO)_2(PPh_3)_2\}^+\{Mo(CO)_3Cp\}^-$.[45] A more quantitative study was reported by Burkett et al.[46] who studied ionic cleavage of the metal–metal bond as shown in reaction 10.28 (X = SCN, Cl, Br, I).

$$[Mo(CO)_3\,Cp]_2 + X^- \xrightarrow{\;h\nu\;} X\text{-}Mo(CO)_3Cp + [Mo(CO)_3Cp]^- \quad (10.28)$$

In the reaction with $Cl^-$, the net quantum yield for reaction at 546 nm was 0.36, but at 366 nm it was only 0.07, which is unusual in that it is less efficient at higher energy. The reactions were studied in acetone, acetonitrile, and THF and solvent dependence was observed. Wrighton and Ginley[47] reported quantum yields for the Cl abstraction shown in Equation 10.29 and proposed that $\sigma$ to $\sigma^*$ transition cleaves the metal–metal bond to radicals in the fundamental photochemical step.

$$[Mo(CO)_3Cp]_2 + CCl_4 \xrightarrow{\;h\nu\;} Cl\text{-}Mo(CO)_3Cp \quad (10.29)$$

Hughey et al.[48] provided proof of the production of radicals as the primary photoproduct in flash photolysis studies of reaction 10.30.

$$[Mo(CO)_3Cp]_2 \xrightarrow{\;h\nu\;} 2^{\bullet}Mo(CO)_3Cp \quad (10.30)$$

The rate of radical recombination was also reported in their seminal paper as being $2 \times 10^9\,M^{-1}\,s^{-1}$ in THF $3 \times 10^9\,M^{-1}\,s^{-1}$ in acetonitrile and $5 \times 10^9\,M^{-1}\,s^{-1}$ in cyclohexane. The generation of $Mo_2(CO)_5Cp_2$ was also noted and it was observed that

The fact that either UV or visible photolysis gives both metal–metal bond cleavage and loss of CO implies that the two intermediates have a common origin, but it is not clear that the intermediate from which CO has been lost is a primary photoproduct. Metal–metal bond cleavage is expected following UV excitation into the σ to σ* (Mo–Mo bond since the excited state should be antibonding with regard to the Mo–Mo bond, and, when thermally equilibrated, most likely consists of $(\eta^5\text{-}C_5H_5)Mo(CO)_3$ fragments in a solvent cage.

The statement quoted above is insightful, and the role of the solvent cage will be discussed later. Work by Laine and Ford[49] also emphasized at an early stage the importance of solvent effects. In study of the rates of Cl atom abstraction by the photo-generated $^\bullet W(CO)_3Cp$ radical, it was found that the order of reactivity was $CCl_4 \gg CHCl_3 > PhCH_2Cl > CH_2Cl_2$. The authors also noted that

In addition, it is clear that the quantum yields observed in the different chloromethane solvents do not reflect simply the ease of chlorine abstraction from the solvent trapping agent, and instead indicate solvent effects on the primary quantum yields for the formation of reactive metal radicals.

Hoffman and Brown[50] utilized photogeneration of metal radicals to initiate a radical chain reaction for ligand substitution of $HM(CO)_3Cp$ (M = Mo, W) (Equations 10.31 and 10.32).

$$^\bullet M(CO)_3Cp + PR_3 \rightarrow {}^\bullet M(CO)_2(PR_3)Cp + CO \qquad (10.31)$$

$$^\bullet M(CO)_2(PR_3)Cp + HM(CO)_3Cp \rightarrow HM(CO)_2(PR_3)Cp + {}^\bullet M(CO)_3Cp \quad (10.32)$$

Photolysis was proposed to initiate the chain reaction by generating metal radicals, as shown in reaction 10.30. Rapid associative substitution occurs as shown in Equation 10.31, followed by H-atom transfer as shown in Equation 10.32. Quantum yields >1000 were found for M = W and >50 for M = Mo. Some early reports proposed that $^\bullet M(CO)_3Cp$ radicals might readily lose CO and that this might play a role in substitutions such as that shown in Equation 10.31. Turaki and Huggins[51] investigated the reactions of $^\bullet W(CO)_3Cp$ generated by the reaction of $HW(CO)_3Cp$ and trityl radicals and concluded based on kinetic analysis of CO dependence that

Our results clearly rule out a dissociative substitution pathway for the transient CpW $(CO)_3$ radical. At present, however, we are unable to distinguish between a 19-electron transition state and a stable 19-electron intermediate in these reactions.

Philbin et al.[52] studied in detail the photochemistry of the disproportionation reaction producing $\{Mo(CO)_2(PPh_3)_2Cp\}^+ \{Mo(CO)_3Cp\}^-$. It was proposed that the initial ionic product was $\{Mo(CO)_3(PPh_3)Cp)\}^+ \{Mo(CO)_3Cp\}^-$ but that this was capable of back reaction. The back reaction was proposed to occur in benzene, but to be slowed down in more polar solvents such as methylene chloride. Furthermore, wavelength-dependent photolysis on the $\{Mo(CO)_3(PPh_3)Cp\}^+$ cation to produce

$\{Mo(CO)_2(PPh_3)_2Cp\}^+$ was observed to occur when $\lambda = 290\,nm$ but not when $\lambda = 525\,nm$. The final product $\{Mo(CO)_2(PPh_3)_2Cp\}^+\{Mo(CO)_3Cp\}^-$ is stable toward the back reaction, whereas the initial product $\{Mo(CO)_3(PPh_3)Cp\}^+\{Mo(CO)_3Cp\}^-$ is not—explaining the wavelength dependence of the photochemistry. Three methods were discussed for preventing the back reaction: use of a more polar solvent, further reaction of the cationic or anionic products, and performing the reaction at lower temperatures.

A number of additional photochemical observations have been made by Tyler and coworker,[53] including photochemical production of radical in aqueous solution using $(C_5H_4COOH)_2W_2(CO)_6$[54] where the photochemical reactions and rates appear to be similar to those performed in organic solvents. Highly interesting results were obtained in aqueous studies of photochemically generated radical ${}^\bullet W(CO)_3(C_5H_4COO^-)$. A more powerful reducing agent was formed in the presence of the water-soluble phosphine $PPh_2R$ ($R = C_6H_4SO_3^-$). Methyl viologen, $Fe(CN)_6^{3-}$, and cytochrome $c^{III}$ could all be photoreduced as substrates, as shown in Equation 10.33.

$$(C_5H_4COO^-)_2\,W_2(CO)_6 + 2PPh_2R^- \rightleftharpoons 2[(C_5H_4COO^-)\,W(CO)_3(PPh_2\,R^-)]^+$$
$$+\,2Substrate \qquad\qquad\qquad\qquad +\,2Substrate^-$$

$$(10.33)$$

In addition, small quantities of $H_2$ were observed in these reactions.

Additional quantitative photochemical studies of the photogenerated water-soluble radical ${}^\bullet W(CO)_3(C_5H_4COO^-)$ were performed by Zhu and Espenson.[55] The rate of dimerization in water was found to be $3.0 \times 10^9\,M^{-1}\,s^{-1}$, which was quite similar to results found in organic solvents for radical recombination of ${}^\bullet M(CO)_3Cp$ (M = Mo, W) as measured in earlier studies by Scott et al.[56] A characteristic reaction of these radicals is trapping by $O_2$, as shown in Equation 10.34.

$$^\bullet W(CO)_3Cp + O_2 \rightarrow {}^\bullet O_2W(CO)_3Cp \qquad (10.34)$$

The second-order rate constant for this reaction is $3.3 \times 10^9\,M^{-1}\,s^{-1}$; rates of reaction with organic halides were found to be highly selective and span seven orders of magnitude. An interesting aspect of this paper was also the observation that reaction of $Bu_3SnH$ and ${}^\bullet W(CO)_3Cp$ was not consistent with either hydrogen atom abstraction or electron transfer. It was proposed that oxidative addition occurred as shown in Equation 10.35.

$$^\bullet W(CO)_3Cp + HSnBu_3 \rightarrow {}^\bullet W(H)(SnBu_3)(CO)_3Cp \rightarrow {}^\bullet W(H)(SnBu_3)(CO)_2Cp + CO$$

$$(10.35)$$

The proposed radical complex ${}^\bullet W(H)(SnBu_3)(CO)_2Cp$ was then hypothesized to undergo further reaction. $H–W(CO)_3Cp$ was not formed in the reaction. The proposed oxidative addition reaction at the metal radical, concomitant with CO loss to yield a W(III) radical is novel, and additional investigation of these types of reactions are needed.

Another interesting property of photogenerated metal radicals is their ability to act either as an oxidant or as a reductant. The rate of the reduction shown in Equation 10.36 was measured[57] to be $1.89 \times 10^7$ in acetonitrile.

$$\cdot W(CO)_3Cp + FeCp_2^+ \rightarrow W(CO)_3Cp^+ + FeCp_2 \xrightarrow{CH_3CN} W(CH_3CN)(CO)_3Cp^+$$

(10.36)

In acetonitrile, it is likely that the 16 $e^-$ cation $W(CO)_3Cp^+$ would react further (see later discussion) to form the 18 $e^-$ cation $W(MeCN)(CO)_3Cp^+$. It is not clear if this occurs as part of reaction 10.36 or rapidly after it. In contrast, with decamethyl, ferrocene reduction was found either not to occur or to proceed on an extremely slow timescale:

$$\cdot W(CO)_3 Cp + FeCp*_2^+ \xrightarrow{\quad X \quad} W(CO)_3Cp^+ + FeCp*_2$$

(10.37)

Thus, while $\cdot W(CO)_3Cp$ radicals could not reduce the $Fe(Cp*)_2^+$ cation, they were capable of oxidizing $Fe(Cp*)_2$; $k_{obs} = 2.23 \times 10^8 M^{-1} s^{-1}$ (Equation 10.38).

$$\cdot W(CO)_3 Cp + FeCp*_2 \rightarrow W(CO)_3Cp^- + FeCp*_2^+$$

(10.38)

The susceptibility to oxidation of photogenerated radicals $\cdot M(CO)_3Cp$ (M = Mo, W) is enhanced both kinetically and thermodynamically by the presence of strong coordinating ligands such as phosphines that stabilize the cations $W(PR_3)(CO)_3Cp^+$.

The ability of photogenerated radicals $\cdot M(CO)_3Cp$ (M = Mo, W) to be both oxidized and reduced is illustrated by the result that TMPD (tetramethylphenylene-diamine) catalyzes the formation of $[Mo(MeCN)(CO)_3Cp]^+ [Mo(CO)_3Cp]^-$.[58] In a first step, the reversible equilibrium shown in reaction 10.39 is established from the photogenerated $\cdot Mo(CO)_3Cp$ radical.

$$\cdot Mo(CO)_3 Cp + TMPD \rightleftharpoons Mo(CO)_3Cp^- + \cdot TMPHD^+$$

(10.39)

The $TMPD^{\cdot +}$ radical cation then serves as oxidizing agent for a second $\cdot Mo(CO)_3Cp$ radical, as shown in reaction 10.40.

$$\cdot Mo(CO)_3 Cp + \cdot TMPD^+ + CH_3 CN \rightarrow \cdot Mo(CH_3 CN)(CO)_3 Cp^+ + TMPD$$

(10.40)

Once formed, the ionic product $[Mo(MeCN)(CO)_3Cp]^+ [Mo(CO)_3Cp]^-$ is thermodynamically unstable with respect to collapse to $[Mo(CO)_3Cp]_2$ and MeCN, however it was found to be kinetically stable to this reaction in acetonitrile.

Time-resolved infrared studies[59] of $Me-W(CO)_3Cp$, which ultimately produces $[W(CO)_3Cp]_2$ and $CH_4$ as products, proceeds primarily by initial CO loss as shown in Equation 10.41, and not Me–W bond homolysis, as shown in Equation 10.42.

$$H_3C\text{-}W(CO)_3Cp \xrightarrow{h\nu} H_3C\text{-}W(CO)_2Cp + CO \qquad (10.41)$$

$$H_3C\text{-}W(CO)_3Cp \xrightarrow{h\nu} {}^\bullet W(CO)_3\,Cp + {}^\bullet CH_3 \qquad (10.42)$$

In spite of uncertainties about how $CH_4$ is produced and whether or not a methyl radical is generated as a primary photochemical step, the complex Me–W(CO)$_3$Cp has been utilized by Mohler in photoinduced DNA cleavage.[60] The mechanism shown in Equation 10.43 has been proposed[61] for the production of a formate radical under these conditions:

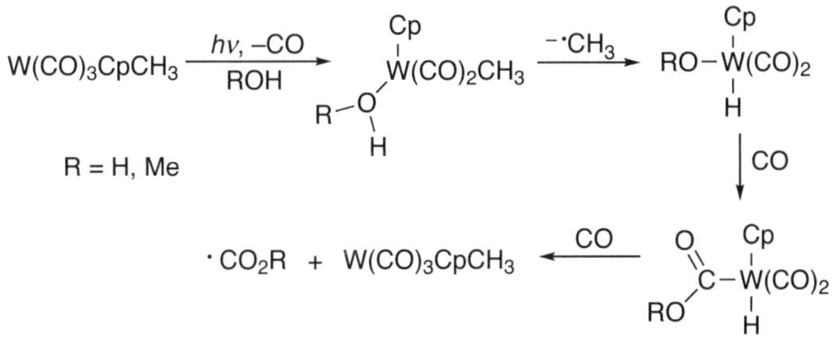

$$(10.43)$$

Recent work in ultrafast IR studies has provided new insight into the mechanisms of metal complex radical generation and reactivity. Photochemical studies[62] of Et–W (CO)$_3$Cp have shown that the primary photochemical pathway was CO loss, as exemplified in Equation 10.43 for the methyl complex Me–W(CO)$_3$Cp. A side reaction did produce radicals, but this may have occurred from vibrationally hot parent molecules. Both $\alpha$ and $\beta$ C–H activation processes were observed. A general scheme elucidated for the ethyl complex is shown in Figure 10.3.

The role of 19 e$^-$ intermediates has been a source of controversy, in particular many mechanisms can be explained without their proposal, and few 19 e$^-$ complexes have been crystallographically characterized. Ultrafast photochemical studies of phosphite substitution in $^\bullet$W(CO)$_3$Cp allowed the direct detection of infrared bands assigned to $^\bullet$W(CO)$_3$(PR$_3$)Cp in Equation 10.44 with decay times on the order of 280 ns.[63]

$$1/2[W(CO)_3\,Cp]_2 \xrightarrow{h\nu} {}^\bullet W(CO)_3Cp \underset{-PR_3}{\overset{+PR_3}{\rightleftharpoons}} {}^\bullet W(CO)_3(PR_3)Cp \qquad (10.44)$$

This leads the authors to conclude as follows:

Numerous studies have shown that the ligand substitution reactions of 17e radicals, such as CpM(CO)$_3$$^\bullet$ (M = Cr, W), CpFe(CO)$_2$$^\bullet$, M(CO)$_5$$^\bullet$ (M = Mn, Re), or V(CO)$_6$$^\bullet$, proceed by an associative mechanism, but this study is the first evidence that a 19e species is an intermediate rather than a transition state in the ligand substitution reaction.

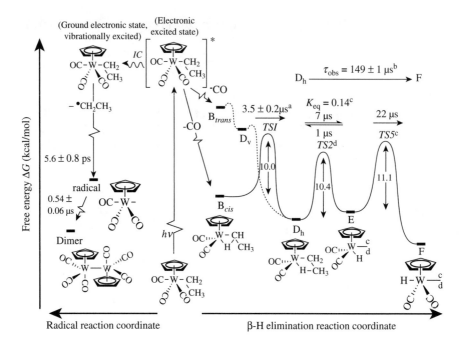

**FIGURE 10.3** Photochemical sequence proposed for β-C–H oxidative addition proceeding by initial CO loss and α- and β-C–H coordination. The radical fragmentation on the left is a side reaction to the major process. Reproduced with permission from reference 63. Copyright 2006. American Chemical Society.

An additional important aspect of this work is that the equilibrium constant for binding was found to be $K_{eq} \approx 3.4 \, M^{-1}$ that implies a near-zero free energy change for the 17/19 e⁻ equilibrium. Thus, the unfavorable entropy of ligand addition is nearly offset by the favorable enthalpy of forming the half bond expected for a 19 e⁻ adduct.

The studies already described were done in relatively dilute solution (85 mM in P(OMe)₃, however the photochemical studies of disproportionation were studied under relatively concentrated solution (1.6 M in P(OMe₃)) as well.[64,65] Under these conditions, disproportionation is observed to occur at a rate even faster than ligand substitution. The reason for this is that high concentrations were used to ensure that at least one P(OMe)₃ was in the first solvation shell of the metal–metal-bonded dimer. In that case, reaction can occur prior to diffusion out of the solvent cage, as shown in Scheme 10.2.

The ratio of disproportionation product to ligand substitution product was controlled by the concentration of P(OMe)₃. At low ligand concentration, escape from the solvent cage of the radical pairs occurred before they could combine with two moles of the photogenerated radical. This is an important result linking together to some extent the chemistry of photogenerated radicals, usually in dilute solution, to that of stable radicals (discussed in the next section) that are generally in more concentrated solution. What occurs within the solvent cage is a third-order reaction in which two

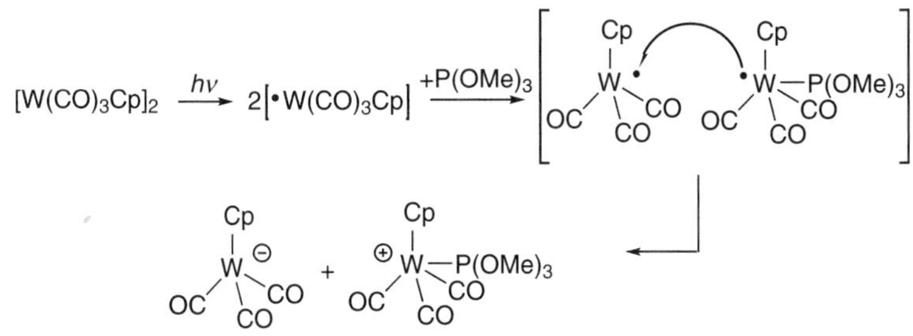

**SCHEME 10.2**  Photochemical sequence proposed for disproportionation within the solvent cage. Adapted with permission from reference 63. Copyright 2006, American Chemical Society.

metal radicals react with a ligand—and this reaction is extremely fast—competitive with escape from the solvent cage. The use of high ligand concentrations allows this electron transfer to occur, and shows that there is a low enthalpy of activation for the ternary transition state. Also bearing on this work—where diffusion is extremely limited—is the work in the area of matrix isolation spectroscopy. Detailed discussion of this topic is beyond the scope of the present study. The reader is referred to an excellent review by Bitterwolf on this topic.[66]

### 10.4.2.3  Reaction Mechanisms Observed for $^{\bullet}Cr(CO)_3C_5R_5$

The radicals $^{\bullet}Cr(CO)_3C_5R_5$ (R = H, Me) have been extensively investigated in solution by several research groups. They are readily generated by dissolution of the metal–metal-bonded dimers. For R = Me, the solutions contain near quantitative conversion to radical monomer at ambient temperatures. For R = H, approximately 1–10% of the complex is dissociated to radicals at typical conditions. Due to the fact that dissociation of the dimeric complexes is rapid (as is radical recombination), most of the chemistry observed originates from the highly reactive 17 e$^{-}$ radical species. While the fundamental reactions in thermal studies in solution are related to the photochemical studies discussed earlier, higher concentrations are generally present in the solution studies and lead more often to concerted reactions involving two metal radicals.

The first extensive investigation of the chemistry of $^{\bullet}Cr(CO)_3Cp^*$ radical was done by Baird.[40] Jaeger and Baird[67] first reported several key reactions such as halogenation that occurs rapidly as shown in Equation 10.45.

$$^{\bullet}Cr(CO)_3(Cp^*) + 1/2\,I_2 \rightarrow I\text{-}Cr(CO)_3(Cp^*) \tag{10.45}$$

While this is a typical reaction of metal–metal-bonded complexes and radicals, rapid atom transfer was also observed between the formed I-Cr(CO)$_3$C$_5$R$_5$ and unreacted $^{\bullet}Cr(CO)_3C_5R_5$:

Interestingly, ($\eta^5$-C$_5$Me$_5$)Cr(CO)$_3$I takes part in an iodine atom exchange process with IV ($^{\bullet}Cr(CO)_3C_5Me_5$), as evidenced by coalescence of the methyl resonances of the two compounds.

Associative ligand substitution reactions such as that shown in Equation 10.46 were also investigated.[68]

$$^{\bullet}Cr(CO)_3 Cp^* + PMe_2Ph \rightleftharpoons {}^{\bullet}Cr(PMe_2 Ph)(CO)_2 Cp^* + CO \qquad (10.46)$$

Phosphine substitution was found to occur by an associative mechanism and the displacement of phosphine by CO (the reverse step in Equation 10.46 by an interchange associative process). Reactions 10.45 (atom transfer) and 10.46 (ligand substitution) are two fundamental steps in metal radical chemistry expected and found to have low barriers in these systems.

The mechanisms of halogen abstraction by $^{\bullet}Cr(CO)_3C_5R_5$ were investigated for a range of organic halides.[69] Some of the most interesting results were for organohalides with β-H-atoms such as PhCHMeBr. Experimental kinetic data were fit to the proposed mechanism in reactions 10.47–10.53.

$$^{\bullet}Cr(CO)_3Cp + PhCHMeBr \rightarrow Br\text{-}Cr(CO)_3Cp + {}^{\bullet}CPhHMe$$
$$k = 0.02 \, M^{-1} \, s^{-1} \qquad (10.47)$$

$$^{\bullet}Cr(CO)_3 Cp + {}^{\bullet}CPhHMe \rightarrow H\text{-}Cr(CO)_3Cp + PhHC = CH_2$$
$$k = 10^9 \, M^{-1} \, s^{-1} \qquad (10.48)$$

$$^{\bullet}Cr(CO)_3 Cp + {}^{\bullet}CHPhHMe \rightarrow PhMeHC\text{-}Cr(CO)_3Cp$$
$$k = 10^9 \, M^{-1} \, s^{-1} \qquad (10.49)$$

$$PhMeHC\text{-}Cr(CO)_3Cp \rightarrow H\text{-}Cr(CO)_3Cp + PhHC = CH_2$$
$$k = 10^3 \, M^{-1} \, s^{-1} \qquad (10.50)$$

$$H\text{-}Cr(CO)_3Cp + {}^{\bullet}CPhHMe \rightarrow {}^{\bullet}Cr(CO)_3Cp + PhCH_2 CH_3$$
$$k = 10^7 \, M^{-1} \, s^{-1} \qquad (10.51)$$

$$H\text{-}Cr(CO)_3Cp + PhHC = CH_2 \rightarrow {}^{\bullet}Cr(CO)_3Cp + {}^{\bullet}CPhHMe$$
$$k = 0.4 \, M^{-1} \, s^{-1} \qquad (10.52)$$

$$PhMeHC\text{-}Cr(CO)_3 Cp \rightarrow {}^{\bullet}Cr(CO)_3 Cp + {}^{\bullet}CPhHMe$$
$$k = 10^3 \, M^{-1} \, s^{-1} \qquad (10.53)$$

The organic radical $(^{\bullet})$CPhHMe generated by a slow atom transfer step in Equation 10.47 can either combine with a second Cr radical as shown in reaction 10.49 or undergo H-atom transfer as shown in reaction 10.48. Reaction 10.50 involves β-H-atom elimination and serves to generate the metal hydride. This can be also involved in radical H-atom transfer reactions such as that shown in Equation 10.51. Finally, H-atom transfer to $PhCH=CH_2$ can generate radical pairs as shown in

Equation 10.52. The alkyl complex can also undergo radical cleavage as shown in Equation 10.53.

It is clear from the discussion above that the "simple" Br-atom transfer in Equation 10.47 brings considerable complexity to the manifold of possible reactions. The reversible nature of these reactions has also been utilized in the study of reversible olefin polymerization. This work has been extensively investigated by Norton and coworkers, and the interested reader is referenced to this work[70] as well as earlier and somewhat related studies of Bullock and Samsel.[71]

The complex manifold of reactions of the "simple" organometallic radical $^\bullet$Cr(CO)$_3$Cp have been elegantly and thoroughly investigated by Goh and Weng, and leading references to this chemistry are cited.[72] Detailed survey of the structural and mechanistic results in this complex chemistry is beyond the scope of this chapter that aims at simpler primary reactions. The interesting complexes formed of interaction with P, S, and other nonmetal chemistry are illustrated by reactivity with $P_4$, $P_4E_3$ ($E = S$, Se) and $Sb_2S_3$ are shown in Figure 10.4.

The enthalpies of hydrogenation of the metal–metal-bonded dimeric complexes of Cr, Mo, and W were measured in 1985[73] (reaction 10.54).

$$[M(CO)_3 Cp]_2 + H_2 \rightarrow H\text{-}M(CO)_3 Cp$$
$$\Delta H(M) = -3(Cr), \; +6(Mo), \; -1(W) kcal/mol \tag{10.54}$$

At that time the radical nature of solutions of $^\bullet$Cr(CO)$_3$Cp was not yet proven. Hydrogenation of solutions of the dimer had been reported by Muetterties and coworkers,[42] but the mechanism of oxidative addition was not known. In addition, the complex [Cr(CO)$_3$Cp]$_2$ is known to catalyze hydrogenation of dienes via H–Cr(CO)$_3$Cp that is regenerated under catalytic conditions. Thermochemical data for reaction 10.54 were obtained by indirect thermochemical cycles, however direct measurement of the enthalpy of reaction of $H_2(g)$ and $^\bullet$Cr(CO)$_3$Cp$^*$ allowed the determination of H–Cr(CO)$_3$Cp$^*$ bond strength as $62.3 \pm 1$ kcal/mol.[74]

Where it is thermodynamically allowed, as in Equation 10.55 ($\Delta H = -2.5 \pm 0.2$ kcal/mol), low kinetic barriers exist ($\Delta H^{\ddagger} = 2.1 \pm 0.2$ kcal/mol, $\Delta S^{\ddagger} = -38.2 \pm 3.8$ cal/(mol K)) for radical H-atom transfer.

$$H\text{-}Cr(PPh_2Me)(CO)_2Cp^* + {}^\bullet Cr(CO)_3 Cp^* \rightarrow H\text{-}Cr(CO)_3 Cp^*$$
$$+ \, {}^\bullet Cr(PPh_2Me)(CO)_2 Cp^* \tag{10.55}$$

That is generally the case when a metal radical attacks a weak bond, such as another metal radical (Equation 10.55) or an organic radical with a weak C–H bond (Equation 10.48). However, with strong bonds such as H–H or R–H, direct attack of a single metal radical is uphill by over 40 kcal/mol and effectively ruled out for thermodynamic reasons.

$$^\bullet Cr(CO)_3 Cp^* + H_2 \rightarrow H\text{-}Cr(CO)_3 Cp^* + H^\bullet \tag{10.56}$$

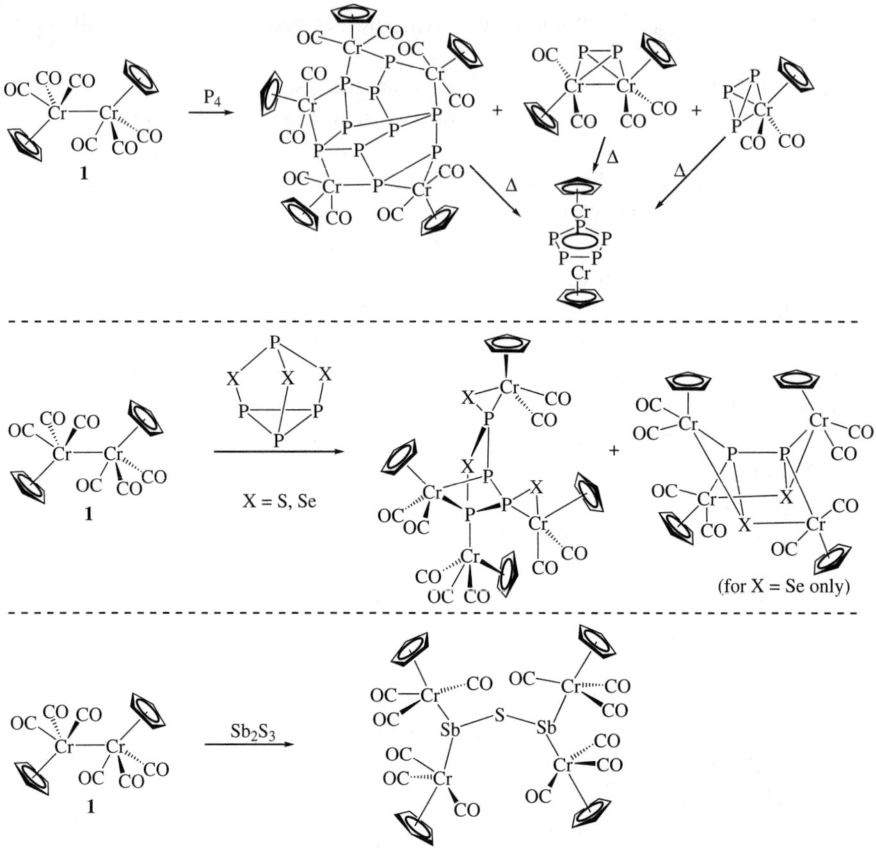

**FIGURE 10.4**    Reaction products assembled from reactions of the $^\bullet Cr(CO)_3Cp$ radical (the reactive species derived from $[Cr(CO)_3Cp]_2$) and $P_4$, $P_4E_3$ ($E = S$, Se), and $Sb_2S_3$. Reproduced with permission from reference 72. Copyright 2004, American Chemical Society.

This situation is typical of metal radicals, the relatively low M–H bond strengths prohibit direct attack on $H_2$ for thermodynamic reasons. Kinetic studies[75] showed that oxidative addition of $H_2$ obeyed an overall third-order rate law $d[P]/dt = k[^\bullet Cr(CO)_3Cp^*]^2[H_2]$ with $\Delta H^\ddagger = 0$ kcal/mol and $\Delta S^\ddagger = -47$ cal/(mol K). The proposed mechanism for the third-order reaction is shown in Equation 10.57.

$$[Cr(CO)_3 Cp^*]_2 + H_2 \rightleftharpoons {}^\bullet H_2\text{-}Cr(CO)_3Cp^* \xrightarrow{{}^\bullet Cr(CO)_3Cp^*} 2H\text{-}Cr(CO)_3Cp^* \quad (10.57)$$

The first step is proposed to be the reversible formation of a 19 e⁻ molecular hydrogen adduct. As discussed above, since the 19 e⁻ adduct would have a half bond between $H_2$ and the metal, its formation is expected to be slightly exothermic. The second step is the attack of a second mole of chromium radical on this adduct to yield the ternary transition state. The enthalpic barrier for this second step is believed to be low and nearly canceled by the exothermic binding of $H_2$ in the first step of Equation 10.57. This results in a zero enthalpy of activation, but a large negative entropy of activation.

One way to reduce the unfavorable entropic barrier would be to use "tethered" radicals designed to "trap" $H_2$ and take advantage of the low enthalpy of activation for dinuclear addition. One system investigated was the fulvalene complex[76] shown in Equation 10.58.

$$\text{(10.58)}$$

The observed enthalpy of activation was higher than expected based on two $^{\bullet}Cr(CO)_3Cp^*$ radicals. That may be due to improper alignment in the transition state of the fulvalene-bridged system in which the Cr radicals are constrained to react along a limited reaction trajectory.

The RS–H bond is weaker than the H–H and R–H bonds and it might be anticipated that single radical attack on this bond might be a viable pathway since the PhS–H bond strength[77] of 79 kcal/mol is only 17 kcal/mol stronger than the Cr–H bond strength. However, detailed kinetic studies of PhSH and BuSH (which has a bond strength $\approx 10$ kcal/mol stronger than PhSH[78]) showed that both reactions occurred at similar rates and followed a third-order reaction pathway.[79] A potential energy diagram for this reaction is shown in Figure 10.5. The measured activation parameters were quite similar to those found for $H_2$ ($\Delta H^{\ddagger} = 0$ kcal/mol and $\Delta S^{\ddagger} = -52$ cal/(mol K) for PhSH, and $\Delta H^{\ddagger} = 0$ kcal/mol and $\Delta S^{\ddagger} = -55$ cal/(mol K) for BuSH).

The activation of S–H bonds is important in both industrial and biological processes.[80] The attack of two moles of $^{\bullet}Cr(CO)_3Cp^*$ on RS–H could be reduced to a process first order in metal radical under two circumstances. The first was reaction with $H_2S$, which was shown[81] to proceed by a relatively rapid process first order in metal radical under Ar and a slower process second order in metal radical under CO atmosphere, as shown in Scheme 10.3.

The upper pathway that is observed under CO pressure is a third-order process similar to the oxidative addition of PhSH and BuSH.

The reaction pathway first order in metal radical had a rate-limiting step ($k_3$ in Scheme 10.3) in which ligand substitution by $H_2S$ of CO produced the substituted radical complex $^{\bullet}Cr(SH_2)(CO)_2Cp^*$. This reaction pathway was suppressed under CO pressure and the reaction was then found to follow the slower upper pathway in Scheme 10.3, which is the "default" pathway—the third-order oxidative addition. The chemistry of the proposed intermediate complex $^{\bullet}Cr(SH_2)(CO)_2Cp^*$ and other such reactive radical species has received little study.

The second example[79] of kinetics first order in chromium radical involved prior coordination to a metal complex that served to activate the RS–H bond. Formation of

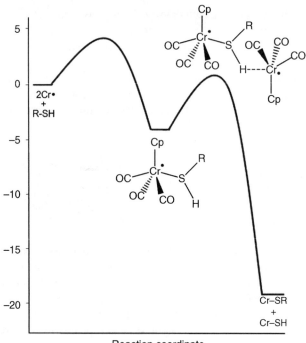

**FIGURE 10.5**  Potential energy diagram (enthalpy of reaction in kcal/mol) for oxidative addition of thiols. The first step involves the formation of a reversible 19 e$^-$ coordinated thiol adduct that has an estimated enthalpy of binding of −3 to −6 kcal/mol. The transition state occurs when this is attacked by a second mole of chromium radical. The measured enthalpy of activation is near 0 kcal/mol since the exothermic enthalpy of binding in the first step is canceled by the activation enthalpy of the second step. Adapted with permission from reference 79. Copyright 1996, American Chemical Society.

W(phen)(BuSH)(CO)$_3$ serves to lower the S–H bond strength for the bound thiol versus free thiol. The mechanism proposed for this rapid reaction is shown in Scheme 10.4. This involves attack of a metal radical on a coordinated ligand at a nonradical metal complex. The coordination of the thiol reduces the S–H bond strength so that the first-order attack by a metalloradical is feasible.

For disulfides,[82] the oxidative addition was found to follow reactions either first or second order in metal radical depending upon the strength of the RS–SR bond that is broken. For PhSSPh, the S–S bond is sufficiently weak ($\approx$46 kcal/mol) so that the rate-limiting step is single metal radical attack. For MeSSMe, the S–S bond ($\approx$60 kcal/mol) is strong enough so that this is no longer viable and the third-order attack occurs. The two pathways are shown in Scheme 10.5. For PhSSPh, the lower pathway is followed in which a free •SPh radical is formed, which then rapidly combines with additional •Cr (CO)$_3$Cp* to form PhS–Cr(CO)$_3$Cp*. For MeSSMe, due to the stronger S–S bond, the third-order kinetics is followed, similar to the activation of H$_2$ and thiols.

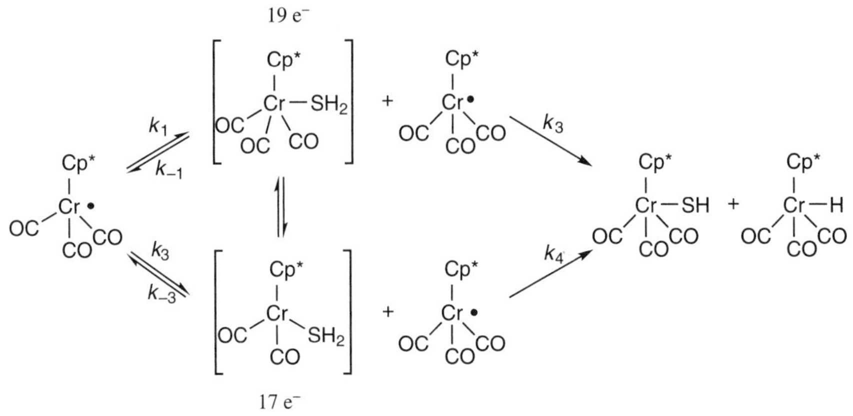

**SCHEME 10.3**  Two mechanistic pathways for oxidative addition of $H_2S$ in reaction with
$^\bullet Cr(CO)_3Cp^*$. The lower pathway occurs under Ar atmosphere and involves rate-limiting CO
substitution by $H_2S$ to produce the substituted radical $^\bullet Cr(SH_2)(CO)_2Cp^*$.

An important difference between the two reaction pathways is whether or not a free
thiyl radical is generated as part of the metal reaction. The free thiyl radical $^\bullet SPh$
generated in reaction of $^\bullet Cr(CO)_3Cp^*$ and PhSSPh was shown to be capable of
reacting with other substrates, notably $HCr(CO)_3Cp^*$ that leads to a radical chain

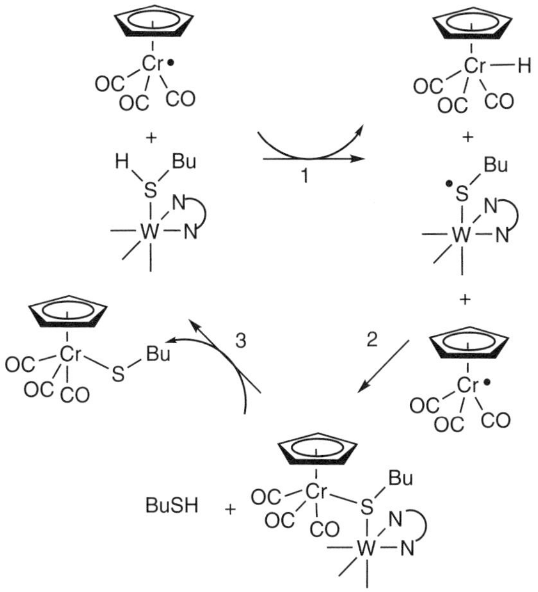

**SCHEME 10.4**  Catalysis of reaction of BuSH and $^\bullet Cr(CO)_3Cp$ by coordination to W(phen)
$(CO)_3$.

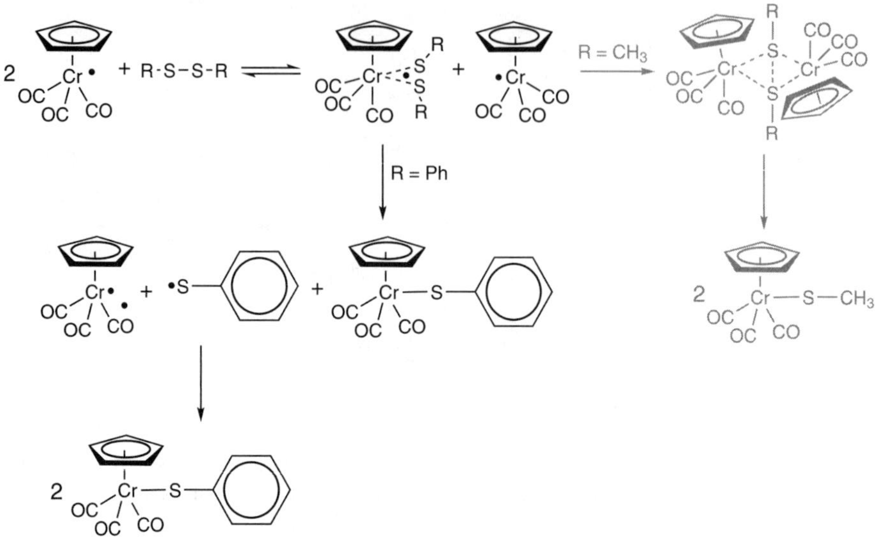

**SCHEME 10.5**    Mechanisms for reaction of ˙Cr(CO)$_3$Cp and disulfides. Both are proposed to proceed through an initial 19 e⁻ fragment (RSSR)˙Cr(CO)$_3$Cp.

reaction for reaction of PhSSPh and HCr(CO)$_3$Cp*. It was found that very pure HCr(CO)$_3$Cp* that contained no residual ˙Cr(CO)$_3$Cp* did not react with PhS-SPh. Injection of small amounts of ˙Cr(CO)$_3$Cp* led to the radical chain reaction shown in Scheme 10.6.

Since MeSSMe does not generate free ˙SMe radicals as shown in Scheme 10.6, injection of small amounts of ˙Cr(CO)$_3$Cp* does not lead to a radical chain reaction for disulfide addition for reaction of MeSSMe and H-Cr(CO)$_3$Cp.

In a later report,[83] stopped-flow kinetic studies were combined with calorimetric studies to generate the data shown in Table 10.1 for PhEEPh (E = S, Se, Te).

Due to the increasingly weaker bonds in descending the chalcogenides, there is a dramatic increase in the rate of oxidative addition and lowering of the enthalpy of activation.

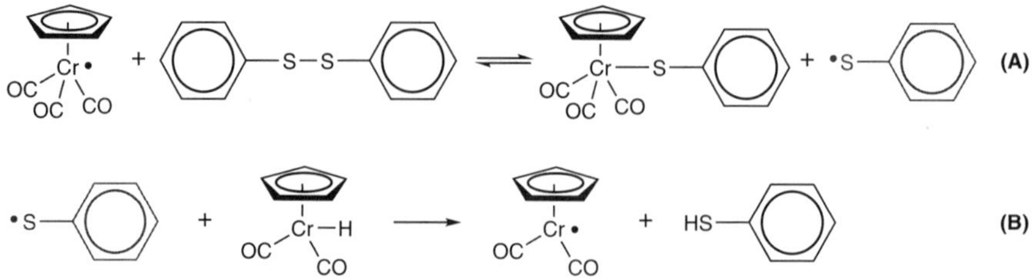

**SCHEME 10.6**    Radical chain process for reaction of PhS-SPh and H-Cr(CO)$_3$Cp.

**TABLE 10.1    Second-Order Rate Constants,[a] Activation Parameters,[b] Enthalpies of Reaction,[c] and Estimated Bond Strength Data[c] for Reactions of PhEEPh and •Cr(CO)₃Cp***

| E | $k$ | $\Delta H^{\ddagger}$ | $\Delta S^{\ddagger}$ | $\Delta H_{rxn}$ | PhE–EPh | PhE–Cr |
|---|---|---|---|---|---|---|
| S | 1.3 | 10.2 | −24.4 | −29.6 | 46 | 38 |
| Se | 1400 | 7 | −22 | −30.8 | 41 | 36 |
| Te | 19,000 | 4 | −26 | −28.9 | 33 | 31 |

[a] $M^{-1}s^{-1}$
[b] $\Delta H^{\ddagger} = kcal/mol$, $\Delta S^{\ddagger} = cal/(mol\,K)$
[c] kcal/mol

Related to the reaction of thiols are pyridine thiones since they exist in an apparent tautomeric equilibrium shown in Equation 10.59 for the 2-pyridinethione/2-pyridinethiol system.

$$(10.59)$$

In toluene solution, the thione tautomer predominates.[84] In spite of that, oxidative addition was found to occur in a manner similar to that of the thiol tautomers, even for 4-pyridine thione,[85] surprisingly it was found to occur much faster than for thiols. This could be explained by a much more favorable enthalpy of formation of the initial 19 e⁻ by coordination of the metal radical to the S=C bond as shown in Scheme 10.7

An alternative explanation for the reactivity would involve electron transfer/proton transfer as shown in Scheme 10.8.

Support for an electron transfer type of mechanism is gained from the observation that 4-N-methyl pyridine thione (incapable of H-atom transfer) yields the ion pair product $[Cr(CO)_3(4\text{-}S\text{-}C_5H_4N\text{-}Me)(C_5Me_5)]^+ [Cr(CO)_3Cp^*]^-$.[85] Mechanistic distinction between H-atom transfer and e⁻/H⁺ transfer is challenging. Other mechanism may be possible as well. The important conclusion is that 4-pyridine thione that contains the good radical receptor S=C[86] is much more reactive than the corresponding thiols in spite of the fact that the thione tautomer is more thermodynamically stable than its thiol counterpart.

Ungváry and coworkers[87] had shown that $Co_2(CO)_8$ catalyzes conversion of diazo compounds to the corresponding ketenes as shown in reaction 10.60.

$$N=N=CHSiMe_3 + CO \rightarrow O=C=CHSiMe_3 + N_2 \qquad (10.60)$$

The •Co(CO)₄ radical was not implicated in this reaction; nevertheless, it was of interest to observe that this reaction occurred in both stoichiometric and catalytic

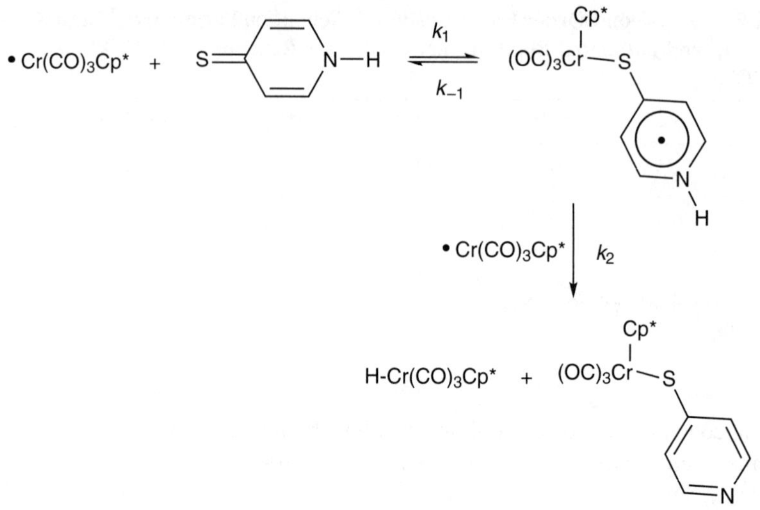

**SCHEME 10.7**    Two-step H-atom transfer mechanism between $Cr(CO)_3Cp^*$ and 4-pyridine thione involving formation of a 19 $e^-$ thione complex followed by H-atom transfer from the 4-position.

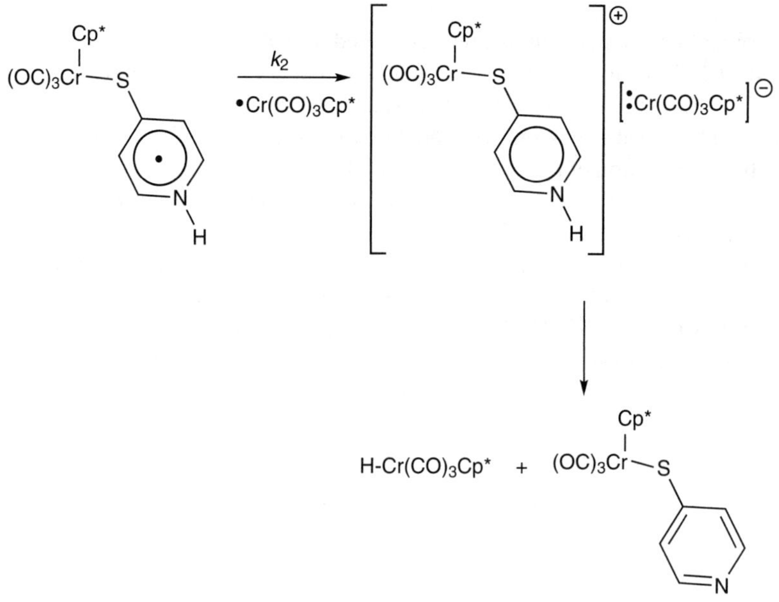

**SCHEME 10.8**    Scheme for reaction of 4-pyridine thione in which the initial 19 $e^-$ thione adduct reacts with $Cr(CO)_3Cp^*$ by a coupled electron transfer/proton transfer mechanism.

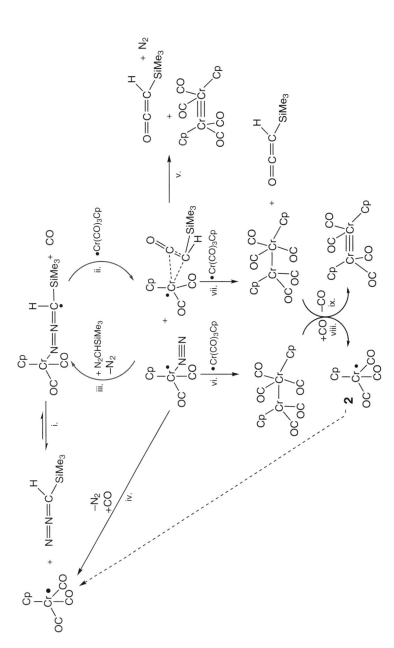

**SCHEME 10.9** Proposed mechanism for carbonylation of diazo compounds by •Cr(CO)$_3$C$_5$R$_5$. Reproduced with permission from reference 88. Copyright 2007, American Chemical Society.

457

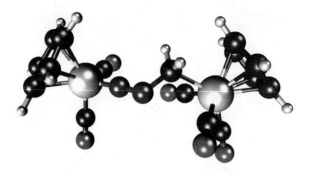

**FIGURE 10.6**    Computed structure of Cp(CO)$_2$Cr=N=N-CH$_2$-Cr(CO)$_3$Cp from sequential addition of two mole of $^\bullet$Cr(CO)$_3$Cp to N=N=CHSiMe$_3$ with loss of CO at the BP86 level of theory. Reproduced with permission from reference 88. Copyright 2007, American Chemical Society.

reactions of the $^\bullet$Cr(CO)$_3$C$_5$R$_5$ radicals. This reaction was proposed to occur by the mechanism shown in Scheme 10.9.

In the overall carbonylation reaction, an induction period was observed as well as partial inhibition by CO.[88] A key step in the mechanism is proposed to be formation of the $^\bullet$Cr(N=N=CHR)(CO)$_2$C$_5$R$_5$ radical in which the diazoalkane has substituted for carbon monoxide—reminiscent of H$_2$S substitution shown above. Attack of a second mole of $^\bullet$Cr(CO)$_3$C$_5$R$_5$ on this intermediate would yield a dinuclear bridging adduct, the computed structure of which is shown in Figure 10.6.

The intermediate bridging structure Cp(CO)$_2$Cr=N=N-CH$_2$-Cr(CO)$_3$Cp was then proposed to undergo carbonyl insertion and fragmentation yielding one mole of $^\bullet$Cr(N$_2$)(CO)$_2$Cp and one mole of $^\bullet$Cr($\eta^2$(O=C=CHSiMe$_3$)(CO)$_2$Cp. The computed transition state structure is shown in Figure 10.7.

The relatively complex mechanism proposed for this reaction arises since two moles of the metalloradical attack a substrate (diazoalkane) in which electron delocalization can readily occur and in which new and highly reactive metal radicals are formed.

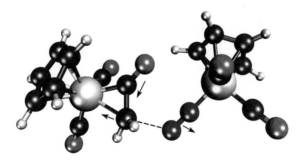

**FIGURE 10.7**    Computed transition state structure in carbonylation of bound diazoalkane at the BP86 level of theory. Reproduced with permission from reference 88. Copyright 2007, American Chemical Society.

This section has attempted to show in some detail reactions of the $^\bullet$Cr(CO)$_3$C$_5$R$_5$ radicals. The fundamental steps—enhanced ligand substitution, formation of 19 e$^-$ intermediates, single metal cleavage of weak bonds to give radicals, concerted cleavage of strong bonds, electron transfer from 19 e$^-$ adducts to form 18 e$^-$ cations, radical attack on coordinated ligands—are all relatively well established. Application of these types of reactions in designed catalytic sequences is a forefront area of inorganic research.

### 10.4.3  Rh(porphyrin) as Representative of Metalloporphyrin Radicals

*10.4.3.1  Introduction*  Typically, the monomeric (por)Rh$^\bullet$ are prepared in solution by thermally or photochemically induced homolytic cleavage of Rh–Rh, Rh–H, or Rh–Me bond of the corresponding precursor.[89] One of the more well-studied Rh porphyrin radical systems, tetramesitylporphyrinrhodium(II), (TMP)Rh$^\bullet$, is prepared by photolyzing benzene or toluene solutions of (TMP)Rh-Me with $\lambda \geq 350$ nm.[90] Figure 10.8 shows common porphyrin Rh$^{II}$ complexes.

Relative stabilities toward dimerization are directly related to the bulkiness of the substituents on the porphyrin ring. In the case of the set of complexes listed in Figure 10.8, bulkiness decreases in the order of (TTiPP)Rh$^\bullet$ > (TTEPP)Rh$^\bullet$ > (TMP)Rh$^\bullet$ > (TXP)Rh$^\bullet$ > (TPP)Rh$^\bullet$ > (OEP)Rh$^\bullet$. Reported bond dissociation enthalpies (BDE) for Rh–Rh bond of [(OEP)Rh]$_2$,[91] [(TXP)Rh]$_2$,[92] and [(TMP)Rh]$_2$[92] are $\approx 15.5$, $\approx 12$, and 0 kcal/mol, respectively. Measurements were made by observations of the effects of temperature on line broadening in NMR spectra. Monomeric (TPP)Rh$^\bullet$ is generated in several ways. Photolysis of [Rh(TPP)]$_2$ at 77K in 2-Me-THF,[93] flash photolysis of μ-TPP[Rh(CO)$_2$]$_2$ in benzene,[93,94] and electrochemical reduction[95] of (TPP)Rh-(NHMe$_2$)$_2$$^+$ in THF create the monomeric (TPP)Rh$^\bullet$. Unfortunately, the monomer is not long lived and dimerizes quickly back to [Rh(TPP)]$_2$ in the absence of other reagents. The more sterically demanding complexes are observed solely as 15 e$^-$ monomeric species. The steric bulk of the porphyrin plays a leading role in both the stability and reactivity of the (por)Rh$^\bullet$ complexes. The following sections are samples of the general reactivities of these intriguing complexes and are in no way a comprehensive list. Many reviews have previously been written[89,96,97] that focus in more detail on these complexes.

*10.4.3.2  Activation of Aliphatic C–H Bonds by (por)Rh$^\bullet$*  (TMP)Rh reversibly reacts with methane to form (TMP)Rh-H and (TMP)Rh-Me, as shown in reaction 10.61. Rh(TXP)$^\bullet$ has also been reported to show a similar reactivity.[92] The reversible nature of the reaction observed in benzene solutions of (por)Rh-H and (por)Rh-Me at 353K yields the reductive elimination products 2 (por)Rh$^\bullet$ and CH$_4$ (por = TMP, TXP).

$$2(\text{TMP})\text{Rh}^\bullet + \text{CH}_4 \rightleftharpoons (\text{TMP})\text{Rh-H} + (\text{TMP})\text{Rh-CH}_3 \qquad (10.61)$$

The reaction is proposed to be third order (first in Me–R and second in Rh(por)$^\bullet$) and to involve a termolecular linear intermediate (Figure 10.9).

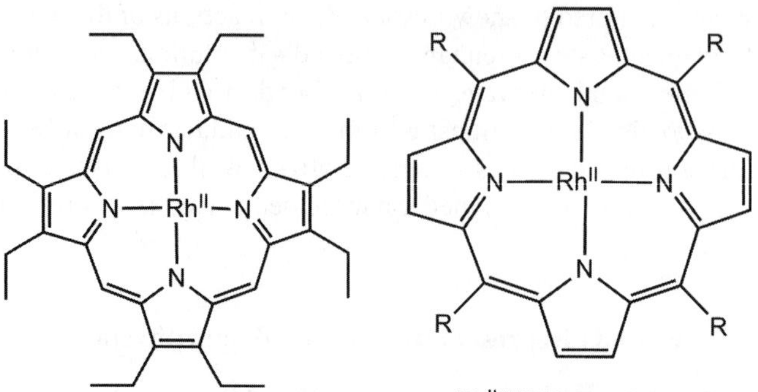

Rh$^{II}$(OEP)

Rh$^{II}$(TPP): R = C$_6$H$_5$
Rh$^{II}$(TXP): R = 3,5-Me$_2$C$_5$H$_3$
Rh$^{II}$(TMP): R = 2,4,6-Me$_3$C$_6$H$_2$
Rh$^{II}$(TTEPP): R = 2,4,6-Et$_3$C$_6$H$_2$
Rh$^{II}$(TTiPP): R = 2,4,6-$^i$Pr$_3$C$_6$H$_2$

**FIGURE 10.8**    15 e$^-$, d$^7$ Square planar Rh$^{II}$ porphyrins and their respective common abbreviations.

The thermodynamic driving force for the reaction is the formation of two weaker bonds, Rh–H (~60 kcal/mol) and Rh–Me (~57 kcal/mol),[92] whose sum is greater than that of the originally broken bond H–Me (104.6 kcal/mol). Measurements of $K_{eq}$ for the reaction of (TMP)Rh$^\bullet$ with methane led, unsurprisingly, to a large activation entropy, $\Delta S^{\ddagger} = -39.2$ cal/(mol K). The large value for $\Delta S^{\ddagger}$ would be expected for a process in which two relatively large molecules and one small molecule must all come together for the reaction to proceed.

Interestingly, no reaction was observed between benzene and (por)Rh$^\bullet$. This was attributed to sterics.[92] The large (por)Rh$^\bullet$ systems cannot form an intermediate analogous to that shown in Figure 10.9 because of the bulkiness of the benzene ring. This unreactivity toward aromatic versus aliphatic C–H bonds was further probed with reactions of (por)Rh with toluene.[92] Both (TXP)Rh$^\bullet$ and (TMP)Rh$^\bullet$ were shown to react readily with toluene to form (por)Rh–H and (por)Rh–CH$_2$(C$_6$H$_5$), as

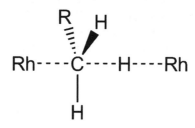

**FIGURE 10.9**    Linear four-centered proposed intermediate in the activation of C–H bonds of Me–R (R = H, C$_6$H$_5$).

shown in reaction 10.61. No evidence was seen for the activation of any aromatic H–C bonds.

$$2(TMP)Rh^{\bullet} + CH_3\text{-}C_6H_5 \rightleftharpoons (TMP)Rh\text{-}H + (TMP)Rh\text{-}CH_2\text{-}C_6H_5 \qquad (10.62)$$

One might suspect that the relative rates of reactions would favor that of the reaction with toluene. Arguments would mostly likely be based on the fact that the H–CH$_2$Ph bond (87.2 kcal/mol) is weaker than that of methane (104.6 kcal/mol). In fact the opposite was observed. The reaction with methane was reported to react ~19 times faster than that of the reaction with toluene. The steric bulk of the Ph group inhibits the formation of the intermediate. This was expressed directly in the observation of a larger activation enthalpy of the toluene reaction ($\Delta H^{\ddagger} = 7.1$ kcal/mol and $\Delta H^{\ddagger} = 17.2$ kcal/mol for methane and toluene, respectively).

**10.4.3.3   *C–C Coupling in the Reaction of CO with (por)Rh*** •   Reactions of (por)Rh$^{\bullet}$ (por = TXP, TMP, and TTiPP) with CO result in the formation of (por)Rh-CO (reaction 10.63), which then dimerizes through C–C bond formation as shown in reaction 10.64.[90]

$$(por)Rh^{\bullet} + CO \rightleftharpoons (por)Rh\text{-}^{\bullet}CO \qquad (10.63)$$

$$2(por)Rh\text{-}^{\bullet}CO \rightleftharpoons (por)Rh\text{-}C(O)\text{-}C(O)\text{-}Rh(por) \qquad (10.64)$$

The inherent steric bulk of the porphyrin directly affects the amount of the bridging α-diketone complex that is formed. At $p_{CO} = 1$ atm and $T = 298$K, formation of the dimer is favored in all but the bulkiest system (TTiPP). In the case of the least sterically demanding porphyrin, (OEP)Rh$^{\bullet}$, reaction with CO produces the dimetal ketone, (OEP)Rh–C(O)–Rh(OEP).[98]

The dimerization of these large systems shows that some of the radical character of the metalloradical is transferred to the CO adduct.[99] ESR spectra of these monomers show that the symmetry of the system is lost upon binding of CO. The one-electron reduction with CO causes the bonding order to decrease from 3 to some intermediate value between 2 and 3. As a result, the CO ligand is no longer linear. This bent confirmation causes the degeneracy in the $\pi^*$ orbitals of the CO ligand to be broken. One orbital preserves its $\pi^*$ character while the other aligns to form a σ-bond with the Rh d$_{z^2}$ orbital. Wayland notes that this occurs at the loss of the overlap between dπ–pπ orbitals.

(TMP)Rh–CO reacts with hydrogen atom sources, HSn($t$-Bu$_3$) and H$_2$,[99] to form the formyl product as shown in reactions 10.65 and 10.66. Reactions with the more bulky (TMP)Rh–H showed no signs of the formyl product.

$$2(TMP)Rh\text{-}^{\bullet}CO + H_2 \rightleftharpoons 2(TMP)Rh\text{-}CHO \qquad (10.65)$$

$$2(TMP)Rh\text{-}^{\bullet}CO + HSn^{t}Bu_3 \rightleftharpoons 2(TMP)Rh\text{-}CHO + {}^{t}Bu_3Sn\text{-}Sn^{t}Bu_3 \qquad (10.66)$$

Studies also showed that no reaction was observed between (TMP)Rh–H and CO or (TMP)Rh–CO.[99] It was speculated that the reaction between (TMP)Rh–CO and H$_2$

involves two molecules of (TMP)Rh–CO and one molecule of $H_2$ through a four-centered transition state.

***10.4.3.4  Tethered (por)Rh• to Increase Reaction Rates***    Reactions of the metal-loradical (por)Rh• with $H_2$ and hydrocarbons proceed through a termolecular transition state that contains two metalloradicals and the substrate in a manner shown in Figure 10.9.[100] As a result, large activation entropies and slow reaction rates are typically associated with these types of reactions. One strategy employed to enhance reaction rates is to tether the two metalloradicals in such a way that both can easily activate the substrate and form the required linear transition state.[100,101]

Use of *m*-xylyl diether tether in conjunction with (TMP)Rh• has shown this increase in reaction rate as a consequence of prearranging the metalloradical centers.[100] A representative drawing for the metalloradical center is shown in Figure 10.10. •Rh(TMP)-*m*-xylyl-(TMP)Rh• is synthesized in a similar way to its single metalloradical counterpart discussed in Section 10.4.3.3.

The reaction of the diradical with various substrates is shown in reaction 10.67.

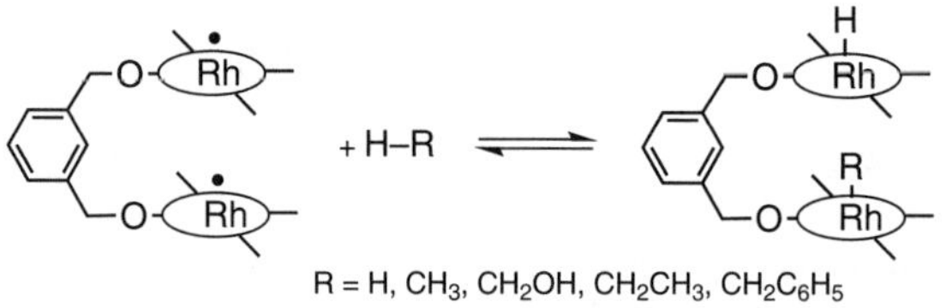

$$+ \, H{-}R \; \rightleftharpoons \tag{10.67}$$

R = H, CH$_3$, CH$_2$OH, CH$_2$CH$_3$, CH$_2$C$_6$H$_5$

Reaction 10.67 was then followed by a moderately fast reaction in which hydrogen atom transfer occurred between H–Rh(TMP)–(*m*-xylyl)–(TMP)Rh–R and •Rh(TMP)–(*m*-xylyl)–(TMP)Rh• to form H–Rh(TMP)–(*m*-xylyl)–(TMP)Rh• and •Rh(TMP)–(*m*-xylyl)–(TMP)Rh–R. Over the course of weeks and months, the dialkyl product R–Rh(TMP)–(*m*-xylyl)–(TMP)Rh–R was ultimately produced. No reactions were observed that involved aromatic H–C bonds.

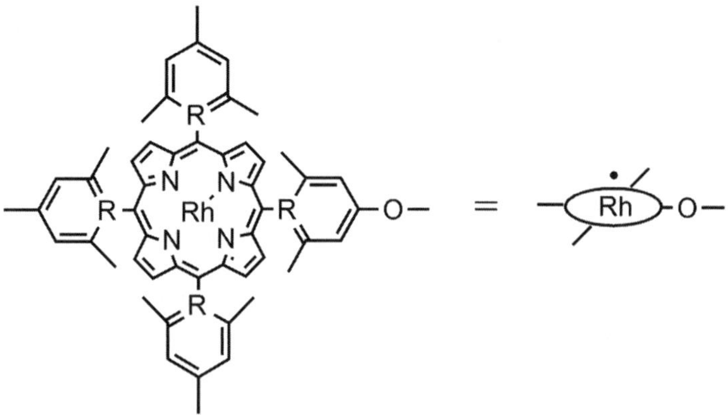

**FIGURE 10.10**    Representative drawing for •Rh(TMP)-*m*-xylyl-(TMP)Rh•.

*10.4.3.5 Other Reactions Involving (por)Rh• Metalloradicals*    Many studies have focused on these selectively reactive (por)Rh• systems. The above sections sample a small portion of these investigations. The versatile reactivities of the (por)Rh$^{II•}$ systems have been demonstrated by its ability to react with numerous substrates. In addition to the reactions stated above, (por)Rh• systems have also been demonstrated to activate and couple ethene;[102] abstract halogen atoms from a variety of organic halide compounds;[95a] form superoxo, peroxo, and hydroperoxo complexes as a result of a reaction with dioxygen;[103] form alkyl adducts via loss of $N_2$ in reactions with diazo compounds;[104] and to activate C–C bonds of a series of nitriles[105] and of TEMPO[106] (TEMPO = 2,2,6,6-tetramethyl-piperdine-1-oxyl).

### 10.4.4    Mo(NRAr)₃ as Representative of Organometallic Radicals with More Than One Unpaired Electron

The ability of Werner-type coordination compounds to exist as low- or high-spin species is well known. The preference between them has been explained as a function of the orbital splitting and pairing energies based on the ligand field theory. 18 e⁻ organometallic species are generally diamagnetic since pairing energies are low for the relative diffuse molecular orbitals involved and the energy gap between bonding and antibonding orbitals is high. However, many transition metal complexes have several lower lying electronic states in a narrow range of energy and are open-shell systems with unpaired electrons in valence space. This occurs particularly often when the ligand field is weak due to the nature or number of the ligands, and when the formal d-electron count on metal is intermediate (e.g., 2–8). When 3d elements are involved, the chance for high-spin ground state is further increased due to the higher exchange interactions between the compact 3d orbitals. Electronic deficient metals with less diffuse orbitals and electron-withdrawing ligands favor high-spin states. As pointed out by Poli, open-shell organometallic compounds can be viewed as bridging the gap between Werner-type and 18 e⁻ organometallic complexes.[107]

Organometallic complexes with two and three unpaired electrons are relatively common. Cr(Cp)₂ and Ni(Cp)₂ are two well-known examples of compounds exhibiting a triplet ground state.[108] Examples of organometallic with 4[109] and 5[108,110] unpaired electrons in the ground state are less common. Frequently, these compounds are better described as an equilibrium between low- and high-spin states where the population of the spin states depends on the temperature.[111] These complexes are better described as Werner-type compounds with an essential ionic interaction. Species with five unpaired electrons have proven to play an important role in biochemistry such as oxygenation processes mediated by cytochrome P450[112] and the activity of hemoproteins.[113] In these cases, solvent effects, protein environment, and H-bonds to amino acid residues are important for the stabilization of high-spin states. Moreover, it has been suggested that the energy difference between the different spin states diminishes along the reaction path. This section provides an introduction to compounds with more than one unpaired electrons, with a central focus on complexes of the formula Mo(NRAr)₃ and related compounds.

The Mo$^{III}$ complex Mo(N[$t$-Bu]Ar)$_3$ (Ar $= 3,5$-Me$_2$C$_6$H$_3$) synthesized by Cummins and coworkers[114] will be considered as an example of a complex with a quartet ground state. $^1$H NMR analysis confirmed the paramagnetism of the complex and magnetic studies revealed a $\mu_{eff} = 3.56\,\mu_B$ indicating the presence of three unpaired electrons. Moreover, its structure was determined by X-ray crystallography. Mo(N[$t$-Bu]Ar)$_3$ is a monomeric species with a trigonal planar geometry. The monomeric nature of the complex contrast with the dimeric Mo$^{III}$ species L$_3$Mo$\equiv$MoL$_3$ (L $=$ alkyl, amide, alkoxide) that are known to contain strong metal–metal triple bonds and are diamagnetic. The steric hindrance imposed by the bulky anilide ligands prevents the formation of the M–M in the former. It also generates a high-spin radical, and many of the characteristic reactions involve at some point spin pairing.

This is illustrated by the cleavage of dinitrogen, one of the most remarkable reactions of this complex.[115] Reaction of two moles of Mo(N[$t$-Bu]Ar)$_3$ ($S = 3/2$) with molecular nitrogen generates two moles of the nitride Mo$^{VI}$ product (N$\equiv$Mo(N[$t$-Bu]Ar)$_3$) ($S = 0$). The mechanism of this process has been widely investigated both experimentally[116] and theoretically.[117] It has also been proved to be catalyzed by the presence of Lewis bases such as pyridine or 2,6-dimethylpyrazine[118] or by addition of NaH.[116b]

Scheme 10.10 shows several of the model complexes that were studied in a recent theoretical study of reaction energetics and spin state effects.

The ground state of Mo(NH$_2$)$_3$ (model species used in the calculations) is computed to be a quartet with three unpaired electrons. Interaction with N$_2$ occurs

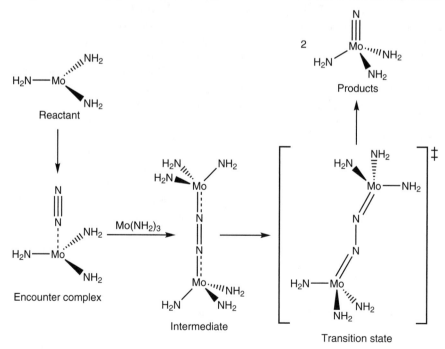

**SCHEME 10.10**  Computed complexes for modeling N$_2$ cleavage catalyzed by Mo(NH$_2$)$_3$.[117a] Adapted with permission from reference 117a. Copyright 2005, American Chemical Society.

via an "encounter complex" for which both doublet and quartet states may exist, but the doublet state was computed to be of lower energy, implying spin state crossover at some point along the reaction coordinate. The intermediate complex $(NH_2)_3Mo=N=N=Mo(NH_2)_3$ is computed to be most stable as a triplet. Cleavage to form two moles of $N\equiv Mo(NH_2)_3$ that exists in a spin-paired singlet state again must go through a transition in spin state. The existence of multiple spin states gives added complexity to these radical reactions. Section 10.6 provides additional introduction to this emerging area of research.

Many of the reactions of $Mo(N[t\text{-}Bu]Ar)_3$, such as ligand addition and oxidative addition, occur rapidly if they occur at all. The high steric demands, and possibly also the quartet ground state, lead to a rich chemistry for this complex. For example, a scheme for N-atom incorporation from $N_2$ into organic nitriles was designed using $Mo(N[t\text{-}Bu]Ar)_3$ as catalyst (Scheme 10.11) where the $N_2$ cleavage is the initial reaction.

In addition to splitting the $N\equiv N$ bond (one of the strongest bonds known), $Mo(N[t\text{-}Bu]Ar)_3$ also reacts rapidly with $N_2O$, albeit in a novel way.[114] Usually nitrous oxide donates an oxygen atom to reducing agents liberating dinitrogen. The bond dissociation energies of the N–O and N–N bonds have been determined as 1.672 and 4.992 eV, respectively.[119] Thus, the N–N bond is about 75 kcal/mol stronger than the N–O bond. Taking into account these values, one might expect that reaction 10.68 would take place.

$$Mo(N(t\text{-}Bu)Ar)_3 + N_2O =\!\!-\!\!X\!\!\rightarrow OMo(N(t\text{-}Bu)Ar)_3 + N_2 \qquad (10.68)$$

However, the expected products $O=Mo(N[t\text{-}Bu]Ar)_3$ and $N_2$ were not observed in the reaction. Surprisingly, reaction of $Mo(N[t\text{-}Bu]Ar)_3$ with $N_2O$ selectively produces a 1:1 mixture of nitride $N\equiv Mo(N[t\text{-}Bu]Ar)_3$ and nitrosyl $ON–Mo(N[t\text{-}Bu]Ar)_3$ as shown in Scheme 10.12.

To understand the striking selectivity of this reaction, calorimetric measurements were carried out. Values of $155.3 \pm 3.3^{[120]}$ and $155.6 \pm 1.6^{[121]}$ kcal/mol for the bond dissociation enthalpies of the $N\equiv Mo$ and $O=Mo$ bonds in $N\equiv Mo(N[t\text{-}Bu]Ar)_3$ and $O=Mo(N[t\text{-}Bu]Ar)_3$, respectively, were estimated. Since deoxygenation of $N_2O$ by $Mo(N[t\text{-}Bu]Ar)_3$ is thermodynamically more favorable than N–N cleavage, it was proposed that the reaction occurs under kinetic control.[120] Again, spin pairing must play a role at some point in Scheme 10.10, since the final products are diamagnetic. In spite of displaying kinetics first order in $Mo(N[t\text{-}Bu]Ar)_3$, a rapid posttransition state reaction with a second mole of $Mo(N[t\text{-}Bu]Ar)_3$ was proposed to occur:

Analysis of the product mixtures resulting from carrying out the $N_2O$ cleavage reactions with $Cr(N^iPr_2)_3$ present as an *in situ* NO scavenger rules out as dominant any mechanism involving the intermediacy of NO. Simplest and consistent with all the available data is a post-rate-determining bimetallic N–N scission process. Kinetic funneling of the reaction as indicated is taken to be governed by the properties of nitrous oxide as a ligand, coupled with the azophilic nature of three-coordinate molybdenum(III) complexes.

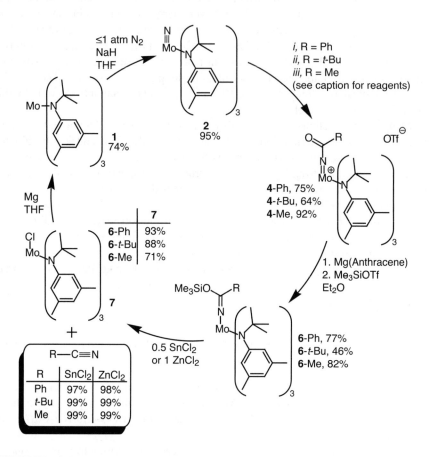

**SCHEME 10.11** A synthetic cycle that incorporates $N_2$ into organic nitriles. (i) (a) $Me_3SiOTf$, (b) 1.25 $PhC(O)Cl$, 0.2 py. (ii) 1.25 $t\text{-}BuC(O)Cl$, $[Me_3Si(py)][OTf]$. (iii) $(i\text{-}Pr)_3 SiOTf$, $MeC(O)Cl$. Yields of RCN were determined by $^1H$ NMR versus an internal standard. The isolated yields of 7 shown were obtained from reactions using $SnCl_2$. Reproduced with permission from reference 116c. Copyright 2008, American Chemical Society.

$Mo(N[t\text{-}Bu]Ar)_3$ has also been demonstrated to activate white phosphorous forming the terminal phosphide $P{\equiv}Mo(N[t\text{-}Bu]Ar)_3$.[122] Additional interesting molecules containing multiple bonds such as the terminal carbide $(C{\equiv}Mo(N[t\text{-}Bu]Ar)_3{}^-)$, selenide $(Se{=}Mo(N[t\text{-}Bu]Ar)_3)$, and telluride $(Te{=}Mo(N[t\text{-}Bu]Ar)_3)$ have also been isolated.[123] Reaction with PhE–EPh complexes have also been investigated and shown to proceed by a radical mechanism.[124]

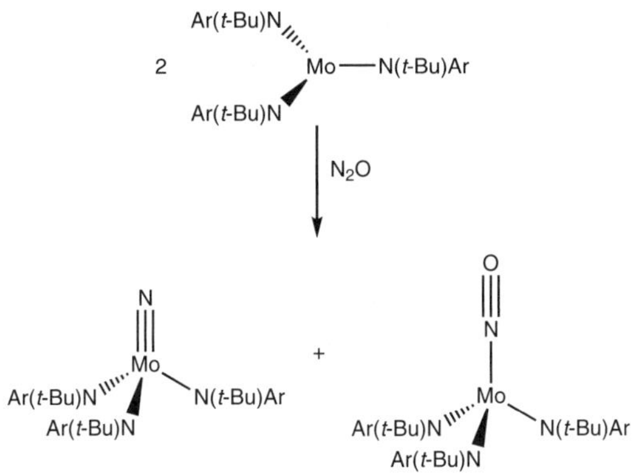

**SCHEME 10.12**   Reaction of Mo(N[*t*-Bu]Ar)₃ with N₂O to form a 1:1 mixture of nitride N≡Mo(N[*t*-Bu]Ar)₃ and nitrosyl ON–Mo(N[*t*-Bu]Ar)₃.

## 10.5   LIGAND-CENTERED REACTIVITY

Organometallic radicals are mostly 17 or 19 e⁻ species with the unpaired electrons located at the metal. Thus, they commonly present reactivity at the metal center. However, alternative descriptions are possible for complexes containing polyhapto or π-accepting ligands in which the spin density resides mainly in the ligand. In these cases, the metal can be viewed as a "normal" 16 or 18 e⁻ species. However, the "metalloradical" and the "ligand radical" are limiting resonance structures of the real electronic structure and normally the former prevails. Radical complexes are generally more reactive at the metal than at the ligand—and this has been more widely studied. However, ligand radical character has proven to be very important in processes catalyzed by enzymes.[125]

Delocalization of electronic density into the ligand, changes in coordination modes in polyhapto ligands, and steric factors are responsible for this kind of reactivity. For instance, the unshared electron in cobaltocene is nearly equally localized between cobalt and the Cp ligand.[126] "Redox non-innocent" ligands are well known to accept electron density from the metal. However, the delocalization of spin density over an extended π-system of a "redox non-innocent ligand" contributes to the relative stability of these radicals. Therefore, this species presents a weak ligand-centered radical reactivity. Delocalization of spin density over multiple metal atoms of multinuclear complexes and clusters is also known to stabilize organometallic radicals.

The highest singly occupied molecular orbital is responsible for the radical reactivity of the complexes. Therefore, theoretical computations to elucidate the nature of this orbital are a valuable tool to characterize the localization of the spin density and, even more important, the reactivity of the organometallic radicals. This topic will be further discussed in a later section.

The ligand-centered reactivity of organometallic radicals has been previously reviewed[127],[128] and will not be discussed extensively. Only several examples will be shown to illustrate the typical reactivity of these systems will be commented.

The most common reaction for ligand-centered radicals in organometallic complexes is dimerization by ligand–ligand coupling where a new C–C bond between the ligands is normally formed. When different hapticities can be achieved due to different ligand bonding modes, several species with different electron counts at the metal can exist in equilibrium. Electron transfer to the rhodocenium cation, depicted in Equation 10.69, promotes radical character from the metal to the cyclopentadienyl ligands and it dimerizes through the Cp ring changing from $\eta^6$ to an $\eta^4$ binding mode.[129]

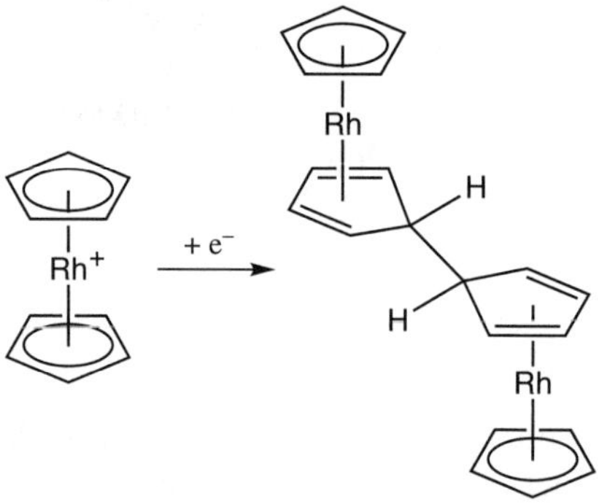

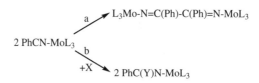

$$(10.69)$$

Furthermore, steric bulk increase around the metal center has been proven useful to promote ligand-centered radical reactivity. Several examples of ligand-centered reactivity in metalloporphyrin systems have been previously discussed. The (TMP)RhCO (TMP = tetramesitylporphyrin) couples through the CO ligand to form the diketone dimer.[90]

Analogously, the steric hindrance around the metal center caused by the three bulky anilide ligands in the nitrile adducts of the molybdenum trisanilide complex discussed previously, RCN–Mo(N[$t$-Bu]Ar)$_3$, prevents the reaction at the metal. In fact, the nitrile adducts have been shown to behave as carbon-based radicals.[124b],[130–132] Reactions of nitriles with Mo(N[$t$-Bu]Ar)$_3$ result in reductive nitrile coupling and the formation of the diiminato product (Scheme 10.13, reaction a). Coupling of the benzonitrile adduct occurs at a much slower rate than that of the MeCN analogue due to a delocalization of

$$
\begin{array}{c}
\text{L}_3\text{Mo-N=C(Ph)-C(Ph)=N-MoL}_3 \\
\nearrow a \\
2\,\text{PhCN-MoL}_3 \\
\searrow b \\
+\text{X} \quad 2\,\text{PhC(Y)N-MoL}_3
\end{array}
$$

**SCHEME 10.13**   Ligand-centered radical reactions typical of the PhCN–MoL$_3$ complex. L = N($t$-Bu)Ar; X = PhC(O)OOC(O)Ph, PhSSPh, PhSeSePh, HMo(CO)$_3$Cp, or H$_2$SnR$_2$; Y = OC(O)Ph, SPh, SePh, H, and H.

the radical into the phenyl ring.[130] Nonetheless, the reactivity of ligand-centered radical systems is not limited to dimerization. Different typical radical reactions such as hydrogen atom abstraction and radical additions to unsaturated substrates are also observed to occur intermolecularly in the presence of suitable species in solution. Reactions of PhCN–Mo(N[$t$-Bu]Ar)$_3$ with radical precursors such as PhC(O)OOC(O)Ph, PhSSPh, PhSeSePh, HMo(CO)$_3$Cp, or H$_2$SnR$_2$ produce the corresponding keti-mides (Scheme 10.13, reaction b). The enthalpies of reaction b in Scheme 10.13 have been determined to be $-22.7$, $-27.1$, $-35.1$, and $-55.3$ kcal/mol for X $=$ PhSeSePh, PhSSPh, HMo(CO)$_3$Cp, and PhC(O)OOC(O)Ph, respectively. Enthal-pies of activation ranging from 2 to 12 kcal/mol and high entropies of activation around 30 cal/(mol deg), as expected for a bimolecular process with bulky frag-ments, were determined. Surprisingly, the reaction of PhCN–Mo(N[$t$-Bu]Ar)$_3$ with HMo(CO)$_3$Cp has almost an entropy of activation of close to zero.[132]

Less common are complexes that present both metal- and ligand-centered radical reactivities.[127] Alkene adducts of paramagnetic Rh and Ir species reveal a diversity of different radical-type reactions, both at the metal and at the alkene moiety.[96] [(Me$_3$tpa)Ir$^{II}$(ethene)]$^{2+}$    (Me$_3$tpa $= N,N,N$-tris(6-methyl-2-pyridylmethyl)amine) complex contains a vacant site at iridium and an ethene fragment, and in principle can present both metal- and ligand-centered reactivities.[133] It does not dimerize in weakly coordination solvents, but react with •NO in acetone to give the nitrosyl complex indicating metalloradical character (Scheme 10.14). However, it spontaneously cou-ples to form an ethylene-bridged species in coordinating solvents such as acetonitrile showing some radical character at the ethene ligand (Scheme 10.14). The reactivity of the Ir$^{II}$ complex is dictated by the coordination ability of the solvent. It behaves as a moderately reactive metalloradical in weakly donor solvents, but it becomes more reactive upon coordination of more coordinative solvent molecules. DFT calculations show that the spin density of [(Me$_3$tpa)Ir$^{II}$(ethene)]$^{2+}$ is essentially localized at the iridium center with little delocalization in the ethene moiety. However, there is a redistribution of the spin density to be localized mainly in the ethene fragment upon coordination of acetonitrile.

## 10.6    SPIN STATE CHANGE EFFECTS

In organometallic chemistry, many reactions involve several electronic states, in particular, states of different spin. Historically, these reactions were referred as "spin forbidden." Later, since reactions with spin state changes were observed to occur, a spin blocking effect that may slow reactions down associated to a hypothetical raise in the activation barriers due to the requirement of spin flip between different state spin multiplicities was proposed.[134] However, several kinetic studies of ligand addition reactions with spin changes showed that they can be as fast as "spin-allowed" processes.[135]

The two-state reactivity concept developed by Schröder et al.[136] contributed to emphasize the importance of spin crossover phenomena in organometallic reactions:

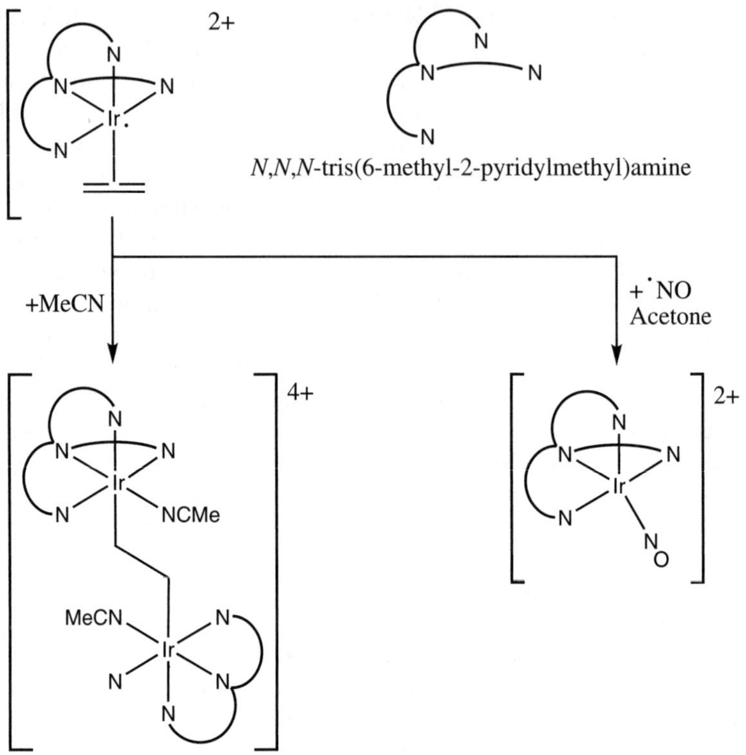

**SCHEME 10.14** Metal and ligand-centered reactivity exhibited by the [(Me₃tpa)Ir^II(ethene)]²⁺ (Me₃tpa = N,N,N-tris(6-methyl-2-pyridylmethyl)amine) complex.

Spin-crossing effects can dramatically affect reaction mechanisms, rate constants, branching ratios, and temperature behaviors of organometallic transformations.

Furthermore, Poli and coworkers have extensively studied the importance of spin crossover and applied to solve real problems in organometallic chemistry[137]:

In the past, mechanistic issues arising for reactions involving changes of spin have often been ignored, glossed over, or even misinterpreted. A much more quantitative approach is now possible.[137b]

The difficulty in recognizing its importance lies in the fact that the reaction intermediates are not easily detected by conventional spectroscopic methods. However, computational chemistry has proven to be a very strong tool to detect and quantify this phenomenon.

The reactions would be true "spin forbidden" in the absence of spin–orbit coupling. However, spin–orbit coupling effects are important for metals, especially for the heavier series. The greater the spin–orbit coupling, the greater the probability that crossover will occur between the two different energy surfaces with different spin states.

Nonetheless, the probability of the hop from potential energy surfaces is not 1. As a consequence, different product distribution and kinetic isotopic effects can also be observed. When very strong spin–orbit coupling between the different states is involved, the reaction behaves like any other, that is, on a single adiabatic potential energy hypersurface whose spin character varies smoothly from reactants to products.

The rate of the reaction is affected by the energy needed to achieve a suitable geometry in the transition state to allow the crossing (classical Arrhenius activation energy) and the transmission factor that determines the transition probability between different potential energy surfaces with different spins. However, the assignment of the amount of the activation energy corresponding to each process is a very difficult task. Moreover, the energy of the barriers varies from system to system.

To understand the reactivity of these processes, knowledge of the spin states of reagents, products, and transition states is needed. Reaction rates depend on the potential energy surfaces and on the topology of their crossing. Conventionally, crossing points were approximately located computationally by the so-called partial optimization method where partial geometry optimizations at fixed values of a selected reaction coordinate are carried out for both spin states. A more accurate and less expensive method to locate the crossover has been developed by Poli and coworkers, the minimum energy crossing point (MECP).[138] The region where the adiabatic wavefunction crosses between different spin states is equivalent to finding the points where the noninteracting corresponding potential energy surfaces cross. As two noninteracting PESs of dimension $3N-6$ (where $N$ is the number of atoms) cross along a hypersurface of dimension $3N-7$,

> The nearest approximation to the adiabatic transition state formed by avoided crossing is the point of minimum energy on this hypersurface, the minimum energy crossing point, or MECP. Of course, where spin–orbit coupling is strong, the adiabatic barrier may be somewhat lower in energy than the MECP.[137b]

Spin crossover can be produced at different points of the reaction coordinate. Therefore, the reaction would be controlled by the spin crossover only if the crossing is located before the saddle point on the reagent's spin surface. When the crossing follows the saddle point, the reaction rate would be affected by the same factors as any other single-state reaction. In some cases, even the reagents and products are in the same spin state, the reaction takes place via a potential energy surface with different spin multiplicity. When the transmission factor is high and the MECP occurs at lower energy than the saddle point in the reagent spin surface, faster rates than those in the absence of crossover would be observed leading to a spin acceleration process.[138a] Thus, as stated by Poli and coworkers,

> Spin-"forbidden" reactions can end up being as fast as spin-allowed ones, or slower, or faster.[137b]

Three qualitative examples are shown in Scheme 10.15. In example (a), the low-spin pathway takes place with a higher barrier than the high-spin one and the low-spin

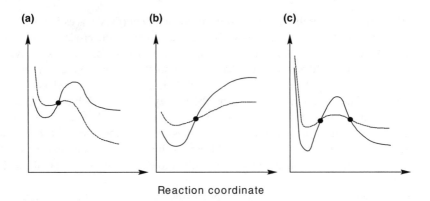

**SCHEME 10.15**  Qualitative examples for reactions taking place with different spin states involved. Dashed curves indicate reaction paths with higher spin state. Black dots indicate minimum energy crossing points (MECPs).

products are also higher in energy. Example (b) presents a typical case for some complexes where barrierless dissociation of a ligand occurs with the coordinatively (more) unsaturated species having a higher spin ground state. These two examples involve only a single crossing between the two PESs. However, there are also cases with a double crossing between the two surfaces of different spin multiplicity. For instance, example (c) depicts a situation with a low-spin pathway having a higher barrier, while the high-spin pathway involves higher energy reactants and products. Such system is expected to undergo spin-state crossover twice, thus resulting in a faster reaction.

A typical example will be presented here with a system that was a focus of considerable experimental and computational interest. The 17 e$^-$ $CpMoCl_2(PR_3)_2$ complex was unexpectedly found to undergo dissociative phosphine exchange in most cases, whereas most low-valent organometallic radicals generally undergo ligand substitution via an associative ligand exchange. The experimental observation that these Mo(III) complexes prefer to react via the unsaturated 15 e$^-$ systems led to the suggestion that a change of spin state occurs during the reaction and the high-spin system favors the dissociative pathway via a coordinative unsaturated $CpMoCl_2(PR_3)$ species. B3LYP and MP2 calculation revealed that this intermediate indeed have a quartet ground state,[139] with the doublet state lying 6.3 kcal/mol higher at B3LYP level (see Figure 10.11).

The partial optimization method is illustrated in Figure 10.12.[138a] The energies of the doublet and quartet states have a different dependence on one particular internal coordinate, namely, the Mo–P distance. Thus, by carrying out partial geometry optimizations on both surfaces at a series of fixed values of this internal coordinate, an approximate crossing point can be found. The real MECP should lie somewhat higher in energy than the isoenergetic points on the two partial optimization curves. An estimated upper limit of the energy of the true MECP can be obtained by computing

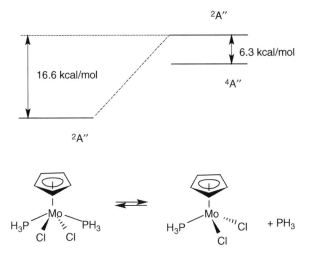

**FIGURE 10.11**   Relative energies for the systems $CpMoCl_2(PH_3)_2$ and $CpMoCl_2(PH_3)$ +
$PH_3$ at the B3LYP level.

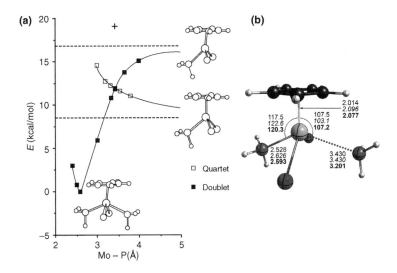

**FIGURE 10.12**   (a) Energies of the partially optimized doublet and quartet $CpMoCl_2(PH_3)_2$
relative to the overall doublet minimum, at various fixed Mo–P distances, including dissociated
$CpMoCl_2(PH_3)$ (dashed lines) and estimated upper limit of the crossing point energy ($+$).
(b) The optimized geometry of the MECP is presented. Selected bond lengths and angles for the
partially optimized doublet (plain text), partially optimized quartet (italics), and MECP
(bold).[138a] Adapted with permission from the American Chemical Society.

single-point energies of each spin state at both partially optimized geometries (see point $(+)$ in Figure 10.12).

The two curves cross near the point with a fixed Mo–P distance of 3.430 Å. The partially optimized geometries of the two PESs and the MECP are also shown in Figure 10.12.

## 10.7 COMPUTATIONAL STUDIES OF RADICAL SYSTEMS

### 10.7.1 General Considerations for Computations on Organometallic Radicals

For a long time, the Hartree–Fock (HF) theory combined with the self-consistent field (SCF) procedure proved arguably to be the most useful method for computational chemists despite its well-known limitations. Within its framework, orbitals can be constructed to reflect paired or unpaired electrons. If the molecular system has a singlet spin state, then the same orbital spatial function can be used for both the $\alpha$- and $\beta$-spin electrons in each pair. This assumption is called the restricted Hartree–Fock method (RHF).

For treating systems with unpaired electron, a different methodology is needed for which there are basically two techniques. One is to construct two completely separate sets of orbitals for the $\alpha$- and $\beta$-electrons, which is called the unrestricted Hartree–Fock (UHF) wavefunction. Here, the paired electrons will not have the same spatial distribution; the wavefunction is no longer an eigenfunction of the total spin $<S^2>$, which may result in an error in the calculation, namely, the spin contamination from the higher spin states. UHF calculations are fairly efficient and easy to implement, therefore this method is usually the default for open-spin systems in most quantum chemical program packages. The other way of constructing open-shell wavefunctions is the restricted open shell (ROHF) method, where the paired electrons share the same spatial orbital, thus avoiding the spin contamination. ROHF calculations are somewhat more computationally intensive. They are used mainly to replace UHF where spin contamination is large. However, adding a spin annihilation step into an UHF calculation often removes the majority of the spin contamination from the wavefunction. The drawback of the ROHF method in some cases is the lack of spin polarization that makes ROHF unreliable in terms of predicting spin densities. For the same reason, ROHF is not particularly useful for predicting ESR spectra.

Spin density calculations analyze the difference between each atom's population of $\alpha$- and $\beta$-electrons using various techniques. Mulliken population analysis[140] is still often used for determining spin densities despite its deficiencies experienced sometimes when large basis sets are used. The Löwdin population analysis Scheme[141] tries to circumvent some of the unreasonable orbital populations predicted by the Mulliken scheme; however it is somewhat still basis set dependent. The natural bond orbital (NBO) methodology[142] is a whole set of analysis techniques, including natural population analysis (NPA), which is less basis set dependent than the Mulliken scheme and often considered one of the most reliable methods for population analysis.

It is well known that the limitation of Hartree–Fock calculations is that electron correlation is not included, that is, HF takes into account the average effect of electron repulsion, but not the electron–electron interaction explicitly. This can be problematic for transition metal complexes, especially for the first-row transition metals due to the much more compact nature of 3d orbitals resulting in strong near-degeneracy effect.[143] Therefore, HF and sometimes even MP2 methods tend to fail to give correct geometries and energies. Density functional theory (DFT), however, includes electron correlation, thus computation with Kohn–Sham orbitals (for which the same SCF method can be applied as for the HF method) became much more reliable for systems containing transition metals. Nowadays computational transition metal chemistry is almost synonymous with DFT. Furthermore, the large size of the organometallic complexes generally prevents the use of higher level correlated *ab initio* calculations. Normally, model complex with simplified ligands is calculated to decrease the computation time. Vast numbers of exchange–correlation functionals have been developed. The ones using the general gradient approximation (GGA) became widely popular. Among the exchange–correlation functionals, the BP86[144] functional tends to be most reliable for transition metal chemistry. Also popular are the hybrid functionals that usually combine Hartree–Fock exchange with exchange functionals. Most robust example in this category is the well-known B3LYP functional.[145] The main limitation of the applicability of the use of DFT is that different functionals can lead to different results. Thus, the functionals employed must be calibrated by comparison to experimental results in the system studied.

Since higher spin states can play an important role in the mechanisms of the reactions as stated previously, it is also necessary to explore these states to understand better the chemistry of the organometallic radicals.

### 10.7.2    Studies on Structure and Reactions of Radical Complexes

*10.7.2.1    Vanadium Carbonyls*    The 17 e$^-$ $^{\bullet}$V(CO)$_6$ complex was among the first radical complexes examined theoretically. Lin and Hall (LH) studied the ligand exchange of the 17 e$^-$ and 18 e$^-$ M(CO)$_6$ complexes (M=V, Nb, Ta, Cr, Mo, W).[146] They optimized the geometries using the Hartree–Fock method combined with double-$\zeta$ basis sets and the corresponding effective core potentials on transition metals. The effect of electron correlation was included by single-point single-reference all-single-and-double-excitation (CISD) calculations at the HF-optimized geometries. To determine the most favorable mode of nucleophilic attack among the associative paths, they optimized the hypervalent $^{\bullet}$Ta(CO)$_7$ and compared the energies of the individual isomers. The results suggested that a face-attacking mode with an adjacent leaving ligand is most favorable. Thus the ligand substitution takes place via a structure similar to **1** in Scheme 10.16, that is, a pseudo-C$_{2v}$ transition state with a three-center five-electron bond.

The activation energies were found to be 16.9, 20.8, and 20.8 kcal/mol for the CO exchange of $^{\bullet}$V(CO)$_6$, $^{\bullet}$Nb(CO)$_6$, and $^{\bullet}$Ta(CO)$_6$, respectively. For the CO substitution reaction of the 18 e$^-$ M(CO)$_6$ complexes (where M = Cr, Mo, W), higher barriers were calculated (19.1, 27.3, and 35.8 kcal/mol, respectively).

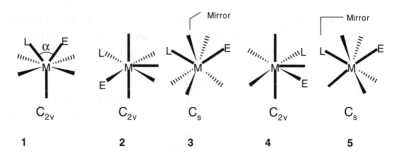

**SCHEME 10.16**    Possible geometries for transition states of ligand exchange of $^{\bullet}M(CO)_6$ (M = V, Nb, Ta). The entering ligand is designated with E, whereas the leaving ligand is designated with L. Reproduced with permission from reference 146. Copyright 1992, American Chemical Society.

We repeated the calculations on the carbonyl exchange reaction of $^{\bullet}V(CO)_6$ within the framework of density functional theory using the BP86 functional with reasonably large basis sets. For vanadium, the (14s11p6d3f)/[8s7p4d1f] all-electron basis from Wachters[147] was applied, whereas the 6-311G(2d,p) basis was used for carbon and oxygen. The lowest energy isomer of $^{\bullet}V(CO)_6$ was found to be a structure with $D_{3d}$ symmetry, whereas the transition state describing the CO exchange process possesses $C_2$ symmetry (Figure 10.13).

The activation energy is 13.6 kcal/mol, somewhat smaller than computed by LH at HF/CISD level. Nevertheless, the barrier computed by these two rather different levels of theory agree quite well.

LH also examined the carbonyl substitution reaction by phosphine. The approximate transition states as well as the barriers were determined by relaxed potential energy surface scans, that is, optimizing a system $L_nMXY$ (where X is the entering

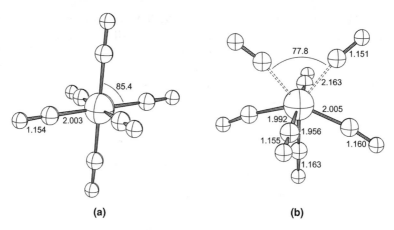

**FIGURE 10.13**    Global minimum of $^{\bullet}V(CO)_6$ (a) and transition state for the carbonyl exchange reaction (b) computed at the BP86 level of theory. Selected bond lengths are given in Å and bond angles are given in degrees.

and Y is the leaving ligand) several times keeping $R$ constant in every individual steps, where $R = \text{M-Y} - \text{M-X}$ from $R = 4.0\,\text{Å}$ to $R = -4.0\,\text{Å}$, and a $C_s$ symmetry was imposed. For the reaction of $Ta(CO)_6 + PH_3$, an activation energy of $21.7\,\text{kcal/mol}$ was calculated. When the same procedure was used for the analogous 18 e$^-$ substitution reaction $W(CO)_6 + PH_3$, a significantly higher activation energy ($32.8\,\text{kcal/mol}$) was estimated.

The structure and molecular properties of $^\bullet V(CO)_6$ were examined by various experimental methods. The investigations revealed a nearly octahedral structure with a distortion caused by Jahn–Teller effect. For the electron diffraction data, the vibrational amplitudes, which are larger than that for $Cr(CO)_6$, are in accordance with a dynamic distortion to $D_{4h}$ or $D_{3d}$ symmetry.[148] The X-ray data at $-30°C$ indicated a small tetragonal distortion that may also be due to crystal packing effects.[149] According to ESR results, $^\bullet V(CO)_6$ exists in a CO matrix at 4K in axial symmetry ($D_{4h}$ or $D_{3d}$).[150]

The structures, relative energies, vibrational spectra, and ESR parameters were computed with DFT and high-level *ab initio* methods using different basis sets.[151] The possible Jahn–Teller distortions from the $O_h$ symmetry according to Scheme 10.17 were investigated.

The tetragonal distortion of $^\bullet V(CO)_6$ from $O_h$ to $D_{4h}$ always lead to a compressed structure with a shorter vanadium–carbon bond in the axial position. The trigonal distortion to $D_{3d}$ changes the C–V–C angle only slightly (not more than $4°$) from the octahedral reference angle of $90°$ while keeping the V–CO moiety almost linear. The geometries showed no pronounced basis set sensitivity. Furthermore, coupled-cluster calculations with single and double excitations (CCSD) resulted in very similar bond angles, albeit the V–C bond distances were longer.

All chosen approaches yielded qualitatively the same energetic order: $D_{3d} < D_{2h} < D_{4h}$. No genuine minima with $O_h$ symmetry were found. Force constant analysis at the UBP86 level confirmed that the $D_{3d}$ and $D_{4h}$ species are minima on the potential surface, whereas the $D_{2h}$ structure is a transition state connecting equivalent $D_{3d}$ species. The infrared spectra of $^\bullet V(CO)_6$ was in reasonable agreement with the computed spectrum of the $D_{3d}$ structure.

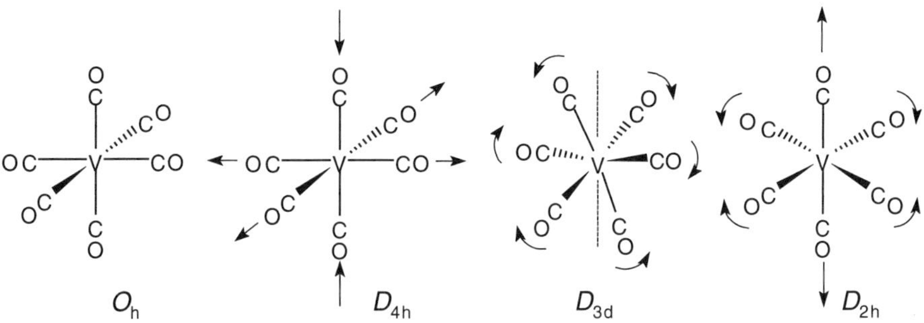

**SCHEME 10.17**  Possible distortions of $^\bullet V(CO)_6$ from $O_h$ symmetry. Reproduced with permission from reference 151. Copyright 2003, American Chemical Society.

Applying the 18 e$^-$ rule to homoleptic vanadium carbonyls would predict a V$_2$(CO)$_{12}$ complex to be the most stable. Instead, the mononuclear radical complex •V(CO)$_6$ was found experimentally. The assumption that •V(CO)$_6$ is in equilibrium with V$_2$(CO)$_{12}$ was proved by IR spectra on a sample prepared with cocondensation of CO with V atoms in rare gas matrices at relatively high V concentrations[152] and with excimer laser photolysis of •V(CO)$_6$.[153] Liu et al. examined the structures and harmonic vibrational frequencies of dinuclear homoleptic vanadium carbonyls V$_2$(CO)$_n$ ($n = 10$–$12$) utilizing B3LYP and BP86 DFT methods.[154] They found an unbridged V$_2$(CO)$_{12}$ isomer with D$_2$ symmetry as global minimum with a V–V bond distance of 3.334 Å (at the BP86 level). The dissociation of this species to two 17 e$^-$ •V (CO)$_6$ complexes is computed to be exothermic by 12.4 kcal/mol using the B3LYP functional but endothermic by 5.7 kcal/mol using the BP86 functional.

Several structures were examined on the singlet and triplet potential energy hypersurfaces of V(CO)$_6$$^+$ using the B3LYP and BP86 functionals and the MP2 level of theory.[155] No D$_{2h}$ or D$_{4h}$ stationary points were found; computations aimed to obtain structures with these symmetries ended up to geometries with higher O$_h$ symmetry. The global minimum for V(CO)$_6$$^+$ was found to be a triplet state with a D$_{3d}$ structure analogous to the global minimum of the neutral doublet state complex.

### 10.7.2.2    Chromium Cyclopentadienyl Carbonyl Complexes

The structure of the radical complex •Cr(CO)$_3$Cp and its reactions such as CO dissociation, dimerization, and reaction with diazo compounds were studied by means of density functional calculations using BP86 and B3LYP functionals.[88] At both levels of theory, the CO dissociation energy of •Cr(CO)$_3$Cp was 38.5 kcal/mol (B3LYP) and 45.9 kcal/mol (BP86), significantly higher than the corresponding experimental value (~14 kcal/mol). At the BP86 level, however, reaction energy of the dimerization of •Cr(CO)$_3$Cp resulting in [Cr(CO)$_3$Cp]$_2$ was 17.1 kcal/mol, being in reasonable accord with experiment. As expected, the rotation of the cyclopentadienyl ring in •Cr(CO)$_3$Cp is a facile process taking place with a tiny free energy barrier, namely, 1.3 kcal/mol. The structures of •Cr(CO)$_3$Cp and •Cr(CO)$_2$Cp, as well as the transition state connecting two equivalent tricarbonyl species, are shown in Figure 10.14.

Replacing one CO group in •Cr(CO)$_3$Cp with diazomethane costs 3.6 kcal/mol in terms of free energy. In the presence of free CO, the formation of a ketene complex was observed experimentally. The structures and harmonic vibrational frequencies of the possible ketene tautomers were also computed (see Figure 10.15). Since ketenes possess C=C as well as C=O π-bonds, both the η$^2$-(C,C) and the η$^2$-(C,O) coordination modes had to be considered. As chromium is of rather oxophilic character, it is not surprising that the η$^2$-(C,O) tautomer (Figure 10.15c) was lower in energy by 1.3 kcal/mol.

The formation of the ketene complex from the diazo complex was proposed to take place via a bridging dinuclear intermediate (see Figure 10.6) formed after a radical recombination of •Cr(CO)$_2$(NNCH$_2$)Cp with •Cr(CO)$_3$Cp. In the second step, the CH$_2$ group migrates toward the carbonyl group in a manner analogous to CO insertion. As this occurs, the driving force to form ketene and N$_2$ provides the needed energy

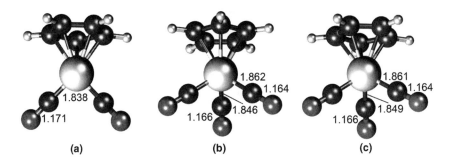

**FIGURE 10.14**   BP86 structures of $^{\bullet}Cr(CO)_2Cp$ (a), $^{\bullet}Cr(CO)_3Cp$ (b), and the transition state connecting two equivalent tricarbonyl structures (c). All species possess $C_s$ symmetry. The bond lengths are given in Å. Reproduced with permission from the American Chemical Society.

to cleave the C–N bond converting the dimer complex to the two radicals $^{\bullet}Cr(CO)_2$ (ketene)Cp and $^{\bullet}Cr(CO)_2(N_2)Cp$. The computed transition state is also shown in Figure 10.7. The reaction is exothermic with a free energy difference of $-9.3\,kcal/mol$ and proceeds with a free energy barrier of $21.3\,kcal/mol$.

The structures and energies of the reactive intermediates formally generated by the removal of a proton, a hydride, or a hydrogen atom from the methyl group of toluene–$Cr(CO)_3$ were investigated by using the B3LYP functional (Figure 10.16).[156] The most stable rotamer of $C_6H_5$–$CH_2^{\bullet}$–$Cr(CO)_3$ is the staggered conformer. The less favored eclipsed structure is only 2.8 kcal/mol higher in energy. The arene ligand is almost planar, however, there is a 3° bending toward the metal. The $C_{aryl}$–$C_{methylene}$ bond is somewhat longer in the eclipsed conformer.

The location of the unpaired electron was supposed to be in between the two extreme cases depicted in the resonance structures shown in Equation 10.70.

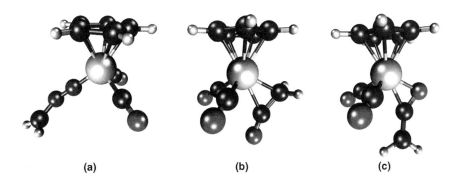

**FIGURE 10.15**   Structures of $^{\bullet}Cr(CO)_2(NNCH_2)Cp$ (a), $^{\bullet}Cr(CO)_2[\eta^2\text{-}(C,C)\text{-}O{=}C{=}CH_2]Cp$ (b), and $^{\bullet}Cr(CO)_2[\eta^2\text{-}(C,O)\text{-}O{=}C{=}CH_2]Cp$ (c). Reproduced with permission from reference 88. Copyright 2007, American Chemical Society.

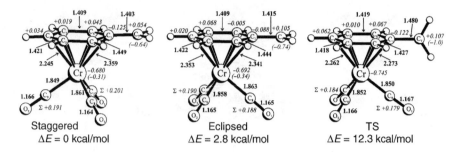

**FIGURE 10.16**   Optimized geometries (bold) and natural charges of the staggered and the eclipsed structure of the $C_6H_5-CH_2{}^\bullet-Cr(CO)_3$ complex and the transition state describing the rotation of the terminal $CH_2$ group. Bond lengths are given in Å and angles in degrees. The charges of the hydrogen atoms are summed into the carbon atoms.[156] Adapted with permission from Wiley–InterScience.

$$\text{(10.70)}$$

The calculation revealed that about two thirds of the excess of spin is located on the terminal $CH_2$ group, whereas 0.31 |e| is localized on the chromium atom. In the rotational transition state, however, the unpaired electron was predicted to be completely localized at the carbon atom of the terminal $CH_2$ group. The computed energy barrier for the $CH_2$ rotation is 12.3 kcal/mol.

### 10.7.2.3   Manganese Carbonyl Complexes

Barckholtz and Bursten studied the bond dissociation energy of binuclear homoleptic carbonyls such as $Mn_2(CO)_{10}$, which made necessary to study the structure and molecular vibrations of the corresponding monomers.[157] For all the calculations, the BP86 functional was used with STO valence basis sets. The structures of the ${}^\bullet Mn(CO)_n$ ($n = 3-5$) species are shown in Figure 10.17. The geometry of ${}^\bullet Mn(CO)_5$ is square pyramidal following $C_{4v}$ symmetry with a ${}^2A_1$ ground state. The 15 e$^-$ complex ${}^\bullet Mn(CO)_4$ has a $C_{2v}$ geometry with a ${}^2B_1$ ground state, while the 13 e$^-$ complex ${}^\bullet Mn(CO)_3$ possesses $C_{3v}$ symmetry with a ${}^2E$ ground state. The computed vibrational frequencies are in good agreement with the experimental values[158] (Table 10.2).

The dissociation energy of the Mn–Mn bond was computed as the energy difference between the unbridged $Mn_2(CO)_{10}$ complex and the two relaxed ${}^\bullet Mn(CO)_5$ radicals. The geometry of the $Mn(CO)_5$ moiety in $Mn_2(CO)_{10}$ changes very little during the bond breaking, thus the energy gain due to relaxation provides a minimal contribution to the Mn–Mn BDE. The Mn–Mn bond energy is rather difficult to measure experimentally, thus the values are in a broad range. Perhaps the most accurate experimental value is $38 \pm 5$ kcal/mol measured by pulsed time-resolved

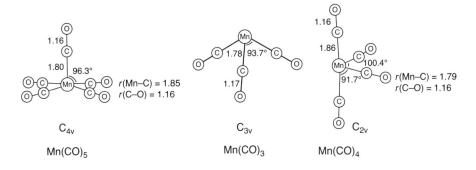

$C_{4v}$          $C_{3v}$        $C_{2v}$

$Mn(CO)_5$          $Mn(CO)_3$     $Mn(CO)_4$

**FIGURE 10.17**   Optimized geometries of $^\bullet Mn(CO)_n$ ($n = 3$–5) at the BP86 level of theory.[157] Adapted with permission from ScienceDirect.

photoacoustic calorimetry,[159] which agrees well with the computed value of 38.9 kcal/mol obtained by the UBP86 level of theory.

***10.7.2.4   The Isoprene–Fe(CO)₃ Radical Complex***   Various acyclic 1,3-diene complexes were investigated using the B3LYP functional.[160] The radical complexes **1** and **2** are formally derived by abstracting a hydrogen atom from the isoprene–Fe $(CO)_3$ complex (Figure 10.18).

Structure **1a**, where the iron coordinates to the butadiene in an $\eta^4$ manner, seems to be a better representation of this complex as spin density is mainly located on the terminal, noncoordinating methylene group. The isomeric structure **2** is favored energetically by 11.6 kcal/mol in comparison with **1**. In this more stable isomer, 83% of the spin density is localized at the iron center.

Both complexes **3** and **4** show an $\eta^3$ coordinating mode of Fe to the organic ligand (Figure 10.19) and the spin density is mainly localized at the iron atom. Therefore,

**TABLE 10.2   Calculated (at the UBP86 Level of Theory) and Experimental CO Stretching Frequencies (cm$^{-1}$) for $^\bullet Mn(CO)_n$ ($n = 3$–5)**

| Molecule | Symmetry | Calculated $\nu_{CO}$ | Experimental $\nu_{CO}$ |
|---|---|---|---|
| $^\bullet Mn(CO)_5 \cdot$ ($C_{4v}$) | $a_1$ | 2094 | 2105[a] |
| | $b_2$ | 2009 | 2018[a] |
| | $a_1$ | 1993 | 1978[b] |
| | e | 1993 | 1988[b] |
| $^\bullet Mn(CO)_4$ ($C_{2v}$) | $a_1$ | 2065 | |
| | $a_1$ | 1971 | |
| | $b_2$ | 1967 | |
| | $b_1$ | 1948 | |
| $^\bullet Mn(CO)_3$ ($C_{3v}$) | $a_1$ | 2005 | |
| | e | 1904 | |

[a] Obtained from force field calculations based on experimental data
[b] Experimental values from Ref. 158.

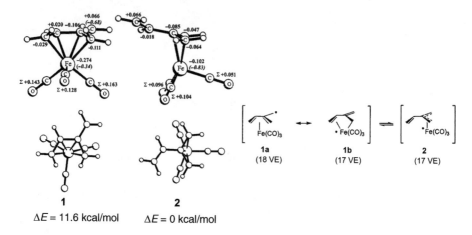

**FIGURE 10.18**    Structure of the radical complexes **1** and **2** formally derived from abstraction of a hydrogen atom from the neutral isoprene–Fe(CO)$_3$ complex with natural charges and the mesomeric forms of **1** and **2**. The charges of the hydrogen atoms are summed into the carbon atoms. Spin densities are given in parentheses.[160] Adapted with permission from Wiley–InterScience.

the mesomeric formulas **3b** and **4b** give a better description for the computed structures.

Complexes **3** and **4** can be considered conformers with an energy difference of 7.7 kcal/mol in favor of complex **3**. The isomer complex **5** with an $\eta^3$ coordinating mode is energetically disfavored as well with an energy of 6.4 kcal/mol higher than complex **3**. Thus, these open-shell systems prefer a 17 valence electron configuration at the Fe center. The formal 18 VE structures **3a** and **4a** and especially the $\eta^3$ 19 VE structure **4c** do not describe these complexes well.

### 10.7.2.5    Reaction of Ruthenium Carbonyls with Alkyl Radicals

Boese and Goldman reported that in the presence of aryl ketones, $d^8$ metal carbonyls such as Ru(CO)$_3$(dmpe) mediate photocatalytic carbonylation of alkanes via a free radical mechanism.[161] The activity was proposed to be initiated by the addition of an alkyl radical to the metal carbonyl and the formation of a metal–acyl radical intermediate. The transition states and the products of the reaction between alkyl radicals and ruthenium carbonyls were studied utilizing the B3LYP level of theory.[162] The methyl addition to a carbonyl of Ru(CO)$_5$ or Ru(CO)$_3$(dmpe) was computed to be about 6 kcal/mol more exothermic than addition to free CO.

Basically, two possibilities for the attack of one of the carbonyl groups by the methyl radical can exist: attack to an axial or attack to an equatorial carbonyl group. The two pathways are depicted in Figure 10.20. The geometry of the transition states differs substantially from the initial trigonal bipyramidal geometry. **TS2a** leads to an acyl product having a square pyramidal geometry in which the acetyl ligand takes the apical position (**2a**). **TS2b**, however, connects the reactants to **2b**, which is a square

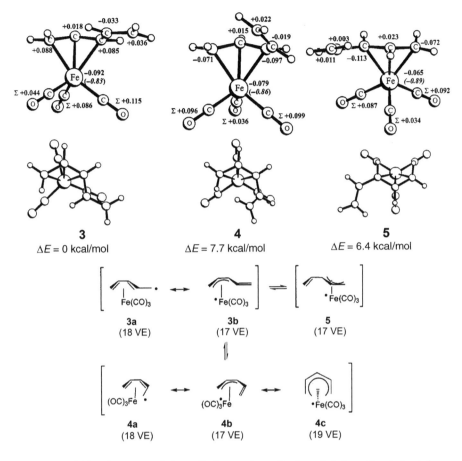

**FIGURE 10.19** Structure of the radical complexes **3**, **4**, and **5** formally derived from abstraction of a hydrogen atom from the neutral pentadiene–Fe(CO)$_3$ complex and the mesomeric forms of **3**, **4**, and **5**. The charges of the hydrogen atoms are summed into the carbon atoms. Spin densities are given in parentheses.[160] Adapted with permission from Wiley–InterScience.

pyramidal product with the acetyl ligand at the basal site. Both transition states have virtually the same energy, but the intermediate **2a** is 3.6 kcal/mol more stable than **2b**. However, conformer **2c**, which is obtained by a 180° rotation of the acyl group, is also found to be a minimum and its energy is almost similar to that of **2a**.

The six-coordinate 19 e$^-$ metal–alkyl complex was also considered, but all attempts to locate a true minimum for this species were unsuccessful, as all the calculations resulted in a dissociation of one of the carbonyl groups giving the square pyramidal Ru(CO)$_4$Me complex. For the net transformation from **1** and free Me to Ru(CO)$_4$Me and free CO, the free energy of the reaction was −2.8 kcal/mol, that is, much less favorable than the formation of the acyl product **2a**.

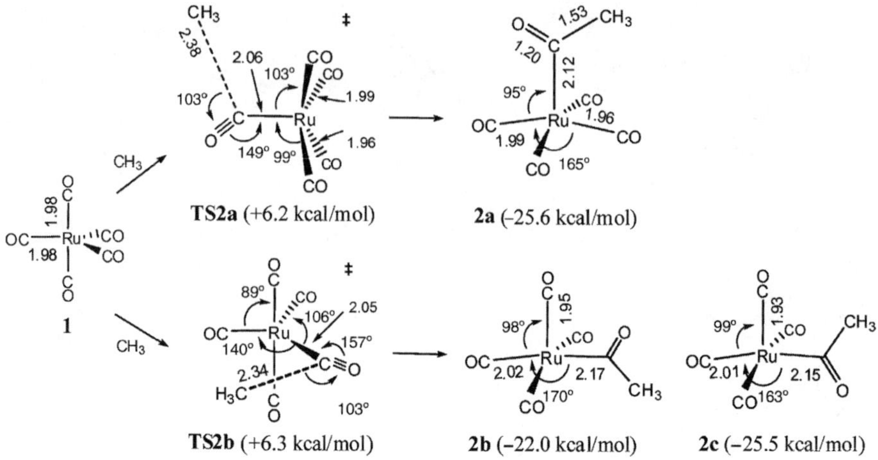

**FIGURE 10.20**   Selected structural parameters, in degrees and Å, of the transition states and products of methyl addition to **1** and their electronic energy ($E_{B3LYP}$) relative to the separate reactants.[162] Reproduced with permission from reference 162. Copyright 2008, American Chemical Society.

*10.7.2.6   Radical Carbonyl Complexes of the Cobalt Group*   The molecular structure and the CO stretch frequencies of the cobalt carbonyls $^{\bullet}Co(CO)_n$ ($n = 1$–4) were studied using the BP86 functional.[163] As no significant improvement was gained by going to a TZVP (triple-$\zeta$ with polarization function) basis, DZVP Slater-type orbital (STO) valence basis sets were used. Calculations for quartet multiplicities were also performed to find the electronic ground state for each system. The geometries of all the binary carbonyls are compared in Table 10.3, and their CO stretching frequencies are summarized in Table 10.4.

**TABLE 10.3   Symmetry, Equilibrium Distances ($R$) in Å, angles ($A$) in degrees, relative energy for different isomers ($E$) in kcal/mol, and average binding energy per carbonyl group E(CoC) for $^{\bullet}Co(CO)_n$, $n = 1$–4 in kcal/mol**

| | $^{\bullet}CoCO$ | $^{\bullet}Co(CO)_2$ | | $^{\bullet}Co(CO)_3$ | $^{\bullet}Co(CO)_4$ | |
| --- | --- | --- | --- | --- | --- | --- |
| Symmetry | $D_{\infty v}$ | $C_{2v}(B)$ | $D_{\infty h}(A)$ | $D_{3h}$ | $C_{3v}$ | $D_{2d}$ |
| $E$ | | 0 | 7 | | 0 | 3 |
| $E(CoC)$ | 57.5 | 53.8 | | 47.3 | 44.1 | |
| $R(C-O)$ | 1.170 | 1.157 | 1.154 | 1.156 | 1.151 | 1.151 |
| $R(Co-C)$ | 1.667 | 1.784 | 1.796 | 1.818 | | 1.813 |
| $R(Co-C)_{ax}$ | | | | | 1.825 | |
| $R(Co-C)_{eq}$ | | | | | 1.794 | |
| $A(Co-C-O)$ | | 175 | | | 177 | 174 |
| $A(C-Co-C)$ | | 152 | | | | 61 |
| $A(C_{ax}-Co-C_{eq})$ | | | | | 98 | |

**TABLE 10.4    Calculated and Experimental CO Stretching Frequencies and Calculated Absorption Intensities for $Co(CO)_n$, $n = 1$–$4$**

| System | Symmetry | $\Gamma$ | $\omega$ (cm$^{-1}$) | Relative Intensity | Symmetry | $\omega$ (cm$^{-1}$) |
|--------|----------|----------|---------------------|--------------------|----------|---------------------|
| | | | Calculated | | Experimental | |
| CO | $C_{\infty v}$ | $\Sigma$ | 2107 | 64 | $C_{\infty v}$ | 2133 |
| ˙CoCO | $C_{\infty v}$ | $\Sigma$ | 1979 | 686 | $C_{\infty v}$ | 1959 |
| ˙$Co(CO)_2$ | $C_{2v}$ | $A_1$ | 2052 | 59 | | |
| | | $B_2$ | 1956 | 2075 | $D_{\infty h}$ | 1925 |
| ˙$Co(CO)_3$ | $D_{3h}$ | $A'$ | 2083 | 0 | | |
| | | $E'$ | 1994 | 1290 | $C_{3v}$ | 1989 |
| ˙$Co(CO)_4$ | $C_{3v}$ | $A_1$ | 2075 | 6 | $C_{3v}$ | 2107 |
| | | $A_1$ | 2006 | 636 | | 2029 |
| | | $E_1$ | 1998 | 1062 | | 2011 |

For ˙CoCO, a linear ground-state geometry $C_{\infty v}$ was found. The computed CO stretch frequency was 20 cm$^{-1}$ higher than the experimental value. For ˙$Co(CO)_2$, two minima were located (bent $C_{2v}$ and linear $D_{\infty h}$), which are nearly degenerate and separated by a very small energy barrier. The bent structure is more stable by 7 kcal/mol, and has a C–Co–C bond angle of 152°. The CO stretch modes have vibration frequencies of 1956 and 2052 cm$^{-1}$ of $B_2$ and $A_1$ symmetry, respectively. Only one isomer of ˙$Co(CO)_3$ was found with $D_{3h}$ symmetry. Calculations started with $C_{3v}$ symmetry constraints also resulted in the $D_{3h}$ structure. This agreed with experimental IR results of Hanlan et al.,[31c] in which only one CO mode was observed. Their ESR results, however, indicated a $C_{3v}$ geometry, and they claimed that small differences from a planar geometry is not easily detected by IR, hence ˙$Co(CO)_3$ might have $C_{3v}$ symmetry in noble gas matrix even if the gas-phase ground state is of $D_{3h}$ symmetry. The calculated asymmetric stretch frequency of 1994 cm$^{-1}$ is only 5 cm$^{-1}$ higher than the experimental value.

In principle, there are many different possible symmetries for ˙$Co(CO)_4$. Geometry optimizations were performed with $T_d$, $D_{2d}$, $C_{3v}$, and $C_{2v}$ symmetry restrictions. Their relative energies were 14, 2, 0, and 6 kcal/mol, indicating the $C_{3v}$ as the ground state. Furthermore, a geometry optimization without any symmetry restrictions starting from a $T_d$ geometry converged to $C_{3v}$. The structure is tetragonal pyramidal, with the apical Co–C bond 0.03 Å longer than the basal ones. This system has three IR-active CO vibrational modes, which is in line with the experimental spectrum.[31] The two $A_1$ and the $E_1$ CO stretch frequencies are 37, 23, and 30 cm$^{-1}$ lower than the experimental ones, respectively.

For the analogous carbonyl radicals ˙$M(CO)_n$ (M = Rh, Ir; $n = 1$–$4$), Zhou and Andrews found different symmetries for the tricarbonyls and tetracarbonyls species both with rhodium and iridium.[164] Unlike ˙$Co(CO)_3$, which is a trigonal $D_{3h}$ structure, the Rh and Ir tricarbonyls have T-shaped structures with $C_{2v}$ symmetry. The tetracarbonyls of Rh and Ir also reveal a different order of thermodynamic stability. While for ˙$Co(CO)_4$, the $C_{3v}$ structure is the ground state and the $D_{2d}$ complex is higher in energy, for Rh and Ir, a reverse relationship was indicated.

Huo et al. studied the structure and energies of cobalt–carbonyl radicals as well as the cationic and anionic homoleptic mononuclear cobalt–carbonlys and the cobalt–carbonyl hydrides.[35] The B3LYP functional was applied with all-electron triple-$\zeta$ quality basis. They found that most of these complexes prefer less symmetrical structures and minor structural deformations may result in large energetic differences.

The structures of $^\bullet Co(CO)_n$ complexes ($n = 1$–4) are shown in Figure 10.21. The neutral monocarbonyl species also has a bent structure. The linear structure has one imaginary frequency (thus, it is not a genuine minimum) and is 21.6 kcal/mol higher in energy. The calculated dissociation energy is 23.0 kcal/mol, much smaller than the value of 57.5 kcal/mol by Ryeng[163] computed at the BP86 level.

For $^\bullet Co(CO)_2$, the very slightly bent $C_s$ structure was found to be most stable, and the linear isomer was higher in energy by 19.6 kcal/mol. These results significantly differ from those by Ryeng et al. obtained at BP86 level, as they found a bent $C_{2v}$ structure with a bond angle of 152° being more stable than the linear one (see above).

$^\bullet Co(CO)_3$ was found to have a trigonal planar structure with $C_s$ symmetry, but the expected $D_{3h}$ conformation is almost degenerate, it is less stable by only 0.1 kcal/mol. In contrast to the BP86 studies, only an isomer of $^\bullet Co(CO)_4$, namely, the $C_{3v}$ structure was found. The paramount difference, compared to the BP86 results,[163] is that the axial Co–C bond length is longer than the equatorial Co–C distances.

The first CO dissociation energies were 37.9, 23.9, and 22.0 kcal/mol for the $^\bullet Co(CO)_2$, $^\bullet Co(CO)_3$, and $^\bullet Co(CO)_4$ ground-state structures, respectively. The atomic natural charge analysis showed that the central cobalt atom is slightly positive in most

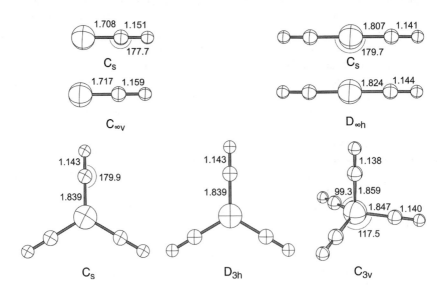

**FIGURE 10.21**  B3LYP structures of the most stable isomers of $^\bullet Co(CO)_n$ ($n = 1$–4). Bond lengths are given in Å and angles in degrees.[35] Reproduced with permission from reference 35. Copyright 2002, American Chemical Society.

cases, being highest for the $C_s$ tricarbonyl structure and close to neutral in the dicarbonyl complexes.

Barckholtz and Bursten[157] studied the homolytic bond dissociation energies of the $Co_2(CO)_8$ and $Co_2(CO)_7$. While there is no experimental results for the latter species, for the octacarbonyl complexes, the BDE was significantly overestimated by the BP86 functional as it is computed to be 29.9 kcal/mol for the $D_{3d}$ isomer of $Co_2(CO)_8$, whereas the experimental BDE for the $Co_2(CO)_8 \rightarrow 2^\bullet Co(CO)_4$ reaction is 19 kcal/mol. As a check on this discrepancy, the BDE was computed for the Co–H bond in $HCo(CO)_4$ as well and it was found that the BP86 calculations yielded a value of 70.1 kcal/mol, whereas the corresponding experimental values were 54 and 59 kcal/mol in two studies.[38,165] The authors came to the conclusion that the energy of $^\bullet Co(CO)_4$ is probably not given correctly by the BP86 functional.

The analogous $^\bullet Rh(CO)_4$ complex as well as its derivatives obtained by the substitution of carbonyl groups by various ligands were studied by de Bruin et al.[29] using also the BP86 functional. As expected, $^\bullet Rh(CO)_4$ had no structure with $T_d$ symmetry due to Jahn–Teller distortion. Thus, for $^\bullet Rh(CO)_4$, a global minimum with $D_{2d}$ symmetry was found with a Rh–C bond length of 1.943 Å and with 52.4° as a smallest angle between planes. Similarly, $^\bullet Rh(cod)_2$ and $^\bullet Rh(dppe)_2$ [cod = cyclooctadiene; dppe = bis(diphenylphosphino-ethane)] also possesses $D_{2d}$ symmetry.

The $^\bullet Rh(trop_2 dach)$ complex ($trop_2 dach$ = 1,4-bis($5H$-dibenzo[$a,d$]-cyclohepten-5-yl)-1,2-diamino-cyclohexane) was synthesized and its structure was computed[29] and the spin density was found to be highly delocalized over the two trop moieties. The spin population was about 36% at the rhodium atom. The structure of the complex and the continuous-wave EPR spectrum of its frozen solution at X-band is depicted in

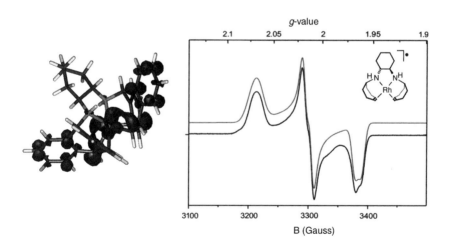

**FIGURE 10.22**    Spin density distribution and the experimental and simulated EPR spectrum of the $^\bullet Rh(trop_2 dach)$ complex at X-band (9.0303044 GHz) at 68K.[29] Adapted with permission from ScienceDirect.

Figure 10.22. The computed $g$-values ($g_1 = 2.054$, $g_2 = 2.018$, and $g_3 = 1.964$) agree very well with the experimental ones ($g_1 = 2.069$, $g_2 = 2.014$, and $g_3 = 1.964$). The calculated absolute hyperfine couplings are slightly smaller than the experimental values.

## 10.8   CONCLUSION AND FUTURE DIRECTIONS

This chapter has tried to present an introduction to the broader aspects of organo-metallic radical chemistry. This is an area of active development in inorganic chemistry. In particular, the development of paramagnetic complexes can be expected to lead to new understanding and advancements. The statement of Astruc[166] quoted below was made about 20 years ago and it remains a challenge to all scientists working in the area of inorganic radical chemistry.

> It is probable that the recognition of the existence and role of 19e species in mechanisms will lead in the near future to the finding of new processes that may involve multicatalytic components with sophisticated technological devices.

## REFERENCES

1. Collman, J. P.; Hegedus, L. S.; Norton, J. R.; Finke, R. G. *Principles and Applications of Organotransition Metal Chemistry*; University Science Books: Sausalito, CA, 1987.
2. Gomberg, M. *J. Am. Chem. Soc.* **1900**, *22*, 757.
3. Paneth, F. A.; Hofeditz, W. *Berichte der Deutschen Chemischen Gesellschaft*, **1929**, *62*, 1335.
4. A special issue dedicated to one-electron processes as appeared. See Poli, R. *J. Organometal. Chem.* **2007**. *692*, 3109 for a preface to this area.
5. Taken from the preface to *Free Radical Chain Reactions in Organic Synthesis*, Motherwell, W. P.; Crich, D.; Academic Press, New York: 1992.
6. Schrauzer, G. N. *Acc. Chem. Res.* **1968**, *1*, 97.
7. Tschugaeff, L. *Berichte der Deutschen Chemischen Gesellschaft* **1905**, *38*, 2520. It is not clear whether the nature of the metal radical were identified at that time.
8. Calvin, M.; Bailes, R. H.; Wilmarth, W. K. *J. Am. Chem. Soc.* **1946**, *68*, 2254.
9. (a) Natta, G.; Ercoli, R.; Bernaldi, G.; Calderazzo, F. *Gazz. Chim.* **1959**, *89*, 809; (b) Ercoli, R.; Calderazzo, R.; Alberola, A. *J. Am. Chem. Soc.* **1960**, *82*, 2966.
10. Sando, G. M.; Spears, K. G. *J. Phys. Chem. A.* **2004**, *108*, 1290.
11. Byers, B. H.; Brown, T. L. *J. Am. Chem. Soc.* **1975**, *97*, 947.
12. King, R. B. *Organomet. Syn.* **1965**, *1*, 70.
13. Geiger, W. E. *Organometallics* **2007**, *26*, 5738.
14. Fischer, E. O.; Wawersik, H. *J. Organomet. Chem.* **1966**, *5*, 559.
15. King, R. B.; Bisnette, M. B. *Inorg. Chem.* **1967**, *6*, 469.
16. Baoslo, F.; Pearson, R. G. *Mechanisms of Inorganic Reactions*; Wiley: New York, 1967.
17. This work is summarized in Halpern, J. *Acc. Chem. Res.* **1970**, *3*, 386.

18. Lappert, M. F.; Lednor, P. W. *Adv. Organometal. Chem.* **1976**, *14*, 345.

19. Crich, D.; Grant, D.; Krishnamurthy, V.; Patel, M. *Acc. Chem. Res.* **2007**, *40*, 453.

20. Pariya, C.; Theopold, K. H. *Curr. Sci.* **2000**, *78*, 1345.

21. Veciana, J.; Carilla, J.; Miravitlles, C.; Molins, E. *J. Chem. Soc. Chem. Commun.* **1987**, 812.

22. Lee, V. Y.; Sekiguchi, A. *Acc. Chem. Res.* **2007**, *40*, 410.

23. Connelly, N. G.; Freeman, M. J., Orpen, A. G.; Sheehan, A. R.; Sheridan, J. B.; Sweigart D. A. *J. Chem. Soc., Dalton Trans.* **1985**, 1019.

24. Fortier, S.; Baird, M. C.; Preston, K. F.; Morton, J. R.; Ziegler, T.; Jaeger, T. J.; Watkins, W. C.; MacNeil, J. H.; Watson, K. A. *J. Am. Chem. Soc.* **1991**, *113*, 542.

25. Hoobler, R. J.; Hutton, M. A.; Dillard, M. M.; Castellani, M. P.; Rheingold, A. L.; Rieger, A. L.; Rieger, P. H.; Richards, T. C.; Geiger, W. E. *Organometallics* **1993**, *12*, 116.

26. Kubas, G. J. *Metal Dihydrogen and σ-Bond Complexes: Structure, Theory, and Reactivity*; Kluwer Academic Publishers Group: Netherlands, 2001.

27. Ward, H. R. *Acc. Chem. Res.* **1972**, *5*, 18.

28. For an excellent summary of quantum methods EPR spectroscopy, see Neese, F. *Coord. Chem. Rev.* **2008**, *253*, 526–563. See also Kababya, S.; Nelson, J.; Calle, C.; Neese, F.; Goldfarb, J. *J. Am. Chem. Soc.* **2006**, *128*, 2017.

29. de Bruin, B.; Russcger, J. C.; Grützmacher, H. *J. Organomet. Chem.* **2007**, *692*, 3167.

30. Adams, R. D.; Miao, S.; Smith, M. D.; Farach, H.; Webster, C. E.; Manson, J.; Hall, M. B. *Inorg. Chem.* **2004**, *43*, 2515.

31. (a) Sweaney, R. L.; Brown, T. L. *Inorg. Chem.* **1977**. *16*, 421; (b) Crichton, O.; Poliakoff, M.; Rest, A. J.; Turner, J. J. *J. Chem. Soc., Dalton Trans.* **1973**, 1321; (c) Hanlan, L. A.; Huber, H.; Kündig, E. P.; McGarvey, B. R.; Ozin, G. A. *J. Am. Chem. Soc.* **1975**, *97*, 7054.

32. Bor, G.; Dietler, U. K.; Noack, K. *J. Chem. Soc. Chem. Commun.* **1976**, 914.

33. For an interesting review of this topic, see Pályi, G.; Ungváry, F.; Galamb, V.; Markó, L. *Coord. Chem. Rev.* **1984**, *53*, 37.

34. Wegman, R. W.; Brown, T. L. *J. Am. Chem. Soc.* **1980**, *102*, 2494.

35. Huo, C. F.; Li, Y. W.; Wu, G. S.; Beller, M.; Jiao, H. *J. Phys. Chem. A.* **2002**, *106*, 12169.

36. Nalesnik, T. E.; Orchin, M. *Organometallics* **1982**, *1*, 222.

37. (a) Klingler, R. J.; Chen, M. J.; Rathke, J. W.; Kramarz, K. W. *Organometallics* **2007**, *26*, 352; (b) Rathke, J. W.; Klingler, R. J.; Krause, T. R. *Organometallics* **1992**, *11*, 585; (c) Klingler, R. J.; Rathke, J. W. *Inorg. Chem.* **1992**, *31*, 804; (d) Rathke, J. W.; Klingler, R. J.; Krause, T. R. *Organometallics* **1991**, *10*, 1350.

38. Klingler, R. J.; Rathke, J. W. *J. Am. Chem. Soc.* **1994**, *116*, 4772.

39. Hackett, P.; O'Neil, P. S.; Manning, A. R. *J. Chem. Soc., Dalton Trans.* **1974**, 1625.

40. Baird, M. C. *Chem. Rev.* **1988**, *88*, 1217.

41. Adams, R. D.; Collins, D. E.; Cotton, F. A. *J. Am. Chem. Soc.* **1974**, *96*, 749.

42. Goh, L.-Y.; D'Aniello, M. J., Jr.; Slater, S.; Muetterties, E. L.; Tavanaiepour, I.; Chang, M. I.; Fredrich, M. F.; Daz, V. W. *Inorg. Chem.* **1979**, *18*, 192.

43. McLain, S. J. *J. Am. Chem. Soc.* **1988**, *110*, 643.

44. Fei, M.; Sur, S. K.; Tyler, D. R. *Organometallics* **1991**, *10*, 419.

45. Haines, R. J.; Nyholm, R. S.; Stiddard, M. H. B. *J. Chem. Soc. A.* **1968**, 43.

46. Burkett, A. R.; Meyer, T. J.; Whitten, D. G. *J. Organomet. Chem.* **1974**, *67*, 67.

47. Wrighton, M. S.; Ginley, D. S. *J. Am. Chem. Soc.* **1975**, *97*, 4246.

48. Hughey, J. L., IV; Bock, C. R.; Meyer, T. J. *J. Am. Chem. Soc.* **1975**, *97*, 4440.

49. Laine, R. M.; Ford, P. C. *Inorg. Chem.* **1977**, *16*, 388.

50. Hoffman, N. W.; Brown, T. L. *Inorg. Chem.* **1978**, *17*, 613.

51. Turaki, N. N.; Huggins, J. M. *Organometallics* **1986**, *5*, 1703.

52. Philbin, C. E.; Goldman, A. S.; Tyler, D. R. *Inorg. Chem.* **1986**, *25*, 4434.

53. Stiegman, A. E.; Tyler, D. R. *Acc. Chem. Res.* **1984**, *17*, 61.

54. Avey, A.; Tenhaeff, S. C.; Weakley, T. J. R.; Tyler, D. R. *Organometallics* **1991**, *10*, 3607.

55. Zhu, Z.; Espenson, J. H. *Organometallics* **1994**, *13*, 1893.

56. Scott, S. L.; Espenson, J. H.; Zhu, Z. *J. Am. Chem. Soc.* **1993**, *115*, 1789.

57. (a) Scott, S. L.; Espenson, J. H.; Chen, W. J. *Organometallics* **1993**, *12*, 4077; (b) Yao, W.; Bakac, A.; Espenson, J. H. *Organometallics* **1993**, *12*, 2010.

58. Balla, J.; Bakac, A.; Espenson, J. H. *Organometallics* **1994**, *13*, 1073.

59. Virrels, I. G.; George, M. W.; Johnson, F. P. A.; Turner, J. J.; Westwell, J. R. *Organometallics* **1995**, *14*, 5203.

60. Mohler, D. L.; Barnhardt, E. K.; Hurley, A. L. *J. Org. Chem.* **2002**, *67*, 4982.

61. Mohler, D. L.; Downs, J. R.; Hurley-Predecki, A. L.; Sallman, J. R.; Gannett, P. M.; Shi, X. *J. Org. Chem.* **2005**, *70*, 9093.

62. Glascoe, E. A.; Kling, M. F.; Cahoon, J. F.; Shanoski, J. E.; DiStassio, R. A.; Payne, C. K., Jr.; Mork, B. V.; Tilley, T. D.; Harris, C. B. *Organometallics* **2007**, *26*, 1424.

63. Cahoon, J. F.; Kling, M. F.; Sawyer, K. R.; Frei, H.; Harris, C. B. *J. Am. Chem. Soc.* **2006**, *128*, 3152.

64. Kling, M. F.; Cahoon, J. F.; Glascoe, E. A.; Shanoski, J. E.; Harris, C. B. *J. Am. Chem. Soc.* **2004**, *126*, 11414.

65. Cahoon, J. F.; Kling, M. F.; Schmatz, S.; Harris, C. B. *J. Am. Chem. Soc.* **2005**, *127*, 12555.

66. Bitterwolf, T. E. *Coord. Chem. Rev.* **2001**, *211*, 235.

67. Jaeger, T. J.; Baird, M. C. *Organometallics* **1988**, *7*, 2074.

68. Watkins, W. C.; Hensel, K.; Fortier, S.; Macartney, D. H.; Baird, M. C.; McLain, S. J. *Organometallics* **1992**, *11*, 2418.

69. (a) MacConnachie, C. A.; Nelson, J. M.; Baird, M. C. *Organometallics* **1992**, *11*, 2521; (b) Huber, T. A.; Macartney, D. H.; Baird, M. C. *Organometallics* **1993**, *12*, 4715; (c) Huber, T. A.; Macartney, D. H.; Baird, M. C. *Organometallics* **1995**, *14*, 592.

70. Choi, J.; Tang, L.; Norton, J. R. *J. Am. Chem. Soc.* **2007**, *129*, 234.

71. Bullock, R. M.; Samsel, E. G. *J. Am. Chem. Soc.* **1990**, *112*, 6886.

72. Weng, Z.; Goh, L.-Y. *Acc. Chem. Res.* **2004**, *37*, 187.

73. Landrum, J. T.; Hoff, C. D. *J. Organomet. Chem.* **1985**, *282*, 215.

74. Kiss, G.; Zhang, K.; Mukerjee, S. L.; Hoff, C. D.; Roper, G. C. *J. Am. Chem. Soc.* **1990**, *112*, 5657.

75. Capps, K. B.; Bauer, A.; Kiss, G.; Hoff, C. D. *J. Organomet. Chem.* **1999**, *586*, 23.

76. Vollhardt, K. P. C.; Cammack, J. K.; Matzger, A. J.; Bauer, A.; Capps, K. B.; Hoff, C. D. *Inorg. Chem.* **1999**, *38*, 2624.

77. Bordwell, F. G.; Zhang, X.-M.; Satish, A. V.; Cheng, J.-P. *J. Am. Chem. Soc.* **1994**, *116*, 6605.

78. Benson, S. W. *Chem. Rev.* **1978**, *78*, 23.

79. Ju, T. D.; Lang, R. F.; Roper, G. C.; Hoff, C. D. *J. Am. Chem. Soc.* **1996**, *118*, 5328.

80. Stieffel, E. I.; Matsumoto, K., Eds. *Transition Metal Sulfur Chemistry: Biological and Industrial Significance*, ACS Symposium Series No. 653, American Chemical Society Publication, 1996.

81. Capps, K. B.; Bauer, A.; Ju, T. D.; Hoff, C. D. *Inorg. Chem.* **1999**, *38*, 6130.

82. Ju, T. D.; Capps, K. B.; Lang, R. F.; Roper, G. C.; Hoff, C. D. *Inorg. Chem.* **1997**, *36*, 614.

83. McDonough, J. E.; Weir, J. J.; Carlson, M. J.; Hoff, C. D.; Kryatova, O. P.; Rybak-Akimova, E. V.; Clough, C. R.; Cummins, C. C. *Inorg. Chem.* **2005**, *44*, 3127.

84. Moran, D.; Sukcharoenphon, K.; Puchta, R.; Schaefer, H. F., III; Schleyer, P. v. R.; Hoff, C. D. *J. Org. Chem.* **2002**, *67*, 9061.

85. Sukcharoenphon, K.; Moran, D.; Schleyer, P. v. R.; McDonough, J. E.; Abboud, K. A.; Hoff, C. D. *Inorg. Chem.* **2003**, *42*, 8494.

86. Crich, D.; Quintero, L. *Chem. Rev.* **1989**, *89*, 1413.

87. (a) Ungvári, N.; Kégl, T.; Ungváry, F. *J. Mol. Catal. A* **2004**, *219*, 7; (b) Kégl, T.; Ungváry, F. *J. Organomet. Chem.* **2007**, *692*, 1825.

88. Fortman, G. C.; Kégl, T.; Li, Q.-S.; Zhang, X.; Schaefer, H. F., III; Xie, Y.; King, R. B.; Telser, J.; Hoff, C. D. *J. Am. Chem. Soc.* **2007**, *129*, 14388.

89. DeWit, D. G. *Coord. Chem. Rev.* **1996**, *147*, 209.

90. Sherry, A. E.; Wayland, B. B. *J. Am. Chem. Soc.* **1989**, *111*, 5010.

91. Wayland, B. B.; Farnos, M. D.; Coffin, V. L. *Inorg. Chem.* **1988**, *27*, 2745.

92. Wayland, B. B.; Ba, S.; Sherry, A. E. *J. Am. Chem. Soc.* **1991**, *113*, 5305.

93. Hoshino, M.; Yasufuku, K.; Konishi, S.; Imamura, M. *Inorg. Chem.* **1984**, *23*, 1982.

94. Yamamoto, S.; Hoshino, M.; Yasufuku, K.; Imamura, M. *Inorg. Chem.* **1984**, *23*, 195.

95. (a) Anderson, J. E.; Yao, C.-L.; Kadish, K. M. *Inorg. Chem.* **1986**, *25*, 718; (b) Anderson, J. E.; Yao, C.-L.; Kadish, K. M. *Organometallics* **1987**, *6*, 706.

96. de Bruin, B.; Hetterscheid, D. G. H. *Eur. J. Inorg. Chem.* **2007**, 211.

97. de Bruin, B.; Hetterscheid, D. H. G.; Koekkoek, A. J. J.; Grützmacher, H. *Prog. Inorg. Chem.* **2007**, *55*, 247.

98. Wayland, B. B.; Woods, B. A.; Coffin, V. L. *Organometallics* **1986**, *5*, 1059.

99. Wayland, B. B.; Sherry, A. E.; Poszmik, G.; Bunn, A. G. *J. Am. Chem. Soc.* **1992**, *114*, 1673.

100. Cui, W.; Wayland, B. B. *J. Am. Chem. Soc.* **2004**, *126*, 8266.

101. Zhang, X.-X.; Wayland, B. B. *J. Am. Chem. Soc.* **1994**, *116*, 7897.

102. Bunn, A. G.; Wayland, B. B. *J. Am. Chem. Soc.* **1992**, *114*, 6917.

103. Cui, W.; Wayland, B. B. *J. Am. Chem. Soc.* **2006**, *128*, 10350.

104. Zhang, L.; Chan, K. S. *Organometallics* **2007**, *26*, 679.

105. Chan, K. S.; Li, X. Z.; Zhang, L.; Fung, C. W. *Organometallics* **2007**, *26*, 2679.

106. Chan, K. S.; Li, X. Z.; Dzik, W. I.; de Bruin, B. *J. Am. Chem. Soc.* **2008**, *130*, 2051.

107. (a) Poli, R. *Chem. Rev.* **1996**, *96*, 2135; (b) Poli, R.; Cacelli, I. *Eur. J. Inorg. Chem.* **2005**, 2324.

108. Haaland, A. *Acc. Chem. Res.* **1979**, *12*, 415.

109. (a) Pitarch López, J.; Heinemann, F. W.; Prakash, R.; Hess, B. A.; Horner, O.; Jeandey, C.; Oddou, J.-L.; Latour, J.-M.; Grohmann, A. *Chem. Eur. J.* **2002**, *8*, 5709; (b) Overby, J. S.; Hanusa, T. P.; Sellers, S. P.; Yee, G. T. *Organometallics* **1999**, *18*, 3561; (c) Voges, M. H.; Rømming, C.; Tilset, M. *Organometallics* **1999**, *18*, 529; (d) Hao, S.; Gambarotta, S.; Bensimon, C. *J. Am. Chem. Soc.* **1992**, *114*, 3556.

110. (a) Tabard, A.; Cocolios, P.; Lagrange, G.; Gerardin, R.; Hubsch, J.; Lecomte, C.; Zarembowitch, J.; Guilard, R. *Inorg. Chem.* **1988**, *27*, 110; (b) Gaudry, J.-B.; Capes, L.; Langot, P.; Marcén, S.; Kollmannsberger, M.; Lavastre, O.; Freysz, E.; Létard, J.-F.; Kahn, O. *Chem. Phys. Lett.* **2000**, *324*, 321; (d) Sheng, T.; Dechert, S.; Hyla-Kryspin, I.; Winter, R. F.; Meyer, F. *Inorg. Chem.* **2005**, *44*, 3863.

111. Ellison, M. K.; Nasri, H.; Xia, Y.-M.; Marchon, J.-C.; Schulz, C. E.; Debrunner, P. G.; Scheidt, W. R. *Inorg. Chem.* **1997**, *36*, 4804.

112. (a) Meunier, B.; de Viser, S. P.; Shaik, S. *Chem. Rev.* **2004**, *104*, 3947; (b) Shaik, S.; Kumar, D.; de Viser, S. P.; Altun, A.; Thiel, W. *Chem. Rev.* **2005**, *105*, 2279.

113. (a) Kelleher, M. J. *Acc. Chem. Res.* **1993**, *26*, 154; (b) McCoy, S.; Caughey, W. S. *Biochemistry* **1970**, *9*, 2387.

114. Laplaza, C. E.; Odom, A. L.; Davis, W. M.; Cummins, C. C. *J. Am. Chem. Soc.* **1995**, *117*, 4999.

115. Laplaza, C. E.; Cummins, C. C. *Science* **1995**, *268*, 861.

116. (a) Laplaza, C. E.; Johnson, M. J. A.; Peters, J. C.; Odom, A. L.; Kim, E.; Cummins, C. C.; George, G. N.; Pickering, I. J. *J. Am. Chem. Soc.* **1996**, *118*, 8623; (b) Peters, J. C.; Cherry, J.-P. F.; Thomas, J. C.; Baraldo, L. M.; Mindiola, D. J.; Davis, W. M.; Cummins, C. C. *J. Am. Chem. Soc.* **1999**, *121*, 10053; (c) Curley, J. J.; Cook, T. R.; Reece, S. Y.; Müller, P.; Cummins, C. C. *J. Am. Chem. Soc.* **2008**, *130*, 9394.

117. (a) Graham, D. C.; Beran, G. J. O.; Head-Gordon, M.; Christian, G.; Stranger, R.; Yates, B. F. *J. Phys. Chem. A* **2005**, *109*, 6762; (b) Marcus, C. D. *Chem. Eur. J.* **2007**, *13*, 3406; (c) Cui, Q.; Musaev, D. G.; Svensson, M.; Sieber, S.; Morokuma, K. *J. Am. Chem. Soc.* **1995**, *117*, 12366; (d) Hahn, J.; Landis, C. R.; Nasluzov, V. A.; Neyman, K. M.; Rosch, N. *Inorg. Chem.* **1997**, *36*, 3947; (e) Neyman, K. M.; Nasluzov, V. A.; Hahn, J.; Landis, C. R.; Rosch, N. *Organometallics* **1997**, *16*, 995; (f) Christian, G.; Stranger, R.; Yates, B. F.; Graham, D. C. *Dalton Trans.* **2005**, 962; (g) Christian, G.; Stranger, R. *Dalton Trans.* **2004**, 2492; (h) Christian, G.; Driver, J.; Stranger, R. *Faraday Discuss.* **2003**, *124*, 331.

118. Tsai, Y. C.; Cummins, C. C. *Inorg. Chim. Acta* **2003**, *345*, 63.

119. Okabe, H. *Photochemistry of Small Molecules*; Wiley: New York, 1978; p 219.

120. Cherry, J.-P. F.; Johnson, A. R.; Baraldo, L. M.; Tsai, Y.-C.; Cummins, C. C.; Kryatov, S. V.; Rybak-Akimova, E. V.; Capps, K. B.; Hoff, C. D.; Haar, C. M.; Nolan, S. P. *J. Am. Chem. Soc.* **2001**, *123*, 7271.

121. Johnson, A. R.; Davis, W. M.; Cummins, C. C.; Serron, S.; Nolan, S. P.; Musaev, D. G.; Morokuma, K. *J. Am. Chem. Soc.* **1998**, *120*, 2071.

122. (a) Laplaza, C. E.; Davis, W. M.; Cummins, C. C. *Angew. Chem., Int. Ed. Engl.* **1995**, *34*, 2042; (b) Stephens, F. H.; Johnson, M. J. A.; Cummins, C. C.; Kryatova, O. P.; Kryatov, S.

V.; Rybak-Akimova, E. V.; McDonough, J. E.; Hoff, C. D. *J. Am. Chem. Soc.* **2005**, *127*, 15191.

123. Cummins, C. C. *Chem. Commun.* **1998**, 1777.

124. (a) McDonough, J. E.; Weir, J. J.; Sukcharoenphon, K.; Hoff, C. D.; Kryatova, O. P.; Rybak-Akimova, E. V.; Scott, B. L.; Kubas, G. J.; Mendiratta, A.; Cummins, C. C. *J. Am. Chem. Soc.* **2006**, *128*, 10295; (b) Mendiratta, A.; Cummins, C. C.; Kryatova, O. P.; Rybak-Akimova, E. V.; McDonough, J. E.; Hoff, C. D. *Inorg. Chem.* **2003**, *42*, 8621.

125. (a) Collman, J. P.; Boulatov, R.; Sunderland, C. J.; Fu, L. *Chem. Rev.* **2004**, *104*, 561; (b) Kim, E.; Chufán, E. E.; Kamaraj, K.; Karlin, K. D. *Chem. Rev.* **2004**, *104*, 1077; (c) Costas, M.; Mehn, M. P.; Jensen, M. P.; Que, L. Jr. *Chem. Rev.* **2004**, *104*, 939; (d) Enemark, J. H.; Cooney, J. J. A. *Chem. Rev.* **2004**, *104*, 1175.

126. Ammeter, J. H.; Swalen, J. D. *J. Chem. Phys.* **1972**, *57*, 678.

127. Torraca, K. E.; McElwee-White, L. *Coord. Chem. Rev.* **2000**, *206-207* 469.

128. Astruc, D. *Electron-Transfer and Radical Processes in Transition-Metal Chemistry*; Wiley-VCH: New York, 1995.

129. Elmurr, N.; Sheats, J. E.; Geiger, W. E.; Holloway, J. D. L. *Inorg. Chem.* **1979**, *18*, 1443.

130. Tsai, Y.-C.; Stephens, F. H.; Meyer, K.; Mendiratta, A.; Gheorghiu, M. D.; Cummins, C. C. *Organometallics* **2003**, *22*, 2902.

131. Mendiratta, A.; Cummins, C. C. *Inorg. Chem.* **2005**, *44*, 7319.

132. Temprado, M.; McDonough, J. E.; Mendiratta, A.; Tsai, Y.-C.; Fortman, G. C.; Cummins, C. C.; Rybak-Akimova, E. V.; Hoff, C. D. *Inorg. Chem.* **2008**, *47*, 9380.

133. Hetterscheid, D. G. H.; Kaiser, J.; Reijerse, E.; Peters, T. P. J.; Thewissen, S.; Blok, A. N. J.; Smits, J. M. M.; de Gelder, R.; de Bruin, B. *J. Am. Chem. Soc.* **2005**, *127*, 1895.

134. Collman, J. P.; Hegedus, L. S. *Principles and Applications of Organotransition Metal Chemistry*; University Science Books: Mill Valley, CA, 1980; p 280.

135. (a) Bengali, A. A.; Bergman, R. G.; Moore, C. B. *J. Am. Chem. Soc.* **1995**, *117*, 3879; (b) Detrich, J. L.; Reinaud, O. M.; Rheingold, A. L.; Theopold, K. H. *J. Am. Chem. Soc.* **1995**, *117*, 11745.

136. (a) Schröder, D.; Shaik, S.; Schwarz, H. *Acc. Chem. Res.* **2000**, *33*, 139; (b) Shaik, S.; Danovich, D.; Fiedler, A.; Schröder, D.; Schwarz, H. *Helv. Chim. Acta* **1995**, *78*, 1393.

137. (a) Poli, R.; Harvey, J. N. *Chem. Soc. Rev.* **2003**, *32*, 1; (b) Harvey, J. N.; Poli, R.; Smith, K. M. *Coord. Chem. Rev.* **2003**, *238-239*. 347; (c) Poli, R. *J. Organomet. Chem.* **2004**, *689*, 4291.

138. (a) Smith, K. M.; Poli, R.; Harvey, J. N. *New J. Chem.* **2000**, *24*, 77; (b) Carreón-Macedo, J. L.; Harvey, J. N. *J. Am. Chem. Soc.* **2004**, *126*, 5789.

139. Cacelli, I.; Keogh, D. W.; Poli, R.; Rizzo, A. *J. Phys. Chem. A* **1997**, *101*, 9801.

140. Mulliken, R. S. *J. Chem. Phys.* **1955**, *23*, 1833.

141. Löwdin, P. O. *Adv. Quantum Chem.* **1970**, *5*, 185.

142. Foster, J. P.; Weinhold, F. *J. Am. Chem. Soc.* **1980**, *102*, 7211.

143. Siegbahn, P. E. M.In *Advances in Chemical Physics*; Rice, S. A.; Prigogine, I., Eds.; Wiley: New York, **1996**; Vol. *XCIII*, p 333.

144. (a) Becke, A. D. *Phys. Rev. A* **1988**, *38*, 3098; (b) Perdew, J. P. *Phys. Rev. B* **1986**, *33*, 8822.

145. (a) Becke, A. D. *J. Chem. Phys.* **1993**, *98*, 5648; (b) Lee, C.; Yang, W.; Parr, R. G. *Phys. Rev. B* **1988**, *37*, 785.

146. Lin, Z.; Hall, M. B. *Inorg. Chem.* **1992**, *31*, 2791.

147. (a) Wachters, A. J. H. *J. Chem. Phys.* **1970**, *52*, 1033; (b) Bauschlicher, C. W.; Langhoff, S. R. Jr.; Barnes, L. A. *J. Chem. Phys.* **1989**, *91*, 2399.

148. Schmidling, D. G. *J. Mol. Struct.* **1975**, *24*, 1.

149. Bellard, S.; Rubinson, K. A.; Sheldrick, G. M. *Acta Crystallogr. B* **1979**, *35*, 271.

150. Parrish, S. H.; Van Zee, R. V.; Weltner, W. J. *J. Phys. Chem. A* **1999**, *103*, 1025.

151. Bernhardt, E.; Willner, H.; Kornath, A.; Breidung, J.; Bühl, M.; Jonas, V.; Thiel, W. *J. Phys. Chem. A* **2003**, *107*, 859.

152. Ford, T. A.; Huber, H.; Klotzbücher, W.; Moskovits, M.; Ozin, G. A. *Inorg. Chem.* **1976**, *15*, 1666.

153. Ishikawa, Y.; Hackett, P. A.; Rayner, D. M. *J. Am. Chem. Soc.* **1987**, *109*, 6644.

154. Liu, Z.; Li, Q.; Xie, Y.; King, R. B.; Schaefer, H. F., III *Inorg. Chem.* **2007**, *46*, 1803.

155. Dicke, J. W.; Stibrich, N. J.; Schaefer, H. F., III *Chem. Phys. Lett.* **2008**, *456*, 13.

156. Pfletschinger, A.; Dargel, T. K.; Bats, J. W.; Schmalz, H.; Koch, W. *Chem. Eur. J.* **1999**, *5*, 537.

157. Barckholtz, T. A.; Bursten, B. E. *J. Organomet. Chem.* **2000**, *596*, 212.

158. (a) Perutz, R. N.; Turner, J. J. *Inorg Chem.* **1975**, *14*, 262; (b) Church, S. P.; Poliakoff, M.; Timney, J. A.; Turner, J. J. *J. Am. Chem. Soc.* **1981**, *103*, 7515.

159. Goodman, J. L.; Peters, K. S.; Vaida, V. *Organometallics* **1986**, *5*, 815.

160. Pfletschinger, A.; Schmalz, H.; Koch, W. *Eur. J. Inorg. Chem.* **1999**, 1869.

161. Boese, E. T.; Goldman, A. S. *J. Am. Chem. Soc.* **1992**, *114*, 350.

162. Hasanayn, F.; Nsouli, N. H.; Al-Ayoubi, A.; Goldman, A. S. *J. Am. Chem. Soc.* **2008**, *130*, 511.

163. Ryeng, H.; Gropen, O.; Swang, O. *J. Phys. Chem. A* **1997**, *101*, 8956.

164. Zhou, M.; Andrews, L. *J. Phys. Chem. A* **1999**, *103*, 7773.

165. Daniel, C. *J. Phys. Chem.* **1991**, *95*, 2394.

166. Astruc, D. *Chem. Rev.* **1988**, *88*, 1189.

# 11  Metal-Mediated Carbon–Hydrogen Bond Activation

THOMAS BRENT GUNNOE

## 11.1  INTRODUCTION

Along with the carbon–carbon bond, the carbon–hydrogen bond forms the foundation of organic chemistry. The combination of the typical strength of C–H bonds (C–H bond dissociation energies (BDEs) are often in the range of 95–110 kcal/mol) and their nonpolar nature often renders C–H bonds chemically inert. These features make it difficult to selectively transform C–H bonds into new functionalities, which is a substantial impediment to the synthesis of complex organic molecules as well as the production of materials on the commodity scale. In the past several decades, the emergence of metal-mediated activation of C–H bonds has resulted in the expectation that homogeneous catalysts for the selective and efficient transformation of C–H bonds can be developed.

Since the demonstration that C–H groups of a ligand on a transition metal can bond with the metal in an intramolecular fashion (i.e., intramolecular agostic interactions; Chart 11.1),[1,2] much effort has been directed toward understanding transition metal/C–H bonding and subsequent C–H bond cleavage by the metal center. Although both fundamental and applied challenges remain, a substantial understanding of how metals coordinate and break C–H bonds has been developed. It is the aim of this chapter to provide an overview of the mechanisms by which metals activate and cleave C–H bonds. This chapter is not intended to present a comprehensive review of all contributions to this field, but rather to provide a representative sampling of seminal studies and a perspective on the history and scope of the studies that have led to the state of the art in metal-mediated C–H activation.

*Physical Inorganic Chemistry: Reactions, Processes, and Applications*   Edited by Andreja Bakac
Copyright © 2010 by John Wiley & Sons, Inc.

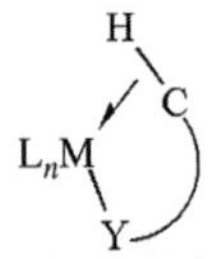

**Chart 11.1**  Generic depiction of an intramolecular agostic bond. (*Note*: Intramolecular agostic interactions are commonly drawn using half arrows to denote the metal/CH bond.)

## 11.2  OVERVIEW OF BONDING BETWEEN A METAL AND C–H BOND AND EARLY STUDIES OF METAL-MEDIATED C–H ACTIVATION

It is generally accepted that metal-mediated rupture of C–H bonds is typically preceded by direct coordination of the C–H unit to the metal center. Similar to the coordination of dihydrogen to metal centers, which was first reported in 1984 by Kubas,[3,4] the metal/C–H bonding interaction involves the donation of electrons from the C–H σ-bonding molecular orbital to an empty metal orbital. Complexes in which the ligand-to-metal electron donation comes from an electron pair that originates from a σ-bonding molecular orbital have been termed σ-complexes (Chart 11.2).[5] In some regards, the coordination of a C–H unit is similar to the coordination of classic Lewis bases (e.g., amines and phosphines) in that the primary bonding interaction often involves the donation of two electrons from the ligand to an empty metal orbital; however, for C–H coordination, the origin of the donated electrons (σ-bonding molecular orbital versus a lone pair or, in some cases, a π molecular orbital) provides an important distinction from typical Lewis bases. Due to the donation of low-energy electrons from a σ-bonding molecular orbital, C–H units coordinate weakly to the metal center relative to the coordination of amines, phosphines, olefins, and so on. In fact, the unique nature of metal/C–H bonding prompted designation of the term "agostic" to describe "situations in which a hydrogen atom is covalently bonded simultaneously to both a carbon atom and a transition metal atom."[1,2] Table 11.1 provides a list of bond dissociation energies of alkanes coordinated to group 6 pentacarbonyl systems. The bond enthalpies range from 8 to 17 kcal/mol, which is typical of metal–alkane BDEs, and are only slightly greater than the strength of typical intermolecular hydrogen bonding. Consistent with the weak metal–alkane bond energies, coordinated alkanes have only been observed using special spectroscopy techniques (see Section 11.4).

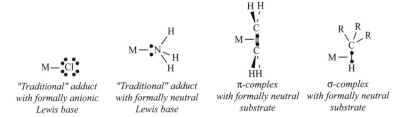

|  |  |  |  |
|---|---|---|---|
| *"Traditional" adduct with formally anionic Lewis base* | *"Traditional" adduct with formally neutral Lewis base* | *π-complex with formally neutral substrate* | *σ-complex with formally neutral substrate* |

**Chart 11.2**  Different classes of Lewis acid/base adducts.

**TABLE 11.1   Representative Bond Enthalpies for Metal–Alkane Bonds for Substrates Coordinated to Metal Pentacarbonyl Fragments, $M(CO)_5$[6]**

| Metal | Alkane | Bond Enthalpy (kcal/mol) |
|---|---|---|
| Cr | Pentane | 8(3) |
| Cr | Heptane | 10(2) |
| Cr | Cyclohexane | 11(2) |
| Mo | Heptane | 17[a] |
| W | Heptane | 13(3) |

[a]Error not reported.

Although many examples of intramolecular agostic interactions have been carefully studied by spectroscopy and X-ray and neutron diffraction, the direct observation of intermolecular coordination (i.e., a metal reacting with and coordinating a "free" molecule) of C–H units is scarce. Based on studies of complexes with intramolecular agostic interactions, theoretical studies, structures of related systems (e.g., coordinated silanes and $BH_4^-$, see below), and spectroscopic data, several possible coordination modes of C–H groups have been proposed (Scheme 11.1).[6] The preferred binding mode likely has a subtle dependence on the metal and the organic substrate. The most commonly invoked coordination mode is $\eta^2$-C,H in which the metal center directly interacts with the carbon atom and a single hydrogen atom. In this coordination mode, two bonds are possible between the metal and C–H unit. As discussed above, the electrons in the C–H σ-bonding molecular orbital are donated into an empty metal orbital of σ symmetry (Scheme 11.2). Thus, the C–H

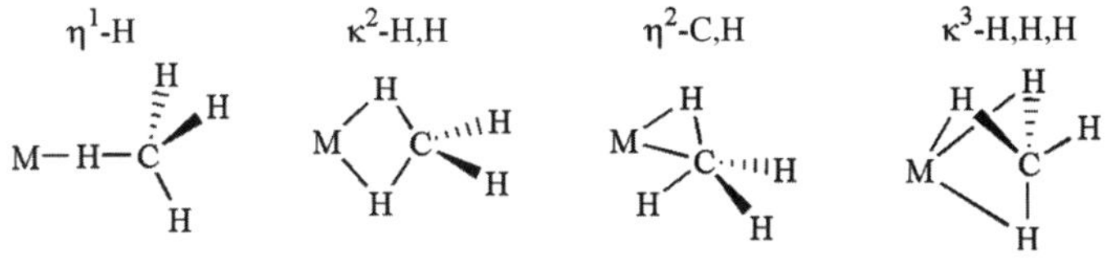

**SCHEME 11.1**   Commonly invoked modes of C–H coordination using methane.

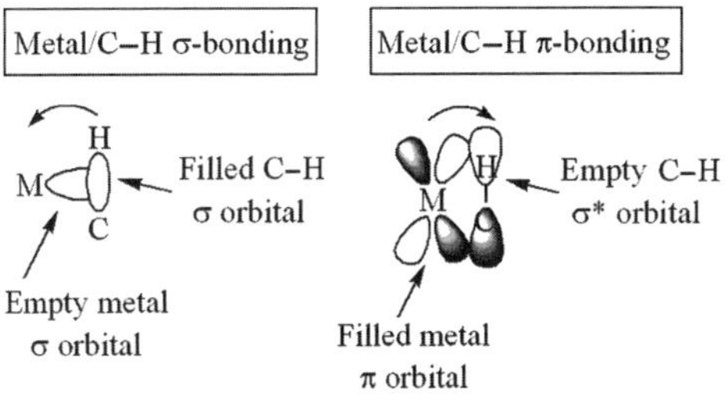

**SCHEME 11.2**   Metal–CH bonding has σ and π components (curved arrows denote the origin and "flow" of electrons involved in both types of bonds).

unit acts as a neutral two-electron donor to the metal center. However, if the metal possesses an occupied orbital that is $\pi$-symmetric with respect to the M–(CH) bond axis, a metal-to-CH $\sigma^*$ backbonding interaction occurs (Scheme 11.2). This metal-to-CH $\pi$-bonding interaction serves to weaken and elongate the C–H bond, and for electron-rich metal centers the metal-to-CH $\sigma^*$ $\pi$-backbonding is often invoked as the primary means of activating of the C–H bond toward cleavage.

Other coordination modes involve variation of the atoms directly bonded to the metal center. Scheme 11.1 illustrates these modes using methane as the hydrocarbon. In addition to the $\eta^2$-C,H coordination mode, coordination of C–H bonds is possible via a linear $\eta^1$-H linkage, $\kappa^2$-H,H, or $\kappa^3$-H,H,H. The $\eta^1$-H coordination mode with relatively long M–H bond distances and large M–H–C bond angles ($\sim$110–170°) has been labeled an "anagostic" interaction.

Although there are no examples of isolable and fully characterized complexes with alkane ligands, other systems have been used as models for alkane coordination. For example, in the absence of metal-to-hydrocarbon backbonding (shown on the right-hand side in Scheme 11.2), the interaction of a metal with two atoms of an alkane gives a three-center/two-electron bonding scheme, which is akin to the electronic structure of $H_3^+$. Given the known preference of $H_3^+$ for a bent geometry, an $\eta^1$-H coordination mode with a truly linear M–H–C unit is anticipated to be unfavorable in most cases. Borohydride anion, $BH_4^-$, is isoelectronic to methane but more basic and forms more stable complexes than alkanes. Several borohydride complexes have been reported, and coordination modes vary among $\eta^1$-H, $\kappa^2$-H,H, $\eta^2$-B,H, and $\kappa^3$-H, H,H.[7–9] The observation of the various coordination geometries for $BH_4^-$ suggests the possibility that analogous coordination patterns are accessible for alkane complexes. In addition, complexes with coordinated silane molecules have been isolated and fully characterized, and these systems exhibit $\eta^2$-Si,H coordination.[10,11]

The elucidation of intramolecular agostic interactions in the 1960s and early examples of C–H activation hinted at the possibility that metals can coordinate alkanes or arenes via C–H bonds to form intermediates.[12] For example, in 1965, the structure of $Ru(Cl)_2(PPh_3)_3$ was reported to have a close Ru–H contact with an *ortho* C–H group of a phosphine ligand (Chart 11.3).[13] Using typical geometric parameters for a phenyl ring, the Ru–H distance was calculated to be $\sim$2.59 Å. While the short Ru–H contact indicated a possible intramolecular Ru/CH bond, due to the lack of precedent, the authors were hesitant to reach this conclusion. On this subject, they stated, ". . . there is no geometrical basis for postulating that this is a weak metal–hydrogen interaction similar to those postulated to account for various spectroscopic

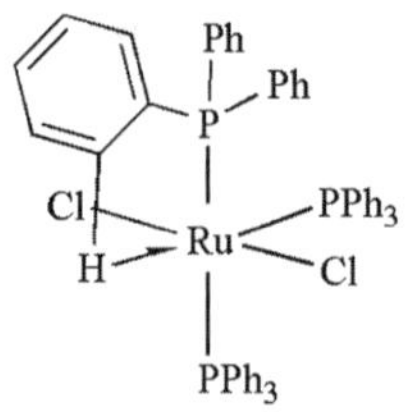

**Chart 11.3** Early example of intramolecular agostic interaction (Ru–H bond distance calculated, based on typical geometry of phenyl ring, to be 2.59 Å).

$$\text{ReH}_7(\text{PEt}_2\text{Ph})_2 + \text{C}_6\text{D}_6 \xrightleftharpoons[\quad]{100\,°\text{C}} \text{ReH}_{7-n}\text{D}_n(\text{PEt}_2\text{Ph})_2 + \text{C}_6\text{D}_{6-n}\text{H}_n$$

$$\text{Cp}_2\text{TaH}_3 + \text{C}_6\text{D}_6 \rightleftharpoons \text{Cp}_2\text{TaH}_{3-n}\text{D}_n + \text{C}_6\text{D}_{6-n}\text{H}_n$$

$$\text{H}_2 + \text{Cp}_2\text{TaH}_3 + \text{C}_6\text{D}_6 \rightleftharpoons \text{Cp}_2\text{TaH}_{3-n}\text{D}_n + \text{C}_6\text{D}_{6-m}\text{H}_m + \text{HD} + \text{D}_2$$

$$\text{H}_2 + (\text{PhPEt}_2)_2\text{IrH}_5 + \text{C}_6\text{D}_6 \rightleftharpoons (\text{PhPEt}_2)_2\text{IrH}_{5-n}\text{D}_n + \text{C}_6\text{D}_{6-m}\text{H}_m + \text{HD} + \text{D}_2$$

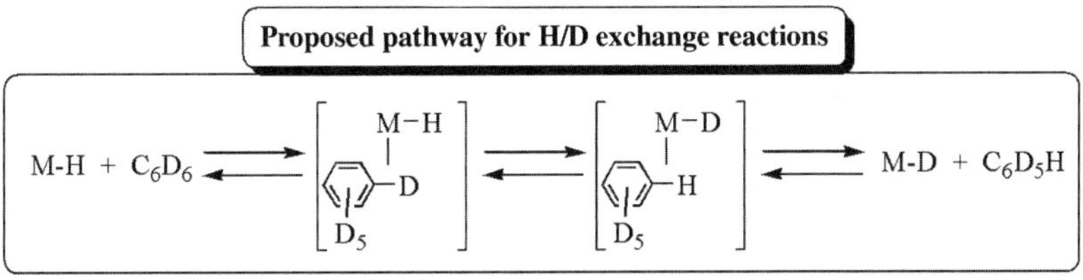

**Proposed pathway for H/D exchange reactions**

**SCHEME 11.3** Early examples of H/D exchange between metal hydrides and deuterated hydrocarbons and possible mechanism that involves C−D(H) coordination to the metal center.

anomalies . . .. The evidence of such weak metal–hydrogen interactions is extremely tenuous . . .." In ensuing work, combined structural and spectroscopic data unambiguously revealed the presence of intramolecular agostic bonds for several complexes.[12,14] Additional early evidence for metal-mediated C−H activation came from isotopic labeling studies. For example, in the late 1960s and early 1970s, several groups reported on H/D exchange reactions between metal hydrides and deuterated hydrocarbons, and a few examples are shown in Scheme 11.3.[15–17] These results are best rationalized by intermediates that involve C−H(D) units coordinated to the metal center.

Chatt and Davidson, in combination with subsequent work by Cotton et al., demonstrated that Ru with two bisphosphine ligands, Ru($\kappa^2$-$P$,$P$-Me$_2$PCH$_2$CH$_2$PMe$_2$)$_2$, reacts to break a phosphine methyl C−H bond to give a dimeric Ru(II) cyclometalated product (Equation 11.1).[18] In addition, the combination of Ru($\kappa^2$-$P$,$P$-Me$_2$PCH$_2$CH$_2$PMe$_2$)$_2$ with naphthalene provides the Ru(II) naphthyl hydride complex.[19] These reactions were among the first examples of well-defined metal-mediated C−H bond cleavage.

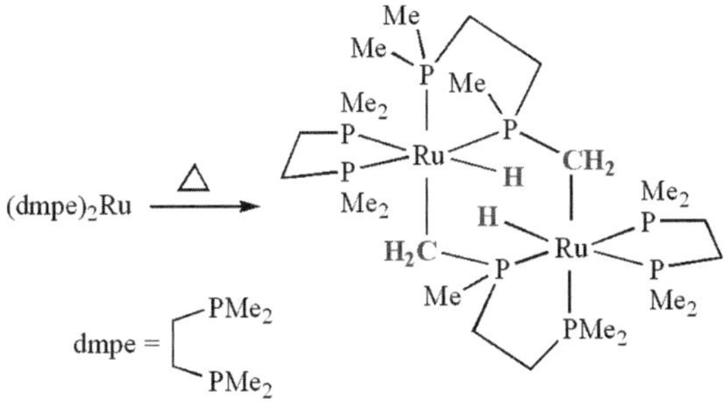

(11.1)

While the observation of intramolecular agostic interactions and intermolecular H/D exchange reactions suggested the possibility of C—H coordinated alkane/arene intermediates in the C—H(D) activation reactions, well-defined experiments based on intramolecular isotope exchange provided more definitive results.[20] Some of the earliest mechanistic studies of hydrocarbon C—H activation were centered on reactions of Cp*M(PMe$_3$)(R)(H) complexes (Cp* = pentamethylcyclopentadienyl; M = Rh or Ir; R = hydrocarbyl or H).[21,22] For example, heating Cp*M(PMe$_3$)(alkyl) (H) in benzene results in the reductive elimination of alkane and production of Cp*M (PMe$_3$)(Ph)(H). Isotopic labeling of the precursor alkyl hydride complexes provided interesting results. Bergman et al. found that heating (130°C) Cp*Ir(PMe$_3$)(cyclo-hexyl)(D) in benzene not only produces free cyclohexane-$d_1$ and Cp*Ir(PMe$_3$)(Ph)(H) but also gives the isotopomer Cp*Ir(PMe$_3$)(cyclohexyl-$d_1$)(H).[23] Thus, the deuteride ligand of the starting material is exchanged with the α-H atom of the cyclohexyl ligand. The isotopic scrambling is best explained by reversible C—D(H) bond formation between the cyclohexyl and deuteride ligands to give a cyclohexane-$d_1$ coordinated intermediate (Scheme 11.4). From the coordinated cyclohexane-$d_1$ intermediate, insertion of Ir into a C—H bond gives Cp*Ir(PMe$_3$)(cylcohexyl-$d_1$) (H), while cyclohexane-$d_1$ dissociation and benzene C—H activation provide

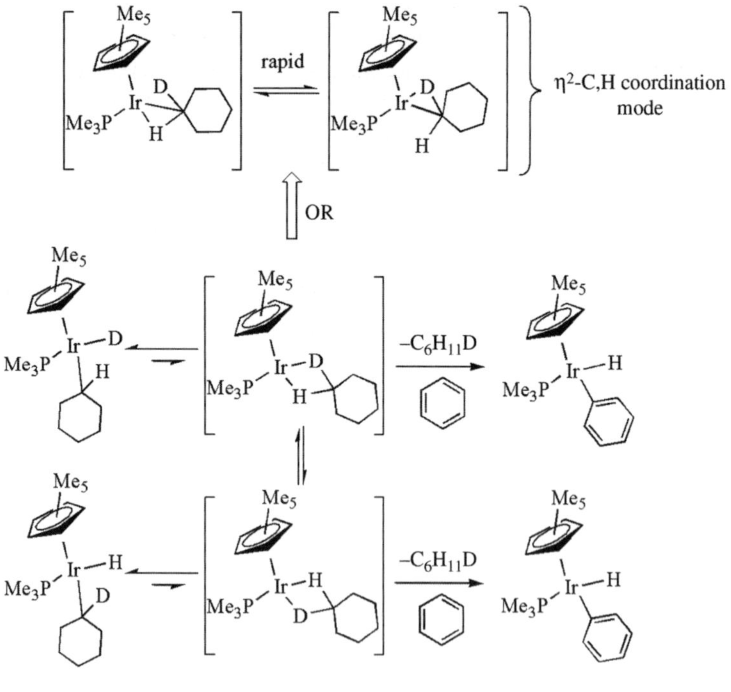

**SCHEME 11.4**    Observation of H/D change with Ir(III) complexes can be explained by intermediacy of Ir/cyclohexane adduct.

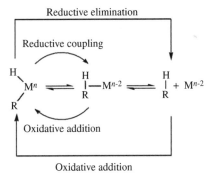

**Chart 11.4** Nomenclature of C—H activation via metal insertion into C—H bond and the microscopic reverse reaction. (*Note*: The term oxidative addition is used for the single step of metal insertion into a coordinated C—H bond as well as the overall two-step process of C—H coordination followed by metal insertion into the C—H bond.)

Cp*Ir(PMe₃)(Ph)(H). Thus, the isotopic scrambling provides evidence for an *alkane adduct* as an intermediate during the conversion of Cp*Ir(PMe₃)(R)(H) and benzene to free RH and Cp*Ir(PMe₃)(Ph)(H).

The conversion of the Ir(III) cyclohexyl hydride complex to an Ir/cyclohexane system involves a change in the formal oxidation state of Ir from $+3$ to $+1$ (i.e., a formal two-electron reduction). As a result, this elementary reaction step is generally called a reductive coupling (Chart 11.4). From a metal hydrocarbyl hydride complex (i.e., M(R)(H)), the overall process of C—H bond formation *and* dissociation of free hydrocarbon (or related functionalized molecule) is called reductive elimination (Chart 11.4). The reverse process, metal coordination of a C—H bond and insertion into the C—H bond, is called oxidative addition. (*Note*: Oxidative addition and reductive elimination reactions are not limited to reactions involving C and H.)

Extension of the isotopic labeling studies to Rh gives similar results to the Cp*Ir systems.[24] At $-80°$C in toluene-$d_8$, Cp*Rh(PMe₃)($^{13}$CH₂CH₃)(D) rapidly isomerizes to give Cp*Rh(PMe₃)($^{13}$CDHCH₃)(H) in equilibrium with the starting isotopomer. At elevated temperature ($-30°$C), the incorporation of deuterium and $^{13}$C labels into the β-position of the ethyl ligand competes with ethane elimination and toluene C—D activation (Scheme 11.5). These results suggest that the migration of the D and $^{13}$C isotopes to the β-ethyl position (at $-30°$C) occurs via the formation of an ethane complex with migration of Rh between the C units of the coordinated ethane (Scheme 11.5). Thus, the dissociation of ethane has a sufficiently large $\Delta G^{\ddagger}$ to render migration along the carbon chain and subsequent C—H activation to form a Rh ethyl hydride complex competitive with ethane dissociation. Importantly, monitoring the deuterated complex Cp*Rh(PMe₃)(CD₂CD₃)(D) in toluene-$d_8$ in the presence of protioethane (C₂H₆) does not result in the incorporation of protioethyl/hydride into the Rh complex. Thus, the H/D scrambling of Cp*Rh(PMe₃)($^{13}$CH₂CH₃)(D) does not likely occur via reversible elimination of ethane from the Rh coordination sphere, which provides additional evidence for the formation of ethane-coordinated Rh intermediates from the Rh ethyl hydride starting material. Furthermore, a mixture of Cp*Rh(PMe₃)(C₂H₅)(H) and

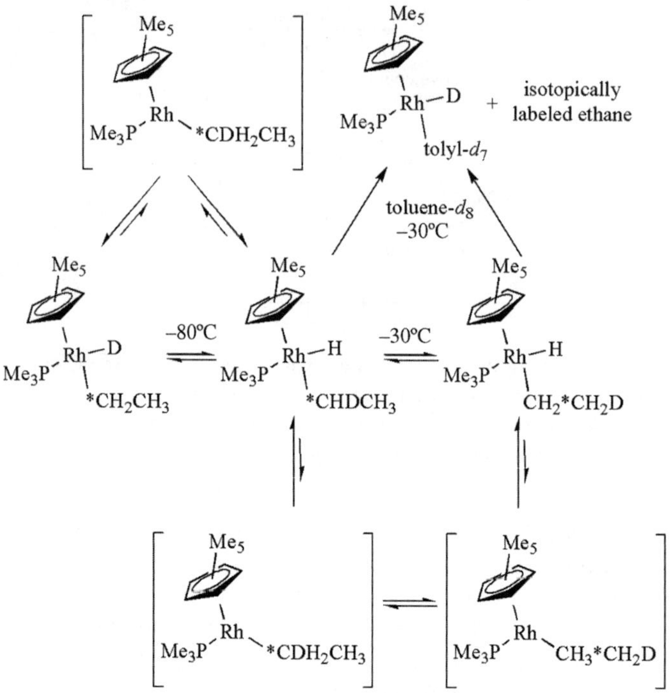

**SCHEME 11.5** H/D exchange reactions for isotopically labeled Cp*Rh complexes suggest a coordinated ethane intermediate (*C = $^{13}$C).

Cp*Rh(PMe$_3$)(C$_2$D$_5$)(D) at $-20°$C in toluene-$d_8$ gives Cp*Rh(PMe$_3$)(C$_7$D$_7$)(D), C$_2$H$_6$, and C$_2$D$_6$. Thus, under the conditions studied, the H/D scrambling occurs via an intramolecular pathway and does not likely involve binuclear Rh complexes.

Studies of other metals give results consistent with alkane (or arene) coordinated intermediates in C–H activation processes. Methane elimination occurs from [Cp$_2$Re (CH$_3$)(H)]$^+$[X]$^-$ (Cp = cyclopentadienyl) to give Cp$_2$Re(X) (X = formally anionic ligand such as chloride) and free CH$_4$. Monitoring the reaction of [Cp$_2$Re(CD$_3$)(D)]$^+$ under CH$_4$ pressure reveals no proton incorporation into the starting material, which suggests that the reductive elimination of methane is irreversible.[25] At $-55°$C, [Cp$_2$Re(CD$_3$)(H)]$^+$ produces [Cp$_2$Re(CD$_2$H)(D)]$^+$. Heinekey and Gould were able to separately monitor the rate of H/D exchange between the methyl and hydride positions of [Cp$_2$Re(CD$_3$)(H)]$^+$, which gives the rate of C–H reductive coupling, and the rate of methane reductive elimination, which is C–H reductive coupling followed by methane dissociation, as a function of temperature. From the Eyring plots, activation parameters for these two processes were determined: $\Delta H^{\ddagger} = 28.1$ (1.2) kcal/mol and $\Delta S^{\ddagger} = 27(6)$ eu for the reductive elimination of methane and $\Delta H^{\ddagger} = 22.3(1.9)$ kcal/mol and $\Delta S^{\ddagger} = 24(9)$ eu for the H/D exchange via the formation of a methane adduct by reversible C–H reductive coupling (Scheme 11.6). It is interesting that the difference in activation barrier between H/D exchange and

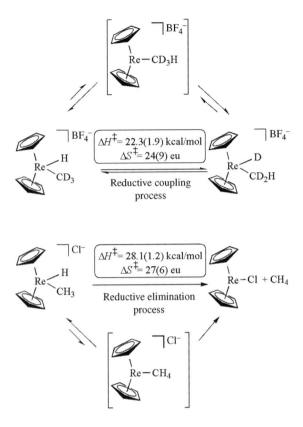

**SCHEME 11.6**   The rates of H/D exchange for a Re(V) perdeuteriomethyl hydride complex and of methane reductive elimination as a function of temperature allow comparison of activation parameters for the two processes.

methane reductive elimination is almost entirely due to differences in the enthalpy of activation with $\Delta\Delta H^{\ddagger} \sim 6\,kcal/mol$, which is within reasonable proximity of most estimated binding energies of alkanes (see Table 11.1 for example).

Parkin and Bercaw reported that $Cp^*_2W(Me)(H)$ eliminates methane to form $Cp^*(\eta^5,\eta^1\text{-}C_5Me_4CH_2)WH$.[26] For the mixed isotopomer, $Cp^*_2W(CH_3)(D)$, H/D scrambling to give $Cp^*_2W(CH_2D)(H)$ is competitive with the methane elimination process (Scheme 11.7). Although the authors point out that the H/D exchange process could occur by pathways other than formation of a methane-coordinated intermediate, the observation of an inverse kinetic isotope effect (KIE) for the methane reductive elimination (see bottom of Scheme 11.7) provides additional support for the reversible formation of coordinated alkane (see below for a more detailed discussion of KIEs for reductive elimination of C−H bonds). Furthermore, at relatively low concentrations, heating a mixture of $Cp^*_2W(CH_3)(H)$ and $Cp^*_2W(CD_3)(D)$ produces only $CH_4$ and $CD_4$ with no observation of H/D crossover, which is consistent with intramolecular C−H(D) processes. Similar results have been obtained for

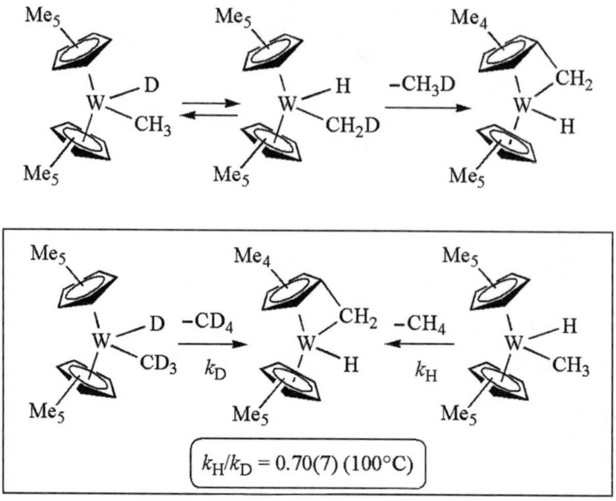

**SCHEME 11.7** H/D exchange reactions of $Cp^*_2W(H)(Me)$ and isotopomers are consistent with methane-coordinated intermediates.

$Cp_2W(CH_3)(H)$ complexes.[27] Interestingly, at higher concentrations, $Cp_2W(D)(CH_3)$ (14.0 mM) and $Cp_2W(H)(CD_3)$ (9.4 mM) eliminate a mixture of all possible methane isotopomers, which suggests that methane elimination occurs via an intermolecular process at the elevated concentrations of W complex. These results point to the subtle energetic balance among the mechanistic possibilities for C–H(D) elimination/addition and hydrogen scrambling pathways. This underscores the importance of detailed studies to accurately elucidate mechanistic pathways and provides a caveat against extrapolating results under one set of conditions to general conclusions for C–H activation/elimination under different conditions, even for processes that involve closely related or identical metal systems.

These experiments, based predominantly on isotopic labeling and kinetics, and related observations cemented the notion of complexes with coordinated C–H bonds as intermediates (i.e., metal/CH σ-complexes) during C–H activation and C–H elimination processes, and they set the stage for more detailed studies of C–H activation mechanisms. The indirect inference of σ-complexes based on isotopic labeling experiments has been supported by spectroscopy including fast IR experiments and specialized NMR spectroscopy experiments. Section 11.4 provides an overview of these experiments.

## 11.3 MECHANISMS FOR C–H ACTIVATION

### 11.3.1 General Comments

Several mechanisms for metal-mediated activation of carbon–hydrogen bonds have now been substantiated (Scheme 11.8).[28] The most commonly accepted pathways

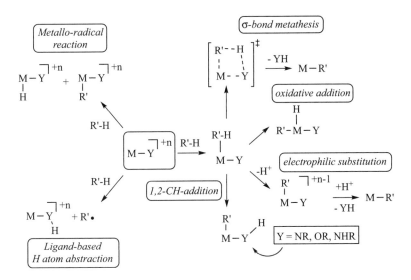

**SCHEME 11.8** Common mechanisms of C–H activation (after the initial complex, the $+n$ charge is indicated only in cases where overall charge of the complex is altered during the reaction pathway). The metalloradical and ligand-based H-atom abstraction reactions (both shown on the left-hand side) are not discussed herein.

include oxidative addition, σ-bond metathesis, electrophilic substitution, 1,2-addition of C–H bonds across metal–heteroatom bonds, and metalloradical reactions. Each mechanism and examples of seminal studies that elucidated the reaction pathways are discussed below. Metalloradical pathways will not be covered herein.[29] In addition, another class of metal-promoted C–H bond scission involves *ligand-centered* hydrogen atom abstraction pathways (also called proton-coupled electron transfer reactions, depending on the specifics of the reaction mechanism). Such reactions typically involve high oxidation state complexes of the middle to late transition metal elements with formally anionic or dianionic heteroatomic ligands.[30,31] In these reactions, although the identity and nature of the metal play an obviously important role, the metal center is not typically thought to directly interact with the C–H bond; hence, this class of reaction along with metalloradical pathways will not be covered in this chapter.

### 11.3.2 Oxidative Addition

Oxidative addition occurs when the metal inserts into the C–H bond to give new hydride and hydrocarbyl ligands (Scheme 11.8). The formation of two new metal–ligand bonds requires four electrons. Two electrons are derived from the C–H bond and two electrons are donated by the metal center. Thus, this process results in a formal two-electron oxidation of the metal center (Scheme 11.9). The C–H oxidative addition with a single metal center requires that the metal precursor have access to two vacant coordination sites and possess at least one lone pair of electrons. Such reactions

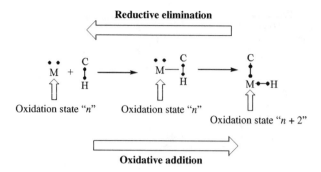

**SCHEME 11.9** C–H oxidative addition with a single metal center results in increase in oxidation state (by $+2$) at the metal.

are commonly observed for electron-rich metals that have a predilection toward losing electrons, and oxidative addition reactions are typical for (though not limited to) complexes of Ru, Os, Re, Rh, Ir, and Pt in low oxidation states. However, caution should be taken with general extrapolation of the notion that more electron-rich metal centers exhibit an enhanced penchant toward oxidative addition processes, since there are other less obvious factors that can impact the energetics of such reactions (including coordination number, coordination geometry, and impact on metal–ligand bonding).[32] Variations of oxidative addition to a single metal that involve C–H addition across two metal centers, which results in a $+1$ change in oxidation state at each metal, and related transformations will not be discussed herein.

Early examples of oxidative addition reactions involved the insertion of transition metals into carbon–halide bonds, and such reactions were typically observed for metals in the middle to late portion of the periodic table in low oxidation states. Likewise, some of the first examples of well-defined oxidative addition involving C–H bonds were centered on Ir(I) and Rh(I) complexes. In a series of seminal papers,[33–38] the groups of Bergman, Graham, and Jones reported that Ir and Rh systems supported by the Cp* ligand initiate oxidative addition of C–H bonds including elusive examples of alkane reactivity. For example, irradiation of Cp*Ir(PMe$_3$)(H)$_2$ results in the extrusion of dihydrogen to generate the unsaturated Ir(I) intermediate Cp*Ir(PMe$_3$), which reacts with benzene, cyclohexane, and neopentane to yield Cp*Ir(PMe$_3$)(Ph)(H), Cp*Ir(PMe$_3$)(cyclohexyl)(H), and Cp*Ir(PMe$_3$)(CH$_2$CMe$_3$)(H), respectively (Scheme 11.10).[33] In their efforts to probe the mechanism of the net oxidative addition reactions, Janowicz and Bergman considered two possible radical pathways (Scheme 11.11).[34] The reaction of Cp*Ir(PMe$_3$)(H)$_2$ with C$_6$D$_{12}$ produces only Cp*Ir(PMe$_3$)(C$_6$D$_{11}$)(D) (Scheme 11.12), which eliminates from consideration the mechanism involving initial Ir–H bond homolysis (mechanism 1 in Scheme 11.11), since this would produce the mixed isotopomer Cp*Ir(PMe$_3$)(C$_6$D$_{11}$)(H). The second radical pathway (mechanism 2 in Scheme 11.11) should favor the activation of weaker C–H bonds. However, the unobserved intermediate Cp*Ir(PMe$_3$) reacts 3.7 times more rapidly with the stronger aromatic C–H bonds of p-xylene than with the benzylic position

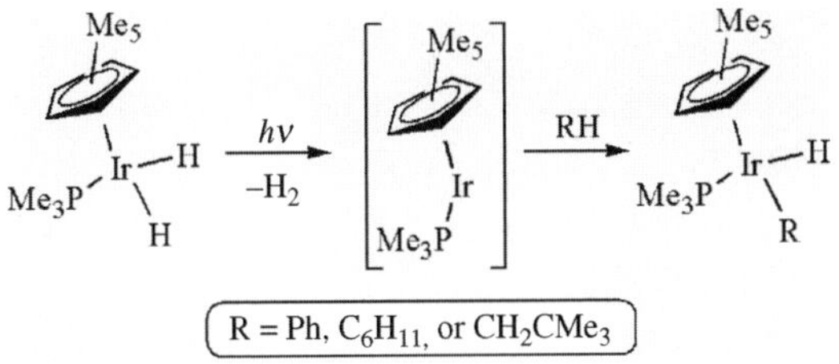

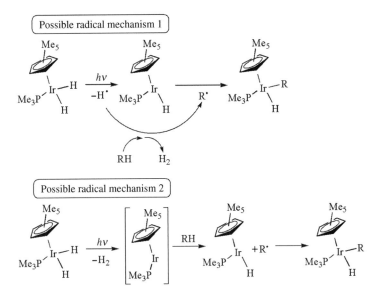

R = Ph, C₆H₁₁, or CH₂CMe₃

**SCHEME 11.10** *In situ* generated Cp*Ir(PMe₃) initiates C—H activation via oxidative addition.

(Scheme 11.12). A labeled crossover experiment provided additional evidence against the second radical reaction (mechanism 2 in Scheme 11.11). The irradiation of Cp*Ir(PMe₃)H₂ in a mixture of perprotioneopentane and deuterated cyclohexane gave the products anticipated for direct insertion of the metal into C—H(D) bonds: Cp*Ir(PMe₃)(CH₂CMe₃)(H) and Cp*Ir(PMe₃)(C₆D₁₁)(D) (Scheme 11.12), whereas the radical mechanism 2 would also produce the crossover products Cp*Ir(PMe₃)(CH₂CMe₃)(D) and Cp*Ir(PMe₃)(C₆D₁₁)(H). These and other experiments are most consistent with a mechanism for overall C—H oxidative addition that occurs through the coordination of the hydrocarbon to Cp*Ir(PMe₃) followed by concerted insertion of the metal into the C—H bond. The Graham group reported analogous results with an Ir carbonyl analogue: Cp*Ir(CO)₂.[35,36] For example, the

**SCHEME 11.11** Possible radical pathways for C—H activation by Cp*Ir(PMe₃), which were shown to be unlikely by the experiments in Scheme 11.12.

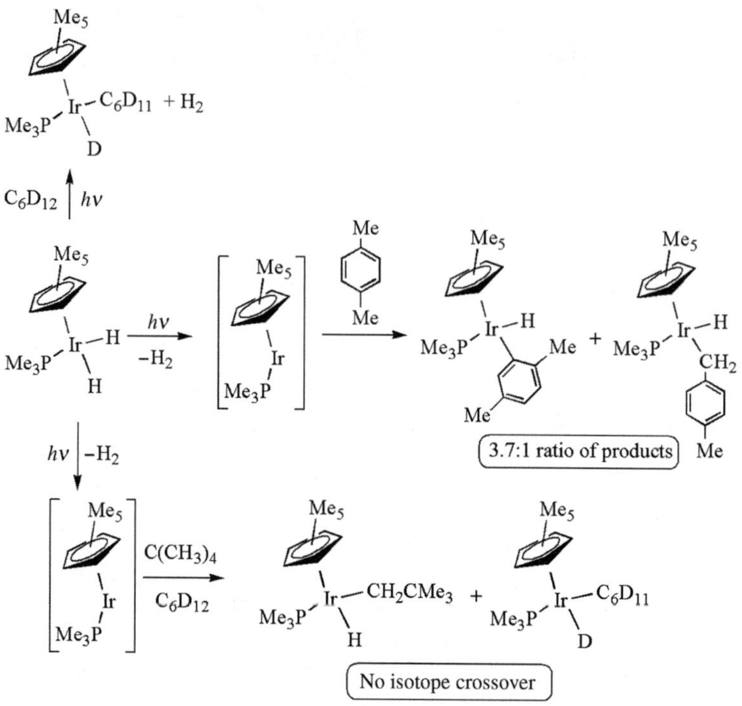

**SCHEME 11.12** Experiments that are consistent with a nonradical mechanism for C–H activation by Cp*Ir(PMe₃)H₂ precursor.

irradiation of Cp*Ir(CO)₂ in neopentane or perfluorohexane under methane pressure produces Cp*Ir(CO)(R)(H) (R = CH₂CMe₃ or Me).

Through a detailed series of studies, Jones and Feher mapped the thermodynamics and kinetics of oxidative addition/reductive elimination of C–H bonds by Cp*Rh(PMe₃), which is generated by photoinduced elimination of dihydrogen from Cp*Rh(PMe₃)H₂.[22] Using a combination of inter- and intramolecular KIEs, the conversion of Cp*Rh(PMe₃) and benzene to Cp*Rh(PMe₃)(Ph)(H) was demonstrated to occur through rate-limiting coordination of benzene with a subsequent fast C–H activation step. Photolysis of Cp*Rh(PMe₃)H₂ in a 1:1 molar ratio of C₆H₆/C₆D₆ resulted in a 1.05:1 molar ratio of Cp*Rh(PMe₃)(C₆H₅)(H) and Cp*Rh(PMe₃)(C₆D₅)(D), which reflects the absence of a primary intermolecular KIE. The absence of a primary intermolecular KIE indicates that benzene C–H(D) activation is not likely the rate-determining step in the overall conversion of Cp*Rh(PMe₃)H₂ and benzene to Cp*Rh(PMe₃)(Ph)(H) and H₂. In contrast, irradiation of Cp*Rh(PMe₃)H₂ in 1,3,5-trideuteriobenzene gives $k_H/k_D = 1.4$ (Scheme 11.13), which is consistent with a primary KIE from a transition state with a nonlinear Rh–H–C linkage. These results can be rationalized by rate-determining coordination of benzene, for which deuteration has minimal impact on the reaction rate, followed by a fast C–H(D) activation step (Scheme 11.14). For η²-C,C coordination of 1,3,5-trideuteriobenzene, only a

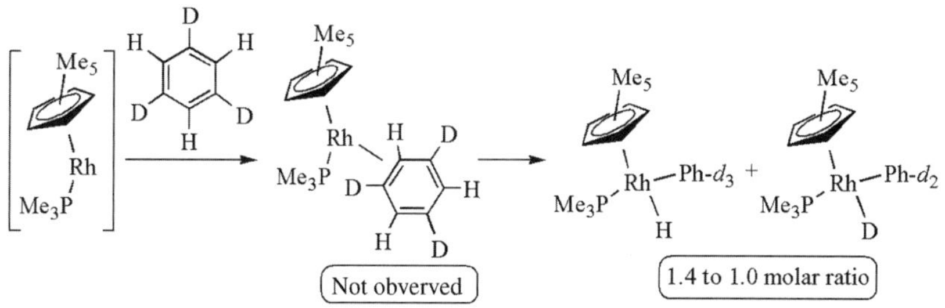

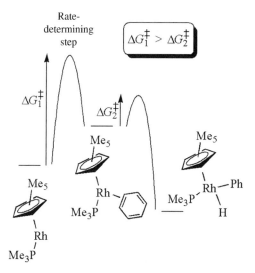

**SCHEME 11.13**   Reaction of Cp*Rh(PMe₃) with 1,3,5-trideuteriobenzene reveals a kinetic isotope effect of $k_H/k_D = 1.4$.

single isomer is possible, and the primary KIE reflects the rate difference between C−H and C−D activation in the second and fast step of the transformation.

Jones, Perutz, and coworkers studied the thermodynamics of equilibria between Cp*Rh(PMe₃)($\eta^2$-aromatic) and Cp*Rh(PMe₃)(aryl)(H) systems (parent Cp systems were also studied) (Equation 11.2).[39] Through a comparison of equilibrium constants for an array of substrates (see Equation 11.2), it was concluded that the equilibria were not controlled by the strength of C−H, M−C, and M−H bonds or by $\Delta S$, but rather by the energy of the $\eta^2$-aromatic complexes. Thus, the details of the Rh-$\eta^2$-aromatic bonding are key to understanding the penchant of the Cp*Rh(PMe₃) and aromatic systems to undergo aromatic C−H oxidative addition to give Cp*Rh(PMe₃)(Ar)(H) products.

**SCHEME 11.14**   Qualitative energy diagram that explains kinetic isotope effects observed for benzene C−H activation by Cp*Rh(PMe₃).

$$
\begin{array}{ccc}
\text{Me}_5 & K_{eq} \; ? & \text{Me}_5 \\
\text{Rh} & \rightleftharpoons & \text{Rh}-\text{Ar} \\
\text{Me}_3\text{P} & & \text{Me}_3\text{P} \quad \text{H}
\end{array}
\qquad (11.2)
$$

benzene, naphthalene, perylene, triphenylene,
phenanthrene, anthracene, 2-methoxynaphthalene,
2,6-dimethoxynaphthalene, pyrene, and fluoranthene

For metal centers with a d-electron count greater than zero, dihapto-C,C coordination of aromatic substrates results in metal-to-aromatic π-backbonding, an interaction that involves electron donation from a filled dπ orbital to a C=C π* molecular orbital (Chart 11.5). If this metal-to-ligand π-interaction is strong, the C–C bond length is increased at the expense of aromatic resonance energy.[40] The C–H oxidative addition step removes the $\eta^2$-C,C bond with Rh and restores aromaticity (Chart 11.5). As a result, aromatic substrates with greater resonance stabilization exhibit an enhanced propensity toward the oxidative addition with equilibria shifted away from the $\eta^2$-aromatic substrates. In contrast, for organic substrates with less aromaticity, $\eta^2$-C,C coordination is correspondingly more favorable, and $K_{eq}$ favors the $\eta^2$-aromatic complex relative to substrates with more aromaticity. Calculated resonance energies for free and bound aromatic substrates are consistent with the proposed equilibrium being dictated by the energy of the $\eta^2$-aromatic complex.

One of the intriguing attributes of many systems that initiate C–H oxidative addition is the commonly observed selectivity for stronger C–H bonds, which can be divided into kinetic and thermodynamic selectivity. For metal-mediated C–H activation, kinetic and thermodynamic selectivities are often identical. For example, arenes often undergo reaction more rapidly than alkanes that possess weaker C–H bonds, and aryl hydride complexes (plus free alkane) are commonly favored thermodynamically over alkyl hydride systems (plus free aromatic substrate). Assuming that

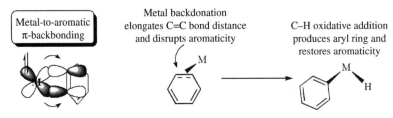

**Chart 11.5**   Generic depiction (on the left-hand side) of metal-to-aromatic π-backbonding with benzene used as the aromatic substrate (arrows denote flow of electron density from the metal dπ orbital to the aromatic C=C π* molecular orbital). On the right-hand side, conversion of $\eta^2$-benzene complex to phenyl hydride complex, via C–H oxidative addition, restores aromaticity.

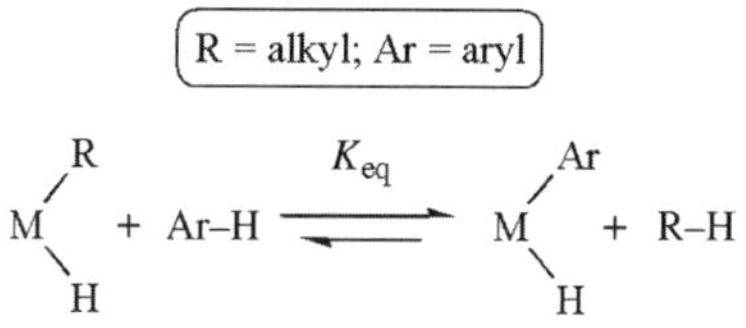

$$\boxed{R = \text{alkyl; } Ar = \text{aryl}}$$

$$\underset{H}{\overset{R}{M}} + Ar\text{–}H \;\underset{}{\overset{K_{eq}}{\rightleftharpoons}}\; \underset{H}{\overset{Ar}{M}} + R\text{–}H$$

$$\Delta H \sim BDE_{M(R)} + BDE_{Ar(H)} - BDE_{M\text{-}Ar} - BDE_{R(H)}$$

$$\text{If } K_{eq} > 1, \left| \Delta BDE_{M(Ar)-M(R)} \right| > \left| \Delta BDE_{Ar(H)-R(H)} \right|$$

**Chart 11.6** Equilibrium favoring metal aryl hydride and alkane over the alkyl hydride complexes and free aromatic substrates likely indicates that the difference of M–Ar and M–R BDEs is greater than difference in BDE for Ar–H and R–H (assuming that $\Delta S \sim 0$ and that the two M–H BDEs are approximately equivalent).

M–H bond energies are approximately constant and negligible difference in $\Delta S$, the thermodynamic preference of metal aryl hydride complexes and free alkane over metal alkyl hydride complexes and free aromatic substrate suggests that the $\Delta$BDE for M–Ar versus M–R is greater than the $\Delta$BDE for Ar–H versus R–H (Chart 11.6).

The kinetic selectivity for aromatic C–H activation over alkane substrates may derive from the more facile coordination of aromatic substrates through the aromatic $\pi$-system compared with the $\sigma$-coordination available to alkanes. Typically, coordination to metal centers through $\pi$-electrons results in stronger bonds than coordination using electrons in C–H $\sigma$-bonds. As a consequence, although not routinely isolated, examples of isolable and fully characterized $\eta^2$-C,C coordinated aromatic substrates are known,[40] while isolated and fully characterized alkane adducts are elusive. Thus, given a choice between aromatic and alkane substrates, a metal center is anticipated to coordinate the $\pi$-system of the aromatic more rapidly than a C–H bond of the alkane in addition to forming a thermally more robust bond with the aromatic substrate (Scheme 11.15). This rationale is perhaps especially suitable for systems in which coordination of the hydrocarbon is the rate-determining step.

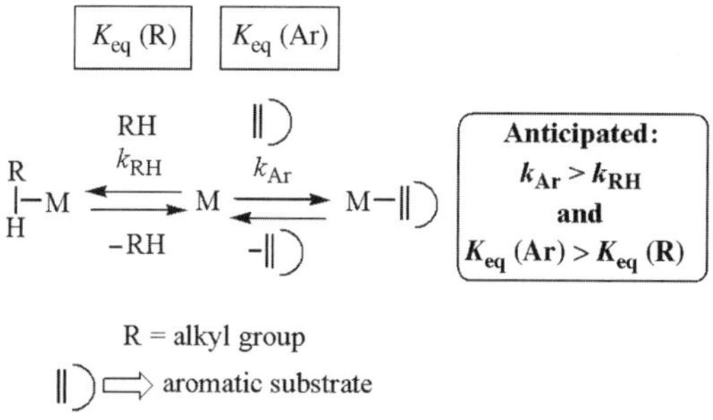

$$\boxed{K_{eq}\,(R)} \quad \boxed{K_{eq}\,(Ar)}$$

| | Anticipated: |
|---|---|
| | $k_{Ar} > k_{RH}$ |
| | and |
| | $K_{eq}\,(Ar) > K_{eq}\,(R)$ |

R = alkyl group

⫼⟩ ⟹ aromatic substrate

**SCHEME 11.15** $\eta^2$-C,C aromatic coordination is typically more facile than C–H coordination.

Selectivity among various sp$^3$ hybrid C–H bonds is more difficult to rationalize. For example, the reaction of *in situ* generated Cp*Rh(PMe$_3$) with alkanes generates the product of activation of the stronger primary C–H bonds in preference to weaker secondary (or tertiary) C–H bonds. Other systems have been demonstrated to possess similar selectivity. A possible explanation of such selectivity is a steric inhibition against activation of internal C–H bonds of a linear alkane. That is, on the timescale of the C–H oxidative addition event, the metal center might access only the terminal C–H bonds. However, recent detailed mechanistic studies have revealed this notion to be incorrect, at least for some systems.

Heating [Cn*Rh(PMe$_3$)(CH$_3$)(D)][BAr$'_4$] (Cn* = 1,4,7-trimethyl-1,4,7-triazacyclononane; Ar' = 3,5-(CF$_3$)$_2$C$_6$H$_3$) to 50°C in a mixture of C$_6$F$_6$ and C$_6$D$_6$ results in H/D scrambling to give an equilibrium with [Cn*Rh(PMe$_3$)(CH$_2$D)(H)][BAr$'_4$] (Equation 11.3) with the ultimate production of free methane and [Cn*Rh(PMe$_3$)(Ph-$d_5$)(D)][BAr$'_4$].[41] In an interesting extension of the H/D exchange of methyl hydride, Flood et al. prepared the complexes [Cn*Rh{P(OMe)$_3$}(R)(H)][BAr$'_4$] where R = Et, Bu, hexyl, and decyl.[42] Initial synthesis at low temperature of the deuterated hexyl complex [Cn*Rh{P(OMe)$_3$}(hexyl)(D)][BAr$'_4$] results in minimal deuterium incorporation into the hexyl ligand; however, at 4°C the deuterium migrates to the α-CH$_2$ group with $k = 4.3 \times 10^{-4}\,\text{s}^{-1}$ (Scheme 11.16). Although deuterium does not migrate to any of the "internal" methylene groups, H/D exchange with the *terminal* methyl group is also observed (Scheme 11.16), albeit at slower rates than for the α-CH$_2$ group. The H/D exchange at the terminal methyl position of "R" is even observed for the decyl complex. The mechanism of H/D exchange between the Rh–D and the terminal methyl position of the alkyl ligands is unlikely to proceed through free alkane, since reactions in benzene would produce the phenyl hydride complex. Dissociation of a nitrogen of the Cn* ligand followed by Rh migration via a

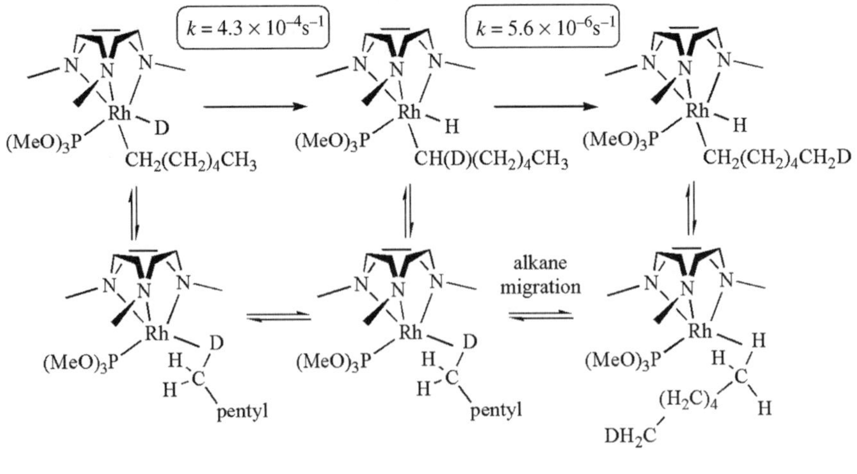

**SCHEME 11.16**  H/D exchange for Rh–hexyl hydride complex with proposed mechanism that involves migration along alkane chain.

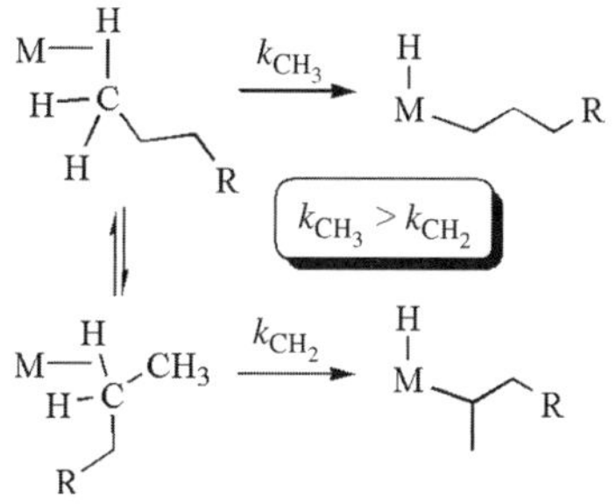

**Chart 11.7**  Oxidative addition of methyl C–H bonds is kinetically more facile than $CH_2$ activation for many-transition metal systems.

sequence of β-hydride elimination/reinsertion reactions was discounted, since D incorporation into the β-position of the alkyl ligands is not observed. Thus, the results point to the possibility of C–H reductive coupling, followed by migration along the carbon chain of the coordinated alkane, and C–H activation of the terminal methyl group. Such a mechanism implies that C–H activation of methyl groups is intrinsically more facile than methylene C–H units (Chart 11.7).

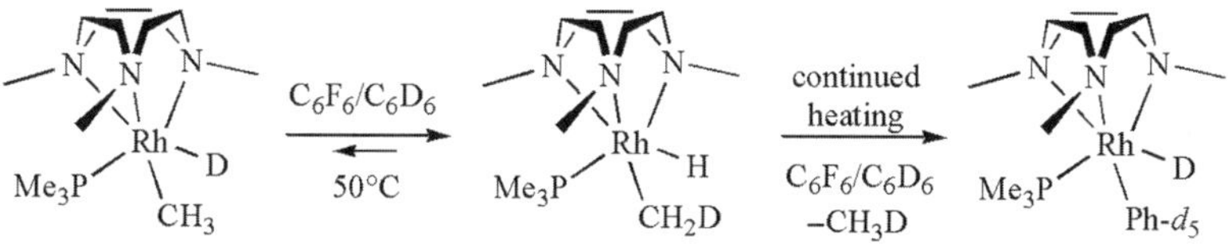

$$(11.3)$$

Detailed studies of C–H reductive elimination and oxidative addition at $Tp^*Rh(CNR')(R)(H)$ systems ($Tp^*$ = hydridotris-3,5-dimethylpyrazolylborate; $R'$ = neopentyl; $R$ = alkyl) provide additional evidence that Rh(I) complexes activate methyl C–H bonds more rapidly than $CH_2$ or CH fragments.[43] In general, the direct reaction of alkanes with Rh(I) systems gives the product of methyl C–H activation. The synthesis of Rh–Cl precursors ($Tp^*Rh(CNR')(R)Cl$) followed by hydride/chloride metathesis provided secondary alkyl hydride complexes. In $C_6D_6$, $Tp^*Rh(CNR')(i\text{-}Pr)(H)$ rearranges to $Tp^*Rh(CNR')(Pr)(H)$ in competition with propane reductive elimination and Rh reaction with $C_6D_6$ to give $Tp^*Rh(CNR')(Ph\text{-}d_5)(D)$. The isotopomer of the $i$-propyl complex, $Tp^*Rh(CNR')(CHMe_2)(D)$, rearranges to produce the linear $n$-propyl complex $Tp^*Rh(CNR')(CH_2CHDCH_3)(H)$ and 2-deuteriopropane/$Tp^*Rh(CNR')(Ph\text{-}d_5)(D)$. No evidence of reversible reductive coupling/oxidative C–H cleavage to give $Tp^*Rh(CNR')(CDMe_2)(H)$ was obtained (Scheme 11.17). Similar results are observed for the *sec*-butyl isomer.

For the Rh alkyl deuteride complexes, migration of the deuterium is observed only at the α-$CH_2$ group and the terminal methyl group. The extent of deuterium incorporation into the terminal methyl group (versus alkane reductive elimination) decreases as a function of increasing alkyl chain length. Similar to the studies of $[Cn^*Rh\{P(OMe)_3\}(R)(H)][BAr'_4]$ complexes, these results suggest that C–H

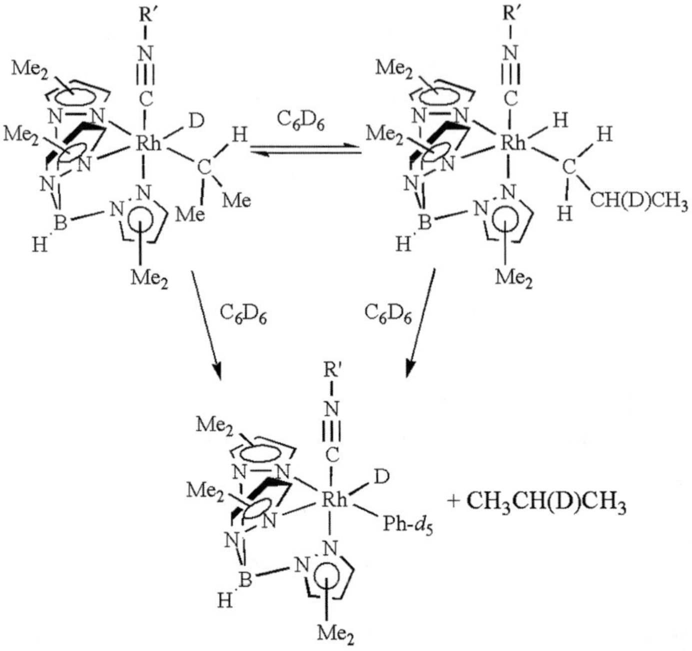

**SCHEME 11.17**   H/D exchange observed with Tp*Rh complexes (R' = neopentyl).

reductive coupling occurs from Tp*Rh(CNR')(R)(H) and that C–H oxidative addition at a methyl group competes with alkane dissociation from Tp*Rh(CNR')(RH) σ-complexes. Importantly, the C–H activation of primary C–H bonds is both kinetically and thermodynamically favored over secondary C–H bonds. The ability to synthesize a wide array of Rh alkyl hydride complexes, including isotopically labeled systems, has allowed the kinetics and thermodynamics of C–H elimination/ addition sequences to be mapped in remarkable detail.[20]

In addition to the indirect evidence of σ-complexes as intermediates in metal-mediated reductive elimination/oxidative addition transformations from isotopic scrambling, KIEs for C–H reductive elimination reactions lend credence to the intermediacy of coordinated alkane intermediates.[20] Assuming coordinated alkane intermediates, the equilibria and rates of reductive elimination and oxidative addition can be described with four rate constants that correspond to reversible C–H reductive coupling/oxidative addition and dissociation/association of the hydrocarbon substrate (Scheme 11.18). Several groups have reported that the rates of C–H reductive elimination of alkanes of perprotio versus perdeuterio variants yield inverse KIEs (i.e., $k_H/k_D < 1$).[20]

For the equilibrium between an alkyl hydride complex and a C–H coordinated substrate, consideration of the differences in zero-point energies suggests an inverse equilibrium isotope effect, which has been used to explain the inverse KIEs observed for C–H(D) reductive elimination processes. Thus, the observation of an inverse KIE

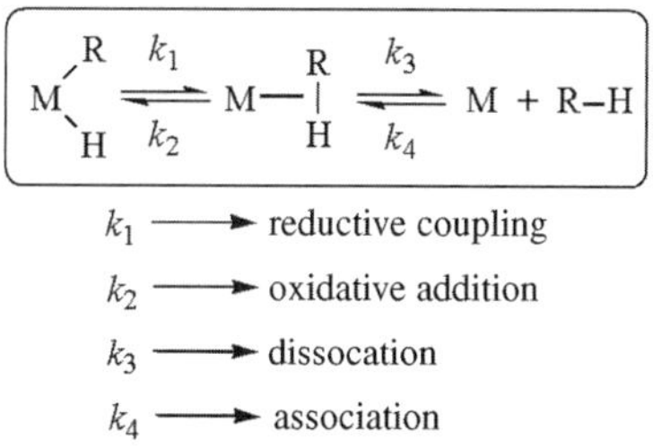

$$M\overset{R}{\underset{H}{\diagdown}}\ \underset{k_2}{\overset{k_1}{\rightleftharpoons}}\ \underset{H}{\overset{R}{M-|}}\ \underset{k_4}{\overset{k_3}{\rightleftharpoons}}\ M\ +\ R-H$$

$k_1 \longrightarrow$ reductive coupling
$k_2 \longrightarrow$ oxidative addition
$k_3 \longrightarrow$ dissocation
$k_4 \longrightarrow$ association

**SCHEME 11.18**  Net C–H reductive elimination and the microscopic reverse, oxidative addition equilibria are typically governed by four rate constants.

for reductive elimination of C–H(D) bonds provides additional support for the intermediacy of coordinated hydrocarbons. However, two different scenarios can lead to the overall inverse KIE for the C–H(D) reductive elimination reaction (Scheme 11.19).[20] If the transition state results in a negligible difference in the zero-point energy between the H and D isotopomers, then both the forward and reverse reactions (i.e., reductive coupling and oxidative addition) would exhibit primary KIEs (as shown on the right-hand side of Scheme 11.19) with the magnitude of the KIE of the oxidative addition greater than the KIE of the reductive coupling. In the second scenario, the difference in zero-point energies in the transition state is intermediate between the difference in the starting alkyl hydride complex and the coordinated alkane. In this case (shown on the left-hand side of Scheme 11.19), the forward process (i.e., C–H reductive coupling) would give an inverse KIE, while the oxidative addition step would give a normal KIE. In both cases shown in Scheme 11.19, the net result for R–H(D) reductive elimination is an inverse KIE.

Jones used the complex $Tp^*Rh(CNR')(i\text{-}Pr)(H)$ to determine the KIE for C–H reductive coupling.[20] In $C_6D_6$, $Tp^*Rh(CNR')(i\text{-}Pr)(H)$ converts to $Tp^*Rh(CNR')(Pr)(H)$ (Pr = $n$-propyl) in addition to the thermodynamic products, free propane and $Tp^*Rh(CNR')(C_6D_5)(D)$. Monitoring the conversion of $Tp^*Rh(CNR')(i\text{-}Pr)(H)$ to $Tp^*Rh(CNR')(Pr)(H)$ and kinetic modeling allowed determination of the rate constant for the C–H reductive coupling step, which is $k_1$ in Schemes 11.18 and 11.19, and comparison to the rate of the same process for $Tp^*Rh(CNR')(i\text{-}Pr)(D)$ provides the KIE for the discrete reductive coupling step with $k_{1H}/k_{1D} = 2.1$. In order to determine the KIE for oxidative addition step from the $Tp^*Rh(CNR')(RH)$ adduct, which is $k_2$ in Schemes 11.17 and 11.18, $CH_2D_2$ was utilized. Irradiation of $Tp^*Rh(CNR')(R'N=C=NPh)$ in $CH_2D_2$ provides a mixture of $Tp^*Rh(CNR')(CH_2D)(D)$ and $Tp^*Rh(CNR')(CHD_2)(H)$, and the ratio of isotopomers in the kinetic products gives the $k_{2H}/k_{2D}$ for the discrete oxidative addition step, which is 4.3 (Equation 11.4). Thus, Jones was able to demonstrate that the inverse KIE for overall reductive elimination of alkanes from $Tp^*Rh(CNR')(R)(H)$ is likely due to two normal isotope effects for the reductive coupling and oxidative addition steps, which corresponds to the scenario on the right-hand side of Scheme 11.19. The generality of these results is unknown.

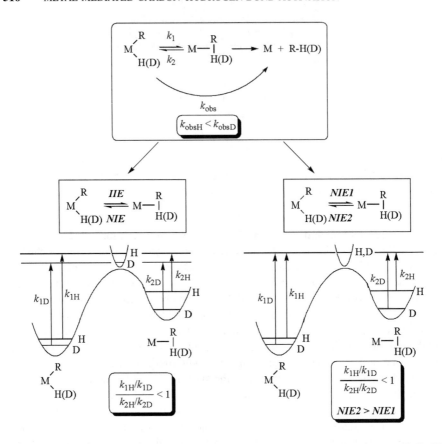

**SCHEME 11.19**  Two scenarios that lead to inverse kinetic isotope effect for overall C–H reductive elimination (IIE = inverse isotope effect; NIE = normal isotope effect).

Oxidative addition and the microscopic reverse, reductive elimination, involving formal Pt(0)/Pt(II) as well as Pt(II)/Pt(IV) redox couples, have been of long-standing interest.[44–47] Using bisphosphine platinum systems, *ab initio* calculations provided insight into the thermodynamics and activation barriers for oxidative addition reactions as a function of the substrate being activated (Scheme 11.20). The calculated

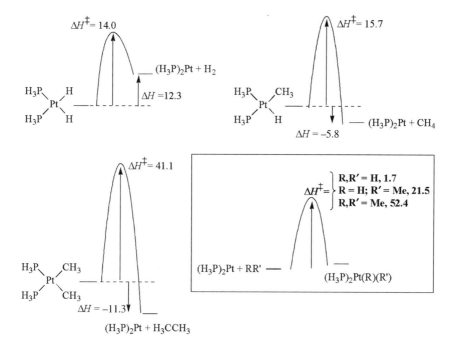

**SCHEME 11.20** Calculated enthalpies (kcal/mol) for oxidative addition and reductive elimination processes involve Pt(II)/Pt(0) cycles with two phosphine ligands.

activation energies for oxidative addition of $H_2$, $CH_4$, and $H_3C-CH_3$ (C–C cleavage) to $Pt(PH_3)_2$ are 1.7, 21.5, and 52.4 kcal/mol, respectively. These reveal a substantial difference in the activation energy for cleavage of bonds that have similar BDEs (i.e., $H_2$ and $CH_4$, 104 and 105 kcal/mol, respectively) and a large increase in activation energy for cleavage of the weaker C–C bond (~90 kcal/mol) of ethane. The source of the difference in kinetics has been attributed, at least in part, to the shape of the orbitals involved in the processes. In the transition from H–H, C–H, or C–C bonds to Pt(H) (H), Pt(H)(CH$_3$), or Pt(CH$_3$)(CH$_3$), the orbitals on the activated atoms transition from overlap with each other (i.e., H–H, C–H, and C–C orbital overlap) to overlap with a directional hybrid d-orbital. Scheme 11.21 illustrates the impact of shape and directionality of the frontier orbitals involved in the reductive coupling step. Using the concept of microscopic reversibility, the same factors should influence the energetics of the reverse reaction, which is oxidative addition. In the transition state, the spherical 1s orbital of hydrogen can bond efficiently with its partner orbital (i.e., another 1s orbital in $H_2$ or an $sp^3$ hybrid orbital for $H_3C-H$), and, due to its spherical nature, the hydrogen 1s atomic orbital can also efficiently overlap with the metal hybrid orbital. In contrast, the directional $sp^3$ hybrid orbitals of the methyl group cannot efficiently overlap simultaneously with the metal orbital and the partner orbital of another H atom or a C atom, thus decreasing overlap of orbitals in the transition state as more methyl groups are introduced. A similar explanation has been given for

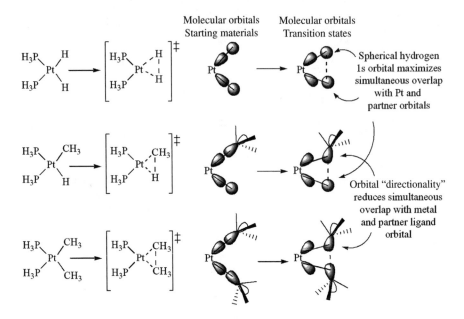

**SCHEME 11.21**   Illustration of orbital shape and impact on energetics of oxidative addition and reductive elimination processes. Directed nature of carbon-based orbitals leads to reduced orbital overlap in the transition state relative to overlap of spherical hydrogen 1s orbital.

relative predilection of substrates for C–H activation by σ-bond metathesis (see below).

In the late 1960s and early 1970s, Shilov and Shul'pin demonstrated that the Pt(II) salt $K_2[PtCl_4]$ could catalytically activate C–H bonds including methane.[48] The production of methanol from methane, thermal stability of the catalyst, and the tolerance for water have prompted substantial efforts to understand the details of the catalysis including the C–H bond-breaking step.[49] Two mechanisms have been considered for C–H activation by this system: electrophilic substitution (see below) and oxidative addition (Scheme 11.22). Mechanistic studies of Pt(II) models of the Shilov system have implicated the oxidative addition route.[50–52] Since study of the forward reaction (i.e., the C–H activation process) has been challenging, monitoring the microscopic reverse reaction, via protonation of Pt(II) alkyl or aryl complexes, has

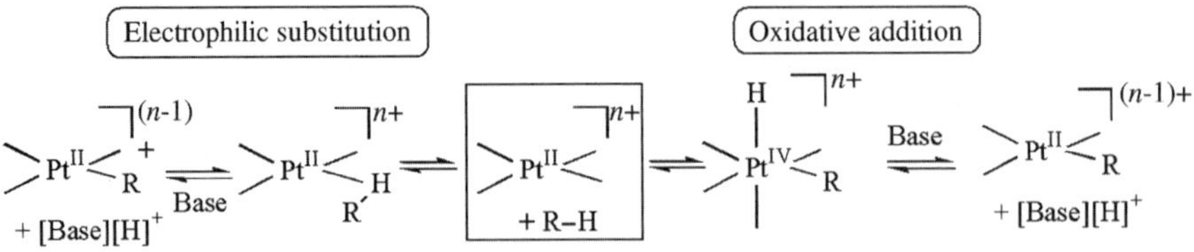

**SCHEME 11.22**   Two proposed mechanisms for C–H activation by the Pt(II) Shilov system.

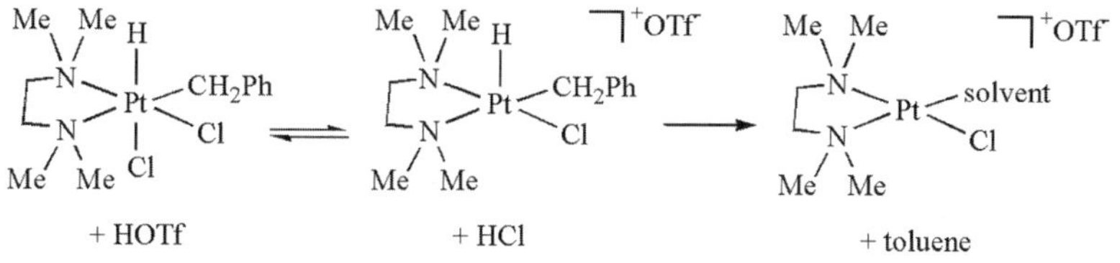

+ HOTf                    + HCl                              + toluene

**SCHEME 11.23**  Proposed mechanism for C–H reductive elimination from observable Pt(IV) intermediate.

been pursued. For example, the addition of HCl to (tmeda)Pt(CH$_2$Ph)(Cl) (tmeda = $\kappa^2$-$N,N,N',N'$-tetramethylethylenediamine) at $-78°$C produces the Pt(IV) complex (tmeda)Pt(CH$_2$Ph)(H)(Cl)$_2$.[50] At $-30°$C, (tmeda)Pt(CH$_2$Ph)(H)(Cl)$_2$ eliminates toluene to form the Pt(II) complex (tmeda)PtCl$_2$. The addition of triflic acid (HOTf) accelerates the reductive elimination of toluene from (tmeda) Pt(CH$_2$Ph)(H)(Cl)$_2$. At constant concentration of HOTf, increasing concentration of HCl slows the rate of toluene elimination. A mechanism consistent with these data is shown in Scheme 11.23. The formation of a Pt(IV) alkyl hydride intermediate and subsequent C–H reductive elimination implicates an oxidative addition mechanism for the reverse process.

Even though C–H oxidative addition from Pt(II) to give Pt(IV) is viable, additional distinctions can be made for these transformations. For example, beginning from four-coordinate Pt(II) complexes of the type [L$_3$Pt(R)]$^+$, C–H activation could proceed directly via C–H coordination and oxidative addition or by initial loss of ligand "L" (Scheme 11.24). To probe these reactions, several groups have studied the reductive elimination of C–H bonds from octahedral Pt(IV) complexes, which by the principle of microscopic reversibility can confer information about the oxidative addition reactions.[52–55]

The Pt(II) cation [(diimine)Pt(CH$_3$)(H$_2$O)]$^+$ (diimine = $\kappa^2$-ArN=C(Me)-C(Me)NAr; Ar = 3,5-(CF$_3$)$_2$C$_6$H$_3$) initiates methane activation in 2,2,2-trifluoroethanol (TFE) as indicated by reaction with the isotopomer CD$_4$ to give H/D scrambling and

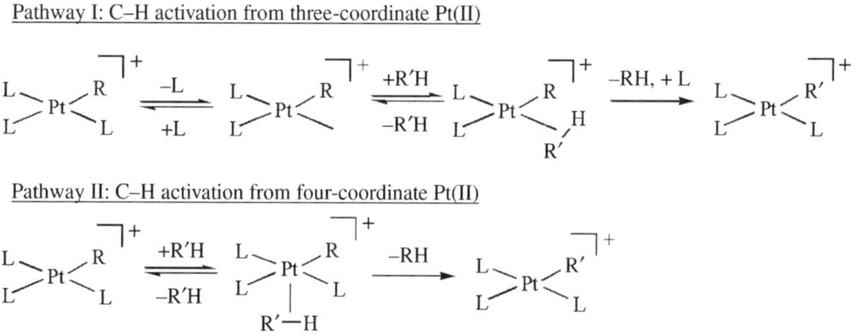

Pathway I: C–H activation from three-coordinate Pt(II)

Pathway II: C–H activation from four-coordinate Pt(II)

**SCHEME 11.24**  Two pathways for C–H activation by four-coordinate Pt(II) precursors.

$CH_nD_{4-n}$ isotopes.[56] The H/D exchange reaction is inhibited by addition of water, suggesting either (1) a pathway that involves reversible loss of water followed by methane coordination and activation by a three-coordinate cationic Pt(II) system (consistent with pathway I in Scheme 11.24) or (2) TFE/water exchange to give methane activation from the four-coordinate Pt(II) systems $[(\text{diimine})\text{Pt}(CH_3)(\text{TFE})]^+[\text{OTf}]^-$, which resembles pathway II in Scheme 11.24. The mechanism of C–H activation was probed by studying the microscopic reverse reaction, reductive elimination from a Pt(IV) precursor. The addition of DOTf to $(\text{diimine})\text{Pt}(CH_3)_2$ results in the formation of $[(\text{diimine})\text{Pt}(D)(CH_3)_2]^+$, which will give $CH_3D$ via C–D reductive coupling and methane elimination. However, if methane C–H(D) activation is competitive with methane dissociation, then a mixture of $CH_4$ and $CH_3D$ will be produced (Scheme 11.25). If methane loss occurs via coordination of L (in this case, NCMe was used as the ligand "L"), through conversion of $[(\text{diimine})\text{Pt}(CH_3D)(CH_3)]^+$ to $[(\text{diimine})\text{Pt}(CH_3D)(CH_3)(\text{NCMe})]^+$, which is the reverse of direct C–H activation by four-coordinate Pt(II), then the ratio of $CH_4$ to $CH_3D$ should depend on the concentration of NCMe. In contrast, loss of methane from $[(\text{diimine})\text{Pt}(CH_3D)(CH_3)]^+$ without coordination of NCMe would give a $CH_4/CH_3D$ ratio that is independent of the concentration of NCMe. The latter result would imply that C–H activation occurs via loss of ligand to give a three-coordinate Pt(II) intermediate. The ratio of $CH_3D$ to $CH_4$ was shown to increase with increasing concentration of NCMe, which is consistent with methane C–H activation directly from a four-coordinate Pt(II) complex (pathway II in Scheme 11.24).

Other studies of Pt(II) systems have directly probed the oxidative addition of C–H bonds and have demonstrated the viability of this transformation. For example, the

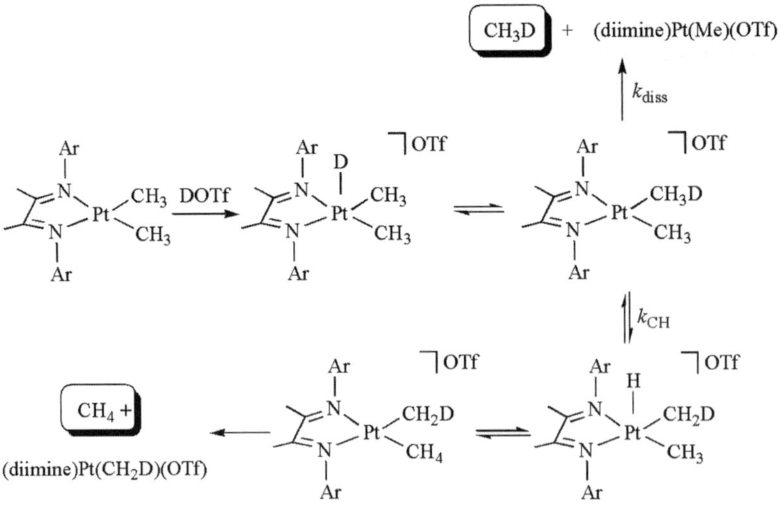

**SCHEME 11.25** Isotopic labeling study to probe competition between methane dissociation ($k_{\text{diss}}$) and methane C–H(D) activation ($k_{\text{CH}}$) (Ar = 3,5-$(CF_3)_2C_6H_3$).

reaction of $K[\kappa^2\text{-}Tp^*Pt(CH_3)_2]$ with $B(C_6F_5)_3$ in benzene, cyclopentane, or cyclohexane gives $K[MeB(C_6F_5)_3]$, via abstraction of a methyl ligand from the Pt starting material, and $\kappa^3\text{-}Tp^*Pt(Me)(R)(H)$ (R = Ph, cyclopentyl or cyclohexyl) (Equation 11.5).[57] The $\kappa^3\text{-}Tp^*Pt(Me)(R)(H)$ complexes form as a result of R–H oxidative addition to Pt(II) followed by conversion of the $Tp^*$ ligand from bidentate to tridentate. In related chemistry, the protonation of $Tp^*Pt(Ph)(H)_2$ gives the four-coordinate complex $[(\kappa^2\text{-}Tp^*H)Pt(H)(\eta^2\text{-}C_6H_6)]^+$.[58] At 252 K, broadened resonances are observed in the $^1H$ NMR spectrum for the coordinated benzene and Pt–H moieties. The line broadening at 252 K is consistent with the onset of an exchange process, which is likely due to a reversible oxidative addition to give the Pt (IV) system $[(\kappa^2\text{-}Tp^*H)Pt(Ph)(H)_2]^+$ (Equation 11.6). Using line broadening techniques,[58] the exchange rate at 252 K is $k = 47 \, s^{-1}$, which gives $\Delta G^{\ddagger} = 12.7 \, kcal/mol$.

$$(11.5)$$

R = phenyl, cyclopentyl, cyclohexyl

$$(11.6)$$

### 11.3.3  σ-Bond Metathesis

The activation of C–H bonds by a σ-bond metathesis pathway involves coordination of the C–H group followed by concerted C–H bond breaking of the activated substrate and C–H bond formation with a hydrocarbyl ligand. Thus, the reaction proceeds through a four-centered transition state (Scheme 11.8).

In 1983, Watson reported that Lu(III) and Y(III) methyl complexes mediate C–H activation of benzene and methane.[59] For benzene activation, $Cp^*_2Lu(Me)$, which is in equilibrium with the species $Cp^*_2Lu(\mu\text{-}Me)Lu(Me)Cp^*_2$, reacts to release methane and produce $Cp^*_2Lu(Ph)$ (Equation 11.7). Kinetic studies revealed a rate equation $-d[Lu\text{-}Me] = (k_1 + k_2[C_6H_6])[Lu\text{-}Me]$ that is consistent with two

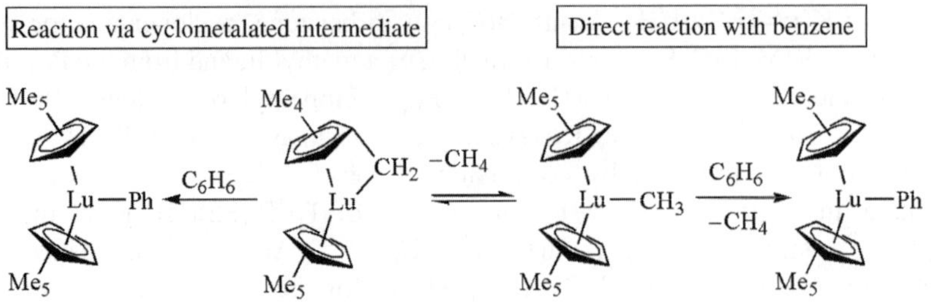

**SCHEME 11.26** Kinetics indicate two pathways are operable for the conversion of $Cp^*_2Lu(Me)$ and benzene to $Cp^*_2Lu(Ph)$ and methane.

pathways, including one pathway that is proposed to involve cyclometalation of the $Cp^*$ ring and release of methane, followed by rapid reaction with benzene to generate the Lu-Ph product (Scheme 11.26). Interestingly, the rate of reaction in sealed NMR tubes was found to be slower than reactions in open vessels, which suggests that the release of methane is reversible. Indeed, reacting $Cp^*_2M(CH_3)$ (M = Lu or Y) with isotopically labeled $^{13}CH_4$ results in methyl exchange to give $Cp^*_2M(^{13}CH_3)$. Since these systems are $d^0$, oxidative addition is not a viable mechanism for the activation of C–H bonds. The four-centered transition state shown in Scheme 11.27 was suggested for the methane activation by $Cp^*_2Lu(Me)$, and the Lu- and Y-based reactions were proposed to initiate C–H bond breaking and bond formation in a concerted single-step manner without altering the metal's formal oxidation state, a reaction that has been termed $\sigma$-bond metathesis.

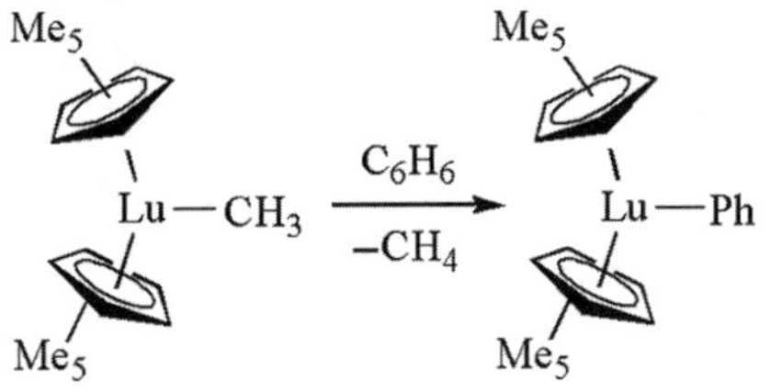

$$(11.7)$$

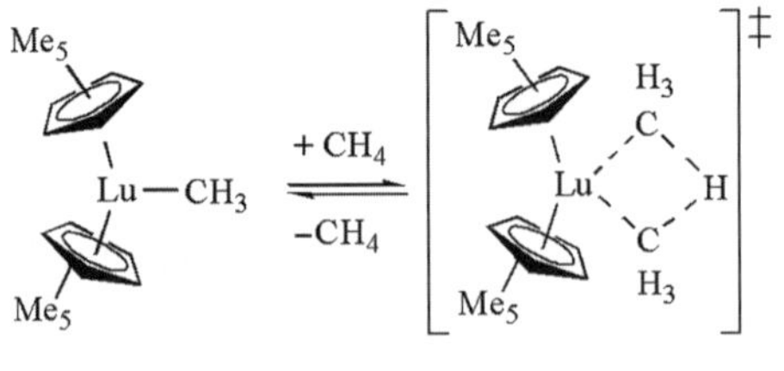

**SCHEME 11.27** Proposed mechanism for Lu-mediated C–H activation, termed $\sigma$-bond metathesis.

Since the report by Watson, many examples of metal-mediated C–H activation by the σ-bond metathesis mechanism have been reported. Predominantly, examples include early transition metal and lanthanide systems that lack access to d-electrons and, hence, cannot undergo oxidative addition reactions.[60] Notable examples include $Cp^*_2Sc(R)$,[61,62] $Cp^*_2Th(R)_2$ systems,[63] and $[Cp_2Zr(R)]^+$.[64] An interesting feature of the σ-bond metathesis reaction is the predilection toward participation of H atoms versus the involvement of alkyl groups. For example, Bercaw and coworkers reported that the relative rate of reactivity for R–H bonds with $Cp^*_2Sc$-R′ systems is R = R′ H ≫ R = H, R′ = alkyl ≫ R–H = sp  C–H, R′ = alkyl > R–H = $sp^2$, R′ = alkyl R–H = $sp^3$, R′ = alkyl,[61] concluding that the rate decreases with decreasing s character of the reacting bonds. This suggests that the geometry of the four-centered transition state results in maximized bonding with nondirectional s-orbitals, and this is especially true of the position β to the metal in the four-member metalacyclic transition state. A similar rationalization for relative rates of oxidative addition to Pt was discussed above (see Scheme 11.21).

Of particular importance to transition metal-mediated chemistry in general, and the σ-bond metathesis reaction specifically, is the forbidden nature of the $[2_\sigma + 2_\sigma]$ reaction for organic substrates, yet the transformations often occur in a facile manner with transition metal complexes. This is due to the presence of the d-orbitals on the metal center.[65,66] Understanding the role of the orbitals on the metal is important to gaining predictive power for organometallic reactions, which is illustrated in the observation for $d^0$ metal-mediated σ-bond metathesis by Steigerwald and Goddard that "the more metal d character in the M–Z bond, the lower the activation barrier for the exchange reaction and analogous insertions,"[65] which results from the necessity to remove metal s or p character in the transition state. It follows that the calculated activation barrier for σ-bond metathesis of $D_2$ with $[Cl_2MH]^n$ (M = Sc, n = 0; M = Ti, n = +1) was much smaller for Ti than for Sc.

### 11.3.4  Oxidative Addition or σ-Bond Metathesis?

Recent studies have suggested that C–H activation by the σ-bond metathesis pathway is not necessarily limited to $d^0$ metal centers. For example, benzene C–H activation by the Ir(III) fragment $(acac)_2Ir(CH_2CH_2Ph)$ (acac = $\kappa^2$-O,O-acetylacetonate) to give $(acac)_2Ir(Ph)$ and ethylbenzene is calculated to proceed from $(acac)_2Ir(CH_2CH_2Ph)(C_6H_6)$ via a concerted σ-bond metathesis-type pathway (Scheme 11.28).[67] The activation of carbon–hydrogen bonds by TpRu(L)R systems (L = CO, $PPh_3$, $P(OCH_2)_3CEt$, and P(N-pyrrolyl)$_3$) is calculated to proceed by closely related transformations (Scheme 11.28).[68,69] The influence of metal identity on the transition state for C–H activation by $TpM(PH_3)(R)$ (M = Fe, Ru, or Os) systems has been probed using density functional theory (DFT) calculations.[70] Among other transformations, transition states for methane C–H activation by $TpM(PH_3)(Me)(\eta^2$-$CH_4)$ were probed. The Os(II) complex is calculated to proceed via an oxidative addition reaction with an Os(IV) intermediate and a small $\Delta G^\ddagger$ of 3.3 kcal/mol (Scheme 11.29). In contrast, the Fe system is calculated to undergo a single-step reaction with a Fe–H bond distance of 1.568 Å in the calculated

| L | X |
|---|---|
| CO | 1.643 |
| $PMe_3$ | 1.583 |
| $P(OCH_2)_3CEt$ | 1.580 |

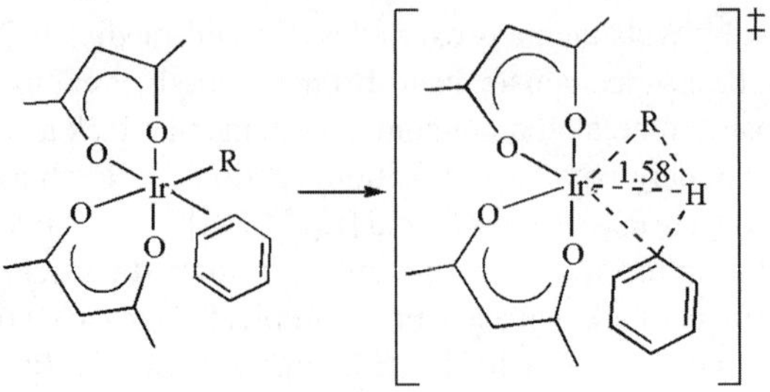

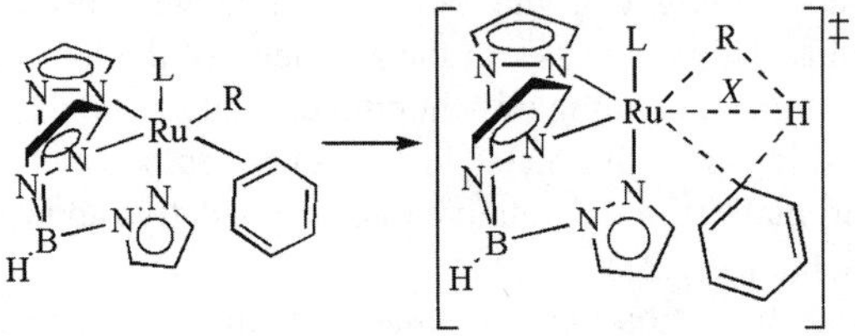

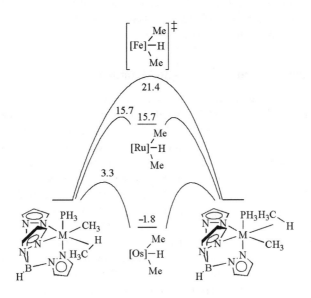

**SCHEME 11.28** Calculated transition states for C−H activation by six-coordinate $d^6$ complexes that proceed by a $\sigma$-bond metathesis-type pathway ($R = CH_2CH_2Ph$; calculated bond distances are given in Å).

**SCHEME 11.29** Comparison of calculated energetics (Gibbs free energy) for methane C−H activation by $TpM(PH_3)(Me)(\eta^2\text{-}CH_4)$.

transition state. For Ru, the calculations are more ambiguous, and although a Ru(IV) intermediate was calculated, the energy difference between it and the transition state was minimal. Similar to calculations for TpRu(L)(CH$_2$CH$_2$Ph) (see above), it was noted that the Ru reaction is probably best considered a concerted process. Of note, calculations on methane C–H activation by [CpM(PH$_3$)(Me)]$^+$ (M = Co, Rh, or Ir) systems revealed quite similar results to the calculated methane C–H activation by TpM(PH$_3$)Me (M = Fe, Ru, or Os).[71]

The recognition that electron-rich late transition metals can activate C–H bonds via oxidative addition *or* by σ-bond metathesis reactions poses a challenging mechanistic question. That is, for late transition metals with d-electron counts greater than zero, overall metathesis between M–R and R′–H substrates to give M–R′ and R–H could proceed via concerted σ-bond metathesis or through stepwise oxidative addition and reductive elimination sequences (e.g., see Scheme 11.29). If an oxidative addition intermediate is not observed, the experimental distinction between the two pathways is quite subtle and challenging.

The report of remarkably facile C–H activation by the Ir(III) complex [Cp*Ir(PMe$_3$)(Me)(ClCH$_2$Cl)]$^+$ sparked substantial interest in the distinction between σ-bond metathesis and oxidative addition/reductive elimination routes. As noted by Bergman and Arndtsen,[72] an oxidative addition process would require an Ir(V) intermediate, which is an unusually high oxidation state for Ir organometallic complexes. Conversely, at the time of this report σ-bond metathesis was generally considered to fall under the purview of d$^0$ metals.

Mass spectrometry provides a convenient method to monitor isotope distributions, and gas-phase mass spectrometry was used to study benzene C–H activation by [Cp*Ir(PMe$_3$)(Me)(NCMe)]$^+$.[73] Given differences between gas- and solution-phase reactivity, extension of mechanistic studies in the gas-phase to solution-phase reactions should be viewed with caution. These studies elucidated the intermediacy of a metallaphosphacyclopropane (in the gas state) that forms via loss of methane.[73] For example, using the isotopically labeled complex [Cp*Ir{P(CD$_3$)$_3$}(CH$_3$)NCMe]$^+$ as the precursor produces CH$_3$D, which suggests that methane production occurs exclusively through cyclometalation of the phosphine ligand rather than the Cp* ligand (Scheme 11.30). Furthermore, reacting the isotopically labeled metallaphosphacyclopropane complex with C$_6$H$_6$ produces [Cp*Ir{P(CD$_3$)$_2$(CHD$_2$)}(Ph)]$^+$.

Is the formation of a metallaphosphacyclopropane a viable mechanism for C–H activation by [Cp*Ir(PMe$_3$)(Me)(ClCH$_2$Cl)]$^+$ in solution? In solution-phase studies, deuterium incorporation into the phosphine ligand is not observed upon reaction of [Cp*Ir(PMe$_3$)(Me)(ClCH$_2$Cl)]$^+$ with C$_6$D$_6$ to produce [Cp*Ir{P(CH$_3$)$_3$}(Ph-$d_5$)(ClCH$_2$Cl)]$^+$ (Scheme 11.31).[74] Furthermore, generation of the cyclometalated triflate complex Cp*Ir{κ$^2$-$P,C$-PMe$_2$(CH$_2$)}OTf in the presence of benzene does not result in the production of [Cp*Ir(PMe$_3$)(Ph)(OTf)] but rather gives decomposition. These results, in combination with kinetic studies, suggest that cyclometalation of the PMe$_3$ ligand is not involved in the solution-phase C–H activation mediated by [Cp*Ir(PMe$_3$)(Me)(ClCH$_2$Cl)]$^+$. Starting from Cp*Ir(Me)$_4$, the Ir(V) complex [Cp*Ir(Me)$_3$(PMe$_3$)][OTf] has been synthesized, which models the intermedi-ates of oxidative addition starting from Cp*Ir(PMe$_3$)(Me)OTf. In

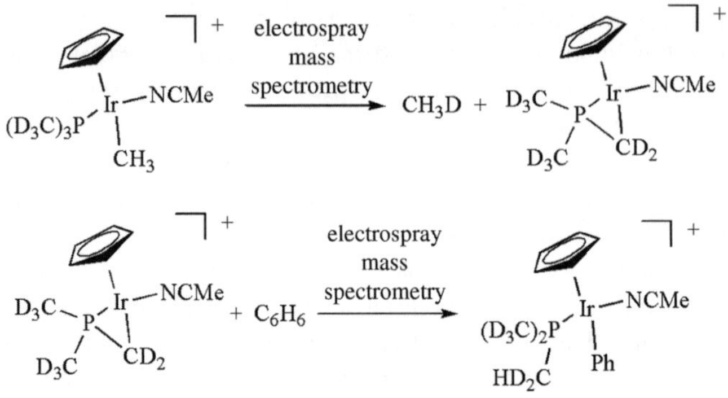

**SCHEME 11.30** Electrospray mass spectrometry studies indicate that C–H activation by $[CpIr(PMe_3)(NCMe)(Me)]^+$ occurs via a cyclometalation pathway in the gas phase.

addition, $[Cp^*Ir(PMe_3)(Me)(ClCH_2Cl)]^+$ was found to react with triphenylsilane to give the Ir(V) complex $[Cp^*Ir(PMe_3)(H)(\kappa^2\text{-}C,Si\text{-}Si(Ph)_2(C_6H_4)]^+$ (Scheme 11.31). These results lend credence to the viability of an Ir(V) oxidative addition intermediate.[75]

The presence of d-electrons for the σ-bond metathesis reactions with late transition metal systems may result in an important difference compared to analogous reactions with $d^0$ metal centers. The calculated metal–hydrogen contacts in the transition states for $d^6$ complexes (e.g., Ru(II), Fe(II), and Ir(III)) suggest that these reactions may possess oxidative character that results from metal-to-hydrogen electron donation. Several groups have suggested nomenclature to distinguish σ-bond metathesis reactions with $d^0$ systems from those with $d^n$ $(n > 0)$ electron counts. Labels for the latter reactions include metal-assisted σ-bond metathesis, oxidatively added transition state, oxidative hydrogen migration, and σ-complex-assisted metathesis.[76]

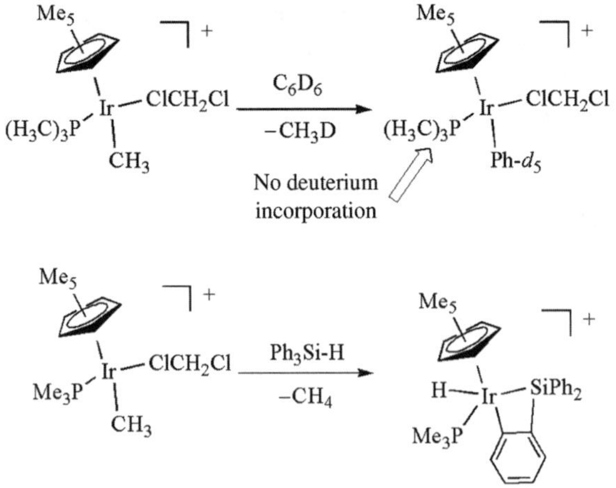

**SCHEME 11.31** Solution-phase experiments that are inconsistent with PMe$_3$ cyclometalation along the pathway for C–H activation by $[Cp^*Ir(PMe_3)(Me)(ClCH_2Cl)]^+$.

The studies of C–H activation by $d^6$ transition metal complexes demonstrate that the mechanism of C–H bond cleavage is likely a subtle balance that depends on the ancillary ligands, metal identity, and metal oxidation state as well as the substrates being activated. Furthermore, these calculations suggest the possibility of a continuum of mechanisms ranging from "classic" σ-bond metathesis to σ-bond metathesis in which the metal–hydrogen distance is short (implying a metal–hydrogen bond) to pathways that involve an oxidative addition intermediate (Scheme 11.32). The impact of the metal center is clearly indicated by the calculated changes in mechanism as a function of variation among Fe/Ru/Os (see Scheme 11.29) as well as Co/Rh/Ir for $CpM(PH_3)(Me)]^+$ systems (see above). In addition, the influence of ancillary ligand is indicated by studies of TpRu(L)(R) systems, in which more donating phosphine and phosphite ligands shorten the Ru–H bond distance in the calculated transition states, implying stronger Ru-to-H electron donation, relative to the less donating CO ligand (Scheme 11.28). Furthermore, calculations that compare methane C–H activation by $CpRe(CO)_2$ and $TpRe(CO)_2$ systems highlight the impact of ancillary ligand Cp versus Tp on the predilection toward oxidative addition.[77] Methane oxidative addition reactions with $CpRe(CO)_2$ and $TpRe(CO)_2$ to give the Re(III) products $CpRe(CO)_2(Me)(H)$ and $TpRe(CO)_2(Me)(H)$, respectively, have been studied using DFT calculations (Scheme 11.33). Oxidative addition to $CpRe(CO)_2$ is calculated to be exothermic with $\Delta H = -7.9$ kcal/mol, while the same reaction with $TpRe(CO)_2$ is calculated to be endothermic with $\Delta H = 6.4$ kcal/mol. Starting from methane adducts, the calculated activation energy for methane oxidative addition by the

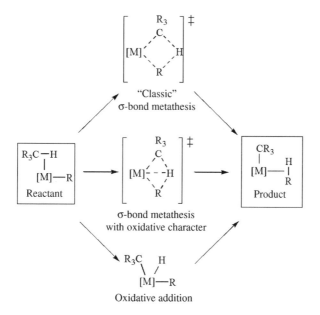

**SCHEME 11.32**   Some possible mechanisms for C–H activation by transition metal complexes with d-electron counts greater than zero.

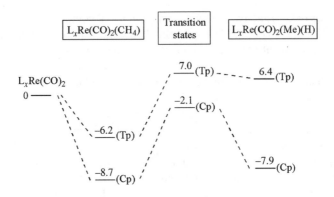

**SCHEME 11.33** Comparison of calculated energetics for methane oxidative addition to $CpRe(CO)_2$ and $TpRe(CO)_2$ (energies given in kcal/mol).

$CpRe(CO)_2(CH_4)$ system is smaller than that for the $TpRe(CO)_2(CH_4)$ complex with $\Delta\Delta H^{\ddagger} = 6.6$ kcal/mol.

### 11.3.5    Electrophilic Substitution

Electrophilic substitution involves coordination and activation of a C–H bond with subsequent loss of a proton. The proton is typically transferred to a weakly basic counteranion (Scheme 11.8). Since Lewis acidity of the metal center plays a key role, electrophilic substitution reactions are generally invoked for late transition metal complexes that are relatively electronegative. For many systems that appear to initiate C–H activation by electrophilic substitution, whether the proton is transferred in an intermolecular (to an uncoordinated base) or intramolecular (to a coordinated basic ligand) fashion is not entirely clear (Scheme 11.34 and Equation 11.8). Since the intramolecular variant of electrophilic substitution is analogous to the σ-bond metathesis reaction and the distinction between intermolecular electrophilic substitution and σ-bond metathesis processes is often difficult, the labels electrophilic substitution and σ-bond metathesis often have been used interchangeably. It is likely the case that, at least for Pt(II) complexes, the two pathways often have similar energetic profiles and that subtle changes in ligand environment, hydrocarbon substrates, solvent, and so on can potentially alter the preferred route.[78]

$$L_nM-R + R'H \xrightarrow[\text{or electrophilic substitution?}]{\text{oxidative addition}} L_nM-R' + RH \qquad (11.8)$$

Electrophilic substitution can be categorized into two subsets. One class of electrophilic substitution reaction involves electrophilic addition of the metal to the π system of an aromatic substrate, to form a "Wheland-type" intermediate, followed by proton loss (Scheme 11.35). This is the metal analogue of the classic organic

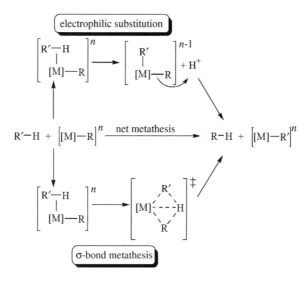

**SCHEME 11.34**  Experimental differentiation of intermolecular electrophilic substitution and $\sigma$-bond metathesis pathways for overall conversion of M(R) and R′H to M(R′) and RH can be challenging.

electrophilic aromatic substitution, and since the reaction does not involve a direct metal interaction with the C−H bond, it will not be considered further here.

The second pathway for electrophilic substitution requires coordination of the C−H bond to a Lewis acidic metal center, which depletes electron density of the C−H moiety and enhances its acidity and, hence, increases the propensity toward proton transfer. This reaction can occur with both aromatic substrates and alkanes. It is known that the coordination of dihydrogen can substantially increase the acidity of $H_2$,[4] and it is reasonable to assume that a similar effect can occur through metal coordination of C−H bonds, particularly with highly Lewis acidic metals. Although the $pK_a$ values of hydrocarbons coordinated to metal centers have not been directly measured, the acidity of intramolecular agostic bonds has been reported.[2]

The Shilov Pt(II) system for methane functionalization has been studied using computations. For example, using $MCl_2(H_2O)_2$ (M = Pt or Pd) systems as models for Shilov-type reactions, the overall catalytic cycle was studied using a combination of

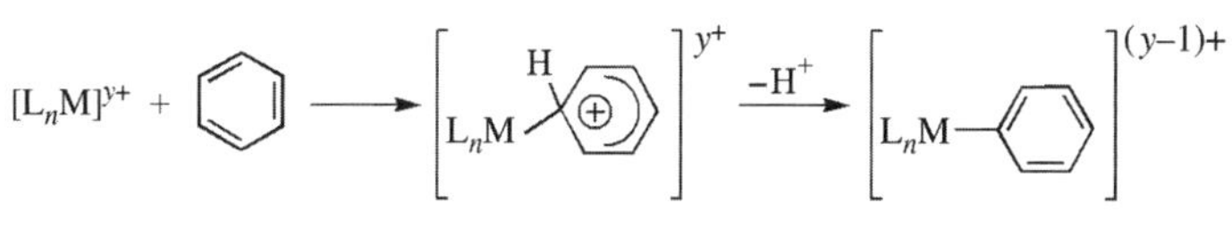

$$[L_nM]^{y+} + \text{⬡} \longrightarrow \left[ L_nM \overset{H}{⬡^{\oplus}} \right]^{y+} \xrightarrow{-H^+} \left[ L_nM\text{—⬡} \right]^{(y-1)+}$$

**SCHEME 11.35**  Net electrophilic substitution of C−H bond via a Wheland-type intermediate.

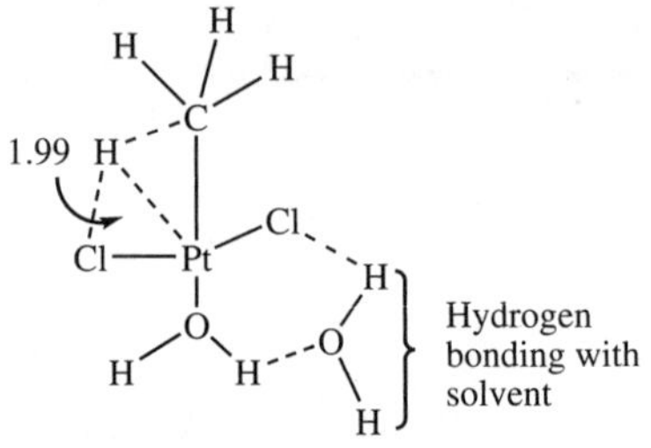

**FIGURE 11.1** Model of calculated transition state for methane C–H activation by $Pt(Cl)_2(H_2O)$ (bond distance between Pt and H given in Å).

DFT and *ab initio* methods.[78] For the C–H activation step using a *trans*-dichloride geometry (calculations on the *cis* isomer were also studied), the calculations suggested a four-center transition step that is akin to a σ-bond metathesis reaction (Figure 11.1), which was calculated to be 20.5 kcal/mol higher in energy than the precursor methane complex. The calculated distance between Pt and the activated hydrogen is 1.99 Å. The Mulliken charge on the activated hydrogen undergoes a negligible change from the methane adduct ($+0.26$) to the transition state ($+0.23$), which suggests little protic character. Comparison of the σ-bond metathesis transition state with that for an oxidative addition to give the Pt(IV) complex, $Pt(H)(Me)(OH_2)_2Cl_2$, reveals that the activation energy for the oxidative addition reaction is 23.8 kcal/mol, only slightly greater than the 20.5 kcal/mol calculated for the σ-bond metathesis transition state. From these studies, it was concluded that for the Pt(II) system, the σ-bond metathesis pathway is most likely, although not definitive, while for Pd(II) systems the oxidative addition reaction is unlikely since it would form an unusual Pd(IV) complex. However, recent examples of oxidative addition from Pd(II) complexes to give Pd(IV) products suggest that high oxidation state Pd(IV) complexes may be more accessible than previously thought.[79,80]

Periana et al. have reported a mercury system that catalyzes the partial oxidation of methane to methanol.[81] Hg(II) is typically considered to be a soft electrophile and is known to initiate electrophilic substitution of protons from aromatic substrates. The catalytic reaction employs mercuric triflate in sulfuric acid, and a key step in the catalytic cycle is Hg(II)-mediated methane C–H activation. For methane C–H activation by Hg(II), an oxidative addition reaction pathway via the formation of Hg(IV) is unlikely. Thus, an electrophilic substitution pathway has been proposed, although differentiation between proton transfer to an uncoordinated anion versus intramolecular proton transfer to a coordinated anion (i.e., σ-bond metathesis) has not been established. Hg(II)-based methane C–H activation was confirmed by the observation of H/D exchange between $CH_4$ and $D_2SO_4$ (Equation 11.9).

$$Hg(OSO_3H)_2 + CH_4 \xrightarrow[\quad]{-H_2SO_4} (CH_3)Hg(OSO_3H) \xrightarrow[-CH_3D]{+D_2SO_4} Hg\{OSO_3H(D)\}_2 \quad (11.9)$$

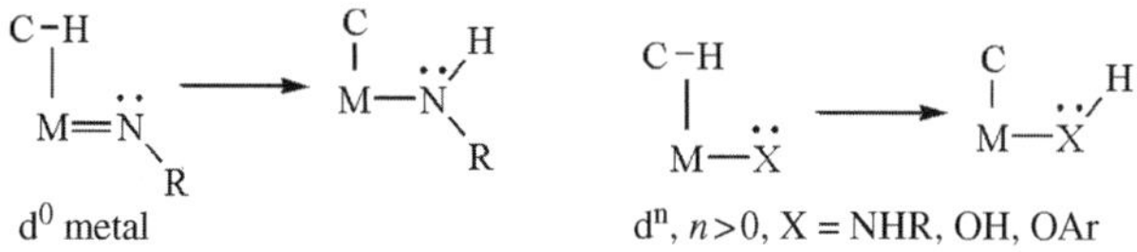

**SCHEME 11.36**  1,2-CH addition reactions have been observed with $d^0$ imido complexes and late transition metal amido, hydroxo, and aryloxo systems.

### 11.3.6  1,2-Addition Across Metal–Heteroatom Bonds

The 1,2-addition of C—H bonds across metal–heteroatom bonds has been reported for two different classes of complexes: early transition metal $d^0$ complexes with imido ligands and late transition metal complexes with amido, hydroxo, and aryloxo ligands (Scheme 11.36). These transformations are potentially related to σ-bond metathesis reactions discussed above; however, the presence of a lone pair on the heteroatom that receives the activated hydrogen may impart important differences.

Thermolysis of $(t\text{-Bu}_3\text{SiNH})_3\text{Zr(Me)}$ in benzene results in the elimination of methane and formation of $(t\text{-Bu}_3\text{SiNH})_3\text{Zr(Ph)}$.[82,83] In $C_6D_6$, $CH_4$ and $(t\text{-Bu}_3\text{SiND})_3\text{Zr}(C_6D_5)$ are produced with a reaction rate that is zero order in benzene concentration. The deuterium incorporation into all of the amido ligands is a result of reversible benzene elimination from $\{t\text{-Bu}_3\text{SiN(H/D)}\}_3\text{Zr}(C_6D_5)$. Heating the isotopomer with deuterium-labeled amido groups, $(t\text{-Bu}_3\text{SiND})_3\text{Zr(Me)}$, in $C_6H_6$ gives $CH_3D$ and $(t\text{-Bu}_3\text{SiNH})_3\text{Zr}(C_6H_5)$. These results are consistent with a rate-limiting formation of methane via abstraction of an amido hydrogen atom by the methyl ligand to generate an unobserved imido complex, $(t\text{-Bu}_3\text{SiNH})_2\text{Zr}=\text{N}\{\text{Si}(t\text{-Bu})_3\}$, which coordinates and activates benzene in a fast step (Scheme 11.37). A large KIE, $k_H/k_D = 7.3(4)$, for the rate of reaction of $(t\text{-Bu}_3\text{SiNH})_3\text{Zr(Me)}$ versus

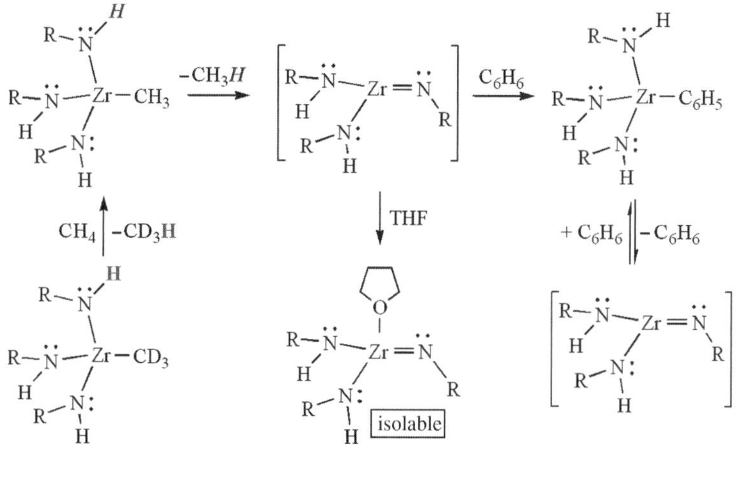

**SCHEME 11.37**  The 1,2-addition of C—H bonds across Zr imido bond (R = $t\text{-Bu}_3\text{Si}$).

($t$-Bu$_3$SiND)$_3$Zr(Me) is consistent with rate-determining methane extrusion. Additional evidence for a Zr imido intermediate is obtained from heating ($t$-Bu$_3$SiNH)$_3$Zr(Ph) in THF, which affords the Zr imido complex as the THF adduct. In addition to benzene C–H activation, the intermediate Zr imido complex can also initiate methane activation, as indicated by the reaction of ($t$-Bu$_3$SiNH)$_3$Zr(CD$_3$) with CH$_4$ to generate ($t$-Bu$_3$SiNH)$_3$Zr(CH$_3$) and CD$_3$H.

Similar to the Zr systems, heating ($t$-Bu$_3$SiNH)($t$-Bu$_3$SiO)$_2$Ti(CH$_3$) provides access to a Ti imido complex, ($t$-Bu$_3$SiO)$_2$Ti=N{Si($t$-Bu)$_3$}, via elimination of methane.[84] The Ti imido complex undergoes reversible dimerization (via bridging imido groups) in addition to reactions with a variety of hydrocarbons to yield ($t$-Bu$_3$SiNH)($t$-Bu$_3$SiO)$_2$Ti(R) (R = CH$_2$CH$_3$, $c$-C$_3$H$_5$, $c$-C$_5$H$_9$, C$_6$H$_4$Me, or C$_6$H$_5$). A detailed series of mechanistic studies of C–H activation by Zr, Ti, and Ta systems has revealed important details.[85] The C–H activation likely occurs in a single-step reaction with the formation of a coordinated hydrocarbon intermediate, and the observation of large kinetic isotope effects indicates that the N$\cdots$H$\cdots$C linkage is likely close to linear in the transition state. Kinetic selectivities for C–H activation by the titanium imido intermediate are influenced more substantially by the strengths of incipient Ti–C bonds rather than C–H bond dissociation energies. Using this suggestion, the kinetic selectivity for C–H activation between two hydrocarbons is given by Equation 11.10, where the slope "$m$" gives the extent of selectivity. A plot of the relative Ti–R bond dissociation energies for a series of ($t$-Bu$_3$SiO)$_2$($t$-Bu$_3$SiNH)Ti(R) complexes versus C–H bond dissociation energies (of the organic "RH") gives a reasonable linear fit. In a scenario where the strength of Ti–C bonds controls kinetic selectivity, the slope of this line dictates the magnitude of selectivity among various C–H bonds as a function of the thermodynamics. For the Ti system, for the activation of sp$^3$ C–H bonds, $m = 0.77$.

$$[(t\text{-Bu}_3\text{SiO})_2\text{Ti}=\text{N}(\text{Si-}t\text{-Bu}_3)] + \text{R-H} + \text{R'-H} \longrightarrow \begin{matrix} (t\text{-Bu}_3\text{SiO})_2(t\text{-Bu}_3\text{SiNH})\text{Ti}(\text{R}) \\ + \\ (t\text{-Bu}_3\text{SiO})_2(t\text{-Bu}_3\text{SiNH})\text{Ti}(\text{R'}) \end{matrix} \quad (11.10)$$

$$\boxed{\Delta\Delta G^{\ddagger} = \Delta G^{\ddagger}_{\text{RH}} - \Delta G^{\ddagger}_{\text{R'H}} \sim m[\Delta G\Upsilon_{\text{Ti-R}} - \Delta G\Upsilon_{\text{Ti-R'}}]}$$

Biscyclopentadienyl Zr(IV) imido complexes also initiate C–H activation via 1,2-CH addition across metal–imido bonds.[86,87] For example, heating Cp$_2$Zr(Me){N(H)-$t$-Bu} in benzene releases methane and gives the product of benzene C–H activation, Cp$_2$Zr(Ph){N(H)-$t$-Bu}. Extending this chemistry to the Zr imido complex supported by ethylenebis(tetrahydro)indenyl allows access to C–H activation of saturated hydrocarbons (Scheme 11.38). For this system, regioselectivity for the terminal methyl position is observed for reactions with linear alkanes. The selectivity for stronger C–H bonds is consistent with the results for Ti imido complexes discussed above. Heating the ethylenebis(tetrahydro)indenyl zirconium complex in a mixture of isotopes $n$-pentane-$d_0$ and $n$-pentane-$d_{12}$ generates the Zr(amido)(pentyl) products

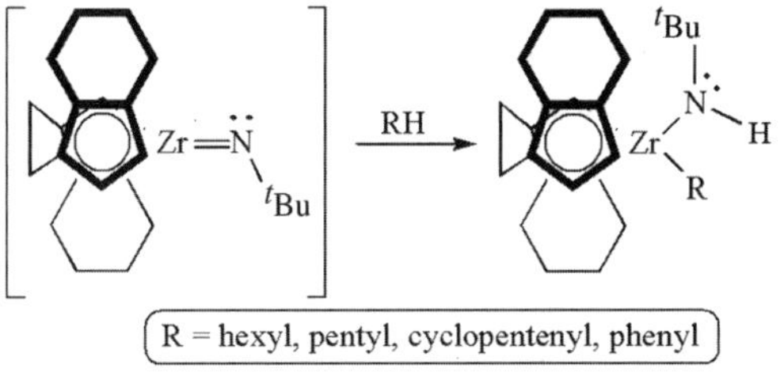

R = hexyl, pentyl, cyclopentenyl, phenyl

**SCHEME 11.38**    Activation of C–H bonds by Zr(IV) imido complex.

without isotope scrambling. These results are consistent with mononuclear reactions, in which the proposed Zr imido complex coordinates the C–H(D) bond and initiates an intramolecular 1,2-CH addition rather than intermolecular chemistry via two Zr centers.

With the observation of 1,2-addition of C–H bonds across $d^0$ metal–imido bonds for several different transition metals, it is possible that this reaction is reasonably general. But, the nature of the metal–imido bonding is likely a critical determinant for the energetics of the transformations. For example, the minimization of metal–imido multiple bonding may be a key aspect of these reactions. In a simple viewpoint, the four-center transition state for C–H activation delivers a proton from the activated C–H bond to the lone pair of the imido nitrogen (Scheme 11.39). Imido ligands can function as four-electron to six-electron donors through the formation of one to two metal–imido $\pi$-bonds (Chart 11.8). But, the presence of multiple $\pi$-donor ligands, termed "$\pi$-loaded" complexes,[88] can disrupt imido-to-metal $\pi$-donation through $p\pi$–$d\pi$ competition. Using the model of C–H activation and the transition state in Scheme 11.39, it might be anticipated that reducing imido $\pi$-donation and,

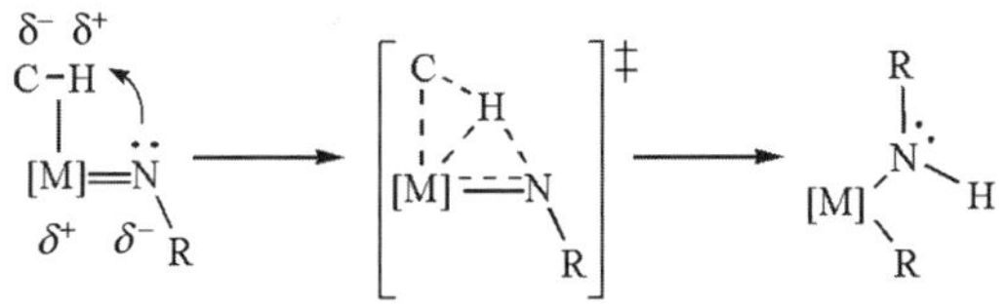

**SCHEME 11.39**    Four-center transition state for 1,2-CH addition across metal–imido bond.

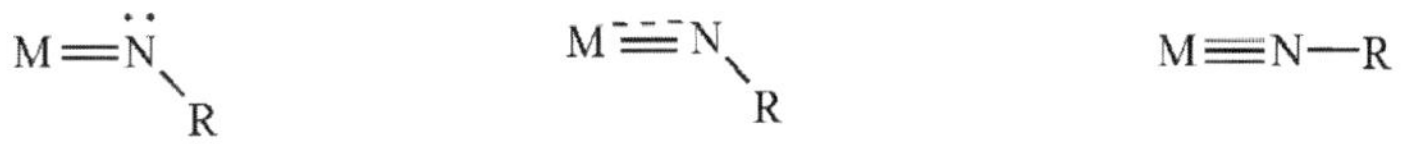

| Double bond, imido is four-electron donor | Intermediate between double and triple bonds, imido is formally five-electron donor | Triple bond, imido is six-electron donor |

**Chart 11.8**    Metal–imido multiple bonding.

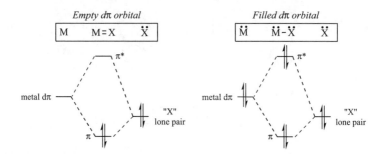

**SCHEME 11.40**   Impact of d-electron count on metal–heteroatom multiple bonding (X = OR, NHR, etc.).

hence, increasing nitrogen-based electron density would facilitate the C–H activation. In fact, more electron-rich systems, in which imido π-donation is minimized, may lead to early transition states for the 1,2-CH addition and provide both kinetic and thermodynamic advantages to C–H activation.

The observation that reduction of imido-to-metal π-bonding (via "π-loading" of the complex) suggests that late transition metal complexes in low oxidation states could be ideally suited for C–H activation via 1,2-addition across metal–heteroatom bonds. Such systems can disrupt ligand to metal π-donation due to *filled* dπ manifolds. Thus, in contrast to $d^0$ metals for which ligand–metal π-interaction must be manipulated through π-competition, for late transition metals ligand–metal π-bonding can formally be "shut off" since metal–ligand π-bonding and π-antibonding molecular orbitals are both occupied (Scheme 11.40). In fact, late transition metal complexes in low oxidation states with amido, hydroxo, and related ligands have been shown to be highly reactive with basic/nucleophilic heteroatomic ligands.[89–92] In 2004, in a report on 1,2-addition of dihydrogen and intramolecular C–H addition across a Ru(II) amido bond, it was stated, "Thus, by accessing Ru(II) complexes that possess an open coordination site and a nondative nitrogen-based ligand, it might be feasible to transiently bind nonpolar X–H (e.g., H–H or C–H) bonds to the metal center, thereby activating the substrate toward *intra*molecular deprotonation."[93] Subsequently, such reactions have been shown to be accessible with amido, hydroxo, and aryloxo ligands coordinated to Ru(II), Ir(III), and Rh(I).

Heating TpRu(PMe$_3$)$_2$OH (Tp = hydridotris(pyrazolyl)borate) in C$_6$D$_6$ results in H/D exchange between the hydroxo ligand and C$_6$D$_6$.[94,95] Mechanistic studies are consistent with a metal-mediated pathway that involves initial dissociation of PMe$_3$ followed by reversible 1,2-addition of C–D across the Ru–OH bond (Scheme 11.41). The regioselectivity for H/D exchange with toluene-$d_8$ was monitored by $^1$H NMR spectroscopy, which revealed the following rates of reaction: *para* *meta* > *ortho* > benzylic (Equation 11.11). The slow reaction at the benzylic position (relative to aromatic positions) indicates that a radical pathway is not likely operative. Furthermore, the selectivity for *para* and *meta* positions over the *ortho* position is consistent with a metal-mediated C–H activation pathway, in which the position *ortho* to the methyl group is sterically protected from reaction with the relatively

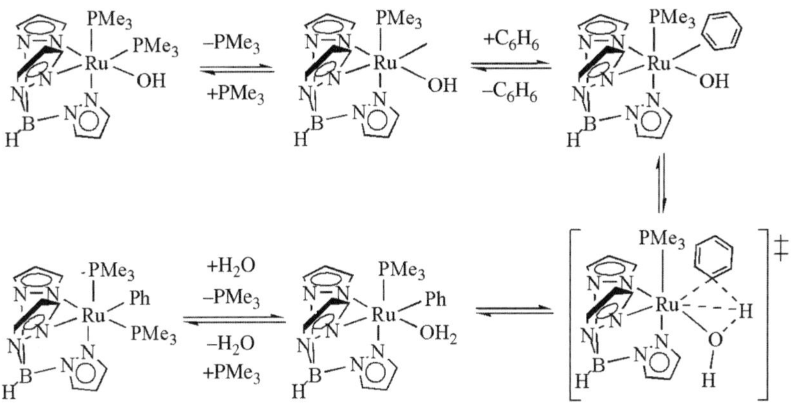

**SCHEME 11.41**    Proposed mechanism for H/D exchange between $C_6D_6$ and Ru–OH bond.

bulky Ru complex. Similar to C–H activation by $TpRu(L)(R)(C_6H_6)$ systems (see above), DFT calculations are consistent with a σ-bond metathesis-type process in which there is a Ru–H bond in the transition state (Scheme 11.42).

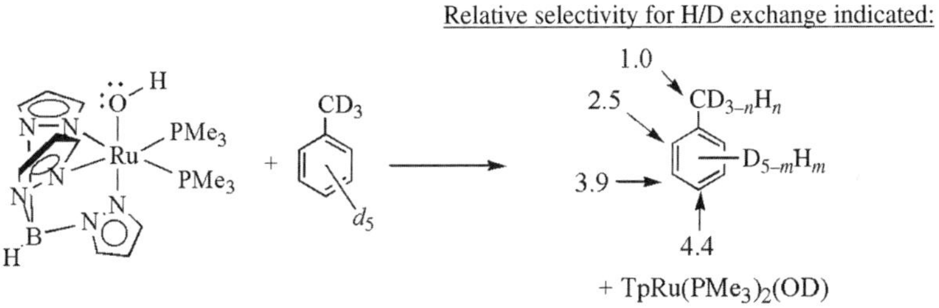

Relative selectivity for H/D exchange indicated:

$$(11.11)$$

$+ \ TpRu(PMe_3)_2(OD)$

A series of complexes of the type $TpRu(PMe_3)_2(X)$ (X = OH, OPh, NHPh, SH, Me, Ph, Cl, OTf) was studied for the 1,2-CH addition reactions. Of the heteroatomic systems for which isotope exchange can be monitored for H/D exchange with $C_6D_6$ (i.e., X = OH, NHPh, and SH), H/D scrambling reactions occur for X = OH and NHPh but not for X = SH. Elucidation of the rates of $PMe_3$ dissociation from $TpRu(PMe_3)_2(X)$ revealed the relative rates for these systems to be $k_{OH} = 2.8k_{NHPh} = 23k_{SH}$, which is consistent with the requirement of phosphine dissociation to initiate C–H activation (Scheme 11.43). While the rates of phosphine dissociation from $TpRu(PMe_3)_2Me$ and $TpRu(PMe_3)_2Ph$ are of similar magnitude to $TpRu(PMe_3)_2NHPh$ ($k_{NHPh} = 2.6k_{Me} = 1.2k_{Ph}$), H/D exchange reactions are not observed for the Ru–hydrocarbyl complexes. At *prolonged* reaction times, $TpRu(PMe_3)_2Me$ in $C_6D_6$ begins to produce $CH_3D$ and $TpRu(PMe_3)_2(Ph-d_5)$. These results suggest the possibility that C–H activation by a σ-bond metathesis-type pathway might be intrinsically favored kinetically when the ligand receiving the activated

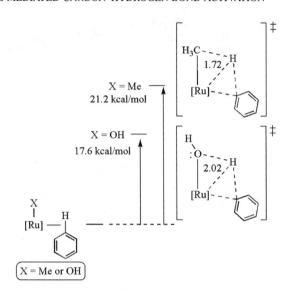

**SCHEME 11.42**    Calculated transition states for benzene C–H activation by Ru hydroxo and methyl complexes ([Ru] — (Tab)Ru(PH₃), Tab = tris(azo)borate; calculated bond distances are given in Å.

hydrogen has a lone pair (e.g., NHR, OR, NR, etc.; see Section 11.3.7 for more discussion). Calculations using the tris(azo)borate and PH₃ ligands (as models for the full Tp and PMe₃ ligands) are consistent with this notion (Scheme 11.42), although more extensive and thorough studies are required to establish whether this is a general trend. The calculated geometries of the transition states might provide an explanation for the relative stabilities. In the calculated transition states, the Ru–H bond distance is short indicating a possible donation of electron density to the activated hydrogen atom. For X = OH, this bond distance (2.02 Å) is calculated to be longer than the Ru–H bond distance for X = Me (1.72 Å), which might reflect increased O · · · H$_{activated}$ bonding and, as a result, reduced Ru–H bonding, due to the presence of a lone pair on oxygen.

Similar C–H activation reactions have been observed for Ir(III) and Rh(I) complexes, suggesting that net 1,2-CH addition across late transition metal

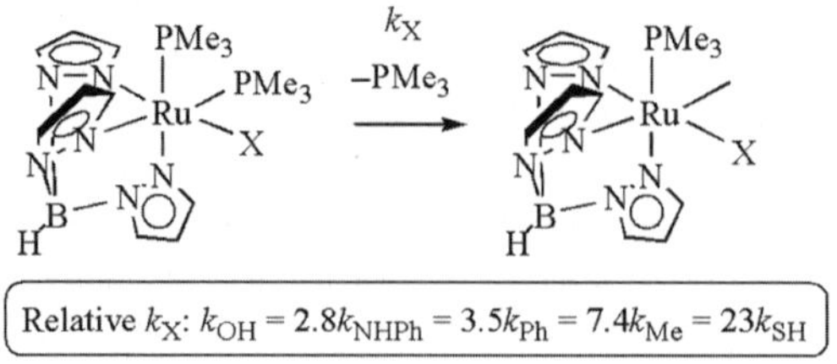

**SCHEME 11.43**    Relative rates of PMe₃ dissociation for TpRu(PMe₃)₂X systems (determined by PMe₃ exchange with PMe₃-$d_{18}$).

heteroatom bonds could be a general reaction, although mechanistic details may vary.[96,97] Heating (acac)$_2$Ir(OMe)(MeOH) in benzene with pyridine (160°C) produces two equivalents of free methanol (one equivalent from the coordination methanol and one equivalent from the methoxy ligand) and (acac)$_2$Ir(Ph)(pyridine) (Equation 11.12). Interestingly, the reaction of the Ir–OMe complex with benzene to give (acac)$_2$Ir(Ph)(pyridine) and free methanol is thermodynamically favored. In contrast, while the 1,2-CH addition reactions with TpRu(PMe$_3$)$_2$X (X = OH, NHPh) complexes are more kinetically facile than the Ir(III) complex, the 1,2-CH addition reactions with the TpRu(II) systems are not thermodynamically favorable.

$$\text{(11.12)}$$

Calculations on the Ir(III) system suggest that it is best considered as an intramolecular proton transfer (labeled an internal electrophilic substitution).[98] An important distinction exists between a σ-bond metathesis-type reaction where the ligand receiving the activated hydrogen is a hydrocarbyl ligand and a nondative heteroatomic ligand. When the receiving ligand is a hydrocarbyl group, a four-electron system results with two electrons originating from the C–H bond that is coordinated to the metal center and two electrons from the metal–carbon bond (Scheme 11.44). In the transition state, the electron pair that originates from the C–H bond is delocalized between both carbon atoms and the activated hydrogen atom. When the receiving ligand is a heteroatom with a lone pair, such as a M–OR complex, a six-electron system results with two electrons each from the C–H bond, the M–heteroatom bond, and the lone pair on the heteroatom. Based on calculations of a model Ir(III) complex,[98] in the transition state, the lone pair from the heteroatom is predominantly X–H in character (Scheme 11.44) while the M/CH electron pair remains delocalized over these three atoms.

The observation of 1,2-CH addition across late transition metal heteroatom bonds is a newly observed reaction, and little is known about factors that control the energetics of these transformations. An important consideration is the impact of electronic structure on the activation barrier, especially modulation of the electron density at the metal center (via tuning of ancillary ligands) and changes in basicity of the heteroatomic ligand. DFT calculations on benzene C–H activation by the Ir (III) systems (acac)$_2$Ir(X)(C$_6$H$_6$) (X = OMe, OCF$_3$, or NH$_2$), which probe the impact of ligand basicity, indicate that the activation barriers change by less than 2 kcal/mol.[98] In contrast, comparing the calculated activation energies for 1,2-CH addition of coordinated methane across the Ir–OH bond of Ir(Me)$_2$(NX$_3$)$_2$(OH)(CH$_4$), which probes the influence of metal electron density, reveals that replacing X = H

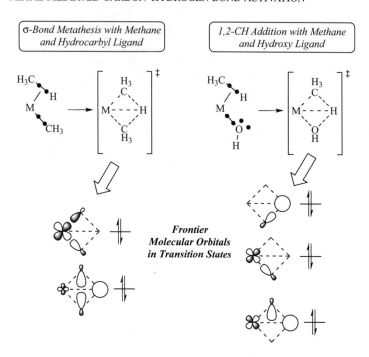

**SCHEME 11.44** Comparison of $\sigma$-bond metathesis-type transition states, including qualitative molecular orbitals, in which the activated hydrogen atom is transferred to a hydrocarbyl group or a nondative heteroatomic ligand.

with X = F lowers the barrier by 6.8 kcal/mol (Scheme 11.45). These results suggest that altering the metal's electrophilicity to enhance C−H coordination and activation will have a more profound impact on activation barriers than changing the identity of the heteroatomic ligand. Using DFT calculations, similar results have

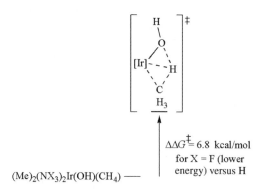

**SCHEME 11.45** Calculated activation barriers for 1,2-CH addition across Ir−OH bond as a function of donor ability of ancillary ligands ([Ir] = (Me)$_2$(NX$_3$)Ir).

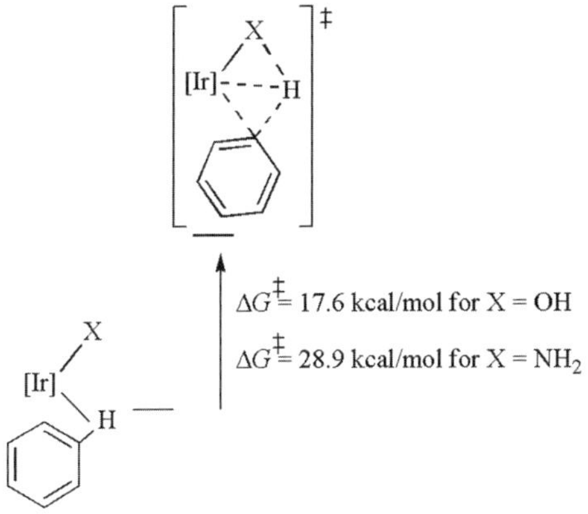

$$\Delta G^{\ddagger} = 17.6 \text{ kcal/mol for } X = OH$$
$$\Delta G^{\ddagger} = 28.9 \text{ kcal/mol for } X = NH_2$$

**SCHEME 11.46**    Calculated activation barriers for 1,2-CH addition across Ir–X (X = OH or NH$_2$) bonds of [(Tab)Ir(PH$_3$)(C$_6$H$_6$)(X)]$^+$ ([Ir] = [(Tab)Ir(PH$_3$)]$^+$).

been obtained for benzene C–H activation by a series of complexes [(Tab)M (PH$_3$)$_2$(X)]$^{m+}$ (X = OH or NH$_2$).[99] That is, the variation of X between OH and NH$_2$ is calculated to have a relatively small influence on the *overall* activation barrier for benzene C–H activation. However, these calculations are for overall benzene C–H activation, which involves PH$_3$ dissociation, benzene coordination, and 1,2-CH addition. In contrast, comparison of the calculations for the elementary reaction step of benzene C–H activation reveals a substantial impact upon variation of the ligand "X". For example, the calculated $\Delta G^{\ddagger}$ for 1,2-CH addition from [(Tab)Ir(PH$_3$)(C$_6$H$_6$)(OH)]$^+$ is 17.6 kcal/mol, while the same reaction for [(Tab)Ir(PH$_3$)(C$_6$H$_6$)(NH$_2$)]$^+$ is calculated to have $\Delta G^{\ddagger} = 28.9$ kcal/mol (Scheme 11.46). The parameters that control the 1,2-CH addition reactions with late transition metals are not yet clearly defined.

Four-coordinate Rh(I) hydroxide and related complexes have been demonstrated to initiate aromatic C–H activation (Equation (11.13)).[97,100] In contrast to the proposed mechanisms for the Ru(II) and Ir(III) reactions, mechanistic studies for the Rh(I) systems suggest the possibility of an initial exchange between RO$^-$ (via Rh–OR *hetero*lytic cleavage), coordination of the substrate undergoing C–H activation, and deprotonation by free RO$^-$ (Scheme 11.47). Alkane C–H activation by late transition metal complexes via 1,2-CH addition across metal–heteroatom bonds has yet to be demonstrated.

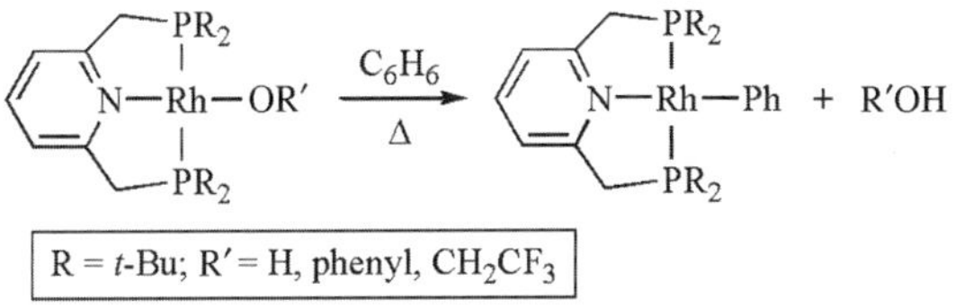

$$\text{R} = t\text{-Bu; R}' = \text{H, phenyl, CH}_2\text{CF}_3$$

(11.13)

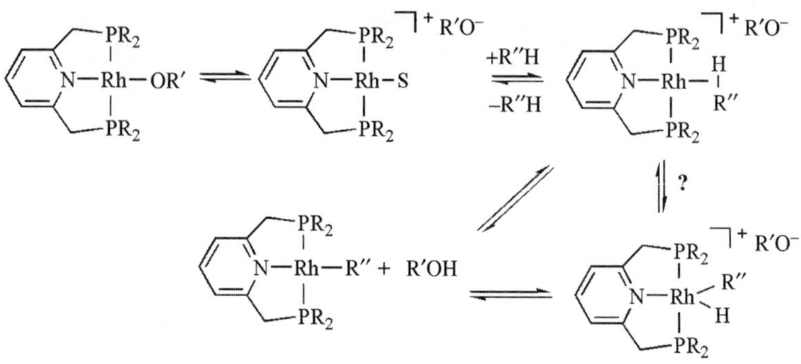

**SCHEME 11.47** Proposed mechanism for Rh(I)-mediated C–H activation with hydroxo, alkoxo, and aryloxo ligands (R = $t$-Bu; S = solvent).

### 11.3.7 Model of Transition for a Nonoxidative Addition C–H Activation by Late Transition Metal Complexes

A detailed series of experimental and computational studies of C–H activation by TpRu(L)(X)(RH) (L = neutral, two-electron donating ligands; X = hydrocarbyl, OR, or NHR; RH = substrate undergoing C–H activation) complexes has resulted in a model for the transition states for C–H activation (Figure 11.2). In this model, coordination of the C–H bond to Ru(II) activates the hydrogen toward an intramolecular proton transfer to the ligand X. In the transition state, Ru backdonates electron density to stabilize the activated hydrogen atom. Hence, it may be anticipated that any changes that stabilize the "protic" hydrogen or the buildup of negative charge density in the moiety R will lower the activation barrier. Possible factors include: 1) Increased electron density at the metal center (e.g., by increasing the donating ability of the ligand L) can enhance the Ru-H bonding interaction in the transition state. Conversely, this feature may also reduce the electrophilicity of the metal center, which implies a competing balance between metal electronic structure that is not fully mapped and understood. 2) Electron withdrawing groups on R will enhance the acidity of the C–H bond and facilitate the reaction. 3) More basic ligands "X" should

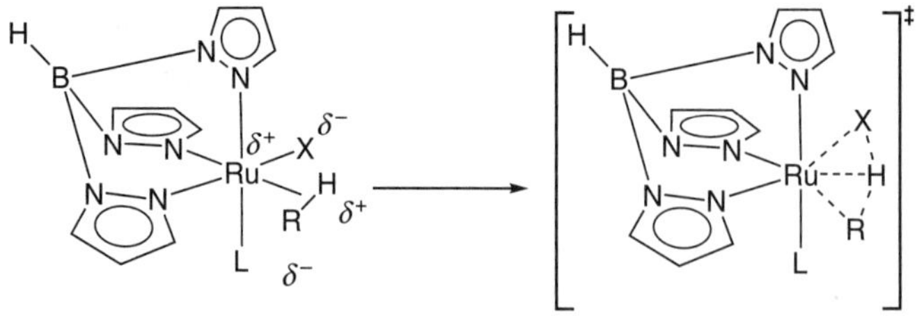

**FIGURE 11.2** Model of C–H activation by TpRu(L)(X)(RH) systems (L = neutral, two-electron donating ligands; X = hydrocarbyl, OR, or NHR; RH = substrate undergoing C–H activation).

increase the aptitude toward the reaction, especially the presence of lone pairs on heteroatomic ligands such as hydroxo, amido and related ligands.

## 11.4    STUDIES OF ALKANE COORDINATION

No isolable and fully characterized examples of a transition metal complex with a coordinated alkane have been reported. With typical metal–alkane bond energies <15 kcal/mol, isolation of these complexes is a substantial challenge. However, spectroscopic methods have been incorporated to directly observe alkane coordination. Initially, fast IR methods were utilized to study transient alkane coordination. More recently, novel NMR techniques have been developed to directly observe the coordination of alkanes to transition metals. Both techniques have afforded valuable insight into metal–hydrocarbon bonding.

### 11.4.1    Fast IR Spectroscopy

There are two predominant challenges to direct observation of alkanes coordinated to transition metals: (1) the short-lived nature of metal/alkane complexes and (2) competition for coordination of the alkane to the metal center. Because of the weak binding energy, alkane coordination is typically short-lived. Thus, fast spectroscopy techniques are required, and these techniques are often coupled with low temperatures in order to slow processes that result in alkane dissociation. In addition to the rapid dissociation of alkanes, most organic substrates will effectively compete (kinetically and thermodynamically) with alkanes for coordination to metals. Thus, the reaction medium is an important consideration since most common solvents are better ligands than alkanes, and attempts to observe alkane coordination have been commonly performed in the gas phase, in hydrocarbon matrices, or in liquid krypton or xenon. Finally, photolysis is generally required to dissociate a ligand at low temperature to create a transient coordination site for the alkane.

   Early efforts to observe alkane coordination involved photolytic dissociation of carbonyl ligands in hydrocarbon matrices and monitoring by IR spectroscopy. For example, gas-phase studies of alkane coordination to the 16-electron complex $W(CO)_5$, which is generated upon photolysis of $W(CO)_6$, have been reported.[101] Using time-resolved IR spectroscopy, free $W(CO)_5$ has been shown to exhibit CO absorptions at 1980 and 1942 cm$^{-1}$. In the presence of a variety of alkanes, new CO absorptions appear, generally in the range of 1973–1969 and 1944–1947 cm$^{-1}$, which are assigned as $W(CO)_5$(alkane) complexes. Alkane coordination occurs in the μs time regime. A variety of alkanes were found to coordinate to $W(CO)_5$ including ethane, propane, butane, isobutane, hexane, and cyclohexane. The generation of $W(CO)_5$ in the presence of methane provides no evidence of methane coordination. From van't Hoff plots, alkane binding energies were determined (Table 11.2). The binding energies correlate with alkane ionization energies (i.e., more electron-rich

**TABLE 11.2  Alkane Binding Energies (kcal/mol) for W(CO)₅(alkane) Complexes**

| Alkane | Binding Energy |
|---|---|
| Methane | <5 |
| Ethane | 7.4(2) |
| Propane | 8.1(2) |
| Butane | 9.1(3) |
| Isobutane | 8.6(2) |
| Pentane | 10.6(3) |
| Cyclopentane | 10.2(3) |

alkanes form stronger bonds with metals) consistent with alkane-to-metal σ-donation dominating the coordination energy.

CpM(L) complexes (M = Rh, Ir; L = neutral, two-electron donor ligands) are known to coordinate and activate alkanes toward oxidative addition. In the gas phase, the irradiation of CpRh(CO)₂ results in CO dissociation to produce CpRh(CO), which exhibits a CO absorption at 1985 cm⁻¹.[102] In the absence of alkane, CpRh(CO) decomposes via reaction with starting Rh complex, CpRh(CO)₂, to form Cp₂Rh₂(CO)₂(μ-CO) with a half-life of approximately 1 μs. In neopentane, CpRh(CO) reacts to form the product of C–H oxidative addition, CpRh(CO)(neopentyl)(H), which exhibits a CO absorption at 2037 cm⁻¹. The increase in CO absorption energy upon conversion of CpRh(CO) to CpRh(CO)(neopentyl)(H) is consistent with the formal increase in Rh oxidation state from +1 to +3, and, hence, a decrease in the extent of Rh to CO π-backbonding.

Analogous studies of Cp*Rh(CO)₂ were used to compare reaction with various alkanes.[103] Photolysis of Cp*Rh(CO)₂ in liquid Kr initially produces Cp*Rh (CO)(Kr), with $\nu_{CO}$ = 1947 cm⁻¹. In addition, CO absorptions in the range of 2000–2008 cm⁻¹ due to the formation of the products of C–H oxidative addition, Cp*Rh(CO)(R)(H), are observed. Using the mechanism for alkane oxidative addition shown in Scheme 11.48 and the expression for $k_{obs}$, monitoring the rate of formation of Cp*Rh(CO)(R)(H) using time-resolved IR spectroscopy allowed extraction of $k_2$ and $K_{eq}$. These data were acquired at temperatures between −80 and −110°C.

The most obvious influence of alkane structure on the reaction is that larger alkane size increases the $K_{eq}$ for Rh–Kr/Rh–alkane exchange (Scheme 11.48). As discussed above, this is consistent with studies of alkane coordination to W(CO)₅ that suggest increasing alkane size results in more strongly coordinated alkanes. Also consistent with the studies of W(CO)₅, the Cp*Rh(CO)(Kr) system does not coordinate methane. These results underpin the especially challenging nature of coordination and activation of methane, even among a series of seemingly closely related alkane substrates. The trend in relative affinity toward the C–H oxidative addition reaction, as indicated by the magnitude of $k_2$, is roughly opposite that of $K_{eq}$. That is, more tightly coordinating alkanes, as indicated by large $K_{eq}$, undergo the oxidative addition step more slowly (Scheme 11.48), which suggests that ground-state stabilization plays a key role in the $\Delta G^{\ddagger}$ for the C–H oxidative addition step.

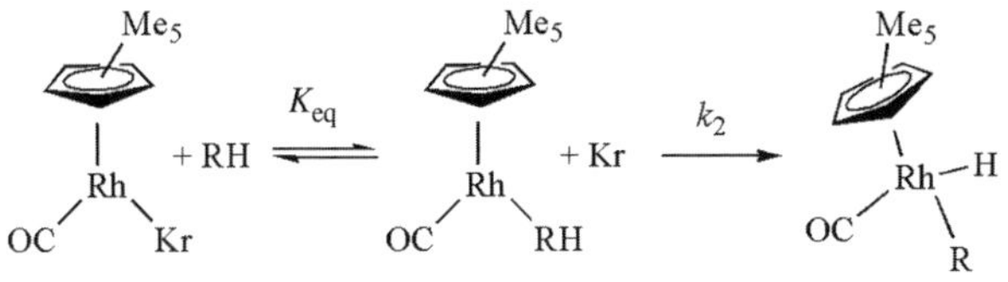

| RH | $K_{eq}$ | $k_2$ |
|---|---|---|
| ethane | 14(1) | 28(2) |
| propane | 140(10) | 11(1) |
| hexane | 350(50) | 11(2) |
| octane | 400(100) | 10(1) |
| cyclopentane | 1200(300) | 8.5(1) |
| cyclooctane | 15000(1000) | 1.0(.2) |

$$k_{obs} = \frac{k_2[RH]}{[RH] + [Kr]} \cdot \frac{1}{\overline{K_{eq}}}$$

**SCHEME 11.48**   Selected data for equilibria between Rh–Kr and Rh–alkane complexes (data acquired at $-80°C$).

One of the potential values of the fast IR studies is to identify trends in alkane coordination. For example, through direct observation of alkane complexes and their decomposition, complexes that most strongly bind alkanes can be identified. The direct observation of a series of heptane complexes with supporting aromatic and carbonyl ligands revealed that $CpRe(CO)_2$(alkane) complexes are relatively stable (Table 11.3).[104] This was determined by monitoring the rate of disappearance of metal–alkane complexes, given as $k$ in Table 11.3. These results provided an important lead to the first observation of an alkane complex using NMR spectroscopy (see Section 11.4.2).

## 11.4.2  NMR Spectroscopy

Observation of an alkane complex by NMR spectroscopy is complicated by several factors, including (1) the complex must be longer lived than is required by fast IR spectroscopy, and (2) generation of a coordinatively unsaturated intermediate

**TABLE 11.3   Rate Constants ($M^{-1}s^{-1}$) for the Exchange of Heptane with CO for a Series of Complexes**

| Complex | $k$ |
|---|---|
| $CpV(CO)_3$(heptane) | $1 \times 10^8$ |
| $CpNb(CO)_3$(heptane) | $7 \times 10^6$ |
| $CpTa(CO)_3$(heptane) | $5 \times 10^6$ |
| $(\eta^6\text{-Benzene})Cr(CO)_2$(heptane) | $2 \times 10^6$ |
| $CpMn(CO)_2$(heptane) | $8 \times 10^5$ |
| $CpRe(CO)_2$(heptane) | $2 \times 10^3$ |

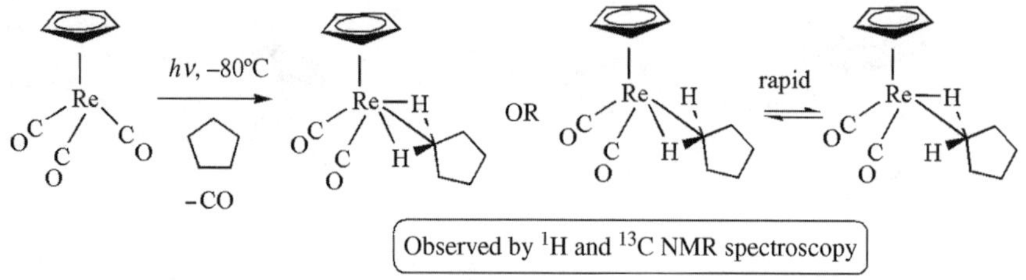

**SCHEME 11.49** Coordination of cyclopentane by CpRe(CO)$_2$ has been directly observed using low-temperature NMR spectroscopy.

and subsequent alkane adduct outside of the NMR probe would result in alkane dissociation by the time the sample is transferred to the NMR probe and prepared for NMR data acquisition. In an elegant series of experiments that overcame these challenges, a fiber optic cable was used to continuously irradiate CpRe(CO)$_3$ in various hydrocarbon solvents in an NMR probe.[105,106]

The irradiation of CpRe(CO)$_3$ in cyclopentane at $-80°$C results in the formation of CpRe(CO)$_2$(cyclopentane) (Scheme 11.49). $^1$H NMR features consistent with the coordination of cyclopentane include a new resonance due to the Cp ligand at 4.92 ppm and an upfield quintet ($^3J_{HH} = 6.6$ Hz) at $-2.32$ ppm, with an integrated ratio of the peaks at 4.92 and $-2.32$ ppm of 5:2, consistent with a Cp ring and a methylene group. A single resonance for the methylene group that bonds with the metal center is consistent with either a $\kappa^2$-H,H coordination mode or rapidly interconverting $\eta^2$-C,H coordination isomers (Scheme 11.49). Performing the same experiment in deuterated cyclopentane results in the new resonance for the Cp ring, but the upfield quintet is not observed suggesting that it indeed arises from coordinated cyclopentane. The reaction of CpRe(CO)$_3$ in a 1:1 ratio of perprotiocyclopentane and perdeuteriocyclopentane reveals an equilibrium isotope effect with $K_{eq} = 1.33$ in favor of coordination of the protio isotopomer. The coordination of $^{13}$C-labeled cyclopentane revealed the chemical shift of the coordinate methylene group at $-31.2$ ppm with $^1J_{CH}$ of 112.9 Hz (cf. 129 Hz for free cyclopentane). At $-80°$C, the $\Delta G^{\ddagger}$ for dissociation of the coordinated methylene group is 10.3(5) kcal/mol.

Photolysis of ($i$-PrCp)Re(CO)$_3$ ($i$-PrCp = (isopropyl)cyclopentadienyl, used to enhance solubility) in pentane at $-110°$C allowed the direct observation of Re-coordinated pentane.[106] In contrast to cyclopentane, which offers a single methylene group, pentane presents three chemically unique CH$_n$ ($n = 2$ or 3) units to coordinate to Re. Although transition metals that initiate oxidative addition are typically selective for terminal methyl groups, isotopic labeling studies (see above) suggest that the metal center can rapidly migrate among various CH$_n$ moieties in linear alkanes. Consistent with these observations, for ($i$-PrCp)Re(CO)$_2$(pentane) the coordination of all three carbon units of pentane was observed with a slight preference (approximately 0.13(2) kcal/mol) for binding to methylene groups over the terminal methyl position. ROESY NMR experiments confirm that these coordination isomers

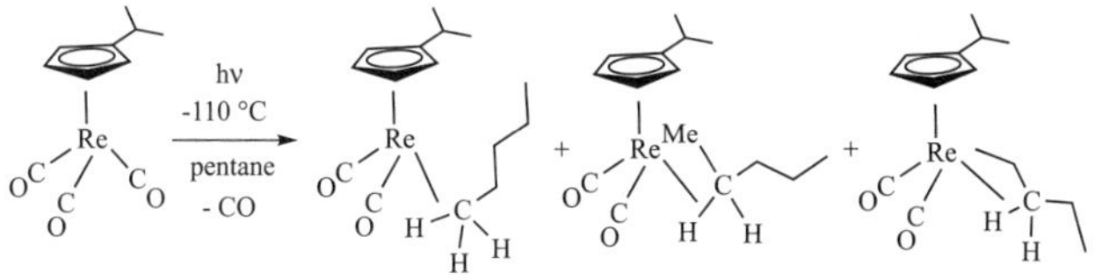

**SCHEME 11.50**   Direct observation of pentane coordination to CpRe(CO)$_2$ allows the identification of an equilibrium between coordination of the methyl and both methylene units.

undergo rapid exchange, which is consistent with the ratios reflecting a thermodynamic distribution rather than a kinetic preference for coordination (Scheme 11.50). The resonance ($^1$H NMR) due to the coordinated methylene groups is shifted farther upfield than the coordinated methyl group. This result is consistent with an $\eta^2$-C,H coordination mode since each hydrogen of a CH$_2$ would exhibit 50% occupation of the metal-coordinated site, while each hydrogen of a CH$_3$ group would interact with the metal 33% of the time.

## 11.5   SUMMARY

Over the past several decades, substantial progress has been made elucidating the details and mechanisms of metal-mediated C−H activation. While a greater understanding of these reactions has been achieved, with each development, a number of new questions arise. The ability to rationally develop catalysts for C−H functionalization and control synthetic processes of hydrocarbons and C−H bonds of functionalized materials depends on an increasingly sophisticated understanding of these fundamental details.

## ACKNOWLEDGMENTS

T. Brent Gunnoe extends gratitude to Professor William Jones (University of Rochester) and Professor William Goddard III (California Institute of Technology) for their time and helpful suggestions.

## REFERENCES

1. Brookhart, M.; Green, M. L. H. *J. Organomet. Chem.* **1983**, *250*, 395–408.
2. Brookhart, M.; Green, M. L. H.; Wong, L. -L. *Prog. Inorg. Chem.* **1988**, *36*, 1–124.
3. Kubas, G. J.; Ryan, R. R.; Swanson, B. I.; Vergamini, P. J.; Wasserman, H. J. *J. Am. Chem. Soc.* **1984**, *106*, 451–452.

4. Kubas, G. J. *Metal Dihydrogen and σ-Bond Complexes*. Kluwer Academic/Plenum Publishers: New York, 2001.

5. Crabtree, R. H. *Angew. Chem., Int. Ed. Engl.* **1993**, *32*, 789–805.

6. Hall, C.; Perutz, R. N. *Chem. Rev.* **1996**, *96*, 3125–3146.

7. Jensen, J. A.; Wilson, S. R.; Schultz, A. J.; Girolami, G. S. *J. Am. Chem. Soc.* **1987**, *109*, 8094–8096.

8. Baker, M. V.; Field, L. D. *Chem. Commun.* **1984**, 996–997.

9. Jensen, J. A.; Girolami, G. S. *Chem. Commun.* **1986**, 1160–1162.

10. Luo, X. -L.; Kubas, G. J.; Burns, C. J.; Bryan, J. C.; Unkefer, C. J. *J. Am. Chem. Soc.* **1995**, *117*, 1159–1160.

11. Schubert, U. *Adv. Organomet. Chem.* **1990**, *30*, 151–187.

12. Brookhart, M.; Green, M. L. H.; Parkin, G. *Proc. Natl. Acad. Sci.* **2007**, *104*, 6908–6914.

13. La Placa, S. J.; Ibers, J. A. *Inorg. Chem.* **1965**, *4*, 778–783.

14. Brookhart, B.; Green, M. L. H.; Wong, L. -L. *Prog. Inorg. Chem.* **1988**, *36*, 1–124.

15. Barefield, E. K.; Parshall, G. W.; Tebbe, F. N. *J. Am. Chem. Soc.* **1970**, *92*, 5234–5235.

16. Chatt, J.; Coffey, R. S. *J. Chem. Soc. A* **1969**, 1963–1972.

17. Garnett, J. L.; Hodges, R. J. *J. Am. Chem. Soc.* **1967**, *89*, 4546–4547.

18. Cotton, F. A.; Frenz, B. A.; Hunter, D. L. *Chem. Commun.* **1974**, 755–756.

19. Chatt, J.; Davidson, J. M. *J. Chem. Soc. A* **1965**, 843–855.

20. Jones, W. D. *Acc. Chem. Res.* **2003**, *36*, 140–146.

21. Arndtsen, B. A.; Bergman, R. G.; Mobley, T. A.; Peterson, T. H. *Acc. Chem. Res.* **1995**, *28*, 154–162.

22. Jones, W. D.; Feher, F. J. *Acc. Chem. Res.* **1989**, *22*, 91–100.

23. Buchanan, J. M.; Stryker, J. M.; Bergman, R. G. *J. Am. Chem. Soc.* **1986**, *108*, 1537–1550.

24. Periana, R. A.; Bergman, R. G. *J. Am. Chem. Soc.* **1986**, *108*, 7332–7346.

25. Gould, G. L.; Heinekey, D. M. *J. Am. Chem. Soc.* **1989**, *111*, 5502–5504.

26. Parkin, G.; Bercaw, J. E. *Organometallics* **1989**, *8*, 1172–1179.

27. Bullock, R. M.; Headford, C. E. L.; Hennessy, K. M.; Kegley, S. E.; Norton, J.R *J. Am. Chem. Soc.* **1989**, *111*, 3897–3908.

28. Labinger, J. A.; Bercaw, J. E. *Nature* **2002**, *417*, 507–514.

29. Sherry, A. E.; Wayland, B. B. *J. Am. Chem. Soc.* **1990**, *112*, 1259–1261.

30. Mayer, J. M. *Acc. Chem. Res.* **1998**, *31*, 441–450.

31. Mayer, J. M. *Annu. Rev. Phys. Chem.* **2004**, *55*, 363–390.

32. Krogh-Jespersen, K.; Czerw, M.; Zhu, K.; Singh, B.; Kanzelberger, M.; Darji, N.; Achord, P. D.; Renkema, K. B.; Goldman, A. S. *J. Am. Chem. Soc.* **2002**, 10797–10809.

33. Janowicz, A. H.; Bergman, R. G. *J. Am. Chem. Soc.* **1982**, *104*, 352–354.

34. Janowicz, A. H.; Bergman, R. G. *J. Am. Chem. Soc.* **1983**, *105*, 3929–3939.

35. Hoyano, J. K.; Graham, W. A. G. *J. Am. Chem. Soc.* **1982**, *104*, 3723–3725.

36. Hoyano, J. K.; McMaster, A. D.; Graham, W. A. G. *J. Am. Chem. Soc.* **1983**, *105*, 7190–7191.

37. Jones, W. D.; Feher, F. J. *J. Am. Chem. Soc.* **1982**, *104*, 4240–4242.

38. Jones, W. D.; Feher, F. J. *J. Am. Chem. Soc.* **1984**, *106*, 1650–1663.

39. Chin, R. M.; Dong, L. Z.; Duckett, S. B.; Partridge, M. G.; Jones, W. D.; Perutz, R. N. *J. Am. Chem. Soc.* **1993**, *115*, 7685–7695.

40. Harman, W. D. *Chem. Rev.* **1997**, *97*, 1953–1978.

41. Wang, C.; Ziller, J. W.; Flood, T. C. *J. Am. Chem. Soc.* **1995**, *117*, 1647–1648.

42. Flood, T. C.; Janak, K. E.; Iimura, M.; Zhen, H. *J. Am. Chem. Soc.* **2000**, *122*, 6783–6784.

43. Northcutt, T. O.; Wick, D. D.; Vetter, A. J.; Jones, W. D. *J. Am. Chem. Soc.* **2001**, *123*, 7257–7270.

44. Low, J. J.; Goddard, W. A., III. *J. Am. Chem. Soc.* **1984**, *106*, 6928–6937.

45. Low, J. J.; Goddard, W. A., III. *J. Am. Chem. Soc.* **1986**, *108*, 6115–6128.

46. Low, J. J.; Goddard, W. A., III. *Organometallics* **1986**, *5*, 609–622.

47. Halpern, J. *Acc. Chem. Res.* **1982**, *15*, 332–338.

48. Shilov, A. E.; Shul'pin, G. B. *Chem. Rev.* **1997**, *97*, 2879–2932.

49. Stahl, S. S.; Labinger, J. A.; Bercaw, J. E. *Angew. Chem., Int. Ed.* **1998**, *37*, 2180–2192.

50. Stahl, S. S.; Labinger, J. A.; Bercaw, J. E. *J. Am. Chem. Soc.* **1995**, *117*, 9371–9372.

51. Johansson, L.; Tilset, M.; Labinger, J. A.; Bercaw, J. E. *J. Am. Chem. Soc.* **2000**, *122*, 10846–10855.

52. Stahl, S. S.; Labinger, J. A.; Bercaw, J. E. *J. Am. Chem. Soc.* **1996**, *118*, 5961–5976.

53. Fekl, U.; Zahl, A.; van Eldik, R. *Organometallics* **1999**, *18*, 4156–4164.

54. Hill, G. S.; Rendina, L. M.; Puddephatt, R. J. *Organometallics* **1995**, *14*, 4966–4968.

55. Bartlett, K. L.; Goldberg, K. I.; Borden, W. T. *J. Am. Chem. Soc.* **2000**, *122*, 1456–1465.

56. Johansson, L.; Tilset, M. *J. Am. Chem. Soc.* **2001**, *123*, 739–740.

57. Wick, D. D.; Goldberg, K. I. *J. Am. Chem. Soc.* **1997**, *119*, 10235–10236.

58. Reinartz, S.; White, P. S.; Brookhart, M.; Templeton, J. L. *J. Am. Chem. Soc.* **2001**, *123*, 12724–12725.

59. Watson, P. L. *J. Am. Chem. Soc.* **1983**, *105*, 6491–6493.

60. Lin, Z. *Coord. Chem. Rev.* **2007**, *251*, 2280–2291.

61. Thompson, M. E.; Baxter, S. M.; Bulls, A. R.; Burger, B. J.; Nolan, M. C.; Santarsiero, B. D.; Schaefer, W. P.; Bercaw, J. E. *J. Am. Chem. Soc.* **1987**, *109*, 203–219.

62. Sadow, A. D.; Tilley, T. D. *J. Am. Chem. Soc.* **2003**, *125*, 7971–7977.

63. Fendrick, C. M.; Marks, T. J. *J. Am. Chem. Soc.* **1984**, *106*, 2214–2216.

64. Jordan, R. F.; Taylor, D. F. *J. Am. Chem. Soc.* **1989**, *111*, 778–779.

65. Steigerwald, M. L.; Goddard, W. A., III. *J. Am. Chem. Soc.* **1984**, *106*, 308–311.

66. Ziegler, T.; Folga, E.; Berces, A. *J. Am. Chem. Soc.* **1993**, *115*, 636–646.

67. Oxgaard, J.; Muller, R. P.; Goddard, W. A., III; Periana, R. A. *J. Am. Chem. Soc.* **2004**, *126*, 352–363.

68. Foley, N. A.; Lail, M.; Lee, J. P.; Gunnoe, T. B.; Cundari, T. R.; Petersen, J. L. *J. Am. Chem. Soc.* **2007**, *129*, 6765–6781.

69. Foley, N. A.; Ke, Z.; Gunnoe, T. B.; Cundari, T. R.; Petersen, J. L. *Organometallics* **2008**, *27*, 3007–3017.

70. Lam, W. H.; Jia, G.; Lin, Z.; Lau, C. P.; Eisenstein, O. *Chem. Eur. J.* **2003**, *9*, 2775–2782.

71. Niu, S.; Hall, M. B. *Chem. Rev.* **2000**, *100*, 353–405.

72. Arndtsen, B. A.; Bergman, R. G. *Science* **1995**, *270*, 1970–1972.

73. Hinderling, C.; Feichtinger, D.; Plattner, D. A.; Chen, P. *J. Am. Chem. Soc.* **1997**, *119*, 10793–10804.

74. Luecke, H. F.; Bergman, R. G. *J. Am. Chem. Soc.* **1997**, *119*, 11538–11539.

75. Klei, S. R.; Tilley, T. D.; Bergman, R. G. *J. Am. Chem. Soc.* **2000**, *122*, 1816–1817.

76. Vastine, B. A.; Hall, M. B. *J. Am. Chem. Soc.* **2007**, *129*, 12068–12069.

77. Bergman, R. G.; Cundari, T. R.; Gillespie, A. M.; Gunnoe, T. B.; Harman, W. D.; Klinckman, T. R.; Temple, M. D.; White, D. P. *Organometallics* **2003**, *22*, 2331–2337.

78. Siegbahn, P. E. M.; Crabtree, R. H. *J. Am. Chem. Soc.* **1996**, *118*, 4442–4450.

79. Whitfield, S. R.; Sanford, M. S. *J. Am. Chem. Soc.* **2007**, *129*, 15142–15143.

80. Dick, A. R.; Hull, K. L.; Sanford, M. S. *J. Am. Chem. Soc.* **2004**, *126*, 2300–2301.

81. Periana, R. A.; Taube, D. J.; Evitt, E. R.; Löffler, D. G.; Wentrcek, P. R.; Voss, G.; Masuda, T. *Science* **1993**, *259*, 340–343.

82. Cummins, C. C.; Baxter, S. M.; Wolczanski, P. T. *J. Am. Chem. Soc.* **1988**, *110*, 8731–8733.

83. Schaller, C. P.; Cummins, C. C.; Wolczanski, P. T. *J. Am. Chem. Soc.* **1996**, *118*, 591–611.

84. Bennett, J. L.; Wolczanski, P. T. *J. Am. Chem. Soc.* **1994**, *116*, 2179–2180.

85. Bennett, J. L.; Wolczanski, P. T. *J. Am. Chem. Soc.* **1997**, *119*, 10696–10719.

86. Walsh, P. J.; Hollander, F. J.; Bergman, R. G. *J. Am. Chem. Soc.* **1988**, *110*, 8729–8731.

87. Hoyt, H. M.; Michael, F. E.; Bergman, R. G. *J. Am. Chem. Soc.* **2004**, *126*, 1018–1019.

88. Chao, Y. -W.; Rodgers, P. M.; Wigley, D. E. *J. Am. Chem. Soc.* **1991**, *113*, 6326–6328.

89. Fulton, J. R.; Holland, A. W.; Fox, D. J.; Bergman, R. G. *Acc. Chem. Res.* **2002**, *35*, 44–56.

90. Gunnoe, T. B. *Eur. J. Inorg. Chem.* **2007**, 1185–1203.

91. Caulton, K. G. *New J. Chem.* **1994**, *18*, 25–41.

92. Mayer, J. M. *Comments Inorg. Chem.* **1988**, *8*, 125–135.

93. Conner, D.; Jayaprakash, K. N.; Cundari, T. R.; Gunnoe, T. B. *Organometallics* **2004**, *23*, 2724–2733.

94. Feng, Y.; Lail, M.; Barakat, K. A.; Cundari, T. R.; Gunnoe, T. B.; Petersen, J. L. *J. Am. Chem. Soc.* **2005**, *127*, 14174–14175.

95. Feng, Y.; Lail, M.; Foley, N. A.; Gunnoe, T. B.; Barakat, K. A.; Cundari, T. R.; Petersen, J. L. *J. Am. Chem. Soc.* **2006**, *128*, 7982–7994.

96. Tenn, W. J., III; Young, K. J. H.; Bhalla, G.; Oxgaard, J.; Goddard, W. A., III; Periana, R. A. *J. Am. Chem. Soc.* **2005**, *127*, 14172–14173.

97. Kloek, S. M.; Heinekey, D. M.; Goldberg, K. I. *Angew. Chem., Int. Ed.* **2007**, *46*, 4736–4738.

98. Oxgaard, J.; Tenn, W. J., III; Nielsen, R. J.; Periana, R. A.; Goddard, W. A., III. *Organometallics* **2007**, *26*, 1565–1567.

99. Cundari, T. R.; Grimes, T.; Gunnoe, T. B. *J. Am. Chem. Soc.* **2007**, *129*, 13172–13182.

100. Kloek, S. M.; Heinekey, D. M.; Goldberg, K. I. *Organometallics* **2008**, *27*, 1454–1463.

101. Brown, C. E.; Ishikawa, Y. -I.; Hackett, P. A.; Rayner, D. M. *J. Am. Chem. Soc.* **1990**, *112*, 2530–2536.

102. Wasserman, E. P.; Moore, C. B.; Bergman, R. G. *Science* **1992**, *255*, 315–318.

103. McNamara, B. K.; Yeston, J. S.; Bergman, R. G.; Moore, C. B. *J. Am. Chem. Soc.* **1999**, *121*, 6437–6443.

104. Sun, X. -Z.; Grills, D. C.; Nikiforov, S. M.; Poliakoff, M.; George, M. W. *J. Am. Chem. Soc.* **1997**, *119*, 7521–7525.

105. Geftakis, S.; Ball, G. E. *J. Am. Chem. Soc.* **1998**, *120*, 9953–9954.

106. Lawes, D. J.; Geftakis, S.; Ball, G. E. *J. Am. Chem. Soc.* **2005**, *127*, 4134–4135.

# 12 Solar Photochemistry with Transition Metal Compounds Anchored to Semiconductor Surfaces

GERALD J. MEYER

## 12.1 INTRODUCTION

There exists a critical need for sustainable power on a terawatt (TW $= 10^{12}$ W) scale.[1,2] As the world's need for energy is related to the number of people on Earth, a staggering 45% population growth over the last quarter-century equates to roughly 2 billion people and a 6 TW ($\sim$63%) increase in needed power.[3] In addition, the urbanism of third-world and industrialized nations and cities has led to an increase in the demand for fuel that has driven gas and oil prices to record highs.[3] Regardless of price, the continued use of fossil fuels cannot be a long-term solution as they come from a limited stock and the deleterious environmental consequences of their combustion have become self-evident.[4,5] The increased average global temperature and rates of glacial melting measured over the past few decades are telling signs.[6–9] Concern should be elicited as ice-core data correlate temperature with greenhouse gas concentrations over the past half-a-million years. The present 380 ppm atmospheric $CO_2$ levels exceed any values attained over the same time period.[5,6] Further, other than natural photosynthesis, there exist no obvious means by which we can decrease today's level. Thus, population, energy demand, and fuel prices do not convey the severity of the need for sustainable energy.

It is our belief that molecular assemblies, particularly molecules arranged at semiconductor interfaces, will one day efficiently harvest and convert energy from the Sun, thereby providing sustainable, carbon-neutral power for future generations. The Sun is in fact the only source that on its own can provide the TWs of power needed. It has been shown that the amount of solar energy reaching the Earth *in one day* could power the planet for *an entire year*.[7,8] Remaining is the important challenge of harvesting, converting, and storing this energy in a cost-effective way. Coordination

*Physical Inorganic Chemistry: Reactions, Processes, and Applications*   Edited by Andreja Bakac
Copyright © 2010 by John Wiley & Sons, Inc.

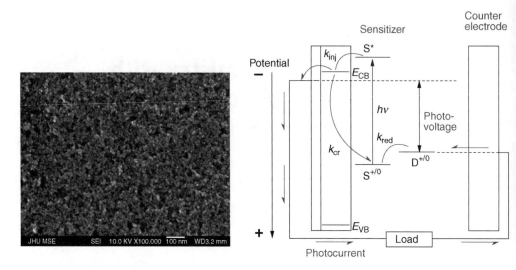

**FIGURE 12.1**   Mechanism for light to electrical energy conversion in a regenerative dye-sensitized solar cell. A photoexcited sensitizer S* injects an electron into the conduction band of a wide-bandgap semiconductor with a rate constant $k_{inj}$. The oxidized sensitizer S$^+$ is regenerated by an electron donor D present in the external electrolyte with a rate constant $k_{red}$. The oxidized donor D$^+$ is reduced at a dark counter electrode. Charge recombination can occur to the oxidized sensitizer or donor represents unwanted loss processes. An SEM image of a mesoporous nanocrystalline (anatase) TiO$_2$ thin film commonly utilized for this application is also shown.

compounds already exist that can effectively harvest large fractions of sunlight and initiate redox reactions, ultimately producing electrical power.[10,11] The class of compounds that have proven to be most robust and useful are Ru$^{II}$ polypyridyl compounds. The metal-to-ligand charge transfer (MLCT) excited states of such compounds can quantitatively store >1.5 eV of free energy.[11] An important next step is to identify methods by which such energy can be permanently stored and/or converted to electrical power. In this regard, these excited states can quantitatively inject electrons into anatase TiO$_2$ nanocrystallites assembled in the mesoporous thin films first described by O'Regan and Grätzel (Figure 12.1). When these sensitized thin films are utilized in regenerative solar cells, power conversion efficiencies greater than 11% have been confirmed.[12]

In this chapter, we describe the excited-state and electron transfer properties of transition metal compounds anchored to nanocrystalline TiO$_2$ (anatase) particles. Emphasis is given to interfacial charge transfer processes relevant to the conversion of light into other forms of energy. We discuss new advances that have enabled practical applications in electrical power generation and the production of useful fuels. The discussion is not meant to be comprehensive but details key advances that represent new opportunities for further research and application.

## 12.2   METAL-TO-LIGAND CHARGE TRANSFER EXCITED STATES

The metal-to-ligand charge transfer excited states of transition metal coordination compounds have emerged as most efficient for solar harvesting and conversion. When one takes into account the need for high stability in a variety of formal oxidation states, $Ru^{II}$ and $Os^{II}$ with coordinated diimine ligands such as those shown in Figure 12.2 are by far the most promising. The pioneering works of Crosby, Demas, Watts, Meyer, Balzani, and others have provided keen insights into the nature of metal-to-ligand charge transfer excited states.[13–27] As the name implies, light absorption promotes an electron from the metal d-orbitals to the $\pi^*$ orbitals of the bipyridine ligand, $d(\pi) \rightarrow \pi^*$.[19–22] Many overlapping charge transfer transitions are observed that give rise to intense broad absorption bands in the visible region with moderate extinction coefficients. There is no formal spin for each excited state due to heavy-atom spin–orbit coupling from the transition metal center (especially for 4d and 5d metals).[23,24] Crosby has proposed that the excited state is accurately described by solely the symmetry label, corresponding to an irreducible representation, of the molecular point group and not the spin and orbital individually.[23] However, the effects of spin–orbit coupling must be introduced in order to rationalize the relative oscillator strengths and absorption spectra of $M(bpy)_3^{2+}$ (M = Fe, Ru, and Os) compounds.

The classical example of a compound with MLCT excited states is $Ru(bpy)_3^{2+}$, where bpy is 2,2′-bipyridine, which is arguably the most well-studied coordination compound in existence. The bpy ligands chelate to the $d\pi^6$ ruthenium metal center in an octahedral arrangement. The ground state is thus threefold symmetric and, based

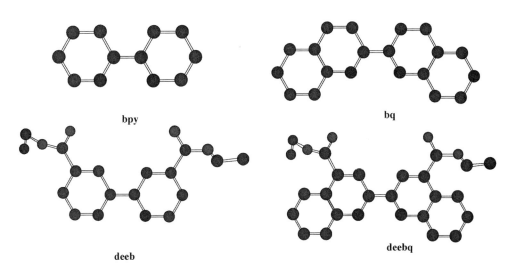

**FIGURE 12.2**   The diimine ligands commonly used for molecular sensitizers: 2,2′-bipyridine (bpy), 4,4′-$(CO_2CH_2CH_3)_2$-bpy (deeb), 2,2′-biquinoline (bq), and 4,4′-$(CO_2CH_2CH_3)_2$-bq (deebq). Note that the ethyl esters are often converted to the corresponding carboxylic acids prior to surface binding and are abbreviated dcb and dcbq.

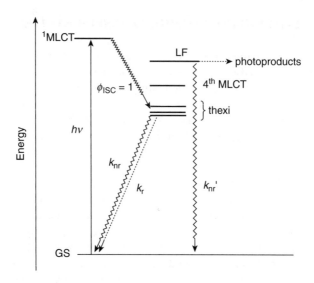

**FIGURE 12.3**    A simplified Jablonski-type diagram for Ru(bpy)$_3$$^{2+}$. Quantitative intersystem crossing from the initially formed singlet excited state to a manifold of thermally equilibrated excited (thexi) MLCT states that have significant Boltzmann population, and behave as a single state, near room temperature. Radiative and nonradiative decays occur from the thexi state. Higher in energy are ligand field excited states that are antibonding with respect to metal–ligand bonds.

on the Franck–Condon principle, the initially formed MLCT excited state possesses the same structural symmetry as the ground state, reasonably formulated as $[\mathrm{Ru^{III}(bpy^{-1/3})_3}]^{2+*}$.

It is now well established that the long-lived excited state observed in fluid solution has an electron localized on a single ligand, $[\mathrm{Ru^{III}(bpy^-)(bpy)_2}]^{2*+}$, with reduced symmetry and an estimated $\sim$10 Debye dipole moment.[28,29] The excited states of Ru(bpy)$_3$$^{2+*}$ have been summarized in the simplified Jablonski diagram (Figure 12.3).[13–18] Note that the long-lived thermally equilibrated excited (thexi) state is actually a manifold of three energetically proximate states with a significant Boltzmann population that behaves as a single state near room temperature. A fourth MLCT state and a ligand field (LF) state are higher in energy. The timescales for excited-state D$_3$ → C$_2$ localization are relevant to interfacial charge transfer as both processes have been found to occur on a femtosecond timescale. The ultrafast excited-state dynamics of MLCT excited states has recently been reviewed by McCusker.[30]

The lifetime of the thexi state of Ru(bpy)$_3$$^{2+}$ is $\sim$1 µs in water at room temperature.[31] The radiative rate constant ($k_r \sim 10^5$ s$^{-1}$) is typically about two orders of magnitude smaller than the nonradiative rate constant ($k_{nr} \sim 10^7$ s$^{-1}$) and hence the excited-state lifetime is controlled by the latter.[31] Ru(II) and Os(II) polypyridyl excited states have been shown to follow Jortner's energy gap law, wherein the nonradiative rate constant increases exponentially with decreased energy separation between the ground and thexi states.[32–36] For this reason, it has proven to be difficult to prepare compounds that both emit in the infrared region and have long-lived excited states. As the energy gap is well approximated by the difference in the $E°(M^{III/II})$ and $E°(bpy^{-/0})$ formal reduction potentials, a simple electrochemical measurement allows one to estimate the excited-state lifetime with considerable accuracy. Furthermore, Lever has reported an empirical model that accurately predicts the redox properties of ruthenium compounds with a wide variety of ligands.[37] Therefore, one can design desired properties such as color, emission energy, excited-state lifetime, and quantum yields into a $Ru^{II}$ compound before any synthesis occurs.

In heteroleptic $Ru^{II}$ compounds, it is found that the electron localizes on the diimine ligand that is reduced most easily.[38] For example, in compounds such as $Ru(dcb)(bpy)_2^{2+}$, where dcb is $4,4'$-$(CO_2H)_2$-bpy, the electron-withdrawing carboxylic acid groups lower the energy of the bipyridine $\pi^*$ orbitals resulting in a redshifted MLCT absorption. The dcb ligand is more easily reduced relative to bpy due to the same inductive effects by the carboxylic acid groups. Thus, the excited state is expected (and found) to be localized on the dcb ligand, $[Ru^{III}(dcb^-)(bpy)_2]^{2*+}$. This is important when considering whether interfacial electron transfer occurs from a ligand that interacts with the semiconductor surface (*adjacent* injection) or a bipyridine ligand that is *remote* to the surface.

The energy gap defines the color and hence the solar light harvesting properties of the compound and can be tuned through synthetic chemistry. This is generally accomplished with substituents on the diimine ligand that inductively tune the $\pi^*$ levels and by controlling the extent of $d\pi$–$\pi^*$ backbonding donation to nonchromophoric ligands such as $Cl^-$, $I^-$, $Br^-$, $NCS^-$, $CN^-$, amines, phosphines, and others. The compound that has emerged as the most efficient sensitizer for regenerative dye-sensitized solar cell applications is $cis$-$Ru(dcb)_2(NCS)_2$, which is often called N3.[39] N3 gains red absorption over $Ru(dcb)(bpy)_2^{2+}$ at the expense of a more negative metal-based reduction potential, $E°(Ru^{III/II})$. Although not generally critical to regenerative solar cells, this loss precludes its use in the "holy grail" of solar photochemistry: splitting water into hydrogen and oxygen gas.[39,40]

## 12.2.1  Behavior on Nanocrystalline $TiO_2$ Thin Films

The most common and successful functional groups for excited-state sensitization are carboxylic acids, usually placed in the 4- and $4'$-positions of bpy.[41] Other functional groups based on phosphonates, siloxanes, acetyl acetonates, ethers, phenols, and cyanides have also been reported to result in binding to $TiO_2$.[41] The surface binding chemistry is usually performed by overnight reactions in organic solvents with freshly prepared $TiO_2$ thin films. In some reports, the $TiO_2$ surface is pretreated with aqueous solutions of known pH or aqueous $TiCl_4$ solutions.

Goodenough first proposed that carboxylic acids would dehydratively couple with titanol groups to form surface ester linkages (Figure 12.4).[42] We have reported spectroscopic evidence for this and for the more commonly observed "carboxylate" linkage in reactions of $Ru(dcb)(bpy)_2(PF_6)_2$ and $Ru(bpy)_2(ina)_2(PF_6)_2$, where dcb is $4,4'$-$(CO_2H)_2$-bpy and ina is isonicotinic acid, with nanocrystalline $TiO_2$ and colloidal

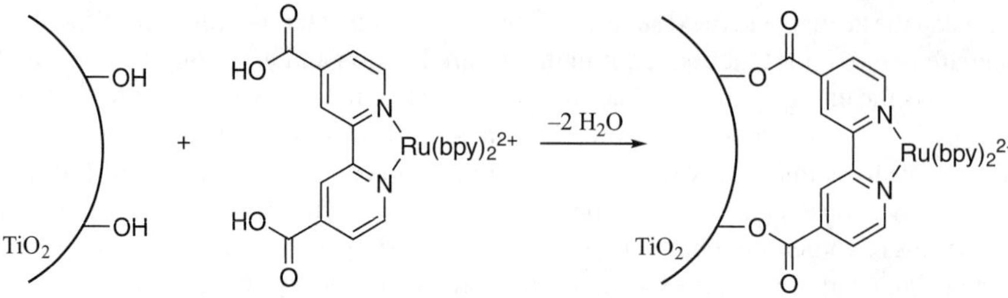

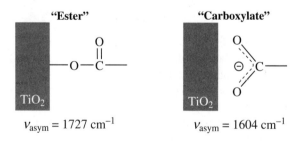

**FIGURE 12.4** Idealized surface chemistry first proposed by Goodenough: dehydrative coupling of surface titanol and carboxylic acid groups to form ester linkages. Note that for mesoporous nanocrystalline $TiO_2$ thin films, the most common surface linkage identified is based on interactions of the carboxylate groups with $TiO_2$.

$ZrO_2$ films in acetonitrile at room temperature.[43] The interfacial proton concentration was intentionally varied by equilibration of the films with aqueous solutions of known pH prior to sensitizer binding. The visible absorption and IR spectral data indicate that a high surface proton concentration yields an ester-type linkage(s) where low proton concentrations favor "carboxylate" type binding mode(s) (Figure 12.5). The $TiO_2$ surface saponified the esters in $Ru(deeb)(bpy)_2(PF_6)_2$, where deeb is $4,4'-(CO_2Et)_2$-bpy, and the carboxylates bound to the surface.[43] For a variety of sensitizers with carboxylic acid functional groups reacted with untreated $TiO_2$, the concentration-dependent binding generally follows the Langmuir adsorption isotherm model with adduct formation constants of $10^4-10^5\,M^{-1}$ and limiting surface coverages of $10^{-8}$ $mol/cm^2$. This corresponds roughly to monolayer coverage and about 700 Ru(II) sensitizers anchored to each $\sim$20 nm $TiO_2$ particle.

Vibrational spectroscopic data reported to date have not directly identified the surface site(s) involved in the sensitizer–semiconductor bond.[44–49] Deacon and Phillips tabulated vibrational data of metal carboxylate compounds with known

|  "Ester" | "Carboxylate" |
|---|---|
| $\nu_{asym} = 1727\ cm^{-1}$ | $\nu_{asym} = 1604\ cm^{-1}$ |

**FIGURE 12.5** Asymmetric CO stretches for $TiO_2$ anchored Ru(II) sensitizers based on the $4,4'-(COOH)_2-2,2'$-bipyridine (dcb) ligand have been observed at $\sim$1730 and $\sim$1600 cm$^{-1}$. The former has only been reported on acidified surfaces. The latter is by far the most commonly observed, although the nature and number of $TiO_2$ sites involved in the carboxylate binding remain speculative.

crystal structures and found an empirical relation between the frequency difference of the asymmetric and symmetric CO stretches and the carboxylate–metal coordination mode.[48] The same approach has been used to predict the carboxylate binding modes to presumed $Ti^{IV}$ sites on the $TiO_2$ surface.[45–48] The analysis is most consistent with the carboxylate oxygens binding to separate $Ti^{IV}$ centers on the anatase surface, in agreement with theoretical studies.[50]

### 12.2.1.1 MLCT Excited States on TiO₂

*12.2.1.1 MLCT Excited States on TiO₂*  Experimental studies of MLCT states on $TiO_2$ (and other semiconductors) are few mainly because of rapid interfacial charge separation that shortens their lifetimes considerably. Some aspects of MLCT excited states anchored to nanocrystalline $TiO_2$ thin films are now becoming available through studies where the semiconductor acceptor states lie above (toward the vacuum level) the excited-state reduction potential of the sensitizer such that excited-state electron transfer from the thexi state is unfavorable.

In many important aspects, the room-temperature photophysical properties of $Ru(dcb)(bpy)_2$ anchored to $TiO_2$ are remarkably similar to those observed in fluid solution particularly when one takes into account the protonation state of the carboxylic acid groups.[51] The most significant difference was that excited-state relaxation on $TiO_2$ (and $ZrO_2$) occurred by a parallel first- and second-order kinetic model where excited-state relaxation in solution is first order. The second-order component was attributed to triplet–triplet annihilation reactions that occur in parallel with radiative and nonradiative decay. Excited-state annihilation is facilitated by the close proximity of the surface-bound compounds that affords rapid, lateral, isoenergetic energy transfer across the semiconductor surface (Figure 12.6).

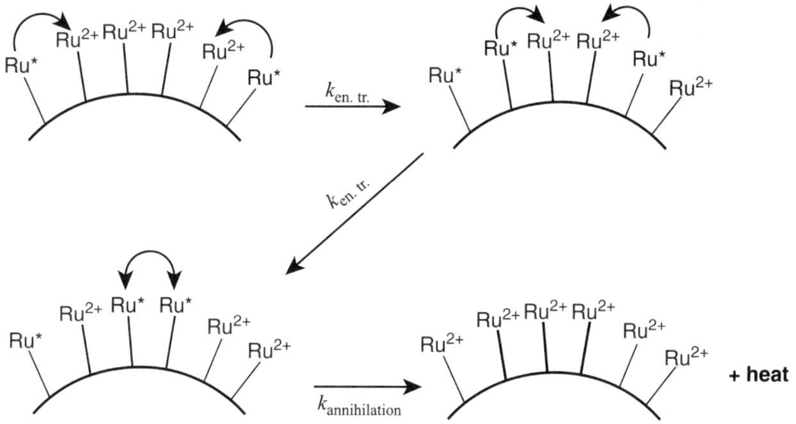

**FIGURE 12.6**  Schematic of how lateral intermolecular energy transfer across the semiconductor surface can lead to a second-order excited-state annihilation reaction. First-order excited-state relaxation was observed for sensitizers with short excited-state lifetimes, $\tau < 50$ ns, and was predominant at low irradiances and surface coverages for all sensitizers.

Direct evidence for energy transfer came from studies where both $Ru(dcb)(bpy)_2(PF_6)_2$ and $Os(dcb)(bpy)_2(PF_6)_2$ were anchored to the same nanocrystalline $TiO_2$ film.[52] The Os compound acts as an energy transfer trap as reaction 12.1 is thermodynamically downhill.

$$Ru(dcb)(bpy)_2^*/TiO_2 + Os(dcb)(bpy)_2/TiO_2 \xrightarrow{\;k_{en}\;} Ru(dcb)(bpy)_2/TiO_2$$

$$+ Os(dcb)(bpy)_2^*/TiO_2$$

$$(12.1)$$

The yields and dynamics for energy transfer from the MLCT excited states of $Ru(dcb)(bpy)_2(PF_6)_2$, $Ru^{2+}$, and $Os(dcb)(bpy)_2(PF_6)_2$, $Os^{2+}$, anchored to mesoporous nanocrystalline (anatase) $TiO_2$ thin films were quantified.[53] Lateral energy transfer from $Ru^{2+*}$ to $Os^{2+}$ was observed and the yields were measured as a function of the relative surface coverage and the external solvent environment ($CH_3CN$, THF, $CCl_4$, and hexane). Excited-state decay of $Ru^{2+*}/TiO_2$ was well described by the parallel first- and second-order kinetic model described previously, whereas $Os^{2+*}/TiO_2$ decayed with first-order kinetics. The first-order component was assigned to the usual sum of the radiative and nonradiative rate constants as is observed in fluid solution ($\tau = 1\,\mu s$ for $Ru^{2+*}/TiO_2$ and $\tau = 50\,ns$ for $Os^{2+*}/TiO_2$). The second-order component was attributed to intermolecular energy transfer followed by triplet–triplet annihilation. An analytical model was derived that allowed determination of the fraction of excited states that follow the two pathways. A trend emerged from systematic surface coverage and excitation irradiance studies. The fraction of $Ru^{2+*}/TiO_2$ that decayed through the second-order pathway increased with irradiance and surface coverage.

Monte Carlo simulations were performed with sensitizers arranged in a $32 \times 32$ grid of either a primitive square geometry with four nearest neighbor sites or a hexagonal arrangement with six nearest neighbors (Figure 12.7).[53] This corresponds to $\sim 1000$ sensitizers anchored to each anatase nanocrystallite at saturation surface coverage. Initially, a set of random numbers was drawn that determined which of the sensitizers was placed in the excited state. With one sun of AM 1.5 irradiation, about three excited states were present on each nanocrystallite. After a 1 ps time step, random numbers determined which of the nearest neighbors accepted the excited-state energy. If two excited states were located adjacent to each other, one annihilation event was counted and one excited state was converted to a ground state and the other to an excited state. This was consistent with known triplet–triplet annihilation reactions that form a singlet excited state and a ground state. The total simulation time was 5 $\mu s$, which corresponded to five lifetimes of $Ru^{2+*}$ and accounted for more than 99% total excited-state decay. The calculations included a circular boundary condition such that excited states that hop off the grid reappear on the other side. After each time step, the whole process was repeated. Simulated excited-state decays were generated from these data. It was found that the hexagonal sensitizer arrangement gave better fits to the experimental decays than did the square geometry.

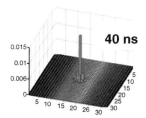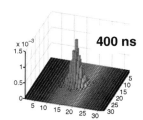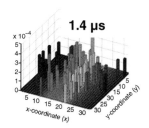

**FIGURE 12.7**   Monte Carlo simulations of lateral intermolecular energy transfer across an anatase $TiO_2$ nanocrystallite. A $32 \times 32$ grid of close-packed sensitizer grid with a continuity condition that allowed excited states that hopped off the grid to reappear on the opposite side was used to approximate the $\sim$700 sensitizers found on a $\sim$20 nm spherical crystal. The probability of where a $Ru(dcb)(bpy)_2^*$ excited state would be found at 40 ns, 400 ns, and 1.4 μs that underwent random $(30\,ns)^{-1}$ intermolecular energy hops is shown. (See the color version of this figure in Color Plates section.)

These simulations also provided an estimation of how far an excited state can migrate by isoenergetic intermolecular energy transfer across a $TiO_2$ nanocrystallite.[53] For long-lived excited states such as $Ru^*/TiO_2$, there was a significant probability that the excited state would migrate to every sensitizer on the surface. This suggests that the sensitized mesoporous films could function like naturally occurring antenna and transfer their energy to specific sites or catalysts. On the other hand, sensitizers with excited-state lifetimes of 50 ns or less did not move far from the initial excitation even after four lifetimes. This is shown graphically in Figure 12.7 and is consistent with the experimental finding that such short-lived excited states did not undergo triplet–triplet annihilation reactions at saturated surface coverages and any of the irradiances studied. Lateral energy transfer may be used to sensitize specific catalytic sites on the semiconductor surface. More fundamentally, energy transfer dynamics can provide direct information on the distance between the surface-bound sensitizers.

## 12.2.2   Ligand Field and Ligand Localized Excited States

The intervention of thexi $\rightarrow$ LF surface crossing in fluid solution may be inferred from the appearance of ligand loss photochemistry.[14–20] The presence of low-lying ligand field states can also deactivate the MLCT excited states and decrease excited-state lifetimes. A classical example of this is $Fe(bpy)_3^{2+}$, which, until recently, was thought to be completely nonemissive due to rapid and quantitative conversion to ligand field states.

The presence of low-lying ligand field states in $Ru^{II}$ polypyridyl compounds may be inferred from the appearance of ligand loss photochemistry and temperature-dependent excited-state lifetimes. A classical example is $cis$-$Ru(bpy)_2(py)_2^{2+}$, where py is pyridine.[54] This compound is nonemissive in room-temperature acetonitrile electrolytes and photolysis is accompanied by significant ligand loss photochemistry.

Another well-studied example is the aqueous photochemistry of $Ru^{II}$ pyridyl ammine compounds that displayed significant photoaquation yields with MLCT excitation.

We recently reported that while $cis$-Ru(bpy)$_2$(ina)$_2$(PF$_6$)$_2$ was also nonemissive in fluid solution, $\tau < 10$ ns, when anchored to nanocrystalline TiO$_2$ films the compound becomes highly photoluminescent with a lifetime of 60 ns.[55] The sensitizer was much more photostable than in fluid solution. The apparent MLCT $\rightarrow$ LF internal conversion activation energy increases to 2500 cm$^{-1}$ upon surface binding. Similar increases in this activation energy have been observed for Ru(bpy)$_3{}^{2+}$ in solid-state media and attributed to destabilization of the ligand field states by the solid.[56] Interestingly, the fraction of $cis$-Ru(bpy)$_2$(ina)$_2$/TiO$_2$ excited states formed was also temperature dependent, behavior indicative of nonunity intersystem crossing yields.[55]

In studies with compounds of the type Ru(ina)(NH$_3$)$_5{}^{2+}$ and Ru(dcb)(NH$_3$)$_4{}^{2+}$, enhanced photostability after attachment to TiO$_2$ was also observed. Isotopic substitution studies suggested that high-frequency N–H modes also play a role in MLCT excited-state relaxation.[57,58] Ru(bpy)$_3{}^{2+}$ and most other tris-heteropleptic Ru(II) compounds have redox and optical properties that are fairly insensitive to their environments[59]. This was not the case for Ru(ina)(NH$_3$)$_5{}^{2+}$ and Ru(dcb)(NH$_3$)$_4{}^{2+}$ in solutions or when attached to TiO$_2$ thin films and immersed in different solutions. Ammine or cyano compounds of the type M(bpy$'$)(X)$_4{}^{2-/2+}$ or $cis$-M(bpy)$_2$(X)$_2{}^{0/2+}$, X $=$ CN$^-$ or NH$_3$, are in fact well known to be highly solvatochromic.[60–63] Outer-sphere interactions with the ammine (or cyano) ligands have a profound influence on the Ru$^{III/II}$ reduction potential and hence MLCT energy gap and the color of the compound. Indeed, these compounds are among the most highly solvatochromic compounds known.[59]

A shortcoming of solar light harvesting by Ru$^{II}$ compounds is their relatively low extinction coefficients compared to $\pi \rightarrow \pi^*$ transitions often found in organic sensitizers. Thus, 6–10 µm thick films of nanocrystallite TiO$_2$ are required for efficient solar harvesting and increased LHE with Ru$^{II}$-based coordination compounds. This precludes the use of many classes of semiconductor materials that have inherently low surface areas. Ru(bpy)$_3{}^{2+}$ has a molar extinction coefficient of about 15,000 M$^{-1}$ cm$^{-1}$ for its MLCT-based electronic transitions.[63] In contrast, natural and synthetic organic pigments also absorb solar photons but with extinction coefficients that are often in excess of 200,000 M$^{-1}$ cm$^{-1}$.[64] It is well established that substituents with low-lying $\pi^*$ orbitals (such as aromatic groups, esters, or carboxylic acids) present on the bpy ligand can enhance MLCT extinction coefficients relative to unsubstituted bpy. Interestingly, 4 and 4$'$ disubstitution of bpy has been found to increase these extinction coefficients more effectively than does disubstitution in the 5- and 5$'$-positions.[62,63] The preparation of high extinction coefficient, heteroleptic N3 derivatives, where one of the dcb ligands is replaced by a 4,4$'$-disubstituted bpy, is an extremely active area of research.[64–68]

We recently found that Ru$^{II}$ compounds with bpy ligands bridged in the 3- and 3$'$-positions by dithiolene had notably high extinction coefficients.[69] Substituent effects in the 3 and 3$'$ bipyridine positions are not well documented as they sterically force the two pyridyl rings out of planarity, behavior that can decrease the stability of the

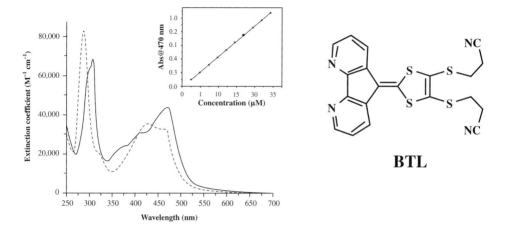

**FIGURE 12.8**   The absorption spectra of high extinction coefficient sensitizers Ru(deeb)$_2$ (BTL)(PF$_6$)$_2$ (solid line) and Ru(bpy)$_2$(BTL)(PF$_6$)$_2$ (dashed line) in neat acetonitrile. The inset shows the concentration dependence of the Ru(deeb)$_2$(BTL)(PF$_6$)$_2$ absorbance at 470 nm, from which an extinction coefficient of 44,300 M$^{-1}$ cm$^{-1}$ was obtained. The BTL ligand shown is somewhat unique in that the extended conjugation on the bipyridine ligand is in the 3- and 3′-positions rather than the 4- and 4′-positions that are normally utilized.

compound. This issue is circumvented with bridging ligands but at the expense of opening up the N–Ru–N bite angle, thereby stabilizing ligand field states. Nevertheless, it was notable that these first-derivative, MLCT dithiolene compounds have extinction coefficients for their lowest energy transitions that are comparable to the highest ever reported based on Ru$^{II}$(4,4′-disubstituted bpy) compounds, $4.4 \times 10^4$ M$^{-1}$ cm$^{-1}$ (Figure 12.8). Part of this success was that the dithiolene-bpy ligands themselves have intraligand absorption bands, in addition to the MLCT absorption bands, in the visible region.[69]

An alternative strategy for increasing the LHE is to use nature's antenna effect.[70] Multiple Ru$^{II}$-bpy units that are suitably arranged can absorb light and vectorially transfer energy to a terminal group that can then inject an electron into the semiconductor. If the additional pigments do not increase the footprint of the sensitizer on the semiconductor surface, this too is a method for enhancing solar light harvesting. Indeed, the trinuclear Ru$^{II}$ sensitizer utilized in the celebrated 1991 Nature paper had been previously designed in Italy to function as an antenna.[10,71] An issue with the Ru(dcb)$_2$(CN)$_2$ group used as the energy transfer acceptor and surface anchor is the *cis*-geometry of the ambidentate cyano ligands. This results in a surface footprint that increases with the number of Ru$^{II}$-bpy units (Figure 12.9). In this regard, a *trans*-geometry is more preferred.[72] Stabilization of a *trans*-geometry generally requires covalently linking two diimine ligands so as to prevent *trans* → *cis* photoisomerization. The synthesis of molecules that function as antenna for use in dye-sensitized solar cells continues to be an active area of research that may one day enable the efficient sensitization of semiconductor materials.

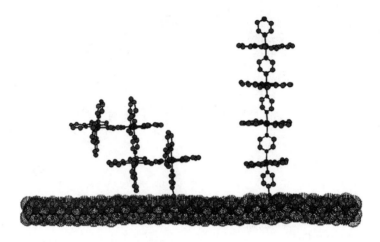

**FIGURE 12.9**    An array of sensitizers anchored to $TiO_2$ surface that consists of *cis-* (left) and *trans-* (right) $[(Ru(LL)_2(pz))_4(ina)]^{8+}$, where pz is ambidentate pyrazine and ina is isonicotinic acid. Note that the *trans*-geometry allows the number of Ru(LL) units to be increased without increasing the occupied surface area.

## 12.3   CHARGE SEPARATION

$Ru^{II}$ polypyridyl compounds can effectively harvest large fractions of sunlight and initiate redox reactions, ultimately producing electrical power or useful fuels such as hydrogen gas. A key initial step is photoinduced charge separation, a generic example of which is given in Figure 12.10. Light absorption creates an excited state $C^*$ that transfers an electron to an acceptor to yield what is hereby referred to as a "charge-separated state" that is comprised of an oxidized donor and a reduced acceptor, $D^+$ and $A^-$. Such an excited-state electron transfer reaction is often referred to as *oxidative quenching*. Since the excited state is both a stronger oxidant and stronger

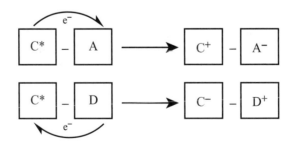

**FIGURE 12.10**    Mechanisms of excited-state electron transfer. In the upper reaction, the excited state acts as a reductant, a process commonly termed *oxidative* quenching. In the lower reaction, the excited state is an oxidant, termed *reductive* quenching.

reductant than the ground state, electron transfer from a donor to the excited state can form a $C^-$, $D^+$ charge-separated state, a process termed *reductive quenching*. The importance of these reactions is that they provide a molecular basis for the conversion of light into potential energy in the form of redox equivalents. The energy is only transiently stored as recombination to ground-state products; that is, $C^+ + A^- \rightarrow C + A$ or $C^- + D^+ \rightarrow C + D$ is thermodynamically downhill. The rate constants reported for charge recombination in fluid solution often approach the diffusion limit.

The thermodynamics for excited-state electron transfer can be estimated from knowledge of ground- and excited-state reduction potentials.[73] The ground-state potentials are often conveniently measured by cyclic voltammetry. Excited-state reduction potentials have been estimated with thermochemical cycles (Equations 12.2 and 12.3), where $\Delta G_{ES}$ is the free energy stored in the thexi state. This energy can be estimated by the photoluminescence onset or through a Franck–Condon line shape analysis of the corrected spectrum. The reduction potential of the initially formed Franck–Condon excited state can be calculated by substituting the excitation energy for $\Delta G_{ES}$. For many studies of sensitized $TiO_2$ thin films, the spectroscopic and electrochemical data needed for such calculations can be measured *in situ*, that is, for the sensitizer anchored to the semiconductor film. Previous studies have shown that $Ru^{II}$ and other sensitizers anchored to mesoporous nanocrystalline $TiO_2$ thin films can be reversibly oxidized in standard electrochemical cells provided that the sensitizer surface coverage exceeds a percolation threshold.[74–76] Cyclic voltammetry and spectroelectrochemistry are thus powerful *in situ* tools for determining formal reduction potentials and absorption spectra of relevant redox states.

$$E^\circ(Ru^{III/II}*) = E^\circ(Ru^{III/II}) - \Delta G_{ES} \qquad (12.2)$$

$$E^\circ(Ru^{II*/+}) = E^\circ(Ru^{II/+}) + \Delta G_{ES} \qquad (12.3)$$

The classical inorganic example of charge separation is electron transfer from the MLCT excited states of $Ru(bpy)_3^{2+}$ to methyl viologen, $MV^{2+}$, where $Ru(bpy)_3^{2+}$ is the chromophoric donor* and $MV^{2+}$ is the electron acceptor.[77,78] Excited-state electron transfer yields a $Ru(bpy)_3^{3+}$ and $MV^{+\bullet}$ charge-separated state that stored about 1.6 eV of free energy (Equation 12.4). The oxidative quenching yields were high, but rapid recombination in the solvent cage lowers the yield of long-lived charge-separated states significantly in aqueous solution (Equation 12.5). It has been over 30 years since Whitten and Meyer first reported the direct observation of the $Ru(bpy)_3^{3+}$, $MV^{+\bullet}$ charge-separated state, and now the details are understood at molecular level.[77,78] The use of $Ru(bpy)_3^{2+*}$ excited states to drive electron transfer processes of fundamental importance to biology has been pioneered by Gray and Winkler.[79] Studies of compounds with covalent bonds between the donor and the acceptor have provided considerable insights into how spin, distance, and thermodynamic driving force influence electron transfer rate constants.[80] In addition, solid-state materials have been successfully employed to spatially arrange and isolate chromophores, donors, and/or acceptors in solid-state materials for long-lived charge separation.[81]

$$Ru^{III}(bpy^{-\bullet})(bpy)_2{}^{2+}* + MV^{2+} \rightarrow Ru(bpy)_3{}^{3+} + MV^{+\bullet} \tag{12.4}$$

$$Ru(bpy)_3{}^{3+} + MV^{+\bullet} \rightarrow Ru(bpy)_3{}^{2+} + MV^{2+} \tag{12.5}$$

An interesting example of intramolecular, photoinduced electron transfer was reported with donor–acceptor compounds based on copper.[82,83] Not only can $Cu^I$ diimine compounds be utilized for solar energy conversion, but they also possess unique structural properties not found in the $d\pi^6$ chemistry of $Ru^{II}$, $Os^{II}$, or $Re^I$. For bis-phenanthroline compounds, the $Cu^I$ state is known to be pseudotetrahedral in the solid state and in solution. In contrast, $Cu^{II}$ compounds are generally five-coordinate trigonal bipyramidal or square pyramidal, with solvent or anion occupying the fifth coordination site as is shown.[95] Therefore, when charge recombination involves reduction of $Cu^{II}$ to $Cu^I$, a large structural, and hence inner-sphere reorganizational energy, change is expected.[84]

The first donor–acceptor compounds based on $Cu^I$ have a viologen tethered to 2,2'-bipyridine, abbreviated bpy-MV$^{2+}$ (Figure 12.11).[82] Visible light excitation of $Cu(bpy-MV^{2+})_2{}^{5+}$ or $Cu(bpy-MV^{2+})(PPh_3)_2{}^{3+}$ produced a charge-separated state with an electron localized on the viologen and a $Cu^{II}$ center. The charge separation rate constant could not be time resolved, $k_{cs} > 10^8 \text{ s}^{-1}$. In contrast, the first-order rate constants for charge recombination were remarkably slow. For $Cu(bpy-MV^{2+})(PPh_3)_2{}^{3+}$, the lifetime of the charge-separated state was 20 ns in dichloromethane, $CH_2Cl_2$, and increased to 1.8 μs in dimethyl sulfoxide, DMSO, for example. The Gibbs free energy stored in the copper charge-separated states is 0.45 eV for the bis-chelate complex and approximately 1.1 eV for the bisphosphine complex in acetonitrile, $CH_3CN$. The lifetimes of the charge-separated states were over three orders of magnitude longer lived than those reported for analogous donor–acceptor compounds based on $Ru^{II}$, where inner-sphere contributions to the reorganizational energy are known to be minimal. To account for the long-lived nature and the remarkable solvent dependence of electron transfer, it was speculated that solvent coordinated to the $Cu^{II}$ center in the charge-separated state.[82] Therefore, incorporated within the total reorganization energy for charge recombination was the enthalpic contribution of Cu–S bond rupture. Temperature-dependent charge recombination kinetics for $Cu(bpy-MV^{2+})(PPh_3)_2{}^{3+}$ were consistent with activated electron transfer in the Marcus normal region.[83] For example, in DMSO an enthalpy of activation $\Delta H = 17(3) \text{ kJ/mol}$ and an entropy of activation $\Delta S = 45(5) \text{ J/(mol K)}$ were measured.

Direct evidence for coordination number changes to the copper center of $[Cu(dmp)_2]^{+*}$, where dmp is 2,9-dimethyl-1,10-phenanthroline, was obtained by XANES and XAFS (Figure 12.12).[85–87] A fifth nearest neighbor was identified, presumably derived from solvent or counterion, and a distorted square-based pyramidal geometry. Such spectroscopic changes were expected as the electronic

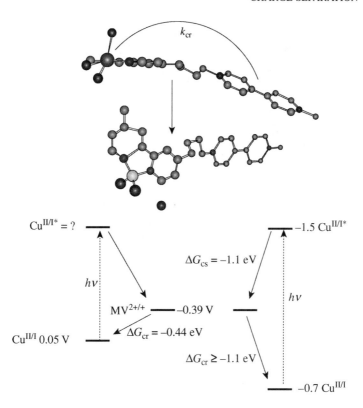

**FIGURE 12.11** Intramolecular charge recombination, $MV^{+\bullet}$-Cu(II) $\rightarrow$ $MV^{2+}$-Cu(I), showing the expected 5 to 4 change in coordination number. The energetics for two dyads based on the bpy-$MV^{2+}$ ligand are shown: one that had two bpy-$MV^{2+}$ ligands and one that had one bpy-$MV^{2+}$ ligand (left) and two triphenyl phosphine ligands (right). Note that the driving force for charge recombination is smaller than that for separation, behavior that is somewhat novel.

configuration changes from $Cu^{I}$ ($3d^{10}$) to formally $Cu^{II}$ ($3d^{9}$) in the MLCT excited state. The geometry changes that may accompany excited-state transfer to an electron acceptor, such as a viologen, will be the subject of future studies.

### 12.3.1   Interfacial Charge Separation

A promising idea is to convert the energy stored as redox equivalents in charge-separated states directly into electrical power. The celebrated "photogalvanic cells" were designed to do just this by selective collection of photogenerated D$^{+}$' and A$^{-}$' at opposite dark electrodes.[88] Photogalvanic cells based on $Ru(bpy)_3^{2+}$ and $Fe^{3+}$ (aq) have been characterized, although solar conversion efficiencies were and remain quite low.[88,89] When a photocurrent is generated from light absorbed by a dye, the process is known as sensitization and the light absorbing dye is referred to as sensitizer.

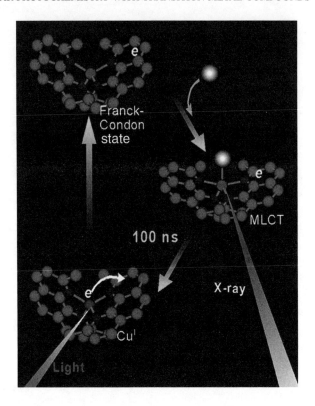

**FIGURE 12.12** X-ray characterization of the equilibrated MLCT excited states of $Cu(dmp)_2^+$ in toluene solution. A pulsed laser was utilized to create the excited state that was subsequently interrogated by a series of X-ray pulses from the advanced light source at Argonne National Laboratory. The results showed that the emissive excited state had five nearest neighbors. (See the color version of this figure in Color Plates section.)

Polypyridyl compounds of $Ru^{II}$ were among the first sensitizers to be studied and continue to be the most promising for real-world applications in regenerative solar cells. The mesoporous nanocrystalline $TiO_2$ thin films developed by Grätzel and O'Regan have resulted in an order of magnitude increase in solar energy conversion efficiencies when employed in such cells.[10] The film thickness and high transparency allow interfacial charge transfer processes to be spectroscopically characterized with signal-to-noise ratios approaching those obtained in fluid solution.

***12.3.1.1 Theory*** Gerischer has described a theory for excited-state electron transfer to semiconductors.[90–92] The rate constant for interfacial electron transfer is proportional to the overlap of occupied donor levels of the excited state, $W_{don}(E)$, with unoccupied acceptor states in the semiconductor $D(E)$ (Equation 12.6):

$$k_{inj} \sim \int \kappa(E) D(E) W_{don}(E) \, dE \qquad (12.6)$$

where $\kappa(E)$ is the transfer frequency. Fluctuations in the solvation of the sensitizer give rise to a distribution of excited-state energies. Gerischer defined the Gaussian donor excited-state distribution function, $W_{don}(E)$:

$$W_{don}(E) = \frac{1}{\sqrt{4\pi\lambda k_B T}} \exp\left(-\frac{(E-{}^\circ E)^2}{4\lambda k_B T}\right) \qquad (12.7)$$

where $\lambda$ is the reorganization energy of interfacial electron transfer, $k_B$ is Boltzmann's constant, $T$ is temperature, $E$ is energy, and ${}^\circ E$ is the energy of the most probable solvation state.[91] Thus, the rate and efficiency of electron injection from the sensitizer excited state depend on the overlap of the sensitizer excited-state distribution function with the density of semiconductor acceptor states (Figure 12.13). This has been exploited to quantify the reorganization energy for excited-state injection by Ru(II) polypyridyl compounds at rutile single crystals. Sensitized photocurrents were measured as a function of pH, and a value of $\lambda = 0.25$ eV was obtained.[93]

***12.3.1.2   TiO₂ DOS***   The nature of the $TiO_2$ electron-acceptor states remains poorly understood and the molecular descriptions a chemist would like are lacking. Distinctions are often made between delocalized and trapped electrons. A commonly held view is that electron injection occurs to the unfilled conduction band states and is rapidly trapped as a $Ti^{III}$-like state. Charge recombination is thus from electrons "trapped" in Ti(III) states within the forbidden energy gap. While this is an appealing idea that makes good chemical sense in many regards, the distinction between trapped and conduction band electrons in $TiO_2$ thin films immersed in acetonitrile electrolytes is somewhat arbitrary. In aqueous electrolytes, there does exist some evidence for discrete trapped and free electronic states.[94] Since most of the studies described herein were performed with sensitized $TiO_2$ thin films in organic solvents such as acetonitrile, we simply refer to reduced $TiO_2$ as $TiO_2(e^-)$.

It is well known that $TiO_2$ can be reduced electrochemically, chemically, or photoelectrochemically and that the resultant blue-black colored material is stable in the absence of dioxygen or other electron acceptors.[95] The $TiO_2(e^-)$ has characteristic EPR and UV-Vis spectra.[95–97] A broad absorption that increases with wavelength through the visible region is observed that is accompanied by a blue shift of the fundamental valence-to-conduction band absorption edge. The extinction coefficient of reduced $TiO_2$ thin films in 1.0 M LiClO₄–acetonitrile is about 1800 $M^{-1} cm^{-1}$ at 800 nm. The electronic spectra of electrochemically reduced $TiO_2$ are within experimental error, the same as that generated by dye sensitization. Therefore, the $TiO_2(e^-)$'s can be generated photo- and electrochemically for thermal electron transfer studies.

Determination of the $TiO_2(e^-)$ "redox potential" is necessary for estimation of the free energy stored in interfacial charge-separated states. Under all conditions we have studied, injected electrons in $TiO_2$ were able to reduce $MV^{2+}$, indicating that the free energy stored is larger than that in the classical $Ru(bpy)_3^{3+}$, $MV^{+\bullet}$ charge-separated state, $\Delta G > 1.6$ eV. The density of unoccupied acceptor states (i.e., DOS) in nanocrystalline anatase $TiO_2$ thin films is, however, somewhat unresolved. The

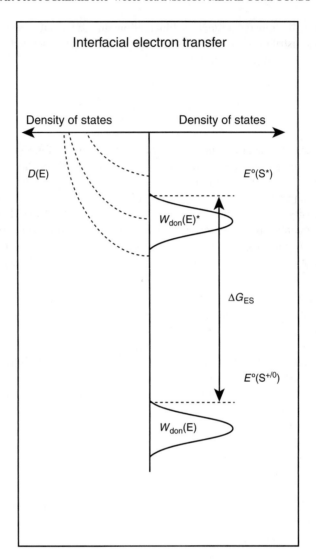

**FIGURE 12.13**    Gerischer-type diagram for interfacial electron transfer. The rate constants for interfacial electron transfer are dependent on the overlap of the sensitizer and the semiconductor density of states. Note that the density of states of the semiconductor is not a singular parameter and can shift with a change in environment, that is, pH, ionic strength, solvent, and so on.

classical method for determining this in the solid state is via photoelectron spectroscopy. Hagfeldt and coworkers have reported such data for a nanocrystalline $TiO_2$ thin film sensitized with N3 in the presence and absence of $Li^+$ salts.[98] These data show a broad distribution of trap states centered at $\sim 1$ eV below the energy of the conduction band edge ($E_{cb}$). However, it is well known that the flatband potentials of the semiconductors are very sensitive to environment. Therefore, the absolute and relative energies in vacuum may not be as relevant to solution electrolytes. In photoelectrochemistry, the standard approach for determining the flatband potentials of semiconductor electrodes is Mott–Schottky analysis of capacitance data.[99] The analysis is based on the potential-dependent capacitance of a depletion layer at the semiconductor surface, behavior that is not likely observed for $\sim$20 nm anatase crystals that are expected to be fully depleted near $kT$.[100]

Rothenberger et al. have proposed an accumulation layer model to describe the potential distribution within the TiO$_2$ particles at negative applied potentials[101] that provides literature estimates of the conduction band edge energy in organic and aqueous solvents with common electrolytes.[102–104] These values give the impression that the nanocrystalline TiO$_2$ thin films have a well-defined conduction band edge. Even if this is the case, there is a tremendous compilation of data supporting the notion that the acceptor states relevant to interfacial charge separation and recombination are more localized and are more easily reduced than literature values indicate. Many electrochemical, photochemical, and spectroscopic techniques have supported the suggestion that mesoporous nanocrystalline TiO$_2$ thin films possess an exponential DOS rather than an abrupt onset from an ideal $E_{cb}$.[105] These states are believed to be unsaturated Ti$^{IV}$ surface states where oxygen vacancies reside. The energetics of such states was shown to be affected by surface chelation from various molecules due to the Lewis acid–base characteristics of the unsaturated Ti$^{IV}$ and surface-bound molecules.[106–108]

A final comment with regard to the semiconductor DOS is that they are not singular material parameters. The most well-known example is the nearly Nernstian shift, that is, 59 mV/pH unit, in aqueous solution over the pH range $H_0 = -8$ to $H_- = +23$ due to protonation/deprotonation of surface titanol groups on TiO$_2$.[109,110] It has also been known for quite some time that the flatband (and conduction band edge) potential of mesoporous nanocrystalline TiO$_2$ (anatase) can be widely tuned by the presence of cations in nonaqueous supporting electrolyte. This effect is greatest with cations possessing a large charge-to-radius ratio in the order Mg$^{2+}$ > Li$^+$ > Na$^+$ > K$^+$ > TBA$^+$.[167,168] For example, $E_{cb}$ has been reported to be $-1.0$ V versus SCE ($-0.76$ V versus NHE) in 0.1 M LiClO$_4$–acetonitrile electrolyte and $\sim-2.0$ V ($-1.76$ V) when Li$^+$ was replaced by TBA$^+$. The direction of the band edge shifts has been confirmed by excited-state quenching data described below. Interestingly, the same order has been seen for the equilibrium of cation adsorption onto TiO$_2$ in aqueous solutions.[111] Although this shift is non-Nernstian, the behavior has been shown to be logarithmic in LiClO$_4$ activity in acetonitrile and other aprotic mixed solvent systems. Similar behavior was not observed in protic solvents hypothesized to be due to selective solvation of Li$^+$ by the protic solvent molecules. This same cation-dependent shift in $E_{cb}$ can be used to promote photoinduced electron injection from surface-bound sensitizers.

### 12.3.1.3 Interfacial Charge Separation Mechanisms

There are three interfacial charge separation mechanisms for electron transfer from a molecular donor to a semiconductor such as TiO$_2$: (1) excited-state transfer, that is, Ru$^{III}$(dcb$^-$)(bpy)$_2^{2+*}$/

$TiO_2 \rightarrow Ru^{III}(dcb)(bpy)_2^{3+}/TiO_2(e^-)$; (2) reduced-state transfer, $[Ru^{II}(dcb^-)(bpy)_2]^+/TiO_2 \rightarrow [Ru^{II}(dcb)(bpy)_2]^{2+}/TiO_2(e^-)$; or (3) molecule-to-particle charge transfer complex, that is, $[(NC)_5Ru^{II}\text{-}CN]^{4-}/TiO_2 \rightarrow [(NC)_5Ru^{III}(CN)]^{3-}/TiO_2(e^-)$. Below we discuss some key observations and unresolved issues for each of these three sensitization mechanisms.

There now exists a large body of experimental data supporting ultrafast electron transfer from MLCT excited states to the acceptor states of anatase $TiO_2$.[112] Most, but not all, of these studies have focused on the famous "N3" dye first prepared by Nazeeruddin, $cis$-Ru(dcb)$_2$(NCS)$_2$.[39] In a recent study, an excitation wavelength dependence of the injection process was time resolved.[113] Femtosecond injection was attributed to the singlet state and a slower picosecond process from the thermally equilibrated triplet state.

Evidence for ultrafast injection has also come from photoluminescence quenching of Ru(dcb)(bpy)$_2$*/TiO$_2$ tuning the nature and concentration of cations at the TiO$_2$ surface.[114] The quantum yield for electron injection was reversibly tuned from below detection limits, ~0, to near unity simply by altering the [Li$^+$] concentration in an external acetonitrile bath. A Gerischer-type model was proposed to account for this behavior wherein cation adsorption to the TiO$_2$ surface shifts the semiconductor acceptor states positive on an electrochemical scale (i.e., away from the vacuum level) resulting in better overlap with the chromophoric donor's excited state (Figure 12.14).

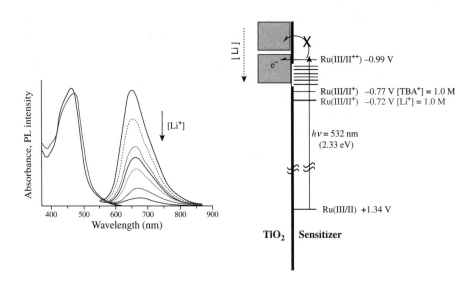

**FIGURE 12.14** Shown are the absorption and photoluminescence from Ru(dcb)(bpy)$_2$/TiO$_2$ immersed in acetonitrile. The addition of LiClO$_4$ to the acetonitrile resulted in a red shift in the absorption spectrum (shown by dotted line) and a quenching of the photoluminescence intensity that was shown to result from oxidative quenching by the conduction band. Time-resolved data were most consistent with the model shown: Li$^+$ adsorption to TiO$_2$ promotes rapid excited-state injection.

Interestingly, the quenching data were most consistent with injection from vibrationally hot excited states, $Ru^{III/II}**$. At low $Li^+$ concentrations, excited-state quenching was found to be static on a nanosecond timescale. In other words, the concentration of excited states generated with light decreased with $Li^+$ addition, while the excited-state lifetime did not change appreciably. If ultrafast injection occurred from the thexi state, the excited-state lifetime would have decreased to one over the injection rate constant, contrary to what was observed.[112] Cation-induced increases in the excited-state injection yields have been reported for other Lewis acid alkali and alkaline earth metals and correlated with the size-to-charge ratio of the cation,[114] behavior exploited for ratiometric photoluminescence sensing applications.[115]

The appearance of subpicosecond injection rate constants is often attributed to strong electronic coupling between the carboxylic acid groups in the dcb ligand and the semiconductor surface. It is therefore of interest to quantify semiconductor–sensitizer electronic coupling effects. Since the radiative and nonradiative rate constants for excited-state decay are relatively small, injection rate constants on the $10^{-8}$ s timescale would be expected to occur with near-unity quantum yields provided that other quenching pathways (such as those with the redox mediator) do not take place. In fact, there may be an as-of-yet undetermined sensitizer–semiconductor electronic interaction, where charge injection occurs quantitatively and recombination is slowed considerably. Such behavior would be expected to increase the photocurrent efficiency from sensitizers with more negative $Ru^{III/II}$ reduction potentials.

Tuning sensitizer–$TiO_2$ electronic interactions was initially accomplished by introducing flexible methylene spacers between the carboxylic acid binding groups and the bipyridine ligand or with bimetallic sensitizers.[75,116,117] In both cases efficient charge injection was realized, and in one report, a dependence on the injection rate constant was reported. More recently, elegant tripodal compounds with phenyl ethyne spacers between the chromophore and three ester (or carboxylic acid) groups have been designed (Figure 12.15).[68,69]

Remarkably, subpicosecond injection rates were measured with Ru sensitizers anchored to a surface about 24 Å from the semiconductor surface if the idealized Eiffel Tower-like orientation shown was achieved.[118–120] Of course delocalization of the excited state onto the phenyl ethyne substituents of the bpy ligands would decrease the injection distance. Indeed, this bridge may act as a conduit for excited-state electron transfer to $TiO_2$.[120]

The quantum yield for electron injection from a bipyridine ligand that was not directly anchored to the semiconductor surface can be unity (Figure 12.16).[121,122] For example, light excitation of $Re(bpy)(CO)_3(ina)^+$ results in quantitative injection from the $Re^{II}(bpy^-)(CO)_3(ina)^{+*}$ state with $k_{inj} > 10^8 s^{-1}$.[121] The injection yield after light excitation of cis-$Ru(bpy)_2(ina)_2/TiO_2$ was lower than that expected based on the excited-state reduction potential. It was found that the injection yield increased to near unity when the solution was cooled. Such behavior is exactly the opposite of what one would expect for activated interfacial charge separation based on Gerischer theory.[91] This sensitizer has low-lying ligand field excited states and it was proposed that rapid internal conversion processes underlie the temperature dependence. To our

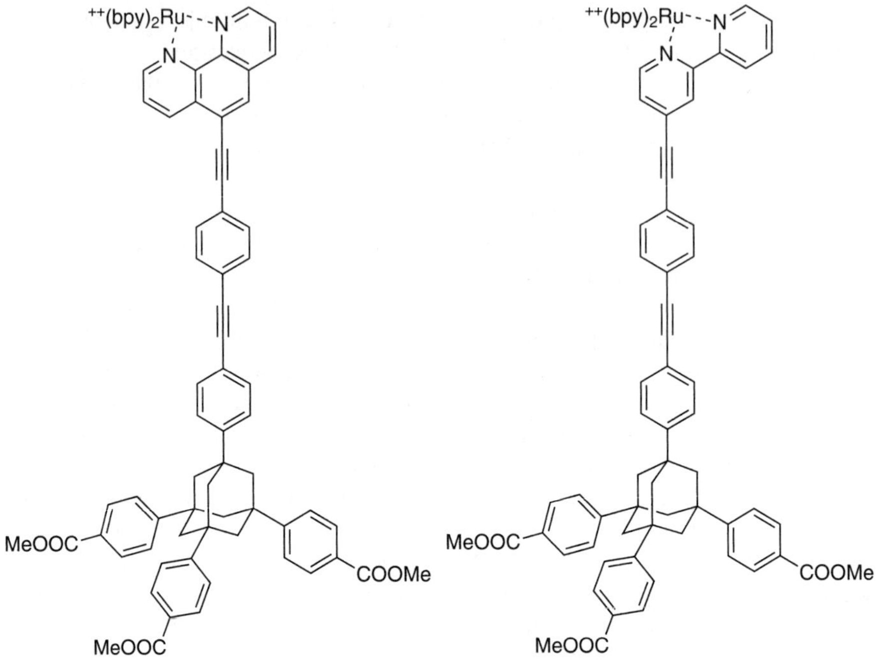

**FIGURE 12.15**    Tripodal sensitizers designed to place the ruthenium at a fixed distance from the semiconductor surface. Ideally, the three ester or carboxylic acid (not shown) groups bind to the surface and the adamantyl group orients the phenyl ethyne spacers approximately normal to the $TiO_2$ surface.

knowledge, this remains the sole example of temperature-dependent electron injection yields at sensitized semiconductor interfaces.

### 12.3.1.4   Is Ultrafast Electron Injection Useful for Solar Energy Conversion?

While ultrafast injection is necessary for sensitizers with inherently short excited-state lifetimes, such as those based on iron polypyridyl compounds, it remains unclear whether ultrafast injection is necessary, or even desirable, for solar energy conversion. A recent example of "trapping" hot carriers provides some clues as to how ultrafast electron transfer might be exploited for enhanced energy conversion efficiency.[123,124]

Anatase $TiO_2$ thin films were functionalized with [Ru(bpy)$_2$(deebq)](PF$_6$)$_2$, [Ru(bq)$_2$deeb](PF$_6$)$_2$, [Ru(deebq)$_2$bpy](PF$_6$)$_2$, [Ru(bpy)(deebq)(NCS)$_2$], or [Os(bpy)$_2$(deebq)](PF$_6$)$_2$, where bpy is 2,2′-bipyridine and deebq is 4,4′-diethyles-ter-2,2′-biquinoline in their carboxylate forms with limiting surface coverages of $8 (\pm 2) \times 10^{-8}$ mol/cm$^2$. Electrochemical measurements show that the first biquino-line-based reduction of these compounds ($-0.70$ V versus SCE) occurred prior to $TiO_2$ reduction. The thermally equilibrated MLCT excited state and the reduced form of these compounds did not inject electrons into $TiO_2$. However, after ultrafast injection, back electron transfer to form the thexi or reduced state was energetically

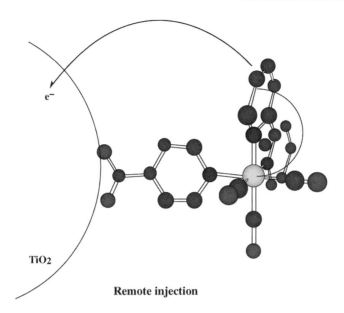

**FIGURE 12.16**    Excited-state injection by *fac*-Re(bpy)(CO)$_3$(ina)$^+$, where ina is isonico-tinic acid. Light excitation promotes an electron to the bpy ligand that was shown to quantitatively inject electrons into the semiconductor even though it is not directly bound.

favored and spectroscopic data for the latter acceptor were clearly observed (Figure 12.17). The quantum yields for this process were found to increase with excitation energy, behavior attributed to stronger overlap between the excited sensitizer and the semiconductor acceptor states. For example, the quantum yields for Os(bpy)$_2$dcbq/TiO$_2$ were $\phi(417\,\text{nm}) = 0.18 \pm 0.02$, $\phi(532.5\,\text{nm}) = 0.08 \pm 0.02$, and $\phi(683\,\text{nm}) = 0.05 \pm 0.01$. Electron transfer to yield ground-state products occurred by lateral intermolecular charge transfer. The driving force for charge recombination was in excess of that stored in the photoluminescent excited state.[124] Chronoabsorption measurements indicate that ligand-based intermolecular electron transfer was about three orders of magnitude faster than metal-centered intermole-cular hole transfer.[125]

   Electrons injected into the semiconductor from upper vibrational excited states were therefore found to yield long-lived charge-separated intermediates that store ~2 eV of free energy.[124] A key to realization of this behavior was to prepare sensitizers where the first reduction potential was below that of the TiO$_2$ acceptor states. These observations open the door toward fundamental studies of charge trapping at semiconductor interfaces where well-defined molecular compounds trap and store charge. The number of charge-separated states could be increased with continued illumination. It is likely that other examples of this behavior will emerge as solar energy researchers utilize ligands with low-lying $\pi^*$ orbitals and tune the conduction band edge position to optimize the spectral sensitivity and power conversion efficiencies of dye-sensitized solar cells.

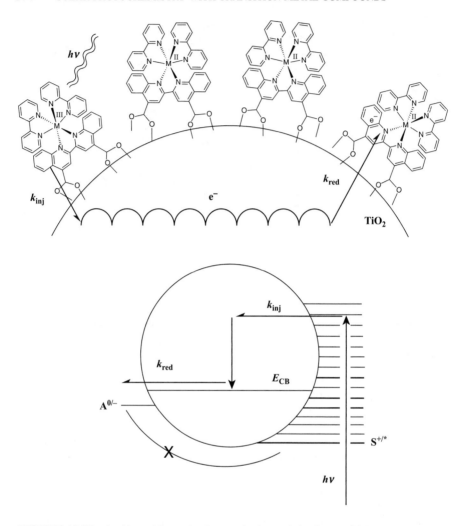

**FIGURE 12.17**    An idea of how ultrafast excited-state injection could be used to reduce electron acceptors that would be energetically unfavorable for the thexi state. An example of this was realized with Ru and Os sensitizers with 4,4'-(COOH)$_2$-2,2'-biquinoline ligands. These ligands were reduced prior to TiO$_2$, a necessary condition for realization of this behavior.

A class of futuristic solar cells, often called "hot carrier" solar cells, seeks to harvest the full energy of solar photons. Such cells would utilize the additional energy content of a blue photon relative to a red one.[126] In present-day solar cells, equilibrated carriers are collected and hence all absorbed photons with energy greater than the bandgap contribute equally to the measured efficiency. The realization of such "hot carrier" solar cells therefore requires electron transfer processes that are competitive with nonradiative decay of molecules or phonon relaxation in solids.[126] Literature data indicate that such relaxation occurs on a femtosecond timescale. The ultrafast injection realized at $TiO_2$ is indeed one of the few molecular interfaces that can allow fundamental studies toward exceeding the well-known Shockley–Queisser limit.[127]

### 12.3.2    Reduced Sensitizer Interfacial Charge Separation

When a donor reductively quenches the excited sensitizer, the reduced state formed, $S^-$, may transfer an electron to the semiconductor, thereby separating charge. Many early dye sensitization studies utilized this approach and the electron donors were termed "super-sensitizers." Photoelectrochemical measurements performed with planar electrodes often did not unambiguously distinguish whether the excited-state or the reduced-state charge separation mechanism was operative.[88] In special cases, the photoelectrochemistry data quite convincingly demonstrated a reduced sensitizer injection mechanism.[128] The observation of ultrafast electron injection coupled with the weak oxidizing power of the excited sensitizers currently in use strongly suggests that an excited-state injection mechanism is operative in regenerative solar cells based on these materials. Nevertheless, absolute proof is lacking as ultrafast spectroscopic studies are rarely done on the fully assembled dye-sensitized solar cell.[112]

Solar cells based on this mechanism are often referred to as photogalvanic cells.[88] A potential advantage of this mechanism is that the reduced sensitizer is a stronger reductant than the MLCT excited state by 300–500 meV (Figure 12.18). Thus, sensitizers that are weak photoreductants may sensitize $TiO_2$ efficiently after reductive quenching. This could be exploited to produce large open-circuit photovoltages or enhanced light harvesting in the near-IR regions. Clear spectroscopic evidence of reduced sensitizer electron transfer to $TiO_2$ was reported.[129] Such interfacial electron transfer was shown to be rate limited by reductive quenching of the sensitizer excited state. The sensitizer used in this work was $Ru(dcbH_2)(bpy)_2(PF_6)_2$ and the electron donor was phenothiazine, PTZ. A drawback of the PTZ donors was that they produce

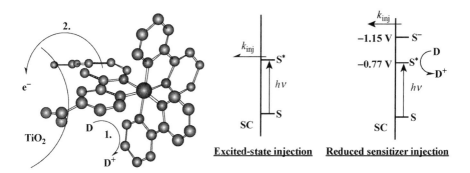

**FIGURE 12.18**    An alternative mechanism for excited-state sensitization of $TiO_2$ by $Ru(dcb)(bpy)_2^{2+}$ wherein a donor first reductively quenches the excited state, followed by thermal injection by the reduced sensitizer. A potential advantage of this mechanism is that the reduced sensitizer is a stronger reductant than the excited state, 380 mV in this case.

negligible photocurrents in the dye-sensitized nanocrystalline solar cells. Iodide is the sole electron donor identified that yields solar energy conversion efficiencies >10% under one sun of AM 1.5 solar irradiation.[12] Potent photooxidants based on $Ru^{II}$ bipyrazine compounds that rapidly oxidize iodide were recently characterized.[130,131] Efficient reductive quenching was observed by a variety of electron donors, and oxidation of iodide was found to produce an iodine atom. Cage escape yields were poor; however, long-lived charge separation was realized and the mechanism for I–I bond formation was elucidated.[131] Unfortunately, the reduced states of compounds such as $[Ru^{II}(dcb)(bpz^-)(bpz)]^+$ did not transfer electrons to $TiO_2$ or $SnO_2$ semiconductors.[130]

### 12.3.3 Molecule-to-Particle Charge Transfer

There exists a class of inorganic and organic compounds that form molecule-to-particle charge transfer (MPCT) donor–acceptor adducts with the $TiO_2$ surface.[132] This interfacial chemistry produces color changes that cannot be explained by trivial acid–base chemistry, decomposition, or aggregation. Such absorption bands were first observed with metal cyano compounds anchored to $TiO_2$ nanocrystals.[133,134] There now exist a large number of theoretical and experimental studies that support this assignment.[132] MPCT interactions raise the interesting issue of where the molecule stops and the extended solid begins. Experimentally, they are less complicated than excited- or reduced-state sensitization as one is assured that each absorbed photon is converted to an interfacial charge-separated state.

The classical inorganic example of MLCT is metal cyano compounds, $[M(CN)_x]^{m-}$ ($M = Fe^{II}$, $Ru^{II}$, $Os^{II}$, $Re^{III}$, $Mo^{IV}$, or $W^{IV}$; $x = 6$, 7, or 8), such as ferrocyanide, $Fe^{II}(CN)_6^{4-}$, anchored to acidic $TiO_2$ particles through the ambidentate cyano ligands. For example, $Fe^{II}(CN)_6^{4-}$ does not absorb light appreciably at wavelengths longer than 380 nm, but a deep orange color with an absorption maximum centered at 420 nm was observed when it forms an adduct with $TiO_2$, $Fe^{II}(CN)_6^{4-}/TiO_2$.[131] The visible absorption was attributed to a metal-to-particle charge transfer complex formed between $Fe^{II}(CN)_6^{4-}$ and surface $Ti^{4+}$ ions, $Fe(II) \rightarrow Ti(IV)$. Sensitized photocurrents and transient absorption studies support this assignment.[133–136]

A series of coordination compounds of the type $Fe(LL)(CN)_4^{2-}$, where LL is a bpy or phenanthroline ligand, were designed to sensitize $TiO_2$ to visible light by two distinct charge transfer pathways (Figure 12.19).[137] The absorption spectra of $Fe(bpy)(CN)_4^{2-}$ compound anchored to $TiO_2$ were well modeled by a sum of metal-to-ligand ($Fe \rightarrow bpy$) and metal-to-particle ($Fe(II) \rightarrow Ti(IV)$) charge transfer bands. Charge separation could not be time resolved while recombination was well described by a second-order kinetic model. An ionic strength-dependent quantum yield for charge separation measured after $Fe \rightarrow bpy$ excitation was taken as evidence for the MLCT excited-state or "indirect" sensitization pathway. The metal-to-particle pathway gave the expected ionic strength-independent yield of unity. The MLCT bands were solvatochromic while the MPCT bands were not. The total reorganization energy for the MPCT of $Fe(CN)_6^{4-}/TiO_2$ was estimated to be $\sim0.6$ eV, which

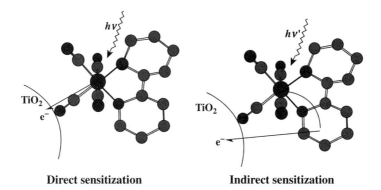

**Direct sensitization**          **Indirect sensitization**

**FIGURE 12.19**    Mechanisms of excited-state sensitization of $TiO_2$ by $Fe(bpy)(CN)_4^{2-}$. In the mechanism termed direct sensitization, an electron is promoted directly from the Fe d-orbitals to $TiO_2$. In the indirect sensitization mechanism, an MLCT excited state is formed followed by injection into $TiO_2$.

compared well with values obtained from MLCT sensitization of single-crystal rutile electrodes in aqueous solution, $\sim 0.3\,eV$.[137]

## 12.4    INTERFACIAL CHARGE RECOMBINATION

Recombination of the injected electron with the oxidized metal center generates ground-state products and wastes the energy stored in the interfacial charge-separated state. For efficient photocurrent generation in regenerative solar cells, iodide oxidation must be faster than charge recombination. It has been known for sometime that recombination occurs on a micro- to millisecond timescale, while injection is orders of magnitude faster. The origin of this fortuitous difference in interfacial electron transfer rates has been the subject of much discussion.[11] Recent studies indicated that charge recombination was not slow because of inherently sluggish rate constants, but because the process was second order in nature[64] and/or limited by transport of the injected electron back to the oxidized sensitizer.[138–141] Systematic charge recombination measurements where the concentration of charge-separated states was independently varied with irradiance or cation adsorption were reported. The obtained second-order rate constants were found to be independent of the number of interfacial charge-separated pairs photocreated. These data represent the strongest evidence to date that interfacial charge recombination is a second-order process.[141]

This behavior was attributed to activated transport of the injected electron back to the oxidized sensitizer as the rate-determining step for charge recombination. Charge transport and recombination in sensitized $TiO_2$ have both been shown to be second order. Nelson has modeled recombination data with the Kohlrausch–Williams–Watts model that is a paradigm for charge transport in disordered materials.[138–140] The rates increased significantly when additional electrons were electrochemically introduced into the $TiO_2$. It therefore appears likely that a high density of trap states and a relatively small number of injected electrons will give rise to this behavior. Since the number of trap states at colloidal semiconductor interfaces is generally expected to be quite high, this observation implies that long-lived charge separation will occur at most sensitized nanocrystalline semiconductor interfaces provided that charge separation occurs.

Interfacial recombination rates have been correlated with the physical location of the oxidized sensitizer's lowest unoccupied molecular orbitals. The rates were found to decrease exponentially with distance over the first few angstroms and a $\beta$ value of $0.95 \pm 0.2 \, \text{Å}^{-1}$ was estimated.[141] Attempts to slow recombination by fixing the distance between the sensitizer and the semiconductor surface with conjugated spacers have been less successful. Rate constants that showed very little distance dependence or did not vary in a systematic way with the expected distance have been reported. This research deserves further study and complements the studies of supramolecular sensitizers described below.

An intriguing approach for tuning electronic coupling and interfacial charge-separated state lifetimes has been the deposition of a thin layer of a second metal oxide material between the sensitizer and $TiO_2$.[142–145] Two classes of materials have been investigated in this regard, insulators and semiconductors. The conduction band edge of the semiconductor layers was designed to lie energetically above that of $TiO_2$, thereby promoting vectorial charge transfer away from the oxidized sensitizer. With insulators, the strategy was to have the excited electron tunnel through the insulating barrier and lower the electronic coupling for back electron transfer. These materials have indeed influenced charge-separated state lifetimes, although they have not enhanced energy conversion efficiencies over what has been realized with sensitized $TiO_2$ films that did not contain the layer.

In a recent study, the effects of insulating layers of $ZrO_2$, $Al_2O_3$, or $ZrO_2$ between $cis$-$Ru(dcb)_2(NCS)_2$ and $TiO_2$ were reported.[145] A good correlation was found between the recombination dynamics and the point-of-zero charge (pzc) of the metal oxide. The material with the most basic pzc, $Al_2O_3$ (pzc = 9.2), was found to inhibit charge recombination most effectively. In a novel extension of the same approach, a $cis$-$Ru(dcb)_2(CN)_2/TiO_2$ thin film was coated with a layer of $Al_2O_3$ to which a Ru phthalocyanine complex was attached to yield $RuPc/Al_2O_3$–$Ru(dcb)_2(CN)_2/TiO_2$. Light excitation was found to initiate electron transfer reactions that ultimately produced an electron in $TiO_2$ and an oxidized phthalocyanine. About half of these states lived for 5 ms prior to recombination. It is noteworthy that the sol–gel synthesis of the core–shell metal oxide materials was sufficiently mild that it allowed these complex molecular heterostructures to be fabricated without significant sensitizer degradation.

## 12.5  SUPRAMOLECULAR SENSITIZERS

Studies of "supramolecular" sensitizers anchored to $TiO_2$ nanocrystallites provide examples of how the principles of stepwise charge separation and hole transfer, originally developed in the field of photochemistry, can be applied to solid-state materials.[146,147] The sensitizers studied were designed as proof-of-principle examples, without any pretension to compete with the sensitizers commonly used in regenerative solar cells.

With binuclear Rh–Ru sensitizers shown, the rhodium center was bound directly to $TiO_2$ and the chromophoric ruthenium donor was fixed away from the semiconductor, $TiO_2$/Rh-L-Ru (Figure 12.20).[148] MLCT excitation resulted in an unprecedented electron "hopping" from bpy to Rh to the semiconductor nanocrystallite. Under the experimental condition studies, about 40% of the electrons that arrive at Rh were found to transfer to $TiO_2$, the rest recombined to Ru(III). The 60/40 branching ratio presumably reflected different orientations of the compound on the $TiO_2$ surface. When employed in regenerative solar cells, the photocurrent efficiency was rather low, mainly because of low charge injection yields. Nevertheless, the results suggest a general strategy to slow down recombination between the injected electron and oxidized sensitizer.[148]

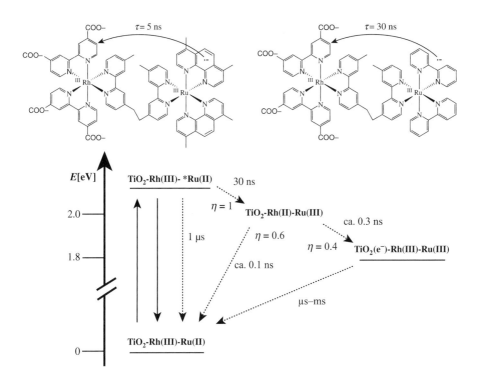

**FIGURE 12.20**    The chemical structure of two Rh–Ru dyads anchored to $TiO_2$ for interfacial electron transfer studies. The Rh(dcb) groups oxidatively quenched the ruthenium MLCT excited state by electron transfer. About 40% of the reduced rhodium units injected electrons to form long-lived charge-separated states, $TiO_2(e^-)$/Rh(III)-Ru(III), that decayed by back electron transfer to ground-state products on a microsecond to millisecond timescale.

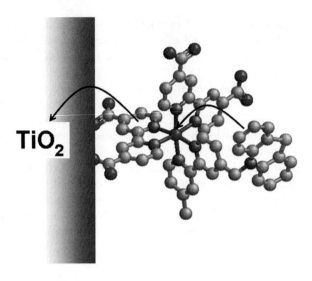

**FIGURE 12.21**    A Ru polypyridyl compound with a covalently bound phenothiazine group. The compound was designed to promote rapid hole transfer away from the Ru(III) center that is generated after excited-state electron injection into $TiO_2$.

Intramolecular "hole" transfer has been used to regenerate the ground state of the Ru(II) chromophore at sensitized $TiO_2$ interfaces.[149,150] The first compound reported to perform this function was $Ru(dcb)_2(4\text{-}CH_3,4'\text{-}CH_2\text{-}PTZ,\text{-}2,2'\text{-bipyridine})^{2+}$, where PTZ is the electron donor phenothiazine, shown in Figure 12.21.

In fluid solution, electron transfer from the PTZ group was moderately exergonic ($<0.25\,eV$) and had an approximate rate constant of $\sim 2.5 \times 10^8\,s^{-1}$ in methanol. The corresponding charge recombination step was faster than the forward one so that there was again, like the Ru–Rh compounds, no appreciable transient accumulation of the electron transfer product.

When attached to $TiO_2$, MLCT excitation resulted in a new charge-separated state with an electron in $TiO_2$ and an oxidized PTZ group, abbreviated $PTZ^+\text{-}Ru/TiO_2(e^-)$. Mechanistically, there were two possible electron transfer pathways available to reach this charge-separated state: (1) charge injection followed by oxidation of phenothiazine by the oxidized sensitizer unit, $PTZ\text{-}Ru(II)^*/TiO_2 \rightarrow PTZ\text{-}Ru(III)/TiO_2(e^-) \rightarrow PTZ^+\text{-}Ru/TiO_2(e^-)$; or (2) reductive quenching by PTZ followed by charge injection into the semiconductor, $TiO_2\text{-}Ru(II)^*\text{-}PTZ \rightarrow TiO_2\text{-}Ru(II)(dcb^-)^+\text{-}PTZ^+ \rightarrow PTZ^+\text{-}Ru/TiO_2(e^-)$. Note that both pathways yield the same electron transfer product.

Recombination of the electron in $TiO_2$ with the oxidized PTZ to yield the ground state occurred with a rate constant of $3.6 \times 10^3\,s^{-1}$. Excitation of a model compound that did not contain the PTZ donor, $Ru(dcb)_2(dmb)_2^{2+}$ where dmb is $4,4'\text{-}(CH_3)_2\text{-bpy}$, under otherwise identical conditions gave rise to the immediate formation of a charge-separated state, $TiO_2(e^-)/Ru(III)$, that recombined with an average rate constant of $3.9 \times 10^6\,s^{-1}$. Therefore, translating the "hole" from the Ru center to the pendant PTZ moiety slowed recombination rate constants by almost three orders of magnitude.

The Ru–PTZ and model compounds were tested in regenerative solar cells, with iodide as an electron donor. The charge separation yields were within experimental error, the same for the two sensitizers. However, the open-circuit photovoltage, $V_{oc}$, was observed to be about $100\,mV$ larger for the Ru–PTZ/TiO$_2$. The enhancement was even more pronounced in the absence of iodide, $180\,mV$ larger over five decades of irradiance. The diode equation, Equation 12.8, predicts a $59\,mV$ increase in $V_{oc}$ for each order of magnitude decrease in the rate constant for charge recombination of the injected electrons with acceptors, $k_i[A]_i$, provided that the electron injection flux into the semiconductor, $I_{inj}$, is constant. Applying the spectroscopically measured rate constants to Equation 12.6 gave a predicted increase in $V_{oc}$ of $200\,mV$, which was in close agreement with the experimentally determined value of $180\,mV$. It was remarkable that these molecular interfaces behaved like ideal diodes over five decades of irradiance with charge separation rate constants at least six orders of magnitude faster than charge recombination.[150]

$$V_{oc} = \left(\frac{kT}{e}\right)\ln\left(\frac{I_{inj}}{n\sum k_i[A]_i}\right) \qquad (12.8)$$

The bimetallic sensitizer [Ru(dcb)$_2$(Cl)-bpa-Os(bpy)$_2$(Cl)](PF$_6$)$_2$, abbreviated Ru-bpa-Os, where bpa is 1,2-bis(4-pyridyl)ethane, was anchored to TiO$_2$ for interfacial electron transfer studies (Figure 12.22).[151]

Pulsed light excitation of a Os-bpa-Ru/TiO$_2$ material immersed in a $1.0\,M$ LiClO$_4$– acetonitrile bath at $25\,°C$ resulted in rapid interfacial electron transfer and

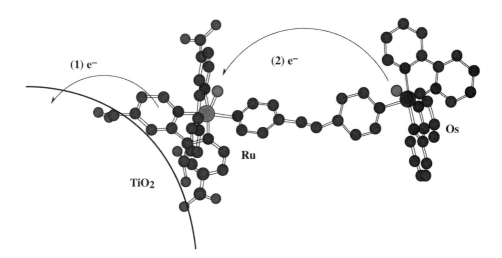

**FIGURE 12.22**   As an extension of previous work with inorganic–organic dyads (Figure 12.21), the bimetallic Ru–Os compound shown was designed to promote rapid intramolecular electron transfer (2), after electron injection (1). This process was demonstrated as well as a novel remote injection from the osmium MLCT excited state.

intramolecular electron transfer ($Os^{II} \rightarrow Ru^{III}$) to ultimately form an interfacial charge-separated state with an electron in $TiO_2$ and an $Os^{III}$ center, abbreviated $Os^{III}$-bpa-Ru/$TiO_2(e^-)$.[94]

The same state could also be generated after selective excitation of the $Os^{II}$ group with red light. The rates of intramolecular and interfacial electron transfer are fast, $k > 10^8\,s^{-1}$, while interfacial charge recombination, $Os^{III}$-bpa-Ru/$TiO_2(e^-) \rightarrow Os$-bpa-Ru/$TiO_2$, required milliseconds for completion. The results show a general strategy for promoting rapid intramolecular electron transfer ($Os^{II} \rightarrow Ru^{III}$) after interfacial electron injection and a "remote" electron injection process that occurs after direct excitation of the Os(II) chromophore. Unfortunately, there was no evidence for an enhanced lifetime of the charge-separated state, presumably because the $Os^{III}$ center was proximate to the semiconductor surface.[151] Grätzel and coworkers have reported related studies with supramolecular sensitizers and have emphasized their potential application in photochromic devices.[152] Interestingly, these workers found long-lived charge separation, like that described for the PTZ-Ru/$TiO_2$ interface described above, in some cases while not in others. Presumably, driving force and semiconductor molecular orientations are two key parameters. Additional experiments have been reported in this area, particularly with amine donors,[153–155] and more are required before this interesting interfacial behavior can be fully understood.

## 12.6   CONCLUSION

It has been about 40 years since chemists first observed photoinduced charge separation.[78] These studies have progressed and naturally evolved toward solid-state materials that are more applicable to solar energy conversion in the real world. The mechanisms for interfacial charge separation and recombination at $Ru^{II}$ sensitized $TiO_2$ thin film interfaces are now well understood in considerable molecular detail. It is now relatively easy to generate interfacial charge-separated states that live for milliseconds and store $>1.5\,eV$ of free energy. Aside from electrical power generation in regenerative solar cells, a significant future challenge is to intercept these states to drive reactions that produce useful fuels as photocatalytic materials. The mesoporous nanocrystalline thin films that are under active investigation provide exciting new opportunities for fundamental research in this area. The ability to store and transport a high density of redox equivalents in a semiconductor material offers the real and near-term possibility of multielectron transfer photocatalysis relevant to water splitting and environmental remediation. Future prospects for solar energy conversion with coordination compounds anchored to semiconductor surfaces appear to be very bright.

## ACKNOWLEDGMENTS

The Division of Chemical Sciences, Office of Basic Energy Sciences, Office of Energy Research, U.S. Department of Energy is gratefully acknowledged for research support. We thank the National Science Foundation for support of the environmental chemistry aspects of this work.

## REFERENCES

1. Hoffert, M. I.; Caldeira, K.; Jain, A. K.; Haites, E. F.; Harvey, L. D. D.; Potter, S. D.; Schlesinger, M. E.; Schneider, S. H.; Watts, R. G.; Wigley, T. M. L.; Wuebbles, D. J. *Nature* **1998**, *395*, 881–884.

2. Caldeira, K.; Jain, A. K.; Hoffert, M. I. *Science* **2003**, *299*, 2052–2054.

3. U.S. Department of Energy, Energy Information Administration, http://www.eia.doe. gov/.

4. Lewis, N. S.; Nocera, D. G. *Proc. Natl. Acad. Sci. USA* **2006**, *103*, 15729–15735.

5. Petit, J. R.; Jouzel, J.; Raynaud, D.; Barkov, N. I.; Barnola, J. M.; Basile, I.; Bender, M.; Chappellaz, J.; Davis, M.; Delaygue, G.; Delmotte, M.; Kotlyakov, V. M.; Legrand, M.; Lipenkov, V. Y.; Lorius, C.; Pepin, L.; Ritz, C.; Saltzman, E.; Stievenard, M. *Nature* **1999**, *399*, 429–436.

6. Siegenthaler, U.; Stocker, T. F.; Monnin, E.; Luthi, D.; Schwander, J.; Stauffer, B.; Raynaud, D.; Barnola, J.-M.; Fischer, H.; Masson-Delmotte, V.; Jouzel, J. *Science* **2005**, *310*, 1313–1317.

7. Lewis, N. S. *Powering the Planet: Global Energy Perspective.* http://nsl.caltech.edu/files/ energy.pdf.

8. U.S. Department of Energy. Report of the Basic Energy Sciences Workshop on Solar Energy Utilization. In *Basic Research Needs for Solar Energy Utilization*; U.S. Department of Energy: Washington, DC, 2005.

9. United Nations. World Energy Assessment Report: Energy and the Challenge of Sustainability. In *United Nations Development Program*; United Nations: New York, 2003.

10. O'Regan, B.; Grätzel, M. *Nature* **1991**, *353*, 737–740.

11. Meyer, G. J. *Inorg. Chem.* **2005**, *44*, 6852–6864.

12. Green, M. A.; Emery, K.; Hishikawa, Y.; Warta, W. *Prog. Photovolt. Res. Appl.* **2008**, *16*, 61–67.

13. Kober, E. M.; Meyer, T. J. *Inorg. Chem.* **1982**, *21*, 3967–3977.

14. Adamson, A. W.; Demas, J. N. *J. Am. Chem. Soc.* **1971**, *93*, 1800–1801.

15. Crosby, G. A.; Demas, J. N. *J. Am. Chem. Soc.* **1971**, *93*, 2841–2847.

16. Demas, J. N.; Taylor, D. G. *Inorg. Chem.* **1979**, *18*, 3177–3179.

17. Yersin, H.; Gallhuber, E. *J. Am. Chem. Soc.* **1984**, *106*, 6582–6586.

18. Adamson, A. W. *J. Chem. Educ.* **1983**, *60*, 797–802.

19. Hager, G. D.; Crosby, G. A. *J. Am. Chem. Soc.* **1975**, *97*, 7031–7037.

20. Hager, G. D.; Watts, R. J.; Crosby, G. A. *J. Am. Chem. Soc.* **1975**, *97*, 7037–7042.

21. Harrigan, R. W.; Crosby, G. A. *J. Chem. Phys.* **1973**, *59*, 3468–3476.

22. Hipps, K. W.; Crosby, G. A. *J. Am. Chem. Soc.* **1975**, *97*, 7042–7048.

23. Crosby, G. A.; Hipps, K. W.; Elfring, W. H. *J. Am. Chem. Soc.* **1974**, *96*, 629–630.

24. Daul, C.; Baerends, E. J.; Vernooijs, P. *J. Am. Chem. Soc.* **1994**, *33*, 3538–3543.

25. Juris, A.; Balzani, V.; Barigelletti, F.; Campagna, S.; Belser, P.; Von Zelewsky, A. *Coord. Chem. Rev.* **1988**, *84*, 85.

26. Kalyanasundaram, K. *Photochemistry of Polypyridine and Porphyrin Complexes*; Academic Press: London, 1992.

27. Meyer, T. J. *Acc. Chem. Res.* **1989**, *22*, 163.

28. Kober, E. M.; Sullivan, B. P.; Meyer, T. J. *Inorg. Chem.* **1984**, *23*, 2098–2104.

29. Dallinger, R. F.; Woodruff, W. H. *J. Am. Chem. Soc.* **1979**, *101*, 4391–4393.

30. McCusker, J. K. *Acc. Chem. Res.* **2003**, *36*, 876–887.

31. Durham, B.; Caspar, J. V.; Nagle, J. K.; Meyer, T. J. *J. Am. Chem. Soc.* **1982**, *104*, 4803–4810.

32. Englman, R.; Jortner, J. *Mol. Phys.* **1970**, *18*, 145–164.

33. Freed, K. F.; Jortner, J. *J. Chem. Phys.* **1970**, *52*, 6272–6291.

34. Bixon, M.; Jortner, J. *J. Chem. Phys.* **1968**, *48*, 715–726.

35. Claude, J. P.; Meyer, T. J. *J. Phys. Chem.* **1995**, *99*, 51–54.

36. Lumpkin, R. S.; Meyer, T. J. *J. Phys. Chem.* **1986**, *90*, 5307–5312.

37. Lever, A. B. P. Ligand electrochemical parameters and electrochemical-optical relationships. In *Comprehensive Coordination Chemistry II*; Elsevier Science, 2003; Vol. 2, pp 251–268.

38. DeArmond, M. K.; Carlin, C. M. *Coord. Chem. Rev.* **1981**, *36*, 325–355.

39. Nazeeruddin, M. K.; Kay, A.; Rodicio, I.; Humphry-Baker, R.; Mueller, E.; Liska, P.; Vlachopoulos, N.; Grätzel, M. *J. Am. Chem. Soc.* **1993**, *115*, 6382–6390.

40. Bard, A. J.; Fox, M. A. *Acc. Chem. Res.* **1995**, *28*, 141–145.

41. Galoppini, E. *Coord. Chem. Rev.* **2004**, *248*, 1283–1297.

42. Anderson, S.; Constable, E. C.; Dare-Edwards, M. P.; Goodenough, J. B.; Hamnett, A.; Seddon, K. R.; Wright, R. D. *Nature* **1979**, *280*, 571–573.

43. Qu, P.; Meyer, G. J. *Langmuir* **2001**, *17*, 6720–6728.

44. Finnie, K. S.; Bartlett, J. R.; Woolfrey, J. L. *Langmuir* **1998**, *14*, 2744–2749.

45. Kilsa, K.; Mayo, E. I.; Brunschwig, B. S.; Gray, H. B.; Lewis, N. S.; Winkler, J. R. *J. Phys. Chem. B* **2004**, *108*, 15640–15651.

46. Meyer, T. J.; Meyer, G. J.; Pfennig, B. W.; Schoonover, J. R.; Timpson, C. J.; Wall, J. F.; Kobusch, C.; Chen, X.; Peek, B. M.; Wall, C. G.; Ou, W.; Erickson, B. W.; Bignozzi, C. A. *Inorg. Chem.* **1994**, *33*, 3952–3964.

47. Dobson, K. D.; McQuillan, A. J. *Spectrochim. Acta A* **2000**, *56*, 557–565.

48. Deacon, G. B.; Phillips, R. J. *Coord. Chem. Rev.* **1980**, *33*, 227–250.

49. Heimer, T. A.; D'Arcangelis, S. T.; Farzad, F.; Stipkala, J. M.; Meyer, G. J. *Inorg. Chem.* **1996**, *35*, 5319–5324.

50. Persson, P.; Bergstrom, R.; Lunell, S. *J. Phys. Chem. B* **2000**, *104*, 10348–10351.

51. Kelly, C. A.; Farzad, F.; Thompson, D. W.; Meyer, G. J. *Langmuir* **1999**, *15*, 731–737.

52. Farzad, F.; Thompson, D. W.; Kelly, C. A.; Meyer, G. J. *J. Am. Chem. Soc.* **1999**, *121*, 5577–5578.

53. Higgins, G. T.; Bergeron, B. V.; Hasselmann, G. M.; Farzad, F.; Meyer, G. J. *J. Phys. Chem. B* **2006**, *110*, 2598–2605.

54. Leasure, R. M.; Oy, W.; Moss, J. A.; Linton, R. W.; Meyer, T. J. *Chem. Mater.* **1996**, *8*, 264–272.

55. Qu, P.; Thompson, D. W.; Meyer, G. J. *Langmuir* **2000**, *16*, 4662–4671.

56. Maruszewski, K.; Strommen, D. P.; Kincaid, J. R. *J. Am. Chem. Soc.* **1993**, *115*, 8345.

57. Liu, F.; Meyer, G. J. *J. Am. Chem. Soc.* **2005**, *127*, 824–825.

58. Liu, F.; Meyer, G. J. *Inorg. Chem.* **2003**, *42*, 7351–7353.

59. Chen, P.; Meyer, T. J. *Chem. Rev.* **1998**, *98*, 1439–1478.

60. Timpson, C. J.; Bignozzi, C. A.; Sullivan, B. P.; Kober, E. M.; Meyer, T. J. *J. Phys. Chem.* **1996**, *100*, 2915–2925.

61. Hasselman, G. M.; Watson, D. F.; Stromberg, J. R.; Bocian, D. F.; Holten, D.; Lindsey, J. S.; Meyer, G. J. *J. Phys. Chem. B* **2006**, *110*, 25430–25440.

62. Liu, Y.; De Nicola, A.; Reiff, O.; Ziessel, R.; Schanze, K. S. *J. Phys. Chem. A* **2003**, *107*, 3476.

63. Argazzi, R.; Bignozzi, C. A.; Heimer, T. A.; Castellano, F. N.; Meyer, G. J. *Inorg. Chem.* **1994**, *33*, 5741–5749.

64. Wang, P.; Klein, C.; Humphy-Baker, R.; Zakeeruddin, S. M.; Grätzel, M. *Appl. Phys. Lett.* **2005**, *86*, 123508–123510.

65. Wang, P.; Klein, C.; Humphry-Baker, R.; Zakeeruddin, S. M.; Grätzel, M. *J. Am. Chem. Soc.* **2005**, *127*, 808–809.

66. Kuang, D.; Ito, S.; Wenger, B.; Klein, C.; Moser, J. E.; Humphry-Baker, R.; Zakeeruddin, S. M.; Grätzel, M. *J. Am. Chem. Soc.* **2006**, *128*, 4146–4154.

67. Snaith, H. J.; Karthikeyan, C. S.; Petrozza, A.; Teuscher, J.; Moser, J. E.; Nazeeruddin, M. K.; Thelakkat, M.; Grätzel, M. *J. Phys. Chem. C* **2008**, *112*, 7562–7566.

68. Aranyos, V.; Hjelm, J.; Hagfeldt, A.; Grennberg, H. *J. Chem. Soc., Dalton Trans.* **2001**, 1319–1325.

69. Staniszewski, A.; Heuer, W. B.; Meyer, G. J. *Inorg. Chem.* **2008**, *47*, 7062–7064.

70. Holten, D.; Bocian, D. F.; Lindsey, J. S. *Acc. Chem. Res.* **2002**, *35*, 57–69.

71. Amadelli, R.; Argazzi, R.; Bignozzi, C. A.; Scandola, F. *J. Am. Chem. Soc.* **1990**, *112*, 7099–7103.

72. Gajardo, F.; Leiva, A. M.; Loeb, B.; Delgadillo, A.; Stromberg, J. R.; Meyer, G. J. *Inorg. Chim. Acta* **2008**, *361*, 613–619.

73. Rehm, D.; Weller, A. *Isr. J. Chem.* **1970**, *8*, 259–267.

74. Bonhote, P.; Gogniat, E.; Tingry, S.; Barbe, C.; Vlachopoulos, N.; Lenzmann, F.; Comte, P.; Grätzel, M. *J. Phys. Chem. B* **1998**, *102*, 1498–1507.

75. Heimer, T. A.; D'Arcangelis, S. T.; Farzad, F.; Stipkala, J. M.; Meyer, G. J. *Inorg. Chem.* **1996**, *35*, 5319–5324.

76. Trammell, S. A.; Meyer, T. J. *J. Phys. Chem. B* **1999**, *103*, 104–107.

77. Young, R. C.; Meyer, T. J.; Whitten, D. G. *J. Am. Chem. Soc.* **1975**, *97*, 4781–4782.

78. Bock, C. R.; Meyer, T. J.; Whitten, D. G. *J. Am. Chem. Soc.* **1975**, *97*, 2909–2911.

79. Gray, H. B.; Winkler, J. R. *Proc. Natl. Acad. Sci. USA* **2005**, *102*, 3534–3539.

80. Schanze, K. S.; Walters, K. A. Photoinduced electron transfer in metal-organic dyads. In *Molecular and Supramolecular Photochemistry*; Ramamurthy, V.; Schanze, K. S., Eds.; Marcel Dekker: New York, 1999, Vol. *8*, Chapter 3, pp 75–126, and references therein.

81. Castellano, F. N.; Meyer, G. J. *Prog. Inorg. Chem.* **1997**, *44*, 167–209.

82. Ruthkosky, M.; Kelly, C. A.; Castellano, F. N.; Meyer, G. J. *J. Am. Chem. Soc.* **1997**, *119*, 12004–12005.

83. Scaltrito, D. V.; Kelly, C. A.; Ruthkosky, M.; Thompson, D. W.; Meyer, G. J. *Inorg. Chem.* **2000**, *39*, 3777–3783.

84. Scaltrito, D. V.; Thompson, D. W.; O'Callahan, J. A.; Meyer, G. J. *Coord. Chem. Rev.* **2000**, *208*, 243–267.

85. Chen, L. X.; Jennings, G.; Liu, T.; Gosztola, D. J.; Hessler, J. P.; Scaltrito, D. V.; Meyer, G. J. *J. Am. Chem. Soc.* **2002**, *124*, 10861–10867.

86. Chen, L. X.; Shaw, G. B.; Liu, T.; Jennings, G.; Attenkofer, K.; Meyer, G. J.; Coppens, P. *J. Am. Chem. Soc.* **2003**, *125*, 7022–7034.

87. Shaw, G. B.; Grant, C. D.; Castner, E. W.; Meyer, G. J.; Chen, L. C. *J. Am. Chem. Soc.* **2007**, *129*, 2147–2160.

88. Albery, W. J. *Acc. Chem. Res.* **1982**, *15*, 142.

89. Gomer, R. *Electrochim. Acta* **1975**, *20*, 13.

90. Gerischer, H.; Willig, F. *Top. Curr. Chem.* **1976**, *61*, 31–84.

91. Gerischer, H. *Photochem. Photobiol.* **1972**, *16*, 243–260.

92. Gerischer, H. *Surf. Sci.* **1969**, *18*, 97–122.

93. Clark, W. D. K.; Sutin, N. *J. Am. Chem. Soc.* **1977**, *99*, 4676 4682.

94. Boschloo, G.; Fitzmaurice, D. *J. Phys. Chem. B* **1999**, *103*, 7860–7868.

95. Enright, B.; Redmond, G.; Fitzmaurice, D. *J. Phys. Chem.* **1994**, *98*, 6195–6200.

96. Rothenberger, G.; Fitzmaurice, D.; Grätzel, M. *J. Phys. Chem.* **1992**, *96*, 5983–5986.

97. O'Regan, B.; Grätzel, M.; Fitzmaurice, D. *J. Phys. Chem.* **1991**, *95*, 10525–10528.

98. Rensmo, H.; Södergren, S.; Patthey, L.; Westermark, K.; Vayssieres, L.; Kohle, O.; Brühwiler, P. A.; Hagfeldt, A.; Siegbahn, H. *Chem. Phys. Lett.* **1997**, *274*, 51–57.

99. Cardon, F.; Gomes, W. P. *J. Phys. D: Appl. Phys.* **1978**, *11*, L63–L67.

100. Albery, W. J.; Bartlett, P. N. *J. Electrochem. Soc.* **1984**, *131*, 315–325.

101. Rothenberger, G.; Fitzmaurice, D.; Grätzel, M. *J. Phys. Chem.* **1992**, *96*, 5983–5986.

102. Enright, B.; Redmond, G.; Fitzmaurice, D. *J. Phys. Chem.* **1994**, *98*, 6195–6200.

103. Redmond, G.; Fitzmaurice, D. *J. Phys. Chem.* **1993**, *97*, 1426–1430.

104. Fitzmaurice, D. *Solar Energy Mater.* **1994**, *32*, 289–305.

105. Bisquert, J.; Fabregat-Santiago, F.; Mora-Sero, I.; Garcia-Belmonte, G.; Barea, E. M.; Palomares, E. *Inorg. Chim. Acta* **2008**, *361*, 684–698.

106. Finklea, H. O. *Semiconductor Electrodes*; Elsevier: Amsterdam, 1988; Vol. 55.

107. Moser, J.; Punchihewa, S.; Infelta, P. P.; Grätzel, M. *Langmuir* **1991**, *7*, 3012–3018.

108. Redmond, G.; Fitzmaurice, D.; Grätzel, M. *J. Phys. Chem.* **1993**, *97*, 6951–6954.

109. Gerischer, H. Semiconductor electrochemistry. In *Physical Chemistry: An Advanced Treatise*; Eyring, H.; Henderson, D.; Jost, W., Eds.; Academic Press: New York, 1970; Vol. 9A, Chapter 5, p 463.

110. Lyon, L. A.; Hupp, J. T. *J. Phys. Chem. B* **1999**, *103*, 4623–4628.

111. Berube, Y. G.; de Bruyn, P. L. *J. Colloid Interface Sci.* **1968**, *28*, 92–105.

112. Watson, D. F.; Meyer, G. J. *Annu. Rev. Phys. Chem.* **2005**, *56*, 119–156.

113. Benko, G.; Kallioinen, J.; Korppi-Tommola, J. E. I.; Yartsev, A. P.; Sundstrom, V. *J. Am. Chem. Soc.* **2002**, *124*, 489–493.

114. Kelly, C. A.; Thompson, D. W.; Farzad, F.; Stipkala, J. M.; Meyer, G. J. *Langmuir* **1999**, *15*, 7047–7054.

115. Stux, A. M.; Meyer, G. J. *J. Fluorescence* **2002**, *12*, 419–423.

116. Anderson, N. A.; Ai, X.; Chen, D.; Mohler, D. L.; Lian, T. *J. Phys. Chem. B* **2003**, *107*, 14231–14239.

117. Argazzi, R.; Bignozzi, C. A.; Heimer, T. A.; Meyer, G. J. *Inorg. Chem.* **1997**, *36*, 2–3.

118. Galoppini, E.; Guo, W.; Qu, P.; Meyer, G. J. *J. Am. Chem. Soc.* **2001**, *123*, 4342–4343.

119. Galoppini, E.; Guo, W.; Hoertz, P.; Qu, P.; Meyer, G. J. *J. Am. Chem. Soc.* **2002**, *124*, 7801–7811.

120. Piotrowiak, P.; Galoppini, E.; Wei, Q.; Meyer, G. J.; Wiewior, P. *J. Am. Chem. Soc.* **2003**, *125*, 5278–5279.

121. Hasselmann, G. M.; Meyer, G. J. *Z. Phys. Chem.* **1999**, *212*, 39–44.

122. Liu, F.; Meyer, G. J. *Inorg. Chem.* **2005**, *44*, 9305–9313.

123. Hoertz, P. G.; Staniszewski, A.; Marton, A.; Higgins, G. T.; Incarvito, C. D.; Rheingold, A. L.; Meyer, G. J. *J. Am. Chem. Soc.* **2006**, *128*, 8234–8245.

124. Hoertz, P. G.; Thompson, D. W.; Friedman, L. A.; Meyer, G. J. *J. Am. Chem. Soc.* **2002**, *124*, 9690–9691.

125. Staniszewski, A.; Morris, A. J.; Ito, T.; Meyer, G. J. *J. Phys. Chem. B* **2007**, *111*, 6822–6828.

126. Ross, R. T.; Nozik, A. J. *J. Appl. Phys.* **1982**, *53*, 3813–3818.

127. Shockley, W.; Queisser, H. J. *J. Appl. Phys.* **1961**, *32*, 510–519.

128. Ortmans, I.; Moucheron, C.; Kirsch-De Mesmaeker, A. *Coord. Chem. Rev.* **1998**, *168*, 233–271.

129. Thompson, D. W.; Kelly, C. A.; Farzad, F.; Meyer, G. J. *Langmuir* **1999**, *15*, 650–653.

130. Bergeron, B. V.; Meyer, G. J. *J. Phys. Chem. B* **2003**, *107*, 245–254.

131. Gardner, J. M.; Giaimuccio, J. M.; Meyer, G. J. *J. Am. Chem. Soc.* **2008**, *130*, 17252–17253.

132. Liu, F.; Yang, M.; Meyer, G. J. Molecule-to-particle charge transfer in sol-gel materials. In *Handbook of Sol-Gel Science and Technology: Processing Characterization and Application; Volume II: Characterization of Sol-Gel Materials and Products*; Almeida, R. M.,Ed.; Kluwer Academic Publishers: Dordrecht, The Netherlands, 2005; Vol. 2, pp 400–428.

133. Vrachnou, E.; Grätzel, M.; McEvoy, A. J. *J. Electroanal. Chem.* **1989**, *258*, 193–205.

134. Vrachnou, E.; Vlachopoulos, N.; Grätzel, M. *Chem. Commun.* **1987**, 868–870.

135. Blackbourn, R. L.; Johnson, C. S.; Hupp, J. T. *J. Am. Chem. Soc.* **1991**, *113*, 1060–1062.

136. Lu, H.; Prieskorn, J. N.; Hupp, J. T. *J. Am. Chem. Soc.* **1993**, *115*, 4927–4928.

137. Yang, M.; Thompson, D. W.; Meyer, G. J. *Inorg. Chem.* **2002**, *41*, 1254–1262.

138. Barzykin, A. V.; Tachiya, M. *J. Phys. Chem. B* **2004**, *108*, 8385–8389.

139. Nelson, J.; Haque, S. A.; Klug, D. R.; Durrant, J. R. *Phys. Rev. B* **2001**, *63*, 205–321.

140. Walker, A. B.; Peter, L. M.; Martínez, D.; Lobato, K. *Chimia* **2007**, *61*, 792–795.

141. Clifford, J.N.; Palomares, E.; Nazeeruddin, M.K.; Gratzel, M.; Nelson, J.; Li, X.; Long, N.J.; Durrant, J.R. J. Am Chem. Soc. **2004**, *126*, 5225–5233.

142. Bedja, I.; Kamat, P. V. *J. Phys. Chem.* **1995**, *99*, 9182–9188.

143. Diamant, Y.; Chappel, S.; Chen, S. G.; Melamed, O.; Zaban, A. *Coord. Chem. Rev.* **2004**, *248*, 1271–1276.

144. Zaban, A.; Chen, S. G.; Chappel, S.; Gregg, B. A. *Chem. Commun.* **2000**, 2231–2232.

145. Palomares, E.; Clifford, J. N.; Haque, S. A.; Lutz, T.; Durrant, J. R. *J. Am. Chem. Soc.* **2003**, *125*, 475–482.

146. Balzani, V.; Moggi, L.; Scandola, F. Towards a supramolecular photochemistry: assembly of molecular components to obtain photochemical molecular devices. In *Supramolecular Photochemistry*; Balzani, V., Ed.; D. Reidel Publishing Co.: Dordrecht, Holland, 1987.

147. Forster, R. J.; Keyes, T. E.; Vos, J. G. *Interfacial Supramolecular Assemblies*; Wiley: Chichester, 2003.

148. Kleverlaan, C. J.; Indelli, M. T.; Bignozzi, C. A.; Pavanin, L.; Scandola, F.; Hasselman, G. M.; Meyer, G. J. *J. Am. Chem. Soc.* **2000**, *122*, 2840–2849.

149. Argazzi, R.; Bignozzi, C. A.; Heimer, T. A.; Castellano, F. N.; Meyer, G. J. *J. Am. Chem. Soc.* **1995**, *117*, 11815–11816.

150. Argazzi, R.; Bignozzi, C. A.; Heimer, T. A.; Castellano, F. N.; Meyer, G. J. *J. Phys. Chem. B* **1997**, *101*, 2591–2597.

151. Kleverlaan, C.; Alebbi, M.; Argazzi, R.; Bignozzi, C. A.; Hasselmann, G. M.; Meyer, G. J. *Inorg. Chem.* **2000**, *39*, 1342–1343.

152. Bonhote, P.; Moser, J. E.; Humphry-Baker, R.; Vlachopoulos, N.; Zakeeruddin, S. M.; Walder, L.; Grätzel, M. *J. Am. Chem. Soc.* **1999**, *121*, 1324–1336.

153. Hirata, N.; Lagref, J.-J.; Palomares, E. J.; Durrant, J. R.; Nazeeruddin, M. K.; Grätzel, M.; Di Censo, D. *Chem.: Eur. J.* **2004**, *10*, 595–602.

154. Haque, S. A.; Handa, S.; Peter, K.; Palomares, E.; Thelakkat, M.; Durrant, J. R. *Angew. Chem., Int. Ed.* **2005**, *44*, 5740–5744.

155. Handa, S.; Wietasch, H.; Thelakkat, M.; Durrant, J. R.; Haque, S. A. *Chem. Commun.* **2007**, 1725–1727.

# INDEX

*Physical Inorganic Chemistry: Reactions, Processes, and Applications*   Edited by Andreja Bakac
Copyright © 2010 by John Wiley & Sons, Inc.